Essentials of Cell Biology

Essentials of Cell Biology

Edited by **Ralph Becker**

Callisto Reference
New York

Published by Callisto Reference,
106 Park Avenue, Suite 200,
New York, NY 10016, USA
www.callistoreference.com

Essentials of Cell Biology
Edited by Ralph Becker

International Standard Book Number: 978-1-63239-719-5 (Hardback)

Printed in China.

Contents

Preface

This book has been a concerted effort by a group of academicians, researchers and scientists, who have contributed their research works for the realization of the book. This book has materialized in the wake of emerging advancements and innovations in this field. Therefore, the need of the hour was to compile all the required researches and disseminate the knowledge to a broad spectrum of people comprising of students, researchers and specialists of the field.

Cell biology is based on the concept that cell is the fundamental unit of life. It is a branch of science which gives detailed information about the structure, properties, metabolic processes, life cycle, signaling pathways and interactions of the cell with the surrounding environment. Cell biology is closely associated with diverse fields such as genetics, molecular biology, immunology, biochemistry, and developmental biology. Cell research work is radical for all biological sciences, for example, biomedical fields such as cancer and other diseases. This book is a detailed and comprehensive medium helping readers to better understand the fundamentals and advances made in this field. It will serve as a great resource for the scientists and professionals working in this domain. It will also prove beneficial for students as it comprises the latest researches in the field.

At the end of the preface, I would like to thank the authors for their brilliant chapters and the publisher for guiding us all-through the making of the book till its final stage. Also, I would like to thank my family for providing the support and encouragement throughout my academic career and research projects.

Editor

1

Are MSCs angiogenic cells? New insights on human nestin-positive bone marrow-derived multipotent cells

Simone Pacini* and Iacopo Petrini

Department of Clinical and Experimental Medicine, University of Pisa, Pisa, Italy

Edited by:
Anthony Scime, York University, Canada

Reviewed by:
Susumu Ikehara, Kansai Medical University, Japan
Robert Paul Rhoads, Virginia Polytechnic Institute and State University, USA

****Correspondence:***
Simone Pacini, Department of Clinical and Experimental Medicine, University of Pisa, Via Roma 56, 56126 Pisa, Italy
e-mail: simone.pacini@do.unipi.it

Recent investigations have made considerable progress in the understanding of tissue regeneration driven by mesenchymal stromal cells (MSCs). Data indicate the anatomical location of MSC as residing in the "perivascular" space of blood vessels dispersed across the whole body. This histological localization suggests that MSCs contribute to the formation of new blood vessels *in vivo*. Indeed, MSCs can release angiogenic factors and protease to facilitate blood vessel formation and *in vitro* are able to promote/support angiogenesis. However, the direct differentiation of MCSs into endothelial cells is still matter of debate. Most of the conflicting data might arise from the presence of multiple subtypes of cells with heterogeneous morpho functional features within the MSC cultures. According to this scenario, we hypothesize that the presence of the recently described Mesodermal Progenitor Cells (MPCs) within the MSCs cultures is responsible for their variable angiogenic potential. Indeed, MPCs are Nestin-positive CD31-positive cells exhibiting angiogenic potential that differentiate in MSC upon proper stimuli. The ISCT criteria do not account for the presence of MPC within MSC culture generating confusion in the interpretation of MSC angiogenic potential. In conclusion, the discovery of MPC gives new insight in defining MSC ancestors in human bone marrow, and indicates the *tunica intima* as a further, and previously overlooked, possible additional source of MSC.

Keywords: mesenchymal stromal cells, endothelial differentiation, angiogenesis, adult stem cells, bone marrow, nestin, neo-vascolarization, *in vivo* MSC

DISCOVER, ISOLATION, AND CHARACTERIZATION OF MSCs

In the late sixties, A. J. Friedenstein and coworkers first described the multipotent mesenchymal stromal cells (MSCs). When human bone marrow (BM) cells were cultured in plastic dishes colonies of adhered fribroblastoid cells proliferate and hematopoietic precursors progressively disappear. These cells have been named *Colony Forming Units of Fibroblastoid cells* (CFU-Fs) because of their ability to form large colonies on plastic surfaces. MSCs are able to differentiate into chondrocytes and osteoblasts, *in vitro* (Friedenstein et al., 1968), and *in vivo* (Friedenstein et al., 1974). By that time, T. M. Dexter and colleagues developed a culture system to study hematopoiesis and demonstrated that bone marrow hematopoietic stem cells (HSCs) were unable to adhere onto the culture flasks but necessitate an underlying layer of adherent cells that mimics the bone marrow stromal compartment (Dexter et al., 1977). After the demonstration that CFU-Fs originate from the bone marrow stroma, their name was changed in bone marrow stromal cells (Lanotte et al., 1981). In 1991, A. I. Caplan suggested the presence of a stem cell population in the adult BM able to differentiate into multiple mature cell lineages sharing a common precursor in the mesodermal layer of the embryo. Therefore, these cells were named "mesenchymal stem cells" (Caplan, 1991). Subsequently, the differentiation potential of MSCs into multiple mature lineages has been confirmed: these cells have a stable phenotype that can be easily expanded in culture and maintain the ability to differentiate into osteoblasts, chondrocytes, adipocytes, tenocytes, myocytes and stromal cells supporting hematopoiesis (Pittenger et al., 1999). MSCs became popular when K. Le Blanc et al. showed that MSCs are "invisible" to the immune system because they express only the class I Major Histocompatibility Complex (MHC-I) but not the class II and co-stimulatory molecules such as CD40, CD80, and CD86. This is relevant in the prospective of allogeneic transplantation with possible therapeutic applications (Le Blanc et al., 2003).

The isolation and the expansion of MSCs are easily feasible from the adult bone marrow in most cellular laboratories. Therefore, MSCs have been largely evaluated and employed in several premature clinical trials. However, a precise characterization of these cells is still missing as well as standardized protocols for their isolation and expansion. Consequently, results are controversial and the biology of MSCs is still unclear.

More recently, the International Society for Cellular Therapy (ISCT) has proposed minimal criteria to define MSCs (Dominici et al., 2006):

(1) MSCs are plastic-adherent cells in standard culture conditions,
(2) MSCs express the surface marker: CD105, CD73, and CD90 and do not to express the hematopoietic markers: CD45, CD34, CD14, and CD11b, and
(3) MSCs must be able to differentiate toward osteogenic, adipogenic and chondrogenic lineages when exposed to the proper conditions *in vitro*.

Finally, MSCs have been re-named "mesenchymal stromal cells" modifying the term "stem" into "stromal" in order to maintain the same acronym and avoid possible over-interpretation of their limited pluripotent potential (Horwitz et al., 2005).

TISSUE ORIGINS AND DISTRIBUTION OF MSCs

Cells matching the ISCT criteria can be isolated not only from the bone marrow but also from other adult and the fetal tissues (De Bari et al., 2001; Zuk et al., 2001; In't Anker et al., 2004; Seo et al., 2004). In these studies the culture condition adopted to establish BM-derived MSCs were applied to cells isolated from other tissues, in order to determine if MSCs reside also in different organs. In mice, L. da Silva Meirelles et al. demonstrated that long-term culture of MSCs could be established from a wide range of different adult tissues including fat, muscles, pancreas, vena cava, kidney glomerulus, aorta, brain and many others alongside bone marrow (da Silva Meirelles et al., 2006). Notably, all the cell populations that L. da Silva Meirelles established, independently from their origin, were long-term culture of adherent cells, with MSC phenotype and able to differentiate into mesenchymal cell lineages. These data suggest the presence of MSCs in virtually in all organs and tissues of the body. Three hypotheses try to explain the tissue distribution of MSCs:

(1) MSCs permanently reside in multiple tissues and organs,
(2) MSCs reside in only in particular tissues but can circulate in blood, and
(3) MSCs are circulating blood cells.

The presence of CFU-Fs in blood of adult mammals was shown in 2007 (He et al., 2007). However, the contamination with fragments of connective tissue could not be ruled out to justify the presence of MSCs in the collected sample. Disputes remain regarding the existence of circulating MSCs (Roufosse et al., 2004; Kuznetsov et al., 2007). L. da Silva Meirelles demonstrated the presence of MSCs within tissue cleaning the organs using intravascular perfusion before their collection. Nonetheless, the possibility that MSCs may circulate locally or systemically under non-physiological conditions, i.e., tissue injury, is not excluded. Although, the features of MSC from different organs are similar, mild differences in differentiation potential and surface markers have been reported. These differences have been related to the influence of a modified local environment (niche) present in different site of the body.

MSC can be isolated from the wall of blood vessels (Doherty et al., 1998; Bianco et al., 2001), and in 2007, B. Sacchetti et al. demonstrated a common phenotype for BM-derived CFU-Fs and Adventitial Reticular cells (ARCs). ARCs populate sinusoids and lay in close contact with the endothelium. Strong evidence indicate that the fibroblastoid colonies described by Friedenstein *in vitro* originate from the ARCs isolated *ex vivo* (Sacchetti et al., 2007). Stromal progenitors in human BM that reside in the sub-endothelial layer of sinusoids strongly express the melanoma-associated adhesion molecule (MCAM/CD146). A. Tormin et al. confirmed that CD271 is an *in vivo* marker of BM-derived MSCs (Quirici et al., 2002) and described a subset of CD271+ cells that express CD146. Cells expressing CD271 and CD146 are the ARCs present in the sub-endothelial layer of sinusoids. The remaining CD271+/CD146− cells maintain the MSC features but *in vivo* reside in the trabecular bone-lining endosteal niche (Tormin et al., 2011). Therefore, there are at least two cells able to generate MSCs from the bone marrow one in the perivascular (CD271+/CD146+) and one in the endosteal niche (CD271+/CD146−).

M. Corselli et al. isolated two distinct MSC progenitors from the stroma vascular fraction (SVF) of the adipose tissues (Corselli et al., 2012). CD34−CD146+ pericytes encircling capillaries and microvessels and CD34+CD146− adventitial cells surrounding larger arteries and veins. MSC-like cultures can be expanded from both these populations, suggesting a vascular origin for the MSCs of the adipose tissue, similarly to what observed in bone marrow. A hierarchical organization of cell differentiation has been proposed for the vascular progenitor of MSCs being the adventitial cells the precursors of pericytes. Indeed, under proper conditions, ARCs differentiate into pericytes *in vitro*. Later, M. Crisan et al., isolating the cells through CD146, demonstrated the perivascular origin of MSCs in multiple organs (Crisan et al., 2008).

A new hypothesis that MSCs are localized *in vivo* in "perivascular" spaces that extend through the whole post-natal organism has been proposed. While this latest hypothesis is gaining consensus among researchers, the term "perivascular" is somehow ambiguous because it include the proximity of vessel and the wall itself. More precisely, two intra-vessel wall compartments, the *adventitia* and *sub-endothelium*, have been indicated as possible locations for these two MSC progenitors. The relationship between these two progenitors remains obscure, even if they have been largely characterized. Their histological localization, suggests a role of MSC in blood vessel formation *in vivo*. MSCs can directly differentiate into vascular cells (endothelial cells and smooth muscle cells) and/or as supporting vascular (re)-generation in response to the paracrine secretion of stimulating factors (Lin and Lue, 2013).

CONTROVERSIES ABOUT THE ANGIOGENIC POTENTIAL OF MSCs

One of the most interesting debates regarding MSCs concerns their angiogenic potential. Due to the possible role of MSC in therapeutic (re)-vascularization, an increasing number of studies *in vitro* and *in vivo* have been performed (reviewed in Vittorio et al., 2013).

The formation of new blood vessel can be divided in:

(1) Vasculogenesis: *de novo* formation of blood vessels from the endothelial precursors or angioblasts,
(2) Angiogenesis: includes sprouting of existing vessels and intussusceptive angiogenesis, and
(3) Arteriogenesis: remodels a pre-existing collateral circulation (Makanya et al., 2009; van Royen et al., 2009; Melero-Martin et al., 2010; Carmeliet and Jain, 2011; Potente et al., 2011).

Although MSCs support these processes through the release of angiogenic factors and protease (reviewed in Watt et al., 2013), the relevance of their differentiation into endothelial lineages is still debated.

The demonstration of MSC commitment toward endothelial lineage is often limited to the detection of the upregulation of typical EC surface molecules including CD31, CD34, VEGF receptors (VEGFR1, VEGFR2) and von Willebrand factor (vWF). As phenotype modification is insufficient to demonstrate differentiation, additional functional tests are often performed. These tests include *in vitro* tube formation on Matrigel® and uptake of acetylated-low density lipoproteins (Ac-LDL). Nonetheless, it might be inaccurate describing these differentiated MSCs as fully mature and functional ECs basing on these *in vitro* assays. For example, Ac-LDL uptake has been described also in macrophages and pericytes (Voyta et al., 1984).

A large effort was spent for the optimization of protocols able to induce endothelial differentiation of MSCs. VEGF stimulates the differentiation of MSCs into ECs. In 2004, J. Oswald et al. demonstrated that confluent human BM-derived MSCs cultured in 2% fetal calf serum (FCS) and 50 ng/ml VEGF for a week, displayed upregulation of endothelial surface markers including VEGFR1, VEGFR2, VE-Cadherin, VCAM-1, and vWF. Moreover, when incubated on Matrigel® *in vitro*, MSCs formed characteristic capillary-like structures (CLS) (Oswald et al., 2004). Similarly, M. Jazayeri et al. cultured human BM-derived MSCs in medium supplemented with 5% FCS, IGF and VEGF and detected CD31, vWF, Tie2, VCAM1, and VE-cadherin on the cell surface. In addiction, applying electron microscopy Authors showed the presence of typical EC morphological features including Weibel-Palade bodies, tight junctions and caveolae (Jazayeri et al., 2008). Similar results were also achieved using "endothelial growth medium-2" (EGM-2, which contains VEGF, EGF, FGF-2, IGF-1, hydrocortisone, heparin, ascorbic acid and 2% FCS) (Liu et al., 2007) and MSCs isolated from the adipose tissue (Cao et al., 2005; Fischer et al., 2009). Conversely, V.D. Roobrouck et al. reported that VEGF treatment of human BM-derived MSCs significantly increased mRNA expression of CD34, VEGFR1, and VEGFR2, but not of Tie-2 and vWF or CD31 that was even decreased. These MSCs also failed forming CLS in Matrigel® assays, (Roobrouck et al., 2011). In parallel, W. Fan et al. demonstrated that human BM-derived MSCs cultured in the presence of different concentrations of VEGF did not show increase in CD31, vWF or VEGFR2 expression (Fan et al., 2011).

In vivo G. V. Silva et al. demonstrated that MSCs applied in a region of myocardial ischemia can differentiate into smooth muscle cells and endothelial cells leading to increased vessel density and an improvement the cardiac function, in a canine model (Silva et al., 2005). Nonetheless, there is a consolidating concept that the angiogenic effect of MSCs is predominantly caused by their paracrine actions rather than their EC trans-differentiation potential. A. Al-Khaldi et al. showed that, in the murine Matrigel® plug assay, more than 99% of the new-formed blood vessels originated from host-derived EC, while a small portion of injected BM-derived MSCs were found in the close proximity of- or within blood vessels (Al-Khaldi et al., 2003). Moreover, the observation that MSCs *in vitro* committed through endothelial lineages were not superior to "*naïve*" MSCs in stimulating *in vivo* angiogenesis, may underline the relevance of the secretion of pro-angiogenic factors that is also sustained by uncommitted MSCs (Liu et al., 2007; Fan et al., 2011). Thus, the up-regulation of endothelial marker, under specific culture conditions, could represent an *in vitro* artifact and not a real differentiation into functional endothelial cells. BM-derived MSCs seems to be an important regulator of neo-vascularization by the secretion of pro-angiogenic factors as well as by differentiating into functional pericytes able to stabilize the new-formed vasculature (Au et al., 2008), rather than a source of endothelial progenitor cell.

Most of the controversial data about endothelial differentiation of MSCs need to be discussed as consequence of the sub-optimal protocols of differentiation (Janeczek Portalska et al., 2012). Moreover, a critical issue is the heterogeneity of the primary MSC cultures used to generate endothelial progenitors.

HETEROGENEITY OF CULTURE EXPANDED MSCs

The anatomic localization and the physiological function of MSCs are not clearly characterized. MSCs are commonly isolated from long-term cultures; therefore, it remains difficult to determine the primary cells of origin. The loose ISCT criteria hamper the identification of unique precursors of MSCs. Indeed, several types of primary cells with different features can fulfill the definition of MSCs *in vitro*. Being the definition permissive, the presence of a unique common precursor for cells with MSC features cannot be hypothesized. In BM, MSCs can originate from both perivascular and endosteal progenitors, therefore, it is difficult to distinguish if there is a unique common precursor or if the loose ISCT definition is unable to identifies two different progenitor populations. However the clinical applications of MSCs are only partially limited by the incomplete characterization of the progenitor cells.

The heterogeneity of MSC cultures, defined according to ISCT criteria, is becoming evident in more recent articles and brings into question the utility of these ambiguous criteria. From the beginning, different terms have described the morphology of plastic-adherent cells: fibroblastoid (Werts et al., 1980), giant fat cells and blanket cells (Allen and Dexter, 1983), spindle shaped flattened cells (Kuznetsov et al., 1997) and very small round cells (Colter et al., 2001). Thus, mesenchymal cell morphology seems to be highly dependent on culture conditions: supplements, seeding density, number of passages and culture time (Wagner and Ho, 2007; Barachini et al., 2009). It is still unclear if there is any relation between these different morphology and cell functions.

There is not a consensus on the surface markers of MSCs, aside from the unspecific CD105, CD90, CD44, and CD73, because different laboratories use different sets of antigens. Therefore, the differentiation into mature cells with a mesengenic ancestor seems to be the more reliable and stringent criteria to define MSCs. However, the differentiation potential of MSC is variable. This variability is observed between different donors (Phinney et al., 1999) and also within different colonies obtained from the same subject (Russell et al., 2010). Indeed, colonies obtained from the same individual could be characterized as mono-, bi- or tri-potent on the basis of their ability to differentiate into, respectively one, two or three types of tissue: osteogenic, chondrogenic and adipogenic lineages. Moreover, it has been clearly demonstrated that repeated passages progressively reduce the multi-lineage differentiation ability, introducing a further element of complexity (Muraglia et al., 2000). It is possible to hypothesize that the angiogenic potential of MSC is subject to a similar variability that is influenced by the same factors.

MSCs are heterogeneous not only among different colonies but also within the same colony (Digirolamo et al., 1999). Cells show variable differentiation potential in relation to their topographic localization inside the colony. Cells from the center and the margins of the colony differ for shape, differentiation potential and surface markers (Ylöstalo et al., 2008; Sengers et al., 2010). Therefore, the term "multipotent mesenchymal stromal cells" does not identify a population of cells with uniform features and unambiguous potential but refers to a highly heterogeneous population that is dramatically affected by donors characteristics (Russell et al., 2013), isolation methods (Wagner and Ho, 2007; Barachini et al., 2009), culture conditions (Bieback et al., 2009).

Several possible mechanisms may explain the basis of the MSC heterogeneity, beside the already described variability introduced by *ex vivo* procedures (Pevsner-Fischer et al., 2011). Hypotheses on the origin of this variability include: stochastic events, occurring during expansion and differentiation and a possible *in vivo* heterogeneity of the isolated cell populations. In this latter hypothesis, specific culture conditions select, or simply promote, particular subpopulations of MSCs giving reason of the observed heterogeneity and the morpho-functional variability. According to this scenario, numerous multipotent cell populations can be described in bone marrow, some of them able to differentiate into lineage from the three germ layers: endoderm, mesoderm and ectoderm (triploblastic differentiation). However, the isolation and successive characterization of these cells is strictly dependent on the application of specific culture conditions. For example, "marrow-isolated adult multipotent inducible (MIAMI)" cells can differentiate into, neural and pancreatic-like cells in addition to skeletal tissue lineage (D'Ippolito et al., 2004). The isolation and expansion of MIAMI cells require specific culture conditions with low oxygen tension.

Recently a unique multipotent sub-population in adult human BM-derived MSCs has been isolated using fluorescent activated cell sorting (FACS) for stage-specific embryonic antigen 3 (SSEA-3). BM-derived MSCs show poluripotency-differentiation properties (Kuroda et al., 2013). Interestingly, stress conditions could enrich the expression of SSEA-3 in cultured MSCs (Kuroda et al., 2010). Y. Kuroda et al. demonstrated that long-term trypsin incubation could increase the recovery of cell clusters containing pluripotency-associated markers and renamed these cells as "multi-lineage differentiating stress-enduring" (MUSE) cells.

Thus, if mild modifications in the culture conditions, or in the culture procedures, can induce/preserve an embryonic-like differentiation potential in BM-derived cells, it is reasonable to suppose that angiogenic potential behaves similarly and is significantly affected by manipulation *in vitro*.

MESODERMAL PROGENITOR CELLS (MPCs) IN ADULT HUMAN BONE MARROW

In 2007, we attempted to optimize MSC culture conditions for clinical application selecting media without supplements of animal origin. When medium was supplemented with autologous serum instead of that of bovine origin, a small population of cells with distinct shape was noticed. These cells presented rounded, fried-egg shape, instead of the usual spindle morphology of MSCs, were highly refractive and remain firmly attached to the plastic during trypsin digestion

In 2008, by replacing fetal bovine serum (FBS) with pooled human AB serum (PhABS) in the culture medium of human BM cells, we were able to characterize this new population of adherent cells (Petrini et al., 2009). These cells are quiescent: Ki-67 negative; with long telomeres and express the pluripotency-associated transcription factors Oct-4 and Nanog instead of RUNX2 and Sox9 typical for MSC-phenotype (Pacini et al., 2010). Phenotypically, these cells share the expression of CD105 with MSCs but lacked expression of CD73, CD90, CD166, CD271 and those other markers typical of the mesenchymal phenotype such as MSCA-1. Interestingly, this cell population rapidly produces mesenchymal offspring when supplemented with FBS or human cord blood serum. Thus, this novel population of cells, isolated from the BM, has *in vitro* characteristics of a progenitor of the mesengenic lineage and therefore has been named "*Mesodermal Progenitor Cells*" (MPCs).

ISOLATION OF MPCs FROM HUMAN BM SAMPLES

Method for the isolation of MPCs from BM samples is feasible, inexpensive and based on selective culture conditions (Trombi et al., 2009). Initially, MPCs were co-isolated together with MSCs applying media supplemented with autologous serum or PhABS, in culture. It became evident that MPCs have different adhesion properties compared to MSCs. In fact, applying standard trypsin-based cell detaching protocols, MSCs were entirely harvested while most of the MPCs remain firmly attached to the plastic surface and required different proteases' solution (TrypLE Select® from LifeTechnologies) to be detached. Therefore, plastic features and coating of the culture surfaces influence the proportion of MPCs and MSCs in the primary cultures. We firstly tested not gas-treated hydrophobic plastics, usually applied for cultures in suspension, and surprisingly MPCs were able to attached on that surface also. Conversely, the hydrophobic conditions resulted not permissive for MSC. Thus, a selective culture could be established using these conditions allowing the recovery of MPCs with a purity of more than 95%. We also noticed that a higher yield of MPC recovery was achieved using higher seeding densities, than that usually applied for MSC isolation. In summary, PhABS supplementation and high density seeding on hydrophobic plastics were the selective culture conditions necessary for MPCs isolation from BM-MNCs. This method has been consolidated and it is highly reproducible allowing the quality screening of the MPC preparations before their employment in the different studies.

The mechanisms behind the difference, in MPC recovery, between culturing cells in FBS or PhABS-containing medium are still unknown. Nonetheless, we demonstrated that the addition of PhABS, even in small percentages, to FBS primary cultures allows MPC isolation, suggesting the presence of undetected agents able to induce MPCs in the human serum (Trombi et al., 2009). Conversely, when FBS is added to cultures grown in PhABS, cells differentiate into MSCs without any significant reduction in the number of MPCs indicating a possible semi-conservative proliferation of MPCs. What characterized FBS against PhABS is the different origin in terms of species (bovine instead of human) and stage of body development (fetal against adult). The differentiation of MPCs into MSCs is induced replacing PhABS with FBS (Petrini et al., 2009), or using human cord blood-derived serum (unpublished data). These preliminary results suggest that media supplementation with fetal sera represent the culture condition for the mesengenic induction of MPCs, independently from the adopted species.

Mesengenic potential of MPCs

Later, we have been able to demonstrate a hierarchical multi-step model of mesenchymal differentiation with at least three different populations of multi-potent cells (Fazzi et al., 2011). Indeed, MPCs can generate exponentially growing MSC cultures after 2 weeks of stimulation with differentiating conditions. The differentiation proceeds through the commitment into an intermediate cell population; we named *early* MSCs. Timing of MPC mesengenic differentiation was definitively clarified and specific morphologies, phenotypes and growing features of the three protagonists described (**Figure 1A**). Studying Wnt signaling activation during MPC differentiation, we showed that non-canonical Wnt5/Calmodulin pathway was involved in the commitment of MPCs into *early* MSCs and demonstrated that Calmidazolium Chloride, a Calmodulin inhibitor, was able to interfere with the differentiation only at this initial step while has no effect on passage from *early* to *late* MSCs.

Angiogenic potential of bone marrow-derived MPCs

From the beginning it has been clear that MPCs have angiogenic potential because they form capillary-like structures (CLS) after a multiple steps of differentiation (Petrini et al., 2009; Trombi et al., 2009; Pacini et al., 2010). Interestingly, the inhibition of the Wnt5/Calmodulin signaling pathway has no effects on the MPC differentiation toward endothelial lineage suggesting that Wnt signaling pathway activation finely regulates MPC fate. On the contrary of mesengenic, all the passage of angiogienic differentiation have not been described to date due to the lack of specific culture protocols. Indeed, to partially induce endothelial differentiation we applied protocols optimized for endothelial progenitor cells (EPCs) with mild modifications (Hill et al., 2003). This was sufficient to demonstrate the angiogenic potential of MPCs and suggested that exist a BM-derived endothelial progenitor cell distinct from EPCs.

The high expression of CD31 and Nestin in MPCs suggest the existence of a primitive progenitor for the endothelial lineages (Petrini et al., 2009; Trombi et al., 2009; Pacini et al., 2010). Nestin is a class VI intermediate filament protein originally described as a marker of neural stem cells that is expressed during the development of central nervous system (CNS) (Lendahl et al., 1990). Although Nestin expression is down regulated during the differentiation into neurons or glial cells (Dahlstrand et al., 1995), it can be detected in adult neural progenitor cells (Reynolds et al., 1992; Morshead et al., 1994) and in some CNS tumors (Tohyama et al., 1992). Frequently, Nestin is not expressed by the cancer cells themselves but can be observed in the endothelial cells of the tumor regardless of malignancy grade or its histotype (Sugawara et al., 2002). This suggests that Nestin can be a marker of proliferating tumor endothelial cells and not only of neuroepithelial elements. Therefore, Nestin expression correlates with angiogenesis because it is expressed in proliferating vascular endothelial cells of the tumor (Kim et al., 2002; Teranishi et al., 2007; Gravdal et al., 2009; Eaton et al., 2010). K. Sugawara et al. demonstrated a high expression of Nestin in bovine aortic proliferating endothelial cells in static culture. This expression rapidly decreases under conditions of laminar shear stress flow, suggesting that Nestin expression is typical of early proliferating endothelial precursors but is loss in mature endothelial cells of normal tissues. In the recent years, the emerging concept of Nestin as a novel early angiogenic marker is gaining consensus in normal ad tumor angiogenesis (reviewed in Matsuda, 2013).

MPCs express CD31/PECAM (Pacini et al., 2013) but rapidly loose these markers during differentiation contextually with the lost of Nestin (**Figure 1A**). Preliminary data show that MPCs can make "sprouting" when directly seeded in Matrigel® 3D-cultures but are not able to efficiently form CLS without a step of pre-differentiation. Therefore, MPCs could represent a very staminal progenitor with angiogenic potential more immature than what is reputed to date.

MPCs DO NOT SHOW FEATURES OF PERICYTES OR ADVENTITIAL PROGENITORS

Pericytes surround blood capillaries, precapillary arterioles, post-capillary venules and collecting venules (Sims, 1986; Allt and Lawrenson, 2001); where they can be identify by the expression of CD146 together with less specific markers such as α-smooth muscle actin (α-SMA), desmin, NG-2, platelet-derived growth factor receptor (PDGFR)-α, aminopeptidase A and N, RGS5, and the promotertrap transgene *XlacZ4* (Gerhardt and Betsholtz, 2003).

Pericytes and Adventitial Progenitor Cells (APCs) belong to the same cell lineage, according to several authors (Tormin et al., 2011; Corselli et al., 2012). Pericytes and APCs differ in their position in the vessel wall morphology, and some surface markers. However, this distinction is not absolute because exist a continuum between the phenotype of the classical APCs and the typical pericytes when they are distributed along small vessels such as arteriole, capillary, and venule. It has been suggested that pericytes may reside under the endothelium even in large vessels and support the endothelial removal and repair after injuries. From this point of view, these cells can be considered a reservoir of MSC-like undifferentiated cells (da Silva Meirelles et al., 2006). Conversely, Corselli's data demonstrated that APCs differentiate into pericytes *in vitro*, suggesting a "centripetal" (from the *adventitia* toward sub-endothelial layer) relationship between these cells (Corselli et al., 2012). Both the "centrifugal" and the "centripetal" model assumed that the *intima* would not be involved in the generation of MSC-like cells.

Interestingly, our MPCs express the early angiogenic markers Nestin and CD31/PECAM, suggesting their plausible location in the *tunica intima*. Although the localization of MPCs in the lumen-facing wall of vessels has been not definitively demonstrated, preliminary histological evaluations revealed the expression of Nestin in the sinusoids, arterioles/venules and larger

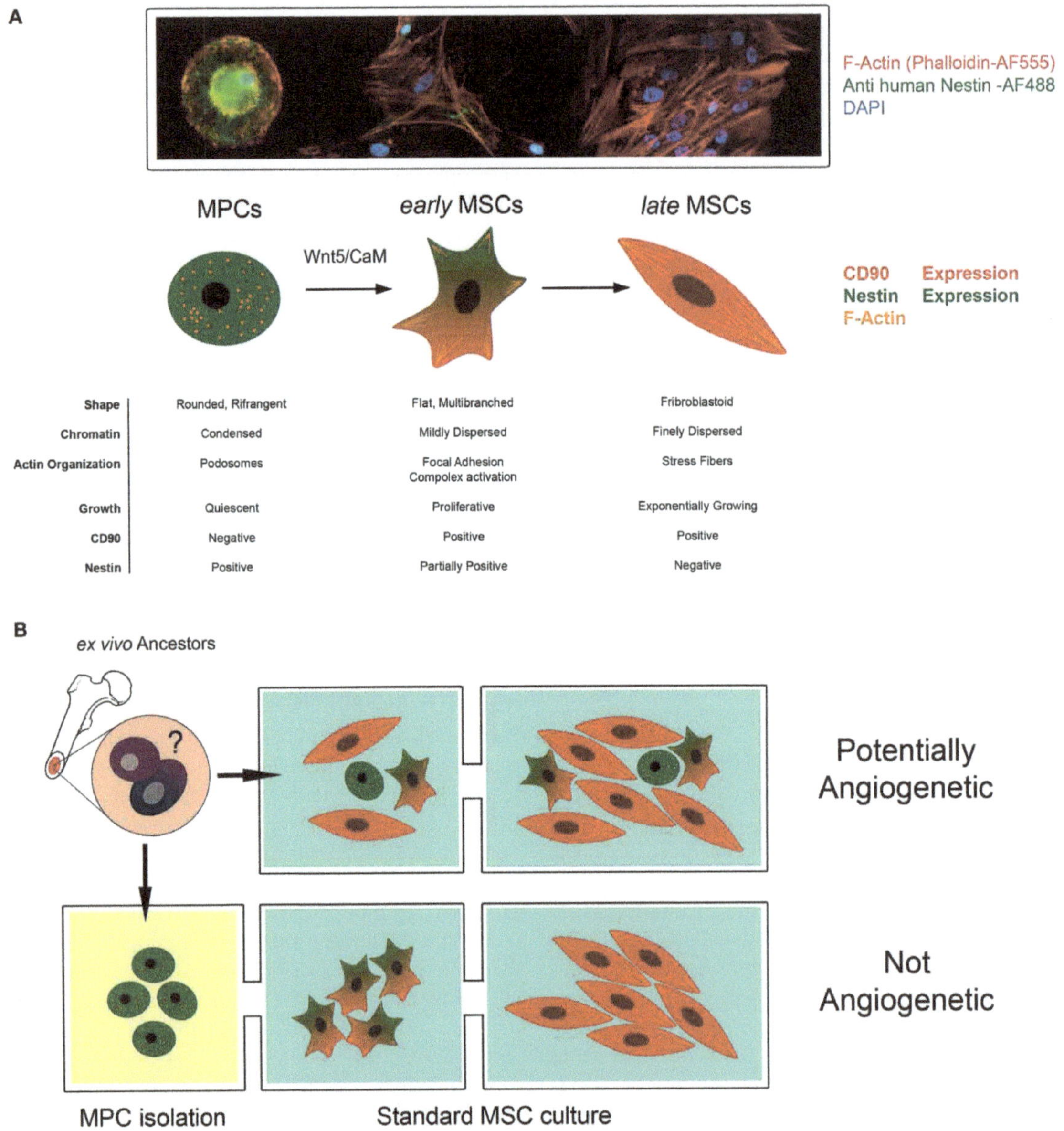

FIGURE 1 | The angiogenic potential of MSC cultures is controversial. (A) Mesodermal Progenitor Cells (MPCs) are CD90-, Nestin+ progenitors of MSCs. These cells remain in a quiescent state presenting a typical fried egg-shape, condensed chromatin and podosomal structures. When mesengenic differentiation is induced, MPCs differentiate into "*early* MSCs." *Early* MSCs slowly proliferate, express Nestin and CD90 (a MSC marker) and modify their shape because of Actin re-organization in focal adhesion complexes. Under persistent stimulation toward mesengenic differentiation, "*early* MSC" become "*late* MSC" showing exponentially growth, fibroblastoid shape and the ability to differentiate into skeletal tissues (fat, bone and cartilage). **(B)** According to this hierarchical model, MPCs can be considered a putative progenitor of the mesenchymal lineage *in vivo*, being present in the bone marrow mononuclear fraction. When supplemented with media containing FBS, these cells rapidly differentiate toward the mesengenic lineage and form asynchronous and heterogeneous cultures fulfilling the ISCT MSC criteria of definition. The undetected and unpredictable presence of sub-populations of MPCs, in culture could explain the variable angiogenic potential described for MSC in the literature. Conversely, the isolation of MPC and their subsequent differentiation using optimized protocol can allow the generation of synchronized and homogeneous mesenchymal stromal cells with a reproducible angiogenic potential.

vessels of bone marrow (**Figure 2**). A higher Nestin expression was observed in sinusoids' endothelium, whereas a lower Nestin expression was localized in only few cells located in the sub-endothelium and *adventitia*.

MPC do not express CD146, a specific markers of pericytes, as well as CD271. Therefore, it is unlikely MPCs reside in the *adventitia* or in the sub-endothelium (Petrini et al., 2009; Trombi et al., 2009; Pacini et al., 2010). Thus, any model of perivascular localization of MSC-like cells will be incomplete without the inclusion of the *tunica intima* among the possible sources of these cells.

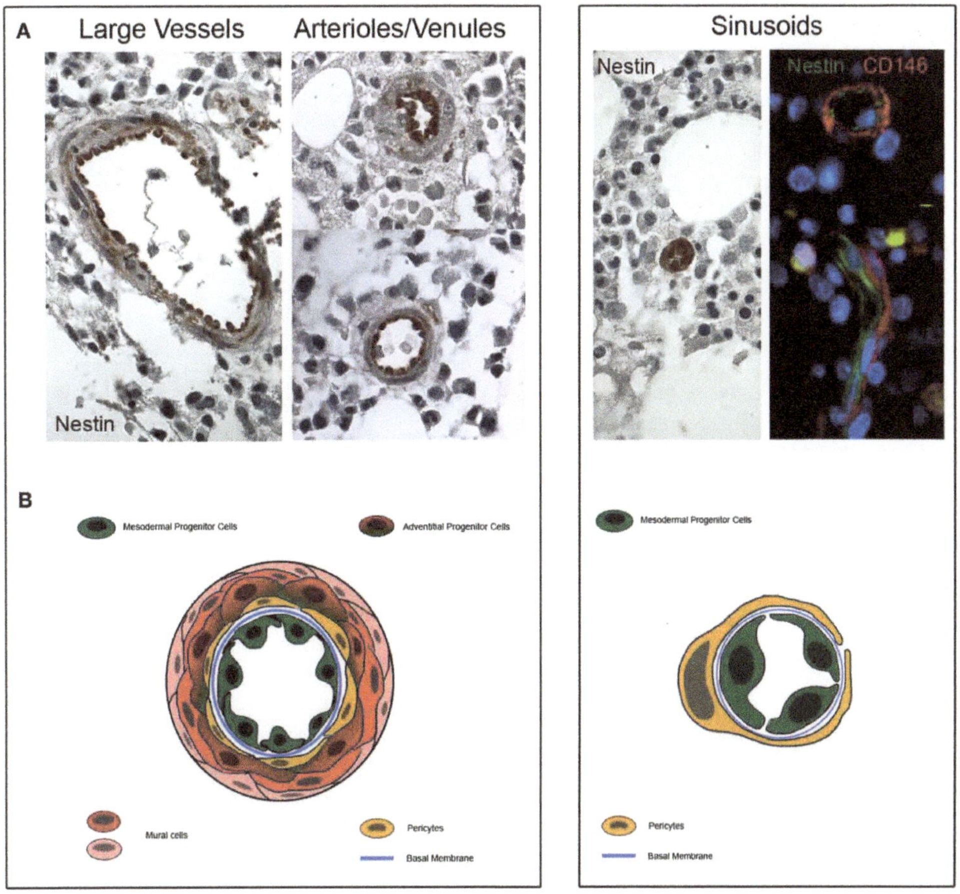

FIGURE 2 | Nestin expression in human bone marrow biopsies. (A) Consistent expression of Nestin is detectable in the *tunica intima* of large vessels, arterioles and venules (dark brown color). Few cells of the sub-endothelium and *adventitia* show a weakly positive Nestin staining (light brownish color). Furthermore, Nestin is highly expressed in the sinusoids and two colors immunofluorescence reveals that Nestin is positive in lumen-facing cells surrounded by CD146+/Nestin− pericytes. **(B)** Schematic representation of possible tissue distribution of MPC (represented in green) and other bone marrow vessels-residing cells.

NEW HYPOTHESIS ON ANGIOGENIC PROPERTIES OF BM-DERIVED MULTIPOTENT STROMAL CELLS

Recently, we have proposed that the yield of MPCs, co-isolated in the standard BM-MSC preparations, is influenced from the host, the batches of the serum and from the density of cell seeding (Petrini et al., 2009). The inter-population variability introduced by different donors and cell isolation protocols, affect the yield of MPCs, *early* MSCs or *late* MSCs. According to the hierarchical model, MPCs are progenitors of the mesenchymal lineage and account for 1–3% of mononuclear cells of the bone marrow (BM-MNCs). When BM-MNCs are seeded in standard FBS-containing media MPCs rapidly differentiate into mesengenic lineages forming the typical MSC culture in few days.

The variable angiogenic potential described in the literature of MSCs is probably related to the heterogeneous composition of the cultures expanded from the bone marrow that includes sub-populations of MPCs and MSCs when defined according to ISCT criteria (**Figure 1B**). Indeed, the expanded or exponentially growing MSCs are Nestin-negative and *de facto* coincide with our *late* MSCs. Late MSCs do not retain any angiogenic potential because these cells are already committed toward other mesengenic lineage (**Figure 1B**). Most of the reports do not apply specific protocols for MPC's isolation; thus, these MSC cultures represent an uncontrolled heterogeneity of multipotent cells with an unpredictable angiogenic potential. Because MPCs show resistance to trypsin digestion, these cells are expected to be lost during subsequent passages reducing progressively the angiogenic potential of the sub-cultures. Not surprisingly, the most successful endothelial differentiation protocols have been obtained from early passages of MSC's cultures (Oswald et al., 2004; Fan et al., 2011; Janeczek Portalska et al., 2012). Conversely, protocols specific for endothelial differentiation may commit a pure population of MPCs into homogeneous clones of MSCs. The clinical utility of BM-derived cells will be improved by a more precise phenotypization able to distinguish MPCs from *early* and *late* MSCs.

Recently, Frenette's group demonstrated in the hematopoietic niche the presence of Nestin-positive bone marrow cells with mesengenic potential of differentiation. Using *Nes-Gfp* transgenic mice, authors identified a highly selected fraction of

Table 1 | *Bona fide* multipotent progenitor cells of bone marrow and adipose tissues.

Acronym		MSCs	MPCs	Pericytes	ARCs	APCs	TBLCs
Phenotype							
	CD90	+	−	+	+	+	+
	CD105	+	+	+	+	+	+
	CD73	+	−				
	Nestin	±	+		±		
	CD31	−	+	−	−	−	−
	CD146	±	−	+	−	−	±
	CD271	+	−	±			+
	CD34	−	−	−	+	+	−
Distribution		Bone marrow (Pittenger et al., 1999), adipose tissue (Zuk et al., 2001) and many others tissue cultures (da Silva Meirelles et al., 2006)	Bone marrow cultures (Petrini et al., 2009)	Sub-endothelium of vessels and microvessels in bone marrow (Tormin et al., 2011) and adipose tissue (Corselli et al., 2012)	Adventitia of vessels and microvessels in bone marrow (Sacchetti et al., 2007)	Adventitia of vessel and microvessel in adipose tissue (Tormin et al., 2011; Corselli et al., 2012)	Adjacent to trabecular bone (Tormin et al., 2011)
Role in angiogenesis		Controversial	Sprouting and direct differentiation into ECs	Stabilization of new formed vasculature	Mural cells and endothelium support	Mural cells and endothelium support	Not involved

MSCs able to form the HSC niche (Méndez-Ferrer et al., 2010). *In vivo*, both in human and mouse, these cells are positive for PDGFRα, CD51 and Nestin expression and negative for CD45 CD31 CD235a (Ter119- in mice). Although the two types of BM stem cell can form a single niche, only a small fraction of Nestin+ cells exhibits MSC activity when tested in mesensphere or CFU-F assays (Méndez-Ferrer et al., 2010). Also for Nestin-positive cells, the protocols of isolation and expansion dramatically affect the composition of cell populations in culture; thus harsh isolation protocols will be needed to avoid the development of heterogeneous populations and allow the characterization and functional definition of these interesting cells. Further the co-expression of PDGFRα and CD51 identifies a subset (about 60%) of Nestin-positive cells with an enriched potential to form HSC niches and to perform mesenchymal differentiation (Pinho et al., 2013). Limited data are available regarding the angiogenic potential of Nestin-positive cells. Because PDGFRα+ CD51+ hematopoietic-supporting stromal cells do not express CD45 (a hematopoietic marker) and CD31 (an endothelial marker), it can be hypothesized the presence of an additional population of Nestin-positive CD31-positive cells not described in the previous analysis.

CONCLUSIONS

Although MSCs have been largely studied for their interesting applications in clinical trials, these cells have not been fully characterized because of the lack of standardized protocols between different laboratories. Controversies remain, and several aspects of MSC biology are still unclear. The heterogeneity and morpho-functional variability of MSC cell preparations could explain most of the conflicting data in the literature. Together with the effects of culture conditions that can indeed select, or simply promote, particular subpopulations of MSC-like cells (**Table 1**), the described possible multiple origins of MSCs contribute to the confusing interpretation of the experimental data. More stringent phenotypization criteria may help to prevent this issue. Recently, perivascular localization of MSC precursors may explain their presence in a wide range of tissues and organs and suggests some angiogenic potential.

Therefore, we hypothesize that the presence of the recently described Mesodermal Progenitor Cells (MPCs) could be responsible for the controversial data regarding angiogenic potential of MSC cultures. Although these cells can be co-isolated with MSC culture, different protocols may determine a different yield of MPCs.

The discovery of Nestin-positive CD31-positive MPCs supports their role as MSC ancestors in human bone marrow and indicates the *tunica intima* as a possible source of MSCs.

Further studies are needed to deeply investigate the MPC biology and confirm their anatomical home in human bone marrow. Nonetheless, the identification of MPCs suggests the opportunity of a revision of the MSC definition in order to achieve their expected clinical utility (Keating, 2012).

Finally, MPCs represent a valuable cell population for the proof of new concepts in tissue engineering, where the neo-vascularization plays a crucial role in the establishment of successful therapies. Future studies evaluating MPC-based therapies will take advantage of their mesengenic and angiogenic potential in order to regenerate skeletal tissues and support their growth with a newly formed vasculature.

AUTHOR CONTRIBUTIONS

Simone Pacini: Literature review and article draft. Iacopo Petrini: revision.

REFERENCES

Al-Khaldi, A., Eliopoulos, N., Martineau, D., Lejeune, L., Lachapelle, K., and Galipeau, J. (2003). Postnatal bone marrow stromal cells elicit a potent VEGF-dependent neoangiogenic response *in vivo*. *Gene Ther.* 10, 621–629. doi: 10.1038/sj.gt.3301934

Allen, T. D., and Dexter, T. M. (1983). Long term bone marrow cultures: an ultrastructural review. *Scan Electron Microsc.* (Pt 4), 1851–1866.

Allt, G., and Lawrenson, J. G. (2001). Pericytes: cell biology and pathology. *Cells Tissues Organs.* 169, 1–11. doi: 10.1159/000047855

Au, P., Tam, J., Fukumura, D., and Jain, R. K. (2008). Bone marrow-derived mesenchymal stem cells facilitate engineering of long-lasting functional vasculature. *Blood* 111, 4551–4558. doi: 10.1182/blood-2007-10-118273

Barachini, S., Trombi, L., Danti, S., D'Alessandro, D., Battolla, B., Legitimo, A., et al. (2009). Morpho-functional characterization of human mesenchymal stem cells from umbilical cord blood for potential uses in regenerative medicine. *Stem Cells Dev.* 18, 293–305. doi: 10.1089/scd.2008.0017

Bianco, P., Riminucci, M., Gronthos, S., and Robey, P. G. (2001). Bone marrow stromal stem cells: nature, biology, and potential applications. *Stem Cells* 19, 180–192. doi: 10.1634/stemcells.19-3-180

Bieback, K., Hecker, A., Kocaömer, A., Lannert, H., Schallmoser, K., Strunk, D., et al. (2009). Human alternatives to fetal bovine serum for the expansion of mesenchymal stromal cells from bone marrow. *Stem Cells* 27, 2331–2341. doi: 10.1002/stem.139

Cao, Y., Sun, Z., Liao, L., Meng, Y., Han, Q., and Zhao, R. C. (2005). Human adipose tissue derived stem cells differentiate into endothelial cells *in vitro* and improve postnatal neovascularization *in vivo*. *Biochem. Biophys. Res. Commun.* 332, 370–379. doi: 10.1016/j.bbrc.2005.04.135

Caplan, A. I. (1991). Mesenchymal stem cells. *J. Orthop. Res.* 9, 641–650. doi: 10.1002/jor.1100090504

Carmeliet, P., and Jain, R. K. (2011). Molecular mechanisms and clinical applications of angiogenesis. *Nature* 473, 298–307. doi: 10.1038/nature10144

Colter, D. C., Sekiya, I., and Prockop, D. J. (2001). Identification of a subpopulation of rapidly self-renewing and multipotential adult stem cells in colonies of human marrow stromal cells. *Proc. Natl. Acad. Sci. U.S.A.* 98, 7841–7845. doi: 10.1073/pnas.141221698

Corselli, M., Chen, C. W., Sun, B., Yap, S., Rubin, J. P., and Péault, B. (2012). The tunica adventitia of human arteries and veins as a source of mesenchymal stem cells. *Stem Cells Dev.* 21, 1299–1308. doi: 10.1089/scd.2011.0200

Crisan, M., Casteilla, L., Lehr, L., Carmona, M., Paoloni-Giacobino, A., Yap, S., et al. (2008). A reservoir of brown adipocyte progenitors in human skeletal muscle. *Stem Cells* 26, 2425–2433. doi: 10.1634/stemcells.

D'Ippolito, G., Diabira, S., Howard, G. A., Menei, P., Roos, B. A., and Schiller, P. C. (2004). Marrow-isolated adult multilineage inducible (MIAMI) cells, a unique population of postnatal young and old human cells with extensive expansion and differentiation potential. *J. Cell Sci.* 117, 2971–2981. doi: 10.1242/jcs.01103

Dahlstrand, J., Lardelli, M., and Lendahl, U. (1995). Nestin mRNA expression correlates with the central nervous system progenitor cell state in many, but not all, regions of developing central nervous system. *Brain Res. Dev. Brain Res.* 84, 109–129. doi: 10.1016/0165-3806(94)00162-S

da Silva Meirelles, L., Chagastelles, P. C., and Nardi, N. B. (2006). Mesenchymal stem cells reside in virtually all post-natal organs and tissues. *J. Cell Sci.* 119, 2204–2211. doi: 10.1242/jcs.02932

De Bari, C., Dell'Accio, F., Tylzanowski, P., and Luyten, F. P. (2001). Multipotent mesenchymal stem cells from adult human synovial membrane. *Arthritis Rheum.* 44, 1928–1942. doi: 10.1002/1529-0131(200108)44:8<1928::AID-ART331>3.0.CO;2-P

Dexter, T. M., Allen, T. D., and Lajtha, L. G. (1977). Conditions controlling the proliferation of haemopoietic stem cells *in vitro*. *J. Cell. Physiol.* 91, 335–344. doi: 10.1002/jcp.1040910303

Digirolamo, C. M., Stokes, D., Colter, D., Phinney, D. G., Class, R., and Prockop, D. J. (1999). Propagation and senescence of human marrow stromal cells in culture: a simple colony-forming assay identifies samples with the greatest potential to propagate and differentiate. *Br. J. Haematol.* 107, 275–281. doi: 10.1046/j.1365-2141.1999.01715.x

Doherty, M. J., Ashton, B. A., Walsh, S., Beresford, J. N., Grant, M. E., and Canfield, A. E. (1998). Vascular pericytes express osteogenic potential *in vitro* and *in vivo*. *J. Bone Miner. Res.* 13, 828–838. doi: 10.1359/jbmr.1998.13.5.828

Dominici, M., Le Blanc, K., Mueller, I., Slaper-Cortenbach, I., Marini, F., Krause, D., et al. (2006). Minimal criteria for defining multipotent mesenchymal stromal cells. The International Society for Cellular Therapy position statement. *Cytotherapy* 8, 315–317. doi: 10.1080/14653240600855905

Eaton, C. L., Colombel, M., van der Pluijm, G., Cecchini, M., Wetterwald, A., Lippitt, J., et al. (2010). Evaluation of the frequency of putative prostate cancer stem cells in primary and metastatic prostate cancer. *Prostate* 70, 875–882. doi: 10.1002/pros.21121

Fan, W., Crawford, R., and Xiao, Y. (2011). The ratio of VEGF/PEDF expression in bone marrow mesenchymal stem cells regulates neovascularization. *Differentiation* 81, 181–191. doi: 10.1016/j.diff.2010.12.003

Fazzi, R., Pacini, S., Carnicelli, V., Trombi, L., Montali, M., Lazzarini, E., et al. (2011). Mesodermal progenitor cells (MPCs) differentiate into mesenchymal stromal cells (MSCs) by activation of Wnt5/calmodulin signalling pathway. *PLoS ONE* 6:e25600 doi: 10.1371/journal.pone.0025600

Fischer, L. J., McIlhenny, S., Tulenko, T., Golesorkhi, N., Zhang, P., Larson, R., et al. (2009). Endothelial differentiation of adipose-derived stem cells: effects of endothelial cell growth supplement and shear force. *J. Surg. Res.* 152, 157–166. doi: 10.1016/j.jss.2008.06.029

Friedenstein, A. J., Chailakhyan, R. K., Latsinik, N. V., Panasyuk, A. F., and Keiliss-Borok, I. V. (1974). Stromal cells responsible for transferring the microenvironment of the hemopoietic tissues. Cloning *in vitro* and retransplantation *in vivo*. *Transplantation* 17, 331–340. doi: 10.1097/00007890-197404000-00001

Friedenstein, A. J., Petrakova, K. V., Kurolesova, A. I., and Frolova, G. P. (1968). Heterotopic of bone marrow. Analysis of precursor cells for osteogenic and hematopoietic tissues. *Transplantation* 6, 230–247. doi: 10.1097/00007890-196803000-00009

Gerhardt, H., and Betsholtz, C. (2003). Endothelial-pericyte interactions in angiogenesis. *Cell Tissue Res.* 314, 15–23. doi: 10.1007/s00441-003-0745-x

Gravdal, K., Halvorsen, O. J., Haukaas, S. A., and Akslen, L. A. (2009). Proliferation of immature tumor vessels is a novel marker of clinical progression in prostate cancer. *Cancer Res.* 69, 4708–4715. doi: 10.1158/0008-5472.CAN-08-4417

He, Q., Wan, C., and Li, G. (2007). Multipotent mesenchymal stromal cells in blood. *Stem Cells* 25, 69–77. doi: 10.1634/stemcells.2006-0335

Hill, J. M., Zalos, G., Halcox, J. P., Schenke, W. H., Waclawiw, M. A., Quyyumi, A. A., et al. (2003). Circulating endothelial progenitor cells, vascular function, and cardiovascular risk. *N. Engl. J. Med.* 348, 593–600. doi: 10.1056/NEJMoa022287

Horwitz, E. M., Le, B. K., Dominici, M., Zetterberg, E., and Ringdén, O. (2005). Clarification of the nomenclature for MSC: the International Society for Cellular Therapy position statement. *Cytotherapy* 7, 393–395. doi: 10.1080/14653240500319234

In't Anker, P. S., Scherjon, S. A., Kleijburg-van der Keur, C., de Groot-Swings, G. M., Claas, F. H., Fibbe, W. E. et al. (2004). Isolation of mesenchymal stem cells of fetal and maternal origin from human placenta. *Stem Cells* 22, 1338–1345. doi: 10.1634/stemcells.2004-0058

Janeczek Portalska, K., Leferink, A., Groen, N., Fernandes, H., Moroni, L., van Blitterswijk, C., et al. (2012). Endothelial differentiation of mesenchymal stromal cells. *PLoS ONE* 7:e46842. doi: 10.1371/journal.pone.0046842

Jazayeri, M., Allameh, A., Soleimani, M., Jazayeri, S. H., Piryaei, A., and Kazemnejad, S. (2008). Molecular and ultrastructural characterization of endothelial cells differentiated from human bone marrow mesenchymal stem cells. *Cell Biol. Int.* 32, 1183–1192. doi: 10.1016/j.cellbi.2008.07.020

Keating, A. (2012). Mesenchymal stromal cells: new directions. *Cell Stem Cell* 10, 709–716. doi: 10.1016/j.stem.2012.05.015

Kim, H. S., Kang, H. S., Messam, C. A., Min, K. W., and Park, C. S. (2002). Comparative evaluation of angiogenesis in gastric adenocarcinoma by nestin and CD34. *Appl. Immunohistochem. Mol. Morphol.* 10, 121–127. doi: 10.1097/00022744-200206000-00005

Kuroda, Y., Kitada, M., Wakao, S., Nishikawa, K., Tanimura, Y., Makinoshima, H., et al. (2010). Unique multipotent cells in adult human mesenchymal cell populations. *Proc. Natl. Acad. Sc.i U.S.A.* 107, 8639–8643. doi: 10.1073/pnas.0911647107

Kuroda, Y., Wakao, S., Kitada, M., Murakami, T., Nojima, M., and Dezawa, M. (2013). Isolation, culture and evaluation of multilineage-differentiating stress-enduring (Muse) cells. *Nat. Protoc.* 8, 1391–1415. doi: 10.1038/nprot.2013.076

Kuznetsov, S. A., Krebsbach, P. H., Satomura, K., Kerr, J., Riminucci, M., Benayahu, D., et al. (1997). Single-colony derived strains of human marrow stromal fibroblasts form bone after transplantation *in vivo*. *J. Bone Miner. Res.* 12, 1335–1347. doi: 10.1359/jbmr.1997.12.9.1335

Kuznetsov, S. A., Mankani, M. H., Leet, A. I., Ziran, N., Gronthos, S., and Robey, P. G. (2007). Circulating connective tissue precursors: extreme rarity in humans and chondrogenic potential in guinea pigs. *Stem Cells* 25, 1830–1839. doi: 10.1634/stemcells.2007-0140

Lanotte, M., Allen, T. D., and Dexter, T. M. (1981). Histochemical and ultrastructural characteristics of a cell line from human bone-marrow stroma. *J. Cell Sci.* 50, 281–297.

Le Blanc, K., Tammik, C., Rosendahl, K., Zetterberg, E., and Ringden, O. (2003). HLA expression and immunologic properties of differentiated and undifferentiated mesenchymal stem cells. *Exp. Hematol.* 31, 890–896. doi: 10.1016/S0301-472X(03)00110-3

Lendahl, U., Zimmerman, L. B., and McKay, R. D. (1990). CNS stem cells express a new class of intermediate filament protein. *Cell* 60, 585–595. doi: 10.1016/0092-8674(90)90662-X

Lin, C. S., and Lue, T. F. (2013). Defining vascular stem cells. *Stem Cells Dev.* 22, 1018–1026. doi: 10.1089/scd.2012.0504

Liu, J. W., Dunoyer-Geindre, S., Serre-Beinier, V., Mai, G., Lambert, J. F., Fish, R. J., et al. (2007). Characterization of endothelial-like cells derived from human mesenchymal stem cells. *J. Thromb. Haemost.* 5, 826–834. doi: 10.1111/j.1538-7836.2007.02381.x

Makanya, A. N., Hlushchuk, R., and Djonov, V. G. (2009). Intussusceptive angiogenesis and its role in vascular morphogenesis, patterning, and remodeling. *Angiogenesis* 12, 113–123. doi: 10.1007/s10456-009-9129-5

Matsuda, Y. (2013). Nestin: a novel angiogenesis marker and possible target for tumor angiogenesis. *World J. Gastroenterol.* 19, 42–48. doi: 10.3748/wjg.v19.i1.42

Melero-Martin, J. M., De Obaldia, M. E., Allen, P., Dudley, A. C., Klagsbrun, M., and Bischoff, J. (2010). Host myeloid cells are necessary for creating bioengineered human vascular networks *in vivo*. *Tissue Eng. Part A* 16, 2457–2466. doi: 10.1089/ten

Méndez-Ferrer, S., Michurina, T. V., Ferraro, F., Mazloom, A. R., Macarthur, B. D., Lira, S. A., et al. (2010). Mesenchymal and haematopoietic stem cells form a unique bone marrow niche. *Nature* 466, 829–834. doi: 10.1038/nature09262

Morshead, C. M., Reynolds, B. A., Craig, C. G., McBurney, M. W., Staines, W. A., Morassutti, D., et al. (1994). Neural stem cells in the adult mammalian forebrain: a relatively quiescent subpopulation of subependymal cells. *Neuron* 13, 1071–1082. doi: 10.1016/0896-6273(94)90046-9

Muraglia, A., Cancedda, R., and Quarto, R. (2000). Clonal mesenchymal progenitors from human bone marrow differentiate *in vitro* according to a hierarchical model. *J. Cell Sci.* 113, 1161–1166.

Oswald, J., Boxberger, S., Jørgensen, B., Feldmann, S., Ehninger, G., Bornhäuser, M., et al. (2004). Mesenchymal stem cells can be differentiated into endothelial cells *in vitro*. *Stem Cell.* 22, 377–384. doi: 10.1634/stemcells.22-3-377

Pacini, S., Carnicelli, V., Trombi, L., Montali, M., Fazzi, R., Lazzarini, E., et al. (2010). Costitutive expression of pluripotency-associated genes in Mesodermal Progenitor Cells (MPCs). *PLoS ONE* 5:e9861. doi: 10.1371/journal.pone.0009861

Pacini, S., Fazzi, R., Montali, M., Carnicelli, V., Lazzarini, E., and Petrini, M. (2013). Specific integrin expression is associated with podosome-like structures on mesodermal progenitor cells. *Stem Cells Dev.* 22, 1830–1838. doi: 10.1089/scd.2012.0423

Petrini, M., Pacini, S., Trombi, L., Fazzi, R., Montali, M., Ikehara, S., et al. (2009). Identification and purification of mesodermal progenitor cells from human adult bone marrow. *Stem Cells Dev.* 18, 857–866. doi: 10.1089/scd.2008.029

Pevsner-Fischer, M., Levin, S., and Zipori, D. (2011). The origins of mesenchymal stromal cell heterogeneity. *Stem Cell Rev.* 7, 560–568. doi: 10.1007/s12015-011-9229-7

Phinney, D. G., Kopen, G., Righter, W., Webster, S., Tremain, N., and Prockop, D. J. (1999). Donor variation in the growth properties and osteogenic potential of human marrow stromal cells. *J. Cell. Biochem.* 75, 424–436. doi: 10.1002/(SICI)1097-4644(19991201)75:3<424::AID-JCB8>3.0.CO;2-8

Pinho, S., Lacombe, J., Hanoun, M., Mizoguchi, T., Bruns, I., Kunisaki, Y., et al. (2013). PDGFRα and CD51 mark human nestin+ sphere-forming mesenchymal stem cells capable of hematopoietic progenitor cell expansion. *J. Exp. Med.* 210, 1351–1367. doi: 10.1084/jem.20122252

Pittenger, M. F., Mackay, A. M., Beck, S. C., Jaiswal, R. K., Douglas, R., Mosca, J. D., et al. (1999). Multilineage potential of adult human mesenchymal stem cells. *Science* 284, 143–147. doi: 10.1126/science.284.5411.143

Potente, M., Gerhardt, H., and Carmeliet, P. (2011). Basic and therapeutic aspects of angiogenesis. *Cell* 146, 873–887. doi: 10.1016/j.cell.2011.08.039

Quirici, N., Soligo, D., Bossolasco, P., Servida, F., Lumini, C., and Deliliers, G. L. (2002). Isolation of bone marrow mesenchymal stem cells by antinerve growth factor receptor antibodies. *Exp. Hematol.* 30, 783–791. doi: 10.1016/S0301-472X(02)00812-3

Reynolds, B. A., Tetzlaff, W., and Weiss, S. (1992). A multipotent EGF-responsive striatal embryonic progenitor cell produces neurons and astrocytes. *J. Neurosci.* 12, 4565–4574.

Roobrouck, V. D., Clavel, C., Jacobs, S. A., Ulloa-Montoya, F., Crippa, S., Sohni, A., et al. (2011). Differentiation potential of human postnatal mesenchymal stem cells, mesoangioblasts, and multipotent adult progenitor cells reflected in their transcriptome and partially influenced by the culture conditions. *Stem Cells* 29, 871–882. doi: 10.1002/stem.633

Roufosse, C. A., Direkze, N. C., Otto, W. R., and Wright, N. A. (2004). Circulating mesenchymal stem cells. *Int. J. Biochem. Cell Biol.* 36, 585–597. doi: 10.1016/j.biocel.2003.10.007

Russell, K. C., Phinney, D. G., Lacey, M. R., Barrilleaux, B. L., Meyertholen, K. E., and O'Connor, K. C. (2010). *In vitro* high-capacity assay to quantify the clonal heterogeneity in trilineage potential of mesenchymal stem cells reveals a complex hierarchy of lineage commitment. *Stem Cells* 28, 788–798. doi: 10.1002/stem.312

Russell, K. C., Tucker, H. A., Bunnell, B. A., Andreeff, M., Schober, W., Gaynor, A. S., et al. (2013). Cell-surface expression of neuron-glial antigen 2 (NG2) and melanoma cell adhesion molecule (CD146) in heterogeneous cultures of marrow-derived mesenchymal stem cells. *Tissue Eng. Part A* 19, 2253–2266. doi: 10.1089/ten

Sacchetti, B., Funari, A., Michienzi, S., Di Cesare, S., Piersanti, S., Saggio, I., et al. (2007). Selfrenewing osteoprogenitors in bone marrow sinusoids can organize a hematopoietic microenvironment. *Cell* 131, 324–336. doi: 10.1016/j.cell.2007.08.025

Sengers, B. G., Dawson, J. I., and Oreffo, R. O. (2010). Characterisation of human bone marrow stromal cell heterogeneity for skeletal regeneration strategies using a two-stage colony assay and computational modelling. *Bone* 46, 496–503. doi: 10.1016/j.bone.2009.10.002

Seo, B. M., Miura, M., Gronthos, S., Bartold, P. M., Batoli, S., Brahim, J., et al. (2004). Investigation of multipotent postnatal stem cells from human periodontal ligament. *Lancet* 364, 149–155. doi: 10.1016/S0140-6736(04)16627-0

Silva, G. V., Litovsky, S., Assad, J. A., Sousa, A. L., Martin, B. J., Vela, D., et al. (2005). Mesenchymal stem cells differentiate into an endothelial phenotype, enhance vascular density, and improve heart function in a canine chronic ischemia model. *Circulation* 111, 150–156. doi: 10.1161/01.CIR.0000151812.86142.45

Sims, D. E. (1986). The pericyte—a review. *Tissue Cell* 18, 153–174. doi: 10.1016/0040-8166(86)90026-1

Sugawara, K., Kurihara, H., Negishi, M., Saito, N., Nakazato, Y., Sasaki, T., et al. (2002). Nestin as a marker for proliferative endothelium in gliomas. *Lab. Invest.* 82, 345–351. doi: 10.1038/labinvest.3780428

Teranishi, N., Naito, Z., Ishiwata, T., Tanaka, N., Furukawa, K., Seya, T., et al. (2007). Identification of neovasculature using nestin in colorectal cancer. *Int. J. Oncol.* 30, 593–603. doi: 10.3892/ijo.30.3.593

Tohyama, T., Lee, V. M., Rorke, L. B., Marvin, M., McKay, R. D., and Trojanowski, J. Q. (1992). Nestin expression in embryonic human neuroepithelium and in human neuroepithelial tumor cells. *Lab. Invest.* 66, 303–313.

Tormin, A., Li, O., Brune, J. C., Walsh, S., Schütz, B., Ehinger, M., et al. (2011). CD146 expression on primary nonhematopoietic bone marrow stem cells is correlated with *in situ* localization. *Blood.* 117, 5067–5077. doi: 10.1182/blood-2010-08-304287

Trombi, L., Pacini, S., Montali, M., Fazzi, R., Chiellini, F., Ikehara, S., et al. (2009). Selective culture of mesodermal progenitor cells. *Stem Cells Dev.* 18, 1227–1234. doi: 10.1089/scd.2009.0054

van Royen, N., Piek, J. J., Schaper, W., and Fulton, W. F. (2009). A critical review of clinical arteriogenesis research. *J. Am. Coll. Cardiol.* 55, 17–25. doi: 10.1016/j.jacc.2009.06.058

Vittorio, O., Jacchetti, E., Pacini, S., and Cecchini, M. (2013). Endothelial differentiation of mesenchymal stromal cells: when traditional biology meets mechanotransduction. *Integr. Biol. (Camb).* 5, 291–299. doi: 10.1039/c2ib20152f

Voyta, J. C., Via, D. P., Butterfield, C. E., and Zetter, B. R. (1984). Identification and isolation of endothelial cells based on their increased uptake of acetylated-low density lipoprotein. *J. Cell Biol.* 99, 2034–2040. doi: 10.1083/jcb.99.6.2034

Wagner, W., and Ho, A. D. (2007). Mesenchymal stem cell preparations–comparing apples and oranges. *Stem Cell Rev.* 3, 239–248. doi: 10.1007/s12015-007-9001-1

Watt, S. M., Gullo, F., van der Garde, M., Markeson, D., Camicia, R., Khoo, C. P., et al. (2013). The angiogenic properties of mesenchymal stem/stromal cells and their therapeutic potential. *Br. Med. Bull.* 108, 25–53. doi: 10.1093/bmb/ldt031

Werts, E. D., DeGowin, R. L., Knapp, S. K., and Gibson, D. P. (1980). Characterization of marrow stromal (fibroblastoid) cells and their association with erythropoiesis. *Exp. Hematol.* 8, 423–433.

Ylöstalo, J., Bazhanov, N., and Prockop, D. J. (2008). Reversible commitment to differentiation by human multipotent stromal cells in single-cell-derived colonies. *Exp. Hematol.* 36, 1390–1402. doi: 10.1016/j.exphem.2008.05.003

Zuk, P. A., Zhu, M., Mizuno, H., Huang, J., Futrell, J. W., Katz, A. J., et al. (2001). Multilineage cells from human adipose tissue: implications for cell-based therapies. *Tissue Eng.* 7, 211–228. doi: 10.1089/107632701300062859

Conflict of Interest Statement: The authors declare that the research was conducted in the absence of any commercial or financial relationships that could be construed as a potential conflict of interest.

2

Generation of a dual-color reporter mouse line to monitor spermatogenesis *in vivo*

Yoshinori Makino*[1,2]*, Erina Inoue*[1]*, Masashi Hada*[1,3]*, Keisuke Aoshima*[4]*, Satsuki Kitano*[5]*, Hitoshi Miyachi*[5] *and Yuki Okada*[1,2,6]*

[1] *Laboratory of Pathology and Development, Institute of Molecular and Cellular Biosciences, The University of Tokyo, Tokyo, Japan*
[2] *Career-Path Promotion Unit for Young Life Scientists, Center for the Promotion of Interdisciplinary Education and Research, Kyoto University, Kyoto, Japan*
[3] *Department of Biotechnology, Graduate School of Agricultural and Life Sciences, The University of Tokyo, Tokyo, Japan*
[4] *Division of Molecular Pathobiology, Center for Zoonosis Control, Hokkaido University, Sapporo, Japan*
[5] *Reproductive Engineering Team, Institute for Virus Research, Kyoto University, Kyoto, Japan*
[6] *PRESTO, Japan Science and Technology Agency, Saitama, Japan*

Edited by:
Guillaume Grenier, Université de Sherbrooke, Canada

Reviewed by:
Liliana Burlibasa, University of Bucharest, Romania
Giulia Ricci, Second University of Naples, Italy
Shree Ram Singh, National Cancer Institute, USA

****Correspondence:***
Yuki Okada, Laboratory of Pathology and Development, Institute of Molecular and Cellular Biosciences, The University of Tokyo, 1-1-1 Yayoi, Bunkyo-Ku, Tokyo 113-0032, Japan
e-mail: ytokada@iam.u-tokyo.ac.jp

In vivo fluorescent imaging technique is a strong tool to visualize the various cellular events such as the proliferation, differentiation, migration, and a lineage tracing in living cells requiring no further experimental procedure such as immunostaining. During spermatogenesis, unique and dynamic histone exchanges occur. Since the timing and types of histone exchanges defines the particular stages of spermatogenesis, visualizing certain types of histones in testes is useful not only for researching specific histone dynamics, but also for monitoring the stages of spermatogenesis *in vivo*. In this study, we report the establishment of two transgenic (Tg) mouse lines expressing histone H4-Venus (H4V) and histone H3.3-mCherry (H33C) fusion proteins in testicular germ cells, and demonstrated their utility for monitoring germ cell development *in vivo*. Because of the choice of promoter as well as the nature of these histones, H4V and H33C were exclusively expressed in the germ cells of the distinct stages, which allowed the determination of spermatogenic stages in real time. In addition, disappearance of H4V and H33C at particular stages of differentiation/fertilization also represented dynamic histone removal. Collectively, these Tg mice are a valuable resource not only for monitoring differentiation stages, but also for studying the chromatin dynamics of post-natal testicular germ cell development *in vivo*.

Keywords: spermatogenesis, histones, reporter mice, transplantation, homologous, spermatogonial stem cells

INTRODUCTION

Spermatogenesis is the process of differentiation from spermatogonia to spermatozoa. Postnatal spermatogenesis originates from spermatogonial/spermatogenic stem cells (SSCs), which continuously undergo self-renewal as well as asymmetrical cell division to produce daughter cells, called spermatogonia. These cells proliferate by mitosis, and become spermatocytes, which enter meiotic stages. Post-meiotic stages are called spermiogenesis, in which haploid germ cells called spermatids dynamically change both their morphological and epigenetic properties. One of the most unique epigenetic changes during spermatogenesis is histone-protamine exchange. During spermiogenesis, most of the histones are gradually eliminated from chromatin, and replaced first by transition proteins (TNPs) and then by protamines, both of which are highly basic male germ cell-specific proteins. These steps make the spermatozoal chromain highly condensed, and are necessary for the acquisition of fertility. However, 1–20% of histones including specific histone variants are still retained in sperm nuclei. A recent study suggests transgenerational effects are caused by the retained histones as well as their modifications (Hammoud et al., 2009).

Like in other types of tissues, pulse chase experiments and live imaging techniques have been utilized for studies of spermatogenesis *in vivo*, particularly for tracing of SSCs/spermatogonial progenitor cells to monitor their lineages (Yoshida, 2012). The former is preferable for monitoring of prolonged period, as the label persists in not only the labeled cells but also in their differentiating/differentiated progenies. However, the label is irreversible, and fixation/staining procedure is sometimes required for visualization. On the other hand, the latter method is more direct to monitor the cell fate in real time. By taking advantage of these two methods, Hara et al successfully demonstrated that entire population of GFRα1-positive undifferentiated spermatogonia contains a single stem cell pool regardless their syncytial state, suggesting the stochastic fates of GFRα1-positive cells to serve the stem cells or differentiating spermatogonia (Hara et al., 2014). Another example is an application of live imaging to Sertoli cells, which are testicular somatic cells adjacent to germ cells in the seminiferous tubules. Nel-Themaat et al established two-color Tg mice, which can visualize the nuclei and plasma membrane of Sertoli cells with distinct colors. This mouse line is useful to monitor not only the cell division and migration of Sertoli cells themselves but also the association with germ cells in spatio-temporal manner (Nel-Themaat et al., 2011).

The entire spermatogenic process described above is reproducible by a transplantation of SSCs to germ cell-deficient recipient testis (Brinster and Zimmermann, 1994). The transplantation assay is useful not only for evaluating the stemness of donor cells, but also for pulse-chase studies for lineage tracing. Recently Sato et al developed *in vitro* testicular organ culture system, that enabled real-time monitoring of *in vitro* spermatogenesis as well as the successful engraftment of transplanted SSCs in *in-vitro*-cultured testes (Sato et al., 2011, 2013). For these experiments, they utilized SSCs isolated from Acr-GFP Tg mice as a reporter, which allowed them to detect the appearance of post-meiotic spermatids in real time. Meanwhile, a bioluminescence imaging, which has been widely utilized for cancer biology, is also reported to be feasible in germ cell study when it's combined with transplantation (Chen et al., 2012).

Since dynamic histone exclusion and retention occur in particular spermatogenic stages, utilizing specific histones as reporters can function as appropriate tools for monitoring spermatogenesis in real time, especially to judge the spermatogenic stages of engrafted cells by transplantation. To this end, here we report the establishment of two Tg mouse lines, which express H4V and H33C in male germ cells, and their possible application for SSC transplantation assays.

MATERIALS AND METHODS

GENERATION OF Tg MICE

Zygotes obtained from C57BL/6 mice (CLEA Japan Inc.) were used to generate Tg mice. For H4-Venus (H4V), Bacterial artificial chromosome (BAC) plasmid modification was carried out as described previously (Copeland et al., 2001). Briefly, an XhoI-NotI-5′ arm-linker-Venus-3xFlag-stop-poly A-XhoI sequence was inserted into a pL452 vector cut with SalI. Subsequently, the 3′ arm sequence was inserted into a pL451 vector with BamHI-NotI; the resultant plasmid was named pL452-H4V3xF. The loxP sequence in the RP23-296J10 BAC plasmid, which includes the H4 gene, was deleted by recombination with the EcoRI-digested fragment of the p23loxZeo plasmid. The Venus-3xFlag sequence was then inserted into the BAC plasmid at the immediate 3′ end of the H4 gene coding sequence (not containing stop codon sequences) by recombination with the NotI-digested fragment of pL452-H4V3xF. The neomycin cassette of the Venus-3xFlag-inserted BAC plasmid was deleted by Cre-recombination; the resultant BAC plasmid was named 296J10-V3F. Recombination and Cre reaction were performed in the E. coli strain EL350. After recombination, the plasmid DNA obtained was purified, and injected into the pronuclei of zygotes without linearization.

For H3.3-mCherry (H33C), a protamine 1 promoter sequence (from −465 to +89 bp, NCBI Gene ID: 19118, TSS: 10796916), an H3.3A coding sequence (NM_008210), mCherry cDNA, and a NotI cutting sequence were inserted into pPBbsr2, a Piggybac system vector with SpeI-XhoI-EcoRI-ClaI (Yusa et al., 2009), designated as pPB-Prm1 pro-H3.3-mCherry. Piggybac mRNA was prepared by *in vitro* transcription using RiboMAX-T7 kit (Promega) from pCMV-PBase, a Piggybac cDNA-encoding plasmid (Yusa et al., 2011). Circular pPB-Prm1 pro-H3.3-mCherry DNA and Piggybac mRNA were mixed at the concentration of 25 and 30 ng/μl in the injected solution, respectively, and were cytoplasmically injected to the zygotes.

The obtained pups were genotyped, and mice carrying the transgenes were heterozygously maintained by crossing with wild-type C57BL/6 mice. To obtain H4V/H33C double Tg mice, animals for H4V were crossed with those heterozygous H33C.

All experimental procedures involving animals were approved by the Animal Experiment Ethics Committees at the Graduate School of Medicine, Kyoto University (MedKyo11094) and the Institute of Molecular and Cellular Biosciences, The University of Tokyo (#23015). All experiments were conducted in accordance with the Guidelines for the Care and Use of Laboratory Animals of Kyoto University and The University of Tokyo. All efforts were made to minimize animal suffering and discomfort and to reduce the number of animals used.

ANTIBODIES

Antibodies used for immunostaining and western blotting were as follows; α-GENA (clone-TRA98; Bio Academia, Cat# 73-003, 1:250), α-γH2A.X (Abcam, Cat# ab22551, 1:250), α-SLA (clone-TRA54; Bio Academia, Cat# 73-001, 1:250), α-Sox9 (Millipore, Cat# AB5535, 1:250), α-DYKDDDDK (= Flag) tag (Clone No. 1E6, WAKO Japan, 1:2000), α-histone H3 (Abcam, Cat# ab1791, 1:1000), and α-histone H4 (CST, Cat# 2592S, 1:1000).

MORPHOLOGICAL AND HISTOLOGICAL ANALYSES OF TESTES

Gross images of testes were obtained by a stereoscopic microscope (SZX16, Olympus) with a long-path filter for GFP (SZX2-FGFP, Olympus) and RFP (SZX2-FRFP1,Olympus). The histological images were taken using an inverted fluorescent microscope (IX-73, Olympus). For histological analysis, testes were fixed with 4% paraformaldehyde, and embedded with OCT compound (SAKURA Finetek Japan Co., Ltd) to make frozen sections. After permeabilization with 0.2% Triton X-100, primary antibodies were applied overnight at 4°C. Alexa Fluor 488 or 568 secondary antibodies (Life Technologies) were used for detection, followed by nuclear staining with Hoechst 33342 (Life Technologies). Staging of spermatogenesis was performed based on the morphological structure or the pattern of γH2A.X staining of testicular germ cells according to definitions (Russell et al., 1990; Hamer et al., 2003).

WESTERN BLOT

Nuclear extracts were prepared from equal weight of testicular pieces isolated from adult wildtype and H33C/H4V double Tg male. 20 mg of testicular tissues were first suspended with low salt buffer (10 mM Hepes-KOH, pH 7.8, 10 mM KCl, 0.1 mM EDTA, pH 8.0) containing 0.1% nonidet P-40 to eliminate cytoplasmic proteins. The pellets were then suspended with 200 μl of RIPA buffer (25 mM Tris-HCl pH7.6, 150 mM NaCl, 1% NP-40, 1% sodium deoxycholate, 0.1% SDS) followed by brief sonications to ensure the release of chromatin proteins. After centrifuge, the supernatants were subjected to 15% SDS-PAGE. After transferring to the PVDF membrane, the signals were detected by using IRDye secondary antibodies and ODYSSEY imaging system (LICOR).

FLOW CYTOMETRY

Single-cell-suspensions of whole testes were prepared from adult H4V/H33C double Tg mice by trypsin-DNase I treatment. Cells were stained with DRAQ5 (Cell Signaling Technology) to distinguish post-meiotic germ cells (1n) based on their DNA content. A BD FACS Aria II cell sorter was used to detect the presence of Venus, mCherry, and DRAQ5 simultaneously. Only cells expressing either H4V or H33C were used for negative controls and fluorescent compensation.

ISOLATION, CULTURE, AND TRANSPLANTATION OF SSCs

Isolation and *in vitro* culture of SSCs from P9 pups were performed as recently described by Aoshima et al. (2013). Cells were subject to transplantation assays at 3–4 weeks after isolation. Approximately 4 week-old WBB6F1/Kit-KitW/KitW-v/Slc male mice (Japan SLC, Inc.) were used as recipients of transplantation, as they lacked endogenous germ cells. 1–2 × 10^5 SSCs were transplanted per testis, and the recipients were analyzed every 2 weeks to evaluate the efficiency of engraftment.

INTRA-CYTOPLASMIC SPERM INJECTION (ICSI)

Epididymal sperm were isolated from adult H33C mice. Metaphase II (MII) oocytes were obtained from wild type C57BL/6 female mice by hyperovulation. Tail-removed sperm heads were injected into the cytoplasm of MII oocytes using a micromanipulation system (Narishige Japan) combined with a Piezo micro manipulator PMM-150HJ (PRIMETECH, Ltd.) and a fluorescent system (IX-73, Olympus). Fertilized oocytes injected with H33C sperm were monitored using the same fluorescent microscope.

RESULTS

GENERATION OF TRANSGENIC (Tg) MOUSE LINES

To generate H4V transgenic mice, we utilized a BAC system to maintain the endogenous expression pattern and level of the H4 histone, as ectopic and excess expression of transgene might be toxic to cells. To this end, we used BAC clone RP23-296J10 containing an entire promoter region of H4, and an open reading frame of H4 fused with Venus fluorescence-3xFlag tags as shown in **Figure 1A**, designated as 296J10-V3F. The DNA was injected to the pronuclei of zygotes, and two independent Tg lines were obtained. Interestingly, one of the Tg lines exhibited significantly higher expression of the H4V transgene in germ cells compared with other types of cells, although it was designed to be expressed by the endogenous H4 promoter and thus systemic ubiquitous expression was expected (data not shown). For the expression of the H33C transgene, we chose a Prm1 promoter, as H3.3 is known to be incorporated in chromatin of post-meiotic germ cells. The DNA fragment encoding a Prm-H3.3-mCherry sequence was subcloned into the Piggybac system vector pBbsr2 (**Figure 1B**), which enabled us to generate a transgenic line in a transposase-dependent manner (Sumiyama et al., 2010; Yusa et al., 2011). As a result, four independent Tg lines were obtained, and all of them exhibited similar properties (data not shown). Thus, line#2 was used for further analysis.

EXPRESSION PATTERN OF H4V AND H33C IN TRANSGENIC MOUSE TESTES

Both H4V and H33C Tg lines were healthy and fertile (data not shown). H4V signals were detectable in postnatal day (P) P0.5 neonatal testes as numerous dots in the seminiferous tubules (**Figure 2A**). These dotted pattern represented nuclear expression of H4V in gonocytes, since the perinatal germ cells which express GENA in their nuclei (**Figure 2B**). In the heterozygous adult testes, H4V and H33C signals, respectively, were strong enough to be detected macroscopically (**Figures 2Ca–f**). In the testes of double Tg mice obtained by crossing H4V and H33C, both transgenes were properly expressed (**Figures 2Cg,h**). In western blotting, the expression of H33C fusion protein was detected around the expected size (~44 KDa) using αH3 antibody (**Figure 2D**). On the other hand, H4V failed to be detected using αH4 antibody presumably due to its lower expression level (data not shown). Alternatively, H4V was detected around the expected size (~40 KDa) by using αFlag antibody (**Figures 1A, 2D**).

Histologically, H4V was prominently expressed in the basal layer of seminiferous tubules including γH2A.X-positive spermatocytes (i.e., leptotene, zygotene, pachytene, and diplotene) (**Figures 3Aa,b**). Smaller cells attached to the basement membrane, which were likely spermatogonia, also possessed intense H4V (**Figures 3Aa,b**). Post-meiotic haploid spermatids marked by SLA also exhibited weaker H4V signals in certain stages in seminiferous tubules including Stage VII, but not in later

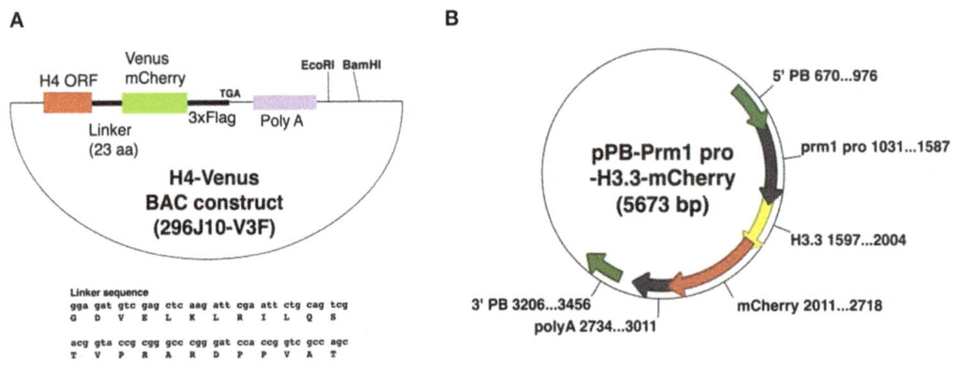

FIGURE 1 | Plasmid maps to generate H4V (A) and H33C (B) Tg mice.

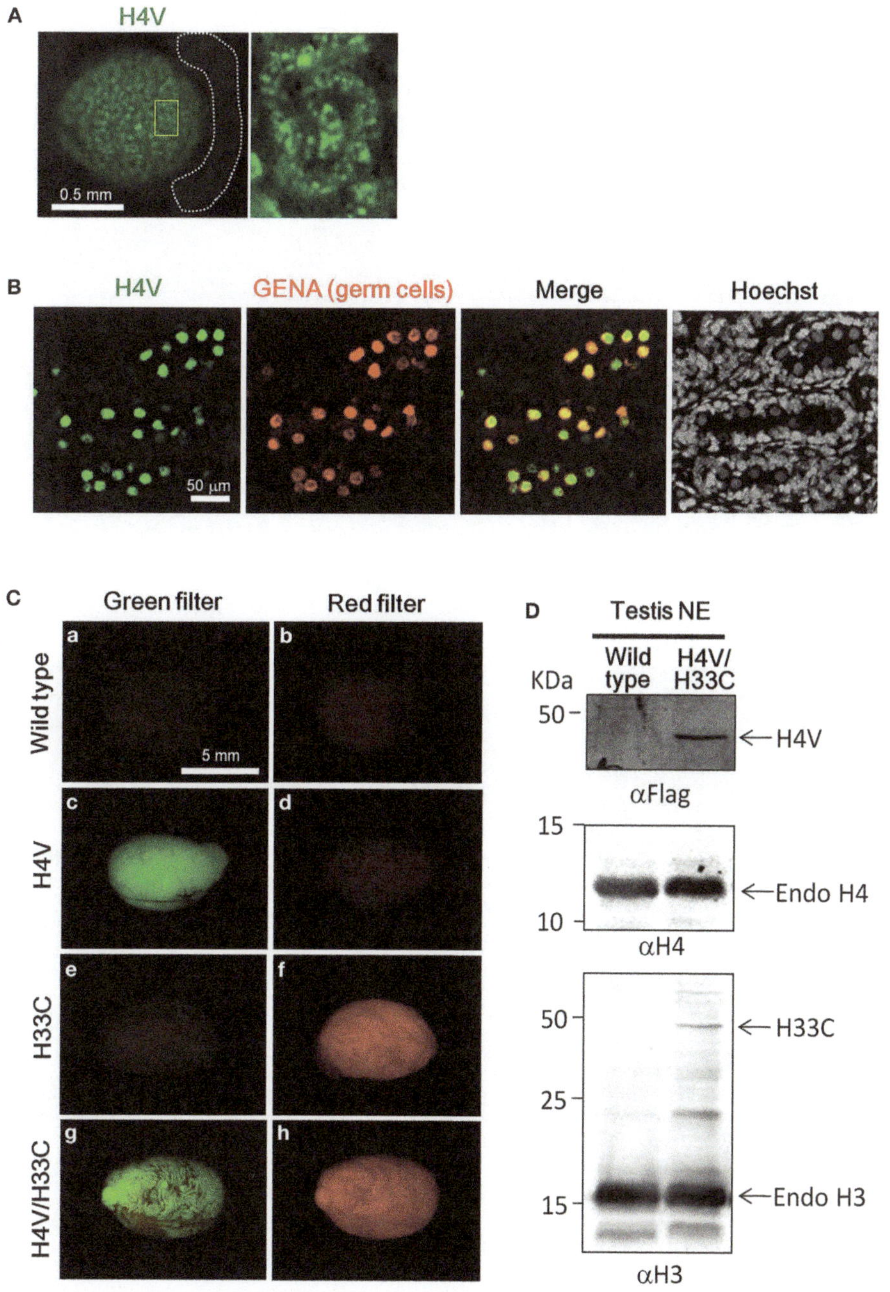

FIGURE 2 | Expression of H4V and H33C in mouse testes. (A) Gross view of P0.5 testis of H4V Tg mouse (left panel). White dotted lines indicate the epididymis. Dotted distribution of H4V-positive cells is exhibited under higher magnification of an area surrounded by yellow line in the left panel (right panel). **(B)** Histological features of P0.5 H4V Tg testis. H4V expression was derived from the transgene (green), whereas GENA, a marker of germ cells (red), was visualized by immunostaining. **(C)** Gross view of adult testes of wild type **(a,b)**, H4V **(c,d)**, H33C **(e,f)**, and H4V/H33C double **(g,h)** Tg testes. Images of left panels were visualized using a green filter, and those of right panels were visualized using a red filter. Scale bars are as indicated. **(D)** Western blot analysis to confirm the expression of H4V and H33C in the double Tg testis. Primary antibodies used for each panel were indicated. Middle panel exhibiting the expression of endogenous H4 was served as a loading control of αFlag western blot (upper panel). NE, nuclear extract.

stages such as Stage XI-XII (**Figures 3Ac,d, 4C**). Sertoli cells, the intra-tubular testicular somatic cells marked by Sox9, did not express H4V (**Figures 3Ae,f**). In contrast, expression of H33C existed mainly in the lumen side of seminiferous tubules. These cells were γH2A.X-negative, suggesting that these were post-meiotic spermatids (**Figures 3Ba,b**). Elongating spermatids in Stage XI–XII seminiferous tubules exhibited modest H33C signals compared with those of elongated spermatids in Stage VIII (**Figures 3Ba,b**), suggesting that the expression of H33C increased as the differentiation of spermatids proceeded. Interestingly, in H4V/H33C double Tg testes, germ cells with both signals were hardly observed in any spermatogenic stages

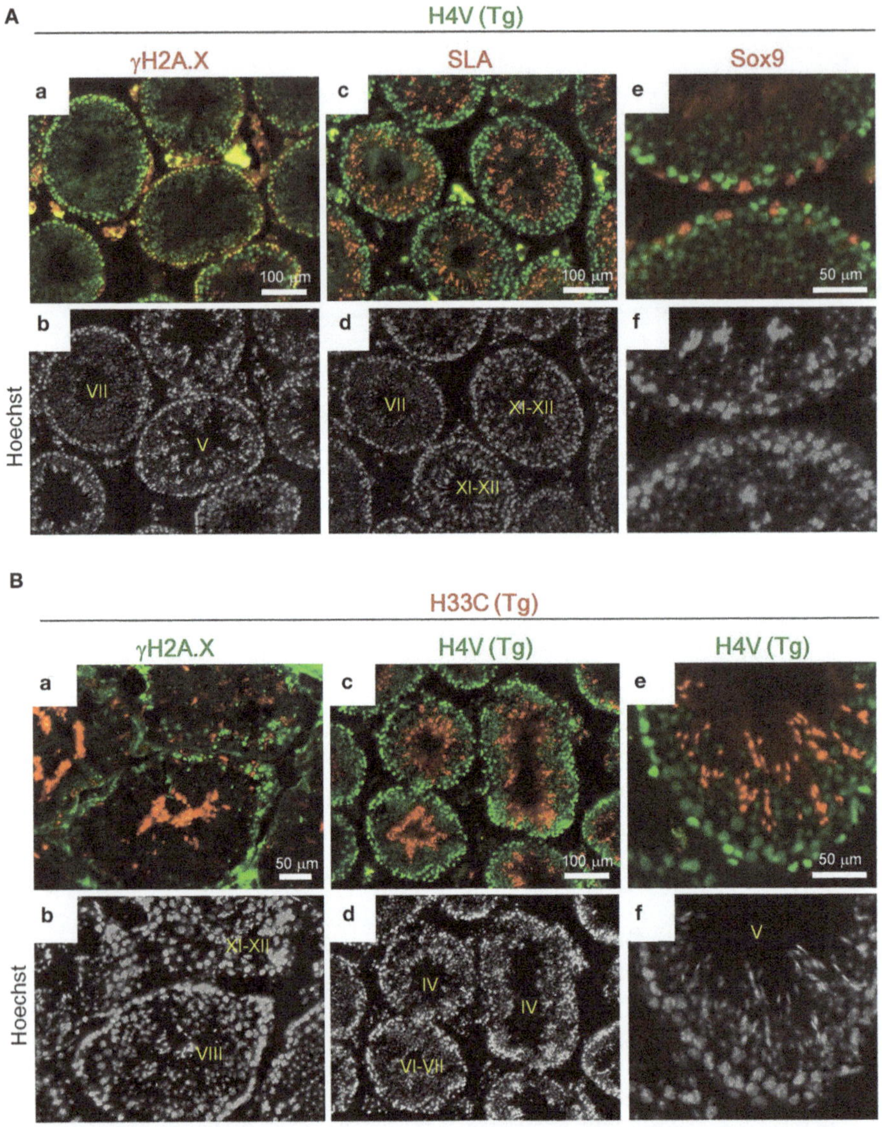

FIGURE 3 | Immunohistological analyses of the adult Tg testes. (A) Expression of γH2A.X **(a,b)**, SLA **(c,d)**, and Sox9 **(e,f)** in testes of adult H4V Tg mice. Roman numerals indicate stages of spermatogenesis of the seminiferous tubules. **(B)** Expression of γH2A.X **(a,b)** and H4V **(c–f)** in testes of adult H33C Tg mice. H4V signals in **(c)** and **(e)** were derived from the transgene (i.e., the double Tg mice). Roman numerals indicate stages of spermatogenesis of the seminiferous tubules. Scale bars are as indicated.

(**Figures 3Bc–f**), implying mutually exclusive expression of H4V and H33C.

Flow cytometrical analysis was performed to confirm the mutual exclusivity between H4V and H33C. Post-meiotic haploid spermatids (1n) were isolated by DNA staining, and their H4V and H33C intensities were examined. We observed that the majority of haploid cells expressed only H33C, and consistent with this histological observation, H4V/H33C double-positive cells were hardly detected (**Figures 4A,B**). Taken together, these results suggest that H4V is intensely expressed in spermatogonia through approximately Step 8 of spermatids, whereas H33C starts to be detected around Step 11, and remains prominent until the final stage of testicular spermatogenesis (**Figure 4C**).

APPLICATION OF H4V/H33C DOUBLE Tg CELLS FOR SSC TRANSPLANTATION ASSAY

Since the results above demonstrated an apparent switch of H4V and H33C during spermatogenesis, the double Tg cells seemed useful to monitor spermatogenesis *in vivo*. To test this possibility, we isolated SSCs from P9 pups of H4V/H33C double Tg mice, and subjected them to transplantation analysis. The SSCs cultured *in vitro* for 3–4 weeks exhibited "grape-shaped" colonies, which are typical for SSCs (**Figure 5A**). Consistent to the expression pattern in testes, the SSCs expressed only H4V, whereas H33C expression remained at background levels (**Figure 5A**). 1–2 × 10^5 SSCs were transplanted to a testis of W/Wv mice, and the engrafted cells were examined every

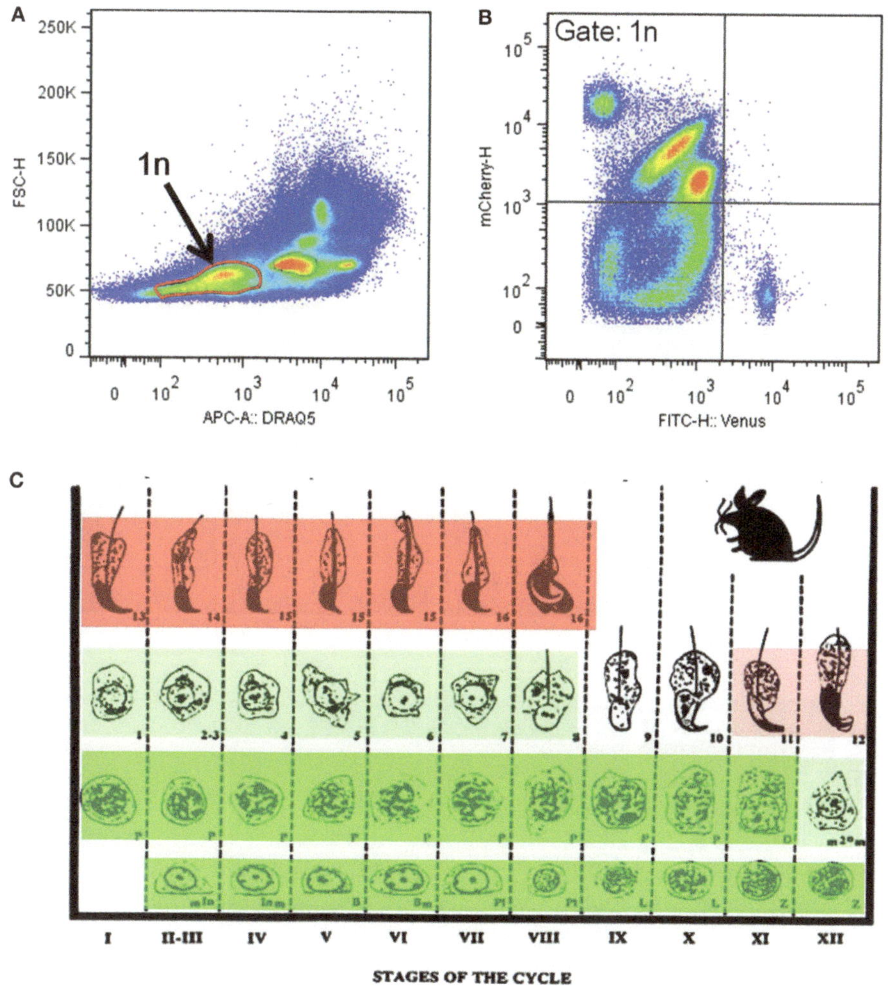

FIGURE 4 | Flow cytometry analysis of the H4V/H33C double Tg testes. **(A)** Separation of testicular cells by the DNA contents. Approximately 30% of cells with low DRAQ5 fluorescence are assessed as post-meiotic haploid cells indicated as 1n. **(B)** Expression of Venus and mCherry in haploid cells gated in **(A)**. **(C)** Scheme of mouse spermatogenesis with expression patterns and the intensities of Venus (green) and mCherry (red). Roman numerals represent spermatogenic stages, and Arabic numerals indicate spermiogenic stages, respectively. This scheme is modified from Russell et al. (1990).

2 weeks. At the second weeks post-transplantation, small clusters of H4V-positive cells were already detected in the recipient testes, indicating a successful engraftment (**Figure 5B**, see a panel of 2w). On the fourth week, H4V-positive clusters increased the number, but H33C expression had not been detected (**Figures 5B,C**, see 4w panels). On the sixth week, a small number of H33C-expressing cells started to form clusters in the seminiferous tubules (**Figures 5B,C**, see 6w panels). Finally at the eighth week, H33C-positive clusters were largely expanded (**Figure 5B**, see 8w panels), and exhibited a discontinuous cluster formation representing spermiogenic step-dependent expression (**Figure 5C**, see 8w panels). Consistently, histological examination revealed H4V-only expression at the fourth week, as engrafted cells had not entered the post-meiotic stages yet (**Figure 5D**, see the upper 4w panels), whereas at the eighth week, the engrafted cells proceeded to the later stages of post-meiosis, where H33C was predominantly expressed (**Figure 5D**, see the lower 8w panels). These results confirmed our hypothesis that the H4V/H33C double Tg cells were useful to monitor spermatogenesis *in vivo*.

Finally, we isolated epididymal sperm from H33C Tg mice to see whether H33C persists in spermatozoa, and found that almost all spermatozoa exhibited H33C prominently in their nuclei (**Figure 6A**). Interestingly, the fluorescence in sperm heads disappeared within 20 min, when they were injected into the cytoplasm of unfertilized oocytes (i.e., ICSI), indicating the dynamic remodeling of male chromatin after fertilization (**Figures 6B,C**).

DISCUSSION

In this study, we established H4V and H33C transgenic mouse lines, and demonstrated that their distinct expression patterns during spermatogenesis can be utilized for monitoring spermatogenesis *in vivo*. Recently, Sato et al developed an *in vitro* testicular organ culture system, which enables the real-time monitoring of spermatogenesis *in vitro* (Sato et al., 2011, 2013). Applying the testes of our Tg mice to this *in vitro* testicular organ culture system

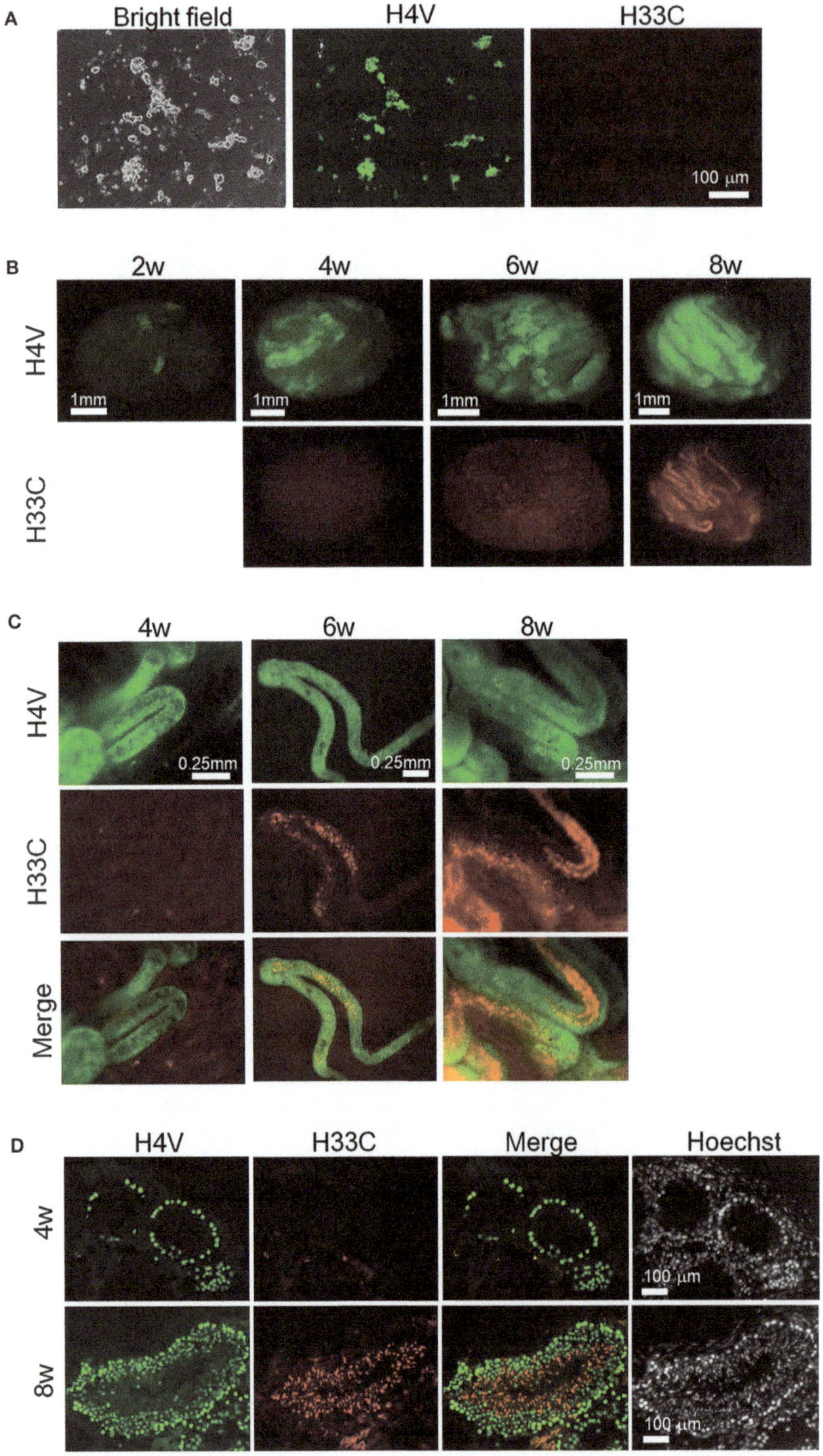

FIGURE 5 | Transplantation assay using H4V/H33C double Tg SSCs. (A) Isolation of SSCs (= donor cells) from H4V/H33C double Tg mice. **(B)** Gross view of the recipient testes representing efficiency of the engraftment after transplantation. The examined timepoints post-transplantation are indicated. **(C)** Higher magnification of seminiferous tubules shown in **(B)**. **(D)** Histological features of the recipient testes at the indicated timepoints. Scale bars are as indicated.

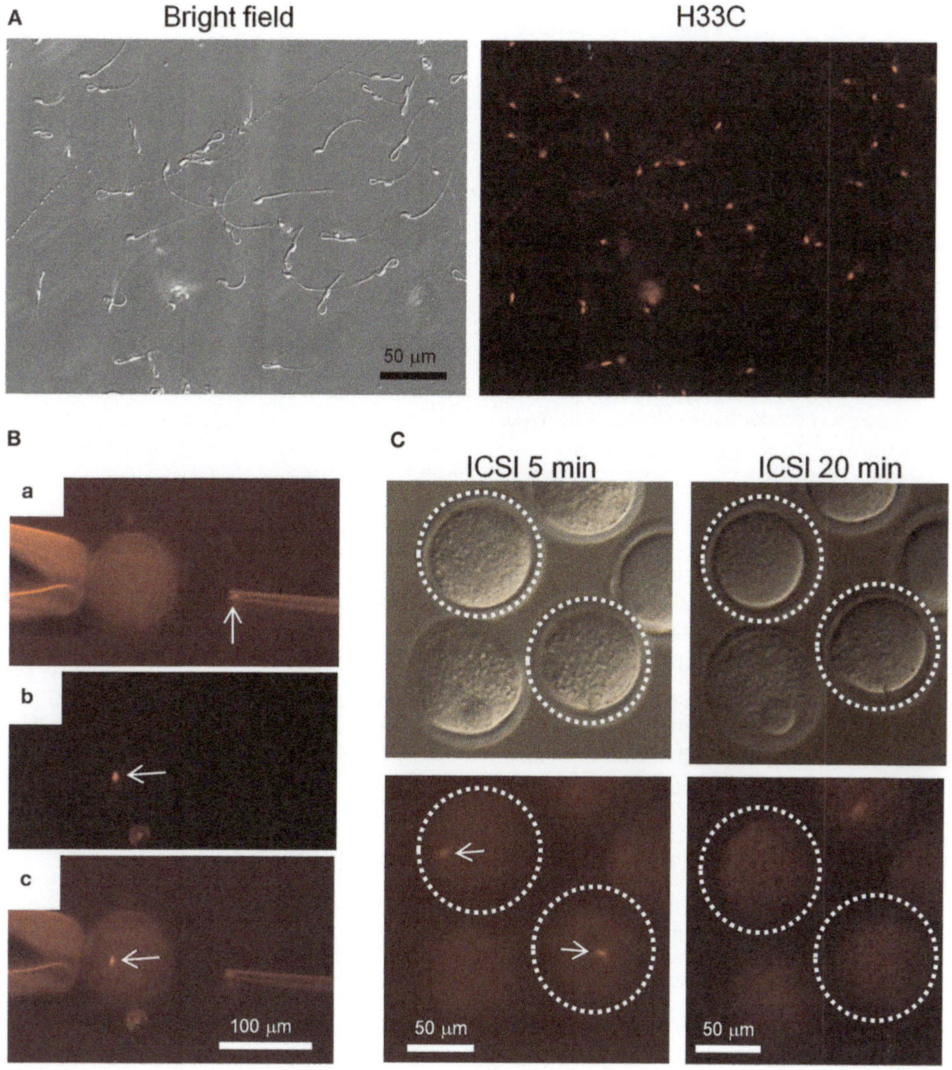

FIGURE 6 | Intra-cytoplasmic sperm injection (ICSI) using H33C sperm. **(A)** Expression of H33C in the head of epididymal sperm. **(B)** A representative image of performing ICSI using H33C sperm. **(a)** An H33C-positive sperm head in an injection capillary before injection. **(b,c)** An H33C-positive sperm (white arrows) in ooplasm immediately after injection. **(C)** H33C-positive sperm-injected oocytes at the indicated timepoints. White dotted lines denote oocytes containing H33C-positive sperm (white arrows). Scale bars are as indicated.

will allow us not only to determine the stages of spermatogenesis without further immunostaining, but also to perform real-time imaging of spermatogenesis. Particularly, as shown in **Figure 5**, using the double Tg SSCs for a transplantation assay is a convenient way to monitor engraftment efficiency, and to evaluate the differentiation stages of transplanted cells in real time.

In addition, these Tg lines can be a strong tool to study histone-protamine replacement during spermiogenesis. It is well known that in the nuclei of spermatids, histones are removed from chromatin en masse, and subjected to degradation in the course of spermiogenesis through their exchange with TNPs and protamines. Recently, Qian et al. demonstrated that a spermatid/sperm-specific proteasomal complex is responsible for histone degradation, and that acetylation of histones including H4 lysine 16 (H4K16) are required for binding histones to the proteasomal complex (Qian et al., 2013). On the other hand, it has also been reported that small amount of histones escape this removal, and are retained in spermatozoa (Hammoud et al., 2009). Because H3.3 is one of the histones selectively retained in spermatozoa (Erkek et al., 2013), our finding that H4V and H33C are mutually exclusive distributed in spermatids is likely to reflect the fine-tuned exchange of histone removal and deposition during spermiogenesis, although the relatively large size of the Venus tag for H4V, and the promoter choice for H33C expression may have had some artificial effects.

Similarly, we also observed a rapid removal of sperm H33C after fertilization, consistent with previous studies that reported that protamine withdrawal occurs 1–60 min after fertilization in different animal species due to protamine association to nucleoplasmin rather than their degradation (Ohsumi and Katagiri,

1991; Jones et al., 2011). In contrast, maternal-derived H3.3 is quickly incorporated into male pronuclei soon after fertilization (Torres-Padilla et al., 2006). Since the localization of H3.3 in the sperm genome was recently reported, it is a challenging but intrinsic task to determine the distribution of H3.3 in male pronuclei to see why the H3.3 needs to be exchanged by another H3.3 in such a short period of time.

Taken above together, we propose that our developed H4V and H33C Tg mice can serve as useful tools for further researches into spermatogenesis.

ACKNOWLEDGMENTS

We would like to thank Dr. N. G. Copeland at The Methodist Hospital Research Institute for kindly providing the EL350 strain and pL452, and Drs. J. Takeda at Osaka University, and K. Yusa and A. Bradley at The Wellcome Trust Sanger Institute for providing the pPBbsr2 and pCMV-PBase plasmids. We also thank Ms. Koga and Ohtaki at the Cell Sorting Facility of the Institute of Molecular and Cellular Biosciences, The University of Tokyo. This work was supported by The Naito Foundation Subsidy for Female Researchers after Maternity Leave (Yuki Okada), the JST PRESTO program (Yuki Okada), a Grant-in-Aid for Scientific Research from the Ministry of Education, Culture, Sports, Science and Technology (MEXT) in Japan (22125007 to Yuki Okada), the Special Coordination Fund for Promoting Science and Technology of MEXT (Yuki Okada). Masashi Hada and Keisuke Aoshima are Research Fellows (DC2) for Young Scientists, JSPS.

REFERENCES

Aoshima, K., Baba, A., Makino, Y., and Okada, Y. (2013). Establishment of alternative culture method for spermatogonial stem cells using knockout serum replacement. *PLoS ONE* 8:e77715. doi: 10.1371/journal.pone.0077715

Brinster, R. L., and Zimmermann, J. W. (1994). Spermatogenesis following male germ-cell transplantation. *Proc. Natl. Acad. Sci. U.S.A.* 91, 11298–11302. doi: 10.1073/pnas.91.24.11298

Chen, C. H., Wang, C. W., Hsu, M. I., Huang, Y. H., Lai, W. F., and Tzeng, C. R. (2012). Bioluminescence imaging as a tool to evaluate germ cells *in vitro* and transplantation *in vivo* as fertility preservation of prepubertal male mice. *Fertil. Steril.* 97, 1192–1198. doi: 10.1016/j.fertnstert.2012.02.003

Copeland, N. G., Jenkins, N. A., and Court, D. L. (2001). Recombineering: a powerful new tool for mouse functional genomics. *Nat. Rev. Genet.* 2, 769–779. doi: 10.1038/35093556

Erkek, S., Hisano, M., Liang, C. Y., Gill, M., Murr, R., Dieker, J., et al. (2013). Molecular determinants of nucleosome retention at CpG-rich sequences in mouse spermatozoa. *Nat. Struct. Mol. Biol.* 20, 868–875. doi: 10.1038/nsmb.2599

Hamer, G., Roepers-Gajadien, H. L., van Duyn-Goedhart, A., Gademan, I. S., Kal, H. B., van Buul, P. P., et al. (2003). DNA double-strand breaks and gamma-H2AX signaling in the testis. *Biol. Reprod.* 68, 628–634. doi: 10.1095/biolreprod.102.008672

Hammoud, S. S., Nix, D. A., Zhang, H., Purwar, J., Carrell, D. T., and Cairns, B. R. (2009). Distinctive chromatin in human sperm packages genes for embryo development. *Nature* 460, 473–478. doi: 10.1038/nature08162

Hara, K., Nakagawa, T., Enomoto, H., Suzuki, M., Yamamoto, M., Simons, B. D., et al. (2014). Mouse spermatogenic stem cells continually interconvert between equipotent singly isolated and syncytial states. *Cell Stem Cell* 14, 658–672. doi: 10.1016/j.stem.2014.01.019

Jones, E. L., Zalensky, A. O., and Zalenskaya, I. A. (2011). Protamine withdrawal from human sperm nuclei following heterologous ICSI into hamster oocytes. *Protein Pept. Lett.* 18, 811–816. doi: 10.2174/092986611795713925

Nel-Themaat, L., Jang, C. W., Stewart, M. D., Akiyama, H., Viger, R. S., and Behringer, R. R. (2011). Sertoli cell behaviors in developing testis cords and postnatal seminiferous tubules of the mouse. *Biol. Reprod.* 84, 342–350. doi: 10.1095/biolreprod.110.086900

Ohsumi, K., and Katagiri, C. (1991). Characterization of the ooplasmic factor inducing decondensation of and protamine removal from toad sperm nuclei: involvement of nucleoplasmin. *Dev. Biol.* 148, 295–305. doi: 10.1016/0012-1606(91)90338-4

Qian, M. X., Pang, Y., Liu, C. H., Haratake, K., Du, B. Y., Ji, D. Y., et al. (2013). Acetylation-mediated proteasomal degradation of core histones during DNA repair and spermatogenesis. *Cell* 153, 1012–1024. doi: 10.1016/j.cell.2013.04.032

Russell, L. D., Ettlin, R. A., Sinha Hikim, A. P., and Clegg, E. D. (1990). *Histological and Histopathological Evaluation of the Testis.* St Louis, MO: Cache River Press.

Sato, T., Katagiri, K., Gohbara, A., Inoue, K., Ogonuki, N., Ogura, A., et al. (2011). *In vitro* production of functional sperm in cultured neonatal mouse testes. *Nature* 471, 504–507. doi: 10.1038/nature09850

Sato, T., Katagiri, K., Kubota, Y., and Ogawa, T. (2013). *In vitro* sperm production from mouse spermatogonial stem cell lines using an organ culture method. *Nat. Protoc.* 8, 2098–2104. doi: 10.1038/nprot.2013.138

Sumiyama, K., Kawakami, K., and Yagita, K. (2010). A simple and highly efficient transgenesis method in mice with the Tol2 transposon system and cytoplasmic microinjection. *Genomics* 95, 306–311. doi: 10.1016/j.ygeno.2010.02.006

Torres-Padilla, M. E., Bannister, A. J., Hurd, P. J., Kouzarides, T., and Zernicka-Goetz, M. (2006). Dynamic distribution of the replacement histone variant H3.3 in the mouse oocyte and preimplantation embryos. *Int. J. Dev. Biol.* 50, 455–461. doi: 10.1387/ijdb.052073mt

Yoshida, S. (2012). Elucidating the identity and behavior of spermatogenic stem cells in the mouse testis. *Reproduction* 144, 293–302. doi: 10.1530/REP-11-0320

Yusa, K., Rad, R., Takeda, J., and Bradley, A. (2009). Generation of transgene-free induced pluripotent mouse stem cells by the piggyBac transposon. *Nat. Methods* 6, 363–369. doi: 10.1038/nmeth.1323

Yusa, K., Zhou, L., Li, M. A., Bradley, A., and Craig, N. L. (2011). A hyperactive piggyBac transposase for mammalian applications. *Proc. Natl. Acad. Sci. U.S.A.* 108, 1531–1536. doi: 10.1073/pnas.1008322108

Conflict of Interest Statement: The authors declare that the research was conducted in the absence of any commercial or financial relationships that could be construed as a potential conflict of interest.

3

A novel function of RIP1 in postnatal development and immune homeostasis by protecting against RIP3-dependent necroptosis and FADD-mediated apoptosis

John P. Dowling, Anirudh Nair and Jianke Zhang*

Department of Microbiology and Immunology, Sidney Kimmel Cancer Center, Sidney Kimmel Medical College, Thomas Jefferson University, Philadelphia, PA, USA

Edited by:
Francis Kaming Chan, University of Massachusetts Medical School, USA

Reviewed by:
Han-Ming Shen, National University of Singapore, Singapore
Adrian T. Ting, Icahn School of Medicine at Mount Sinai, USA

***Correspondence:**
Jianke Zhang, Department of Microbiology and Immunology, Sidney Kimmel Cancer Center, Sidney Kimmel Medical College, Thomas Jefferson University, 233 S. 10th St., Room 731 BLSB, Philadelphia, PA 19107, USA
e-mail: jianke.zhang@jefferson.edu

RIP1 is an adaptor kinase originally identified as being able to associate with TNFR1 and Fas, and is later shown to be involved in signaling induced by TLRs. Major signaling pathways regulated by RIP1 include necroptosis, apoptosis, and pro-survival/inflammation NF-κB activation. Previous studies show that RIP1 deficiency has no effect on mouse embryogenesis, but blocks postnatal development. This phenotype could not readily be explained, since mice lacking TNFR1, Fas, or TLRs show no apparent developmental defect. Certain types of RIP1-deficient cells are hypersensitive to TNF-induced apoptosis. However, in our previous study, deletion of the apoptotic adaptor protein, FADD, provides marginal improvement of postnatal development of *rip1*$^{-/-}$ mice. Remarkably, the current data shows that haploid insufficiency of RIP3, a known mediator of necroptosis, allowed survival of *rip1*$^{-/-}$*fadd*$^{-/-}$ mice beyond weaning age, although the resulting *rip1*$^{-/-}$*fadd*$^{-/-}$ *rip3*$^{+/-}$ mice were significant smaller in size and weight. Moreover, complete absence of RIP3 further improved postnatal development of the resulting *rip1*$^{-/-}$*fadd*$^{-/-}$*rip3*$^{-/-}$ mice, which display normal size and weight. In such triple knockout (TKO) mice, lymphocytes underwent normal development, but progressively accumulated as mice age. This lymphoproliferative (*lpr*) disease in TKO mice is, however, less severe than that of *fadd*$^{-/-}$*rip3*$^{-/-}$ double knockout mice. In total, the data show that the postnatal developmental defect in *rip1*$^{-/-}$ mice is due in part to FADD-mediated apoptosis as well as RIP3-dependent necroptosis. Moreover, the function of RIP1 contributes to development of *lpr* diseases.

Keywords: mouse development, apoptosis, necroptosis, programmed cell death (PCD), immune homeostasis

INTRODUCTION

Apoptosis and necroptosis are programmed cell death (PCD) pathways that play an essential role during animal development and immune responses (Moriwaki and Chan, 2013; Walsh, 2014). Although these two PCD pathways have distinct mechanisms and outcomes, they are intricately connected and, therefore, must be tightly regulated. PCD can be initiated through the death receptors (DRs) like Fas and tumor necrosis factor receptor-1 (TNFR1) (Ashkenazi and Dixit, 1998). Apoptotic death induced by DRs is mediated by the Fas-associated death domain (FADD or MORT1) adaptor protein (Boldin et al., 1995; Chinnaiyan et al., 1995; Zhang and Winoto, 1996). FADD binds to and activates the apical Caspase 8, which then cleaves and activates the effector Caspases, and triggers the apoptotic program (Boldin et al., 1996; Muzio et al., 1996). When apoptotic pathway is blocked, such as through viral defense mechanisms or Caspase inhibitory conditions, DRs signaling could lead to necroptosis which is mediated through the receptor interacting protein (RIP)1-RIP3-mixed lineage kinase domain-like (MLKL) protein axis (Holler et al., 2000; Degterev and Yuan, 2008; Cho et al., 2009; Sun et al., 2012; Zhao et al., 2012).

One striking example of the importance of these PCD pathways in immune homeostasis is illustrated by a mutant mouse strain with lymphoproliferative (*lpr*) disease caused by mutations in the *fas* gene (Strasser et al., 2009). In *lpr* mice, lymphocytes are unable to undergo Fas-mediated cell death, leading to accumulation of a unique subset of T cells that lose expression of CD4 and CD8 and gain expression of B220, a traditional B cell marker (Strasser et al., 2009). Moreover, in certain genetic backgrounds, the *lpr* Fas mutation facilitates autoimmunity including glomerulonephritis, arthritis, and autoantibody production, and diseases sharing symptoms of systemic lupus erythematosus (SLE) (Cohen and Eisenberg, 1992; Shi et al., 2002). In human patients containing mutations in the *fas* gene, a similar disease has been identified as autoimmune-lymphoproliferative symdrome (ALPS) (Fisher et al., 1995). Deletion of TNFR1 alone does not affect lymphoid homeostasis (Pfeffer et al., 1993; Rothe et al., 1993), but exacerbates the *lpr* disease of Fas mutant mice (Zhou et al., 1996).

Besides abnormality in certain immune functions, mice lacking each individual DR undergo normal embryogenesis and postnatal development in mice (Pfeffer et al., 1993; Rothe et al., 1993; Adachi et al., 1996; Diehl et al., 2004). Surprisingly, inactivation of FADD, the common signaling adaptor of the DRs, results in death around embryonic day (E)11.5 (Yeh et al., 1998; Zhang et al., 1998). It has been demonstrated that *fadd*$^{-/-}$ embryos contain elevated levels of RIP1 and undergo massive necrosis (Zhang et al., 2011). Lack of RIP1 completely corrects the embryonic defect in *fadd*$^{-/-}$ embryos. Moreover, RIP3 deletion prevents T cell death induced by expression of a FADD dominant-negative (DN) mutant (Lu et al., 2011). Unlike RIP1, RIP3 does not play a role in NF-κB activation (Newton et al., 2004). Deletion of RIP3 completely restores normal development of mice lacking FADD or Caspase 8 (Ch'en et al., 2011; Kaiser et al., 2011; Oberst et al., 2011), indicating that RIP1/RIP3-dependent necroptosis is unleashed in the absence of FADD or Caspase 8.

Though RIP1 was initially identified as a Fas-binding protein (Stanger et al., 1995), apoptosis through Fas is not affected in *rip1*$^{-/-}$ cells (Ting et al., 1996; Kelliher et al., 1998). However, RIP1 deficiency leads to hypersensitivity to TNF-induced apoptosis, which has been attributed in part to a defect in NF-κB activation in certain cell types. When signaling through receptors like TNFR1 or toll like receptors (TLRs), the ubiquitination status of RIP1 can determine whether a cell lives or dies (Vucic et al., 2011; Schmukle and Walczak, 2012). For example, upon activation of TNFR1, RIP1 promotes NF-κB activation when ubiquitinated, but mediates apoptosis when deubiquitinated and associated with FADD and Caspase 8. Furthermore, if the function of FADD or Caspase 8 is interrupted, RIP1 kinase activity signals through RIP3, leading to necroptosis (Moriwaki and Chan, 2013; Vanden Berghe et al., 2014). These multiple functions of RIP1 participate in regulating inflammatory signaling and have been a target of interest for inflammatory disease (Moriwaki and Chan, 2013; Vanden Berghe et al., 2014; Walsh, 2014).

The importance of RIP1 during development is manifested in *rip1*$^{-/-}$ mice, which die at birth with extensive cell death in the lymphoid and adipose tissues (Kelliher et al., 1998) Since without RIP1, TNFR1 initiates FADD-mediated apoptosis (Cusson et al., 2002; O'Donnell and Ting, 2011), it was surprising that deletion of FADD does not rescue *rip1*$^{-/-}$ mice from perinatal lethality (Zhang et al., 2011). This suggests that RIP1 could play different roles during embryonic development, where it may initiate necroptosis in the absence of FADD, and in perinatal mice, where RIP1 is essential for survival. Unlike perinatal lethality in *rip1*$^{-/-}$ or *rip1*$^{-/-}$*fadd*$^{-/-}$ mice, normal embryonic and postnatal development is observed in *fadd*$^{-/-}$*rip3*$^{-/-}$ mice (Dillon et al., 2012). Moreover, under certain circumstances, RIP3 is able to signal necroptosis in the absence of RIP1 (Upton et al., 2010; Moujalled et al., 2013). In this study, we performed mouse genetic studies and show that lack of RIP3 is insufficient to correct the defect in RIP1 mice. However, the resulting *rip1*$^{-/-}$*rip3*$^{-/-}$ mice could survive to adulthood if FADD is also inactivated. Therefore, our data indicate that RIP1 has a novel function, which appears to provide protection against RIP3-dependent necroptosis and FADD-mediated apoptosis during postnatal development.

MATERIALS AND METHODS

MICE

fadd$^{+/-}$ mice have been previously described (Zhang et al., 1998). *rip1*$^{+/-}$ mice were provided by Dr. Michelle Kelliher (Kelliher et al., 1998). *rip3*$^{-/-}$ mice were provided by Drs. Kim Newton and Vishva Dixit (Genentech) (Newton et al., 2004). All animal studies were approved by the Institutional Animal Care and Use Committee (IACUC) at Thomas Jefferson University.

MOUSE GENETIC ANALYSIS

rip1$^{+/-}$*fadd*$^{+/-}$ mice were crossed with *rip3*$^{-/-}$ mice. The resulting *rip1*$^{+/-}$*fadd*$^{+/-}$*rip3*$^{+/-}$ triple heterozygous (THZ) mice were intercrossed. From the offspring, *rip1*$^{+/-}$*fadd*$^{-/-}$*rip3*$^{-/-}$ mice were mated with THZ mice, producing viable and fertile *rip1*$^{-/-}$*fadd*$^{-/-}$*rip3*$^{-/-}$ triple knockout (TKO) mice, which were mated with *rip1*$^{+/-}$*fadd*$^{-/-}$*rip3*$^{-/-}$ mice. For genotyping, genomic DNA was isolated from the tails of newborns to weaning-age mice using proteinase K and sodium dodecyl sulfate (SDS)-containing lysis buffer followed by phenol extraction and ethanol precipitation. Allele-specific primers were used to detect the presence of wild type and knockout alleles of *rip1* and *fadd* as previously described (Zhang et al., 2011). *rip3*$^{-/-}$ mice were typed using the forward primers 5′-GCCTGCCC ATCAGCAACT-3′ and 5′-CCAGAGGCCACTTGTGTAGCG-3′ and reverse primer 5′-CGCTTTAGAAGCCTTCAGGTTGAC-3′ (Newton et al., 2004). PCR produces a 320 bp wild type band and a 485 bp knockout band. For postnatal survival analysis, neonates were monitored for survival from birth to weaning age; deceased pups were immediately collected and genotyped. Survival curves were generated using Prism software (Graphpad Software, Inc.). For genetic analysis, the numbers of mice containing the most numbers of wild type alleles of the genes of concern were used to predict the numbers of other genotypes among the offspring of the indicated crosses, based on Mendelian ratios.

WESTERN BLOTTING ANALYSIS

The presence or absence of FADD, RIP3, and RIP1 protein was confirmed by western blotting. Total splenocytes were isolated from mice of the indicated genotypes. Cell lysates were prepared in cold RIPA lysis buffer containing 50 mM Tris pH 8.0, 150 mM NaCl, 1% Nonidet P-40, 0.5% deoxycholate, 0.1% SDS, 1 mM phenylmethyl sulphonyl fluoride (PMSF), and a proteinase inhibitor cocktail (Roche). Proteins (30 μg) were separated on a 10% SDS-PAGE gel and transferred to a nitrocellulose membrane. Proteins on the membrane were stained with Ponceau S (Sigma-Aldrich) as a loading/transfer control. Antibodies specific for FADD (generated in house), RIP3 (ProSci, Catalog #2283), and RIP1 (BD Biosciences, Catalog #610459) were incubated with the membrane overnight at 4°C followed by HRP-conjugated goat anti-rabbit antibodies (1/10,000). Chemiluminescent signals were detected on the ProteinSimple FluorChem M Western Blot Imaging machine.

FLOW CYTOMETRIC ANALYSIS

Spleen, lymph nodes, and thymus of mice were isolated during mouse dissection. Red blood cell lysis of the spleen was performed using ACK buffer and single cell suspensions were prepared. Cells (10^6) were stained with fluorochrome-conjugated antibodies in PBS containing 3% BSA, 0.5 mM EDTA, and 0.05% sodium azide, and samples were incubated on ice for 30 min. Antibodies used for flow cytometry were anti-CD3 (BD Biosciences), anti-CD4 (Caltag), anti-CD8 (Invitrogen), and anti-B220 (BD Biosciences).

Data were acquired on an LSR II cytometric analyzer (BD Biosciences) and analyzed using FlowJo software (TreeStar).

CELL CULTURE

Thymocytes and peripheral T cells purified from the spleen and lymph nodes were cultured in complete RPMI media (cRPMI) containing 10% FBS, 2 mM L-Glutamine, penicillin (100 U/mL), streptomycin (100 μg/mL), and β-mercaptoethanol (50 μM).

TOTAL CELLULARITY CALCULATIONS

Total cell number was obtained using single cell suspensions spleen and lymph nodes counted using a Countess Automated Cell Counter (Invitrogen). Cell number of various lymphocyte subsets was obtained by multiplying the total cell number with percentages of subsets of $CD3^+$ and $B220^+$ cells obtained by flow cytometry.

CELL DEATH ASSAYS

Thymocytes (10^5) were seeded in a 96-well flat-bottom plate in 100 μL of cRPMI and incubated at 37°C with 5% $CO_2$for 16 h with indicated concentrations of anti-Fas antibodies, TNFα, or dexamethasone. Thymocytes were also incubated with no treatment, staurosporine (1 μg/mL), and ionomycin (1 μg/mL) for indicated time points. Cells were collected, each well was washed with PBS, and 1 μg/mL propidium iodide was added to the sample. Cells were analyzed by flow cytometry on an LSR II cytometric analyzer (BD Biosciences). In some cases, Percent Specific Death was calculated by: (percent cell killing with treatment)—(percent cell killing without treatment).

PROLIFERATION ASSAYS

T cells were purified by pooling total spleen and lymph node cells of five-week-old mice and using antibodies against Thy1.2 and B220. $Thy1.2^+B220^-$ T cells were sorting using a FACS Aria (BD Biosciences). Purified T cells were labeled with CellTrace Violet (Life Technologies), as per the manufacturer's instructions. Cells were placed in a 96 well round-bottom plate (2 × 10^5 cells/well) and stimulated with 1 μg/mL anti-CD3 antibody and 0.2 μg/mL anti-CD28 antibody in 100 μL cRPMI for various timepoints. Cells were stained with 1 μg/mL propidium iodide and run on an LSR II cytometric analyzer (BD Biosciences). For ^{3}H-thymidine experiments, cells were placed in a 96 well round-bottom plate (10^5 cells/well) and stimulated with 1 μg/mL anti-CD3 antibody and 0.2 μg/mL anti-CD28 antibody for 40 h. Two microcuries of tridiated thymidine were added to each well and cells were

A

THZ x THZ	*rip1 (+/+,+/-) fadd (+/+,+/-) rip3 (+/+,+/-)*	*rip1 (-/-) fadd (+/+,+/-) rip3 (+/+,+/-)*	*rip1 (+/+,+/-) fadd (+/+,+/-) rip3 (-/-)*	*rip1 (-/-) fadd (+/+, +/-) rip3 (-/-)*	*rip1 (+/+,+/-) fadd (-/-) rip3 (+/+,+/-)*	*rip1 (-/-) fadd (-/-) rip3 (+/+)*	*rip1 (-/-) fadd (-/-) rip3 (+/-)*	*rip1 (+/+, +/-) fadd (-/-) rip3 (-/-)*	*rip1 (-/-) fadd (-/-) rip3 (-/-)*	Total
Probability	27/64	9/64	9/64	3/64	9/64	1/64	2/64	3/64	1/64	64/64
Predicted	108	36	36	12	36	4	8	12	4	256
Actual	74	0	19	0	0	0	2	5	3	103

B

rip1+/- fadd-/- rip3-/- x THZ	*rip1(+/+, +/-) fadd(+/-) rip3(+/-)*	*rip1(-/-) fadd(+/-) rip3(+/-)*	*rip1 (+/+, +/-) fadd(+/-) rip3 (-/-)*	*rip1(-/-) fadd(+/-) rip3(-/-)*	*rip1(+/+, +/-) fadd(-/-) rip3(+/-)*	*rip1(-/-) fadd(-/-) rip3(+/-)*	*rip1(+/+, +/-) fadd(-/-) rip3(-/-)*	*rip1(-/-) fadd(-/-) rip3 (-/-)*	Total
Probability	3/16	1/16	3/16	1/16	3/16	1/16	3/16	1/16	16/16
Predicted	24	8	24	8	24	8	24	8	60
Actual	21	0	16	0	0	3	17	3	128

C

rip1+/- fadd-/- rip3-/- x TKO	*rip1+/- fadd-/- rip3-/-*	*rip1-/- fadd-/- rip3-/-*
Probability	1/2	1/2
Predicted	16	16
Actual	16	6

D

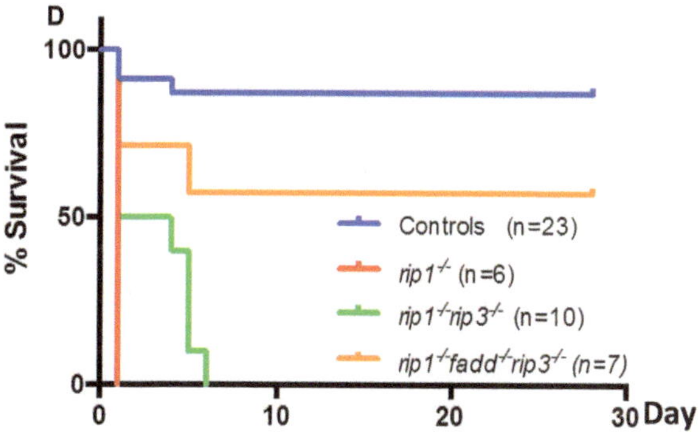

FIGURE 1 | *rip1*$^{-/-}$ perinatal lethality is mediated by FADD and RIP3. (A–C) Predicted and actual frequency of indicated genotypes of weaning age mice resulting from intercrosses of *rip1*$^{+/-}$ *fadd*$^{+/-}$ *rip3*$^{+/-}$ THZ mice **(A)**, crosses of *rip1*$^{+/-}$ *fadd*$^{+/-}$ *rip3*$^{+/-}$ and *rip1*$^{+/-}$ *fadd*$^{-/-}$ *rip3*$^{-/-}$ mice **(B)**, and crosses of *rip1*$^{+/-}$ *fadd*$^{-/-}$ *rip3*$^{-/-}$ and *rip1*$^{-/-}$ *fadd*$^{-/-}$ *rip3*$^{-/-}$ mice **(C)**. **(D)** Survival of newborn mice from various intercrosses. For this and following figures, control mice are littermates that have at least one wild type allele of FADD and RIP1. The *fadd*$^{-/-}$ *rip3*$^{-/-}$ mice shown contain at least one wild type allele of *rip1*.

incubated for an additional 8 h. Samples were placed onto a glass fiber filter using a Mach Harvester 96 (Tomtec) and analyzed using a Wallac 1205 Betaplate Counter (Perkin Elmer).

STATISTICAL ANALYSIS

Data is expressed as mean ± standard deviation. Student's *t*-tests were performed using Prism software (Graphpad Software, Inc).

RESULTS

POSTNATAL DEVELOPMENT OF *rip1*$^{-/-}$ MICE IN THE ABSENCE OF RIP3 AND/OR FADD

To determine whether RIP3 plays a role in the postnatal defect of *rip1*$^{-/-}$ mice, *rip1*$^{+/-}$*rip3*$^{+/-}$ or *rip1*$^{+/-}$*rip3*$^{-/-}$ mice were intercrossed, which produced no viable *rip1*$^{-/-}$*rip3*$^{-/-}$ double knockout (DKO) or *rip1*$^{-/-}$*rip3*$^{+/-}$ mice at weaning age (data not shown). This observation was confirmed when *rip1*$^{+/-}$*fadd*$^{+/-}$*rip3*$^{+/-}$ triple heterozygous (THZ) mutant mice were intercrossed, which produced no viable *rip1*$^{-/-}$*rip3*$^{-/-}$ mice containing one or both alleles of FADD, though Mendelian genetics predicts 12 mice of such genotypes (column 5, **Figure 1A**). Nonetheless, *rip1*$^{-/-}$*rip3*$^{-/-}$ newborns were detected, and showed slightly improved survival up to postpartum day (p)6, as compared to *rip1*$^{-/-}$ newborns which died at p1 (**Figure 1D**). In contrast, if one or both alleles of the *rip1* gene were present, *fadd*$^{-/-}$*rip3*$^{-/-}$ DKO mice were viable from birth to adulthood (column 9, **Figure 1A**).

Similar to the previous study (Zhang et al., 2011), no viable *rip1*$^{-/-}$*fadd*$^{-/-}$ mice were present at weaning age among the offspring of THZ intercrosses (column 7, **Figure 1A**). Strikingly, however, absence of one allele of RIP3 allowed survival of *rip1*$^{-/-}$*fadd*$^{-/-}$ mice beyond weaning ages up to adulthood (column 8, **Figures 1A, 2A**). Presence or absence of the FADD, RIP1, and RIP3 proteins was analyzed by western blotting (**Figure 2B**). These observations were also confirmed by analysis of the offspring from crosses of *rip1*$^{+/-}$*fadd*$^{-/-}$*rip3*$^{-/-}$ mice with *rip1*$^{+/-}$*fadd*$^{+/-}$*rip3*$^{+/-}$ THZ mice (**Figure 1B**), which produce higher ratios of *rip1*$^{-/-}$*fadd*$^{-/-}$*rip3*$^{+/-}$ mice.

Further examination revealed that although *rip1*$^{-/-}$*fadd*$^{-/-}$*rip3*$^{+/-}$ mice were viable, they were much smaller in size (**Figure 2A**), and also weighed less than wild type controls or *fadd*$^{-/-}$*rip3*$^{-/-}$ mice during the indicated period of analysis (**Figure 2C**). Moreover, *rip1*$^{-/-}$*fadd*$^{-/-}$*rip3*$^{+/-}$ mice appeared to be present at lower numbers (2) at weaning age than the predicted Mendelian frequencies (8) (column 8, **Figure 1A**).

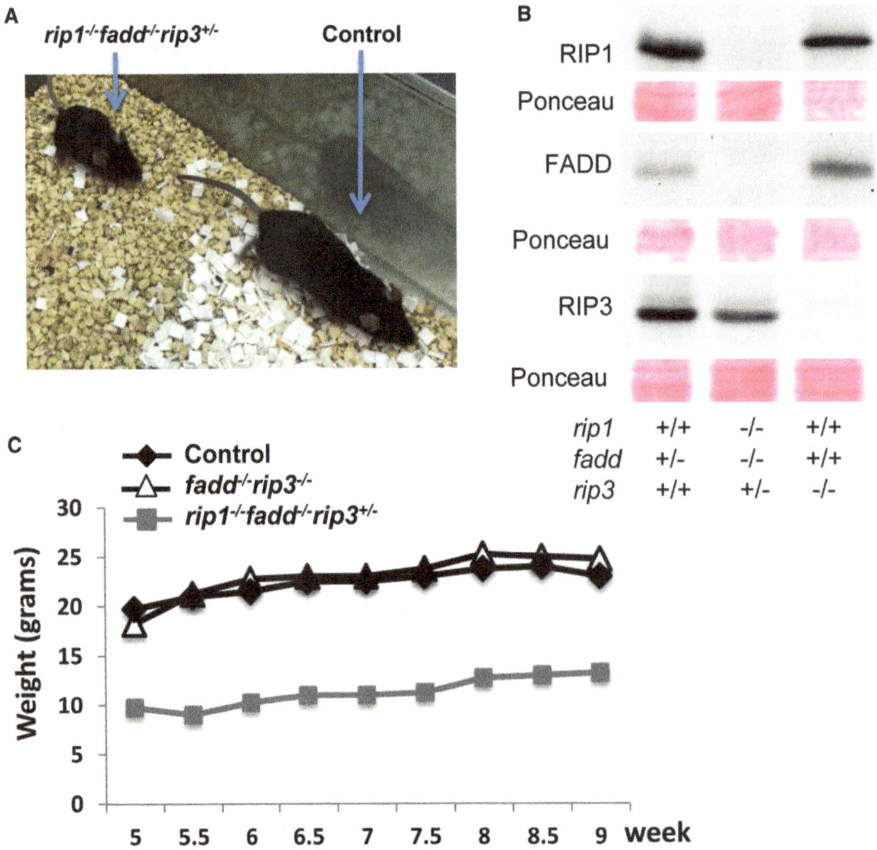

FIGURE 2 | Haploid insufficiency of *rip3* improves survival of *rip1*$^{-/-}$*fadd*$^{-/-}$ mice, which are viable but weigh less than wild type controls. (A) Adult *rip1*$^{-/-}$*fadd*$^{-/-}$*rip3*$^{+/-}$ mouse and littermate control. **(B)** Western blot analysis of total splenocytes confirming presence or absence of FADD, RIP3, and RIP1 protein. **(C)** Weight curves showing the *rip1*$^{-/-}$*fadd*$^{-/-}$*rip3*$^{+/-}$ mouse with age-matched controls. Data shown is representative of three mice of each genotype.

Western blotting analysis indicated that ablation of one allele of RIP3 resulted in lower RIP3 protein levels (**Figure 2B**), which helped improve survival of *rip1*$^{-/-}$*fadd*$^{-/-}$ mice.

Interestingly, when both alleles of RIP3 were absent, the resulting *rip1*$^{-/-}$*fadd*$^{-/-}$*rip3*$^{-/-}$ triple knockout (TKO) mice gained weight normally, as compared to wild type control or *fadd*$^{-/-}$*rip3*$^{-/-}$ DKO mice (**Figures 1, 3A,C,D**). Nonetheless, the numbers of the TKO mice appeared to be lower than that predicted by Mendelian ratios, particularly among the offspring of the mouse cross in **Figure 1B** (column 10). This defect is more pronounced through analyzing the offspring of crosses between *rip1*$^{+/-}$*fadd*$^{-/-}$*rip3*$^{-/-}$ mice and TKO mice (**Figure 1C**). This correlates with decreased survival of *rip1*$^{-/-}$*fadd*$^{-/-}$*rip3*$^{-/-}$mice from birth to weaning age (**Figure 1D**). Knockout of these genes was confirmed by western blotting (**Figure 3B**). In total, these data showed that while *rip1*$^{-/-}$*rip3*$^{-/-}$ mice did survive several days longer than *rip1*$^{-/-}$, knockout of both FADD and RIP3 was necessary for survival of *rip1*$^{-/-}$ mice to adulthood.

THE ROLE OF RIP1, FADD, AND RIP3 IN LYMPHOCYTE RESPONSES

The role of these proteins in lymphocyte development and function has been previously investigated through gene deletion in germ cells or in specific lineages. The lymphoid system of *rip3*$^{-/-}$ mice is normal (Newton et al., 2004). Without FADD function, peripheral T cell number decreases (Walsh et al., 1998; Zhang et al., 2005), but the B cell compartment is normal (Imtiyaz et al., 2006). Lastly, inducible deletion of RIP1 results in greatly reduced lymphocyte numbers (Roderick et al., 2014). We initially analyzed young (4 week old) mice, due to concerns of age-dependent diseases (see below). The *fadd*$^{-/-}$*rip3*$^{-/-}$ DKO and *rip1*$^{-/-}$*fadd*$^{-/-}$*rip3*$^{-/-}$ TKO mutant mice contained normal T cell populations in the thymus, similar to that in control mice (**Figure 4B**). This indicates that RIP1 deficiency does not affect T lymphocyte development, as long as FADD and RIP3 are absent. The lymphoid organs of DKO and TKO mice at 5 weeks of age were not significantly larger than control mice, and these mutant mice appeared to contain higher numbers of B cells in the lymph nodes (**Figures 4A,C,D**).

The extrinsic cell death pathways mediated by the DRs are essential for immune homeostasis, as indicated by the *lpr* mouse model. Disruption of FADD function blocks Fas-induced apoptosis in T cells (Walsh, 2014). We previously have shown that absence of RIP1 has no effect on Fas-induced death, but leads to hypersensitivity to TNF-induced death in thymocytes (Zhang et al., 2011). In contrast, *rip1*$^{-/-}$*fadd*$^{-/-}$ thymocytes were resistant to TNF, similar to wild type thymocytes (Zhang et al., 2011). In order to determine the status of these pathways

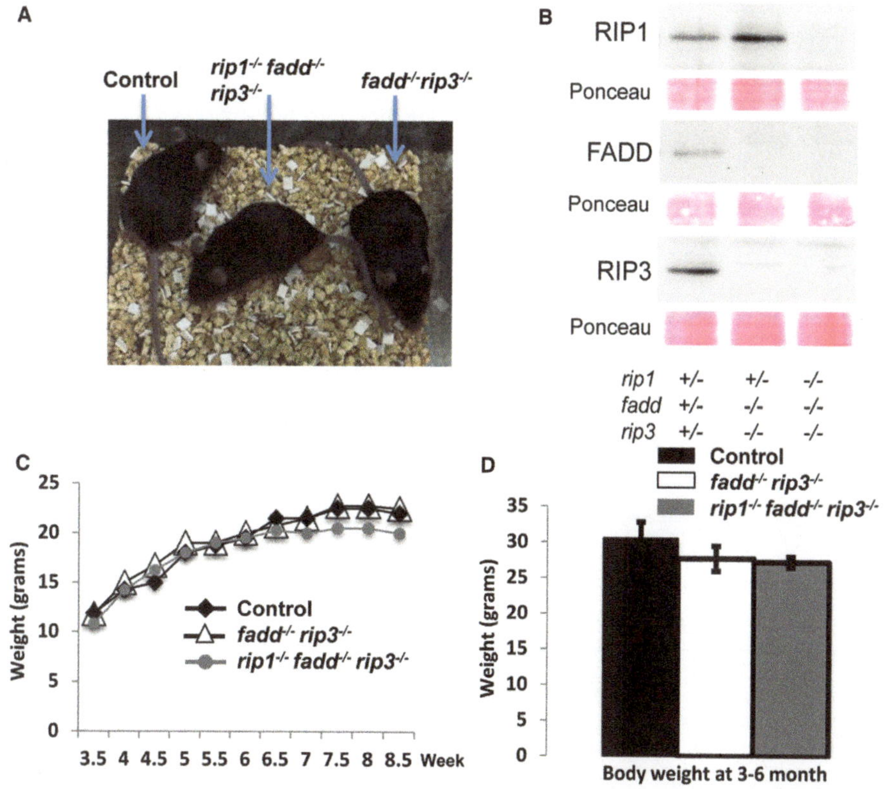

FIGURE 3 | Complete ablation of both RIP3 and FADD rescued *rip1*$^{-/-}$ perinatal lethality. **(A)** *rip1*$^{-/-}$*fadd*$^{-/-}$*rip3*$^{-/-}$ mouse with *fadd*$^{-/-}$*rip3*$^{-/-}$ and littermate control. **(B)** Western blot of total splenocytes confirming presence or absence of FADD, RIP3, and RIP1 protein. **(C)** Weight curves of *rip1*$^{-/-}$*fadd*$^{-/-}$*rip3*$^{-/-}$ with age-matched controls, representative of at least three mice of each genotype. **(D)** Body weight of 3- to 6-month-old mice at time of analysis. Data shown is representative of four mice of each genotype.

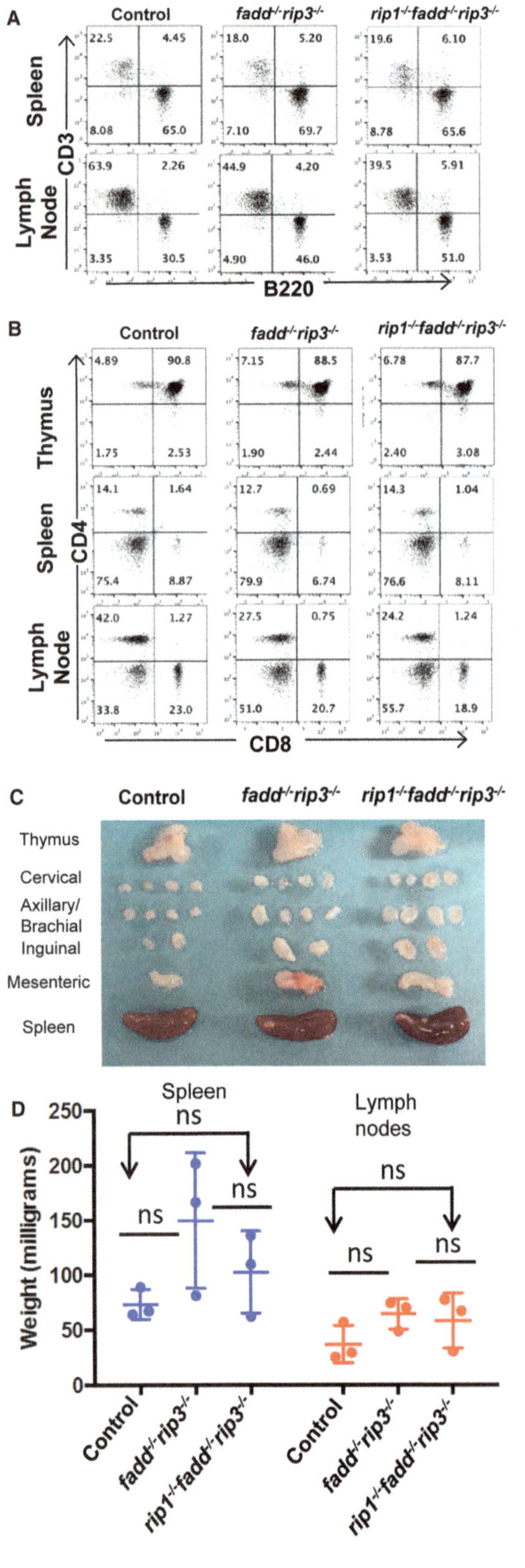

FIGURE 4 | Young $rip1^{-/-}fadd^{-/-}rip3^{-/-}$ **mice contain normal T and B cell populations. (A,B)** Spleen, lymph node, and thymus of 5 week old mice of the indicated genotypes were stained with antibodies for CD3 and *(Continued)*

FIGURE 4 | Continued
B220 **(A)** or CD4 and CD8 **(B)**. Numbers indicated are percentages of T cell and B cell subsets. **(C)** Lymph nodes, spleen, thymus of 5 week-old mice. **(D)** Weight of spleen and lymph node of mice 4–6 week old mice. ns, not significant. Data shown is representative of at least three independent experiments with one mouse of each genotype.

in $rip1^{-/-}fadd^{-/-}rip3^{-/-}$ mice, thymocytes were isolated and treated with various stimuli to initiate extrinsic or intrinsic cell death. Similar to $fadd^{-/-}rip3^{-/-}$ DKO thymocytes, TKO thymocytes were highly resistant to cell death responses induced by Fas (**Figure 5A**). Unlike $rip1^{-/-}$ thymocytes (Zhang et al., 2011), TKO and DKO mutant thymocytes were not sensitive to TNFα-induced cell death (**Figure 5B**). However, they were able to undergo intrinsic apoptosis in response to dexamethasone, similar to wild type thymocytes (**Figure 5C**). Furthermore, DKO and TKO thymocytes were sensitive to staurosporine and ionomycin-induced intrinsic apoptosis, as compared with untreated cells (**Figures 5D–F**). These results imply that resistance to extrinsic cell death is essential for the survival of $rip1^{-/-}$ mice.

Previous studies indicated that FADD is required during lymphocyte proliferative responses, to sequester RIP1/RIP3-dependent necrosis (Lu et al., 2011; Zhang et al., 2011). To analyze TCR-induced proliferation, peripheral T cells were isolated and stimulated with anti-CD3 antibodies. In tritiated thymidine incorporation assays, there appeared to be more proliferation in $fadd^{-/-}rip3^{-/-}$ T cells and $rip1^{-/-}fadd^{-/-}rip3^{-/-}$ T cells than control T cells (**Figure 6A**). In further analysis, T cell division kinetics and death responses were analyzed through two-color flow cytometry. T cells were pre-labeled with the CellTrace dye, and then stimulated with anti-CD3/CD28 antibodies. T cell division is indicated by dilution of CellTrace, and cell death measured through propidium iodide dye staining. As shown in **Figure 6B**, accelerated proliferative responses were seen in both $fadd^{-/-}rip3^{-/-}$ and $rip1^{-/-}fadd^{-/-}rip3^{-/-}$ T cells, when compared to control T cells. Moreover, there was less cell death in DKO and TKO mutant T cells than in control T cells (**Figure 6B**).

THE ROLE OF RIP1 IN AGE-DEPENDENT *lpr* DISEASES

Previous studies have shown that disruption of the function of FADD by using a FADD-DN mutant results in an *lpr*-like symptom in aged $rip3^{-/-}$ mice (Lu et al., 2011). Similarly, $fadd^{-/-}rip3^{-/-}$ DKO mutant mice developed *lpr*-like symptoms including enlarged lymph nodes and spleen in as early as 5-week old mice (**Figure 7** and data not shown). As mice age, the lymphadenopathy and splenomegaly phenotype become more pronounced, as indicated by greatly increased size, weight, and total cellularity of the secondary lymphoid organs of the $fadd^{-/-}rip3^{-/-}$ DKO mice (**Figures 7A–C**). Interestingly, these signature *lpr* symptoms were diminished in TKO mice, as compared to DKO mice (**Figure 7**).

Flow cytometric analyses showed that DKO mice contained the unique subset of $CD3^{+}$ T cells, which express the B cell marker $B220^{+}$, a signature phenotype of the Fas mutant mice (**Figure 8A**). Furthermore, these T cells lose the expression of CD4 and CD8 (**Figure 8B**). When absolute numbers of distinct

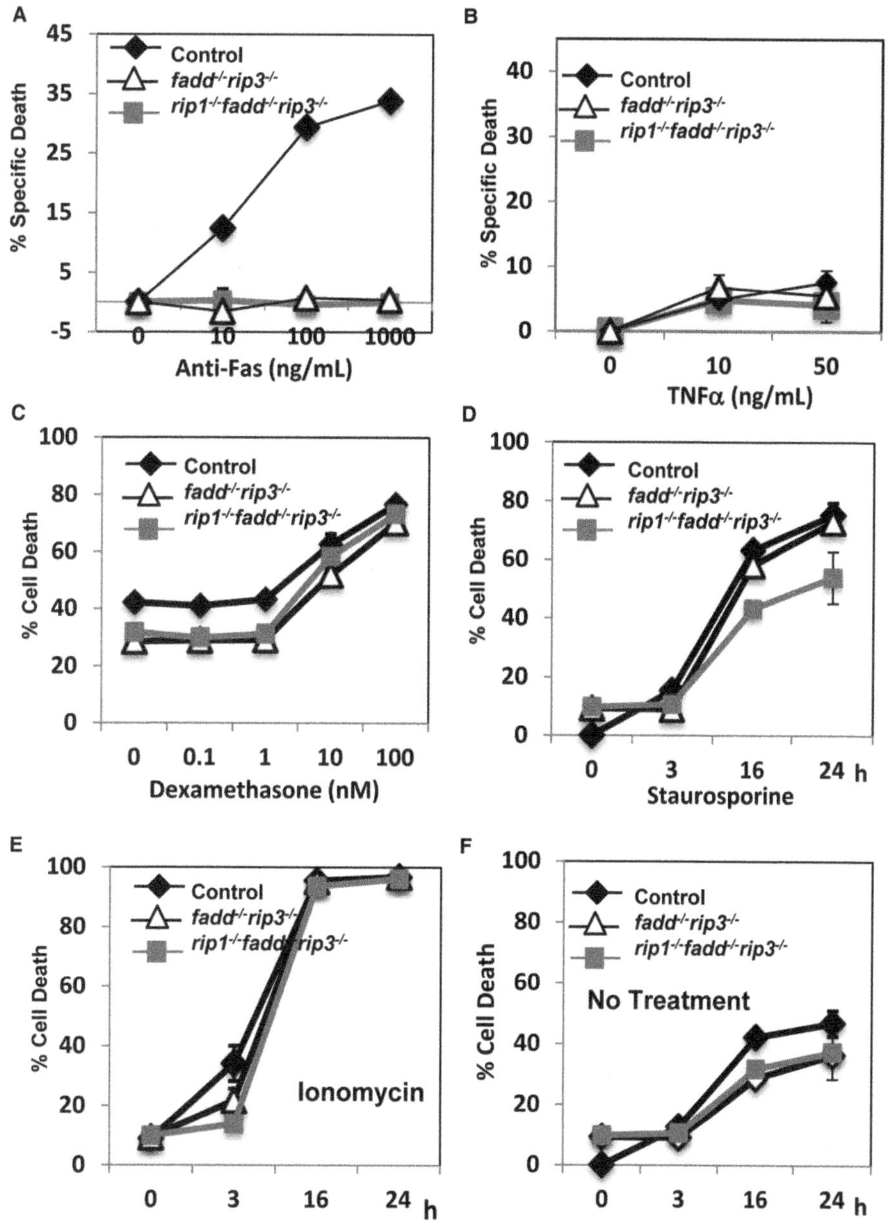

FIGURE 5 | *rip1*$^{-/-}$*fadd*$^{-/-}$*rip3*$^{-/-}$ mice had impairment of extrinsic, but not intrinsic, cell death pathways. (A–D) Thymocytes of indicated genotypes were treated with various doses of **(A)** anti-Fas antibody, **(B)** TNFα, **(C)** Dexamethasone and cell death was determined by PI uptake after 16 h incubation. **(D–F)** Thymocytes were treated with **(D)** staurosporin (1 μg/ml), **(E)** ionomycin (1 μg/mL) or **(F)** no treatment for indicated time and cell death was determined by PI uptake. Experiments were performed in triplicate.

lymphocyte populations were determined, DKO mice aged 3–6 months were shown to accumulate dramatically higher numbers of CD3$^+$B220$^+$ T cells (**Figure 8C**). Moreover, B cells and conventional T cells (CD3$^+$B220$^-$) number also increased in DKO mice (**Figures 8D,E**). Such *lpr*-like phenotypes were significantly less severe when RIP1 was deleted (**Figures 7, 8**).

DISCUSSION

PCD is tightly controlled in order to ensure proper embryonic and postnatal development as well as to maintain tissue homeostasis. Two major PCD pathways, intrinsic and extrinsic apoptosis, have been well-studied. Intrinsic apoptosis is mediated by pro-apoptotic members of the Bcl-2 family proteins including Bax and Bak, which are counteracted by Bcl-2 and Bcl-$X_{S/L}$ (Marsden and Strasser, 2003; Danial and Korsmeyer, 2004). Absence of Bcl-$X_{S/L}$ blocks embryogenesis at early-midgestation stages due to massive cell death, and deletion of Bcl-2 results in lymphopenic conditions due to greatly increased apoptosis in lymphocytes. In contrast, when proapoptotic Bim is inactivated, mice develop autoimmune and lymphoproliferative diseases

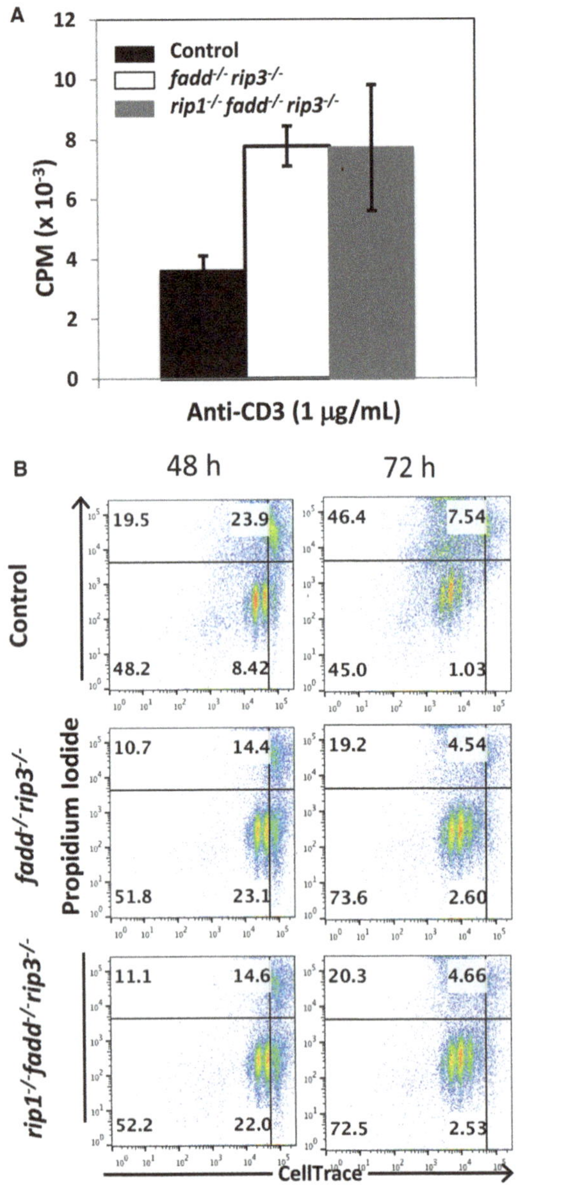

FIGURE 6 | Proliferative responses of TKO and DKO T cells are more robust than control T cells. (A) T cells of indicated genotypes were stimulated with anti-CD3, anti-CD28 antibodies. Proliferation was measured by [^{3}H]-thymidine incorporation (cpm, counts per minute). **(B)** T cells labeled with CellTrace dye were stimulated with anti-CD3, anti-CD28 antibodies. Dilution of CellTrace dye and uptake of propidium iodide was monitored by flow cytometry at various time points. Data shown is representative of at least three independent experiments with one mouse of each genotype.

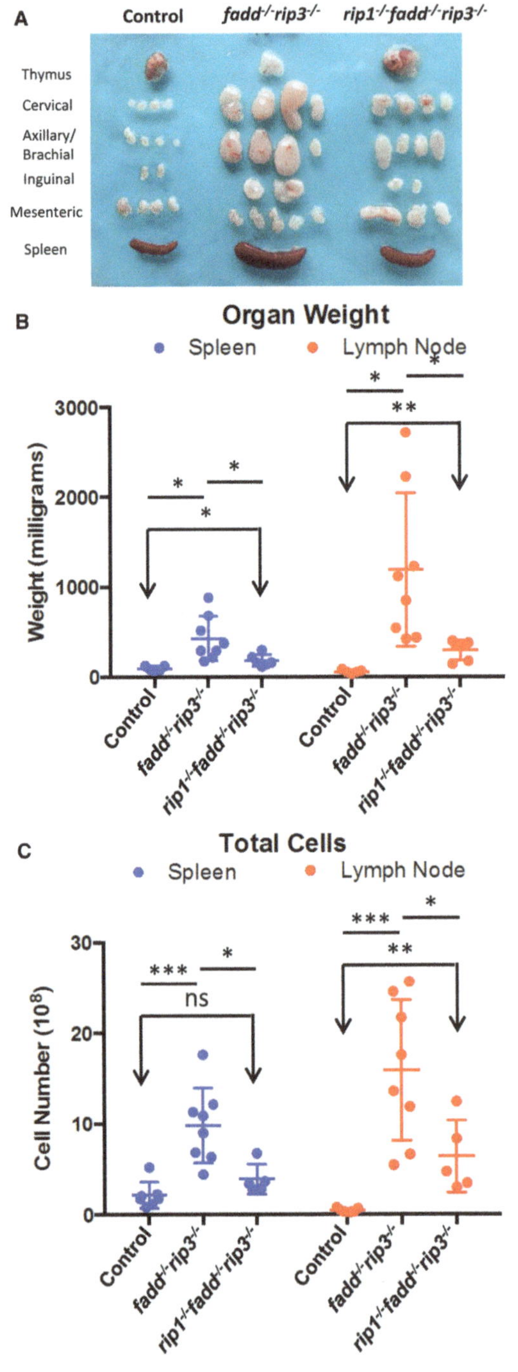

FIGURE 7 | Lymphadenopathy and splenomegaly is less severe in TKO mice, compared to DKO mice. (A) Thymus, lymph nodes, and spleen of 6-month-old mice **(B)** Weight of spleen and lymph nodes of mice >3 months old **(C)** Total cell number of spleen and lymph node of mice 3–6.5 months old. At least five mice from each group were used to generate given data. $^*p < 0.05$; $^{**}p < 0.01$; $^{***}p < 0.001$; ns, not significant.

(Bouillet et al., 2002). The significance of extrinsic PCD pathways during animal development mediated by DRs or TLRs has been underappreciated, mainly because mice lacking these receptors undergo normal embryonic and postnatal development. As such, developmental defects in mice without the downstream signaling proteins of the DRs, including RIP1, FADD, Caspase 8, or cFLIP, remained largely a conundrum for over a decade, until a latent, RIP1/RIP3-dependent necrosis-like death pathway has been revealed recently (Moriwaki and Chan, 2013; Walsh, 2014).

The fact that RIP1, as implicated in studies based on cell systems, is involved in the regulation of NF-κB activation, apoptosis,

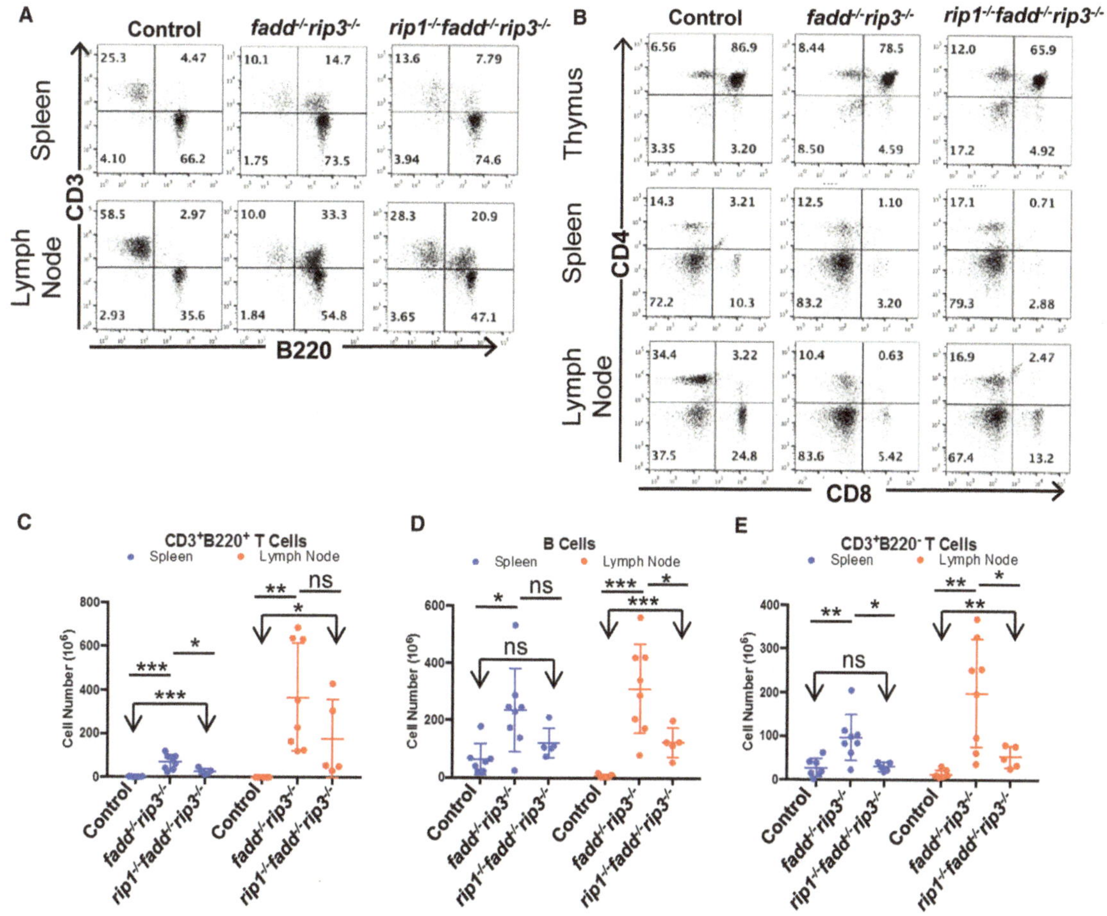

FIGURE 8 | Knockout of RIP1 reduced severity of the *lpr* symptoms of lymphadenopathy and splenomegaly in *fadd*$^{-/-}$ *rip3*$^{-/-}$. (A,B) Spleen, lymph node, and thymus of six-month-old mice of the indicated genotypes were stained with antibodies for CD3 and B220 **(A)** or for CD4 and CD8 **(B)**. Numbers indicated are percentages of T cell and B cell subsets. **(C–E)** Scatter plots of total number of **(C)** CD3$^+$B220$^+$ T Cells, **(D)** B cells, and **(E)** conventional CD3$^+$B220$^-$ T cells. At least five mice from each group were used to generate given data. $^*p < 0.05$; $^{**}p < 0.01$; $^{***}p < 0.001$; ns, not significant.

and necroptosis, poses as a challenge in understanding the perinatal lethality phenotype of *rip1*$^{-/-}$ mice (Kelliher et al., 1998). One possible mechanism is that *rip1*$^{-/-}$ cells become hypersensitive to extrinsic cell death, as observed previously(Ting et al., 1996; Kelliher et al., 1998). However, our previous study shows that blocking extrinsic apoptosis pathways alone through deletion of FADD is insufficient, as postnatal survival of *rip1*$^{-/-}$*fadd*$^{-/-}$ mice were only marginally improved (Zhang et al., 2011). Alternatively, evidence exists, which implies that RIP3 mediates necroptosis in the absence of RIP1 (Upton et al., 2010; Moujalled et al., 2013). Indeed, we found that deletion of RIP3 helped improve the survival of *rip1*$^{-/-}$ mice for several additional days (**Figure 1** and data not shown). Importantly, the current study demonstrates that postnatal lethality in *rip1*$^{-/-}$ mice is due to the functions of not only FADD but also RIP3, because *rip1*$^{-/-}$ mice could reach adulthood only if both the FADD and RIP3 genes are ablated simultaneously (**Figures 1–3**). Firstly, decreasing the RIP3 protein level through deletion of just one allele of the RIP3 gene significantly improved survival of *rip1*$^{-/-}$*fadd*$^{-/-}$, although the resulting *rip1*$^{-/-}$*fadd*$^{-/-}$*rip3*$^{+/-}$ mice had trouble gaining weight (**Figures 1, 2**). This could suggest that the initial pathology of cell death in the adipose tissue *rip1*$^{-/-}$ mice is partially mediated by RIP3, which may explain the systemic inflammation displayed in *rip1*$^{-/-}$*fadd*$^{-/-}$*rip3*$^{+/-}$ mice, as indicated in a previous independent study (Kaiser et al., 2014). Secondly, when both alleles of RIP3 were deleted, the resulting TKO mice show no defect in body sizes and weights as compared to DKO or control littermates (**Figure 3**). This data implies that during postnatal development, *rip1*$^{-/-}$ cells may undergo both FADD-mediated apoptosis and RIP3-dependent necroptosis. This new finding is counterintuitive because RIP1 appears to signal through RIP3 in promoting necroptosis in embryonic cells lacking FADD or Caspase 8 or cFLIP (Kaiser et al., 2011; Oberst et al., 2011; Dillon et al., 2012). In particular, our previous study showed that deletion of the RIP1 adaptor kinase is able to correct the embryonic defect in *fadd*$^{-/-}$ mice (Zhang et al., 2011). Moreover, *fadd*$^{-/-}$ or *caspase 8*$^{-/-}$ mice could reach adulthood when RIP3 is deleted (Kaiser et al., 2011; Oberst et al., 2011; Dillon et al., 2012). It appears that RIP1 plays distinct roles at different stages during development. While dispensable in embryos, RIP1 is required to restrict

FADD-mediated apoptosis and RIP3-dependent necroptosis in order to ensure proper postnatal development.

The precise mechanism involved in RIP1-mediated sequestration of apoptosis and necroptosis remains to be further defined. One scenario involves cFLIP, an inhibitor of both apoptosis and necroptosis. cFLIP is a target of NF-κB (Kreuz et al., 2001), which could be activated through RIP1 under certain circumstances. Likely, lack of RIP1 impairs the expression of cFLIP. It was previously thought that the defect in *rip1*$^{-/-}$ mice is due, in part, to TNFR1-triggered apoptosis. Recently, deletion of TNFR1 or caspase 8 in addition to RIP1 was attempted to mitigate uncontrolled apoptosis, but provided no survival advantage (Rickard et al., 2014). A RIP1-independent necroptotic pathway is induced by TLR3 and TLR4 through its adaptor TIR-domain-containing adaptor-inducing interferon-β (TRIF), which can bind RIP3 via the RIP homotypic interaction motif (RHIM) (Kaiser et al., 2013; Dillon et al., 2014). These two TLRs could also trigger activation of NF-κB through RIP1 (Meylan et al., 2004; Zhang et al., 2011). The type I and type II interferon (IFN) receptors are also capable of inducing necroptosis (Balachandran et al., 2004; Kaiser et al., 2014). In concordance with these models, knockout of TRIF or the type I IFN receptor, in addition to knockout of RIP1 and TNFR, modestly prolonged survival, allowing a small number of these mice to survive to weaning age (Dillon et al., 2014). The perinatal lethality of *rip1*$^{-/-}$ mice was partially rescued when apoptosis or necroptosis mediated by TNFR, TLRs, or IFN receptors was blunted (Dillon et al., 2014; Kaiser et al., 2014; Rickard et al., 2014).

RIP3 and MLKL have been shown to execute necroptosis in a RIP1-independent manner through Toll-like Receptor-3 (TLR-3) (Upton et al., 2010; He et al., 2011; Kaiser et al., 2013). Recent independent study has shown that ablation of either RIP3 or MLKL in addition to RIP1 resulted in lower levels of cleaved caspase 3 in various organs in newborns compared with *rip1*$^{-/-}$ newborns as well as lower levels of inflammatory cytokines, potentially due to lower levels of damage-associated molecular patterns (DAMPs) released from necrotic cells (Rickard et al., 2014). However, these mice did not survive past day p4. Another study found that a small percentage of *rip1*$^{-/-}$*rip3*$^{-/-}$ mice survived up to 21 days, however, we see no survival of *rip1*$^{-/-}$*rip3*$^{-/-}$ mice beyond day p6 (**Figure 1D**). Furthermore, some *rip1*$^{-/-}$*tnfr*$^{-/-}$ survived for 18 days (Dillon et al., 2014), whereas such mice die right at birth in a separate study (Rickard et al., 2014). Combined knockout of RIP3 and TNFR1 did allow survival of *rip1*$^{-/-}$ mice into adulthood, but these mice ultimately succumbed to blood bacteremia, sepsis, and prevalent apoptosis in the intestines (Dillon et al., 2014). This lethal intestinal defect caused by RIP1 ablation that was recently characterized using cell-specific Cre-mediated deletion of RIP1 (Dannappel et al., 2014; Takahashi et al., 2014). In another study, it was noted that *rip1*$^{-/-}$*rip3*$^{-/-}$*tnf*$^{-/-}$ mice did not survive past day 28 (Kaiser et al., 2014). The discrepancy in survival of mice in these studies could be due to differences in genetic backgrounds of the mice used. The lethality of *rip1*$^{-/-}$*rip3*$^{-/-}$*tnfr*$^{-/-}$ mice also suggests a key role for FADD/caspase 8-mediated apoptosis in RIP1 perinatal lethality. Therefore, lethality of RIP1 knockout mice is due to multiple factors.

Thymocytes isolated from DKO and TKO mice were resistant to death induced by the extrinsic pathways triggered through stimulations with anti-Fas antibodies or TNFα (**Figures 5A,B**). However, these mutant T cells were sensitive to cell death induced by dexamethasone, staurosporine, or ionomycin, similar to that of the wild-type control mice (**Figures 5C,D,E**). Extrinsic cell death has been shown to play a key role in lymphocyte homeostasis and peripheral tolerance (Strasser et al., 2009). When lymphocytes are not capable of being eliminated in the periphery, progressive accumulation of lymphocytes can occur in the peripheral lymphoid organs. This is the case in *lpr* mice, displaying splenomegaly and lymphadenopathy as a result of a mutation in the Fas gene. Whereas disruption of the FADD function alone through gene deletion or using FADD-DN results in a lymphopenic condition (Newton et al., 1998; Walsh et al., 1998; Zhang et al., 2005), double deletion of both FADD and RIP3 leads to *lpr*-like symptoms (**Figures 7, 8**), which is similar to that in *caspase8*$^{-/-}$*rip3*$^{-/-}$ mice (Kaiser et al., 2011; Oberst et al., 2011), but is more severe that in *fas*$^{-/-}$ mutant mice. Notably, however, a diminished *lpr* disease was present in *rip1*$^{-/-}$*fadd*$^{-/-}$*rip3*$^{-/-}$ mice (**Figures 7, 8**). This indicates that RIP1 has additional functions besides blocking FADD-mediated apoptosis and RIP3-dependent necroptosis. Interestingly, *rip1*$^{-/-}$*fadd*$^{-/-}$*rip3*$^{+/-}$ mice develop splenomegaly, but not lymphadenopathy, after several months of age (data not shown), indicating a role for RIP3-dependent necroptosis in the clearance of lymphocytes in the lymph nodes, but not the spleen. These *rip1*$^{-/-}$*fadd*$^{-/-}$*rip3*$^{+/-}$, as well as DKO and TKO mice, contain varying numbers of the peculiar $CD3^{+}B220^{+}$ T cells, which lose CD4 and CD8 expression (**Figure 8** and data not shown). These data, in total, illustrate that in the immune system, chronic activation of lymphocytes disrupts homeostasis and leads to autoimmunity, which can be prevented by FADD-mediated apoptosis, a sterile PCD. RIP3-deficient T cells undergo normal activation-induced cell death (AICD), indicating that FADD-mediated apoptosis is sufficient for maintaining homeostasis. Intuitively, apoptosis is a preferred pathway for AICD as necroptosis may potentially lead to inflammation which needs to be suppressed by FADD. Therefore, although pro-inflammatory, RIP1/RIP3-dependent necroptosis serves as a backup, when the FADD pathway is compromised. It is interesting to note that there appeared to be more cell division in activated *fadd*$^{-/-}$*rip3*$^{-/-}$ T cells than in wild type T cells (**Figure 6**). This may due in part to decreased cell death in activated *fadd*$^{-/-}$*rip3*$^{-/-}$ T cells, and/or a role for FADD in negatively regulating T cell proliferative responses. As such, both of these functions of FADD likely contribute to the greatly accelerated *lpr* disease in *fadd*$^{-/-}$*rip3*$^{-/-}$ mice.

Clearly, RIP1 plays an essential role in the life of an organism and the absence of this protein can only be compensated for by removing both apoptosis and necroptosis pathways. Due to its essential function of regulating not only cell death, but also cell survival, there may be further important consequences for the loss of such a key protein. Future investigations will drive the discoveries of the essential functions of RIP1 and the consequences for deleting RIP1. These studies will broaden our understanding of RIP1 function and could lead to the development of new strategies to target inflammatory conditions as well as cancer and autoimmune diseases, that may be exacerbated due to unregulated necrosis.

ACKNOWLEDGMENTS

This study is supported in part by NIH grant R21 AI105794 to JZ. We thank Dr. Michelle Kelliher for providing $rip1^{+/-}$ mice and Drs. Kim Newton and Vishva Dixit for providing $rip3^{-/-}$ mice, Jennifer Wilson for critical reading of the manuscript, technical contribution by Ben Davis, the technical staff for their help at the Flow Cytometry Facility and Research Animal Facility of the Sidney Kimmel Cancer Center, which is supported in part by NCI Cancer Center support grant P30CA56036.

REFERENCES

Adachi, M., Suematsu, S., Suda, T., Watanabe, D., Fukuyama, H., Ogasawara, J., et al. (1996). Enhanced and accelerated lymphoproliferation in Fas-null mice. *Proc. Natl. Acad. Sci. U.S.A.* 93, 2131–2136. doi: 10.1073/pnas.93.5.2131

Ashkenazi, A., and Dixit, V. M. (1998). Death receptors: signaling and modulation. *Science* 281, 1305–1308. doi: 10.1126/science.281.5381.1305

Balachandran, S., Thomas, E., and Barber, G. N. (2004). A FADD-dependent innate immune mechanism in mammalian cells. *Nature* 432, 401–405. doi: 10.1038/nature03124

Boldin, M. P., Goncharov, T. M., Goltsev, Y. V., and Wallach, D. (1996). Involvement of MACH, a novel MORT1/FADD-interacting protease, in Fas/APO-1-and TNF receptor-induced cell death. *Cell* 85, 803–815. doi: 10.1016/S0092-8674(00)81265-9

Boldin, M. P., Varfolomeev, E. E., Pancer, Z., Mett, I. L., Camonis, J. H., and Wallach, D. (1995). A novel protein that interacts with the death domain of Fas/APO1 contains a sequence motif related to the death domain. *J. Biol. Chem.* 270, 7795–7798. doi: 10.1074/jbc.270.14.7795

Bouillet, P., Purton, J. F., Godfrey, D. I., Zhang, L. C., Coultas, L., Puthalakath, H., et al. (2002). BH3-only Bcl-2 family member Bim is required for apoptosis of autoreactive thymocytes. *Nature* 415, 922–926. doi: 10.1038/415922a

Ch'en, I. L., Tsau, J. S., Molkentin, J. D., Komatsu, M., and Hedrick, S. M. (2011). Mechanisms of necroptosis in T cells. *J. Exp. Med.* 208, 633–641. doi: 10.1084/jem.20110251

Chinnaiyan, A. M., O'Rourke, K., Tewari, M., and Dixit, V. M. (1995). FADD, a novel death domain-containing protein, interacts with the death domain of Fas and initiates apoptosis. *Cell* 81, 505–512. doi: 10.1016/0092-8674(95)90071-3

Cho, Y. S., Challa, S., Moquin, D., Genga, R., Ray, T. D., Guildford, M., et al. (2009). Phosphorylation-driven assembly of the RIP1-RIP3 complex regulates programmed necrosis and virus-induced inflammation. *Cell* 137, 1112–1123. doi: 10.1016/j.cell.2009.05.037

Cohen, P. L., and Eisenberg, R. A. (1992). The lpr and gld genes in systemic autoimmunity: life and death in the Fas lane. *Immunol. Today* 13, 427–428. doi: 10.1016/0167-5699(92)90066-G

Cusson, N., Oikemus, S., Kilpatrick, E. D., Cunningham, L., and Kelliher, M. (2002). The death domain kinase RIP protects thymocytes from tumor necrosis factor receptor type 2-induced cell death. *J. Exp. Med.* 196, 15–26. doi: 10.1084/jem.20011470

Danial, N. N., and Korsmeyer, S. J. (2004). Cell death: critical control points. *Cell* 116, 205–219. doi: 10.1016/S0092-8674(04)00046-7

Dannappel, M., Vlantis, K., Kumari, S., Polykratis, A., Kim, C., Wachsmuth, L., et al. (2014). RIPK1 maintains epithelial homeostasis by inhibiting apoptosis and necroptosis. *Nature* 513, 90–94. doi: 10.1038/nature13608

Degterev, A., and Yuan, J. (2008). Expansion and evolution of cell death programmes. *Nat. Rev. Mol. Cell Biol.* 9, 378–390. doi: 10.1038/nrm2393

Diehl, G. E., Yue, H. H., Hsieh, K., Kuang, A. A., Ho, M., Morici, L. A., et al. (2004). TRAIL-R as a negative regulator of innate immune cell responses. *Immunity* 21, 877–889. doi: 10.1016/j.immuni.2004.11.008

Dillon, C. P., Weinlich, R., Rodriguez, D. A., Cripps, J. G., Quarato, G., Gurung, P., et al. (2014). RIPK1 blocks early postnatal lethality mediated by caspase-8 and RIPK3. *Cell* 157, 1189–1202. doi: 10.1016/j.cell.2014.04.018

Dillon, C. P., Oberst, A., Weinlich, R., Janke, L. J., Kang, T. B., Ben-Moshe, T., et al. (2012). Survival function of the FADD-CASPASE-8-cFLIP(L) complex. *Cell Rep.* 1, 401–407. doi: 10.1016/j.celrep.2012.03.010

Fisher, G. H., Rosenberg, F. J., Straus, S. E., Dale, J. K., Middleton, L. A., Lin, A. Y., et al. (1995). Dominant interfering Fas gene mutations impair apoptosis in a human autoimmune lymphoproliferative syndrome. *Cell* 81, 935–946. doi: 10.1016/0092-8674(95)90013-6

He, S., Liang, Y., Shao, F., and Wang, X. (2011). Toll-like receptors activate programmed necrosis in macrophages through a receptor-interacting kinase-3-mediated pathway. *Proc. Natl. Acad. Sci. U.S.A.* 108, 20054–20059. doi: 10.1073/pnas.1116302108

Holler, N., Zaru, R., Micheau, O., Thome, M., Attinger, A., Valitutti, S., et al. (2000). Fas triggers an alternative, caspase-8-independent cell death pathway using the kinase RIP as effector molecule. *Nat. Immunol.* 1, 489–495. doi: 10.1038/82732

Imtiyaz, H. Z., Rosenberg, S., Zhang, Y., Rahman, Z. S., Hou, Y. J., Manser, T., et al. (2006). The Fas-associated death domain protein is required in apoptosis and TLR-induced proliferative responses in B cells. *J. Immunol.* 176, 6852–6861. doi: 10.4049/jimmunol.176.11.6852

Kaiser, W. J., Daley-Bauer, L. P., Thapa, R. J., Mandal, P., Berger, S. B., Huang, C., et al. (2014). RIP1 suppresses innate immune necrotic as well as apoptotic cell death during mammalian parturition. *Proc. Natl. Acad. Sci. U.S.A.* 111, 7753–7758. doi: 10.1073/pnas.1401857111

Kaiser, W. J., Sridharan, H., Huang, C., Mandal, P., Upton, J. W., Gough, P. J., et al. (2013). Toll-like receptor 3-mediated necrosis via TRIF, RIP3, and MLKL. *J. Biol. Chem.* 288, 31268–31279. doi: 10.1074/jbc.M113.462341

Kaiser, W. J., Upton, J. W., Long, A. B., Livingston-Rosanoff, D., Daley-Bauer, L. P., Hakem, R., et al. (2011). RIP3 mediates the embryonic lethality of caspase-8-deficient mice. *Nature* 471, 368–372. doi: 10.1038/nature09857

Kelliher, M. A., Grimm, S., Ishida, Y., Kuo, F., Stanger, B. Z., and Leder, P. (1998). The death domain kinase RIP mediates the TNF-induced NF-κB signal. *Immunity* 8, 297–303. doi: 10.1016/S1074-7613(00)80535-X

Kreuz, S., Siegmund, D., Scheurich, P., and Wajant, H. (2001). NF-kappaB inducers upregulate cFLIP, a cycloheximide-sensitive inhibitor of death receptor signaling. *Mol. Cell. Biol.* 21, 3964–3973. doi: 10.1128/MCB.21.12.3964-3973.2001

Lu, J. V., Weist, B. M., van Raam, B. J., Marro, B. S., Nguyen, L. V., Srinivas, P., et al. (2011). Complementary roles of Fas-associated death domain (FADD) and receptor interacting protein kinase-3 (RIPK3) in T-cell homeostasis and antiviral immunity. *Proc. Natl. Acad. Sci. U.S.A.* 108, 15312–15317. doi: 10.1073/pnas.1102779108

Marsden, V. S., and Strasser, A. (2003). Control of apoptosis in the immune system: Bcl-2, BH3-only proteins and more. *Annu. Rev. Immunol.* 21, 71–105. doi: 10.1146/annurev.immunol.21.120601.141029

Meylan, E., Burns, K., Hofmann, K., Blancheteau, V., Martinon, F., Kelliher, M., et al. (2004). RIP1 is an essential mediator of Toll-like receptor 3-induced NF-kappa B activation. *Nat. Immunol.* 5, 503–507. doi: 10.1038/ni1061

Moriwaki, K., and Chan, F. K. (2013). RIP3: a molecular switch for necrosis and inflammation. *Genes Dev.* 27, 1640–1649. doi: 10.1101/gad.223321.113

Moujalled, D. M., Cook, W. D., Okamoto, T., Murphy, J., Lawlor, K. E., Vince, J. E., et al. (2013). TNF can activate RIPK3 and cause programmed necrosis in the absence of RIPK1. *Cell Death Dis.* 4, e465. doi: 10.1038/cddis.2012.201

Muzio, M., Chinnaiyan, A. M., Kischkel, F. C., O'Rourke, K., Shevchenko, A., Ni, J., et al. (1996). FLICE, a novel FADD-homologous ICE/CED-3-like protease, is recruited to the CD95 (Fas/APO-1) death-inducing signaling complex. *Cell* 85, 817–827. doi: 10.1016/S0092-8674(00)81266-0

Newton, K., Harris, A. W., Bath, M. L., Smith, K. G. C., and Strasser, A. (1998). A dominant interfering mutant of FADD/MORT1 enhance deletion of autoreactive thymocytes and inhibits proliferation of mature T lymphocytes. *EMBO J.* 17, 706–718. doi: 10.1093/emboj/17.3.706

Newton, K., Sun, X., and Dixit, V. M. (2004). Kinase RIP3 is dispensable for normal NF-kappa Bs, signaling by the B-cell and T-cell receptors, tumor necrosis factor receptor 1, and Toll-like receptors 2 and 4. *Mol. Cell. Biol.* 24, 1464–1469. doi: 10.1128/MCB.24.4.1464-1469.2004

Oberst, A., Dillon, C. P., Weinlich, R., McCormick, L. L., Fitzgerald, P., Pop, C., et al. (2011). Catalytic activity of the caspase-8-FLIP(L) complex inhibits RIPK3-dependent necrosis. *Nature* 471, 363–367. doi: 10.1038/nature09852

O'Donnell, M. A., and Ting, A. T. (2011). RIP1 comes back to life as a cell death regulator in TNFR1 signaling. *FEBS J.* 278, 877–887. doi: 10.1111/j.1742-4658.2011.08016.x

Pfeffer, K., Matsuyama, T., Kundig, T. M., Wakeham, A., Kishihara, K., Shahinian, A., et al. (1993). Mice deficient for the 55 kd tumor necrosis factor receptor are resistant to endotoxic shock, yet succumb to L. monocytogenes infection. *Cell* 73, 457–467. doi: 10.1016/0092-8674(93)90134-C

Rickard, J. A., O'Donnell, J. A., Evans, J. M., Lalaoui, N., Poh, A. R., Rogers, T., et al. (2014). RIPK1 regulates RIPK3-MLKL-driven systemic inflammation and emergency hematopoiesis. *Cell* 157, 1175–1188. doi: 10.1016/j.cell.2014.04.019

Roderick, J. E., Hermance, N., Zelic, M., Simmons, M. J., Polykratis, A., Pasparakis, M., et al. (2014). Hematopoietic RIPK1 deficiency results in bone marrow failure caused by apoptosis and RIPK3-mediated necroptosis. *Proc. Natl. Acad. Sci. U.S.A.* 111, 14436–14441. doi: 10.1073/pnas.1409389111

Rothe, J., Lesslauer, W., Lotscher, H., Lang, Y., Koebel, P., Kontgen, F., et al. (1993). Mice lacking the tumour necrosis factor receptor 1 are resistant to TNF-mediated toxicity but highly susceptible to infection by Listeria monocytogenes. *Nature* 364, 798–802. doi: 10.1038/364798a0

Schmukle, A. C., and Walczak, H. (2012). No one can whistle a symphony alone—how different ubiquitin linkages cooperate to orchestrate NF-kappaB activity. *J. Cell Sci.* 125, 549–559. doi: 10.1242/jcs.091793

Shi, X., Xie, C., Kreska, D., Richardson, J. A., and Mohan, C. (2002). Genetic dissection of SLE: SLE1 and FAS impact alternate pathways leading to lymphoproliferative autoimmunity. *J. Exp. Med.* 196, 281–292. doi: 10.1084/jem.20010955

Stanger, B. Z., Leder, P., Lee, T.-H., Kim, E., and Seed, B. (1995). RIP: a novel protein containing a death domain that interacts with Fas/APO-1 (CD95) in yeast and causes cell death. *Cell* 81, 513–523. doi: 10.1016/0092-8674(95)90072-1

Strasser, A., Jost, P. J., and Nagata, S. (2009). The many roles of FAS receptor signaling in the immune system. *Immunity* 30, 180–192. doi: 10.1016/j.immuni.2009.01.001

Sun, L., Wang, H., Wang, Z., He, S., Chen, S., Liao, D., et al. (2012). Mixed lineage kinase domain-like protein mediates necrosis signaling downstream of RIP3 kinase. *Cell* 148, 213–227. doi: 10.1016/j.cell.2011.11.031

Takahashi, N., Vereecke, L., Bertrand, M. J. M., Duprez, L., Berger, S. B., Divert, T., et al. (2014). RIPK1 ensures intestinal homeostasis by protecting the epithelium against apoptosis. *Nature* 513, 95–99. doi: 10.1038/nature13706

Ting, A. T., Pimentel-Muinos, F. X., and Seed, B. (1996). RIP mediates tumor necrosis factor receptor 1 activation of NF-κB but not Fas/APO-1-intitiated apoptosis. *EMBO J.* 15, 6189–6196.

Upton, J. W., Kaiser, W. J., and Mocarski, E. S. (2010). Virus inhibition of RIP3-dependent necrosis. *Cell Host Microbe* 7, 302–313. doi: 10.1016/j.chom.2010.03.006

Vanden Berghe, T., Linkermann, A., Jouan-Lanhouet, S., Walczak, H., and Vandenabeele, P. (2014). Regulated necrosis: the expanding network of non-apoptotic cell death pathways. *Nat. Rev. Mol. Cell Biol.* 15, 135–147. doi: 10.1038/nrm3737

Vucic, D., Dixit, V. M., and Wertz, I. E. (2011). Ubiquitylation in apoptosis: a post-translational modification at the edge of life and death. *Nat. Rev. Mol. Cell Biol.* 12, 439–452. doi: 10.1038/nrm3143

Walsh, C. M. (2014). Grand challenges in cell death and survival: apoptosis vs. necroptosis. *Front. Cell Dev. Biol.* 2:3. doi: 10.3389/fcell.2014.00003

Walsh, C. M., Wen, B. G., Chinnaiyan, A. M., O'Rourke, K., Dixit, V. M., and Hedrick, S. M. (1998). A role for FADD in T cell activation and development. *Immunity* 8, 439–449. doi: 10.1016/S1074-7613(00)80549-X

Yeh, W.-C., Pompa, J. L., McCurrach, M. E., Shu, H.-B., Elia, A. J., Shahinian, A., et al. (1998). FADD: essential for embryo development and signaling from some, but not all, inducers of apoptosis. *Science* 279, 1954–1958. doi: 10.1126/science.279.5358.1954

Zhang, H., Zhou, X., McQuade, T., Li, J., Chan, F. K., and Zhang, J. (2011). Functional complementation between FADD and RIP1 in embryos and lymphocytes. *Nature* 471, 373–376. doi: 10.1038/nature09878

Zhang, J., Cado, D., Chen, A., Kabra, N. H., and Winoto, A. (1998). Fas-mediated apoptosis and activation-induced T-cell proliferation are defective in mice lacking FADD/Mort1. *Nature* 392, 296–300. doi: 10.1038/32681

Zhang, J., and Winoto, A. (1996). A mouse Fas-associated protein with homology to the human Mort1/FADD protein is essential for Fas-induced apoptosis. *Mol. Cell. Biol.* 16, 2756–2763.

Zhang, Y., Rosenberg, S., Wang, H., Imtiyaz, H. Z., Hou, Y. J., and Zhang, J. (2005). Conditional Fas-Associated Death Domain Protein (FADD):GFP Knockout mice reveal FADD Is dispensable in thymic development but essential in peripheral T cell homeostasis. *J. Immunol.* 175, 3033–3044. doi: 10.4049/jimmunol.175.5.3033

Zhao, J., Jitkaew, S., Cai, Z., Choksi, S., Li, Q., Luo, J., et al. (2012). Mixed lineage kinase domain-like is a key receptor interacting protein 3 downstream component of TNF-induced necrosis. *Proc. Natl. Acad. Sci. U.S.A.* 109, 5322–5327. doi: 10.1073/pnas.1200012109

Zhou, T., Edwards, C. K. R., Yang, P., Wang, Z., Bluethmann, H., and Mountz, J. D. (1996). Greatly accelerated lymphadenopathy and autoimmune disease in lpr mice lacking tumor necrosis factor receptor I. *J. Immunol.* 156, 2661–2665.

Conflict of Interest Statement: The authors declare that the research was conducted in the absence of any commercial or financial relationships that could be construed as a potential conflict of interest.

4

GRAF1 forms a complex with MICAL-L1 and EHD1 to cooperate in tubular recycling endosome vesiculation

Bishuang Cai†, Shuwei Xie, Steve Caplan* and Naava Naslavsky*

Department of Biochemistry and Molecular Biology and the Fred and Pamela Buffett Cancer Center, The University of Nebraska Medical Center, Omaha, NE, USA

Edited by:
Roberto Weigert, National Institute Dental And Craniofacial Research - National Institutes of Health, USA

Reviewed by:
Julie G. Donaldson, National Heart, Lung and Blood Institute, National Institutes of Health, USA
Peter Scott McPherson, McGill University, Canada

****Correspondence:***
Steve Caplan and Naava Naslavsky, Department of Biochemistry and Molecular Biology and the Fred and Pamela Buffett Cancer Center, The University of Nebraska Medical Center, Omaha, NE 68198, USA
e-mail: scaplan@unmc.edu; nnaslavsky@unmc.edu

†Present address:
Bishuang Cai, Department of Medicine, Columbia University, New York, NY, USA

The biogenesis of tubular recycling endosomes (TREs) and their subsequent vesiculation after cargo-sorting has occurred, is essential for receptor and lipid recycling to the plasma membrane. Although recent studies have implicated the C-terminal Eps15 Homology Domain (EHD) protein, EHD1, as a key regulator of TRE vesiculation, additional proteins involved in this process have been largely uncharacterized. In the present study, we identify the GTPase Regulator Associated with Focal adhesion kinase-1 (GRAF1) protein in a complex with EHD1 and the TRE hub protein, Molecules Interacting with CasL-Like1 (MICAL-L1). Over-expression of GRAF1 caused vesiculation of MICAL-L1-containing TRE, whereas GRAF1-depletion led to impaired TRE vesiculation and delayed receptor recycling. Moreover, co-addition of purified EHD1 and GRAF1 in a semi-permeabilized cell vesiculation assay produced synergistic TRE vesiculation. Overall, based on our data, we suggest that in addition to its roles in clathrin-independent endocytosis, GRAF1 synergizes with EHD1 to support TRE vesiculation.

Keywords: BAR domain, PH domain, tubulation, EHD1, MICAL-L1, GRAF1, tubular recycling endosome, vesiculation

INTRODUCTION

Endosomes are a primary "vehicle" for protein transport within the cell (Jovic et al., 2010). This transportation includes the sorting and delivering of cargo molecules to their corresponding destinations, which depends upon a series of well-coordinated membrane fission and fusion events. A large group of cytosolic proteins assemble onto membranes to influence their shape, and facilitate the formation of a budding vesicle. These vesicles/carriers are then transported along microtubule tracks to a specific destination where they eventually fuse with a target membrane and release their cargo. Such proteins often assemble as a complex and can act as an "operational unit."

We recently identified a protein that serves as a membrane-hub on tubular recycling endosomes (TREs) (Rahajeng et al., 2012), and coordinates multiple events that shape the membranes during endocytic transport (Giridharan et al., 2013). Molecules Interacting with CAsL-Like1 (MICAL-L1) is a novel TRE marker (Sharma et al., 2009), which recruits and stabilizes a battery of proteins that leads to a direct impact on membrane shaping (Giridharan et al., 2013). MICAL-L1 recruits many essential membrane modulators to TREs, including Rab8 (Sharma et al., 2009) and proteins that possess Bin–Amphiphysin–Rvs (BAR) domains (summarized in **Figure 8A**). Such BAR domain-containing proteins induce membranes into highly curved shapes (McMahon and Gallop, 2005; Zimmerberg and Kozlov, 2006) and include the N-BAR protein Amphiphysin/Bin1 (Pant et al., 2009), and the F-BAR protein Syndapin2, also known as PACSIN2 (Braun et al., 2005; Giridharan et al., 2013).

MICAL-L1 also interacts with the C-terminal Eps15 Homology Domain (EHD) proteins, EHD3 and EHD1 (Sharma et al., 2009; Kieken et al., 2010), which are involved in membrane tubulation and vesiculation, respectively (Cai et al., 2013). Also crucial for TRE biogenesis is the generation of phosphatidic acid (PA), an essential lipid component of TRE tubules that binds and recruits MICAL-L1 and Syndapin2, thus promoting F-BAR-induced tubulation (Giridharan et al., 2013). EHD1 subsequently joins this complex on TRE where it binds to both MICAL-L1 and Syndapin2 and performs scission, giving rise to newly formed recycling endosomes (Cai et al., 2013; Giridharan et al., 2013).

Along with the growing consensus for EHD1 as a vesiculator of TRE (Jakobsson et al., 2011; Cai et al., 2012, 2013), there is also evidence suggesting that additional proteins are involved in the scission process. For example, EHD1 directly interacts and cooperates with the lipid-modifier enzyme, cPLA2α, to initiate and facilitate vesiculation, thus supporting the generation of transport vesicles (Cai et al., 2012). Another protein with potential to be involved in TRE vesiculation is the GTPase Regulator Associated with Focal adhesion kinase-1 (GRAF1; also known as Oligophrenin-1-Like). GRAF1 is a protein that cooperates with other proteins to remodel tubulo-vesicular clathrin-independent carriers (CLICs) derived from the plasma membrane (Lundmark et al., 2008), thus regulating endocytosis.

GRAF1 belongs to a family of proteins, all of which are comprised of a similar domain-configuration (see scheme in **Figure 1A**) consisting of: an N-terminal BAR domain, a pleckstrin

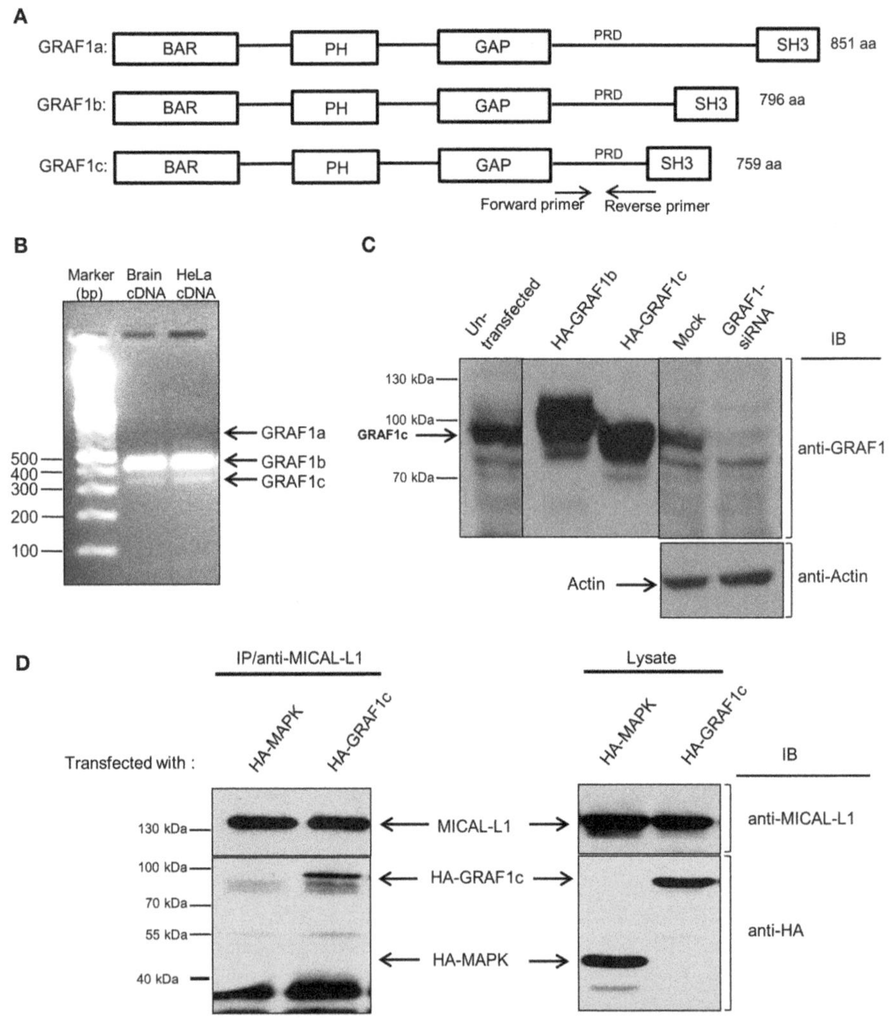

FIGURE 1 | HeLa cells predominantly express the GRAF1c isoform, which interacts with the hub-protein, MICAL-L1. (A) The domain architecture of the three GRAF1 isoforms. **(B)** GRAF1 isoforms (a, b, and c) were amplified from either human brain or HeLa cDNA libraries with two primers as indicated in **(A)**. **(C)** Cells were either left in culture, or transfected with either HA-GRAF1b, or HA-GRAF1c for 16 h. In the left panel, the cells were subsequently mock-treated or treated with GRAF1-siRNA for 72 h before being lysed on ice. Lysates were subjected to 8% SDS-PAGE, followed by immunoblotting with anti-GRAF1 and anti-Actin antibodies (loading control). **(D)** Cells were either transfected with HA-MAPK as a negative control or HA-GRAF1c. After 48 h, cells were lysed for 1 h in buffer containing 50 mM Tris pH 7.4, 150 mM NaCl, 0.5% Triton X-100 and protease inhibitors. Cell lysates were incubated with mouse anti-MICAL-L1 antibody overnight. Protein L beads were added to the mixture of cell lysate and MICAL-L1 antibody for 3 h at 4°C. Beads were washed in buffer containing 50 mM Tris pH 7.4, 150 mM NaCl, 0.1% Triton. Proteins were eluted by adding SDS loading buffer. Samples were subjected to 8% SDS-PAGE, followed by blotting with anti-HA and anti-MICAL-L1 antibodies.

homology (PH) domain, a Rho-GAP, a proline-rich domain (PRD) and a Src homology 3 (SH3) domain (Doherty and Lundmark, 2009). The PH domain is found in a variety of proteins, and it directly binds to Phosphatidylinositol-4,5-bisphosphate (PIP2) (Harlan et al., 1995). The BAR and PH domains of GRAF1 work cooperatively to bind to highly curved PIP2-containing tubular and vesicular membranes (Lundmark et al., 2008). Via its Rho-GAP domain, GRAF1 can also affect certain small G proteins such as RhoA and Cdc42 (Taylor et al., 1999; Shibata et al., 2001; Jelen et al., 2009).

Due to its multi-domain architecture, GRAF1 has numerous functions that include (1) localizing onto membranes (2) interacting with membrane-shaping proteins, such as Dynamin, and (3) activating RhoA/Cdc42 small G-proteins, thereby indirectly impacting cytoskeleton-related processes. In addition, GRAF1 interacts with Focal Adhesion Kinase (FAK) to regulate cell spreading by remodeling microdomains and the cytoskeleton at podosomes, leading to membrane turnover during morphological adjustments (summarized in Model **Figure 8A**) (Hildebrand et al., 1996; Doherty et al., 2011).

We propose a novel function for GRAF1 and hypothesize that in addition to its previously described regulatory roles, GRAF1 might be involved in TRE vesiculation. Our study indicates that GRAF1 interacts with both MICAL-L1 and EHD1 to regulate TRE vesiculation and mediate endocytic recycling.

METHODS AND MATERIALS

ANTIBODIES

The affinity-purified rabbit polyclonal peptide antibodies directed against the C-terminus of human EHD1 (DLPPHLVPPSKRRHE) were previously described (Caplan et al., 2002). Mouse monoclonal MEM-43 antibody against CD59 was a generous gift of Dr. V. Horejsi (Praha, Czech Republic) (Cai et al., 2011, 2012). The following commercial antibodies were used in this study: mouse anti-GRAF1 (Abcam), mouse anti-HA (Covance), rabbit anti-HA (Signalway), mouse anti-Actin and mouse anti-MICAL-L1 (Novus Biologicals, Inc.), mouse anti-Rab5 (Transduction Laboratories), rabbit anti-Rab11 (US Biologicals), mouse anti-β1 integrin (AbD Serotec), goat anti-mouse horseradish-peroxidase (HRP) (Jackson ImmunoResearch Laboratories, Inc.), donkey anti-rabbit HRP (GE Healthcare), Alexa 568 goat anti-mouse, Alexa 488 goat anti-rabbit and Alexa 568 goat anti-rabbit (Invitrogen). The transfection reagents X-treme GENE 9, GeneExpresso and Oligofectamine were obtained from Roche Applied Science Innovita and Invitrogen, respectively.

cDNA

GRAF1b and GRAF1c were amplified from HeLa cDNA and human brain cDNA, respectively, using the primers CCGGAATTCGGATGGGGCTCCCAGCGCTCGAGTTCAG and AAGGAAAAAAGCGGCCGCTTAGAGGAACTCCACGTAATTC TCAGG. Amplified GRAF1b and GRAF1c were then cloned into the pHA-CMV vector. HA-GRAF1c (BAR) (1-250aa), HA-GRAF1c (BAR-PH) (1-383aa), HA-GRAF1c (BAR-PH-GAP) (1-663aa) and HA-GRAF1c (R412D) mutants were engineered using PCR directed mutagenesis (Stratagene). GST-GRAF1c was generated as follows: GRAF1c was amplified from HA-GRAF1c by using primers CGCGGATCCATGGGGCTCCCAGCGCTCGAGTTCAG and AA GGAAAAAAGCGGCCGCTTAGAGGAACTCCACGTAATTCTC AGG and cloned into pGEX-6P-2 vector. GST-GRAF1c (BAR-PH) was created by introducing a stop codon following aa 383 in the wild-type GST-GRAF1c.

PROTEIN PURIFICATION

GST-GRAF1c and GST-GRAF1c (BAR-PH) were transformed into BL21 bacterial, which was cultured in Super Broth media at 37°C. When the OD reached 0.5, cells were induced with 0.2 mM IPTG and cultured at 18°C overnight. Cells were then lysed by French Press. GST-GRAF1c and GST-GRAF1c (BAR-PH) were purified from the lysates using Glutathione S-transferase beads. EHD1 was cloned into a pCOLD-GST vector and purified as described in Hayashi and Kojima (2008).

GENE KNOCKDOWN BY SILENCING RNA (siRNA) AND RESCUE EXPERIMENTS

Four specific oligonucleotides directed at GRAF1 (GCACTAC TGTACATATCAA, CAAGGGCTGTATCGAATTG, GAACGGA TACGGATGATTG, GGAGAAACGTCACTTTGGA), using On-Target SMART pool (Dhamacon), were transfected into HeLa cells with Oligofectamine (Invitrogen) for 72 h to knockdown endogenous human GRAF1. Mock-SiRNA was essentially the same treatment as for SiRNA-treated cells, excluding the oligonucleotide(s). The siRNA-resistant HA-GRAF1 was engineered using QuickChange site directed mutagenesis. The specificity of the knockdown was confirmed by rescue experiments: the siRNA-resistant HA-GRAF1 was transfected into GRAF1-siRNA-treated cells by using X-treme GENE 9, 48 h after the start of siRNA treatment.

IMMUNOBLOTTING

HeLa cells were harvested and lysed on ice for 30 min in buffer containing 50 mM Tris pH7.4, 150 mM NaCl, 0.5% Triton X-100 and protease inhibitor cocktail (Roche Molecular Biochemicals). The lysate supernatants were obtained after centrifugation. Total protein level in the lysate was quantified by Bio-Rad protein assay (Bio-Rad laboratories, Inc.) for equal protein loading on gels in siRNA experiments. Samples were resolved by 8% SDS-PAGE.

CO-IMMUNOPRECIPITATION

HeLa cells growing on 100 mm dishes were transfected with HA-tagged proteins by using GeneExpresso. After 20 h, cells were lysed for 1 h in buffer containing 50 mM Tris pH7.4, 150 mM NaCl, 0.5% Triton X-100 and protease inhibitor cocktail. Cell lysates were incubated with either mouse anti-MICAL-L1 antibody or rabbit anti-EHD1 antibody at 4°C overnight, followed by incubation with protein L agarose beads (Fisher Thermo Scientific) and Rabbit IP Matrix beads (Santa Cruz Biotechnology) for 3 h at 4°C, respectively. Beads were washed in buffer containing 50 mM Tris pH7.4, 150 mM NaCl, 0.1% Triton X-100 4 times, and proteins were eluted using 4×SDS loading buffer. Samples were subjected to 8% SDS-PAGE, followed by blotting with appropriate antibodies.

IMMUNOFLUORESCENCE, CD59 UPTAKE ASSAY AND β1 INTEGRIN PULSE-CHASE ASSAY

HeLa cells were grown on cover slips, occasionally transfected with X-treme GENE 9 or oligofectamine, as noted, and fixed with 4% (v/v) paraformaldehyde in PBS as described previously (Cai et al., 2012). Fixed cells were incubated with primary antibodies prepared in staining solution [0.2% saponin (w/v) and 0.5% (w/v) BSA in PBS] for 1 h at room temperature. After washes with PBS, cells were incubated with the appropriate fluorochrome-conjugated secondary antibody mixture in staining solution for 30 min at room temperature. All images were acquired using a Zeiss LSM 5 Pascal confocal microscope (Carl Zeiss) with a 63x 1.4 numerical aperture objective with appropriate filters. For CD59 uptake assay, cells in complete medium were pulsed with anti-CD59 antibody for 15 min at 37°C, followed by 1 min incubation with acid stripping buffer (0.5 M NaCl and 0.5% acetic acid, pH3.0) to remove surface-bound CD59 antibody. Cells were fixed after stripping and stained with appropriate secondary antibodies. For β1 integrin pulse-chase assay, cells were pulsed with anti-β1 integrin for 1 h in complete media at 37°C, followed by 1 min acid stripping to remove surface-bound β1 integrin antibody. Cells were then either fixed directly (pulse only) or "chased" in complete media for 4 h to allow β1 to recycle back to the plasma membrane (pulse and chase). After chase, cells were acid stripped again to remove β1 integrin which had recycled back to the plasma membrane.

SEMI-PERMEABILIZATION ASSAY

This assay was developed (Cai et al., 2013) and modified from the methods described by Balch (Beckers et al., 1987) and Aridor (Long et al., 2010). Briefly, HeLa cells were grown on coverslips. When the cell confluence reached 80–90%, cytosol was extracted with 20 μg/ml digitonin for 40 s at room temperature. After three times washing in KHM buffer containing 110 mM KOAc, 20 mM HEPES, and 2 mM MgOAc, pH 7.2, semi-permeabilized cells were incubated with GST proteins in the presence of 25 mM HEPES, 115 mM KCl, 2.5 mM $MgCl_2$, 1 mM ATP, 5 mM creatine phosphate and 0.2 IU of creatine phosphokinase for 30 min at 37°C. Cells were then washed with KHM buffer for three times, followed by fixation. Fixed cells were stained with anti-MICAL-L1 antibody.

QUANTIFICATION OF MICAL-L1 TUBULES BY IMAGEJ

In the IMAGEJ software, background from the images was removed by adjusting the threshold. The size of particles measured was set between 2 and 150 μm^2. Every MICAL-L1-containing particle in this range was counted. Ten fields (images) from each treatment were analyzed.

STATISTICAL ANALYSIS

One-Way ANOVA was used to calculate statistical significance, and all graphs show standard error bars.

SELECTIVE YEAST TWO-HYBRID ASSAYS

Yeast two-hybrid studies were performed as previously described (Giridharan et al., 2013). The *Saccharomyces cerevisiae* strain AH109 was co-transformed with mentioned constructs using lithium acetate procedure and allowed to grow at 30 C after streaking on plates lacking leucine and tryptophan. Once the colonies grow large enough, an average of three to four colonies were chosen, suspended in water, equilibrated to same optical density at 600 nm and replated on plates lacking leucine and tryptophan (+His) as well as plates also lacking histidine (−His).

RESULTS

The multi-domain-containing protein, GRAF1, is involved in a variety of processes including membrane-shaping, small G-protein signaling, clathrin-independent endocytosis and cell spreading (Doherty and Lundmark, 2009). Because it contains both a BAR and PH domain, and has a role in membrane bending similar to that of other BAR-containing proteins, and because it localizes to tubular endosomes (Lundmark et al., 2008), we posited that GRAF1 may be involved in the biogenesis of TREs.

Recently, a studies from our lab has demonstrated the role of two key proteins in TRE biogenesis: Syndapin2 and MICAL-L1 (Giridharan et al., 2013), along with the C-Terminal EHD protein, EHD3 (Cai et al., 2013). As does GRAF1, Syndapin2 has a membrane-bending BAR domain, whereas MICAL-L1 has 14 PRDs, short motifs known to interact with proteins containing SH3 domains. Since GRAF1 possesses a SH3 domain (see **Figure 1A**), we hypothesized that GRAF1 and MICAL-L1 might function together on MICAL-L1-enriched TRE. GRAF1 has 3 alternative splicing products, GRAF1a, b, and c, comprised of 851, 796, and 759 amino acids, respectively (**Figure 1A**). We first assessed which alternatively spliced GRAF1 product is expressed endogenously in HeLa cells. As shown in **Figure 1B**, GRAF1b RNA transcripts were abundant in a HeLa cDNA library, whereas we were unable to detect GRAF1a transcripts (**Figure 1B**). However, since it is not known in which cells and/or tissues GRAF1a is expressed (leaving us without a "positive control" for our detection system), we cannot rule out the possibility that GRAF1a RNA might be expressed in HeLa cells. On the other hand, GRAF1c was the only protein detected (**Figure 1C**, left lane). This was verified by comparing overexpressed HA-GRAF1b and c, with endogenously expressed GRAF1 in non-transfected HeLa cells (**Figure 1C**; middle panel). Endogenous GRAF1c protein expression was almost completely knocked-down by siRNA, using specific oligonucleotides (**Figure 1C**; right panel). Actin was used as a control for equal protein loading.

To assess the potential interaction between MICAL-L1 and GRAF1, HeLa cells were transiently transfected with either HA-GRAF1c, or with a negative control, HA-MAPK (Mitogen-Activated Protein Kinase). The latter control was chosen because it was an unrelated protein with an identical HA-tag to HA-GRAF1c, and its expression was similar to that of HA-GRAF1c. As shown in **Figure 1D** (left panel), endogenous MICAL-L1 pulled-down HA-GRAF1c but not HA-MAPK (**Figure 1D**; right panel). Additional controls were done immunoprecipitating with Protein L beads but without the anti-MICAL-L1, further confirming the specificity of this interaction (**Figure S1D**). However, data from selective yeast two-hybrid assays lead us to suggest that this interaction is likely indirect (**Figure S1A**).

MICAL-L1, GRAF1, AND EHD1 ARE COMPONENTS OF A COMPLEX

The apparent indirect interaction between GRAF1 and MICAL-L1 led us to query whether these proteins might be linked by a common interaction partner. We have previously demonstrated that MICAL-L1 interacts with EHD1 (Sharma et al., 2009), and we now asked whether EHD1 was necessary for the MICAL-L1/GRAF1 interaction. To this aim, HeLa cells were either mock-treated or depleted of EHD1 by siRNA. The cells were then transfected with either HA-GRAF1 or HA-MAPK (negative control). Lysates from these cells (**Figure 2A**; right panel) demonstrate that HA-MAPK, HA-GRAF1, endogenous MICAL-L1 and EHD1 were readily detected by their respective antibodies. The lysates also confirmed that EHD1 expression was completely inhibited by EHD1-siRNA. Under these conditions, anti-MICAL-L1 antibody was used for immunoprecipitation, and the presence of HA-GRAF1 or HA-MAPK was tested in the immunoprecipitate. As demonstrated (**Figure 2A**; left panel), GRAF1 (but not the MAPK negative control) was co-immunoprecipitated by anti-MICAL-L1 antibody in *both* mock-treated cells and cells EHD1-depleted cells (red asterisks). Our data promote the notion that EHD1 is *not* required for the interaction between MICAL-L1 and GRAF1.

The possibility that the interaction between MICAL-L1 and GRAF1 represents a multi-protein complex that includes EHD1 was then tested. MICAL-L1-depleted cells were transfected with cDNA coding for HA-MARCKS (Myristoylated alanine-rich C-kinase substrate; negative control) or HA-GRAF1 (**Figure 2B**). HA-MARCKS was selected as a negative control because it

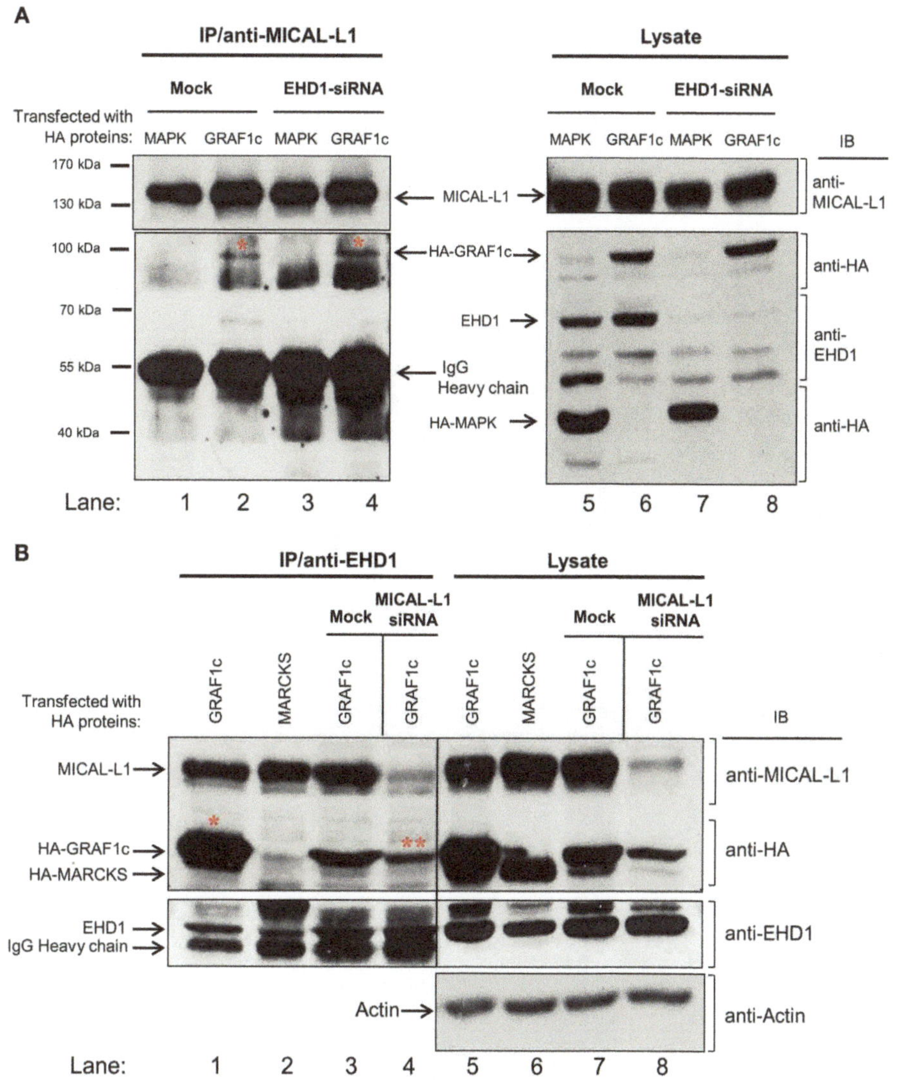

FIGURE 2 | GRAF1c, MICAL-L1, and EHD1 reside in a common complex. **(A)** HeLa cells were either mock-treated or treated with EHD1-siRNA for 72 h. In the final 48 h of the siRNA treatment, cells were transfected with either HA-MAPK (negative control) or HA-GRAF1c. Cells were then lysed with 0.5% Triton X-100 (right panel shows input). Lysates (left panel) were incubated with anti-MICAL-L1 antibody. Protein L beads were added to pull-down MICAL-L1 antibody and elution was done by adding SDS loading buffer. Samples were subjected to 8% SDS-PAGE, followed by blotting with anti-EHD1, anti-HA, and anti-MICAL-L1 antibodies. **(B)** Cells were either transfected with HA-MARCKS (negative control), or HA-GRAF1c, and then either mock-treated or treated with MICAL-L1-siRNA (right panel shows input). Rabbit IP Matrix beads were added to the mixture of cell lysate and EHD1 antibody for 3 h at 4°C. Proteins were eluted and subjected to SDS-PAGE, followed by blotting with anti-HA, anti-MICAL-L1, anti-EHD1, anti-MICAL-L1, and anti-Actin antibodies. Red stars indicate GRAF1c pulled down by either MICAL-L1 or EHD1.

was unrelated to the TRE biogenesis and vesiculation process, because of its HA-tag, and because it has a similar molecular weight to GRAF1c. The efficiency of MICAL-L1 knock-down is demonstrated in lanes 7–8. EHD1, HA-MARCKS, HA-GRAF1, MICAL-L1, and actin (loading control) were detected in the lysate (input) with the corresponding antibodies (**Figure 2B**; right panel, lanes 5–8). As previously shown, anti-EHD1 readily pulled down endogenous MICAL-L1 (**Figure 2B**, lanes 1–3) but not HA-MARCKS (**Figure 2B**, lane 2). In an additional control immunoprecipitation without the specific EHD1 antibody, no HA-GRAF1c was observed (**Figure S1E**). However, EHD1 precipitated HA-GRAF1 even in the absence of MICAL-L1 (**Figure 2B**, lane 4; see double asterisks). This interaction also appeared to be indirect, as suggested by data from yeast two-hybrid assays (**Figure S1A**; right panel) nor is it regulated by the TRE regulator, Syndapin2 (**Figure S1B**; right panel). Taken together, our data support the localization of GRAF1 to a complex containing both MICAL-L1 and EHD1.

GRAF1 SUPPORTS *IN VIVO* VESICULATION OF TUBULAR RECYCLING ENDOSOMES

Endogenous MICAL-L1 decorates TRE and plays an essential role in their biogenesis by recruiting "tubulators" and "vesiculators" (Giridharan et al., 2013). Support for the involvement of GRAF1 in the biogenesis of TRE was obtained using MICAL-L1 as a marker for these structures. As shown in **Figure 3A**, MICAL-L1-decorated TRE are readily detected in untreated HeLa cells. However, upon GRAF1-depletion, there is a striking increase in the length and number of TRE (**Figure 3B**). Quantification of the total area of MICAL-L1-containing TRE indicates a two fold increase per image field upon GRAF1-depletion (**Figure 3C**). While the total level of MICAL-L1, as tested by immunoblot, is not altered by the GRAF1-knockdown (**Figure S1C**), the increased number and overall area of TRE at GRAF1-knockdown suggest a potential role for GRAF1 in vesiculating these endosomes.

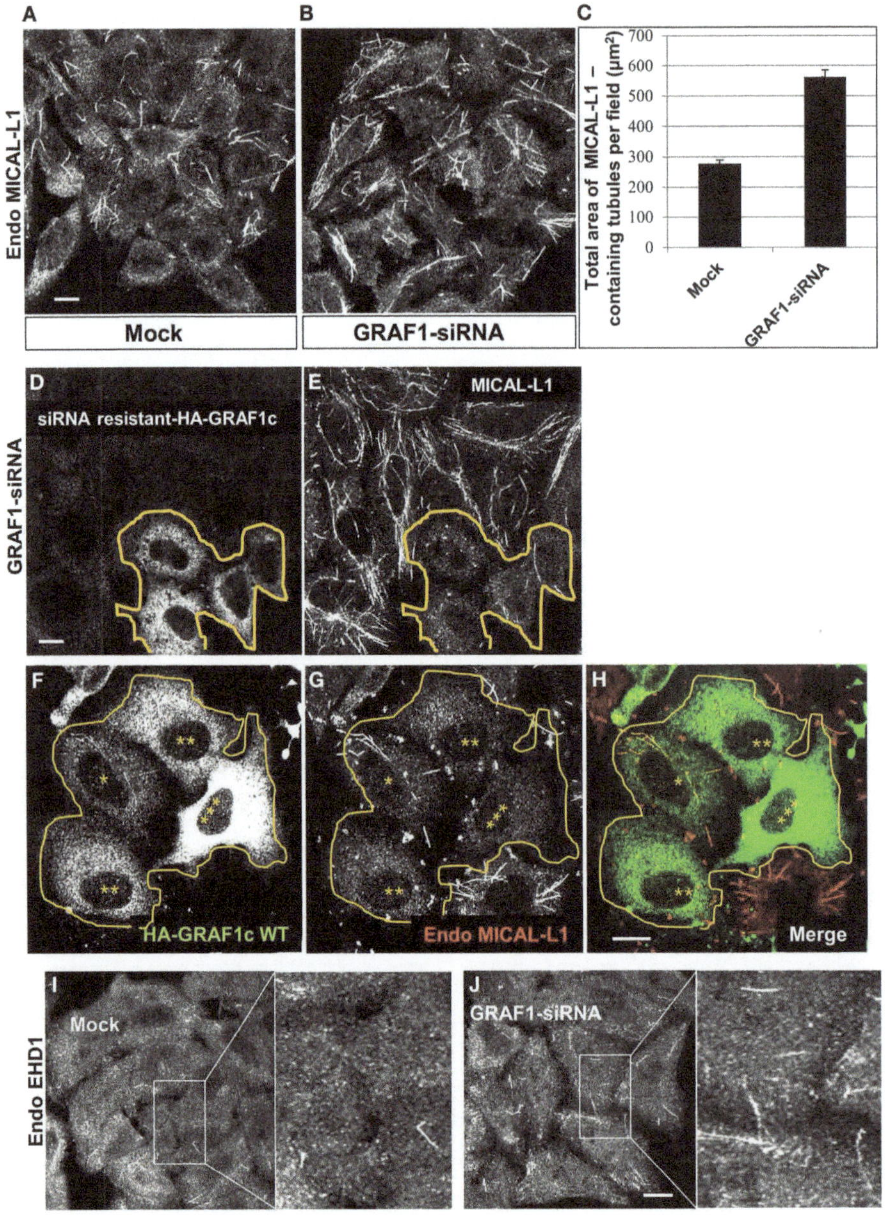

FIGURE 3 | GRAF1 vesiculates MICAL-L1- and EHD1-decorated membrane tubules. **(A)** HeLa cells grown on coverslips were either mock-treated or **(B)** treated with GRAF1-siRNA for 72 h. After fixation, cells were stained with mouse anti-MICAL-L1 antibody for 1 h, followed by Alex 568-goat anti-mouse secondary antibody. **(C)** The total area of MICAL-L1-containing tubules per field in both mock-treated and GRAF1-siRNA-treated cells was quantified by using Image J from three independent experiments (represented in **A,B**). Error bars represent standard error; $p < 0.001$. **(D,E)** siRNA resistant-HA-GRAF1c was re-introduced into GRAF1-depleted cells for 24 h followed by fixation. Fixed cells were then stained with rabbit anti-HA **(D)** and mouse anti-MICAL-L1 antibodies **(E)**, followed by Alexa 488-goat anti-rabbit and Alexa 568-goat anti-mouse secondary antibodies. **(F)** Cells transfected with HA-GRAF1c are shown at variable expression levels (3 asterisks designate the highest expression level). **(G)** These transfected cells (orange line) are stained with anti-MICAL-L1. Note that TRE tend to disappear in the HA-GRAF1c high expressing cells. **(H)** Merged image displaying staining for HA-GRAF1c (green) and endogenous MICAL-L1 (red). **(I,J)** Endogenous EHD1 was detected in either mock-treated **(I)** or GRAF1-depleted cells **(J)**. Bar, 10 μm.

To ascertain that these results can be directly attributed to GRAF1 knock-down (as opposed to off-target effects), we transfected a siRNA-resistant HA-GRAF1 into the depleted cells (in **Figure 3D**, boundary marked with an orange line) and analyzed the levels of TRE. Expression of the resistant HA-GRAF1 construct was confirmed (see **Figure S2D**). As demonstrated, transfected cells in **Figures 3D,E** had fewer and shorter TRE compared to the neighboring GRAF1-depleted cells (**Figure 3E**), confirming that indeed GRAF1 is responsible for the phenotype observed.

Whereas GRAF1-knockdown led to inefficient TRE vesiculation, over-expression of HA-GRAF1 induced the opposite effect, causing extensive vesiculation (**Figures 3F–H**). In **Figure 3F** a group of transfected cells expressing HA-GRAF1 at variable levels is observed (denoted by asterisks, with 3 asterisks representing the highest level of GRAF1 expression). In **Figure 3G**, the degree of TRE vesiculation is observed in these transfected cells; in low-expressing HA-GRAF1 cells, a small number of TRE were still visible, but at medium-to-high expression levels, TRE were absent due to extensive vesiculation.

TRE are enriched in PA, which selectively recruits MICAL-L1 and Syndapin2 (Giridharan et al., 2013). To determine whether GRAF1-decorated membrane tubules also contain PA, we co-transfected HA-GRAF1 (at low expression) along with GFP-Spo20p, a well-characterized PA-binding protein (Zeniou-Meyer et al., 2007), (**Figures S2A–C**; see insets). Their partial co-localization on tubular structures further supports the notion of GRAF1 association with TRE *in vivo*.

Endogenous EHD1 localizes to a wide range of vesicular and tubular membranes within the cell, including TRE (Caplan et al., 2002; Cai et al., 2011), albeit to a lesser degree than MICAL-L1, presumably because it begins to induce vesiculation of these membranes. However, as demonstrated, GRAF1-depletion induced an increase in endogenous EHD-decorated TRE (**Figures 3I,J**; see insets), further supporting the involvement of GRAF1 in the process of vesiculation.

GRAF1 PROMOTES VESICULATION OF INCOMING AND OUTGOING ENDOSOMES

Since GRAF1-depletion affects EHD1-decorated endosomes, this prompted us to further investigate the type of endosomes that serve as target membranes for GRAF1 action. To this aim, in-bound vesicles were visualized by following the uptake of the endogenous GPI-anchored protein (GPI-AP), CD59. Previous studies have demonstrated that GFP-GPI, a transfected GPI-AP model protein, traverses GRAF1-containing CLICs (Lundmark et al., 2008). Endogenous CD59 is a typical cargo of the clathrin-independent pathway that associates with microdomains enriched with cholesterol and sphingolipids (Stulnig et al., 1998). In mock- and GRAF1-siRNA-treated HeLa cells, anti-CD59 antibody was incubated with cells to allow binding to surface-CD59 and internalization of antibody-receptor complexes. Remaining anti-CD59 on the cell surface was stripped with a short acid-wash, to allow examination exclusively of internalized CD59.

We have previously shown that following 15 min. of incubation with anti-CD59 antibody, CD59 is primarily localized to tubular endosomes that are on route to the endocytic recycling compartment (ERC) (Cai et al., 2011, 2012), and therefore used this incubation time to study incoming tubules. Compared to mock-treated cells (**Figure 4A**), incoming endosomes containing CD59 were extensively tubular upon GRAF1-depletion (**Figure 4B**; see insets with higher magnification). Note that the bead-like pattern of CD59 is typically obtained post-fixation (Cai et al., 2011).

We next addressed the potential involvement of GRAF1 in the shaping of two well-established endocytic organelles: sorting endosomes (SE) and recycling endosomes (RE). To this aim, we addressed the distribution of the small GTP-binding proteins,

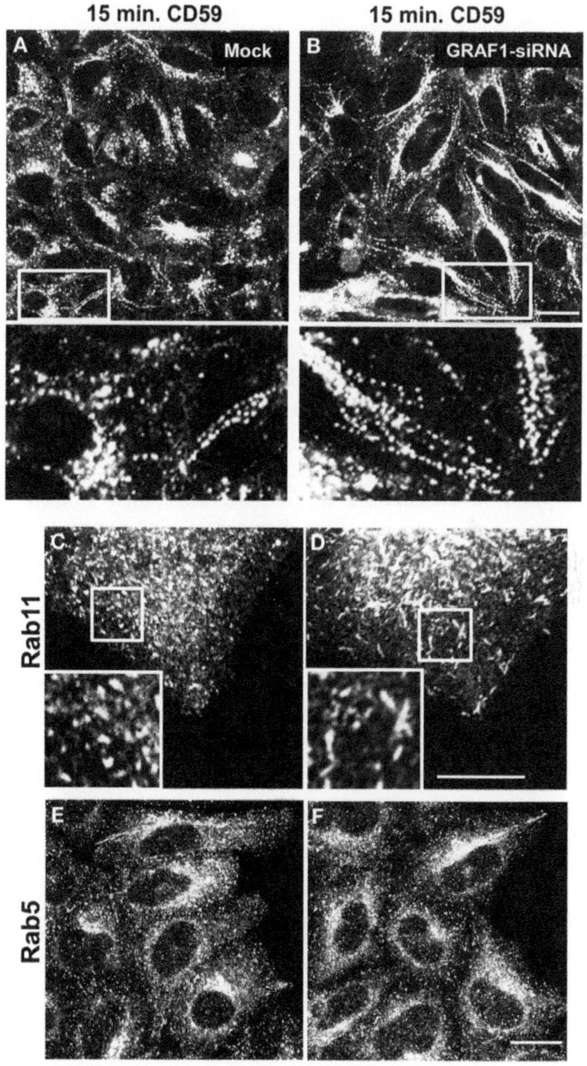

FIGURE 4 | GRAF1-depletion induces hyper-tubulation of CD59-containing endosomes and Rab11-positive recycling endosomes. Cells grown on coverslips were either mock-treated **(A,C,E)** or treated with GRAF1-siRNA **(B,D,F)** for 72 h. **(A,B** and insets**)** Cells were incubated with mouse anti-CD59 antibody for 15 min. At 37°C, followed by a 1 min. acid stripping and fixation. Fixed cells were stained with Alex 568-goat anti mouse secondary antibody. After siRNA-treatment, cells were fixed and stained with either anti-Rab11 antibody **(C,D)** or anti-Rab5 antibody **(E,F)**, followed by appropriate secondary antibodies. Bar, 10 μm.

Rab5 and Rab11 (respectively) in mock- vs. GRAF1-depleted cells. In **Figures 4C,D**, Rab11-associated RE were mostly punctate in mock-treated cells, but displayed a degree of elongation (short tubules) upon the loss of GRAF1. On the other hand, Rab5-decorated SE did not undergo discernible changes (**Figures 4E,F**). Taken together, GRAF1 affects membrane-vesiculation of clathrin-independent incoming endosomes as well as RE, but not Rab5-sorting endosomes.

EXIT OF β1 INTEGRIN FROM THE ENDOCYTIC RECYCLING COMPARTMENT IS DELAYED IN THE ABSENCE OF GRAF1

Vesiculation gives rise to mobile transport vesicles, which move along microfilaments, tethered to motor proteins, and unload their cargo at their target destination, and is thus crucial for the process of recycling. Therefore, since MICAL-L1- and Rab11-containing RE are target-vesicles for GRAF1, we hypothesized that GRAF1-induced vesiculation might directly affect the exit of cargo molecules from the recycling compartment either to the plasma membrane (PM), or to the intracellular degradative pathways. β1 integrins are transmembrane proteins that mediate the attachment of the cells to the extracellular matrix. They can be internalized independently of clathrin (Altankov and Grinnell, 1995; Ylanne et al., 1995), or via clathrin-coated pits (Nishimura and Kaibuchi, 2007; Ezratty et al., 2009), and recycle through a MICAL-L1-mediated pathway (Sharma et al., 2009). Accordingly, we assessed the involvement of GRAF1 in β1 integrin exit from the recycling compartment. Cells were mock-treated, or treated with GRAF1-siRNA, then incubated with β1 integrin antibody for 1 h to allow internalization, followed by a short acid-strip to remove remaining non-internalized antibody. While similar amounts of β1 integrin were internalized upon GRAF1-depletion (compare **Figures 5B** to **5A**; quantified in **5E**, 1 h pulse), a clear delay in β1 integrin exit from the recycling compartment was evident upon GRAF1-depletion (compare **Figures 5D** to **5C**; quantified in **5E**, 4 h chase). Indeed, we observed a ~25% delay in the rate of β1 integrin recycling to the PM and/or degradation in GRAF1-depleted cells (**Figure 5E**).

THE RHOA-GAP DOMAIN OF GRAF1 IS REQUIRED FOR VESICULATION OF TUBULAR RECYCLING ENDOSOMES

To identify the domain(s) of GRAF1 responsible for vesiculation, we transfected a series of truncation mutants or HA-GRAF1 point mutations and used the morphology of MICAL-L1-containing TRE as a readout to measure vesiculation *in vivo*. A truncated HA-GRAF1 form, containing only the BAR domain (**Figure 6A**; quantified in **Figure 6I**), displayed a cytoplasmic distribution and had no effect on MICAL-L1-decorated endosomes (**Figure 6B**). A form of GRAF1 comprised of only the BAR+PH domains (**Figure 6C**), extensively localized to MICAL-L1-decorated TRE (**Figure 6D**, see inset; quantified in **Figure 6I**). These findings support previous observations by Lundmark et al. (2008) characterizing the BAR-PH region of GRAF1 as an inducer of static tubular invaginations, incapable of undergoing fission. Thus, in analogy with the role of the PH domain in linking GRAF1 to PIP2 at the plasma membrane (Lundmark et al., 2008), our results support a role for this domain in recruiting GRAF1 to TRE, which also contain PIP2 (Jovic et al., 2009).

Nevertheless, when the Rho-GAP domain was added to the BAR+PH domains (without the SH3 domain), transfected cells were now devoid of TRE compared to neighboring untransfected cells (BAR+PH+GAP; **Figures 6E,F**; quantified in **Figure 6I**). These experiments led to the notion that the Rho-GAP domain may have the ability to vesiculate membranes independently of its SH3 domain, despite the fact that the SH3 domain interacts with Dynamin1 and 2 *in vitro* (Lundmark et al., 2008). Thus, it is possible that the recruitment of Dynamin via the GRAF1 SH3 domain may not be required for TRE vesiculation *in vivo*.

To further address the ability of the Rho-GAP domain to vesiculate TRE *in vivo*, site-directed mutagenesis was used to inactivate the Rho-GAP domain (R412D) of full-length HA-GRAF1, giving rise to a "dead-GAP" mutant, incapable of stimulating GTP hydrolysis by Cdc42 (Doherty et al., 2011). At both high and low expression levels of the R412D GRAF1 mutant (**Figure 6H**), MICAL-L1-decorated TRE were preserved (**Figure 6I**; quantified in **Figure 6J**) and comparable to those in untransfected cells. Moreover, the association of GRAF1 (BAR+PH) with TRE did not require the TRE hub protein, MICAL-L1 or Syndapin2 (**Figures S3A–E**). Taken together, these data provide evidence for the involvement of the Rho-GAP domain of GRAF1 in TRE vesiculation *in vivo*.

GRAF1 AND EHD1 VESICULATE TUBULAR RECYCLING ENDOSOMES SYNERGISTICALLY IN A SEMI-PERMEABILIZED CELL SYSTEM

In vitro experiments with liposomes have demonstrated that N-BAR-domain-containing proteins shape membranes into tubules, or into small vesicles, depending on the local concentration of the BAR domain, its degree of oligomerization and the membrane density (Peter et al., 2004; Ayton et al., 2009). Our *in vivo* experiments in this study implicate GRAF1 in the vesiculation process; such membrane sculpting activity might be assisted by other proteins that associate with GRAF1. Since TRE nonetheless underwent scission *in vivo* when an over-expressed GRAF1 protein lacking its SH3 domain was introduced into cells (**Figures 6E,F**), we speculated that the SH3-interacting scission protein, Dynamin, might be dispensable for TRE vesiculation. Accordingly, we turned to investigate the participation of other putative "co-vesiculators" in this process.

Members of the mammalian EHD family (which do not contain a BAR domain) were likewise implicated in oligomerizing and remodeling membranes *in vivo* and *in vitro* (Daumke et al., 2007; Moren et al., 2012; Giridharan et al., 2013). Human EHD1, which decorates a wide range of endosomes, including TRE, demonstrated vesiculation-capabilities *in vivo*, upon ATP hydrolysis (Cai et al., 2012, 2013). It is of interest that a common EHD1/EHD3 ortholog in the lamprey (l-EHD), which displays 83% identity to both human EHD, cooperates with dynamin to promote vesicle budding and to prevent the generation of long tubular structures (Jakobsson et al., 2011). However, depletion of dynamin2 did not lead to elongation of TRE (**Figures S3F,G**), as might be expected if dynamin2 played an essential role in TRE vesiculation. Indeed, the loss of dynamin appeared to destabilize the tubules (**Figures S3F,G**). While the reason for this phenotype remains unknown, it is apparent that dynamin2 is not likely to be a major player in the scission of TRE alongside EHD1 and GRAF1.

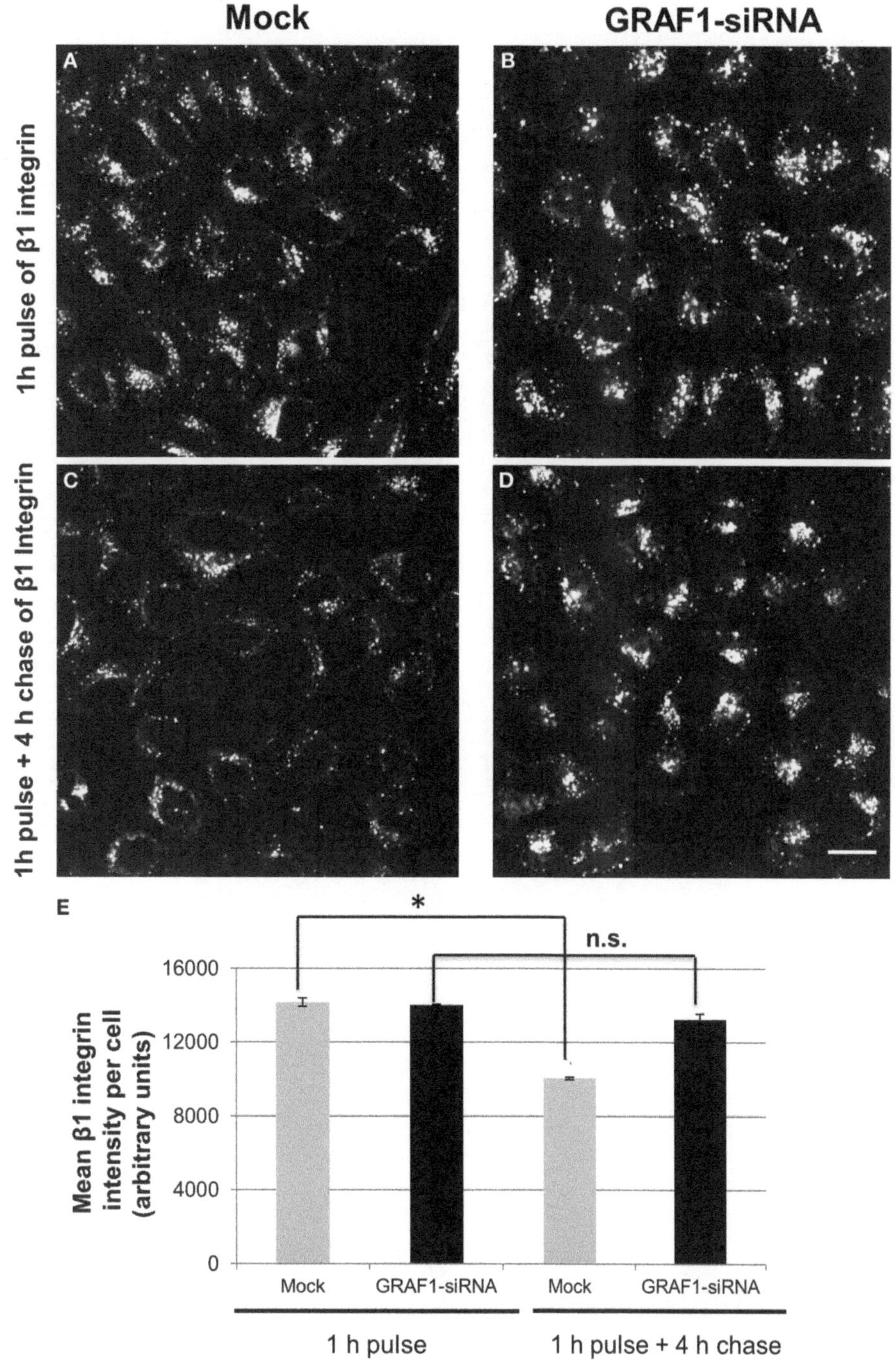

FIGURE 5 | The recycling of β1 integrin is delayed upon GRAF1-depletion. (A) Cells grown on coverslips were either mock-treated or **(B)** treated with GRAF1-siRNA for 72 h. Cells were incubated with mouse anti-β1 integrin for 1 h at 37°C, followed by a short acid strip and either fixed directly **(A,B)** or "chased" in complete media for 4 h, followed by a second acid strip and fixation **(C,D)**. Fixed cells were stained with Alexa 568-goat anti-mouse secondary antibody **(A,D)**. **(E)** The intensity of remaining β1 integrin per cell in both mock-and siRNA treatments was quantified by Image J from three independent experiments **(E)**. $^{*}p < 0.001$. n.s., not significant. Error bars depict standard error. Bar, 10 μm.

Since MICAL-L1 binds to both EHD1 and GRAF1 directly, and indirectly, respectively, and because EHD1 also associates with GRAF1 (**Figure 2B**), we aimed to assess the impact of both GRAF1 and EHD1 on TRE-scission in a newly developed semi-permeabilized cell system (Cai et al., 2013). HeLa cells grown on coverslips were gently perforated with digitonin for 40 s at room temperature and much of the cytosol content was washed away. Upon restoring an ATP-regenerating system, the cells were viable for ~1 h, and MICAL-L1-containing TRE remained partially intact. During this period, purified GST-linked

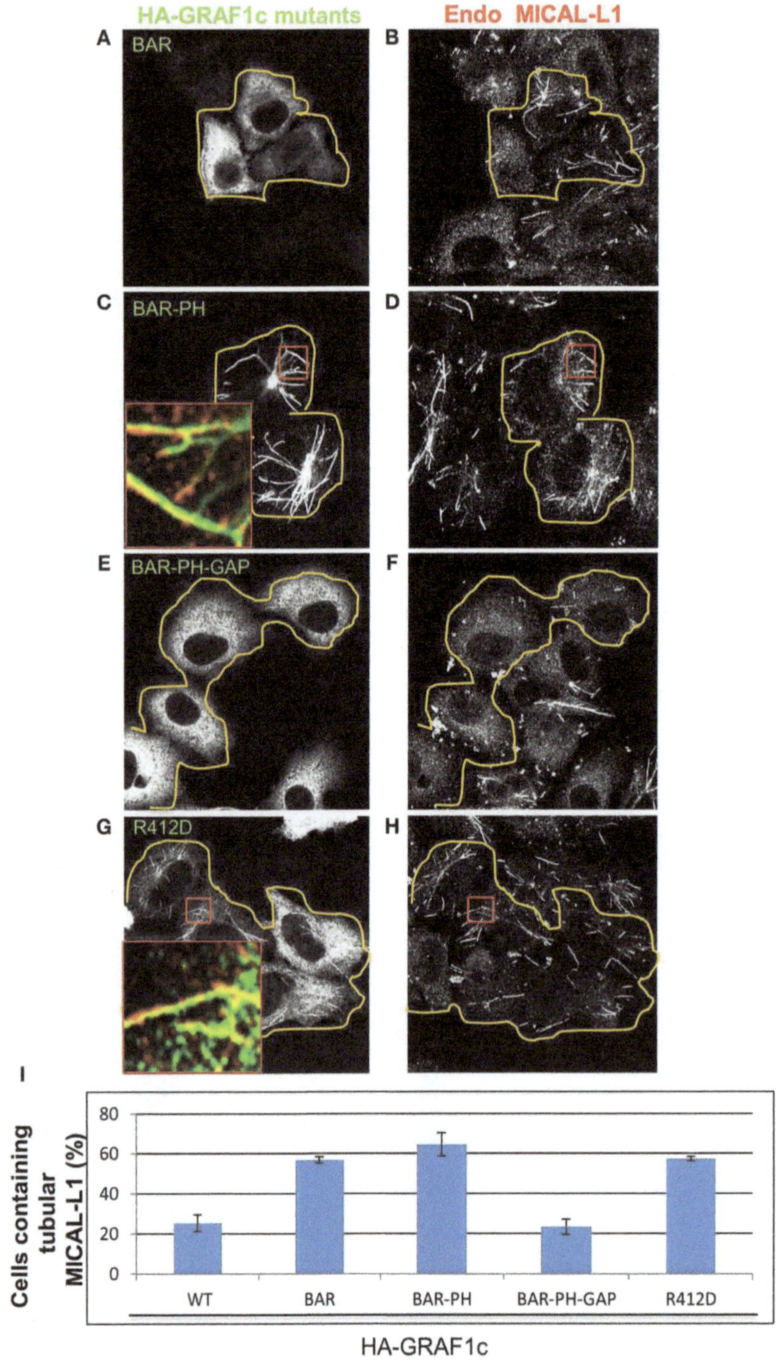

FIGURE 6 | The GAP domain of GRAF1c is required to promote vesiculation of MICAL-L1-containing tubular endosomes. Cells grown on coverslips were either transfected with **(A,B)** the HA-GRAF1c BAR domain (orange lines indicate the transfected cells), **(C,D)** the HA-GRAF1c BAR-PH domains, **(E,F)** the HA-GRAF1c BAR-PH-GAP domains, or **(G,H)** the mutant HA-GRAF1c (R412D) for 12 h. Cells were then fixed and stained with rabbit anti-HA and mouse anti-MICAL-L1 antibodies, followed by Alexa-488 goat anti-rabbit and Alex-568 goat anti- mouse antibodies. **(I)** Quantification of the percentage of cells containing MICAL-L1-decorated tubular recycling endosomes, derived from three independent experiments. Error bars denote standard error. Bar, 10 μm.

proteins (**Figure 7A**) were introduced to the semi-permeabilized cells for 30 min at 37°C, and subsequently fixed. TRE morphology was then visualized by staining endogenous MICAL-L1 (**Figures 7B–E**) and the mean area of MICAL-L1-containing tubules per field was quantified (**Figure 7F**).

The baseline level of TRE was defined by incubating semi-permeabilized cells with GST-only (**Figure 7B**). Under these conditions, GST-EHD1 incubation (**Figure 7C**) decreased the tubulation by 27% (mean square area of TRE from 3 independent experiments is plotted in **Figure 7F**), leaving numerous short,

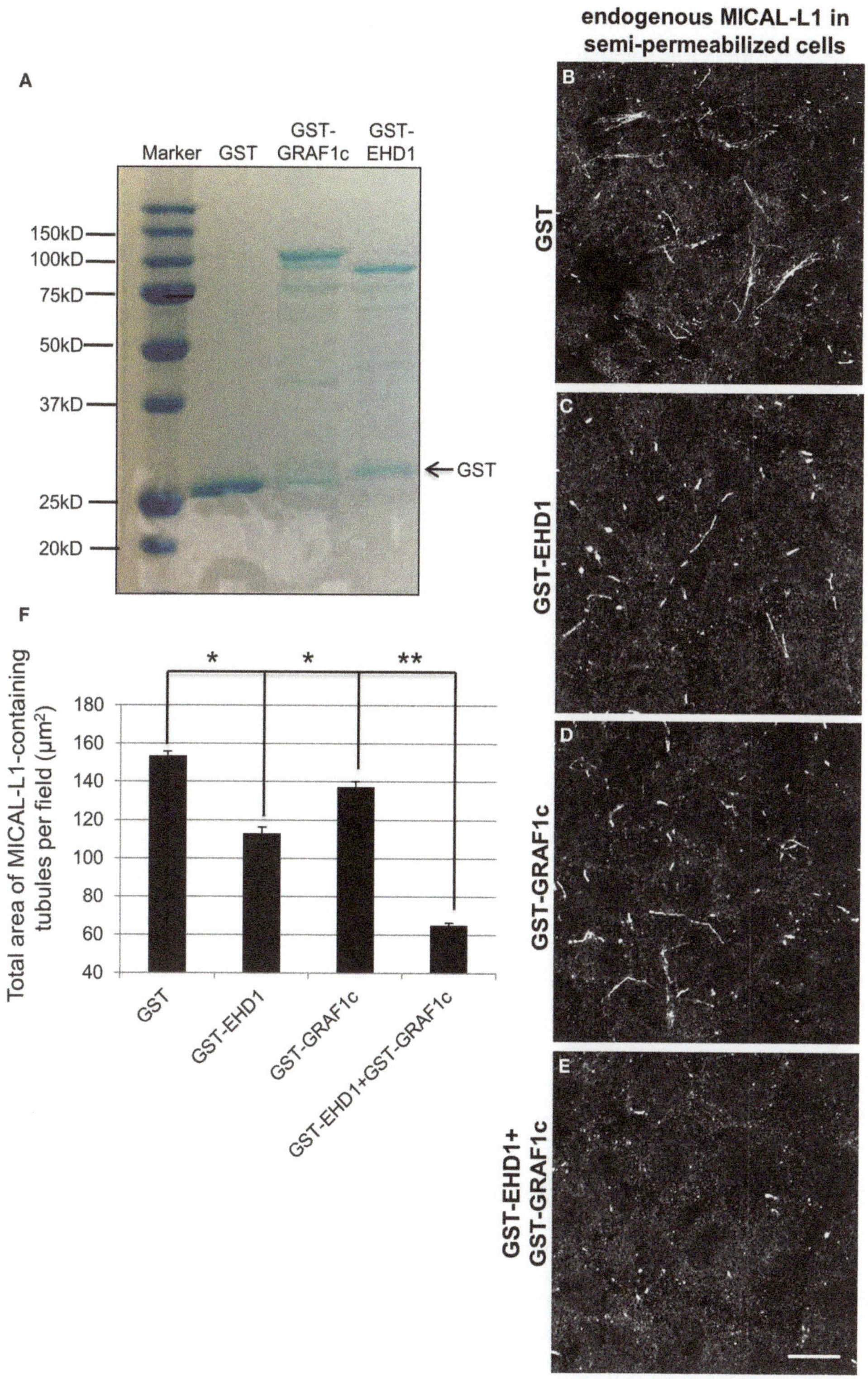

FIGURE 7 | GRAF1c and EHD1 vesiculate MICAL-L1-containing tubules synergistically in a semi-permeabilized cell system. (A) Purified GST only, GST-GRAF1c, and GST-EHD1 were detected by SDS-PAGE and Coomassie Blue staining. **(B–E)** Cells were permeabilized with 20 μg/ml digitonin for 40 s at room temperature. Permeabilized cells were then either incubated for 30 min. at 37°C with **(B)** GST-only, **(C)** GST-EHD1, **(D)** GST-GRAF1c, or **(E)** GST-EHD1 and GST-GRAF1c. After fixation, the remaining TRE were stained with anti-MICAL-L1 antibody **(B–E)**. **(F)** The total area of MICAL-L1-containing tubules per field was quantified by Image J from three independent experiments. $^*p < 0.005$, $^{**}p < 0.001$. Error Bar denotes standard error. Bar, 10 μm.

but somewhat elongated endosomes. This is in line with our previous report (Cai et al., 2013) and our *in vivo* observation attributing a role for EHD1 in TRE vesiculation (Cai et al., 2012). By itself, GST-GRAF1 caused a milder decrease in TRE levels (10%) (**Figure 7D** and quantified in **7F**). However, the addition of both GST-EHD1 and GST GRAF1 *together* brought about a dramatic loss of TRE (57% decrease, **Figure 7E** and quantified in **7F**), suggesting that GRAF1 and EHD1 act synergistically to induce TRE vesiculation in a semi-permeabilized cell system. Interestingly, the truncated GST-GRAF1 (BAR+PH) did not alter TRE morphology (**Figures S4A–C**), possibly due to the absence of as-yet-unidentified cytoplasmic components that are lacking in the semi-permeabilized system, but are present in cells. Overall, these data support a role for the GRAF1 Rho-GAP domain in vesiculation.

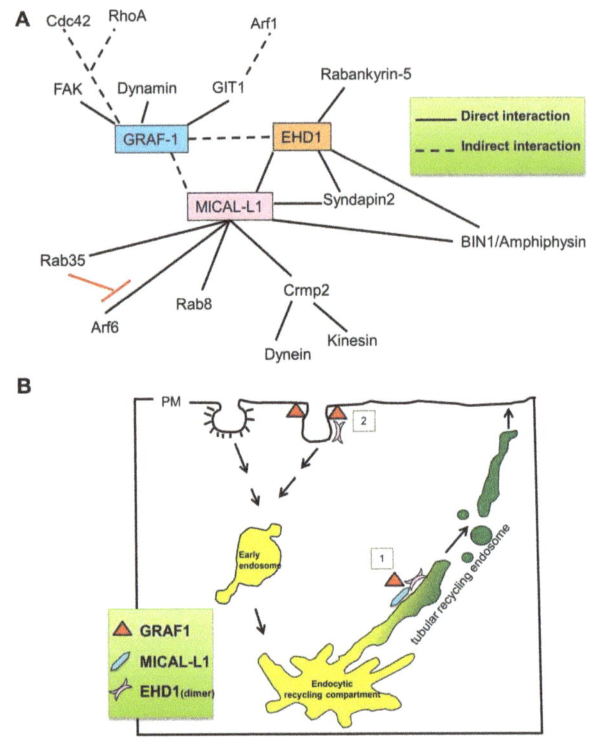

FIGURE 8 | Proposed role for GRAF1 in mediating vesiculation of tubular recycling endosomes. (A) Key interaction partners of GRAF1, MICAL-L1 and EHD1. **(B)** Proposed roles for GRAF1 and EHD1 at the plasma membrane, in the clathrin-independent endocytic pathway [pathway 2] [(Cai et al., 2011) and reviewed in Doherty and Lundmark (2009)], and for GRAF1, MICAL-L1 and EHD1 at tubular recycling endosomes [pathway 1]; MICAL-L1 is recruited onto TRE by directly binding to phosphatidic acid. It then recruits EHD1 by a direct interaction and interacts indirectly with GRAF1. EHD1, and GRAF1 then synergistically vesiculate TRE, giving rise to small and motile recycling endosomes.

DISCUSSION

BIOGENESIS OF TUBULAR RECYCLING ENDOSOMES

The ERC is considered a morphologically distinct series of membranes from early endosomes and comprises a series of vesicular and tubular structures that enable sorting and recycling of internalized receptors and lipids back to the plasma membrane (Gruenberg and Maxfield, 1995). The mechanisms involved in the generation of the ERC and the TREs that comprise this compartment are only partially understood. Recent studies support the role of BAR domain-containing proteins such as BIN1/Amphiphysin (Pant et al., 2009) and Syndapin2 (Braun et al., 2005; Giridharan et al., 2013) in TRE biogenesis, and as well as MICAL-L1 (Giridharan et al., 2013) and the C-terminal EHD protein, EHD3 (Cai et al., 2013). However, once TREs are generated and cargo has been sorted, for recycling to occur there must be efficient vesiculation of these tubular structures.

In previous studies, we and others have identified EHD1 as an ATPase involved in membrane scission and the generation of recycling vesicles (Jakobsson et al., 2011; Cai et al., 2012). While the addition of purified EHD1 and an ATP regenerating system may be sufficient in a semi-permeabilized cell vesiculation assay, *in vivo* other proteins likely contribute to this process. For example, recruitment of PLA2α to tubular membranes may further contribute TRE vesiculation through the generation of lysophosphatidic acid (Cai et al., 2012). We now provide evidence that GRAF1 contributes to TRE vesiculation.

VESICULATION OF MEMBRANES BY GRAF1, EHD1 AND DYNAMIN2

GRAF1 has been implicated in the control of the poorly characterized CLIC/GEEC endocytic pathway through its ability to regulate the activity of Cdc42 and RhoA (Lundmark et al., 2008) (summarized in **Figure 8A**). Via its BAR and PH domains, GRAF1 localizes to incoming tubular and vesicular membrane structures, which carry internalized cargo, including bacterial toxins and fluid phase cargo, as well as the model GPI anchored protein, GFP-GPI (Lundmark et al., 2008). Formation of CLIC vesicles requires membrane deformation followed by fission, yet little is known about the driving force for this process. Although dynamin, a fission protein in clathrin-dependent endocytosis, interacts with the SH3 domain of GRAF1 through its PRD domain (**Figure 8A**), it appears to be dispensable for membrane fission in the CLIC pathway (Doherty and Lundmark, 2009). Consistent with these findings, we also observed that dynamin2 is neither involved in the fission of endosomes containing the GPI-AP, CD59, nor TRE decorated with MICAL-L1. Instead, we have demonstrated that EHD1 mediates this vesiculation process (Cai et al., 2012).

In this study, we show that the GRAF1-depletion and over-expression of GRAF1 induce hypertubulation and vesiculation of recycling tubules, respectively, in a manner analogous to that of EHD1 (Cai et al., 2012). Our data are therefore consistent with the notion that GRAF1 also plays a role in the vesiculation of membrane tubules. Membrane curvature induced by BAR-domain may lead to membrane fission (Peter et al., 2004; Gallop and McMahon, 2005). However, compared with other BAR-containing proteins, the BAR domain of GRAF1 is relatively inefficient in this respect (Doherty and Lundmark, 2009). In agreement with this, the over-expression of the GRAF1-BAR domain did not enhance the vesiculation of recycling tubules (**Figures 6A,B**). Nor did it enhance vesiculation when expressed together with the PH domain (BAR-PH) (**Figures 6C,D**). In contrast, when the GAP domain of GRAF1 was transfected into cells, TRE were extensively vesiculated (**Figures 6E,F**), whereas the GAP-dead mutant was incapable of promoting vesiculation (**Figures 6G,H**). Overall, our study supports the idea that the GAP domain of GRAF1 plays a role in the vesiculation process.

Via its PH domain, GRAF1 was previously shown to predominantly bind to PIP2 (Lundmark et al., 2008; Doherty and Lundmark, 2009), which is a major phosphoinositide in the plasma membrane. On the other hand, MICAL-L1 binds selectively to PA, which is enriched on TRE (Giridharan et al., 2013). However, both phosphatidylinositol-4-phosphate and PIP2 are also localized to TRE (Jovic et al., 2009), potentially mediating docking of GRAF1 on tubular and vesicular endosomes. Although GRAF1 seemingly interacts with both MICAL-L1 and EHD1 indirectly (see **Figure 8A**), MICAL-L1 nevertheless appears to be a major hub that recruits both the components involved in TRE biogenesis (tubulation) as well as vesiculation, potentially allowing synergistic activity between EHD1 and GRAF1 (**Figure 7**).

Based on our findings, we propose that in addition to GRAF1's characterized roles in CLIC/GEEC regulation and its functions at the plasma membrane (Lundmark et al., 2008; Doherty and Lundmark, 2009; Doherty et al., 2011) (see Model in **Figure 8B**; pathway number 2, at clathrin-independent invaginations), GRAF1 is also a component of the vesiculation complex comprised of MICAL-L1 and EHD1 on TRE (**Figure 8B**; pathway number 1, on TRE). Future studies will be needed to determine whether the combined scission activities of EHD1 and GRAF1 are required *in vivo* to overcome energy barriers, or perhaps are needed to effectively regulate the complex process of recycling from the TRE.

ACKNOWLEDGMENTS

The authors gratefully acknowledge support from the National Institutes of Health grants R01GM087455 and R01GM074876. Research reported in this publication, release was also supported by an Institutional Development Award (IDeA) from the National Institute of General Medical Sciences of the National Institutes of Health under grant number 5P20GM103489-10.

SUPPLEMENTARY MATERIAL

The Supplementary Material for this article can be found online at: http://www.frontiersin.org/journal/10.3389/fcell.2014.00022/abstract

Figure S1 | GRAF1c does not interact with MICAL-L1, EHD1, and Syndapin2 by selective yeast two hybrid assays. (A) The *Saccharomyces cerevisiae* yeast strain AH109 was co-transformed with the indicated GAL4-binding domain (GAL4bd) fusion constructs GAL4bd- GRAF1c and GAL4bd-p53 (control), together with the indicated GAL4 transcription activation (GAL4ad) fusion constructs GAL4ad-EHD1, GAL4ad-MICAL-L1 and GAL4ad-SV40 large T-antigen (control). **(B)** The interaction between GAL4ad-GRAF1c and GAL4bd-Syndapin2 was analyzed. Cotransformants were assayed for their growth on non-selective (+HIS) and selective (-HIS) media. **(C)** GRAF1-depletion does not alter the cellular expression levels of either MICAL-L1 or EHD1. Mock-treated or GRAF1-siRNA-treated cells were subject to immunoblotting with anti-MICAL-L1, GRAF1 (control), EHD1 and actin (loading control). **(D,E)** Cells were transfected with HA-GRAF1c. After 48 h, cells were lysed for 1 h in buffer containing 50 mM Tris pH 7.4, 150 mM NaCl, 0.5% Triton X-100 and protease inhibitors. Cell lysates were incubated with mouse anti-MICAL-L1 antibody or rabbit EHD1 antibody overnight. Protein L beads **(D)** or Rabbit IP Matrix beads **(E)** were added to the mixture of cell lysate and MICAL-L1 or EHD1 antibody for 3 h at 4°C. Beads were washed in buffer containing 50 mM Tris pH 7.4, 150 mM NaCl, 0.1% Triton. Proteins were eluted by adding SDS loading buffer. Samples were subjected to 8% SDS-PAGE, followed by blotting with anti-HA and either anti-MICAL-L1 or EHD1 antibodies.

Figure S2 | GRAF1-containing tubular recycling endosomes are positive for phosphatidic acid. (A–C) HeLa cells were cotransfected with HA-GRAF1 and the phosphatidic acid-binding yeast protein, GFP-SPO-20. Insets show that SPO-20 localizes in part to GRAF1-containing tubular endosomes. **(D)** Mock- or GRAF1-depleted cells were transfected after 48 h treatment with either wild-type or siRNA-resistant HA-GRAF1c for an additional 24 h. Lysates were then separated by SDS PAGE and immunoblotted with anti-HA or anti-actin antibodies. Bar, 10 μm.

Figure S3 | The tubular localization of the GRAF1c BAR-PH domains do not require MICAL-L1 or Syndapin2 expression. (A,B) Cells grown in 35 mm dishes were either mock-treated, treated with **(A)** MICAL-L1-siRNA or **(B)** Syndapin2-siRNA for 72 h. After lysis, cell lysates were subjected to 8% SDS-PAGE, followed by immunoblotting with anti-MICAL-L1, Syndapin2 and Actin antibodies. **(C–E)** In the last 20 h of siRNA-treatment, cells were transfected with HA-GRAF1c (BAR-PH). After fixation, cells were stained with rabbit anti-HA antibody, followed by Alexa-488 goat anti-rabbit secondary antibody. **(F,G)** Cells were treated with Mock- or Dynamin2-SiRNA for 48 h, and then fixed and stained with anti-MICAL-L1. Efficacy of Dynamin2 depletion was demonstrated by its impact on transferrin uptake (data not shown). Bars, 10 μm.

Figure S4 | The GRAF1c BAR-PH domains are unable to vesiculate MICAL-L1-containing tubules. (A) Purified GST and GST-GRAF1c (BAR-PH) proteins were detected by SDS-PAGE and Coomassie Blue staining. Cells were permeabilized with 20 μg/ml digitonin for 40 s at room temperature. Permeabilized cells were either incubated with **(B)** GST-only and **(C)** GST-GRAF1c (BAR-PH) for 30 min. at 37°C. Cells were then fixed and stained with anti-MICAL-L1 antibody. Bar, 10 μm.

REFERENCES

Altankov, G., and Grinnell, F. (1995). Fibronectin receptor internalization and AP-2 complex reorganization in potassium-depleted fibroblasts. *Exp. Cell Res.* 216, 299–309. doi: 10.1006/excr.1995.1038

Ayton, G. S., Lyman, E., Krishna, V., Swenson, R. D., Mim, C., Unger, V. M., et al. (2009). New insights into BAR domain-induced membrane remodeling. *Biophys. J.* 97, 1616–1625. doi: 10.1016/j.bpj.2009.06.036

Beckers, C. J., Keller, D. S., and Balch, W. E. (1987). Semi-intact cells permeable to macromolecules: use in reconstitution of protein transport from the endoplasmic reticulum to the Golgi complex. *Cell* 50, 523–534. doi: 10.1016/0092-8674(87)90025-0

Braun, A., Pinyol, R., Dahlhaus, R., Koch, D., Fonarev, P., Grant, B. D., et al. (2005). EHD proteins associate with syndapin I and II and such interactions play a crucial role in endosomal recycling. *Mol. Biol. Cell* 16, 3642–3658. doi: 10.1091/mbc.E05-01-0076

Cai, B., Caplan, S., and Naslavsky, N. (2012). cPLA2alpha and EHD1 interact and regulate the vesiculation of cholesterol-rich, GPI-anchored, protein-containing endosomes. *Mol. Biol. Cell* 23, 1874–1888. doi: 10.1091/mbc.E11-10-0881

Cai, B., Giridharan, S. S., Zhang, J., Saxena, S., Bahl, K., Schmidt, J. A., et al. (2013). Differential roles of C-terminal Eps15 Homology Domain proteins as vesiculators and tubulators of recycling endosomes. *J. Biol. Chem.* 288, 30172–30180. doi: 10.1074/jbc.M113.488627

Cai, B., Katafiasz, D., Horejsi, V., and Naslavsky, N. (2011). Pre-sorting endosomal transport of the GPI-anchored protein, CD59, is regulated by EHD1. *Traffic* 12, 102–120. doi: 10.1111/j.1600-0854.2010.01135.x

Caplan, S., Naslavsky, N., Hartnell, L. M., Lodge, R., Polishchuk, R. S., Donaldson, J. G., et al. (2002). A tubular EHD1-containing compartment involved in the recycling of major histocompatibility complex class I molecules to the plasma membrane. *EMBO J.* 21, 2557–2567. doi: 10.1093/emboj/21.11.2557

Daumke, O., Lundmark, R., Vallis, Y., Martens, S., Butler, P. J., and McMahon, H. T. (2007). Architectural and mechanistic insights into an EHD ATPase involved in membrane remodelling. *Nature* 449, 923–927. doi: 10.1038/nature06173

Doherty, G. J., Ahlund, M. K., Howes, M. T., Moren, B., Parton, R. G., McMahon, H. T., et al. (2011). The endocytic protein GRAF1 is directed to cell-matrix adhesion sites and regulates cell spreading. *Mol. Biol. Cell* 22, 4380–4389. doi: 10.1091/mbc.E10-12-0936

Doherty, G. J., and Lundmark, R. (2009). GRAF1-dependent endocytosis. *Biochem. Soc. Trans.* 37, 1061–1065. doi: 10.1042/BST0371061

Ezratty, E. J., Bertaux, C., Marcantonio, E. E., and Gundersen, G. G. (2009). Clathrin mediates integrin endocytosis for focal adhesion disassembly in migrating cells. *J. Cell Biol.* 187, 733–747. doi: 10.1083/jcb.200904054

Gallop, J. L., and McMahon, H. T. (2005). BAR domains and membrane curvature: bringing your curves to the BAR. *Biochem. Soc. Symp.* 72, 223–231.

Giridharan, S. S., Cai, B., Vitale, N., Naslavsky, N., and Caplan, S. (2013). Cooperation of MICAL-L1, syndapin2, and phosphatidic acid in tubular recycling endosome biogenesis. *Mol. Biol. Cell* 24, 1776–1790, S1771–S1715. doi: 10.1091/mbc.E13-01-0026

Gruenberg, J., and Maxfield, F. R. (1995). Membrane transport in the endocytic pathway. *Curr. Opin. Cell Biol.* 7, 552–563. doi: 10.1016/0955-0674(95)80013-1

Harlan, J. E., Yoon, H. S., Hajduk, P. J., and Fesik, S. W. (1995). Structural characterization of the interaction between a pleckstrin homology domain and phosphatidylinositol 4,5-bisphosphate. *Biochemistry* 34, 9859–9864. doi: 10.1021/bi00031a006

Hayashi, K., and Kojima, C. (2008). pCold-GST vector: a novel cold-shock vector containing GST tag for soluble protein production. *Protein Expr. Purif.* 62, 120–127. doi: 10.1016/j.pep.2008.07.007

Hildebrand, J. D., Taylor, J. M., and Parsons, J. T. (1996). An SH3 domain-containing GTPase-activating protein for Rho and Cdc42 associates with focal adhesion kinase. *Mol. Cell. Biol.* 16, 3169–3178.

Jakobsson, J., Ackermann, F., Andersson, F., Larhammar, D., Low, P., and Brodin, L. (2011). Regulation of synaptic vesicle budding and dynamin function by an EHD ATPase. *J. Neurosci.* 31, 13972–13980. doi: 10.1523/JNEUROSCI.1289-11.2011

Jelen, F., Lachowicz, P., Apostoluk, W., Mateja, A., Derewenda, Z. S., and Otlewski, J. (2009). Dissecting the thermodynamics of GAP-RhoA interactions. *J. Struct. Biol.* 165, 10–18. doi: 10.1016/j.jsb.2008.09.007

Jovic, M., Kieken, F., Naslavsky, N., Sorgen, P. L., and Caplan, S. (2009). Eps15 homology domain 1-associated tubules contain phosphatidylinositol-4-phosphate and phosphatidylinositol-(4,5)-bisphosphate and are required for efficient recycling. *Mol. Biol. Cell* 20, 2731–2743. doi: 10.1091/mbc.E08-11-1102

Jovic, M., Sharma, M., Rahajeng, J., and Caplan, S. (2010). The early endosome: a busy sorting station for proteins at the crossroads. *Histol Histopathol* 25, 99–112.

Kieken, F., Sharma, M., Jovic, M., Giridharan, S. S., Naslavsky, N., Caplan, S., et al. (2010). Mechanism for the selective interaction of C-terminal Eps15 homology domain proteins with specific Asn-Pro-Phe-containing partners. *J. Biol. Chem.* 285, 8687–8694. doi: 10.1074/jbc.M109.045666

Long, K. R., Yamamoto, Y., Baker, A. L., Watkins, S. C., Coyne, C. B., Conway, J. F., et al. (2010). Sar1 assembly regulates membrane constriction and ER export. *J. Cell. Biol.* 190, 115–128. doi: 10.1083/jcb.201004132

Lundmark, R., Doherty, G. J., Howes, M. T., Cortese, K., Vallis, Y., Parton, R. G., et al. (2008). The GTPase-activating protein GRAF1 regulates the CLIC/GEEC endocytic pathway. *Curr. Biol.* 18, 1802–1808. doi: 10.1016/j.cub.2008.10.044

McMahon, H. T., and Gallop, J. L. (2005). Membrane curvature and mechanisms of dynamic cell membrane remodelling. *Nature* 438, 590–596. doi: 10.1038/nature04396

Moren, B., Shah, C., Howes, M. T., Schieber, N. L., McMahon, H. T., Parton, R. G., et al. (2012). EHD2 regulates caveolar dynamics via ATP-driven targeting and oligomerization. *Mol. Biol. Cell* 23, 1316–1329. doi: 10.1091/mbc.E11-09-0787

Nishimura, T., and Kaibuchi, K. (2007). Numb controls integrin endocytosis for directional cell migration with aPKC and PAR-3. *Dev. Cell* 13, 15–28. doi: 10.1016/j.devcel.2007.05.003

Pant, S., Sharma, M., Patel, K., Caplan, S., Carr, C. M., and Grant, B. D. (2009). AMPH-1/Amphiphysin/Bin1 functions with RME-1/Ehd1 in endocytic recycling. *Nat. Cell Biol.* 11, 1399–1410. doi: 10.1038/ncb1986

Peter, B. J., Kent, H. M., Mills, I. G., Vallis, Y., Butler, P. J., Evans, P. R., et al. (2004). BAR domains as sensors of membrane curvature: the amphiphysin BAR structure. *Science* 303, 495–499. doi: 10.1126/science.1092586

Rahajeng, J., Giridharan, S. S., Cai, B., Naslavsky, N., and Caplan, S. (2012). MICAL-L1 is a tubular endosomal membrane hub that connects Rab35 and Arf6 with Rab8a. *Traffic* 13, 82–93. doi: 10.1111/j.1600-0854.2011.01294.x

Sharma, M., Giridharan, S. S., Rahajeng, J., Naslavsky, N., and Caplan, S. (2009). MICAL-L1 links EHD1 to tubular recycling endosomes and regulates receptor recycling. *Mol. Biol. Cell* 20, 5181–5194. doi: 10.1091/mbc.E09-06-0535

Shibata, H., Oishi, K., Yamagiwa, A., Matsumoto, M., Mukai, H., and Ono, Y. (2001). PKNbeta interacts with the SH3 domains of Graf and a novel Graf related protein, Graf2, which are GTPase activating proteins for Rho family. *J. Biochem.* 130, 23–31. doi: 10.1093/oxfordjournals.jbchem.a002958

Stulnig, T. M., Berger, M., Sigmund, T., Raederstorff, D., Stockinger, H., and Waldhausl, W. (1998). Polyunsaturated fatty acids inhibit T cell signal transduction by modification of detergent-insoluble membrane domains. *J. Cell. Biol.* 143, 637–644. doi: 10.1083/jcb.143.3.637

Taylor, J. M., Macklem, M. M., and Parsons, J. T. (1999). Cytoskeletal changes induced by GRAF, the GTPase regulator associated with focal adhesion kinase, are mediated by Rho. *J. Cell Sci.* 112(pt 2), 231–242.

Ylanne, J., Huuskonen, J., O'Toole, T. E., Ginsberg, M. H., Virtanen, I., and Gahmberg, C. G. (1995). Mutation of the cytoplasmic domain of the integrin beta 3 subunit. Differential effects on cell spreading, recruitment to adhesion plaques, endocytosis, and phagocytosis. *J. Biol. Chem.* 270, 9550–9557.

Zeniou-Meyer, M., Zabari, N., Ashery, U., Chasserot-Golaz, S., Haeberle, A. M., Demais, V., et al. (2007). Phospholipase D1 production of phosphatidic acid at the plasma membrane promotes exocytosis of large dense-core granules at a late stage. *J. Biol. Chem.* 282, 21746–21757. doi: 10.1074/jbc.M702968200

Zimmerberg, J., and Kozlov, M. M. (2006). How proteins produce cellular membrane curvature. *Nat. Rev. Mol. Cell Biol.* 7, 9–19. doi: 10.1038/nrm1784

Conflict of Interest Statement: The authors declare that the research was conducted in the absence of any commercial or financial relationships that could be construed as a potential conflict of interest.

5

Intractable diseases treated with intra-bone marrow-bone marrow transplantation

Ming Li[1], Kuquan Guo[2] and Susumu Ikehara[1]*

[1] Department of Stem Cell Disorders, Kansai Medical University, Hirakata City, Japan

[2] Department of Cardiac Surgery, Beijing Institute of Heart, Lung and Blood Vessel Disease, Beijing Anzhen Hospital Affiliated to Capital Medical University, Beijing, China

Edited by:
Atsushi Asakura, University of Minnesota, USA

Reviewed by:
Hong Zhang, Harvard Medical School, USA
Simone Pacini, University of Pisa, Italy

***Correspondence:**
Susumu Ikehara, Department of Stem Cell Disorders, Kansai Medical University, 2-5-1 Shinmachi, Hirakata City, Osaka 570-1010, Japan
e-mail: ikehara@hirakata.kmu.ac.jp

Bone marrow transplantation (BMT) is used to treat hematological disorders, autoimmune diseases (ADs) and lymphoid cancers. Intra bone marrow-BMT (IBM-BMT) has been proven to be a powerful strategy for allogeneic BMT due to the rapid hematopoietic recovery and the complete restoration of T cell functions. IBM-BMT not only replaces hematopoietic stem cells (HSCs) but also mesenchymal stromal cells (MSCs). MSCs are multi-potent stem cells that can be isolated from bone marrow (BM), umbilical cord blood (UCB), and adipose tissue. MSCs play an important role in the support of hematopoiesis, and modify and influence the innate and adaptive immune systems. MSCs also differentiate into mesodermal, endodermal and ectodermal lineage cells to repair tissues. This review aims to summarize the functions of BM-derived-MSCs, and the treatment of intractable diseases such as rheumatoid arthritis (RA) and malignant tumors with IBM-BMT.

Keywords: intra-bone marrow-bone marrow transplantation, mesenchymal stem cell, rheumatoid arthritis, malignant tumors, autoimmune diseases

INTRODUCTION

Mesenchymal stromal cells (MSCs) are multi-potent progenitor cells mainly isolated from bone marrow (BM) (Campagnoli et al., 2001), adipose tissue (Zuk et al., 2001), and the umbilical cord (UC) (Erices et al., 2000). MSCs have been shown to differentiate into osteoblasts, adipocytes (Dominici et al., 2006), cardiomyocytes (Makino et al., 1999), and pancreatic islets (Tang et al., 2004). Furthermore, MSCs have the ability to migrate to injured tissue of liver (van Poll et al., 2008) and heart (Yokokawa et al., 2008). Adipose tissue and BM are the most readily available sources of MSCs because they are easy to harvest, and there are no ethical concerns. BM-derived MSCs (BMMSCs) have a higher degree of commitment to differentiate into chondrogenic and osteogenic lineages than adipose tissue-derived MSCs (Gimble et al., 2007), although there appears to be no difference between adipose tissue-derived MSCs and BM MSCs in terms of immunoregulatory functions and support of hematopoiesis (Poloni et al., 2012). On the other hand, BM MSCs modulate the immune response, suppress allogeneic T cell responses, and prevent the development of graft-versus-host disease (GVHD) (English, 2013).

BM transplantation (BMT) is useful for treating hematopoietic disorders, allogeneic BMT also being used to treat autoimmune diseases (ADs) (Nishimura et al., 1994). Intra-bone marrow-bone marrow transplantation (IBM-BMT) has been proven to be the most effective approach to treating allogeneic BMT, since IBM-BMT can replace not only hematopoietic stem cells (HSCs) but also BMMSCs Thus hematopoietic recovery is rapid, and no GVHD develops even if whole BM cells are injected (Kushida et al., 2001; Ikehara, 2003). In this review, we focus on rheumatoid arthritis (RA) and malignant tumors treated with IBM-BMT.

IMMUNOREGULATORY FUNCTIONS OF BMMSCs

BMMSCs have been reported to have the ability to modify and influence almost all the cells of the innate and adaptive immune systems mediated by BMMSC soluble factors, including IL-6, M-CSF, IL-10, TGFβ, HGF, and PGE2 (Aggarwal and Pittenger, 2005; Beyth et al., 2005; Ramasamy et al., 2007). The adaptive immune system, which is composed of T and B lymphocytes, generates specific immune responses to pathogens with the production of memory cells. BMMSCs modulate the function of dendritic cells (DCs), indirectly regulate T and B cell activity, and delay or prevent the development of acute GVHD (Zhang et al., 2009). BMMSCs have also been shown to suppress the differentiation of DCs and their function during allogeneic islet transplantation (Urban et al., 2008; Aldinucci et al., 2010). BMMSCs strongly inhibited the maturation and functioning of monocyte-derived DCs by interfering selectively with the generation of immature cells via inhibitory mediator of MSC-derived PGE2 (Lee et al., 2006). PGE2 has been identified as one of the candidates responsible for T cell inhibition by BMMSCs, and may have an immunostimulatory role by facilitating Th1 differentiation and expanding the Th17 T cell population (English et al., 2009; Yao et al., 2009). The expression of PGE2 was shown to be upregulated by IFNγ and TNFα in the BMMSCs for immunomodulatory function (English et al., 2007). BMMSCs can inhibit the cytotoxic effects of antigen-primed cytotoxic T cells by suppressing the proliferation and activity (Zhao et al., 2005) via the inhibition of the nuclear translocation of nuclear factor-kappa B (Matsuda-Hashii et al., 2004). BMMSCs have been shown to alter the NK cell phenotype and suppress proliferation of NK cells via the secretion of TGFβ1 and PGE2, and via cytotoxicity against HLA class I-expressing targets (Aggarwal and Pittenger, 2005; Sotiropoulou et al., 2006; Ryan et al., 2007; Uccelli et al., 2008). BMMSCs have also been shown to inhibit the proliferation of B cells when stimulated with anti-CD40L and IL-4 (Glennie et al., 2005). One report has suggested that allogeneic BMMSCs inhibit the activation, proliferation and IgG secretion of B cells in a BXSB mouse model of human systemic lupus erythematosus (Deng et al., 2005).

Allogeneic BMMSCs are effective in the treatment of murine models of human disease (Zappia et al., 2005; Ding et al., 2009; Fiorina et al., 2009). BMMSCs were shown to be able to secrete regulatory cytokines that affect regulatory T cells, and to modulate the immunological dysregulation observed in antibody producing B cells and cytotoxic NK cells in the NOD mouse (Anderson and Bluestone, 2005). BMMSCs promote the endogenous repair of pancreatic islets and renal glomeruli in a streptozotocin-induced diabetic mouse model (Lee et al., 2006). Co-infusion of BMMSCs and BM cells was shown to inhibit the beta cell-specific T cell proliferation and to restore insulin and glucose levels (Urban et al., 2008). BMMSCs secrete many cytokines and growth factors such as HGF, which shows anti-apoptotic activity in hepatocytes and plays an essential part in the regeneration of the liver (Trim et al., 2000; Matsuda-Hashii et al., 2004). BMMSCs have also been shown to protect against experimental liver fibrosis in CCl4-induced rats (Zhao et al., 2005), and to suppress CD3 T-cell proliferation in collagen-induced arthritis (Schurgers et al., 2010).

In mammals, there are seven sirtuin family members, named Sirt1-7. Sirtuins plays a critical role in the regulation of fundamental biological responses to nutritional and environmental stimuli in each subcellular compartment (Blander and Guarente, 2004; Imai and Guarente, 2010). Sirt1 is a class III protein deacetylase, and Sirt1 activity can be regulated through NAD^+. Sirt1 binds to and deacetylates a number of important transcription factors—such as peroxisome proliferator-activated receptor gamma (PPARγ), PPARα, PPAR gamma coactivator 1 alpha (PGC-1α), and the forkhead box, subgroup O (FOXO) family of transcription factors—to drive metabolic responses such as insulin secretion, gluconeogenesis, and fatty acid oxidation (Haigis and Sinclair, 2010). Some reports indicate that Sirt1 promotes osteogenesis and decreases adipogenesis of BMMSCs *in vitro* (Tseng et al., 2011; Peltz et al., 2012; Puri et al., 2012).

Sirt1 deacetylates β-catenin to regulate differentiation of MSCs in MSCs specific Sirt1 knock-out mice (MSC KO) (Simic et al., 2013). Moreover, Sirt1 has been shown to directly downregulate *Sost* gene expression, and promote bone formation in the treatment of osteoporosis (Cohen-Kfir et al., 2011). One report has shown that CD8 T cell differentiation is regulated by basic leucine zipper transcription factor, ATF-like (BATF), which is a member of the AP-1 family, via Sirt1 expression, BATF deficiency inducing high levels of Sirt1 expression in memory CD8 T cells but not in naive CD8 T cells (Kuroda et al., 2011).

IBM-BMT

We reported that MRL/lpr mice possess abnormal radioresistant stem cells and have provided impressive evidence regarding the origin of ADs in this strain (Ikehara et al., 1989). BMT plus bone graft, which can recruit donor stroma cells, can prevent the recurrence of ADs (Ishida et al., 1994). However, allogeneic BMT + bone grafts failed to treat ADs in MRL/lpr mice, because these mice become more radiosensitive after the onset of lupus nephritis. Moreover, our previous reports showed that stroma cells can be trapped in the liver when BM cells are injected via the portal vein. Thus, directly injecting whole BM cells into the BM, as in IBM-BMT, has been shown to be a powerful strategy for the treatment of ADs in MRL/lpr mice. IBM-BMT, which not only replaces HSCs but also MSCs, has been proven to be the best method for allogeneic BMT: (1) hematopoietic recovery is rapid because the MSCs directly home to the bone cavity, (2) the restoration of T cell functions is complete even in donor-recipient combinations across the MHC barriers, and (3) no graft failure occurs even if the radiation dose is reduced (Kushida et al., 2001). Moreover, IBM-BMT of young marrow cells reversed the reduction of pro-B cells and pre-B cells. The frequency of follicular-B cells in the IBM-BMT group was significantly increased compared to the old group (Hida et al., 2010). We have already used IBM-BMT to successfully treat ADs, osteoporosis, diabetes, Alzheimer's disease, and for the induction of tolerance for organ transplantation (Takada et al., 2006; Guo et al., 2008; Kushida et al., 2009; Li et al., 2009, 2010) (**Table 1**).

BM cells mainly include HSCs and MSCs. MSCs are essential for supporting hematopoiesis in the BM. HSCs can normally proliferate in major histocompatibility complex (MHC)-compatible MSCs even in allogeneic microenvironments. Because the BMCs are directly injected into bone, IBM-BMT circumvents the risk of MSCs being trapped in the lung and liver. And because both MSCs and HSCs are transplanted, hematopoiesis can be rapidly restored. Moreover, IBM-BMT can prevent the risk of graft rejection, even with the use of a mild conditioning regimen (Kushida et al., 2001).

IBM-BMT FOR TREATMENT OF RA

RA is an AD that results in a chronic, systemic inflammatory disorder that may affect many tissues and organs. RA primarily affects joints, but it also affects other organs such heart, kidney, and blood vessels (Turesson et al., 2003). Its pathophysiology indicates that TNFα drives synovial inflammation and joint destruction. The synovial cells include both fibroblast-like and macrophage-like synoviocytes. Fibroblast-like synoviocytes show abnormal behavior in RA (Scott et al., 2010). About 50% of RA is caused by genetic abnormalities (van der Woude et al., 2009). The classification criteria for RA by the American College of Rheumatology (2010), and the treatment options, are summarized in the work by Scott et al (Scott et al., 2010). Here we talk about stem cell therapy for the treatment of RA in basic experiments and clinical applications.

SKG/Jcl mice are a murine model for RA. BM cells of C57BL/6J mice were transplanted into SKG/Jcl mice using IBM-BMT, and the hematolymphoid cells in the recipient mice were reconstituted by donor-derived cells. There was no evidence of arthritis

Table 1 | IBM-BMT treatment of various diseases and induction of tolerance for organ transplantation.

Authors	Animal model	Effect of IBM-BMT
Li et al., 2012	Mouse	Improve renal function
Zhang et al., 2012	Mouse	Prevention of leukemia
Shi et al., 2011	Mouse	Diminish risk of GVHD
Feng et al., 2010	Mouse	Prevention of premature ovarian failure
Li et al., 2009	Mouse	Amelioration of cognitive ability
Kushida et al., 2009	Mouse	Prevention of rheumatoid arthritis
Okazaki et al., 2008	Mouse	Liver transplantation
Miyake et al., 2008	Mouse	Prevention of GVHD
Abraham et al., 2008	Mouse	Prevention of type 2 diabetes
Guo et al., 2008	Rat	Long-term donor specific tolerance in cardiac allograft
Feng et al., 2007	Mouse	Prevention of osteoporosis and hypogonadism
Koike et al., 2007	Mouse	Suppression of growth of colon cancer cells
Ikebukuro et al., 2006	Mouse	Tolerance induction in allogeneic pancreatic islets
Kaneda et al., 2005	Rat	Induction of tolerance for lung transplantation
Takada et al., 2006	Mouse	Prevention of senile osteoporosis
Taira et al., 2005	Rat	Prevention of type 1 diabetes
Nakamura et al., 2004	Mouse	Prevention of GVHD
Esumi et al., 2003	Rat	Induction of tolerance for allogeneic leg transplantation
Ichioka et al., 2002	Mouse	Prevention of senile osteoporosis

in the SKG/Jcl mice at 12 months after transplantation. Moreover, IBM-BMT has been shown to normalize the percentages of Treg (Foxp3+/CD4+) cells, the percentages of receptor activator of NF-kB ligand+ cells on the CD4+ T cells and the serum levels of TNFα, IL-1, and IL-6. One report demonstrated that IBM-BMT is a viable method of immunological manipulation that suppresses the severe joint destruction and bone absorption in SKG/Jcl mice and lends further credence to the use of this methodology in humans with intractable RA (Kushida et al., 2009). Human UC-derived MSCs have been discussed as a possible treatment for RA in the clinical setting. TNFα and IL-6 decreased and CD4+ CD25+ Foxp3+ T cells increased, in active RA patients after UC-derived MSCs were infused, and the UC-derived MSCs survived for 3–6 months, suggesting that treatment with MSCs would benefit RA patients (Wang et al., 2013). Expression of IL-17, IL-6, and TNFα were inhibited when allogeneic UC-derived MSCs were cultured with peripheral blood mononuclear cells (PBMCs) from RA patients, suggesting that MSCs can prevent the expression of these cytokines and that they have therapeutic potential in the treatment of RA (Wang et al., 2012).

MALIGNANT TUMORS TREATED WITH IBM-BMT + THYMUS TRANSPLANTATION (TT)

Donor lymphocyte infusion (DLI) is a useful method for the treatment of malignant tumors, but it also induces GVHD. However, IBM-BMT has been shown to prevent not only graft failure but also GVHD in animals, even when the radiation dose is reduced (Nakamura et al., 2004). Thus, IBM-BMT plus DLI were used to treat malignant tumors (fibrosarcomas) induced by a tumor cell line (methA). DLI ($CD4^-$ spleen cells) can prevent GVHD, but the tumor growth was not suppressed, indicating that CD4+ cells play important roles in graft-versus-tumor (GVT) and GVHD. Our previous results showed that IBM-BMT plus DLI ($CD4^-$lymphocytes) suppressed not only GVHD but also tumor growth (Suzuki et al., 2005). Moreover, the combination of DC, IBM-BMT and DLI showed even better results than the combination of IBM-BMT and DLI in the treatment of solid tumors (Mukaide et al., 2007).

The thymus regulates the production, proliferation and functions of T cells. BMT + TT has been shown to be useful in the treatment of ADs in the MRL/Lpr mouse, because the allogeneic T cells newly-developed by TT are naïve T cells, which show less Fas expression and more resistance to apoptosis than the activated memory T cells with their high Fas expression. We found that the combination of allogeneic IBM-BMT + adult TT from the same donor is effective in mice with solid tumors, as it can induce high thymopoiesis, preserving strong GVT effects without inducing a severe graft-versus-host reaction (GVHR). Meth A sarcoma cells were subcutaneously inoculated into mice, and IBM-BMT + adult TT was then used to treat these mice when the tumor had grown to 5 mm. In tumor-bearing mice, tumor growth was more strongly inhibited by IBM-BMT + adult TT than by IBM-BMT alone. The numbers of $CD8^+$ T cells that infiltrated the tumors, and the number of apoptotic tumor cells, both significantly increased in the mice treated with IBM-BMT + adult TT. IBM-BMT + adult TT prevented tumor development with mild GVHR resulting from the induction of high thymopoiesis and a strong GVT effect in the tumor-bearing mice. The number of $CD4^+$ $FoxP3^+$ cells was lower in the mice treated with IBM-BMT + adult TT than in those treated with IBM-BMT alone. Furthermore, the numbers of $CD8^+$ cells infiltrating the tumor and the levels of IFN-γ were higher in the mice treated with IBM-BMT + adult TT than in those treated with IBM-BMT alone (Miyake et al., 2009). Although T regs have been reported to suppress the GVHR induced by $CD4^+$T cells, they did not reduce the GVT induced by $CD8^+$ T cells (Edinger et al., 2003). Tumors were suppressed to a greater extent as a result of the increased $CD4^+$ and $CD8^+$ T cells and decreased number of Gr-1^+/$CD11b^+$ myeloid suppressor cells and $Foxp3^+$/$CD4^+$ T regs, Moreover, the production of $CD62L^-CD44^+$ effector memory T cells and IFN-γ were also higher (Zhang et al., 2011).

IBM-BMT seems to be better than co-transplantation of HSCs and cultured MSCs, mainly because the number of functional MSCs may drop after being cultured *in vitro*, and cultured MSCs also are trapped by the liver and lung in the case of IV-BMT. Umbilical cord blood (UCB) can also be used a source of stem cells for transplantation, although the numbers are generally insufficient to allow this to be used as a general source. IBM thus appears to be the best choice for allogeneic transplantation, despite the limited number of stem cells that can be directly transplanted into the bone cavity. In conclusion, IBM-BMT can efficiently transplant both HSCs and MSCs, is useful to treat intractable diseases such as RA and malignant tumors, and in the future may be useful for treating various intractable diseases.

ACKNOWLEDGMENTS

We would like to thank Mr. Hilary Eastwick-Field and Ms. Keiko Ando for their help in the preparation of the manuscript.

REFERENCES

Abraham, N. G., Li, M., Vanella, L., Peterson, S. J., Ikehara, S., and Asprinio, D. (2008). Bone marrow stem cell transplant into intra-bone cavity prevents type 2 diabetes: role of heme oxygenase-adiponectin. *J. Autoimmun.* 30, 128–135. doi: 10.1016/j.jaut.2007.12.005

Aggarwal, S., and Pittenger, M. F. (2005). Human mesenchymal stem cells modulate allogeneic immune cell responses. *Blood* 105, 1815–1822. doi: 10.1182/blood-2004-04-1559

Aldinucci, A., Rizzetto, L., Pieri, L., Nosi, D., Romagnoli, P., Biagioli, T., et al. (2010). Inhibition of immune synapse by altered dendritic cell actin distribution: a new pathway of mesenchymal stem cell immune regulation. *J. Immunol.* 185, 5102–5110. doi: 10.4049/jimmunol.1001332

Anderson, M. S., and Bluestone, J. A. (2005). The NOD mouse: a model of immune dysregulation. *Annu. Rev. Immunol.* 23, 447–485. doi: 10.1146/annurev.immunol.23.021704.115643

Beyth, S., Borovsky, Z., Mevorach, D., Liebergall, M., Gazit, Z., Aslan, H., et al. (2005). Human mesenchymal stem cells alter antigen-presenting cell maturation and induce T-cell unresponsiveness. *Blood* 105, 2214–2219. doi: 10.1182/blood-2004-07-2921

Blander, G., and Guarente, L. (2004). The Sir2 family of protein deacetylases. *Annu. Rev. Biochem.* 73, 417–435. doi: 10.1146/annurev.biochem.73.011303.073651

Campagnoli, C., Roberts, I. A., Kumar, S., Bennett, P. R., Bellantuono, I., and Fisk, N. M. (2001). Identification of mesenchymal stem/progenitor cells in human first-trimester fetal blood, liver, and bone marrow. *Blood* 98, 2396–2402. doi: 10.1182/blood.V98.8.2396

Cohen-Kfir, E., Artsi, H., Levin, A., Abramowitz, E., Bajayo, A., Gurt, I., et al. (2011). Sirt1 is a regulator of bone mass and a repressor of Sost encoding for sclerostin, a bone formation inhibitor. *Endocrinology* 152, 4514–4524. doi: 10.1210/en.2011-1128

Deng, W., Han, Q., Liao, L., You, S., Deng, H., and Zhao, R. C. (2005). Effects of allogeneic bone marrow-derived mesenchymal stem cells on T and B lymphocytes from BXSB mice. *DNA Cell Biol.* 24, 458–463. doi: 10.1089/dna.2005.24.458

Ding, Y., Xu, D., Feng, G., Bushell, A., Muschel, R. J., and Wood, K. J. (2009). Mesenchymal stem cells prevent the rejection of fully allogenic islet grafts by the immunosuppressive activity of matrix metalloproteinase-2 and -9. *Diabetes* 58, 1797–1806. doi: 10.2337/db09-0317

Dominici, M., Le Blanc, K., Mueller, I., Slaper-Cortenbach, I., Marini, F., Krause, D., et al. (2006). Minimal criteria for defining multipotent mesenchymal stromal cells. The international society for cellular therapy position statement. *Cytotherapy* 8, 315–317. doi: 10.1080/14653240600855905

Edinger, M., Hoffmann, P., Ermann, J., Drago, K., Fathman, C. G., Strober, S., et al. (2003). CD4+CD25+ regulatory T cells preserve graft-versus-tumor activity while inhibiting graft-versus-host disease after bone marrow transplantation. *Nat. Med.* 9, 1144–1150. doi: 10.1038/nm915

English, K. (2013). Mechanisms of mesenchymal stromal cell immunomodulation. *Immunol. Cell Biol.* 91, 19–26. doi: 10.1038/icb.2012.56

English, K., Barry, F. P., Field-Corbett, C. P., and Mahon, B. P. (2007). IFN-gamma and TNF-alpha differentially regulate immunomodulation by murine mesenchymal stem cells. *Immunol. Lett.* 110, 91–100. doi: 10.1016/j.imlet.2007.04.001

English, K., Ryan, J. M., Tobin, L., Murphy, M. J., Barry, F. P., and Mahon, B. P. (2009). Cell contact, prostaglandin E(2) and transforming growth factor beta 1 play non-redundant roles in human mesenchymal stem cell induction of CD4+CD25(High) forkhead box P3+ regulatory T cells. *Clin. Exp. Immunol.* 156, 149–160. doi: 10.1111/j.1365-2249.2009.03874.x

Erices, A., Conget, P., and Minguell, J. J. (2000). Mesenchymal progenitor cells in human umbilical cord blood. *Br. J. Haematol.* 109, 235–242. doi: 10.1046/j.1365-2141.2000.01986.x

Esumi, T., Inaba, M., Ichioka, N., Kushida, T., Iida, H., and Ikehara, S. (2003). Successful allogeneic leg transplantation in rats in conjunction with intra-bone marrow injection of donor bone marrow cells. *Transplantation* 76, 1543–1548. doi: 10.1097/01.TP.0000081945.23698.5D

Feng, W., Cui, Y., Song, C., Zhan, H., Wang, X., Li, Q., et al. (2007). Prevention of osteoporosis and hypogonadism by allogeneic ovarian transplantation in conjunction with intra-bone marrow-bone marrow transplantation. *Transplantation* 84, 1459–1466. doi: 10.1097/01.tp.0000288182.75398.74

Feng, W., Cui, Y., Zhan, H., Shi, M., Cui, W., Guo, K., et al. (2010). Prevention of premature ovarian failure and osteoporosis induced by irradiation using allogeneic ovarian/bone marrow transplantation. *Transplantation* 89, 395–401. doi: 10.1097/TP.0b013e3181ca86bb

Fiorina, P., Jurewicz, M., Augello, A., Vergani, A., Dada, S., La Rosa, S., et al. (2009). Immunomodulatory function of bone marrow-derived mesenchymal stem cells in experimental autoimmune type 1 diabetes. *J. Immunol.* 183, 993–1004. doi: 10.4049/jimmunol.0900803

Gimble, J. M., Katz, A. J., and Bunnell, B. A. (2007). Adipose-derived stem cells for regenerative medicine. *Circ. Res.* 100, 1249–1260. doi: 10.1161/01.RES.0000265074.83288.09

Glennie, S., Soeiro, I., Dyson, P. J., Lam, E. W., and Dazzi, F. (2005). Bone marrow mesenchymal stem cells induce division arrest anergy of activated T cells. *Blood* 105, 2821–2827. doi: 10.1182/blood-2004-09-3696

Guo, K., Inaba, M., Li, M., An, J., Cui, W., Song, C., et al. (2008). Long-term donor-specific tolerance in rat cardiac allografts by intrabone marrow injection of donor bone marrow cells. *Transplantation* 85, 93–101. doi: 10.1097/01.tp.0000296061.71662.76

Haigis, M. C., and Sinclair, D. A. (2010). Mammalian sirtuins: biological insights and disease relevance. *Annu. Rev. Pathol.* 5, 253–295. doi: 10.1146/annurev.pathol.4.110807.092250

Hida, D., Ishiguro, N., Haneda, M., Ishida, Y., Suzuki, H., and Isobe, K. (2010). Intra-bone marrow bone marrow transplantation rejuvenates the B-cell lineage in aged mice. *Immunol. Cell Biol.* 88, 87–94. doi: 10.1038/icb.2009.69

Ichioka, N., Inaba, M., Kushida, T., Esumi, T., Takahara, K., Inaba, K., et al. (2002). Prevention of senile osteoporosis in SAMP6 mice by intrabone marrow injection of allogeneic bone marrow cells. *Stem Cells* 20, 542–551. doi: 10.1634/stemcells.20-6-542

Ikebukuro, K., Adachi, Y., Suzuki, Y., Iwasaki, M., Nakano, K., Koike, Y., et al. (2006). Synergistic effects of injection of bone marrow cells into both portal vein and bone marrow on tolerance induction in transplantation of allogeneic pancreatic islets. *Bone Marrow Transplant.* 38, 657–664. doi: 10.1038/sj.bmt.1705500

Ikehara, S. (2003). A novel strategy for allogeneic stem cell transplantation: perfusion method plus intra-bone marrow injection of stem cells. *Exp. Hematol.* 31, 1142–1146. doi: 10.1016/j.exphem.2003.08.020

Ikehara, S., Yasumizu, R., Inaba, M., Izui, S., Hayakawa, K., Sekita, K., et al. (1989). Long-term observations of autoimmune-prone mice treated for autoimmune disease by allogeneic bone marrow transplantation. *Proc. Natl. Acad. Sci. U.S.A.* 86, 3306–3310.

Imai, S., and Guarente, L. (2010). Ten years of NAD-dependent SIR2 family deacetylases: implications for metabolic diseases. *Trends Pharmacol. Sci.* 31, 212–220. doi: 10.1016/j.tips.2010.02.003

Ishida, T., Inaba, M., Hisha, H., Sugiura, K., Adachi, Y., Nagata, N., et al. (1994). Requirement of donor-derived stromal cells in the bone marrow for successful allogeneic bone marrow transplantation. Complete prevention of recurrence of autoimmune diseases in MRL/MP-Ipr/Ipr mice by transplantation of bone marrow plus bones (stromal cells) from the same donor. *J. Immunol.* 152, 3119–3127.

Kaneda, H., Adachi, Y., Saito, Y., Ikebukuro, K., Machida, H., Suzuki, Y., et al. (2005). Long-term observation after simultaneous lung and intra-bone marrow-bone marrow transplantation. *J. Heart Lung Transplant.* 24, 1415–1423. doi: 10.1016/j.healun.2004.08.015

Koike, Y., Adachi, Y., Suzuki, Y., Iwasaki, M., Koike-Kiriyama, N., Minamino, K., et al. (2007). Allogeneic intrabone marrow-bone marrow transplantation plus donor lymphocyte infusion suppresses growth of colon cancer cells implanted in skin and liver of rats. *Stem Cells* 25, 385–391. doi: 10.1634/stemcells.2006-0227

Kuroda, S., Yamazaki, M., Abe, M., Sakimura, K., Takayanagi, H., and Iwai, Y. (2011). Basic leucine zipper transcription factor, ATF-like (BATF) regulates epigenetically and energetically effector CD8 T-cell differentiation via Sirt1 expression. *Proc. Natl. Acad. Sci. U.S.A.* 108, 14885–14889. doi: 10.1073/pnas.1105133108

Kushida, T., Inaba, M., Hisha, H., Ichioka, N., Esumi, T., Ogawa, R., et al. (2001). Intra-bone marrow injection of allogeneic bone marrow cells: a powerful new strategy for treatment of intractable autoimmune diseases in MRL/lpr mice. *Blood* 97, 3292–3299. doi: 10.1182/blood.V97.10.3292

Kushida, T., Ueda, Y., Umeda, M., Oe, K., Okamoto, N., Iida, H., et al. (2009). Allogeneic intra-bone marrow transplantation prevents rheumatoid arthritis in SKG/Jcl mice. *J. Autoimmun.* 32, 216–222. doi: 10.1016/j.jaut.2009.02.015

Lee, R. H., Seo, M. J., Reger, R. L., Spees, J. L., Pulin, A. A., Olson, S. D., et al. (2006). Multipotent stromal cells from human marrow home to and promote repair of pancreatic islets and renal glomeruli in diabetic NOD/scid mice. *Proc. Natl. Acad. Sci. U.S.A.* 103, 17438–17443. doi: 10.1073/pnas.0608249103

Li, M., Abraham, N. G., Vanella, L., Zhang, Y., Inaba, M., Hosaka, N., et al. (2010). Successful modulation of type 2 diabetes in db/db mice with intra-bone marrow–bone marrow transplantation plus concurrent thymic transplantation. *J. Autoimmun.* 35, 414–423. doi: 10.1016/j.jaut.2010.09.001

Li, M., Inaba, M., Guo, K., Abraham, N. G., and Ikehara, S. (2009). Amelioration of cognitive ability in senescence-accelerated mouse prone 8 (SAMP8) by intra-bone marrow-bone marrow transplantation. *Neurosci. Lett.* 465, 36–40. doi: 10.1016/j.neulet.2009.09.001

Li, M., Vanella, L., Zhang, Y., Shi, M., Takaki, T., Shapiro, J. I., et al. (2012). Stem cell transplantation increases antioxidant effects in diabetic mice. *Int. J. Biol. Sci.* 8, 1335–1344. doi: 10.7150/ijbs.4654

Makino, S., Fukuda, K., Miyoshi, S., Konishi, F., Kodama, H., Pan, J., et al. (1999). Cardiomyocytes can be generated from marrow stromal cells *in vitro*. *J. Clin. Invest.* 103, 697–705. doi: 10.1172/JCI5298

Matsuda-Hashii, Y., Takai, K., Ohta, H., Fujisaki, H., Tokimasa, S., Osugi, Y., et al. (2004). Hepatocyte growth factor plays roles in the induction and autocrine maintenance of bone marrow stromal cell IL-11, SDF-1 alpha, and stem cell factor. *Exp. Hematol.* 32, 955–961. doi: 10.1016/j.exphem.2004.06.012

Miyake, T., Hosaka, N., Cui, W., Nishida, T., Takaki, T., Inaba, M., et al. (2009). Adult thymus transplantation with allogeneic intra-bone marrow-bone marrow transplantation from same donor induces high thymopoiesis, mild graft-versus-host reaction and strong graft-versus-tumour effects. *Immunology* 126, 552–564. doi: 10.1111/j.1365-2567.2008.02920.x

Miyake, T., Inaba, M., Fukui, J., Ueda, Y., Hosaka, N., Kamiyama, Y., et al. (2008). Prevention of graft-versus-host disease by intrabone marrow injection of donor T cells: involvement of bone marrow stromal cells. *Clin. Exp. Immunol.* 152, 153–162. doi: 10.1111/j.1365-2249.2008.03615.x

Mukaide, H., Adachi, Y., Koike-Kiriyama, N., Suzuki, Y., Minamino, K., Iwasaki, M., et al. (2007). Immunotherapy for malignant tumors using combination of allogeneic intra-bone marrow-bone marrow transplantation, donor lymphocyte infusion and dendritic cells. *Int. J. Oncol.* 30, 1309–1315.

Nakamura, K., Inaba, M., Sugiura, K., Yoshimura, T., Kwon, A. H., Kamiyama, Y., et al. (2004). Enhancement of allogeneic hematopoietic stem cell engraftment and prevention of GVHD by intra-bone marrow bone marrow transplantation plus donor lymphocyte infusion. *Stem Cells* 22, 125–134. doi: 10.1634/stemcells.22-2-125

Nishimura, M., Toki, J., Sugiura, K., Hashimoto, F., Tomita, T., Fujishima, H., et al. (1994). Focal segmental glomerular sclerosis, a type of intractable chronic glomerulonephritis, is a stem cell disorder. *J. Exp. Med.* 179, 1053–1058.

Okazaki, S., Hisha, H., Mizokami, T., Takaki, T., Wang, X., Song, C., et al. (2008). Successful acceptance of adult liver allografts by intra-bone marrow-bone marrow transplantation. *Stem Cells Dev.* 17, 629–639. doi: 10.1089/scd.2007.0218

Peltz, L., Gomez, J., Marquez, M., Alencastro, F., Atashpanjeh, N., Quang, T., et al. (2012). Resveratrol exerts dosage and duration dependent effect on human mesenchymal stem cell development. *PLoS ONE* 7:e37162. doi: 10.1371/journal.pone.0037162

Poloni, A., Maurizi, G., Leoni, P., Serrani, F., Mancini, S., Frontini, A., et al. (2012). Human dedifferentiated adipocytes show similar properties to bone marrow-derived mesenchymal stem cells. *Stem Cells* 30, 965–974. doi: 10.1002/stem.1067

Puri, N., Sodhi, K., Haarstad, M., Kim, D. H., Bohinc, S., Foglio, E., et al. (2012). Heme induced oxidative stress attenuates sirtuin1 and enhances adipogenesis in mesenchymal stem cells and mouse pre-adipocytes. *J. Cell. Biochem.* 113, 1926–1935. doi: 10.1002/jcb.24061

Ramasamy, R., Fazekasova, H., Lam, E. W., Soeiro, I., Lombardi, G., and Dazzi, F. (2007). Mesenchymal stem cells inhibit dendritic cell differentiation and function by preventing entry into the cell cycle. *Transplantation* 83, 71–76. doi: 10.1097/01.tp.0000244572.24780.54

Ryan, J. M., Barry, F., Murphy, J. M., and Mahon, B. P. (2007). Interferon-gamma does not break, but promotes the immunosuppressive capacity of adult human mesenchymal stem cells. *Clin. Exp. Immunol.* 149, 353–363. doi: 10.1111/j.1365-2249.2007.03422.x

Schurgers, E., Kelchtermans, H., Mitera, T., Geboes, L., and Matthys, P. (2010). Discrepancy between the *in vitro* and *in vivo* effects of murine mesenchymal stem cells on T-cell proliferation and collagen-induced arthritis. *Arthritis Res. Ther.* 12, R31. doi: 10.1186/ar2939

Scott, D. L., Wolfe, F., and Huizinga, T. W. (2010). Rheumatoid arthritis. *Lancet* 376, 1094–1108. doi: 10.1016/S0140-6736(10)60826-4

Shi, M., Adachi, Y., Cui, Y., Li, M., Lian, Z., Zhang, Y., et al. (2011). Combination of intra-bone marrow-bone marrow transplantation and subcutaneous donor splenocyte injection diminishes risk of graft-versus-host disease and enhances survival rate. *Stem Cells Dev.* 20, 759–768. doi: 10.1089/scd.2010.0232

Simic, P., Zainabadi, K., Bell, E., Sykes, D. B., Saez, B., Lotinun, S., et al. (2013). SIRT1 regulates differentiation of mesenchymal stem cells by deacetylating beta-catenin. *EMBO Mol. Med.* 5, 430–440. doi: 10.1002/emmm.201201606

Sotiropoulou, P. A., Perez, S. A., Gritzapis, A. D., Baxevanis, C. N., and Papamichail, M. (2006). Interactions between human mesenchymal stem cells and natural killer cells. *Stem Cells* 24, 74–85. doi: 10.1634/stemcells.2004-0359

Suzuki, Y., Adachi, Y., Minamino, K., Zhang, Y., Iwasaki, M., Nakano, K., et al. (2005). A new strategy for treatment of malignant tumor: intra-bone marrow-bone marrow transplantation plus CD4- donor lymphocyte infusion. *Stem Cells* 23, 365–370. doi: 10.1634/stemcells.2004-0258

Taira, M., Inaba, M., Takada, K., Baba, S., Fukui, J., Ueda, Y., et al. (2005). Treatment of streptozotocin-induced diabetes mellitus in rats by transplantation of islet cells from two major histocompatibility complex disparate rats in combination with intra bone marrow injection of allogeneic bone marrow cells. *Transplantation* 79, 680–687. doi: 10.1097/01.TP.0000155500.17348.94

Takada, K., Inaba, M., Ichioka, N., Ueda, Y., Taira, M., Baba, S., et al. (2006). Treatment of senile osteoporosis in SAMP6 mice by intra-bone marrow injection of allogeneic bone marrow cells. *Stem Cells* 24, 399–405. doi: 10.1634/stemcells.2005-0068

Tang, D. Q., Cao, L. Z., Burkhardt, B. R., Xia, C. Q., Litherland, S. A., Atkinson, M. A., et al. (2004). *In vivo* and *in vitro* characterization of insulin-producing cells obtained from murine bone marrow. *Diabetes* 53, 1721–1732. doi: 10.2337/diabetes.53.7.1721

Trim, N., Morgan, S., Evans, M., Issa, R., Fine, D., Afford, S., et al. (2000). Hepatic stellate cells express the low affinity nerve growth factor receptor p75 and undergo apoptosis in response to nerve growth factor stimulation. *Am. J. Pathol.* 156, 1235–1243. doi: 10.1016/S0002-9440(10)64994-2

Tseng, P. C., Hou, S. M., Chen, R. J., Peng, H. W., Hsieh, C. F., Kuo, M. L., et al. (2011). Resveratrol promotes osteogenesis of human mesenchymal stem cells by upregulating RUNX2 gene expression via the SIRT1/FOXO3A axis. *J. Bone Miner. Res.* 26, 2552–2563. doi: 10.1002/jbmr.460

Turesson, C., O'Fallon, W. M., Crowson, C. S., Gabriel, S. E., and Matteson, E. L. (2003). Extra-articular disease manifestations in rheumatoid arthritis: incidence trends and risk factors over 46 years. *Ann. Rheum. Dis.* 62, 722–727. doi: 10.1136/ard.62.8.722

Uccelli, A., Moretta, L., and Pistoia, V. (2008). Mesenchymal stem cells in health and disease. *Nat. Rev. Immunol.* 8, 726–736. doi: 10.1038/nri2395

Urban, V. S., Kiss, J., Kovacs, J., Gocza, E., Vas, V., Monostori, E., et al. (2008). Mesenchymal stem cells cooperate with bone marrow cells in therapy of diabetes. *Stem Cells* 26, 244–253. doi: 10.1634/stemcells.2007-0267

van der Woude, D., Houwing-Duistermaat, J. J., Toes, R. E., Huizinga, T. W., Thomson, W., Worthington, J., et al. (2009). Quantitative heritability of anti-citrullinated protein antibody-positive and anti-citrullinated protein antibody-negative rheumatoid arthritis. *Arthritis Rheum.* 60, 916–923. doi: 10.1002/art.24385 van Poll, D., Parekkadan, B., Cho, C. H., Berthiaume, F., Nahmias, Y., Tilles, A. W., et al. (2008). Mesenchymal stem cell-derived molecules directly modulate hepatocellular death and regeneration *in vitro* and *in vivo*. *Hepatology* 47, 1634–1643. doi: 10.1002/hep.22236

Wang, Q., Li, X., Luo, J., Zhang, L., Ma, L., Lv, Z., et al. (2012). The allogeneic umbilical cord mesenchymal stem cells regulate the function of T helper 17 cells from patients with rheumatoid arthritis in an *in vitro* co-culture system. *BMC Musculoskelet. Disord.* 13:249. doi: 10.1186/1471-2474-13-249

Wang, L., Cong, X., Liu, G., Zhou, J., Bai, B., Li, Y., et al. (2013). Human umbilical cord mesenchymal stem cell therapy for patients with active rheumatoid arthritis: safety and efficacy. *Stem Cells Dev.* 22, 3192–3202. doi: 10.1089/scd.2013.0023

Yao, C., Sakata, D., Esaki, Y., Li, Y., Matsuoka, T., Kuroiwa, K., et al. (2009). Prostaglandin E2-EP4 signaling promotes immune inflammation through Th1 cell differentiation and Th17 cell expansion. *Nat. Med.* 15, 633–640. doi: 10.1038/nm.1968

Yokokawa, M., Ohnishi, S., Ishibashi-Ueda, H., Obata, H., Otani, K., Miyahara, Y., et al. (2008). Transplantation of mesenchymal stem cells improves atrioventricular conduction in a rat model of complete atrioventricular block. *Cell Transplant.* 17, 1145–1155. doi: 10.3727/096368908787236594

Zappia, E., Casazza, S., Pedemonte, E., Benvenuto, F., Bonanni, I., Gerdoni, E., et al. (2005). Mesenchymal stem cells ameliorate experimental autoimmune encephalomyelitis inducing T-cell anergy. *Blood* 106, 1755–1761. doi: 10.1182/blood-2005-04-1496

Zhang, B., Liu, R., Shi, D., Liu, X., Chen, Y., Dou, X., et al. (2009). Mesenchymal stem cells induce mature dendritic cells into a novel Jagged-2-dependent regulatory dendritic cell population. *Blood* 113, 46–57. doi: 10.1182/blood-2008-04-154138

Zhang, Y., Hosaka, N., Cui, Y., Shi, M., and Ikehara, S. (2011). Effects of allogeneic hematopoietic stem cell transplantation plus thymus transplantation on malignant tumors: comparison between fetal, newborn, and adult mice. *Stem Cells Dev.* 20, 599–607. doi: 10.1089/scd.2010.0230

Zhang, Y., Hosaka, N., Cui, Y., Shi, M., Li, M., Li, Q., et al. (2012). Effects of intrabone marrow-bone marrow transplantation plus adult thymus transplantation on survival of mice bearing leukemia. *Stem Cells Dev.* 21, 1441–1448. doi: 10.1089/scd.2011.0358

Zhao, D. C., Lei, J. X., Chen, R., Yu, W. H., Zhang, X. M., Li, S. N., et al. (2005). Bone marrow-derived mesenchymal stem cells protect against experimental liver fibrosis in rats. *World J. Gastroenterol.* 11, 3431–3440.

Zuk, P. A., Zhu, M., Mizuno, H., Huang, J., Futrell, J. W., Katz, A. J., et al. (2001). Multilineage cells from human adipose tissue: implications for cell-based therapies. *Tissue Eng.* 7, 211–228. doi: 10.1089/107632701300062859

Conflict of Interest Statement: The authors declare that the research was conducted in the absence of any commercial or financial relationships that could be construed as a potential conflict of interest.

6

An emerging role of the cellular prion protein as a modulator of a morphogenetic program underlying epithelial-to-mesenchymal transition

Mohadeseh Mehrabian[1,2], Sepehr Ehsani[1,2†] and Gerold Schmitt-Ulms[1,2*]

[1] Tanz Centre for Research in Neurodegenerative Diseases, University of Toronto, Toronto, ON, Canada
[2] Department of Laboratory Medicine and Pathobiology, University of Toronto, Toronto, ON, Canada

Edited by:
Sophie Mouillet-Richard, INSERM Unit 1124, France

Reviewed by:
Séverine Martin-Lannerée, Université Paris Descartes, UMR-S1124, France
Edward Málaga-Trillo, University of Konstanz, Germany

****Correspondence:***
Gerold Schmitt-Ulms, Tanz Centre for Research in Neurodegenerative Diseases, University of Toronto, Krembil Discovery Tower, 60 Leonard Avenue, Toronto, ON M5T 2S8, Canada
e-mail: g.schmittulms@utoronto.ca

†Present address:
Sepehr Ehsani, Whitehead Institute for Biomedical Research, Cambridge, MA, USA

Knowledge of phenotypic changes the cellular prion protein (PrP^C) contributes to may provide novel avenues for understanding its function. Here we consider data from functional knockout/down studies and protein–protein interaction analyses from the perspective of PrP's relationship to its ancestral ZIP metal ion transporting proteins. When approached in this manner, a role of PrP^C as a modulator of a complex morphogenetic program that underlies epithelial-to-mesenchymal transition (EMT) emerges. To execute EMT, cells have to master the challenge to shift from cell-cell to cell-substrate modes of adherence. During this process, cell-cell junctions stabilized by E-cadherins are replaced by focal adhesions that mediate cell-substrate contacts. A similar reprogramming occurs during distinct organogenesis events that have been shown to rely on ZIP transporters. A model is presented that sees ZIP transporters, and possibly also PrP^C, affect this balance of adherence modes at both the transcriptional and post-translational levels.

Keywords: epithelial-to-mesenchymal transition, function, phenotype, prion protein, ZIP transporter

INTRODUCTION

To someone new to the prion literature, it might be confusing that on the one hand some researchers lament the absence of a known function of the cellular prion protein (PrP^C), while on the other hand, there seems to be a myriad of roles assigned to this protein (Steele et al., 2007; Aguzzi et al., 2008; Zomosa-Signoret et al., 2008). Indeed, over the past few decades, mouse models harboring a deletion of the prion gene have revealed some interesting and highly diverse phenotypes, including (i) defects in the circadian rhythm (Tobler et al., 1996), (ii) abnormal development of infrapyramidal mossy fibers (Colling et al., 1997), (iii) a reduced capacity to generate hematopoietic stem cells (Zhang et al., 2006), and (iv) a peripheral neuropathy that is most conspicuously characterized by a myelin maintenance defect (Bremer et al., 2010). Due to the complex nature of the biology underlying any of these phenotypic changes, none of them immediately suggest how PrP^C might be involved. It also does not help that each of them requires a unique environment not easily accessible by experimental means. The only phenotype for which a wealth of available data has provided a near-complete molecular explanation constitutes the well-known resistance of PrP-knockout mice to infection with prion disease (Sailer et al., 1994).

This mini-review will add to this body of literature by shining a spotlight on the possibility that the prion protein may, in some cellular contexts, operate in signaling events that influence a shift from cell-cell to cell-substrate modes of adherence and play, for example, a critical role during EMT. In light of the emphasis of this special issue on research that explores the function of the prion protein, a brief clarification of terms and concepts may be helpful.

THE PHYSIOLOGICAL ROLE OF A PROTEIN—PHENOTYPIC CHANGE vs. FUNCTION

Although intuitively appealing, applying the concept of "function" to a protein can be a difficult undertaking. Any protein that emerges from ribosome-mediated translation enters into a unique molecular life-cycle that ultimately comes to an end when the protein is degraded. During this cycle, a protein encounters many other molecules that may impact its own biology. Some of these may, for instance, facilitate its post-translational maturation and folding or influence its sorting and trafficking. The consequences of molecular interactions may be equidirectional, orthogonal or opposite with regard to a specific outcome. They may also possess other characteristics that complicate their

Abbreviations: CRISPR, clustered regularly interspaced short palindromic repeats; EMT, epithelial-to-mesenchymal transition; FOI, Fear-Of-Intimacy; GSK-3β, glycogen synthase kinase 3 beta; MAP, mitogen-activated protein; NCAM, neural cell adhesion molecule; PL, PrP-like; PrP^C, cellular prion protein; SFK, Src family kinase; TGFβ, transforming growth factor beta; ZIP, Zrt-, Irt-like Protein.

study; for example, they can be transient or long-lasting, and can qualitatively or quantitatively change over time. The net effect of these molecular interactions will not only depend on the physicochemical properties of the protein of interest itself, but also on the status of the cell or multicellular entity into which it was born. The latter will, in turn, be influenced by the characteristics and locations of all other molecules present at a given time. For example, there are several proteins (e.g., WNT5A, RACK1, SIRT1 and several F-box proteins) whose expression levels appear to correlate with the risk to acquire certain cancers, yet the same proteins have been observed to confer protection against cancers in other paradigms (McDonald and Silver, 2009; Lin and Fang, 2013; Li and Xie, 2014; Wang et al., 2014).

A phenotypic change caused by a protein can be discerned if its molecular interactions cumulatively change the biology of the experimental paradigm in a way that is detectable (Crusio, 2002). It follows that the description of such a phenotypic change only becomes meaningful if a precise account of the experimental paradigm is also provided.

What is the relationship between phenotypic change and function? The term "function" seems overused in the protein literature in general and the prion literature in particular. Here, the use of this term will be restricted to an immediate and context-independent role of a protein (e.g., the capacity of substrate phosphorylation by a kinase) and will not be employed to capture the contribution of a protein to a broad and indirect phenotypic change (e.g., mitosis).

When viewed in this light, the body of literature dealing with functional assignments of the prion protein begins to look considerably less controversial. It comes as no surprise that a protein studied as extensively as PrP^C has been observed to contribute to several biological processes that give rise to diverse phenotypes. While the phenotypic changes a protein causes are not to be confused with its function, once the function is known, it will hopefully reconcile a majority of these phenotypic changes with a consistent molecular and mechanistic explanation.

EMT, YET ANOTHER PHENOTYPIC CHANGE THE PRION PROTEIN MAY CONTRIBUTE TO—BUT NOT ITS FUNCTION

The term EMT refers to a complex morphogenetic reprogramming of cells observed during their transition from epithelial to mesenchymal phenotypes. EMT is naturally initiated in cells during distinct stages of development and wound healing (Lim and Thiery, 2012). Similar morphogenetic changes also take place during organogenesis and in pathological conditions, including fibrosis and cancers, where their occurrence most often correlates with a higher metastasis propensity (Thiery et al., 2009). During the EMT process, strong epithelial cell-to-cell contacts in the form of adherens junctions gradually give way to more transient cell-to-matrix connections evident in cultured cells (referred to as focal adhesions). As adherens junctions are dissolved, functional levels of E-cadherin, their main molecular constituent, are also reduced. Consequently, cells acquire a more fibroblast-like morphology and gradually rely on a more dynamic molecular network involving integrins for cell–matrix interactions (Lamouille et al., 2014). During the cellular reprogramming that accompanies EMT, the levels and/or activities of many molecules change (**Figure 1**).

The notion of an involvement of PrP^C in the morphogenetic program underlying EMT emerged recently from several independent strands of investigation, including (1) observations in cancer paradigms that predominantly see a role of PrP^C as a promoter of cellular invasiveness and malignancy; (2) morpholino-based manipulations of PrP levels undertaken in zebrafish; (3) a subset of available PrP protein–protein interaction data; and (4) commonalities between PrP and the subbranch of ZIP transporters from which it descended. Finally, pilot data generated in mammalian cells in which the expression of PrP^C was eliminated using CRISPR/Cas gene editing tools (Hsu et al., 2014) are also consistent with a role of PrP^C as a modulator of a cellular program that defines the balance between cell-cell and cell-substrate adhesion (unpublished observations, Schmitt-Ulms laboratory). Therefore, when asked to contribute to this special issue, we decided to take the opportunity to further investigate the merits of an involvement of PrP^C in EMT based on published reports.

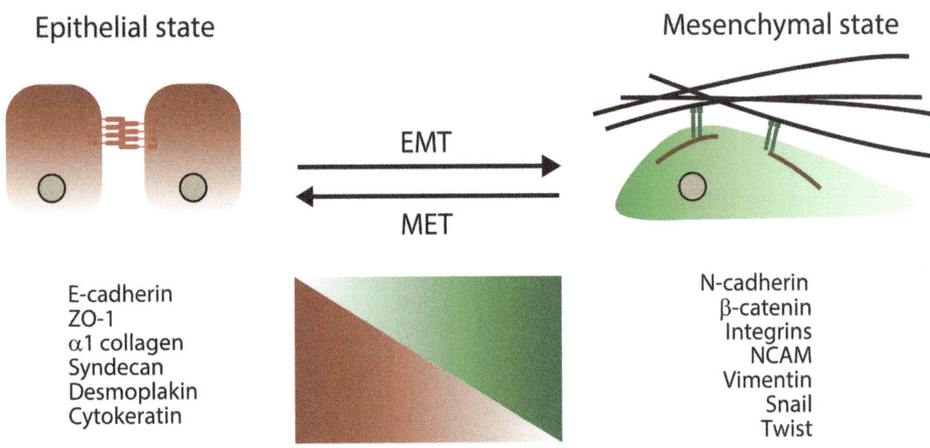

FIGURE 1 | Schematic outlining key morphological and molecular changes that accompany EMT.

Needless to say, no claim is made that an involvement of PrP^C in EMT, even if further substantiated experimentally, would represent its physiological function. Instead, it is hoped that this contribution might further stimulate research that may uncover the function of PrP^C.

EVIDENCE FOR A ROLE OF PrP^C IN EMT IN THE CANCER LITERATURE

In light of the central role that EMT plays in the cellular biology that governs tumor malignancy, it may not be surprising that some support for a possible role of PrP^C in EMT can be found in the cancer literature. Several groups have reported that the prion protein is upregulated in a subset of human cancers ranging from colorectal and gastric cancers to breast cancer and glioblastomas (Mehrpour and Codogno, 2010). Furthermore, a positive correlation has been reported between the upregulation of PrP^C and the tumorigenicity of some cancer cells, suggestive of a more causative involvement. For instance, in one of the first reports that tied PrP^C to cancer biology, the overexpression of PrP^C was observed to promote invasive and metastatic properties of gastric cancer cell lines (Pan et al., 2006), and a more recent report suggested that high PrP^C levels can be predictive of disease recurrence in colorectal cancers (Antonacopoulou et al., 2010). Overall, a broad consensus seems to be emerging that PrP^C can promote key indicators of malignancy in several cancer paradigms. However, less information and agreement exist at this time on the signaling pathways and molecular events involved. Several observations suggest that PrP^C may modulate signaling downstream of TGFβ. For example, it was reported that PrP^C promotes the release of certain matrix metalloproteinases (Pan et al., 2006; Wurm and Wechselberger, 2006) that are activated during TGFβ signaling and are known to promote the breakdown of the extracellular matrix, thereby leading to cell detachment. Also, researchers have proposed that the serine/threonine kinases Erk1/2 may play a critical role in the cancer-related signaling cascade emanating from PrP^C. This conclusion was drawn on the basis of data documenting that small-molecule inhibitors against key components of the mitogen-activated protein (MAP) kinase signaling pathway were able to rescue invasive and metastatic characteristics contributed by PrP^C(Pan et al., 2006; Du et al., 2013). Although an inverse relationship between E-cadherin and PrP^C levels has been observed (Du et al., 2013), whether this relationship is merely correlative or if PrP^C causally influences E-cadherin biology has, to our knowledge, not yet been addressed in the aforementioned cancer paradigms.

ZEBRAFISH PrP KNOCKDOWN MODELS

To date, the most dramatic phenotypic change associated with PrP deficiency was not reported for the mouse prion protein but for one of its two orthologs in zebrafish (Malaga-Trillo et al., 2009). More specifically, the morpholino-based knockdown of PrP1 (but not PrP2) was observed to cause an arrest of zebrafish embryogenesis at the gastrula stage in the course of cellular rearrangements that rely on EMT. Interestingly, the PrP1-deficient embryos do not appear to suffer from an inability to initiate EMT; rather, they are impaired in their ability to fully execute a cell migration program that is limited to a small number of cells and requires re-establishing cellular contacts following their migration (Malaga-Trillo et al., 2009). Whereas in normal embryos PrP1 was shown to activate Src family kinases (SFKs) and modulate the post-transcriptional E-cadherin biology, both of these activities were impaired in PrP1-deficient embryos. A follow-up investigation revealed that several features within PrP1, including the repeat domain, globular domain, N-glycosylation sites and GPI anchor attachment signal, influence the proper positioning of PrP1 at cell-cell contacts (Solis et al., 2013). Importantly, the phenotype does not appear to reflect a functional specialization or idiosyncrasy of zebrafish PrP1 because it could be rescued by the introduction of mammalian PrP^C (Malaga-Trillo and Sempou, 2009; Malaga-Trillo et al., 2009).

PROTEIN–PROTEIN INTERACTIONS OF PrP^C WITH A KNOWN EMT CONNECTION

Although there are divergent views regarding the significance of protein–protein interactions of PrP^C (Watts and Westaway, 2007; Aguzzi et al., 2008; Rubenstein, 2012), there appears to be a broad consensus that PrP^C exerts some of its biological role through its affiliation with caveolae and specialized membrane domains referred to as lipid rafts (Naslavsky et al., 1997; Mouillet-Richard et al., 2000). Furthermore, there is agreement that the molecular microenvironment of PrP^C in neurons is dominated by cell adhesion molecules that include neural cell adhesion molecule (NCAM) (Walsh et al., 1989), L1, integrins and non-integrin laminin receptors known to modulate cell-to-substrate contacts (Gauczynski et al., 2001; Watts et al., 2009). Finally, there is a broad consensus that signals emanating from PrP^C can lead to Fyn activation (Mouillet-Richard et al., 2000; Santuccione et al., 2005; Toni et al., 2006; Pantera et al., 2009; Tomasi, 2010; Um et al., 2012).

Several of these physiological interactors of the prion protein have been implicated in EMT. NCAM, for example, has recently been shown to be a critical regulator of EMT (Lehembre et al., 2008; Evseenko et al., 2010). Specifically, (i) increases in the levels of NCAM and (ii) a redistribution of NCAM that may involve its detachment from fibroblast growth factor (FGF) receptors and recruitment into caveolae and/or raft-like domains (Niethammer et al., 2002; Santuccione et al., 2005), have been recognized as early steps during EMT (Lehembre et al., 2008). Not surprisingly, these changes to the cellular NCAM pool and the concomitant upregulation of functional integrins occurring during EMT lead to increases in signaling through Fyn (Lehembre et al., 2008). Moreover, αV-containing integrins, another family of PrP^C interactors (Watts et al., 2009), are critical for activation of the EMT master regulator TGFβ1 (Munger et al., 1999; Mu et al., 2002). Mice harboring point mutations in the αV integrin "RGD" recognition motif within the TGFβ1 preprotein ($TGF\beta1^{RGE/RGE}$) were shown to phenocopy TGFβ1-null mice (Yang et al., 2007). As an additional example, β1 integrin-L1 complexes have long been known to play a role in controlling cellular migration processes. More recently, L1 has been linked to EMT and a molecular biology that determines cancer invasiveness (Kiefel et al., 2012). Thus, the stage, players and activities that PrP^C is surrounded by are familiar to the research community studying EMT.

FUNCTION OF CLOSEST EVOLUTIONARY RELATIVES OF PrP

Members of the mammalian prion protein family (PrP^C, Dpl and Sho) were recently shown to have evolved from the family of ZIP (Zrt-, Irt-like Protein) metal ion transporters (Schmitt-Ulms et al., 2009), with up to 30% amino acid sequence identity observed in some pair-wise comparisons of PrP and ZIP sequences. Subsequent work established that the founding event of the prion gene family coincided with the speciation of early vertebrates around 500 million years ago. This event relied on a genomic rearrangement that involved the retroinsertion of an ancient ZIP transcript and, probably, was mediated by retrotransposition elements (Ehsani et al., 2011). Amino acid sequence comparisons revealed that PrP is most similar to ZIPs 5, 6 and 10, which constitute a distinct sub-branch within this gene family that consists of 14 paralogs in humans (Schmitt-Ulms et al., 2009). The primary cellular function of ZIP transporters is the import of zinc and other divalent cations into the cytosol (Lichten and Cousins, 2009). The PrP-like (PL) domains within ZIPs 5, 6 and 10 represent their N-terminal ectodomains and resemble PrP with regard to orientation and relative distance to their downstream membrane anchorage sites. The PL domain most likely fulfills a role as a modulator of the cation import channel to which it is attached (Ehsani et al., 2012).

Interestingly, ZIP6 deficiency in zebrafish embryos has been reported to give rise to a gastrulation defect that is similar to the aforementioned defect observed in PrP1-deficient zebrafish (Yamashita et al., 2004). Subsequent work documented that manipulation of ZIP6 expression levels in mammalian cells leads to equidirectional changes in levels of E-cadherin expression (Shen et al., 2009). The close *Drosophila* ZIP ortholog Fear-Of-Intimacy (FOI) (Mathews et al., 2005) has similarly been shown to play a role in EMT-like morphogenetic cell movements underlying both gonad and trachea formation (Godt and Tepass, 2003; Van Doren et al., 2003). More specifically, the zinc transport activity of FOI was shown to be essential for post-transcriptional stabilization of E-cadherin expression, with loss-of-function mutants of FOI leading to strongly reduced levels of functional E-cadherin. This did not preclude initiation of cell migration, but interfered with the coalescence of cells following their migration (Mathews et al., 2006). Additional support for the notion that ZIP proteins may affect EMT-like signaling came from a recent genetic report of a family afflicted with a subtype of myopia, a leading cause of blindness in humans, which identified a nonsense mutation in the ZIP5 gene to co-segregate with the phenotype (Guo et al., 2014). Follow-up biochemical work led the authors to observe that ZIP5 deficiency interfered with signaling downstream of TGFβ.

MODEL OF PrP/ZIP E-CADHERIN MODULATION

Given that the knockdown of PrP1 or its molecular cousin ZIP6 gives rise to similar gastrulation defects in zebrafish, the question arises as to whether a primary role of the prion protein is to modulate ZIP-dependent cation import. The evolutionary relationship of PrP and ZIPs and the ability of ZIPs to interact with PrP^C are consistent with such a model (Schmitt-Ulms et al., 2009). However, close scrutiny of the precise E-cadherin deficiencies in the respective knockdown phenotypes may suggest a more complex scenario. Whereas the zinc import function of the aforementioned ZIPs appears to reduce E-cadherin levels at the transcriptional level (Yamashita et al., 2004), PrP deficiency in zebrafish does not seem to interfere with E-cadherin transcription but may prevent E-cadherin from reaching the plasma membrane (Malaga-Trillo et al., 2009; Solis et al., 2013). Thus, PrP/ZIPs may influence functional E-cadherin levels by a coordinated transcriptional and post-translational regulation. A plausible scenario (**Figure 2**) sees the zinc import function of ZIPs regulate E-cadherin gene transcription, possibly by controlling the zinc-dependent nuclear translocation of certain transcription factors such as the zinc finger EMT master regulator Snail (Yamashita et al., 2004). This signaling pathway may involve glycogen synthase kinase 3 beta (GSK-3β) (Hogstrand et al., 2013). PrP may also act as a modulator of this ZIP-dependent zinc import by acting as a scavenger or sensor of zinc ions (Watt et al., 2013), or on account of its ability to acquire alternative N-terminal folds in response to changes in its surrounding cation milieu (Chattopadhyay et al., 2005). Metal-dependent conformational changes of PrP could, for instance, modulate access of divalent cations to the ion channel present in a nearby ZIP protein, either by sterically blocking the channel or by exerting an influence on the arrangement of the channel transmembrane domains. On the other hand, PrP may serve a role in controlling functional E-cadherin levels post-translationally through the activation of Fyn (Lilien and Balsamo, 2005; Smyth et al., 2012). Assuming that PrP inherited this property from its ZIP ancestor, and that it can still be found in contemporary ZIP transporters equipped with a PL ectodomain, these ZIP molecules may operate as highly specialized regulators of EMT. It will be interesting to explore if the PL ectodomain of ZIPs can influence Fyn activation similar to PrP, and if such dual capability is at play in EMT paradigms that lack the prion protein. An example could be the process of gonad formation in *Drosophila*, where the ZIP paralog FOI is known to play a critical role in modulating E-cadherin expression and stability (Jenkins et al., 2003; Van Doren et al., 2003; Mathews et al., 2006). Indeed, a first indication that FOI might influence E-cadherin biology at the transcriptional and post-translational levels emerged from elegant *in vivo* functional rescue experiments conducted with FOI-knockout flies. In contrast to the aforementioned ZIP6 zebrafish gastrulation paradigm, mutant zinc import-defective derivatives of FOI were, observed to exhibit reduced DE-cadherin expression. Significantly, forced expression of DE-cadherin from an unrelated tubulin promoter still did not lead to a functional rescue unless FOI was also present to ensure the post-translational stabilization of DE-cadherin (Mathews et al., 2006).

CONCLUSIONS

During the course of EMT, cells have to master the feat of gradually shifting their cell-cell and cell-substrate contacts from a stably adherent to a more transient and focal mode of attachment. The cellular program underlying these changes requires the concerted action of many molecules. Here we reviewed evidence consistent with the notion that members of the ZIP/PrP protein family may influence this program by affecting E-cadherin biology at multiple levels. It is likely that the cellular context will play a

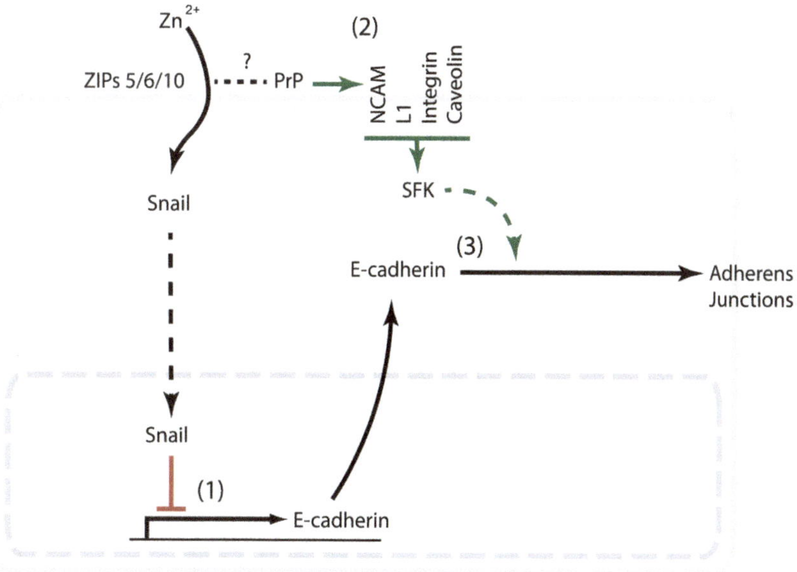

FIGURE 2 | Schematic summarizing evidence consistent with a role of PrP^C in EMT. (1) The closest molecular cousins of PrP from the ZIP zinc transporter family possess a documented role in EMT. Knockout or knockdown phenotypes of the respective ZIPs in zebrafish, *Drosophila* or mammalian cells demonstrate a critical role of ZIP-dependent zinc import in the transcriptional control of the E-cadherin gene. Note that whereas in the zebrafish EMT paradigm this transcriptional control is orchestrated by the EMT master regulator Snail and suppresses E-cadherin expression, the zinc import function of the *Drosophila* FOI gene was observed to promote E-cadherin transcription during gonad organogenesis (not shown). (2) Multiple interactors of PrP^C with a known ability to modulate the activity of Fyn (a Src family kinase) are independently known to play a role in EMT. (3) PrP knockdown in zebrafish is characterized by a gastrulation arrest phenotype with cytoplasmic accumulation of E-cadherin.

critical role for the direction and extent to which the expression of ZIPs/PrP affects this biology. With the advent of novel genome editing technologies and advances in quantitative protein mass spectrometry, the involvement of PrP in EMT and related cellular processes can now be studied much more elegantly. It is to be expected that these efforts will in the not-too-distant future lead to a much more detailed understanding of the cellular pathways that underlie phenotypic changes observed in PrP knockout paradigms and, ultimately, reveal the molecular function of this protein.

ACKNOWLEDGMENTS

Work on this project was funded through support by the Canadian Institutes of Health Research and The W. Garfield Weston Foundation.

REFERENCES

Aguzzi, A., Baumann, F., and Bremer, J. (2008). The Prion's elusive reason for being. *Annu. Rev. Neurosci.* 31, 439–477. doi: 10.1146/annurev.neuro.31.060407.125620

Antonacopoulou, A. G., Palli, M., Marousi, S., Dimitrakopoulos, F. I., Kyriakopoulou, U., Tsamandas, A. C., et al. (2010). Prion protein expression and the M129V polymorphism of the PRNP gene in patients with colorectal cancer. *Mol. Carcinog.* 49, 693–699. doi: 10.1002/mc.20642

Bremer, J., Baumann, F., Tiberi, C., Wessig, C., Fischer, H., Schwarz, P., et al. (2010). Axonal prion protein is required for peripheral myelin maintenance. *Nat. Neurosci.* 13, 310–318. doi: 10.1038/nn.2483

Chattopadhyay, M., Walter, E. D., Newell, D. J., Jackson, P. J., Aronoff-Spencer, E., Peisach, J., et al. (2005). The octarepeat domain of the prion protein binds Cu(II) with three distinct coordination modes at pH 7.4. *J. Am. Chem. Soc.* 127, 12647–12656. doi: 10.1021/ja053254z

Colling, S. B., Khana, M., Collinge, J., and Jefferys, J. G. (1997). Mossy fibre reorganization in the hippocampus of prion protein null mice. *Brain Res.* 755, 28–35. doi: 10.1016/S0006-8993(97)00087-5

Crusio, W. E. (2002). My mouse has no phenotype. *Genes Brain Behav.* 1, 71. doi: 10.1034/j.1601-183X.2002.10201.x

Du, L., Rao, G., Wang, H., Li, B., Tian, W., Cui, J. T., et al. (2013). CD44-positive cancer stem cells expressing cellular prion protein contribute to metastatic capacity in colorectal cancer. *Cancer Res.* 73, 2682–2694. doi: 10.1158/0008-5472.CAN-12-3759

Ehsani, S., Mehrabian, M., Pocanschi, C. L., and Schmitt-Ulms, G. (2012). The ZIP-prion connection. *Prion* 6, 317–321. doi: 10.4161/pri.20196

Ehsani, S., Tao, R., Pocanschi, C. L., Ren, H., Harrison, P. M., and Schmitt-Ulms, G. (2011). Evidence for retrogene origins of the prion gene family. *PLoS ONE* 6:e26800. doi: 10.1371/journal.pone.0026800

Evseenko, D., Zhu, Y., Schenke-Layland, K., Kuo, J., Latour, B., Ge, S., et al. (2010). Mapping the first stages of mesoderm commitment during differentiation of human embryonic stem cells. *Proc. Natl. Acad. Sci. U.S.A.* 107, 13742–13747. doi: 10.1073/pnas.1002077107

Gauczynski, S., Peyrin, J. M., Haik, S., Leucht, C., Hundt, C., Rieger, R., et al. (2001). The 37-kDa/67-kDa laminin receptor acts as the cell-surface receptor for the cellular prion protein. *EMBO J.* 20, 5863–5875. doi: 10.1093/emboj/20.21.5863

Godt, D., and Tepass, U. (2003). Organogenesis: keeping in touch with the germ cells. *Curr. Biol.* 13, R683–R685. doi: 10.1016/S0960-9822(03)00609-2

Guo, H., Jin, X., Zhu, T., Wang, T., Tong, P., Tian, L., et al. (2014). SLC39A5 mutations interfering with the BMP/TGF-beta pathway in non-syndromic high myopia. *J. Med. Genet.* 51, 518–525. doi: 10.1136/jmedgenet-2014-102351

Hogstrand, C., Kille, P., Ackland, M. L., Hiscox, S., and Taylor, K. M. (2013). A mechanism for epithelial-mesenchymal transition and anoikis resistance in breast cancer triggered by Zinc channel zip6 and signal transducer and activator of transcription 3 (STAT3). *Biochem. J.* 455, 229–237. doi: 10.1042/bj20130483

Hsu, P. D., Lander, E. S., and Zhang, F. (2014). Development and applications of CRISPR-Cas9 for genome engineering. *Cell* 157, 1262–1278. doi: 10.1016/j.cell.2014.05.010

Jenkins, A. B., McCaffery, J. M., and Van Doren, M. (2003). Drosophila E-cadherin is essential for proper germ cell-soma interaction during gonad morphogenesis. *Development* 130, 4417–4426. doi: 10.1242/dev.00639

Kiefel, H., Bondong, S., Pfeifer, M., Schirmer, U., Erbe-Hoffmann, N., Schafer, H., et al. (2012). EMT-associated up-regulation of L1CAM provides insights into L1CAM-mediated integrin signalling and NF-kappaB activation. *Carcinogenesis* 33, 1919–1929. doi: 10.1093/carcin/bgs220

Lamouille, S., Xu, J., and Derynck, R. (2014). Molecular mechanisms of epithelial-mesenchymal transition. *Nat. Rev. Mol. Cell Biol.* 15, 178–196. doi: 10.1038/nrm3758

Lehembre, F., Yilmaz, M., Wicki, A., Schomber, T., Strittmatter, K., Ziegler, D., et al. (2008). NCAM-induced focal adhesion assembly: a functional switch upon loss of E-cadherin. EMBO J. 27, 2603–2615. doi: 10.1038/emboj.2008.178

Li, J. J., and Xie, D. (2014). RACK1, a versatile hub in cancer. *Oncogene.* doi: 10.1038/onc.2014.127. [Epub ahead of print].

Lichten, L. A., and Cousins, R. J. (2009). Mammalian zinc transporters: nutritional and physiologic regulation. *Annu. Rev. Nutr.* 29, 153–176. doi: 10.1146/annurev-nutr-033009-083312

Lilien, J., and Balsamo, J. (2005). The regulation of cadherin-mediated adhesion by tyrosine phosphorylation/dephosphorylation of beta-catenin. *Curr. Opin. Cell Biol.* 17, 459–465. doi: 10.1016/j.ceb.2005.08.009

Lim, J., and Thiery, J. P. (2012). Epithelial-mesenchymal transitions: insights from development. *Development* 139, 3471–3486. doi: 10.1242/dev.071209

Lin, Z., and Fang, D. (2013). The Roles of SIRT1 in Cancer. *Genes Cancer* 4, 97–104. doi: 10.1177/1947601912475079

Malaga-Trillo, E., and Sempou, E. (2009). PrPs: proteins with a purpose: lessons from the zebrafish. *Prion* 3, 129–133. doi: 10.4161/pri.3.3.9651

Malaga-Trillo, E., Solis, G. P., Schrock, Y., Geiss, C., Luncz, L., Thomanetz, V., et al. (2009). Regulation of embryonic cell adhesion by the prion protein. *PLoS Biol.* 7:e55. doi: 10.1371/journal.pbio.1000055

Mathews, W. R., Ong, D., Milutinovich, A. B., and Van Doren, M. (2006). Zinc transport activity of fear of Intimacy is essential for proper gonad morphogenesis and DE-cadherin expression. *Development* 133, 1143–1153. doi: 10.1242/dev.02256

Mathews, W. R., Wang, F., Eide, D. J., and Van Doren, M. (2005). Drosophila fear of intimacy encodes a Zrt/IRT-like protein (ZIP) family zinc transporter functionally related to mammalian ZIP proteins. *J. Biol. Chem.* 280, 787–795. doi: 10.1074/jbc.M411308200

McDonald, S. L., and Silver, A. (2009). The opposing roles of Wnt-5a in cancer. *Br. J. Cancer* 101, 209–214. doi: 10.1038/sj.bjc.6605174

Mehrpour, M., and Codogno, P. (2010). Prion protein: from physiology to cancer biology. *Cancer Lett.* 290, 1–23. doi: 10.1016/j.canlet.2009.07.009

Mouillet-Richard, S., Ermonval, M., Chebassier, C., Laplanche, J. L., Lehmann, S., Launay, J. M., et al. (2000). Signal transduction through prion protein. *Science* 289, 1925–1928. doi: 10.1126/science.289.5486.1925

Mu, D., Cambier, S., Fjellbirkeland, L., Baron, J. L., Munger, J. S., Kawakatsu, H., et al. (2002). The integrin alpha(v)beta8 mediates epithelial homeostasis through MT1-MMP-dependent activation of TGF-beta1. *J. Cell Biol.* 157, 493–507. doi: 10.1083/jcb.200109100

Munger, J. S., Huang, X., Kawakatsu, H., Griffiths, M. J., Dalton, S. L., Wu, J., et al. (1999). The integrin alpha v beta 6 binds and activates latent TGF beta 1: a mechanism for regulating pulmonary inflammation and fibrosis. *Cell* 96, 319–328. doi: 10.1016/S0092-8674(00)80545-0

Naslavsky, N., Stein, R., Yanai, A., Friedlander, G., and Taraboulos, A. (1997). Characterization of detergent-insoluble complexes containing the cellular prion protein and its scrapie isoform. *J. Biol. Chem.* 272, 6324–6331. doi: 10.1074/jbc.272.10.6324

Niethammer, P., Delling, M., Sytnyk, V., Dityatev, A., Fukami, K., and Schachner, M. (2002). Cosignaling of NCAM via lipid rafts and the FGF receptor is required for neuritogenesis. *J. Cell Biol.* 157, 521–532. doi: 10.1083/jcb.200109059

Pan, Y., Zhao, L., Liang, J., Liu, J., Shi, Y., Liu, N., et al. (2006). Cellular prion protein promotes invasion and metastasis of gastric cancer. *FASEB J.* 20, 1886–1888. doi: 10.1096/fj.06-6138fje

Pantera, B., Bini, C., Cirri, P., Paoli, P., Camici, G., Manao, G., et al. (2009). PrPc activation induces neurite outgrowth and differentiation in PC12 cells: role for caveolin-1 in the signal transduction pathway. *J. Neurochem.* 110, 194–207. doi: 10.1111/j.1471-4159.2009.06123.x

Rubenstein, R. (2012). Proteomic analysis of prion diseases: creating clarity or causing confusion? *Electrophoresis* 33, 3631–3643. doi: 10.1002/elps.201200310

Sailer, A., Bueler, H., Fischer, M., Aguzzi, A., and Weissmann, C. (1994). No propagation of prions in mice devoid of PrP. *Cell* 77, 967–968. doi: 10.1016/0092-8674(94)90436-7

Santuccione, A., Sytnyk, V., Leshchyns'ka, I., and Schachner, M. (2005). Prion protein recruits its neuronal receptor NCAM to lipid rafts to activate p59fyn and to enhance neurite outgrowth. *J. Cell Biol.* 169, 341–354. doi: 10.1083/jcb.200409127

Schmitt-Ulms, G., Ehsani, S., Watts, J. C., Westaway, D., and Wille, H. (2009). Evolutionary descent of prion genes from the ZIP family of metal ion transporters. *PLoS ONE* 4:e7208. doi: 10.1371/journal.pone.0007208

Shen, H., Qin, H., and Guo, J. (2009). Concordant correlation of LIV-1 and E-cadherin expression in human breast cancer cell MCF-7. *Mol. Biol. Rep.* 36, 653–659. doi: 10.1007/s11033-008-9225-4

Smyth, D., Leung, G., Fernando, M., and McKay, D. M. (2012). Reduced surface expression of epithelial E-cadherin evoked by interferon-gamma is Fyn kinase-dependent. *PLoS ONE* 7:e38441. doi: 10.1371/journal.pone.0038441

Solis, G. P., Radon, Y., Sempou, E., Jechow, K., Stuermer, C. A., and Malaga-Trillo, E. (2013). Conserved roles of the prion protein domains on subcellular localization and cell-cell adhesion. *PLoS ONE* 8:e70327. doi: 10.1371/journal.pone.0070327

Steele, A. D., Lindquist, S., and Aguzzi, A. (2007). The prion protein knockout mouse: a phenotype under challenge. *Prion* 1, 83–93. doi: 10.4161/pri.1.2.4346

Thiery, J. P., Acloque, H., Huang, R. Y., and Nieto, M. A. (2009). Epithelial-mesenchymal transitions in development and disease. *Cell* 139, 871–890. doi: 10.1016/j.cell.2009.11.007

Tobler, I., Gaus, S. E., Deboer, T., Achermann, P., Fischer, M., Rulicke, T., et al. (1996). Altered circadian activity rhythms and sleep in mice devoid of prion protein. *Nature* 380, 639–642. doi: 10.1038/380639a0

Tomasi, V. (2010). Signal transduction in neurons: effects of cellular prion protein on fyn kinase and ERK1/2 kinase. *Immun. Ageing* 7, S5. doi: 10.1186/1742-4933-7-S1-S5

Toni, M., Spisni, E., Griffoni, C., Santi, S., Riccio, M., Lenaz, P., et al. (2006). Cellular prion protein and caveolin-1 interaction in a neuronal cell line precedes Fyn/Erk 1/2 signal transduction. *J. Biomed. Biotechnol.* 2006, 1–13. doi: 10.1155/JBB/2006/69469

Um, J. W., Nygaard, H. B., Heiss, J. K., Kostylev, M. A., Stagi, M., Vortmeyer, A., et al. (2012). Alzheimer amyloid-beta oligomer bound to postsynaptic prion protein activates Fyn to impair neurons. *Nat. Neurosci.* 15, 1227–1235. doi: 10.1038/nn.3178

Van Doren, M., Mathews, W. R., Samuels, M., Moore, L. A., Broihier, H. T., and Lehmann, R. (2003). Fear of intimacy encodes a novel transmembrane protein required for gonad morphogenesis in Drosophila. *Development* 130, 2355–2364. doi: 10.1242/dev.00454

Walsh, F. S., Parekh, R. B., Moore, S. E., Dickson, G., Barton, C. H., Gower, H. J., et al. (1989). Tissue specific O-linked glycosylation of the neural cell adhesion molecule (N-CAM). *Development* 105, 803–811.

Wang, Z., Liu, P., Inuzuka, H., and Wei, W. (2014). Roles of F-box proteins in cancer. *Nat. Rev. Cancer* 14, 233–247. doi: 10.1038/nrc3700

Watt, N. T., Griffiths, H. H., and Hooper, N. M. (2013). Neuronal zinc regulation and the prion protein. *Prion* 7, 203–208. doi: 10.4161/pri.24503

Watts, J. C., Huo, H., Bai, Y., Ehsani, S., Jeon, A. H., Shi, T., et al. (2009). Interactome analyses identify ties of PrP and its mammalian paralogs to oligomannosidic N-glycans and endoplasmic reticulum-derived chaperones. *PLoS Pathog.* 5:e1000608. doi: 10.1371/journal.ppat.1000608

Watts, J. C., and Westaway, D. (2007). The prion protein family: diversity, rivalry, and dysfunction. *Biochim. Biophys. Acta* 1772, 654–672. doi: 10.1016/j.bbadis.2007.05.001

Wurm, S., and Wechselberger, C. (2006). Prion protein modifies TGF-beta induced signal transduction. *Biochem. Biophys. Res. Commun.* 349, 525–532. doi: 10.1016/j.bbrc.2006.08.074

Yamashita, S., Miyagi, C., Fukada, T., Kagara, N., Che, Y. S., and Hirano, T. (2004). Zinc transporter LIVI controls epithelial-mesenchymal transition in zebrafish gastrula organizer. *Nature* 429, 298–302. doi: 10.1038/nature02545

Yang, Z., Mu, Z., Dabovic, B., Jurukovski, V., Yu, D., Sung, J., et al. (2007). Absence of integrin-mediated TGFbeta1 activation *in vivo* recapitulates the phenotype of TGFbeta1-null mice. *J. Cell Biol.* 176, 787–793. doi: 10.1083/jcb.2006 11044

Zhang, C. C., Steele, A. D., Lindquist, S., and Lodish, H. F. (2006). Prion protein is expressed on long-term repopulating hematopoietic stem cells and is important for their self-renewal. *Proc. Natl. Acad. Sci. U.S.A.* 103, 2184–2189. doi: 10.1073/pnas.0510577103

Zomosa-Signoret, V., Arnaud, J. D., Fontes, P., Alvarez-Martinez, M. T., and Liautard, J. P. (2008). Physiological role of the cellular prion protein. *Vet. Res.* 39, 9. doi: 10.1051/vetres:2007048

Conflict of Interest Statement: The authors declare that the research was conducted in the absence of any commercial or financial relationships that could be construed as a potential conflict of interest.

7

PrP^C from stem cells to cancer

Séverine Martin-Lannerée[1,2], Théo Z. Hirsch[1,2], Julia Hernandez-Rapp[1,2,3], Sophie Halliez[4], Jean-Luc Vilotte[5], Jean-Marie Launay[6,7] and Sophie Mouillet-Richard[1,2]*

[1] Toxicology, Pharmacology and Cellular Signaling, INSERM UMR-S1124, Paris, France
[2] Toxicology, Pharmacology and Cellular Signaling, Université Paris Descartes, Sorbonne Paris Cité, UMR-S1124, Paris, France
[3] Université Paris Sud 11, ED419 Biosigne, Orsay, France
[4] U892 Virologie et Immunologie Moléculaires, INRA, Jouy-en-Josas, France
[5] UMR1313 Génétique Animale et Biologie Intégrative, INRA, Jouy-en-Josas, France
[6] AP-HP Service de Biochimie, Fondation FondaMental, INSERM U942 Hôpital Lariboisière, Paris, France
[7] Pharma Research Department, F. Hoffmann-La-Roche Ltd., Basel, Switzerland

Edited by:
Craig Michael Walsh, University of California, Irvine, USA

Reviewed by:
Mario Cioce, New York University Langone Medical Center, USA
Han-Ming Shen, National University of Singapore, Singapore

***Correspondence:**
Sophie Mouillet-Richard, Toxicology, Pharmacology and Cellular Signaling, INSERM U-1124, Université Paris Descartes, 45, rue des Saints Pères, 75006 Paris, France
e-mail: sophie.mouillet-richard@parisdescartes.fr

The cellular prion protein PrP^C was initially discovered as the normal counterpart of the pathological scrapie prion protein PrP^Sc, the main component of the infectious agent of Transmissible Spongiform Encephalopathies. While clues as to the physiological function of this ubiquitous protein were greatly anticipated from the development of knockout animals, PrP-null mice turned out to be viable and to develop without major phenotypic abnormalities. Notwithstanding, the discovery that hematopoietic stem cells from PrP-null mice have impaired long-term repopulating potential has set the stage for investigating into the role of PrP^C in stem cell biology. A wealth of data have now exemplified that PrP^C is expressed in distinct types of stem cells and regulates their self-renewal as well as their differentiation potential. A role for PrP^C in the fate restriction of embryonic stem cells has further been proposed. Paralleling these observations, an overexpression of PrP^C has been documented in various types of tumors. In line with the contribution of PrP^C to stemness and to the proliferation of cancer cells, PrP^C was recently found to be enriched in subpopulations of tumor-initiating cells. In the present review, we summarize the current knowledge of the role played by PrP^C in stem cell biology and discuss how the subversion of its function may contribute to cancer progression.

Keywords: cellular prion protein, stem cell, cancer, self-renewal, cell fate specification, prion infection

INTRODUCTION

The discovery of the cellular prion protein PrP^C dates back to 1985 with the identification that the scrapie prion protein PrP^Sc, the main component of the infectious agent responsible for Transmissible Spongiform Encephalopathies (TSEs) was encoded by a gene of the host, termed *Prnp* (Oesch et al., 1985). PrP^C has been extensively scrutinized as the endogenous substrate for conversion into its pathogenic PrP^Sc counterpart (Aguzzi and Calella, 2009), while studies on its physiological function have long been overlooked. At the molecular and cellular levels, it is well established that PrP^C is anchored to the outer leaflet of the plasma membrane through a glycosyl-phosphatidylinositol (GPI) moiety (Linden et al., 2008). It may exist under a great diversity of isoforms as a result of heterogeneous glycosylation (Ermonval et al., 2003) and proteolytic cleavage (McDonald et al., 2014). Although it is suspected that the wide repertoire of PrP^C species may endow the protein with the capacity to interact with multiple soluble ligands, extracellular matrix components or cell-surface proteins, the specific tissue distribution, and function of each isoform remain elusive (Linden et al., 2008). That research on PrP^C function has lagged behind that of TSE pathophysiology may notably be explained by the lack of major abnormalities in PrP-null mice (Steele et al., 2007), whose most obvious phenotype is their resistance to TSE agents (Bueler et al., 1993). Because PrP is ubiquitously expressed and very much conserved in mammals, with *Prnp* orthologs identified in fish, birds, and reptiles (Premzl and Gamulin, 2007), the apparent normal phenotype of PrP null mice was quite unexpected and proposed to reflect the occurrence of compensatory mechanisms. One major contribution of these mice, however, was the demonstration that PrP^C is mandatory for the long-term repopulating activity of hematopoietic stem cells (HSCs) (see below) (Zhang et al., 2006). This seminal report set the stage for investigating into the role exerted by PrP^C in stem cell biology. Here, we provide an overview of the recent advances regarding the contribution of PrP^C to stem cell biology and their pathophysiological implications.

PrP^C EXPRESSION AND ROLE DURING DEVELOPMENT

Studies on PrP^C have initially focused on the adult central nervous system (CNS), since it is the only target of PrP^Sc-associated toxicity (Aguzzi and Calella, 2009). Further, PrP^C is most abundantly found in neurons (Linden et al., 2008). Notwithstanding, PrP^C is highly expressed during embryonic development, as first shown by Manson et al. over two decades ago (Manson et al., 1992). This *in situ* hybridization analysis revealed widespread expression of *Prnp* transcripts in the developing central and peripheral nervous system at embryonic days E13.5 and E16.5, as well as in other tissues such as the intestine or the dental lamina (Manson et al., 1992). *Prnp* mRNA was also detected in extra-embryonic tissues from E6.5, pointing for the first time to a potential role for PrP^C in the placenta, which has started to be accurately assessed recently (Alfaidy et al., 2012; Passet et al., 2012). These first data were refined with the detection of *Prnp* mRNA starting at E8.5–E9 in the differentiating neuroepithelium (Miele et al., 2003). The induction of *Prnp* expression at this stage in the developing CNS and heart was confirmed in a study using *Prnp*-LacZ reporter mice (Tremblay et al., 2007).

Based on this developmental pattern of expression, transcriptomic analyses were carried out on early *Prnp* knockout vs. wild-type (WT) embryos and revealed prominent alterations, with a total number of 263 genes differentially expressed at day E7.5 (Khalife et al., 2011). The array of genes with altered expression notably includes a set of growth factors and growth factor receptors, supporting the notion that PrP^C plays an important role in the regulation of cascades associated with embryonic development (Khalife et al., 2011). Interestingly, the pattern of pathways affected overlaps with that obtained after early embryonic gene expression profiling of zebrafish PrP2 morphants (Nourizadeh-Lillabadi et al., 2010).

The zebrafish model actually allowed bringing to light a vital function for PrP^C since morpholino-mediated knockdown of the PrP ortholog PrP1 in this species leads to loss of embryonic cell adhesion and gastrulation arrest (Malaga-Trillo et al., 2009). Of note, the defects observed could be rescued with mouse *Prnp* mRNA, indicating that this function is evolutionary conserved (Malaga-Trillo et al., 2009). Thus, the overall data gained at the animal scale argue that PrP^C fulfills an important function during embryogenesis and that its ablation in mice triggers the implementation of yet-to-be-deciphered compensatory mechanisms.

PrP^C REGULATES THE SELF-RENEWAL OF STEM/PROGENITOR CELLS

The link between PrP^C and stem cell biology was first uncovered in HSCs. Investigations of PrP^C in the hematopoietic system were initially prompted by the observation that PrP^{Sc} accumulates in lymphoid organs and by the quest to understand the cellular mechanisms sustaining prion propagation in the periphery (Mabbott and MacPherson, 2006). These studies demonstrated that PrP^C is highly expressed at the surface of various hematopoietic cells, including B and T lymphocytes, monocytes, dendritic cells, megakaryocytes and platelets, but not erythrocytes or granulocytes (Linden et al., 2008). In the human bone marrow, PrP^C was found to be present in the $CD34^+$ stem / progenitor cell population (Dodelet and Cashman, 1998) and to be preferentially expressed on murine $CD43^+$, $B220^-$, IL-$7R^-$ cells, enriched in immature progenitors (Liu et al., 2001). A major advance was the discovery by the team of Lodish that PrP^C is very abundant at the surface of mouse bone marrow $Lin^-Sca^+Endoglin^+$ cells, a population comprising immature HSCs (Zhang et al., 2006). These authors then assessed the ability of bone marrow-derived $Lin^-Sca^+Endoglin^+$ cells from *Prnp* knockout mice to reconstitute the hematopoietic system of lethally irradiated mice in serial transplantation assays, and demonstrated that HSCs from *Prnp* null mice lack long-term repopulating activity (see **Table 1**) (Zhang et al., 2006). These experiments further allowed substantiating that the $Lin^-Sca^+Endoglin^+$ cell population endowed with long-term repopulating activity is PrP^C positive (Zhang et al., 2006).

Much like HSCs, neuroepithelial stem cells have the capacity to proliferate through repeated symmetric divisions and generate radial glial cells, which then undergo asymmetric division and give rise to neurons, oligodendrocytes, and astrocytes (Gotz and Huttner, 2005). The proper proceeding of these expansive and neurogenic phases is crucial for the development of the CNS. By comparing neural stem cells isolated from *Prnp* knockout, WT or PrP overexpressing mice at embryonic day 13.5, Steele et al. documented that PrP^C levels directly increase the differentiation rate of multipotent neural precursor cells (Steele et al., 2006). In the same study, PrP^C expression levels were further found to correlate with the proliferation rate in the two adult neurogenic regions, the subventricular zone (SVZ) or the dentate gyrus (DG) (Steele et al., 2006). The latter finding was corroborated by two independent studies showing that the formation of neurospheres from fetal (Santos et al., 2011) or postnatal (Prodromidou et al., 2014) brains is less efficient with *Prnp* knockout than WT mice.

The notion that PrP^C contributes to the proliferation of stem cells extends beyond hematopoietic and neural stem cells. Indeed, the level of PrP^C was found to serve as an effective cell surface marker for self-renewing mammary gland stem cells in mice (Liao et al., 2007). More recently, PrP^C was further shown to promote the expansion and engraftment of bone marrow-derived human mesenchymal stem cells (MSCs) (Mohanty et al., 2012). Finally, PrP^C was shown to exert either an anti- or a pro-proliferative effect in human embryonic stem (ES) cells, depending on whether they are grown under self-renewing or differentiating conditions, respectively (Lee and Baskakov, 2012).

PrP^C INFLUENCES STEM CELL FATE

The identification of PrP^C as a broad cell surface marker for stem / progenitor cells raises the question as to whether the expression of PrP^C is a determinant of the stem cell fate. In this respect, we (Mouillet-Richard et al., 1999) and others (Peralta et al., 2011) have provided evidence that PrP^C is upregulated following the cell fate restriction of multipotential ES or embryonic carcinoma (EC) cells toward the neuronal lineage. Similarly, PrP^C is induced in ES-derived cardiomyogenic progenitors obtained after embryoid body (EB) formation (Hidaka et al., 2010). In line with this, the expression of PrP^C was found to be increased during spontaneous differentiation of mouse and human ES cells (Lee and Baskakov, 2010; Miranda et al., 2011) and, reciprocally, induction of PrP^C in human ES cells grown under self-renewal conditions was shown to promote their differentiation (Lee and Baskakov, 2012). Intriguingly, exposure of ES cells to recombinant PrP delays their spontaneous differentiation (Lee and Baskakov, 2010). In view of the early expression of PrP^C in extra-embryonic tissues, it is tempting to speculate that placenta-derived PrP^C may serve as a paracrine signal to maintain the self-renewal of inner mass cells, until their appropriate induction toward either of the three lineages.

Table 1 | Summary of effects of PrP^C depletion on stem and progenitor cells.

Cell type/tissue	PrP^C-null/KD vs. wt	References
EMBRYO/POSTNATAL		
ES	Increased apoptosis in embryoid bodies	Miranda et al., 2011
Human ES	Inhibition of ectodermal differentiation Reduced transition from G1 to S phase during spontaneous differentiation	Lee and Baskakov, 2012
Multipotent neural precursors E13.5 (telencephalon)	Delayed neuronal differentiation	Steele et al., 2006
Neurospheres isolated from E14 forebrain	Reduced neurosphere formation	Santos et al., 2011
E16.5 oligodendrocyte precursor cells (OPC) in optic nerve explants	Increased proliferation (BrdU incorporation)	Bribian et al., 2012
E16.5 neocortex	Increased $NG2^+$ $Olig2^+$ OPC	Bribian et al., 2012
P0-P2 cortical OPC primary culture	Delayed differentiation	Bribian et al., 2012
P5 neurospheres isolated from SVZ	Reduced proliferation (BrdU incorporation) and secondary neurosphere formation	Prodromidou et al., 2014
ADULT		
Dentate gyrus	Reduced proliferation (BrdU incorporation)	Steele et al., 2006
	Reduced proliferation of neuroprogenitors and/or neurogenesis ($BrdU^+$ $NeuN^+$)	Bribian et al., 2012
Neocortex	Increased $NG2^+$ $Olig2^+$ OPC	Bribian et al., 2012
SVZ	Reduction of cycling $GFAP^+$ neural stem cells Decreased $Mash^+$ progenitors and accumulation of proliferating neuroblasts DCX^+	Prodromidou et al., 2014
HSC	Loss of long-term repopulating activity in serial transplantation assays	Zhang et al., 2006
Skeletal muscle	Delayed regeneration after injury (retarded exit from the cell cycle of myogenic precursor cells)	Stella et al., 2010
Human MSC	Reduced clonogenic potential, proliferation and differentiation	Mohanty et al., 2012

Beyond lineage specification, the regulation of PrP^C expression also accompanies differentiation toward a given fate. Along the hematopoietic lineage, PrP^C appears to be downregulated upon differentiation of $CD34^+$ progenitors toward a granulocytic fate, while its expression is retained in B and T lymphocytes as well as monocytes (Dodelet and Cashman, 1998). In addition, PrP^C is absent from erythrocytes (Dodelet and Cashman, 1998) and abundant in megakaryocytes and platelets (Starke et al., 2005), suggesting that the expression of PrP^C is switched off with the commitment of megakaryocytic-erythrocytic progenitors toward the erythroid fate, or is decreased along erythroid differentiation, in line with (Panigaj et al., 2011).

As for neural progenitor cells, the expression of PrP^C was reported to be increased along neuronal differentiation, while barely detected in astrocytes or oligodendrocytes (Steele et al., 2006). This high neuronal PrP^C expression is in line with the well-documented contribution of PrP^C to neuronal differentiation, including neurite outgrowth (Chen et al., 2003; Santuccione et al., 2005; Loubet et al., 2012; Santos et al., 2013) or synapse maturation (Kanaani et al., 2005). The lack of PrP^C detection in differentiating oligodendrocytes and astrocytes in the study by Steele et al. (2006) is, however, in contrast with several reports documenting an abundant PrP^C expression in these two cell types in late embryos or in the postnatal brain (Moser et al., 1995; Lima et al., 2007; Bribian et al., 2012). Interestingly, both oligodendrocytic (Bribian et al., 2012) and astrocytic (Arantes et al., 2009) differentiation kinetics appear to be delayed in *Prnp* knockout mice. These observations recall the delay in neuronal differentiation, as initially reported by Steele et al. (2006), as well as the slower regeneration of muscle after injury (Stella et al., 2010) in a PrP null context. Whether PrP^C expression affects the balance from one fate to another remains, however, to be investigated. In this regard, it is worth noting that prion infection in adult neural stem cells (NSCs) favors the differentiation toward the glial lineage at the expense of neuronal differentiation (Relaño-Ginés et al., 2013).

STEM CELLS AND PRION REPLICATION

Whether stem cells are susceptible to prion infection may at first seem a question without relevance, since TSEs are neurodegenerative diseases. However, as rightly underlined in the study by Relaño-Ginés (Relaño-Ginés et al., 2013), exploiting the potential of adult NSCs is currently considered as a promising avenue to mitigate neurodegeneration (Bellenchi et al., 2013). While several studies had reported an efficient replication of PrP^{Sc} in neurospheres isolated from fetal brain (Milhavet et al., 2006; Herva et al., 2010), the susceptibility of adult NSC toward prion infection has been evaluated only recently. In line with the results obtained with embryonic-derived cultures, neurospheres isolated from the SVZ or the DG of adult mice were shown to support prion replication (Relaño-Ginés et al., 2013). The same study further documented the presence of dense PrP^{Sc} deposits in the

DG and the SVZ of prion-infected mice, indicating that prions colonize adult NSCs, the brain's endogenous repair machinery (Relaño-Ginés et al., 2013, 2014). Of note, prion replication of adult NSCs was found to impair neuronal differentiation, both *in vitro* and *in vivo* (Relaño-Ginés et al., 2013). Thus, in addition to constituting a reservoir of PrP^{Sc}, the replication of prions in adult NSCs may also compromise the regeneration of damaged neurons. Finally, because PrP^{Sc} is known to deviate the normal function of PrP^C (Westergard et al., 2007; Pradines et al., 2013), these observations suggest that studying the impact of prion infection on the self-renewal and fate of NSCs may improve our understanding of the physiological role exerted by PrP^C in these processes.

PrP^C-DEPENDENT CONTROL OF STEM CELL SELF-RENEWAL AND FATE: MECHANISTIC INSIGHT

Notwithstanding the well-established involvement of PrP^C in the self-renewal of diverse types of stem /progenitor cells, the molecular mechanisms at play remain obscure. One possible mode of action of PrP^C would be through the interaction with one of its ligands. This view is clearly supported by the demonstration that the binding of PrP^C with STI-1 is critical for the formation and proliferation of neurospheres cultured from fetal forebrain (Santos et al., 2011). While several signaling cascades elicited by the interaction of STI-1 with PrP^C have been described in a neuronal context (Hirsch et al., 2014), the pathways mobilized to sustain neurosphere self-renewal and proliferation have not been analyzed so far (Santos et al., 2011). On another hand, the presence of PrP^C on neurospheres was recently shown to be required for NCAM-induced neuronal differentiation (Prodromidou et al., 2014). These two sets of observations raise the question as to the PrP^C isoforms that respectively bind STI-1 and NCAM, since these two molecules instruct distinct responses. Another PrP^C partner that may have relevance to stem cell biology is the amyloid precursor protein APP, whose functional interaction with PrP in the zebrafish modulates cell adhesion and CNS development (Kaiser et al., 2012). Whether the APP-PrP^C interaction is involved in the regulation of E-cadherin-dependent adhesion in zebrafish embryos deserves further investigation (Malaga-Trillo et al., 2009). Interestingly, our own studies on a neuroectodermal stem cell line also substantiate a disruption of cadherin-mediated cell contacts upon PrP^C depletion (Martin-Lannerée et al., unpublished observations). It is of note that cell adhesion processes are now recognized as major determinants of stem cell biology in relation with their local microenvironment (stem cell niche) (Marthiens et al., 2010). By affecting adhesion properties of stem cells, the depletion of PrP^C may thus in turn impact on their retention, self-renewal or exit from their niche.

A ROLE FOR PrP^C IN CANCER STEM CELLS?

The contribution of PrP^C to cell proliferation appears to apply to many cell types beyond stem/progenitor cells. These notably include cancer cells, as first demonstrated in gastric tumor cell lines (Liang et al., 2007a). In these cells, PrP^C was shown to accelerate the G1 to S phase transition in the cell cycle and to sustain proliferation by inducing the expression of Cyclin D1 through a PI3K/Akt pathway (Liang et al., 2007a). The PrP^C-interacting protein(s) involved in this cascade remain(s), however, to be identified. Beyond proliferation, PrP^C overexpression in cancer cells was further shown to confer resistance to various cytotoxic agents (Mehrpour and Codogno, 2010) as well as invasive properties (Pan et al., 2006). For instance, PrP^C levels were shown to correlate with resistance to TNFα-induced cell death in the MCF-7 breast cancer cell line (Diarra-Mehrpour et al., 2004). Very recently, PrP^C was found to interact with the cell surface protein CD44 in adriamycin-resistant breast cancer cells, and to promote their proliferation and migration (Cheng et al., 2013). Interestingly, CD44 has been reported to be enriched at the cell surface of various types of tumor-initiating cells, which bear similarities with embryonic or adult stem cells and are often referred to as cancer stem cells (CSCs) (Medema, 2013). It is also noteworthy that CSCs have been associated with increased resistance to antitumor treatments (Singh and Settleman, 2010). In line with the above-mentioned role of PrP^C in the self-renewal of stem cells, Du et al. depicted a population of $CD44^+PrP^{C+}$ cells from primary colorectal tumors endowed with enhanced tumor-initiating and metastatic capacity (Du et al., 2013). At a mechanistic level, PrP^C was shown to promote an epithelial to mesenchymal transition (EMT) through the regulation of the Twist transcription factor (Du et al., 2013). These observations are in agreement with the notion that the emergence of CSCs and EMT are intimately connected (Singh and Settleman, 2010).

A still unresolved question concerns the molecular mechanisms sustaining the enhanced expression of PrP^C in cancer cells. PrP^C expression has been shown to be increased in response to hypoxia in gastric cancer cell lines (Liang et al., 2007b). Other PrP^C-inducing signals include oxidative (Sauer et al., 1999) and endoplasmic-reticulum (Dery et al., 2013) stresses. Some deregulation of PrP^C function may also arise with aging. Indeed, PrP^C was recently shown to accumulate in lipid rafts in the mouse aging brain (Agostini et al., 2013). Whether this change in PrP^C distribution also occurs in other tissues with aging is worth considering, since it would potentially impact on the recruitment of downstream signaling cascades. As observed in the context of neurodegeneration (Hirsch et al., 2014), the subversion of PrP^C function may over-activate src kinases and further promote alterations in lipid raft-initiated signaling pathways, known to be detrimental in cancer (Patra, 2008). Such changes may in turn have consequences on the cell local environment, and, in the case of stem cells, deregulate the interactions with their niche. This scenario is considered as a potential cause of CSCs emergence (Rezza et al., 2014), and this may have particular relevance with respect to aging.

OPEN QUESTIONS AND THERAPEUTIC PROSPECTS

Harnessing the self-renewal and differentiation potential of stem cells represents a major challenge for regenerative medicine. The recent accumulation of data regarding the involvement of PrP^C in stem cell biology warrants further studying the molecular and cellular mechanisms sustaining the contribution of this protein to the proliferation of stem cells, their maintenance in an undifferentiated state, their capacity to respond to fate determination inputs and to implement a given differentiation program. Achieving this task is complicated by the multiplicity of PrP^C isoforms and partners, which may fulfill promiscuous functions. That PrP^C is required for efficient tissue repair after injury is clearly indicated in the context of bone-marrow reconstitution (Zhang et al., 2006) or muscle regeneration (Stella et al., 2010), which suggest that the mobilization of PrP^C-dependent cascades via appropriate ligands may provide a fruitful approach to enhance the regeneration of lesionned tissues. As a prerequisite, manipulating conditions would need to be carefully adjusted in order to control activating signals, given the pathological implications that may ensue from PrP^C over-activation.

Finally, the emerging roles of PrP^C in stemness on the one hand and in various aspects of cancer cell biology on the other hand bring new light on this already fascinating molecule. Given the relationship between stem cells and oncogenesis, advance in the understanding of the role played by PrP^C in stem cells is likely to illuminate the issue of its contribution to tumorigenesis and vice-versa. One major remaining challenge is to decipher the mechanisms controlling the expression levels of PrP^C in normal and cancer stem cells. While the cues underlying the induction of PrP^C during embryonic development are elusive, several cancer-associated conditions, including hypoxia (Liang et al., 2007b), oxidative (Sauer et al., 1999) or endoplasmic reticulum (Dery et al., 2013) stresses have been reported to activate PrP^C transcription. More broadly, increasing our knowledge of the regulation of PrP gene expression may help design novel strategies for therapeutic intervention in cancer, beyond directly targeting PrP^C through antisense oligonucleotides (Meslin et al., 2007) or monoclonal antibodies (Du et al., 2013).

To conclude, progress in the stem cell and cancer fields should increase our knowledge of how PrP^C, as a cell surface receptor or co-receptor, connects cells with their environment to drive adaptive, homeostatic responses and how this function is corrupted in disease-associated states.

ACKNOWLEDGMENTS

We apologize to colleagues in the field whose work we were unable to cite owing to space limitations. We acknowledge financial support from the ARC, INSERM, as well as the Region Ile de France (DIM-Stem Pôle).

REFERENCES

Agostini, F., Dotti, C. G., Perez-Canamas, A., Ledesma, M. D., Benetti, F., and Legname, G. (2013). Prion protein accumulation in lipid rafts of mouse aging brain. *PLoS ONE* 8:e74244. doi: 10.1371/journal.pone.0074244

Aguzzi, A., and Calella, A. M. (2009). Prions: protein aggregation and infectious diseases. *Physiol. Rev.* 89, 1105–1152. doi: 10.1152/physrev.00006.2009

Alfaidy, N., Chauvet, S., Andrei, S., Salomon, A., Saoudi, Y., Richaud, P., et al. (2012). Prion protein expression and functional importance in developmental angiogenesis: role in oxidative stress and copper homeostasis. *Antioxid. Redox Signal.* 18, 400–411. doi: 10.1089/ars.2012.4637

Arantes, C., Nomizo, R., Lopes, M. H., Hajj, G. N., Lima, F. R., and Martins, V. R. (2009). Prion protein and its ligand stress inducible protein 1 regulate astrocyte development. *Glia* 57, 1439–1449. doi: 10.1002/glia.20861

Bellenchi, G. C., Volpicelli, F., Piscopo, V., Perrone-Capano, C., and di Porzio, U. (2013). Adult neural stem cells: an endogenous tool to repair brain injury? *J. Neurochem.* 124, 159–167. doi: 10.1111/jnc.12084

Bribian, A., Fontana, X., Llorens, F., Gavin, R., Reina, M., Garcia-Verdugo, J. M., et al. (2012). Role of the cellular prion protein in oligodendrocyte precursor cell proliferation and differentiation in the developing and adult mouse CNS. *PLoS ONE* 7:e33872. doi: 10.1371/journal.pone.0033872

Bueler, H., Aguzzi, A., Sailer, A., Greiner, R. A., Autenried, P., Aguet, M., et al. (1993). Mice devoid of PrP are resistant to scrapie. *Cell* 73, 1339–1347. doi: 10.1016/0092-8674(93)90360-3

Chen, S., Mange, A., Dong, L., Lehmann, S., and Schachner, M. (2003). Prion protein as trans-interacting partner for neurons is involved in neurite outgrowth and neuronal survival. *Mol. Cell. Neurosci.* 22, 227–233. doi: 10.1016/S1044-7431(02)00014-3

Cheng, Y., Tao, L., Xu, J., Li, Q., Yu, J., Jin, Y., et al. (2013). CD44/Cellular prion protein interact in multidrug resistant breast cancer cells and correlate with responses to neoadjuvant chemotherapy in breast cancer patients. *Mol. Carcinog.* 53, 686–697. doi: 10.1002/mc.22021

Dery, M. A., Jodoin, J., Ursini-Siegel, J., Aleynikova, O., Ferrario, C., Hassan, S., et al. (2013). Endoplasmic reticulum stress induces PRNP prion protein gene expression in breast cancer. *Breast Cancer Res.* 15, R22. doi: 10.1186/bcr3398

Diarra-Mehrpour, M., Arrabal, S., Jalil, A., Pinson, X., Gaudin, C., Pietu, G., et al. (2004). Prion protein prevents human breast carcinoma cell line from tumor necrosis factor alpha-induced cell death. *Cancer Res.* 64, 719–727. doi: 10.1158/0008-5472.CAN-03-1735

Dodelet, V. C., and Cashman, N. R. (1998). Prion protein expression in human leukocyte differentiation. *Blood* 91, 1556–1561.

Du, L., Rao, G., Wang, H., Li, B., Tian, W., Cui, J., et al. (2013). CD44-positive cancer stem cells expressing cellular prion protein contribute to metastatic capacity in colorectal cancer. *Cancer Res.* 73, 2682–2694. doi: 10.1158/0008-5472.CAN-12-3759

Ermonval, M., Mouillet-Richard, S., Codogno, P., Kellermann, O., and Botti, J. (2003). Evolving views in prion glycosylation: functional and pathological implications. *Biochimie* 85, 33–45. doi: 10.1016/S0300-9084(03)00040-3

Gotz, M., and Huttner, W. B. (2005). The cell biology of neurogenesis. *Nat. Rev. Mol. Cell Biol.* 6, 777–788. doi: 10.1038/nrm1739

Herva, M. E., Relaño-Ginés, A., Villa, A., and Torres, J. M. (2010). Prion infection of differentiated neurospheres. *J. Neurosci. Methods* 188, 270–275. doi: 10.1016/j.jneumeth.2010.02.022

Hidaka, K., Shirai, M., Lee, J. K., Wakayama, T., Kodama, I., Schneider, M. D., et al. (2010). The cellular prion protein identifies bipotential cardiomyogenic progenitors. *Circ. Res.* 106, 111–119. doi: 10.1161/CIRCRESAHA.109.209478

Hirsch, T. Z., Hernandez-Rapp, J., Martin-Lanneree, S., Launay, J. M., and Mouillet-Richard, S. (2014). PrPC signalling in neurons: from basics to clinical challenges. *Biochimie* 104C, 2–11. doi: 10.1016/j.biochi.2014.06.009

Kaiser, D. M., Acharya, M., Leighton, P. L., Wang, H., Daude, N., Wohlgemuth, S., et al. (2012). Amyloid beta precursor protein and prion protein have a conserved interaction affecting cell adhesion and CNS development. *PLoS ONE* 7:e51305. doi: 10.1371/journal.pone.0051305

Kanaani, J., Prusiner, S. B., Diacovo, J., Baekkeskov, S., and Legname, G. (2005). Recombinant prion protein induces rapid polarization and development of synapses in embryonic rat hippocampal neurons *in vitro*. *J. Neurochem.* 95, 1373–1386. doi: 10.1111/j.1471-4159.2005.03469.x

Khalife, M., Young, R., Passet, B., Halliez, S., Vilotte, M., Jaffrezic, F., et al. (2011). Transcriptomic analysis brings new insight into the biological role of the prion protein during mouse embryogenesis. *PLoS ONE* 6:e23253. doi: 10.1371/journal.pone.0023253

Lee, Y. J., and Baskakov, I. V. (2010). Treatment with normal prion protein delays differentiation and helps to maintain high proliferation activity in human embryonic stem cells. *J. Neurochem.* 114, 362–373. doi: 10.1111/j.1471-4159.2010.06601.x

Lee, Y. J., and Baskakov, I. V. (2012). The cellular form of the prion protein is involved in controlling cell cycle dynamics, self-renewal, and the fate of human embryonic stem cell differentiation. *J. Neurochem.* 124, 310–322. doi: 10.1111/j.1471-4159.2012.07913.x

Liang, J., Bai, F., Luo, G., Wang, J., Liu, J., Ge, F., et al. (2007b). Hypoxia induced overexpression of PrP(C) in gastric cancer cell lines. *Cancer Biol. Ther.* 6, 769–774. doi: 10.4161/cbt.6.5.4001

Liang, J., Pan, Y., Zhang, D., Guo, C., Shi, Y., Wang, J., et al. (2007a). Cellular prion protein promotes proliferation and G1/S transition of human gastric cancer cells SGC7901 and AGS. *FASEB J.* 21, 2247–2256. doi: 10.1096/fj.06-7799com

Liao, M. J., Zhang, C. C., Zhou, B., Zimonjic, D. B., Mani, S. A., Kaba, M., et al. (2007). Enrichment of a population of mammary gland cells that form mammospheres and have *in vivo* repopulating activity. *Cancer Res.* 67, 8131–8138. doi: 10.1158/0008-5472.CAN-06-4493

Lima, F. R., Arantes, C. P., Muras, A. G., Nomizo, R., Brentani, R. R., and Martins, V. R. (2007). Cellular prion protein expression in astrocytes modulates neuronal survival and differentiation. *J. Neurochem.* 103, 2164–2176. doi: 10.1111/j.1471-4159.2007.04904.x

Linden, R., Martins, V. R., Prado, M. A., Cammarota, M., Izquierdo, I., and Brentani, R. R. (2008). Physiology of the prion protein. *Physiol. Rev.* 88, 673–728. doi: 10.1152/physrev.00007.2007

Liu, T., Li, R., Wong, B. S., Liu, D., Pan, T., Petersen, R. B., et al. (2001). Normal cellular prion protein is preferentially expressed on subpopulations of murine hemopoietic cells. *J. Immunol.* 166, 3733–3742. doi: 10.4049/jimmunol.166.6.3733

Loubet, D., Dakowski, C., Pietri, M., Pradines, E., Bernard, S., Callebert, J., et al. (2012). Neuritogenesis: the prion protein controls beta1 integrin signaling activity. *FASEB J.* 26, 678–690. doi: 10.1096/fj.11-185579

Mabbott, N. A., and MacPherson, G. G. (2006). Prions and their lethal journey to the brain. *Nat. Rev. Microbiol.* 4, 201–211. doi: 10.1038/nrmicro1346

Malaga-Trillo, E., Solis, G. P., Schrock, Y., Geiss, C., Luncz, L., Thomanetz, V., et al. (2009). Regulation of embryonic cell adhesion by the prion protein. *PLoS Biol.* 7:e55. doi: 10.1371/journal.pbio.1000055

Manson, J., West, J. D., Thomson, V., McBride, P., Kaufman, M. H., and Hope, J. (1992). The prion protein gene: a role in mouse embryogenesis? *Development* 115, 117–122.

Marthiens, V., Kazanis, I., Moss, L., Long, K., and Ffrench-Constant, C. (2010). Adhesion molecules in the stem cell niche–more than just staying in shape? *J. Cell Sci.* 123, 1613–1622. doi: 10.1242/jcs.054312

McDonald, A. J., Dibble, J. P., Evans, E. G., and Millhauser, G. L. (2014). A new paradigm for enzymatic control of alpha-cleavage and beta-cleavage of the prion protein. *J. Biol. Chem.* 289, 803–813. doi: 10.1074/jbc.M113.502351

Medema, J. P. (2013). Cancer stem cells: the challenges ahead. *Nat. Cell Biol.* 15, 338–344. doi: 10.1038/ncb2717

Mehrpour, M., and Codogno, P. (2010). Prion protein: from physiology to cancer biology. *Cancer Lett.* 290, 1–23. doi: 10.1016/j.canlet.2009.07.009

Meslin, F., Hamai, A., Gao, P., Jalil, A., Cahuzac, N., Chouaib, S., et al. (2007). Silencing of prion protein sensitizes breast adriamycin-resistant carcinoma cells to TRAIL-mediated cell death. *Cancer Res.* 67, 10910–10919. doi: 10.1158/0008-5472.CAN-07-0512

Miele, G., Alejo Blanco, A. R., Baybutt, H., Horvat, S., Manson, J., and Clinton, M. (2003). Embryonic activation and developmental expression of the murine prion protein gene. *Gene Expr.* 11, 1–12. doi: 10.3727/000000003783992324

Milhavet, O., Casanova, D., Chevallier, N., McKay, R. D., and Lehmann, S. (2006). Neural stem cell model for prion propagation. *Stem Cells* 10, 2284–2291. doi: 10.1634/stemcells.2006-0088

Miranda, A., Pericuesta, E., Ramirez, M. A., and Gutierrez-Adan, A. (2011). Prion protein expression regulates embryonic stem cell pluripotency and differentiation. *PLoS ONE* 6:e18422. doi: 10.1371/journal.pone.0018422

Mohanty, S. T., Cairney, C. J., Chantry, A. D., Madan, S., Fernandes, J. A., Howe, S. J., et al. (2012). A small molecule modulator of prion protein increases human mesenchymal stem cell lifespan, *ex vivo* expansion, and engraftment to bone marrow in NOD/SCID mice. *Stem Cells* 30, 1134–1143. doi: 10.1002/stem.1065

Moser, M., Colello, R. J., Pott, U., and Oesch, B. (1995). Developmental expression of the prion protein gene in glial cells. *Neuron* 14, 509–517. doi: 10.1016/0896-6273(95)90307-0

Mouillet-Richard, S., Laurendeau, I., Vidaud, M., Kellermann, O., and Laplanche, J. L. (1999). Prion protein and neuronal differentiation: quantitative analysis of prnp gene expression in a murine inducible neuroectodermal progenitor. *Microbes Infect.* 1, 969–976. doi: 10.1016/S1286-4579(99)80514-0

Nourizadeh-Lillabadi, R., Seilo Torgersen, J., Vestrheim, O., Konig, M., Alestrom, P., and Syed, M. (2010). Early embryonic gene expression profiling of zebrafish prion protein (Prp2) morphants. *PLoS ONE* 5:e13573. doi: 10.1371/journal.pone.0013573

Oesch, B., Westaway, D., Walchli, M., McKinley, M. P., Kent, S. B., Aebersold, R., et al. (1985). A cellular gene encodes scrapie PrP 27-30 protein. *Cell* 40, 735–746. doi: 10.1016/0092-8674(85)90333-2

Pan, Y., Zhao, L., Liang, J., Liu, J., Shi, Y., Liu, N., et al. (2006). Cellular prion protein promotes invasion and metastasis of gastric cancer. *FASEB J.* 20, 1886–1888. doi: 10.1096/fj.06-6138fje

Panigaj, M., Glier, H., Wildova, M., and Holada, K. (2011). Expression of prion protein in mouse erythroid progenitors and differentiating murine erythroleukemia cells. *PLoS ONE* 6:e24599. doi: 10.1371/journal.pone.0024599

Passet, B., Young, R., Makhzami, S., Vilotte, M., Jaffrezic, F., Halliez, S., et al. (2012). Prion protein and Shadoo are involved in overlapping embryonic pathways and trophoblastic development. *PLoS ONE* 7:e41959. doi: 10.1371/journal.pone.0041959

Patra, S. K. (2008). Dissecting lipid raft facilitated cell signaling pathways in cancer. *Biochim. Biophys. Acta* 1785, 182–206. doi: 10.1016/j.bbcan.2007.11.002

Peralta, O. A., Huckle, W. R., and Eyestone, W. H. (2011). Expression and knockdown of cellular prion protein (PrPC) in differentiating mouse embryonic stem cells. *Differentiation* 81, 68–77. doi: 10.1016/j.diff.2010.09.181

Pradines, E., Hernandez-Rapp, J., Villa-Diaz, A., Dakowski, C., Ardila-Osorio, H., Haik, S., et al. (2013). Pathogenic prions deviate PrP(C) signaling in neuronal cells and impair A-beta clearance. *Cell Death Dis.* 4, e456. doi: 10.1038/cddis.2012.195

Premzl, M., and Gamulin, V. (2007). Comparative genomic analysis of prion genes. *BMC Genomics* 8:1. doi: 10.1186/1471-2164-8-1

Prodromidou, K., Papastefanaki, F., Sklaviadis, T., and Matsas, R. (2014). Functional cross-talk between the cellular prion protein and the neural cell adhesion molecule is critical for neuronal differentiation of neural stem/precursor cells. *Stem Cells* 32, 1674–1687. doi: 10.1002/stem.1663

Relaño-Ginés, A., Gabelle, A., Hamela, C., Belondrade, M., Casanova, D., Mourton-Gilles, C., et al. (2013). Prion replication occurs in endogenous adult neural stem cells and alters their neuronal fate: involvement of endogenous neural stem cells in prion diseases. *PLoS Pathog.* 9:e1003485. doi: 10.1371/journal.ppat.1003485

Relaño-Ginés, A., Lehmann, S., and Crozet, C. (2014). Prion diseases and adult neurogenesis: how do prions counteract the brain's endogenous repair machinery? *Prion* 8, 1–7. doi: 10.4161/pri.29021

Rezza, A., Sennett, R., and Rendl, M. (2014). Adult stem cell niches: cellular and molecular components. *Curr. Top. Dev. Biol.* 107, 333–372. doi: 10.1016/B978-0-12-416022-4.00012-3

Santos, T. G., Beraldo, F. H., Hajj, G. N., Lopes, M. H., Roffe, M., Lupinacci, F. C., et al. (2013). Laminin-gamma1 chain and stress inducible protein 1 synergistically mediate PrPC-dependent axonal growth via Ca2+ mobilization in dorsal root ganglia neurons. *J. Neurochem.* 124, 210–223. doi: 10.1111/jnc.12091

Santos, T. G., Silva, I. R., Costa-Silva, B., Lepique, A. P., Martins, V. R., and Lopes, M. H. (2011). Enhanced neural progenitor/stem cells self-renewal via the interaction of stress-inducible protein 1 with the prion protein. *Stem Cells* 29, 1126–1136. doi: 10.1002/stem.664

Santuccione, A., Sytnyk, V., Leshchyns'ka, I., and Schachner, M. (2005). Prion protein recruits its neuronal receptor NCAM to lipid rafts to activate p59fyn and to enhance neurite outgrowth. *J. Cell Biol.* 169, 341–354. doi: 10.1083/jcb.200409127

Sauer, H., Dagdanova, A., Hescheler, J., and Wartenberg, M. (1999). Redox-regulation of intrinsic prion expression in multicellular prostate tumor spheroids. *Free Radic. Biol. Med.* 27, 1276–1283. doi: 10.1016/S0891-5849(99)00164-1

Singh, A., and Settleman, J. (2010). EMT, cancer stem cells and drug resistance: an emerging axis of evil in the war on cancer. *Oncogene* 29, 4741–4751. doi: 10.1038/onc.2010.215

Starke, R., Harrison, P., Mackie, I., Wang, G., Erusalimsky, J. D., Gale, R., et al. (2005). The expression of prion protein (PrP(C)) in the megakaryocyte lineage. *J. Thromb. Haemost.* 3, 1266–1273. doi: 10.1111/j.1538-7836.2005.01343.x

Steele, A. D., Emsley, J. G., Ozdinler, P. H., Lindquist, S., and Macklis, J. D. (2006). Prion protein (PrPc) positively regulates neural precursor proliferation during developmental and adult mammalian neurogenesis. *Proc. Natl. Acad. Sci. U.S.A.* 103, 3416–3421. doi: 10.1073/pnas.0511290103

Steele, A. D., Lindquist, S., and Aguzzi, A. (2007). The prion protein knockout mouse: a phenotype under challenge. *Prion* 1, 83–93. doi: 10.4161/pri.1.2.4346

Stella, R., Massimino, M. L., Sandri, M., Sorgato, M. C., and Bertoli, A. (2010). Cellular prion protein promotes regeneration of adult muscle tissue. *Mol. Cell. Biol.* 30, 4864–4876. doi: 10.1128/MCB.01040-09

Tremblay, P., Bouzamondo-Bernstein, E., Heinrich, C., Prusiner, S. B., and DeArmond, S. J. (2007). Developmental expression of PrP in the post-implantation embryo. *Brain Res.* 1139, 60–67. doi: 10.1016/j.brainres.2006.12.055

Westergard, L., Christensen, H. M., and Harris, D. A. (2007). The cellular prion protein (PrP(C)): its physiological function and role in disease. *Biochim. Biophys. Acta* 1772, 629–644. doi: 10.1016/j.bbadis.2007.02.011

Zhang, C. C., Steele, A. D., Lindquist, S., and Lodish, H. F. (2006). Prion protein is expressed on long-term repopulating hematopoietic stem cells and is important for their self-renewal. *Proc. Natl. Acad. Sci. U.S.A.* 103, 2184–2189. doi: 10.1073/pnas.0510577103

Conflict of Interest Statement: The authors declare that the research was conducted in the absence of any commercial or financial relationships that could be construed as a potential conflict of interest.

8

Role of the prion protein family in the gonads

Aurélie Allais-Bonnet and Eric Pailhoux*

Institut National de la Recherche Agronomique, UMR 1198, Biologie du Développement et Reproduction, Jouy-en-Josas, France

Edited by:
Jean-Luc Vilotte, National Institute of Agronomical Research, France

Reviewed by:
Rafael Linden, Federal University of Rio de Janeiro, Brazil
Reiner A. Veitia, Paris Diderot, France

***Correspondence:**
Eric Pailhoux, Institut National de la Recherche Agronomique, UMR 1198, Biologie du Développement et Reproduction, Bâtiment 440, F-78350 Jouy-en-Josas, France
e-mail: eric.pailhoux@jouy.inra.fr

The prion-gene family comprises four members named *PRNP* (PRP^{C}), *PRND* (Doppel), *PRNT* (PRT), and *SPRN* (Shadoo). According to species, *PRND* is located 16–52 kb downstream from the *PRNP* locus, whereas *SPRN* is located on another chromosome. The fourth prion-family gene, *PRNT*, belongs to the same genomic cluster as *PRNP* and *PRND* in humans and bovidae. *PRNT* and *PRND* possibly resulted from a duplication event of *PRND* and *PRNP*, respectively, that occurred early during eutherian species divergence. Although most of the studies concerning the prion-family has been done on PRP^{C} and its involvement in transmissible neurodegenerative disorders, different works report some potential roles of these proteins in the reproductive function of both sexes. Among them, a clear role of *PRND*, that encodes for the Doppel protein, in male fertility has been demonstrated through gene targeting studies in mice. In other species, Doppel seems to play a role in testis and ovary development but its cellular localization is variable according to the gonadal developmental stage and to the mammalian species considered. For the other three genes, their roles in reproductive function appear ill-defined and/or controversial. The present review aimed to synthesize all the available data on these prion-family members and their relations with reproductive processes, mainly in the gonad of both sexes.

Keywords: prion, Doppel, Shadoo, PRT, reproduction, gonads

INTRODUCTION

In Eutherian mammals, the reproductive system is composed by gonads (testes and ovaries) and the genital tract (male: penis, prostate, seminal vesicle, vas deferens, epididymis; female: vulva, vagina, uterus, and oviduct). Gonads produce sex hormones and gametes (sperms and oocytes), whereas the genital tract provides a suitable environment for the maturation and transport of gametes, the fertilization and implantation of the eggs. The differentiation of reproductive organs follows a specific and variable chronology according to species. In every case, the sex determination occurs immediately at fertilization with the addition of male and female gamete genomes. This step determines the genetic sex of the embryo and induces latter on the differentiation of gonads (arising from mesonephros, a transient embryonic kidney) toward a testicular (XY) or an ovarian (XX) differentiating pathway (DeFalco and Capel, 2009; for review; **Figure 1**). The undifferentiated gonad is composed by a germinal and two somatic cell-lineages. Each somatic line presents a double potentiality and will be turned toward a specific gonadal fate depending of the genes involved in sex determination, with *SRY* (Sex-determining Region of Y) being at the top of them (Kashimada and Koopman, 2010; for review). In somatic cell populations, we distinguish: (1) the supporting cells, which will differentiate into Sertoli cells in males and into follicular (also called granulosa) cells in females (these cells are responsible for the growth and the maturation of the germ line); (2) the steroidogenic cells, which will differentiate in Leydig cells in male and theca cells in female (**Figure 1**). The phenotypic sex, which depends on the gonadal sex and its hormonal production, is set up in many successive steps during development from early gonad differentiation until adulthood. Testes produce androgens and AMH (Anti-Müllerian Hormone) which are responsible for the differentiation of the genital tract toward the male pathway. Without these hormones the genital tract will differentiate into the female pathway.

Most of the major genes involved in gonad differentiation have been discovered through human genetic studies of DSD (Disorders of Sex Differentiation) cases, but many other genes were found to be expressed in gonads following high throughput mRNA sequencing or other expressional studies. Most of the genes of this last category could putatively be involved in gonadal processes but their role remains to be defined. Among these genes are those of the prion-family. Notably, *PRND* has been shown to be essential for testicular function in several species (Behrens et al., 2002; Paisley et al., 2004; Kocer et al., 2007). The most studied gene in the prion family is *PRNP* which encodes a cell surface glycoprotein, the prion protein (PRP^{C}). An infectious isoform of PRP^{C} (PRP^{SC}) has been shown to be the major component of "Prion," the etiological agent of transmissible spongiform encephalopathies (TSEs). These fatal neurodegenerative disorders include Creutzfelt-Jacob disease (CJD) in humans, bovine spongiform encephalopathy and scrapie in bovidae (Prusiner, 1998).

THE PRION GENE FAMILY COMPRISES FOUR MEMBERS

The "prion gene complex" encompasses four members named *PRNP*, *PRND* (downstream prion protein-like gene), *PRNT* (prion protein testis-specific gene), and *SPRN* (shadow of the

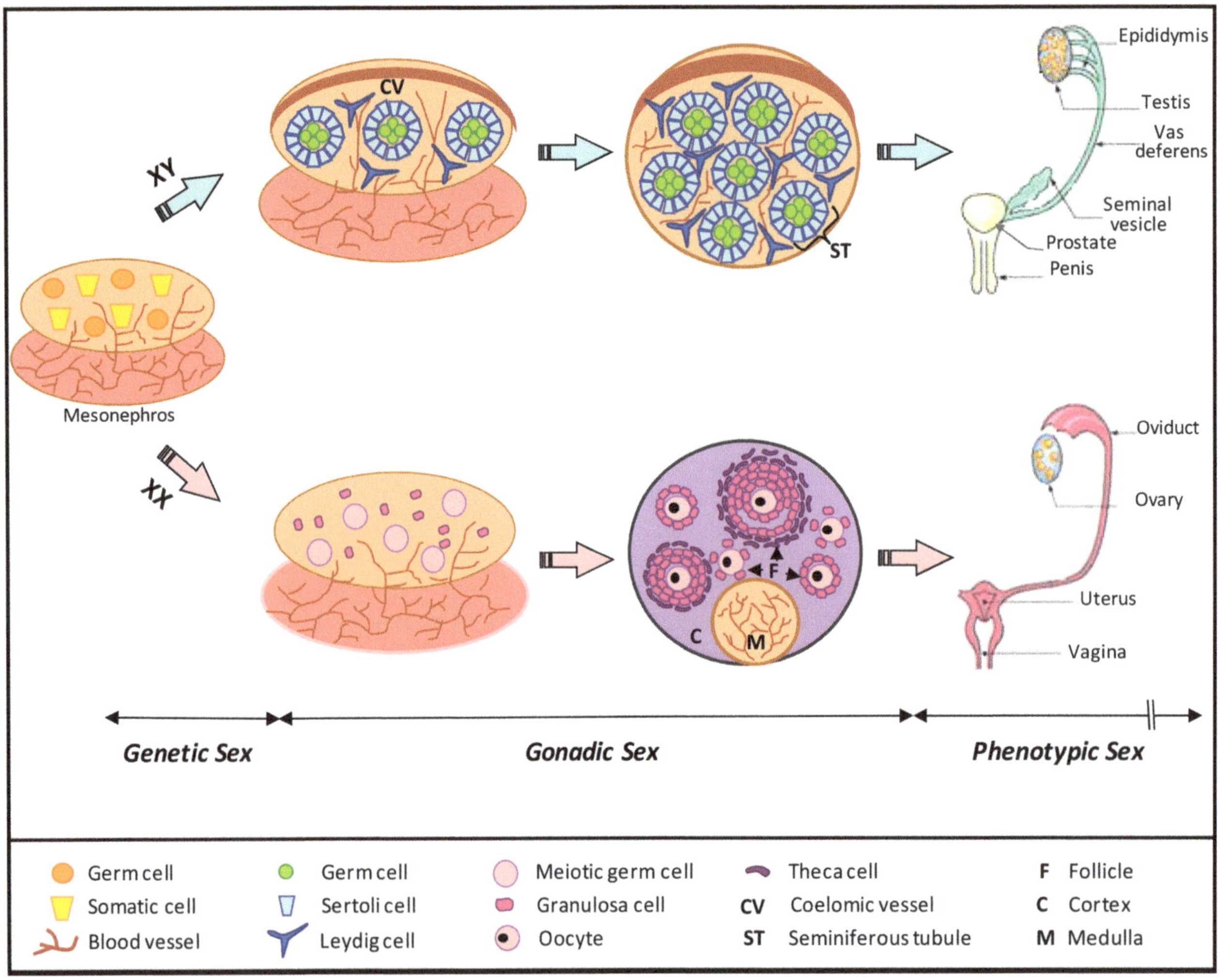

FIGURE 1 | Schematic representation of mammalian gonad differentiation. The genetic sex directs the development of the bipotential gonad toward a male or female gonadal fate. Three cellular types are present in differentiating gonad: somatic supporting cells, somatic steroid-producing cells, and germinal cells. Male and female somatic supporting cells have a common origin and differentiate respectively in Sertoli and granulosa cells. Steroidogenic cells will differentiate in Leydig cells in testis and theca cells in ovary. Testicular-specific features include the formation of the coelomic arterial vessel and of the seminiferous tubules formed by germ and Sertoli cells. Ovarian-specific features are in a chronological order: (i) entry of germ cells into meiosis, (ii) establishment of cortical and medullar compartments, and (iii) formation of follicles, which contain oocytes surrounded by granulosa cells. In agreement with the genetic sex, the development of the phenotypic sex is achieved with the differentiation of the genital tract: epididymis, vas deferens, seminal vesicle, prostate, and penis in males; oviduct, uterus and vagina in female (Adapted from DeFalco and Capel, 2009; for review).

prion protein gene). In mouse, sheep, cattle, and rat, the *PRNP* gene is composed of three exons, whereas only two are present in humans (Yoshimoto et al., 1992; Saeki et al., 1996; Lee et al., 1998; **Figure 2**). Depending on the studied species, *PRND* is located 16–52 kb downstream of *PRNP* and *PRNT* 3 and 6 kb downstream of human and cattle *PRND,* respectively (Moore et al., 1999; Comincini et al., 2001; Essalmani et al., 2002; Makrinou et al., 2002; Kocer et al., 2007). *PRND* and *PRNT* share with *PRNP* the same genomic cluster and possibly result from a duplication, that occurred early during eutherian species divergence, of *PRNP* and *PRND,* respectively. As *PRNP*, *PRND* structure can vary from two to three exons between species, whereas *PRNT* has two exons in humans (Comincini et al., 2001; Makrinou et al., 2002; **Figure 2**). The same organization of *PRNT* was predicted in cow, sheep, horse, dog and primates whereas this gene seems to be absent in rodents (Premzl et al., 2004; Harrison et al., 2010; **Figure 2**). *SPRN* is not part of the *PRNP* genomic locus, and is located on another chromosome. *SPRN* comprises two exons and its structure is conserved in fishes and mammals (Premzl et al., 2003; **Figure 2**). Some of the prion-family genes produce different transcripts of variable compositions and numbers according to the species.

The mammalian *PRNP* encodes the PRP protein (PRP^C) that contains several distinct domains, including an N-terminal signal peptide, an octapeptide repeat domain, a highly conserved hydrophobic segment and a C-terminal hydrophobic region which contains a glycosylphosphatidylinisitol (GPI) anchor (**Figure 3A**). This glycoprotein possesses two N-linked

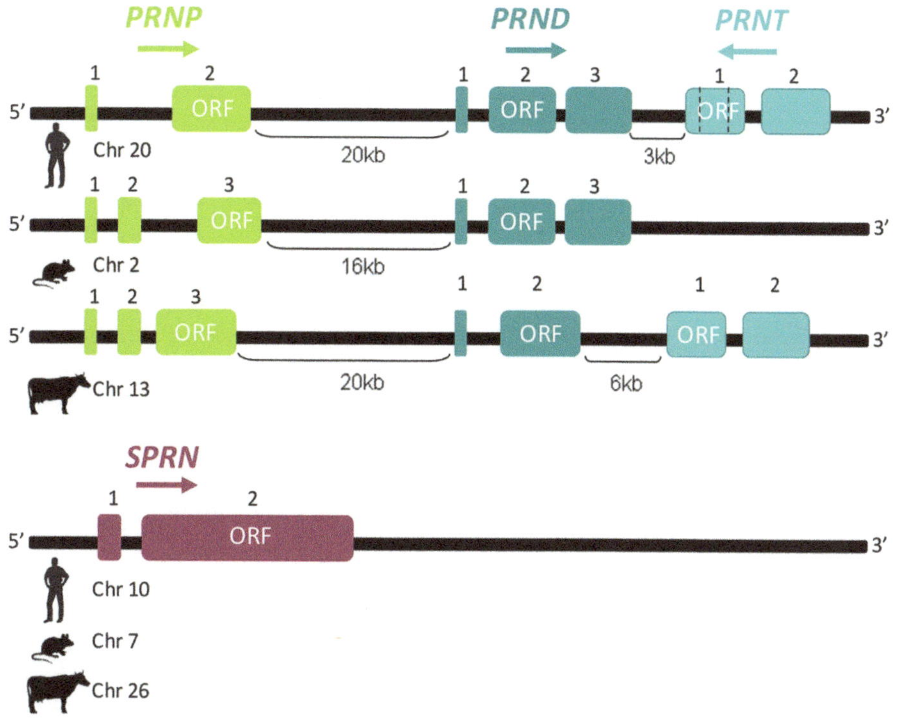

FIGURE 2 | Schematic structural representation of prion genes family members. *PRNP, PRND,* and *PRNT* are clustered on the same genomic locus. The distance between *PRNP, PRND,* and *PRNT* are given in kilobases (kb); according to species, the number of exons could be variable. *SPRN* consist of two exons and is located on another chromosome in all studied species. Arrows indicate the relative orientation of the genes. Chromosomes carrying these genes are annotated near each species symbols. For each gene, the open reading frame (ORF) is indicated on relevant exons.

glycosylation sites and exists in bi-, mono-, and un-glycosylated forms (**Figure 3A**). Its secondary structure is defined by the presence of three α-helices and two β-strands (Harris, 1999; for review; **Figures 3A,B**). The mature Doppel protein (DPL) encoded by *PRND* is a protein which resembles a N-terminally truncated PRP^{C} protein lacking the octamer repeats (**Figure 3**). In contrast, *SPRN* encodes the Shadoo protein (SHO) which shares with PRP^{C} a similar N-terminal region with a basic repeat region and a hydrophobic domain (Watts and Westaway, 2007; for review; **Figure 3**). Few data are available concerning the structure of PRT, the *PRNT*-encoded protein (Makrinou et al., 2002). No signal peptide was predicted for the 94 aa PRT protein, suggesting it could be an intracellular protein (Premzl and Gamulin, 2007). In bovine, *PRNT* encodes for an N-terminally truncated protein of 55 aa in length, sharing 55% identity with its human counterpart (Kocer et al., 2007).

Although PRP^{C} and SHO are mainly expressed in the central nervous system (CNS), these two proteins were also detected in male and female gonads (Bendheim et al., 1992; Tanji et al., 1995; Young et al., 2011). DPL and PRT are described as testis-specific proteins (Moore et al., 1999; Tranulis et al., 2001; Makrinou et al., 2002; Kocer et al., 2007). Nevertheless, a transient expression of DPL has been observed in brain of neonatal mice, but this protein is absent in the CNS of adult healthy animals (Li et al., 2000).

Many studies have been conducted on the role of DPL in testis differentiation but the implication of the other prion proteins in reproduction still remains subjective.

TESTIS-SPECIFIC PRION PROTEINS

In order to comment on the role of a testis-specific protein, it is necessary to introduce the spermatogenesis process allowing the production of male germ cells, spermatozoa (spz). Three different generations of germ cells are visible at any given time among the epithelial cells of the seminiferous tubule: spermatogonia (sg), spermatocytes (spc), and spermatids (std; **Figure 4**). During spermatogenesis, developing germ cells undergo several mitotic divisions and two meiotic divisions after translocation through the blood-testis barrier, to the luminal side of the epithelium, that is defined by tight junctions between non-dividing Sertoli cells. The final stage of spermatogenesis, known as spermiogenesis, consists of the complex differentiation of round spermatids into spermatozoa (cell elongation, nucleus condensation, acrosome formation, cytoplasm reduction; **Figure 4**). Throughout spermatogenesis each cluster of germ cells, derived from a single spermatogonium, is interconnected by cytoplasmic bridges that are important for synchronizing the developing process. Thus, the germ cells can be considered as being isolated cells only after they are released as spermatozoa into the lumen of the seminiferous tubule. Upon release, spermatozoa leave behind excess cytoplasm, in the form of interconnected syncytial chains called cytoplasmic lobes (Sprando and Russell, 1987; Weber and Russell, 1987), which are subsequently engulfed and degraded as residual bodies by the Sertoli cells (Espenes et al., 2006). Then spermatozoa are stored in epididymis and undergo maturation processes necessary to acquire motility and capacity to fertilize. Final maturation is

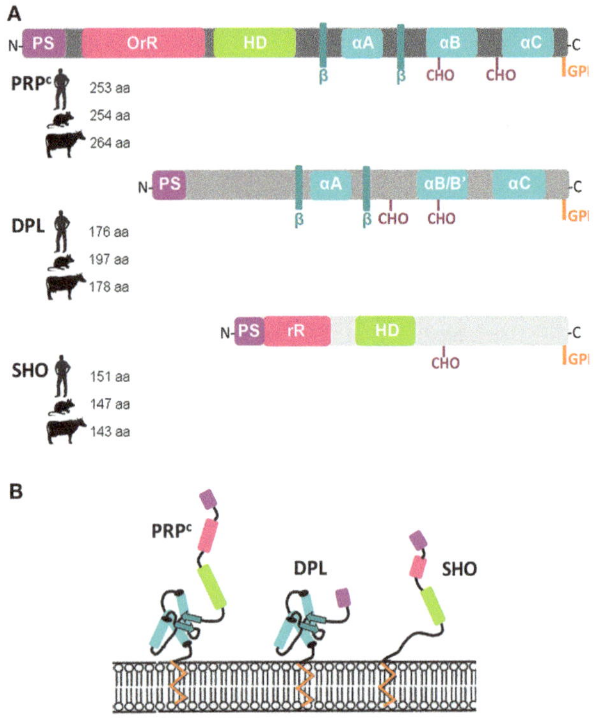

FIGURE 3 | Predicted structural features of Prion (PRPc), Doppel (DPL), and Shadoo (SHO) proteins. (A): Schematic description of these glycophosphatidylinositol (GPI)-anchored glycoproteins structure includes signal peptide, octarepeat or basic repeat region (OrR or rR), hydrophobic domain (HD), alpha (α)-helices, beta (β)-strands, N-glycosylation sites (CHO). The size of the full-length proteins in amino acids (aa) is annotated near each species symbols. **(B)**: Schematic representation of prion family proteins anchored in a membrane. (Adapted from Schmitt-Ulms et al., 2009; Daude and Westaway, 2011; for review).

completed in the female reproductive tract where spermatozoa acquire the capacities to fertilize oocytes. This last step called capacitation ended with acrosome reaction which consists on the fusion of the acrosome and oocyte membranes, allowing fertilization.

Two members of the prion-gene family could be considered as testis-specific proteins, *PRND* and *PRNT*. *PRND*/DPL expression has been studied mainly during spermatogenesis in many species. The first common observation is that DPL is expressed in Sertoli cells at various concentrations according to the species (Westaway, 2004; Rondena et al., 2005; Serres et al., 2006; Kocer et al., 2007; **Figure 5**). Nevertheless, in germ cells, the localization of DPL is less comparable between animals. For example in bovidae, DPL is present early in primordial germ cells in goat fetal gonads (Kocer et al., 2007) and in bovine, this protein is expressed in all stages of male germ cell development during spermatogenesis (spermatogones to ejaculated spermatozoa, Rondena et al., 2005; **Figure 5**). In contrast, DPL is only detected in spermatids at final stages of spermiogenesis in ovine, human and mice, (Behrens et al., 2002; Espenes et al., 2006; Serres et al., 2006; **Figure 5**). Interestingly in ram, DPL is observed in spermatids during the elongation process, in Sertoli cells after sperm release and is completely absent in spermatozoa (Espenes et al., 2006). Authors suggest that DPL is present in cytoplasmic lobes of maturing spermatids that culminate in DPL concentration in residual bodies which are subsequently released by spermatids then engulfed by Sertoli cells. DPL seems not to be detected in spermatozoa. However, sperm supplementation with recombinant ovine DPL protein during *in vitro* capacitation, significantly improves spermatozoa motility, vigor, viability and fertilization rate (Pimenta et al., 2012a). At this step we can proposed that DPL expression in ovine ejaculated sperm may be under the threshold of the detection limit of the method and the antibodies used. Otherwise, DPL could be produced by another cell type in the genital tract thus influencing the behavior of spermatozoa. This hypothesis was also suggested in human by Peoc'h and collaborators which completed previous studies by the localization of DPL on the flagella of epididymal and mature spermatozoa, and in seminal plasma (Peoc'h et al., 2002). As DPL seems to be transiently expressed in spermatids but is not detected in differentiated testicular spermatozoa, these authors considered that DPL could be acquired during the maturation of spermatozoa through the epididymis, as it has been described for other GPI-anchored proteins on spermatozoa (Peoc'h et al., 2002). Serres and collaborators reinforced this hypothesis by the observation of a DPL-staining in epithelial cells of the boar epididymis, suggesting a possible epididymal origin of DPL and a potential role during spermatozoa maturation (Serres et al., 2006). The transient presence of DPL in the final stages of spermiogenesis points an important role of this protein in the final remodeling of spermatids prior to their release into the testicular seminiferous lumen. The role of DPL in this spermiogenesis process was completely demonstrated by the analysis of mouse *Prnd* knock-out lines. Indeed, ablation of *Prnd* (*Prnd*$^{-/-}$) in two different mouse lines lead to infertile males with different sperm phenotypes.

The *Prnd*$^{-/-}$ mouse line with a 129/Ola genetic background produced low numbers of spermatozoa with poor motility and abnormal nuclei and acrosomes, greatly affecting the fertilization process. Indeed, sperm from DPL-deficient mice appears to be unable to undergo the normal acrosome reaction that is necessary to penetrate the zona pellucida of the ovum and *Prnd*$^{-/-}$ males are completely sterile (Behrens et al., 2002).

The second *Prnd*$^{-/-}$ mouse line produced on a mixed C57BL6/CBA genetic background produced a normal number of motile spermatozoa but these spermatozoa had an altered chromatin structure and DNA damages that induce an early arrest of embryo development (Paisley et al., 2004). A common phenotype between both *Prnd*$^{-/-}$ mice is a loss of sperm head integrity.

In conclusion, the localization of DPL on both somatic (Sertoli cells) and germinal cells strongly suggests that this protein plays a major role in male fertility. In most species, its expression in testicular germ cells was detected at late stage of spemiogenesis, principally in spermatids with a transient presence in acrosome. These data show that DPL is implicated in normal acrosome genesis and thus in spermatozoa fertilizing ability. Supplemental roles could be attributing to DPL by its presence at earliest stages of testis development in bovine and goats, suggesting an involvement in germinal cell ontogeny. Moreover, DPL has also been detected in goat fetal Leydig cells (steroidogenic cells) where its role remains to be defined (Kocer et al., 2007). Finally some roles

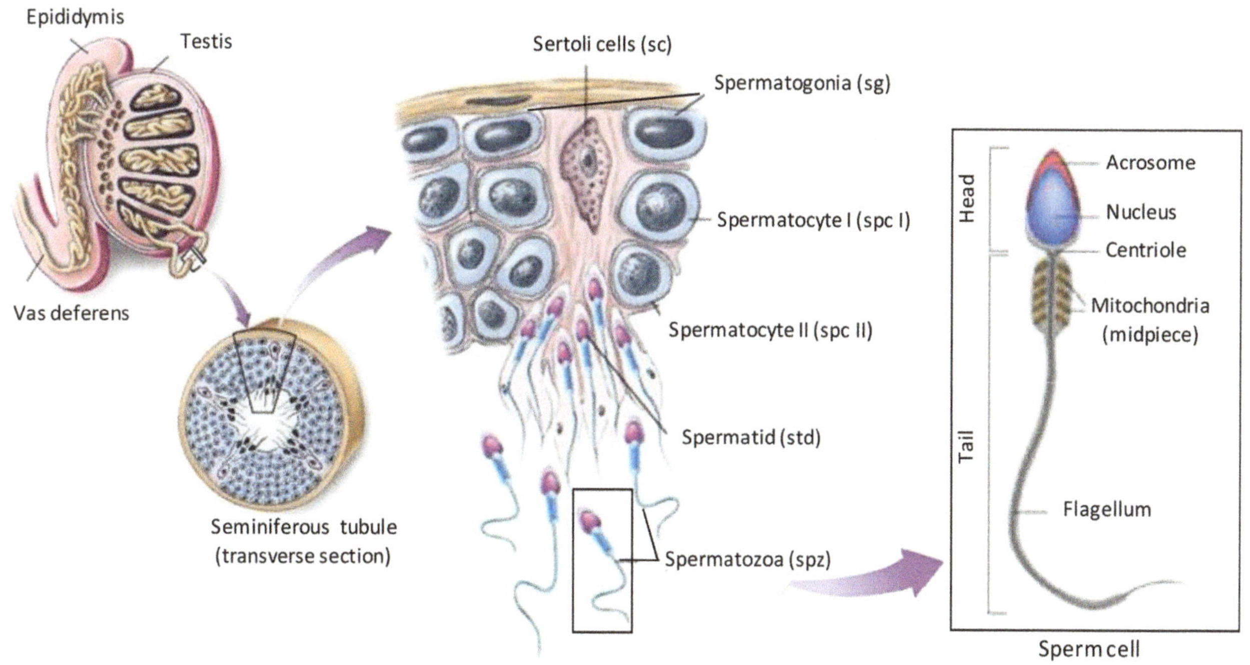

FIGURE 4 | Illustration of spermatogenesis. Spermatogenesis occurs within the seminiferous tubules of the testes of a post pubescent male. Diploid primordial germ cells (also called spermatogonia; sg) near the basal lamina of the seminiferous tubules undergo an initial mitotic division to produce diploid primary spermatocytes (spc I). Nearly half the primary spermatocytes produced remain near the basal lamina to continue to divide mitotically, thus allowing spermatogenesis to be continuous during male's reproductive lifespan. Other primary spermatocytes migrate toward the lumen of the seminiferous tubules and begin to undergo meiosis I, resulting in haploid secondary spermatocytes (spc II). These secondary spermatocytes further divide through meiosis II, producing haploid spermatids (std). Mature sperm cells (spermatozoa; spz) capable of fertilizing an egg develop from spermatids through the final stage called spermiogenesis. In this stage specific regions of the spermatid differentiate into the head, mid-piece, and the tail of the sperm cell. Within the head, an acrosomal space is developed which houses specific enzymes required for fertilization. Specialized acrosomal membranes that are pertinent for fertilization also differentiate in the head of the sperm. A flagellum develops as a means of motility near the posterior aspect which is fuelled by the abundance of mitochondria in the mid-piece of the sperm cell. (Adapted from buffonescience9.wikispaces website).

of DPL in ovarian differentiation couldn't be discarded since DPL has been detected in female germ cells of goat fetuses and in granulosa cells and follicular fluid in bovine (Rondena et al., 2005; Kocer et al., 2007). This last observation reinforces the idea that DPL may contribute to regulate fertility, since follicular fluid has been shown to influence sperm mobility and fertility (Rodriguez et al., 2001).

In order to pinpoint putative other reproductive roles of DPL, it will be of great interest to engineer *PRND* mutant animals in mammalian species other than mice. For the sex-differentiating process, it is know that the mouse species remains less sensible to gene dosage and haplo-insufficiency than humans. Furthermore, *PRND* expression profiles suggest additional roles for DPL in goat testis differentiation compare to mice (Kocer et al., 2007). Accordingly, *PRND* gene ablation in goats is currently under investigation in our laboratory that has recently succeeded in such technologies (Boulanger et al., 2014).

Three isoforms of human *PRNT* have been described and are exclusively expressed in the adult testis, thus absent in fetal tissues including testis (Makrinou et al., 2002). In goats, *PRNT* is weakly and stochastically expressed in both testes and ovaries at various developmental stages, suggesting that the expression pattern of this gene differs between ruminant and human or, most probably, that ruminant *PRNT* is a pseudogene (Kocer et al., 2007). By contrast, recent results demonstrate that PRT is found in the ram germinal cells. Notably, PRT expression is localized in the nuclei of spermatogonia, spermatocytes, spermatids and in the sperm acrosome. These observations suggest that ovine *PRNT* could be a translated protein-coding gene, pointing to a role for PRT in the ram spermatogenesis, throughout spermatogenic cell proliferation and sperm maturation (Pimenta et al., 2012b). However, it is difficult to conclude on a real role of PRT during spermatogenesis. The lack of PRT detection in others species, supports the hypothesis that PRT could be a pseudogene.

PRP^C AND ITS SHADOW

In the CNS, PRP^C, and SHO present a partially reciprocal pattern of expression, suggesting a common function of these two proteins in neuronal cells. This overlapping expression leads to Shadoo protein designation (Shadoo is the Japanese word for shadow) and this protein was considered as the putative host-encoded protein that compensates for the lack of PRP^C. Although PRP^C and SHO are present in gonads, their single and/or common roles are not established in reproductive biology.

PRP^C has been more studied and was detected on spermatozoa of different species including human, cattle and mouse

DPL						
TESTIS						
fetuses germ cells	✓					
Spermatogones		✓				
Spermatocytes		✓				
Spermatids		✓	✓	✓	✓	
Spermatozoa (spz)		✓				
Sertoli cells	✓	✓	✓	✓	✓	✓
Leydig cells	✓					
EPIDIDYMIS						
Epididymal epithelial cells			?	?		✓
Epididymal spz				✓		
EJACULATED SPZ		✓		✓	✓	
SEMINAL PLASMA				✓		
OVARY						
fetuses germ cells	✓					
Oocytes						
Granulosa cells		✓				
Folicular fluid		✓				

FIGURE 5 | Table summarizing the cellular localization of Dopple (DPL) in gonads of various species.

(Shaked et al., 1999) but the nature of PRP^c isoforms on spermatozoa is debated and appears different according to antibodies. A first study showed the presence of PRP^c on epididymal sperm extracts from epididymis of mouse and bovine and from ejaculated spermatozoa from bovine and human (**Figure 6**). The molecular weight of PRP^c in epididymal sperm cells was similar to that of the brain (control tissue) whereas in mature sperm cells, the PRP^c isoform detected is smaller and C-terminally-truncated. Authors suggested that the C-terminal portion of the PRP^c is removed during the process of sperm maturation in epididymis and that the protein is inserted via its N-terminal part in the membrane of ejaculated sperm (Shaked et al., 1999). In contrast, Peoc'h group demonstrated that PRP^c was recognized on sperm membranes by antibodies binding to the C-terminus part of the protein, suggesting that only N-terminally truncated fragments of PRP^c are present in these cells in the human species. In addition in human testes, 3 isoforms of PRP^c, an unglycosylated full-length and two N-terminally truncated proteins, are also detected in spermatocytes and spermatids (Peoc'h et al., 2002; **Figure 6**). Differences in the nature of PRP^c truncated (C- or N-term) could be due to differences in the protocol for preparing spermatozoa extracts that could affect proteolysis differentially, or to the antibodies used. Studies in ram supports the work of Shaked and collaborators by demonstrating that one major glycosylated C-terminally truncated PRP^c isoform is associated with sperm from testis, cauda epididymis and semen and also in sperm cytoplasmic droplets that are released during maturation (Ecroyd et al., 2004; **Figure 6**). Other PRP^c isoforms were compartmentalized within cauda epididymal fluid and semina plasma (Ecroyd et al., 2004). Indeed, Gatti and colleagues demonstrated the synthesis and secretion of soluble PRP^c by the epithelial cells lining the ram epididymis (**Figure 6**). These cells produce large quantities of a specific isoform of PRP^c that seems to be processed post-secretion in different ways during epididymal transit. In ovine spermatozoa, different forms of PRP^c have been found, as reported previously for others species (Shaked et al., 1999), that seem to be inserted into the sperm membrane mainly during ejaculation (Gatti et al., 2002). In summary, both glycosylated and proteolytic isoforms of PRP^c are present in the male reproductive tract. However, the main isoforms differ between the sperm and the reproductive fluid, suggesting only a low extend of exchange between these two compartments. By studying *Prnp* deficient mice, a protective role of PRP^c against copper toxicity has been proposed since sperm cells originating from *Prnp*$^{-/-}$ mice were significantly more susceptible to high copper concentrations than sperm from wild-type mice (Shaked et al., 1999). The presence of an anti-oxidant defense in the sperm-surrounding media is highly important, especially during passage and storage in the epididymis, since these germinal cells lack the molecular machinery to regenerate damaged lipids and proteins. Nevertheless, *Prnp* null mice are fully fertile and the ablation of *Prnp* in *Prnd*$^{-/-}$ mice has no additional effect on the phenotype described for

PRP^c				
TESTIS				
Spermatogonia				
Spermatocytes	✓			
Spermatids	✓			
Spermatozoa (spz)				✓
Sertoli cells				
Leydig cells				
EPIDIDYMIS				
Epididymal epithelial cells				✓
Epididymal spz		✓	✓	✓
EJACULATED SPZ	✓	✓		✓
SEMINAL PLASMA	✓			✓
OVARY				✓
Oocytes				
Granulosa cells		✓		
Theca cells		✓		
Folicular fluid				
OVIDUCT				✓
UTERUS				✓

FIGURE 6 | Table summarizing the cellular localization of Prion protein (PRP^c) in gonads of various species.

Prnd$^{-/-}$ males, suggesting that PRP^c is not involved in fertilizing capacity of mice spermatozoa, at least under normal breeding conditions (Büeler et al., 1992; Manson et al., 1994; Paisley et al., 2004).

PRP^c is also present in the female reproductive tract. This protein has been detected in the ovary, oviduct and uterus of pregnant and cyclic ewes (Moudjou et al., 2001; Tuo et al., 2001; **Figure 6**). In bovine, *PRNP* is expressed in both theca and granulosa cells of ovarian follicles notably in developing follicles suggesting that it could promote the growth of dominant follicles (Forde et al., 2008; **Figure 6**). Again, invalidation of this gene did not induce any noticeable fertility defect in the studied females. However, these animals were kept under control breeding conditions and not challenged, through induction of oxidative stresses for example. Such challenges revealed yet undiscovered function of PrP in other organs, such as placenta (Alfaidy et al., 2013), and such experiments would be of interest to further assess the function of PrP in the gonads.

Although no biological role of SHO has been defined in reproduction, few data are available in mice. Generation of transgenic reporter mice for the gene encoding SHO protein (*Sprn*) has permitted to show a *Sprn*-LacZ expression in the male and female gonads. In both cases, staining was cell-specific, in the interstitial Leydig cells in testis and in granulosa cells of ovarian follicle (Young et al., 2011). Leydig cells are the site of testosterone biosynthesis that is required for the development of the male reproductive system, and the initiation and maintenance of spermatogenesis. Deregulation of some genes expressed in Leydig cells such as proliferin-related protein (PRP) result in decreased testosterone production and has an impact on development of male reproductive system and fertility (Zhao et al., 2011). If we consider the testicular SHO localization, could this protein have the same function in male fertility? Another way to determine the function of a protein is to study the ablation of its coding gene. The lack of *Sprn* in mice has no effect on fertility, as judged by the measurement of the litter size (Daude et al., 2012). Given that DPL is implicated in spermiogenesis and PRP^c is present on spermatozoa, one can imagine that these close proteins can compensate each other. *Sprn* invalidation in *Prnd*$^{-/-}$ mice does not increase the testicular phenotype associated with single *Prnd* knockout (our unpublished data). In the same way, mice deficient for both SHO and PRP^c were also found to be viable and fertile (Daude et al., 2012). By contrast, the *Sprn* knockdown in *Prnp*$^{-/-}$ mice presents an embryonic lethal phenotype at developmental stage E7.5 (Passet et al., 2012). Discordant results between *Sprn*^{KO}/*Prnp*^{KO} and *Sprn*^{KD}/*Prnp*^{KO} could be due to different mice genetic backgrounds used in these studies or to a genetic adaptation of the double knockout animals. Nevertheless, gene-targeting experiments in mice do not allow yet defining a putative role of SHO in reproduction.

CONCLUSION

Among the prion-gene family members, *PRND* is yet the sole to have been clearly linked to reproductive biology with a crucial function during the late steps of spermatogenesis. Indeed its gene ablation in mice leads to male infertility. Although expression of the other three members *PRNP*, *PRNT*, and *SPRN* has been demonstrated at mRNA and protein levels in gonads and/or reproductive tracts, their specific reproductive functions, if any, remain to be elucidated. The fact that these putative functions haven't been pointed on by human genetics or gene-targeting in other mammalian species may suggest that these proteins by alone have no critical reproductive functions or that their functions aren't hugely affected by mutational events. Indeed, even if the physiological role of PRP^c remains unclear in the central nervous system, according to the fact that $Prnp^{-/-}$ mice or goats seem to be unaffected and well-being, it is a mis-folding of PRP^c (PRP^{sc}) that remains highly detrimental for the central nervous system physiology (a sort of gain-of-function mutation). This means that we cannot exclude some detrimental reproductive effects of these proteins without any mutational changes, but with post-translational and/or conformational changes remaining very difficult to pin-point.

ACKNOWLEDGMENTS

The authors want to thank Sophie Mouillet-Richard and Jean-Luc Vilotte for their invitation to submit this manuscript. We also thank Jean-Luc Vilotte for his constructive criticisms. The authors are supported by the French ANR-09-BLAN-0015-01 and ANR-09-GENM-009.

REFERENCES

Alfaidy, N., Chauvet, S., Donadio-Andrei, S., Salomon, A., Saoudi, Y., Richaud, P., et al. (2013). Prion protein expression and functional importance in developmental angiogenesis: role in oxidative stress and copper homeostasis. *Antioxid. Redox Signal.* 18, 400–411. doi: 10.1089/ars.2012.4637

Behrens, A., Genoud, N., Naumann, H., Rülicke, T., Janett, F., Heppner, F. L., et al. (2002). Absence of the prion protein homologue Doppel causes male sterility. *EMBO J.* 21, 3652–3658. doi: 10.1093/emboj/cdf386

Bendheim, P. E., Brown, H. R., Rudelli, R. D., Scala, L. J., Goller, N. L., Wen, G. Y., et al. (1992). Nearly ubiquitous tissue distribution of the scrapie agent precursor protein. *Neurology* 42, 149–156. doi: 10.1212/WNL.42.1.149

Boulanger, L., Pannetier, M., Gall, L., Allais-Bonnet, A., El Zaiat, M., Le Bourhis, D., et al. (2014). FOXL2 is a female sex-determining gene in the goat. *Curr. Biol.* 24, 404–408. doi: 10.1016/j.cub.2013.12.039

Büeler, H., Fischer, M., Lang, Y., Bluethmann, H., Lipp, H. P., DeArmond, S. J., et al. (1992). Normal development and behaviour of mice lacking the neuronal cell-surface PrP protein. *Nature* 356, 577–582. doi: 10.1038/356577a0

Comincini, S., Foti, M. G., Tranulis, M. A., Hills, D., Di Guardo, G., Vaccari, G., et al. (2001). Genomic organization, comparative analysis, and genetic polymorphisms of the bovine and ovine prion Doppel genes (*PRND*). *Mamm. Genome* 12, 729–733. doi: 10.1007/s00335-001-2064-4

Daude, N., and Westaway, D. (2011). Biological properties of the PRP-like Shadoo protein. *Front. Biosci.* 16, 1505–1516. doi: 10.2741/3801

Daude, N., Wohlgemuth, S., Brown, R., Pitstick, R., Gapeshina, H., Yang, J., et al. (2012). Knockout of the prion protein (PrP)-like *Sprn* gene does not produce embryonic lethality in combinaison with PrP^c-deficiency. *Proc. Natl. Acad. Sci. U.S.A.* 109, 9035–9040. doi: 10.1073/pnas.1202130109

DeFalco, T., and Capel, B. (2009). Gonad morphogenesis in vertebrates: divergent means to a convergent end. *Annu. Rev. Cell Dev. Biol.* 25, 457–482. doi: 10.1146/annurev.cellbio.042308.13350

Ecroyd, H., Sarradin, P., Dacheux, J. L., and Gatti, J. L. (2004). Compartmentalization of prion isoforms within the reproductive tract of the ram. *Bio. Reprod.* 71, 993–1001. doi: 10.1095/biolreprod.104.029801

Espenes, A., Harbitz, I., Skogtvedt, S., Fuglestveit, R., Berg, K. A., Dick, G., et al. (2006). Dynamic expression of the prion-like protein Doppel in ovine testicular tissue. *Int. J. Androl.* 29, 400–408. doi: 10.1111/j.1365-2605.2005.00618.x

Essalmani, R., Taourit, S., Besnard, N., and Vilotte, J. L. (2002). Sequence determination and expression of the ovine doppel-encoding gene in transgenic mice. *Gene* 285, 287–290. doi: 10.1016/S0378-1119(02)00391-8

Forde, N., Rogers, M., Canty, M. J., Lonergan, P., Smith, G. W., Coussens, P. M., et al. (2008). Association of the prion protein and its expression with ovarian follicle development in cattle. *Mol. Reprod. Dev.* 75, 243–249. doi: 10.1002/mrd.20807

Gatti, J. L., Métayer, S., Moudjou, M., Andréoletti, O., Lantier, F., Dacheux, J. L., et al. (2002). Prion protein is secreted in soluble forms in the epididymal fluid and proteolytically processed and transported in seminal plasma. *Biol. Reprod.* 67, 393–400. doi: 10.1095/biolreprod67.2.393

Harris, D. A. (1999). Cell biological studies of the prion protein. *Curr. Issues Mol. Biol.* 1, 65–75.

Harrison, P. M., Khachane, A., and Kumar, M. (2010). Genomic assessment of the evolution of the prion protein gene family in vertebrates. *Genomics* 95, 268–277. doi: 10.1016/j.ygeno.2010.02.008

Kashimada, K., and Koopman, P. (2010). Sry: the master switch in mammalian sex determination. *Development* 137:3921–3930 doi: 10.1242/dev.048983

Kocer, A., Gallozzi, M., Renault, L., Tilly, G., Pinheiro, L., Le Provost, F., et al. (2007). Goat *PRND* expression pattern suggests its involvement in early sex differentiation. *Dev. Dyn.* 236, 836–842. doi: 10.1002/dvdy.21066

Lee, I. Y., Westaway, D., Smit, A. F., Wang, K., Seto, J., Chen, L., et al. (1998). Complete genomic sequence and analysis of the prion protein gene region from three mammalian species. *Genome Res.* 8, 1022–1037.

Li, A., Sakaguchi, S., Shigematsu, K., Atarashi, R., Roy, B.C., Nakaoke, R., et al. (2000). Physiological expression of the gene for PrP-like protein, PrPLP/Dpl, by brain endothelial cells and its ectopic expression in neurons of PrP-deficient mice ataxic due to Purkinje cell degeneration. *Am. J. Pathol.* 157, 1447–1452. doi: 10.1016/S0002-9440(10)64782-7

Makrinou, E., Collinge, J., and Antoniou, M. (2002). Genomic characterization of the human protein (PrP) gene locus. *Mamm. Genome* 13, 696–703. doi: 10.1007/s00335-002-3043-0

Manson, J. C., Clarke, A. R., Hopper, M. L., Aitchison, L., McConnell, I., and Hope, J. (1994). 129/Ola mice carrying a null mutation in PrP that abolishes mRNA production are developmentally normal. *Mol. Neurobiol.* 8, 121–127. doi: 10.1007/BF02780662

Moore, R. C., Lee, I. Y., Silverman, G. L., Harrison, P. M., Strome, R., Heinrich, C., et al. (1999). Ataxia in prion protein (PrP)-deficient mice is associated with upregulation of the novel PrP-like protein doppel. *J. Mol. Biol.* 292, 797–817. doi: 10.1006/jmbi.1999.3108

Moudjou, M., Frobert, Y., Grassi, J., and La Bonnardière, C. (2001). Cellular prion protein status in sheep: tissue-specific biochemical signatures. *J. Gen. Virol.* 82, 2017–2024. doi: 10.1099/vir.0.17776-0

Paisley, D., Banks, S., Selfridge, J., McLennan, N. F., Ritchie, A. M., McEwan, C., et al. (2004). Male infertility and DNA damage in Doppel knockout and prion protein/Doppel double-knoukout mice. *Am. J. Pathol.* 164, 2279–2288. doi: 10.1016/S0002-9440(10)63784-4

Passet, B., Young, R., Makhzami, S., Vilotte, M., Jaffrezic, F., Halliez, S., et al. (2012). Prion protein and Shadoo are involved in overlapping embryonic pathways and trophoblastic development. *PLoS ONE* 7:e41959. doi: 10.1371/journal.pone.0041959.

Peoc'h, K., Serres, C., Frobert, Y., Martin, C., Lehmann, S., Chasseigneaux, S., et al. (2002). The human "prion-like" protein Doppel is expressed in both Sertoli cells ans spermatozoa. *J. Biol. Chem.* 277, 43071–43078. doi: 10.1074/jbc.M206357200

Pimenta, J., Dias, F. M., Marques, C. C., Baptista, M. C., Vasques, M. I., Horta, A. E., et al. (2012a). The prion-like protein Doppel enhances ovine spermatozoa fertilizing ability. *Reprod. Domest. Anim.* 47, 196–202. doi: 10.1111/j.1439-0531.2011.01827.x

Pimenta, J., Domingos, A., Santos, P., Marques, C. C., Cantante, C., Santos, A., et al. (2012b). Is prnt a pseudogene? Identification of ram Prt in testis and ejaculated spermatozoa. *PLoS ONE* 7:e42957. doi: 10.1371/journal.pone.0042957

Premzl, M., and Gamulin, V. (2007). Comparative genomic analysis of prion genes. *BMC Genomics* 8:1. doi: 10.1186/1471-2164-8-1

Premzl, M., Gready, J. E., Jermiin, L. S., Simonic, T., and Marshall Graves, J. A. (2004). Evolution of vertebrate genes related to prion and shadoo proteins—clues from comparative genomic analysis. *Mol. Bio. Evol.* 21, 2210–2231. doi: 10.1093/molbev/msh245

Premzl, M., Sangiorgio, L., Strumbo, B., Marshall Graves, J. A., Simonic, T., and Gready, J. E. (2003). Shadoo, a new protein highly conserved from fish to mammals and with similarity to prion protein. *Gene* 314, 89–102. doi: 10.1016/S0378-1119(03)00707-8

Prusiner, S. B. (1998). Prions. *Proc. Natl. Acad. Sci. U.S.A.* 95, 13363–13383. doi: 10.1073/pnas.95.23.13363

Rodriguez, H., Torres, C., Valdes, X., Guerra, H., Pastor, L. M., Maccallini, G., et al. (2001). The acrosomic reaction in stallion spermatozoa: inductive effect of the mare preovulatory follicular fluid. *Biocell.* 25, 115–120.

Rondena, M., Ceciliani, F., Comazzi, S., Pocacqua, V., Bazzocchi, C., Luvoni, C., et al. (2005). Identification of bovine doppel protein in testis, ovary and ejaculated spermatozoa. *Theriogenology* 63, 1195–1206. doi: 10.1016/j.theriogenology.2004.06.009

Saeki, K., Matsumoto, Y., Hirota, Y., Matsumoto, Y., and Onodera, T. (1996). Three-exon structure of the gene encoding the rat prion protein and its expression in tissues. *Virus Genes* 12, 15–20. doi: 10.1007/BF00369996

Schmitt-Ulms, G., Ehsani, S., Watts, J. C., Westaway, D., and Wille, H. (2009). Evolutionary descent of prion genes from the ZIP family of metal ion transporters. *PLoS ONE* 4:e7208. doi: 10.1371/journal.pone.0007208.

Serres, C., Peoc'h, K., Courtot, A. M., Lesaffre, C., Jouannet, P., and Laplanche, J. L. (2006). Spatio-developmental distribution of the prion-like protein doppel in Mammalian testis: a comparative analysis focusing on its presence in the acrosome of spermatids. *Biol. Reprod.* 74, 816–823. doi: 10.1095/biolreprod.105.047829

Shaked, Y., Rosenmann, H., Talmor, G., and Gabizon, R. (1999). A C-terminal-truncated PrP isoform is present in mature sperm. *J. Biol. Chem.* 274, 32153–32158. doi: 10.1074/jbc.274.45.32153

Sprando, R. L., and Russell, L. D. (1987). Comparative study of cytoplasmic elimination in spermatids of selected mammalian species. *Am. J. Anat.* 178, 72–80. doi: 10.1002/aja.1001780109

Tanji, K., Saeki, K., Matsumoto, Y., Takeda, M., Hirasawa, K., Doi, K., et al. (1995). Analysis of PRPc mRNA by *in situ* hybridization in brain, placenta, uterus and testis of rats. *Intervirology* 38, 309–315.

Tranulis, M. A., Espenes, A., Comincini, S., Skretting, G., and Harbitz, I. (2001). The PrP-like protein Doppel gene in sheep and cattle: cDNA sequence and expression. *Mamm. Genome* 12, 376–379. doi: 10.1007/s0033500 10285

Tuo, W., Zhuang, D., Knowles, D. P., Cheevers, W. P., Sy, M. S., and O'Rourke, K. I. (2001). Prp-c and Prp-sc at the fetal maternal interface. *J. Biol. Chem.* 276, 18229–18234. doi: 10.1074/jbc.M008887200

Watts, J. C., and Westaway, D. (2007). The prion protein family: diversity, rivalry, and dysfunction. *Biochim. Biophys. Acta* 1772, 654–672. doi: 10.1016/j.bbadis.2007.05.001

Weber, J. E., and Russell, L. D. (1987). A study of intercellular bridges during spermatogenesis in the rat. *Am. J. Anat.* 180, 1–24. doi: 10.1002/aja.1001800102

Westaway, D. (2004). "Inherited prion diseases," in *Prion Biology and Diseases*, ed S. B. Prusiner (New York, NY: Cold Spring Harbor Laboratory Press), 673–775.

Yoshimoto, J., Iinuma, T., Ishiguro, N., Horiuchi, M., Imamura, M., and Shinagawa M. (1992). Comparative sequence analysis and expression of bovine PrP gene in mouse L-929 cells. *Virus Genes* 6, 343–356. doi: 10.1007/BF01703083

Young, R., Le Guillou, S., Tilly, G., Passet, B., Castille, J., Beringue, V., et al. (2011). Generation of *Sprn*-regulated reporter mice reveals gonadic spatial expression of the prion-like protein Shadoo in mice. *Biochem. Biophys. Res. Commun.* 412, 752–756. doi: 10.1016/j.bbrc.2011.08.049

Zhao, L., Hao, J., Hu, J., Wang, Q., Lü, Z., Wang, L., et al. (2011). Expression of proliferin-related protein in testis and the biological significance in testosterone production. *Mol. Cell. Endocrinol.* 343, 25–31 doi: 10.1016/j.mce.2011.05.046

Conflict of Interest Statement: The authors declare that the research was conducted in the absence of any commercial or financial relationships that could be construed as a potential conflict of interest.

9

Regulation of PrP^C signaling and processing by dimerization

Xavier Roucou *

Department of Biochemistry, Faculty of Medicine, Université de Sherbrooke, Sherbrooke, QC, Canada

Edited by:
Sophie Mouillet-Richard, Institut National de la Santé et de la Recherche Médicale, France

Reviewed by:
Markus Glatzel, University Medical Center Hamburg-Eppendorf, Germany
Joel Watts, University of Toronto, Canada

***Correspondence:**
Xavier Roucou, Département de Biochimie (Z8-2001), Faculté de Médecine, Université de Sherbrooke, 3201 Jean Mignault, Sherbrooke, QC J1E4K8, Canada
e-mail: xavier.roucou@usherbrooke.ca

The cellular prion protein (PrP^C) is a glycosylphosphatidylinositol (GPI)-anchored protein present at the cell surface. PrP^C N-terminal moiety is intrinsically disordered and is able to interact with a variety of ligands. Physiological ligands have neurotrophic activity, whilst others, including protein toxic oligomers, have neurotoxic functions. These two opposite activities involve different interacting partners and result from different PrP^C-activated signaling pathways. Remarkably, PrP^C may be inactivated either by physiological endoproteolysis and release of the N-terminal domain, or by ectodomain shedding. Ligand-induced PrP^C dimerization or enforced dimerization of PrP^C indicate that PrP^C dimerization represents an important molecular switch for both intracellular signaling and inactivation by the release of PrP^C N-terminal domain or shedding. In this review, we summarize evidence that cell surface receptor activity of PrP^C is finely regulated by dimerization.

Keywords: prion protein trafficking, dimerization, signaling, neuroprotection, neurodegeneration

INTRODUCTION

PrP^C is a cell surface protein with a bipartite structure: the N-terminal domain is disordered and the C-terminal domain is structured and contains three α-helices and two short β-strands (Wuthrich and Riek, 2001). The physiological function of PrP^C is unclear, but a large body of evidence indicates that PrP^C is a neuroprotective and neurotrophic protein (Linden et al., 2008). The neuroprotective function of PrP^C against different insults was demonstrated *in vitro* in primary neurons and in cell lines, and *in vivo* (Roucou and LeBlanc, 2005; Lo et al., 2007). In these studies, PrP^C expression was able to slow or halt cell death whilst PrP^C absence did not prevent cell death. In a pioneer investigation, PrP^C expression prevented cell death triggered by serum deprivation of immortalized hippocampal neurons (Kuwahara et al., 1999). Subsequent studies provided significant evidence for the implication of PrP^C in cell survival. In addition to these neuroprotective effects, PrP^C regulates cell proliferation, differenciation, growth, and PrP^C is also important for the expansion of stem cells in culture (Martins et al., 2010; Miranda et al., 2013). Some of these trophic mechanisms have been addressed and involve the assembly of protein complexes at the cell surface.

Most of neuroprotective and neurotrophic activities result from PrP^C-mediated signaling (Martins et al., 2010; Schneider et al., 2011). Thus, a large body of data indicate that GPI-anchored PrP^C is a cell surface receptor or co-receptor and that its engagement with one of its numerous ligands or with antibodies activates different intracellular pathways. Cell surface receptors are generally activated by dimerization (Heldin, 1995), and this may also be valid for PrP^C which forms dimers in native conditions and can be experimentally engaged with cross-linking antibodies (Mouillet-Richard et al., 2000; Rambold et al., 2008).

In prion diseases, PrP^C changes conformation into a pathological conformer termed PrP^{Sc} (Prusiner, 1998). The exact mechanism of this conformational change or prion conversion is unclear but may involve the initial formation of dimers. During the process of prion conversion, PrP^{Sc} oligomerizes and form toxic oligomers that interact with PrP^C and switch its neuroprotective/neurotrophic signaling to a neurotoxic signaling (Rambold et al., 2008; Resenberger et al., 2011).

In this review, I will summarize some of the most important studies on the role of dimerization on the physiological and pathological function of PrP^C and PrP^{Sc}, respectively.

PrP^C DIMERIZATION AND CELL SIGNALING (FIGURE 1)

DETECTION OF PrP^C DIMERS IN NATIVE CONDITIONS AND CYTOPROTECTION

PrP^C dimers were detected in solution in a partially purified fraction from normal bovine brain thalamus (Meyer et al., 2000), and in murine neuroblastoma N2a cells expressing Syrian hamster PrP^C (Priola et al., 1995). Syrian hamster, human and bovine PrP^C expressed in baculovirus and purified under native conditions spontaneously form dimers (Hundt et al., 2003). Dimerization of human PrP^C was confirmed in BHK cells overexpressing PrP^C and in yeast two-hybrid assays (Hundt et al., 2003). More recently, endogenous PrP^C dimers were also detected by blue native PAGE in N2a cells and the dimerization domain mapped to a hydrophobic domain of the protein (amino acids 112-MAGAAAAGAVVGGLGGYMLGSA-133) (Rambold et al., 2008). Finally, PrP^C dimers were detected after chemical crosslinking in crude membranes from human neuroblastoma SH-SY5Y cells and mouse brains (Rambold et al., 2008). These results convincingly demonstrate that PrP^C has an intrinsic tendency to dimerize in native conditions and suggest that dimerization is important for the physiological function of PrP^C.

The assembly of natural PrP^C dimers at the plasma membrane is associated with protective activity against the excitotoxin kainate and altering dimer formation results in cell death (Rambold et al., 2008). Based on their own data and previous data showing PrP^C-mediated signaling using anti-PrP^C antibodies (see below), the authors proposed that cell surface PrP^C dimers induce protective signaling through an unknown transmembrane receptor. This study did not elucidate whether the formation of dimers is constitutive or depends on an unknown ligand. Also, the proportion of PrP^C dimers is unknown.

ANTIBODY-INDUCED PrP^C DIMERIZATION REVEALS THE SIGNALING PATHWAY CONTROLLED BY PrP^C

Antibody-induced dimerization (also termed antibody-induced ligation or -cross-linking) is used to mimic an extracellular signal on cell surface receptors and trigger signal transduction. Although it is unclear if such strategy mimics the interaction with a partner or dimerization of the receptor, antibody-induced dimerization is largely used to engage a receptor in the absence of its ligand and relays intracellular signals. GPI-anchored proteins associate with raft domains in the plasma membrane and activate signal transduction pathways upon engagement with ligands or via antibody-induced dimerization (Robinson, 1991; Suzuki et al., 2012). For GPI-anchored proteins, signal transduction occurs through activation of intracellular tyrosine kinases including the Src-family kinases (Stefanova et al., 1991; Chen et al., 2006). Mouillet-Richard et al were the first to show that engagement of PrP^C using an antibody-induced dimerization approach activates a Fyn-dependent signaling pathway in serotonergic and noradrenergic mouse cells differenciated from the murine neuroectodermal progenitor 1C11 clone (Mouillet-Richard et al., 2000). Similar results were obtained with two different antibodies, 1A8 and SAF61 targeting C-terminal epitopes. Using 4 different antibodies to induce PrP^C dimerization, SAF61, Bar221, and 1A8 that target C-terminal epitopes, and SAF32 which targets epitope 79-92, NADPH oxidase was subsequently identified as the main primary target of PrP^C-mediated signaling. NADPH oxidase-dependent reactive oxygen species production stimulated the phosphorylation of extracellular regulated kinases 1/2 (Erk1/2) in the 1C11 neuroectodermal precursor and its neuronal differentiated progenies, the hypothalamic GT1-7 cells, and the T lymphoid BW5147 cells (Schneider et al., 2003). PrP^C signaling was dependent on Fyn in neuronal cells only, indicating specificity in the control of PrP^C function. PrP^C-mediated phosphorylation of Erk1/2 was independently confirmed in GT1-7 neuronal cells (Monnet et al., 2004) and in human neuroblastoma SH-SY5Y cells (Rambold et al., 2008). PrP^C-induced ROS production and Erk1/2 phosphorylation was confirmed using an inducible dimerization strategy (Beland et al., 2012).

These studies lend support for a role of PrP^C in signal transduction and further investigations provided more insight into the physiological consequence of PrP^C signaling in neuronal 1C11 cells. In 1C11 serotonergic cells expressing 5-HT_{2B}, 5-$HT_{1B/D}$, and 5-HT_{2A} receptor subtypes. PrP^C dimerization interfered with the signaling activity of these three serotonergic receptors belonging to the GPCR family likely by modulating the recruitment of G-proteins (Mouillet-Richard et al., 2005). PrP^C dimerization promoted the recruitment of the cAMP responsive element binding protein (CREB) transcription factor and the transcription of several genes with important function in cellular protection and neuronal plasticity (Pradines et al., 2008). In addition, PrP^C dimerization inactivated the Glycogen Synthase Kinase 3β and activated serotonergic signaling through inhibition of the serotonin 1B receptor (Hernandez-Rapp et al., 2014). CREB recruitment and GSK3β are generally associated with cytoprotection, suggesting an important function of PrP^C in cell survival and homeostasis.

For several years, these data were in contradiction with previous results indicating that antibody-induced PrP^C dimerization is neurotoxic *in vivo* (Solforosi et al., 2004). However, these results were later invalidated with similar and other antibodies (Klohn et al., 2012). This debate is still ongoing since a recent study demonstrated that anti-PrP^C antibodies induce rapid neurotoxicity in mice and cerebellar organotypic cultured slices (Sonati et al., 2012). Importantly, PrP^C dimerization is unlikely to be involved in neuronal toxicity since single-chain antibodies were also toxic.

PrP^C SIGNALING ACTIVATED BY DIFFERENT LIGANDS

At the cell surface, PrP^C interacts directly or indirectly with a variety of ligands as diverse as metals, lipids, nucleic acids, glycosaminoglycans, and other proteins (Linden et al., 2008; Beland and Roucou, 2012). In physiological conditions, it was proposed that PrP^C is a scaffolding protein providing essential molecular interactions and signaling neurotrophic activities (Martins et al., 2010). PrP^C ligands promoting neurotrophic activity include laminin, the 37-kDa/67-kDa laminin receptor precursor/laminin receptor, vitronectin, the neural cell adhesion molecule, and the Stress Inducible Protein 1 (Martins et al., 2010).

In pathological conditions, binding of PrP^{Sc} to cell surface PrP^C corrupts PrP^C signaling and results in cellular toxicity (Rambold et al., 2008; Resenberger et al., 2011). This finding is particularly important as it provides a simple explanation for the observation that PrP^C on the cell surface is critical for the neurotoxicity of PrP^{Sc} in prion diseases (Brandner et al., 1996; Chesebro et al., 2005). PrP^C dimerization is essential for the toxicity of PrP^{Sc} (Rambold et al., 2008). PrP^C is also a receptor for other toxic β-sheet oligomers, including Aβ in Alzheimer's disease (Lauren et al., 2009; Gunther and Strittmatter, 2010; Resenberger et al., 2011).

PrP^C INTEGRITY AT THE CELL SURFACE IS REGULATED BY PROTEOLYSIS AND DIMERIZATION: PrP^C METABOLITES AND NEUROPROTECTION (FIGURE 1)

PrP^C IS A TARGET FOR SEVERAL POSTTRANSLATIONAL ENDOPROTEOLYTIC EVENTS

Following translocation into the endoplasmic reticulum, signal peptidase removes a N-terminal signal peptide, and a C-terminal peptide is removed prior to the attachment of the GPI anchor. Thus, human PrP^C is translated as an immature 253 amino acids protein and mature PrP^C is a 208 residues protein. After trafficking through the secretory pathway, a fraction of PrP^C may undergo three proteolytic cleavages (Altmeppen et al., 2012). An α-cleavage between residues 110–111 and 112 in a late compartment of the secretory pathway produces PrPC1, a 17 kDa GPI-anchored C-terminal polypeptide, and a 11 kDa N-terminal polypeptide released in the extracellular space. The identity of the protease responsible for α-cleavage, termed the α-PrPase (Oliveira-Martins et al., 2010), is still unclear. A β-cleavage at amino acids 89/90 generates PrPC2, a 20 kDa GPI-anchored polypeptide, and the corresponding 8 kDa PrPN2 fragment. β-cleavage occurs at the cell surface mainly in pathological conditions; calpains execute β-cleavage in prion diseases whilst reactive oxygen species perform β-cleavage under conditions of oxidative stress. In addition, a fraction of PrP^C is constitutively shed from the cell surface after proteolytic cleavage close to the GPI anchor. *In vivo*, the main protease responsible for PrP^C shedding is the zinc metalloproteinase ADAM10 (Altmeppen et al., 2012).

NEUROPROTECTIVE PRP^C-DERIVED PRPN1 AND PRPC1 METABOLITES

In recent years, α-cleavage attracted a lot of attention because it results in the production of PrPN1, a natural PrP^C metabolite with a clear neuroprotective activity against different insults *in vivo*, in primary neuronal cultures and in cell lines (Guillot-Sestier et al., 2009, 2012; Resenberger et al., 2011; Beland et al., 2012; Fluharty et al., 2013; Beland and Roucou, 2013a). In particular, the neuroprotection against soluble Aβ oligomers that may be the culprit species in Alzheimer's disease may pave the way for the discovery of a new class of therapeutic molecules (Beland et al., 2012; Fluharty et al., 2013). There is also some evidence that α-cleavage is increased in post-mortem brain tissues of Alzheimer's disease patients, and that PrPN1 traps Aβ into amorphous aggregates unable to transform into soluble and toxic Aβ oligomers, and that α-cleavage decrease promotes neurotoxicity in prion and Alzheimer's diseases (Pietri et al., 2013; Beland et al., 2014). PrPN1 also binds to and antagonizes the toxicity of other β-sheet rich oligomers, including PrP^{Sc} oligomers, and PrPN1-derived therapeutic molecules may help treat different neurodegenerative disorders (Resenberger et al., 2011).

The GPI-anchored PrPC1 fragment after α-cleavage of PrP^C protects against prion infection of neuronal and non-neuronal cell lines and acts as a dominant-negative inhibitor of prion conversion *in vivo* (Lewis et al., 2009; Westergard et al., 2011). The mechanism of action of PrPC1 is unclear, but since PrPC1 is resistant to prion conversion, the authors proposed that PrPC1 competes with PrP^C for binding to infectious PrP^{Sc} (Westergard et al., 2011). Thus, PrPC1-derived peptides may have therapeutic benefits in prion diseases.

PRP^C DIMERIZATION STIMULATES ITS TRAFFICKING TO THE PLASMA MEMBRANE AND THE PRODUCTION OF PRPN1 AND PRPC1

As many experimental data converge to support the proposition that PrPN1 and PrPC1 are neuroprotective metabolites, two therapeutic avenues could be proposed in prion diseases: to provide exogenous PrPN1- or PrPC1-derived molecules, or to increase the natural production of PrPN1 and PrPC1 by stimulating the α-cleavage. The α-cleavage mechanism is nebulous and only two elements are known: it occurs in the late secretory pathway but the enzyme is still unknown, and the hydrophobic domain is essential for this cleavage (Bremer et al., 2010; Oliveira-Martins et al., 2010). This domain is also essential for the physiological dimerization of PrP^C (Rambold et al., 2008), supporting the hypothesis of a possible connection between dimerization and α-cleavage. Using an inducible dimerization strategy with a permeable dimerizer, we were able to show that PrP^C dimerization in cell lines and primary neurons increase PrP^C trafficking to the plasma membrane and largely increase the production of PrPN1, PrPC1, and shed PrP^C (Beland et al., 2012). After dimerization, conditioned medium containing these three metabolites strongly protected cells against toxic Aβ oligomers.

Since levels of the products of both α-cleavage and shedding rose after dimerization, we concluded that the large increase of PrP^C trafficking to the plasma membrane was sufficient to explain the high levels of its metabolites. This effect was fast and occurred 4 h post-dimerization. Deletion of the hydrophobic domain, the natural dimerization domain, does not prevent PrP^C trafficking to the plasma membrane (Winklhofer et al., 2003). Thus, dimerization is not essential for PrP^C trafficking. We proposed a model with a constitutive and dimerization-independent pathway for PrP^C secretion, and a pathway regulated by dimerization. This regulated pathway would allow the cells to quickly respond to toxic insults by increasing the levels of protective PrP^C metabolites (Beland and Roucou, 2013a,b).

THE DARK SIDE OF PrP^C DIMERIZATION REVEALED FROM *IN VITRO* PRION CONVERSION ASSAYS

$PrP^C \rightarrow PrP^{Sc}$ conversion or prion conversion is central to prion diseases; this process is neurotoxic and PrP^{Sc} molecules assemble into infectious particles responsible for the transmission of the disease (Prusiner, 1998; Mallucci et al., 2003). Not surprisingly, numerous mechanistic studies have addressed this process using recombinant PrP (recPrP) and several experimental data have indicated an intrinsic tendency of the protein to form dimers during the initial steps of prion conversion. A fraction of Syrian hamster recPrP(90-231) forms alpha-helical dimers in solution in the presence of submicellar concentrations of SDS (Jansen et al., 2001). These dimers believed to be intermediates in prion conversion were observed by size exclusion chromatography, chemical crosslinking and analytical ultracentrifugation (Kaimann et al., 2008; Stöhr et al., 2008; Jansen et al., 2001). In these studies, non-denaturing concentrations of SDS were used to mimic membrane-like features. Using a reduction-oxidation protocol to induce the fibrillar assembly of Syrian hamster PrP(90–231), Lee and Eisenberg also observed the presence of dimeric intermediates in polyacrylamide native gels (Lee and Eisenberg, 2003). A similar conclusion was obtained with murine recPrP(23–231). During the conversion of murine PrP(23–231), an intermediate water-soluble β-sheet isoform termed PrP^{β} was identified (Luhrs et al., 2006). The kinetics of $PrP^C \rightarrow PrP^{\beta}$ conversion suggest that dimerization is the rate-limiting step for the transition. The dimerization of murine recPrP(23–231) as a key molecular step during the conversion was confirmed in a subsequent study (Hafner-Bratkovic et al., 2011). Additionally, 3D reconstruction of murine recPrP(91–230) amyloid fibrils led to the proposition that dimers represent building units of such fibrils (Tattum et al., 2006).

These experiments were performed with non-posttranslationally modified PrP^C. Yet *in vivo*, PrP^C carries two N-glycosylations and a GPI anchor and these complex posttranslational modifications may play an important role in prion conversion. To address this issue, posttranslationally modified PrP^C was purified from Chinese hamster ovary cells overexpressing Syrian hamster PrP^C. This native PrP^C spontaneously formed dimers stabilized by intermolecular β-sheets after insertion into artificial membranes (Elfrink et al., 2008).

Altogether, these studies support the hypothesis that dimerization is an important step for prion conversion but they did not directly test this hypothesis. To address this issue, two strategies have been used. First, two monomeric mouse PrP (23–231) were covalently linked with a linker and recombinantly purified. This tandem protein oligomerized after purification, and thioflavine T staining indicated that such oligomers were likely on the pathway of amyloid formation, although this was not demonstrated (Simoneau et al., 2007). In a second strategy, we used a conditional dimerization approach to induce chemical dimerization of mouse recPrP(23–231) and human recPrP (23-231). α-helical PrP dimers spontaneously converted into β-sheet oligomers and amyloid fibrils were detected by electron microscopy and thioflavinT staining (Roostaee et al., 2009). Importantly, these experiments were performed in physiological-like conditions in the absence of any detergent or chaotropic agents.

Models of PrP dimers and the role of dimers in prion conversion are available (Warwicker, 2000; Gauczynski et al., 2001; Tompa et al., 2002). However, in all the above studies, prion conversion was assessed by the formation of PrP amyloid fibrils and/or partial resistance to proteinase K rather than by the formation of infectious PrP^{Sc} in animal bioassays. Hence, the biological significance of PrP^C dimerization for the formation of infectious PrP^{Sc} remains speculative.

CONCLUSION

PrP^C forms dimers in native conditions and cell surface dimerization clearly regulates PrP^C-mediated signaling and the resulting physiological neuroprotective/neurotrophic activities (Rambold et al., 2008). Intracellular dimerization also drastically increases its trafficking to the plasma membrane and the production of its natural metabolites PrPN1 and PrPC1 (Beland et al., 2012). The combination of these two effects of PrP^C dimerization likely provides PrP^C with a powerful neuroprotective/neurotrophic function (**Figure 1**). However, the flip side of the coin is that unwanted dimerization may initiate prion conversion and result in neuronal toxicity (Tompa et al., 2002). Regulating PrP^C dimerization may help translate these findings into

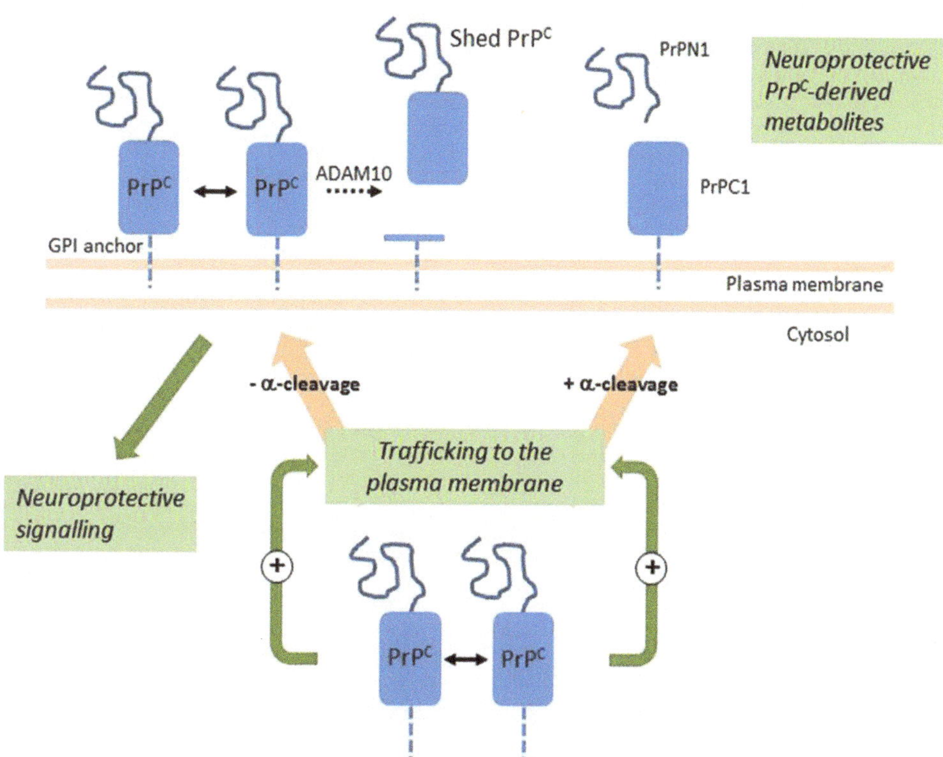

FIGURE 1 | PrP^C Dimerization activates a neuroprotective signaling pathway and the production of neuroprotective PrP^C-derived metabolites. En route to the plasma membrane, a fraction of PrP^C undergoes a physiological cleavage termed α-cleavage. Intracellular PrP^C dimerization stimulates trafficking to the plasma membrane and the release of PrPN1 and PrPC1 at the cell surface. PrP^C dimerization at the cell surface activates an intracellular signaling pathway with a neuroprotective outcome. At the cell surface, a fraction of PrP^C undergoes ADAM10-mediated shedding. Double arrows indicate dimerization; dotted arrow indicates shedding. See text for details.

novel therapeutic interventions in neurodegenerative diseases (Beland and Roucou, 2013b).

REFERENCES

Altmeppen, H. C., Puig, B., Dohler, F., Thurm, D. K., Falker, C., Krasemann, S., et al. (2012). Proteolytic processing of the prion protein in health and disease. *Am. J. Neurodegener. Dis.* 1, 15–31.

Beland, M., Bedard, M., Tremblay, G., Lavigne, P., and Roucou, X. (2014). Abeta induces its own prion protein N-terminal fragment (PrPN1)-mediated neutralization in amorphous aggregates. *Neurobiol. Aging* 35, 1537–1548. doi: 10.1016/j.neurobiolaging.2014.02.001

Beland, M., Motard, J., Barbarin, A., and Roucou, X. (2012). PrP(C) homodimerization stimulates the production of PrPC cleaved fragments PrPN1 and PrPC1. *J. Neurosci.* 32, 13255–13263. doi: 10.1523/JNEUROSCI.2236-12.2012

Beland, M., and Roucou, X. (2012). The prion protein unstructured N-terminal region is a broad-spectrum molecular sensor with diverse and contrasting potential functions. *J. Neurochem.* 120, 853–868. doi: 10.1111/j.1471-4159.2011.07613.x

Beland, M., and Roucou, X. (2013a). Homodimerization as a molecular switch between low and high efficiency PrP C cell surface delivery and neuroprotective activity. *Prion* 7, 170–174. doi: 10.4161/pri.23583

Beland, M., and Roucou, X. (2013b). Taking advantage of physiological proteolytic processing of the prion protein for a therapeutic perspective in prion and Alzheimer diseases. *Prion* 8, 106–110. doi: 10.4161/pri.27438

Brandner, S., Isenmann, S., Raeber, A., Fischer, M., Sailer, A., Kobayashi, Y., et al. (1996). Normal host prion protein necessary for scrapie-induced neurotoxicity. *Nature* 379, 339–343. doi: 10.1038/379339a0

Bremer, J., Baumann, F., Tiberi, C., Wessig, C., Fischer, H., Schwarz, P., et al. (2010). Axonal prion protein is required for peripheral myelin maintenance. *Nat. Neurosci.* 13, 310–318. doi: 10.1038/nn.2483

Chen, Y., Thelin, W. R., Yang, B., Milgram, S. L., and Jacobson, K. (2006). Transient anchorage of cross-linked glycosyl-phosphatidylinositol-anchored proteins depends on cholesterol, Src family kinases, caveolin, and phosphoinositides. *J. Cell Biol.* 175, 169–178. doi: 10.1083/jcb.200512116

Chesebro, B., Trifilo, M., Race, R., Meade-White, K., Teng, C., LaCasse, R., et al. (2005). Anchorless prion protein results in infectious amyloid disease without clinical scrapie. *Science* 308, 1435–1439. doi: 10.1126/science.1110837

Elfrink, K., Ollesch, J., Stohr, J., Willbold, D., Riesner, D., and Gerwert, K. (2008). Structural changes of membrane-anchored native PrP(C). *Proc. Natl. Acad. Sci. U.S.A.* 105, 10815–10819. doi: 10.1073/pnas.0804721105

Fluharty, B. R., Biasini, E., Stravalaci, M., Sclip, A., Diomede, L., Balducci, C., et al. (2013). An N-terminal fragment of the prion protein binds to amyloid-beta oligomers and inhibits their neurotoxicity *in vivo*. *J. Biol. Chem.* 288, 7857–7866. doi: 10.1074/jbc.M112.423954

Gauczynski, S., Hundt, C., Leucht, C., and Weiss, S. (2001). Interaction of prion proteins with cell surface receptors, molecular chaperones, and other molecules. *Adv. Protein Chem.* 57, 229–272. doi: 10.1016/S0065-3233(01)57024-2

Guillot-Sestier, M. V., Sunyach, C., Druon, C., Scarzello, S., and Checler, F. (2009). The alpha-secretase-derived N-terminal product of cellular prion, N1, displays neuroprotective function *in vitro* and *in vivo*. *J. Biol. Chem.* 284, 35973–35986. doi: 10.1074/jbc.M109.051086

Guillot-Sestier, M. V., Sunyach, C., Ferreira, S. T., Marzolo, M. P., Bauer, C., Thevenet, A., et al. (2012). alpha-Secretase-derived fragment of cellular prion, N1, protects against monomeric and oligomeric amyloid beta (Abeta)-associated cell death. *J. Biol. Chem.* 287, 5021–5032. doi: 10.1074/jbc.M111.323626

Gunther, E. C., and Strittmatter, S. M. (2010). Beta-amyloid oligomers and cellular prion protein in Alzheimer's disease. *J. Mol. Med.* 88, 331–338. doi: 10.1007/s00109-009-0568-7

Hafner-Bratkovic, I., Bester, R., Pristovsek, P., Gaedtke, L., Veranic, P., Gaspersic, J., et al. (2011). Globular domain of the prion protein needs to be unlocked by domain swapping to support prion protein conversion. *J. Biol. Chem.* 286, 12149–12156. doi: 10.1074/jbc.M110.213926

Heldin, C. H. (1995). Dimerization of cell surface receptors in signal transduction. *Cell* 80, 213–223. doi: 10.1016/0092-8674(95)90404-2

Hernandez-Rapp, J., Martin-Lanneree, S., Hirsch, T. Z., Pradines, E., Alleaume-Butaux, A., Schneider, B., et al. (2014). A PrP(C)-caveolin-Lyn complex negatively controls neuronal GSK3beta and serotonin 1B receptor. *Sci. Rep.* 4:4881. doi: 10.1038/srep04881

Hundt, C., Gauczynski, S., Leucht, C., Riley, M. L., and Weiss, S. (2003). Intra- and interspecies interactions between prion proteins and effects of mutations and polymorphisms. *Biol. Chem.* 384, 791–803. doi: 10.1515/BC.2003.088

Jansen, K., Schafer, O., Birkmann, E., Post, K., Serban, H., Prusiner, S. B., et al. (2001). Structural intermediates in the putative pathway from the cellular prion protein to the pathogenic form. *Biol. Chem.* 382, 683–691. doi: 10.1515/BC.2001.081

Kaimann, T., Metzger, S., Kuhlmann, K., Brandt, B., Birkmann, E., Höltje, H. D., et al. (2008). Molecular model of an alpha-helical prion protein dimer and its monomeric subunits as derived from chemical cross-linking and molecular modeling calculations. *J. Mol. Biol.* 376, 582–596. doi: 10.1016/j.jmb.2007.11.035

Klohn, P. C., Farmer, M., Linehan, J. M., O'Malley, C., Fernandez de Marco, M., Taylor, W., et al. (2012). PrP antibodies do not trigger mouse hippocampal neuron apoptosis. *Science* 335, 52. doi: 10.1126/science.1215579

Kuwahara, C., Takeuchi, A. M., Nishimura, T., Haraguchi, K., Kubosaki, A., Matsumoto, Y., et al. (1999). Prions prevent neuronal cell-line death. *Nature* 400, 225–226. doi: 10.1038/22241

Lauren, J., Gimbel, D. A., Nygaard, H. B., Gilbert, J. W., and Strittmatter, S. M. (2009). Cellular prion protein mediates impairment of synaptic plasticity by amyloid-beta oligomers. *Nature* 457, 1128–1132. doi: 10.1038/nature07761

Lee, S., and Eisenberg, D. (2003). Seeded conversion of recombinant prion protein to a disulfide-bonded oligomer by a reduction-oxidation process. *Nat. Struct. Biol.* 10, 725–730. doi: 10.1038/nsb961

Lewis, V., Hill, A. F., Haigh, C. L., Klug, G. M., Masters, C. L., Lawson, V. A., et al. (2009). Increased proportions of C1 truncated prion protein protect against cellular M1000 prion infection. *J. Neuropathol. Exp. Neurol.* 68, 1125–1135. doi: 10.1097/NEN.0b013e3181b96981

Linden, R., Martins, V. R., Prado, M. A., Cammarota, M., Izquierdo, I., and Brentani, R. R. (2008). Physiology of the prion protein. *Physiol. Rev.* 88, 673–728. doi: 10.1152/physrev.00007.2007

Lo, R. Y., Shyu, W. C., Lin, S. Z., Wang, H. J., Chen, S. S., and Li, H. (2007). New molecular insights into cellular survival and stress responses: neuroprotective role of cellular prion protein (PrPC). *Mol. Neurobiol.* 35, 236–244. doi: 10.1007/s12035-007-8003-y

Luhrs, T., Zahn, R., and Wuthrich, K. (2006). Amyloid formation by recombinant full-length prion proteins in phospholipid bicelle solutions. *J. Mol. Biol.* 357, 833–841. doi: 10.1016/j.jmb.2006.01.016

Mallucci, G., Dickinson, A., Linehan, J., Klöhn, P. C., Brandner, S., and Collinge, J. (2003). Depleting neuronal PrP in prion infection prevents disease and reverses spongiosis. *Science* 302, 871–874. doi: 10.1126/science.1090187

Martins, V. R., Beraldo, F. H., Hajj, G. N., Lopes, M. H., Lee, K. S., Prado, M. A., et al. (2010). Prion protein: orchestrating neurotrophic activities. *Curr. Issues Mol. Biol.* 12, 63–86.

Meyer, R. K., Lustig, A., Oesch, B., Fatzer, R., Zurbriggen, A., and Vandevelde, M. (2000). A monomer-dimer equilibrium of a cellular prion protein (PrPC) not observed with recombinant PrP. *J. Biol. Chem.* 275, 38081–38087. doi: 10.1074/jbc.M007114200

Miranda, A., Ramos-Ibeas, P., Pericuesta, E., Ramirez, M. A., and Gutierrez-Adan, A. (2013). The role of prion protein in stem cell regulation. *Reproduction* 146, R91–R99. doi: 10.1530/REP-13-0100

Monnet, C., Gavard, J., Mege, R. M., and Sobel, A. (2004). Clustering of cellular prion protein induces ERK1/2 and stathmin phosphorylation in GT1-7 neuronal cells. *FEBS Lett.* 576, 114–118. doi: 10.1016/j.febslet.2004.08.076

Mouillet-Richard, S., Ermonval, M., Chebassier, C., Laplanche, J. L., Lehmann, S., Launay, J. M., et al. (2000). Signal transduction through prion protein. *Science* 289, 1925–1928. doi: 10.1126/science.289.5486.1925

Mouillet-Richard, S., Pietri, M., Schneider, B., Vidal, C., Mutel, V., Launay, J. M., et al. (2005). Modulation of serotonergic receptor signaling and cross-talk by prion protein. *J. Biol. Chem.* 280, 4592–4601. doi: 10.1074/jbc.M406199200

Oliveira-Martins, J. B., Yusa, S., Calella, A. M., Bridel, C., Baumann, F., Dametto, P., et al. (2010). Unexpected tolerance of alpha-cleavage of the prion protein to sequence variations. *PLoS ONE* 5:e9107. doi: 10.1371/journal.pone.0009107

Pietri, M., Dakowski, C., Hannaoui, S., Alleaume-Butaux, A., Hernandez-Rapp, J., Ragagnin, A., et al. (2013). PDK1 decreases TACE-mediated α-secretase activity and promotes disease progression in prion and Alzheimer's diseases. *Nat. Med.* 19, 1124–1131. doi: 10.1038/nm.3302

Pradines, E., Loubet, D., Schneider, B., Launay, J. M., Kellermann, O., and Mouillet-Richard, S. (2008). CREB-dependent gene regulation by prion protein: impact on MMP-9 and beta-dystroglycan. *Cell. Signal.* 20, 2050–2058. doi: 10.1016/j.cellsig.2008.07.016

Priola, S. A., Caughey, B., Wehrly, K., and Chesebro, B. (1995). A 60-kDa prion protein (PrP) with properties of both the normal and scrapie-associated forms of PrP. *J. Biol. Chem.* 270, 3299–3305. doi: 10.1074/jbc.270.7.3299

Prusiner, S. B. (1998). Prions. *Proc. Natl. Acad. Sci. U.S.A.* 95, 13363–13383. doi: 10.1073/pnas.95.23.13363

Rambold, A. S., Muller, V., Ron, U., Ben-Tal, N., Winklhofer, K. F., and Tatzelt, J. (2008). Stress-protective signalling of prion protein is corrupted by scrapie prions. *EMBO J.* 27, 1974–1984. doi: 10.1038/emboj.2008.122

Resenberger, U. K., Harmeier, A., Woerner, A. C., Goodman, J. L., Muller, V., Krishnan, R., et al. (2011). The cellular prion protein mediates neurotoxic signalling of beta-sheet-rich conformers independent of prion replication. *EMBO J.* 30, 2057–2070. doi: 10.1038/emboj.2011.86

Robinson, P. J. (1991). Phosphatidylinositol membrane anchors and T-cell activation. *Immunol. Today* 12, 35–41. doi: 10.1016/0167-5699(91)90110-F

Roostaee, A., Cote, S., and Roucou, X. (2009). Aggregation and amyloid fibril formation induced by chemical dimerization of recombinant prion protein in physiological-like conditions. *J. Biol. Chem.* 284, 30907–30916. doi: 10.1074/jbc.M109.057950

Roucou, X., and LeBlanc, A. C. (2005). Cellular prion protein neuroprotective function: implications in prion diseases. *J. Mol. Med.* 83, 3–11. doi: 10.1007/s00109-004-0605-5

Schneider, B., Mutel, V., Pietri, M., Ermonval, M., Mouillet-Richard, S., and Kellermann, O. (2003). NADPH oxidase and extracellular regulated kinases 1/2 are targets of prion protein signaling in neuronal and nonneuronal cells. *Proc. Natl. Acad. Sci. U.S.A.* 100, 13326–13331. doi: 10.1073/pnas.2235648100

Schneider, B., Pietri, M., Pradines, E., Loubet, D., Launay, J. M., Kellermann, O., et al. (2011). Understanding the neurospecificity of Prion protein signaling. *Front. Biosci. (Landmark Ed.)* 16, 169–186. doi: 10.2741/3682

Simoneau, S., Rezaei, H., Sales, N., Kaiser-Schulz, G., Lefebvre-Roque, M., Vidal, C., et al. (2007). *In vitro* and *in vivo* neurotoxicity of prion protein oligomers. *PLoS Pathog.* 3:e125. doi: 10.1371/journal.ppat.0030125

Solforosi, L., Criado, J. R., McGavern, D. B., Wirz, S., Sanchez-Alavez, M., Sugama, S., et al. (2004). Cross-linking cellular prion protein triggers neuronal apoptosis *in vivo*. *Science* 303, 1514–1516. doi: 10.1126/science.1094273

Sonati, T., Reimann, R. R., Falsig, J., Baral, P. K., O'Connor, T., Hornemann, S., et al. (2012). The toxicity of antiprion antibodies is mediated by the flexible tail of the prion protein. *Nature* 501, 102–106. doi: 10.1038/nature12402

Stefanova, I., Horejsi, V., Ansotegui, I. J., Knapp, W., and Stockinger, H. (1991). GPI-anchored cell-surface molecules complexed to protein tyrosine kinases. *Science* 254, 1016–1019. doi: 10.1126/science.1719635

Stöhr, J., Weinmann, N., Wille, H., Kaimann, T., Nagel-Steger, L., Birkmann, E., et al. (2008). Mechanisms of prion protein assembly into amyloid. *Proc. Natl. Acad. Sci. U.S.A.* 105, 2409–2414. doi: 10.1073/pnas.0712036105

Suzuki, K. G., Kasai, R. S., Hirosawa, K. M., Nemoto, Y. L., Ishibashi, M., Miwa, Y., et al. (2012). Transient GPI-anchored protein homodimers are units for raft organization and function. *Nat. Chem. Biol.* 8, 774–783. doi: 10.1038/nchembio.1028

Tattum, M. H., Cohen-Krausz, S., Thumanu, K., Wharton, C. W., Khalili-Shirazi, A., Jackson, G. S., et al. (2006). Elongated oligomers assemble into mammalian PrP amyloid fibrils. *J. Mol. Biol.* 357, 975–985. doi: 10.1016/j.jmb.2006.01.052

Tompa, P., Tusnady, G. E., Friedrich, P., and Simon, I. (2002). The role of dimerization in prion replication. *Biophys. J.* 82, 1711–1718. doi: 10.1016/S0006-3495(02)75523-9

Warwicker, J. (2000). Modeling a prion protein dimer: predictions for fibril formation. *Biochem. Biophys. Res. Commun.* 278, 646–652. doi: 10.1006/bbrc.2000.3829

Westergard, L., Turnbaugh, J. A., and Harris, D. A. (2011). A naturally occurring C-terminal fragment of the prion protein (PrP) delays disease and acts as a dominant-negative inhibitor of PrPSc formation. *J. Biol. Chem.* 286, 44234–44242. doi: 10.1074/jbc.M111.286195

Winklhofer, K. F., Heske, J., Heller, U., Reintjes, A., Muranyi, W., Moarefi, I., et al. (2003). Determinants of the *in vivo* folding of the prion protein. A bipartite function of helix 1 in folding and aggregation. *J. Biol. Chem.* 278, 14961–14970. doi: 10.1074/jbc.M209942200

Wuthrich, K., and Riek, R. (2001). Three-dimensional structures of prion proteins. *Adv. Protein Chem.* 57, 55–82. doi: 10.1016/S0065-3233(01)57018-7

Conflict of Interest Statement: The author declares that the research was conducted in the absence of any commercial or financial relationships that could be construed as a potential conflict of interest.

10

To develop with or without the prion protein

Sophie Halliez[1]*, Bruno Passet[2], Séverine Martin-Lannerée[3,4], Julia Hernandez-Rapp[3,4], Hubert Laude[1], Sophie Mouillet-Richard[3,4], Jean-Luc Vilotte[2] and Vincent Béringue[1]*

[1] Institut National de la Recherche Agronomique, U892 Virologie et Immunologie Moléculaires, Jouy-en-Josas, France
[2] Institut National de la Recherche Agronomique, UMR1313 Génétique Animale et Biologie Intégrative, Jouy-en-Josas, France
[3] Institut National de la Santé et de la Recherche Médicale, UMR-S1124, Paris, France
[4] Université Paris Descartes, Sorbonne Paris Cité, UMR-S1124, Paris, France

Edited by:
Kim Newton, Genentech, Inc., USA

Reviewed by:
Peter Christian Kloehn, University College London, UK
Joseph W. Lewcock, Genentech, Inc., USA

***Correspondence:**
Sophie Halliez and Vincent Béringue, Institut National de la Recherche Agronomique, U892 Virologie et Immunologie Moléculaires, Equipe MAP2, Bâtiment 440, 78350 Jouy-en-Josas, France
e-mail: shalliez@jouy.inra.fr; vincent.beringue@jouy.inra.fr

The deletion of the cellular form of the prion protein (PrP^C) in mouse, goat, and cattle has no drastic phenotypic consequence. This stands in apparent contradiction with PrP^C quasi-ubiquitous expression and conserved primary and tertiary structures in mammals, and its pivotal role in neurodegenerative diseases such as prion and Alzheimer's diseases. In zebrafish embryos, depletion of PrP ortholog leads to a severe loss-of-function phenotype. This raises the question of a potential role of PrP^C in the development of all vertebrates. This view is further supported by the early expression of the PrP^C encoding gene (*Prnp*) in many tissues of the mouse embryo, the transient disruption of a broad number of cellular pathways in early $Prnp^{-/-}$ mouse embryos, and a growing body of evidence for PrP^C involvement in the regulation of cell proliferation and differentiation in various types of mammalian stem cells and progenitors. Finally, several studies in both zebrafish embryos and in mammalian cells and tissues in formation support a role for PrP^C in cell adhesion, extra-cellular matrix interactions and cytoskeleton. In this review, we summarize and compare the different models used to decipher PrP^C functions at early developmental stages during embryo- and organo-genesis and discuss their relevance.

Keywords: prion protein, development, neural development, stem cells, cell adhesion, extra-cellular matrix, cytoskeleton

INTRODUCTION

Prion diseases are a group of fatal and transmissible neurodegenerative diseases affecting a broad range of mammals including humans. The causative agent (the prion) is primarily composed of abnormally folded and aggregated forms of a host-encoded protein, the cellular prion protein (PrP^C). PrP^C is a glycosyl-phosphatidyl-inositol anchored cell surface sialoglycoprotein associated with lipid rafts (Taylor et al., 2009). PrP primary sequence is highly conserved among mammals (Wopfner et al., 1999) and PrP putative functional domains are structurally conserved between mammals, avians, and fish (Wopfner et al., 1999; Rivera-Milla et al., 2006). PrP^C is widely expressed in nearly all the organism, albeit at highest levels in the adult nervous system (Bendheim et al., 1992). Together, these data lend support for an essential role of PrP^C in mammals and possibly in vertebrates in general. However, the production of mice, goat, and cattle lacking PrP did not lead to any obvious phenotype (Bueler et al., 1992; Manson et al., 1994a; Richt et al., 2007; Yu et al., 2009) except, for mice, a resistance to experimental prion diseases (Bueler et al., 1993; Prusiner et al., 1993; Manson et al., 1994b; Mallucci et al., 2003). Additionally, goats, naturally devoid of PrP^C due to a nonsense mutation, do not seem to present any abnormal phenotype (Benestad et al., 2012). Subtle behavioral and oxidative stress-related alterations have been then reported in adult mice devoid of PrP^C (see **Table 1**) (Tobler et al., 1996; Wong et al., 2001; Roucou et al., 2004; Meotti et al., 2007; Sanchez-Alavez et al., 2008; Le Pichon et al., 2009; Gadotti et al., 2012) although some of them are debated (Steele et al., 2007) and some may be related to *Prnp*-flanking genes rather than to PrP^C absence itself (Nuvolone et al., 2013). However, none of them seems, at first glance, so important to justify the conserved structure and broad expression of PrP^C. The physiological role of PrP^C still remains highly uncertain despite more than two decades of research and numerous proposed functions (Nicolas et al., 2009; Martins et al., 2010). To conciliate these discrepant data, it has been hypothesized that PrP^C function is either dispensable or redundant with that of other proteins. Yet, recent advances, notably in the developmental biology, shed a new light on PrP^C functions and suggest that, perhaps, the quest for PrP^C functions has been made at the wrong place and/or at the wrong period of time.

EARLY DEVELOPMENTAL EXPRESSION OF THE PrP GENES IN SELECTED VERTEBRATES

The apparent lack of a phenotype in mice invalidated for PrP^C sounds at odds with the very early and quasi-ubiquitous expression pattern of the protein at embryonic and postnatal stages, respectively. Expression of the gene encoding PrP—*Prnp*—is detected in post-implantation embryo from embryonic day (E) 6.5 in extra-embryonic regions (Manson et al., 1992) and from E7.5 (late allantoic bud stage) in cardiac mesoderm (Hidaka et al., 2010). *Prnp* expression is then observed in the developing central nervous system and heart around E8 before extending rapidly almost to the entire embryo (Tremblay et al., 2007). *Prnp* may even be expressed earlier on as PrP mRNAs have been

Table 1 | Phenotypes associated to PrP invalidation/ectopic activation.

Life period/developmental stage	Type of manipulation	Phenotype(s)	Comments	References
Early embryo (zebrafish)	*PrP-1* knockdown	**Lethal** – gastrulation arrest due to impaired cell adhesion	– partial rescue of the morphants by the injection of PrP mRNA including mammalian sequence	Malaga-Trillo et al., 2009
Early embryo (mouse)	*Prnp* knockout embryos	**Moderately severe** – transcriptomic analysis shows differential expression of genes from multiple cell pathways	– transient phenotype	Khalife et al., 2011; Passet et al., 2012
Early embryo (mouse)	PrP and Shadoo co-invalidation obtained by: *Sprn* knockdown and *Prnp* knockout Double knockout	**Controversial** – impaired trophoblast development, lethality – no phenotype reported but no assessment at embryonic stage	– discussed in Makhzami et al. (2014)	Young et al., 2009; Passet et al., 2012 Daude et al., 2012
Pharyngula stage (zebrafish)	*PrP-2* knockdown *PrP-2* knockout	**Controversial** – differential expression of genes involved in multiple cell pathways – no obvious developmental abnormalities but no transcriptomic analysis performed	– no phenotype rescue of the morphants	Nourizadeh-Lillabadi et al., 2010 Fleisch et al., 2013
Larva (zebrafish)	*PrP-2* knockdown *PrP-2* knockout	**Controversial** – head malformations, missing neuronal structures – no obvious abnormalities at larval stage	– no phenotype rescue of the morphants	Malaga-Trillo et al., 2009; Nourizadeh-Lillabadi et al., 2010 Fleisch et al., 2013
Late embryo/newborn (mouse)	*Prnp* knockout embryos	**Minor** – increased proliferation and maturation delay of the oligodendrocyte precursor cell population in the brain – earlier formation of dentin and enamel in the developing tooth	– no brain abnormalities or myelin defect – no enamel defect but reduced hardness of dentin at adult stage	Bribian et al., 2012 Zhang et al., 2011
Juvenile (mouse)	*Prnp* knockout mice	**Moderately severe** – functional abnormalities and persisting cell proliferation in the cerebellum, impaired locomotor abilities	– transient phenotype	Prestori et al., 2008
Adult Brain (mouse)	*Prnp* knockout mice	**Minor** – transcriptomic and proteomic analysis revealed no important differences – increased protein oxidation, protein ubiquitination and lipid peroxidation – increased proliferation and maturation delay of the oligodendrocyte precursor cells – delayed maturation of astrocytes	– no brain abnormalities or myelin defect associated	Crecelius et al., 2008; Chadi et al., 2010 Wong et al., 2001 Bribian et al., 2012 Arantes et al., 2009

(Continued)

Table 1 | Continued

Life period/developmental stage	Type of manipulation	Phenotype(s)	Comments	References
		– decreased cell proliferation in the dentate gyrus (adult neurogenic region) – functional abnormalities in the hippocampus		Steele et al., 2006 Collinge et al., 1994
Adult brain (mouse)	*Prnp* overexpressing mice	**Minor** – increased cell proliferation in the subventricular zone (adult neurogenic region) – shorten astrocyte maturation phase	– no brain abnormalities associated	Steele et al., 2006 Hartmann et al., 2013
Adult extraneural tissues (mouse)	*Prnp* knockout mice	**Minor** – delayed mineralization of the continuously erupting incisors – slower regeneration of muscle fibers – shortening of intestinal villi, cell cycle alterations in intestinal crypts and reduced size of desmosomes in intestinal epithelium	– regeneration and renewing tissues	Zhang et al., 2011; Stella et al., 2010 Morel et al., 2008
Adult (mouse)	*Prnp* knockout mice	**Minor and/or partially controversial** – depending of the study, no phenotype observed to minor alterations such as altered olfactory behavior	– few studies were carried out using distinct genetic background	Bueler et al., 1992; Collinge et al., 1994; Tobler et al., 1996; Walz et al., 1999; Nico et al., 2005; Meotti et al., 2007; Sanchez-Alavez et al., 2008; Le Pichon et al., 2009; Gadotti et al., 2012; Rial et al., 2014
Adult (zebrafish)	*PrP-2* knockout	**Minor** – increased susceptibility to a convulsant drug – kinetics alteration of NMDA receptors		Fleisch et al., 2013
Adult (zebrafish)	*PrP-1* knockout	**Unknown**	– no transgenic (inducible) line established	
Aged animal (mouse)	*Prnp* knockout mice	**Minor** – behavior alterations		Rial et al., 2009; Massimino et al., 2013
Aged animal nerves (mouse)	*Prnp* knockout and conditional *Prnp* knockout	**Minor** – increased myelin abnormalities in peripheral nerves	– several genetic background analyzed	Bremer et al., 2010

found maternally expressed in the zygotes of *Xenopus laevis* and zebrafish (*Danio rerio*) (van Rosmalen et al., 2006; Malaga-Trillo et al., 2009).

Two PrP homologs, PrP-1, PrP-2 and a more divergent form, PrP-3, have been identified in zebrafish (Cotto et al., 2005; Rivera-Milla et al., 2006). The expression patterns of the corresponding genes (*PrP-1*, *PrP-2*, and *PrP-3*, respectively) in zebrafish embryos are partially conflicting (Cotto et al., 2005; Malaga-Trillo et al., 2009) although both studies agree on their spatio-temporal complementarity. Both studies describe early PrP gene expression from the mid-blastula stage although it is *PrP-2* according to the Cotto et al. study and *PrP-1* according to Malaga-Trillo et al. At pharyngula stage (24–48 h postfertilization), both studies show a strong and broad expression of *PrP-2* in the developing central nervous system. Additionally, the Cotto et al. study reports the expression of *PrP-1* in the floor plate, an essential organizing center of the central nervous system and in the peripheral nervous system around the same developmental period. The expression of *PrP-2* and *PrP-3* in the developing neuromasts (sensory organs) is also described at the larval stage. PrP-encoding genes are also expressed in developing organs or tissues outside the nervous system such as the heart (*PrP-2* and *PrP-3*), the kidney (*PrP-2*), the pectoral fins (*PrP-2* and *PrP-3*) and the intestinal epithelium (*PrP-2*) (Cotto et al., 2005).

THE EMBRYO SPILLS THE BEANS

In sharp contrast with the situation in mammals, knockdown (KD) of *PrP-1* or *PrP-2* by morpholino injection led to severe biological phenotypes in zebrafish (see **Table 1**): gastrulation arrest for *PrP-1* KD (Malaga-Trillo et al., 2009) and malformations of the brain and the eyes associated to a reduced number of peripheral neurons for *PrP-2* KD (Malaga-Trillo et al., 2009; Nourizadeh-Lillabadi et al., 2010). However, *PrP-2* KD-induced phenotype is subject to caution as *PrP-2* knockout (KO) by zinc finger nuclease-induced targeted mutation did not lead to any obvious developmental phenotype (Fleisch et al., 2013). Additional and aspecific effects due to morpholino injection may have occurred. There is less concern about *PrP-1* KD phenotype as partial rescue was obtained by the injection of *PrP-1* mRNA, *PrP-2* mRNA and, remarkably, mouse *Prnp* mRNA (Malaga-Trillo et al., 2009), strongly suggesting that zebrafish and mammalian PrPs may share conserved functions.

PrP inactivation studies in zebrafish allow investigating PrP^C functions. They also favor a more critical role of PrP^C in early development rather than during postnatal stages (see **Table 1**). Accordingly, transcriptomic analyses (by RNAseq) of $Prnp^{-/-}$ versus wild-type mouse embryos identified numerous genes differentially expressed (73 and 263 genes at E6.5 and E7.5, respectively) (Khalife et al., 2011), while only 1 gene was found differentially expressed in adult brains (Chadi et al., 2010). Proteomic analysis also failed to identify major alterations (Crecelius et al., 2008) although many cellular stress markers such as protein oxidation, protein ubiquitination, and lipid peroxidation were reportedly activated in the brains of $Prnp^{-/-}$ versus wild-type mice (Wong et al., 2001). This suggests either that PrP^C is quite dispensable in the adult brain or that, in a $Prnp^{-/-}$ context, alternative mechanisms are activated during the brain development to compensate the absence of PrP^C. Supporting the second hypothesis, the number of genes differentially expressed in the adult $Prnp^{-/-}$ mouse brain was substantially higher upon postnatal depletion in neurons than embryonically (Chadi et al., 2010), albeit to a less extent than in the embryo. The genes differentially expressed in the embryo covered various cell functions including adhesion, apoptosis and proliferation and are related to heart formation and blood vessel development, vascular diseases, immune response, nervous system development, and prion diseases (Khalife et al., 2011). Interestingly and in line with the idea of evolutionary conserved functions of PrPs, similar biological functions were impacted by *PrP-2* KD in zebrafish, as revealed by transcriptomic analyses: development including that of the cardiovascular and nervous systems specifically, cell death, cellular assembly, cell cycle, immunological, and neurological diseases (Nourizadeh-Lillabadi et al., 2010).

Finally, the KD of the *Sprn* gene encoding a PrP-related protein, Shadoo resulted in embryonic lethality between E8 and E11 specifically in $Prnp^{-/-}$ embryos, supporting a role for PrP^C and Shadoo in mouse embryogenesis, notably cell adhesion and placenta development, and a possible functional redundancy between the two proteins (Young et al., 2009; Passet et al., 2012). However, the recent generation of viable $Prnp^{-/-}$; $Sprn^{-/-}$ animals is challenging this view (Daude et al., 2012). These apparently conflicting observations are discussed in Makhzami et al. (2014).

Thus, studying the consequences of PrP inactivation during developmental stages rather than in adult animals could be a more efficient, although not easier, strategy to decipher PrP^C functions. Additionally, studying PrP^C implications in regenerative processes and in renewing tissues could also be an informative approach (see **Table 1**).

THE CURRENT LIMITS OF *IN VIVO* STUDIES IN MAMMALS

As PrP^C is mediating neuronal dysfunction and degeneration in prion diseases and reportedly in Alzheimer's disease (Lauren et al., 2009; Klyubin et al., 2014), it has been logically suspected to play a role in neuronal homeostasis and during neural development. As PrP^C functions in mature neurons are not under the scope of the present review, we will not discuss them and only focus on developmental aspects. *In vivo* direct evidence supporting a role of PrP^C in neural development is quite elusive and PrP^C expression pattern remains one of the strongest pro arguments. Indeed, PrP^C expression increases as neurons mature, although it is not detected in mitotically active neural progenitor cells (Steele et al., 2006; Peralta et al., 2011). Other experimental evidence was obtained from the comparative study of WT and $Prnp^{-/-}$ and/or *Prnp* overexpressing animals (see **Table 1**). The levels of PrP^C were found to positively regulate the proliferation in neurogenic regions in adult mice (Steele et al., 2006; Bribian et al., 2012) and the maturation of astrocytes in the developing brain (Arantes et al., 2009; Hartmann et al., 2013). Conversely, the absence of PrP^C was associated with an increased proliferation in oligodendrocyte precursor cells both in embryos and adults (Bribian et al., 2012). However, no brain morphology abnormalities or myelin defects could be observed in $Prnp^{-/-}$ animals, casting doubt on a crucial role of PrP^C in cell proliferation and maturation during brain development or, at least, suggesting the occurrence of compensatory mechanisms. Yet, electrophysiological recordings on hippocampus from $Prnp^{-/-}$ mice (Collinge et al., 1994) and on cerebellum from juvenile $Prnp^{-/-}$ mice (Prestori et al., 2008) revealed functional alterations. However, no hippocampus-related behavioral or learning alterations were observed in $Prnp^{-/-}$ mice (Bueler et al., 1992) suggesting the hippocampus functional alterations are minor or at least compensated. Conversely, cerebellum functional alterations were found associated to a failure of juvenile $Prnp^{-/-}$ mice in motor control test and to protracted cell proliferation in the cerebellum granular layer at the third week of postnatal development although all these abnormalities vanished in older animals (Prestori et al., 2008). Taken together, these arguments suggest a delay in the maturation of the cerebellar granule cells in *Prnp*$^{-/-}$ mice.

Further dissecting the underlying mechanisms responsible for the subtle phenotypes in *Prnp*$^{-/-}$ mice is a particularly arduous task *in vivo*. For example, determining whether the effects are cell-autonomous would require the use of several transgenic lines or genetic chimeras to limit the expression of PrP^{C} to a subset of cells. And this question is far from being irrelevant as PrP^{C} is a cell surface protein. Such topology could elicit specific cell responses in neighboring cells. Additionally, PrP^{C} may act at distance since secreted forms of have been described both *in vitro* and *in vivo* (Borchelt et al., 1993; Harris et al., 1993).

CONTRIBUTIONS OF *IN VITRO* AND *EX VIVO* MODELS TO UNDERSTANDING THE ROLE OF PrP^{C} IN NEURAL DEVELOPMENT

The use of *in vitro* or *ex vivo* models has generated a wealth of data with regard to the involvement of PrP^{C} in neural development and, importantly, has comforted the putative functions suggested by the *in vivo* approaches. Moreover they often allow pinpointing more precisely the cellular processes and signaling pathways involving PrP^{C}. Collectively, these experimental data argue that PrP^{C} positively regulates (1) the differentiation of pluripotent progenitors cells toward the neural lineage (Peralta et al., 2011), (2) the proliferation of neural progenitor cells (Prodromidou et al., 2014) (3) the self-renewal of neural progenitor cells (Santos et al., 2011; Prodromidou et al., 2014) (4) the neuronal differentiation in terms of choice of cell fate (Steele et al., 2006) and (5) acquisition of neuronal features (Graner et al., 2000). However, how PrP^{C} fulfills these different actions is only partially understood. The current idea is that PrP^{C} is part of multiprotein signaling complexes, able to bind various partners, and participate to signal transduction events along different cellular pathways depending on the cellular context (Caughey and Baron, 2006; Watts et al., 2009; Stuermer, 2011; Hirsch et al., 2014). Non-exhaustively, the interaction between PrP^{C} and one of its identified ligand, Stress inducible protein 1 (STI1), has been shown to promote the self-renewal of the neural progenitor cells (Santos et al., 2011). Besides, interaction with the neural cell adhesion molecule (NCAM) can induce their neuronal differentiation (Prodromidou et al., 2014). In cultured neurons, STI1 binding to PrP^{C} increases protein synthesis (Roffe et al., 2010), STI1 and Laminin-γ1 binding promotes intracellular Ca^{2+} increase and axonogenesis (Santos et al., 2013), and vitronectin binding, axonal growth (Hajj et al., 2007). Experimental evidence also favors a role of PrP^{C} in glia development since the lack of PrP^{C} promotes the proliferation of oligodendrocyte precursor cells and delays their differentiation in culture (Bribian et al., 2012). Studies in astrocyte primary cultures revealed that the involvement of PrP^{C} in astrocytes maturation is dependent upon its interaction with STI1 and the downstream activation of ERK1/2 (Arantes et al., 2009). Of note, PrP^{C} action does not necessarily rely on PrP^{C} expression solely in the cell population considered. For instance, proper neuritogenesis has been shown to require PrP^{C} expression in both neurons and surrounding astrocytes in co-cultures experiments (Lima et al., 2007). Beyond a role in the interactions between both populations, PrP^{C} serves as a neuronal receptor for laminin (Graner et al., 2000) and is also involved in the organization of the extracellular matrix by the astrocytes, especially through laminin deposition (Lima et al., 2007).

It should also be stressed that PrP^{C} partners and the cellular context could tightly determine the final consequences of PrP^{C} stimulation. For example PrP^{C} interaction with contactin-associated protein (Caspr) protects Caspr from proteolysis allowing it to inhibit more efficiently neurite outgrowth (Devanathan et al., 2010). On the opposite, PrP^{C} and NCAM interactions allow stabilization of NCAM in lipid rafts allowing Fyn kinase activation and promoting neurite outgrowth (Santuccione et al., 2005).

In vitro, PrP^{C} is thereby found involved in all the major steps of neural development. Since it notably regulates the self-renewal and differentiation processes in neural progenitor cells, the question arises as to whether this role can be extended to other cell types.

WHAT THE STUDY OF STEM CELLS REVEALS

The role of PrP^{C} is far from being restricted to neural development and a growing body of evidence support its involvement in stem cell biology see also Martin-Lannerée et al. (2014). Whether PrP^{C} is expressed in embryonic stem cells (ESC) at basal state remains unclear (Lee and Baskakov, 2010; Miranda et al., 2011), possibly due to limitations in the sensitivity of the detection methods, choice of the target (i.e., the gene or the protein itself) and methods of production and culture of the cells. Nevertheless, there is a consensus that PrP^{C} expression progressively increases during stem cell differentiation (Miranda et al., 2011; Lee and Baskakov, 2013). To understand the role played by PrP^{C} during early embryogenesis, the consequences of *Prnp* KD or overexpression were studied in human ESC (Lee and Baskakov, 2013). Forcing PrP^{C} expression under self-renewal conditions was found to alter cell cycle regulation and to promote ESC differentiation. Blocking PrP^{C} expression in differentiating ESC also impacts on cell cycle regulation and inhibits the differentiation toward ectodermal lineages, in line with PrP^{C} functions in neural differentiation (Peralta et al., 2011) and the reduced expression of *Nestin*, an ectodermal marker, observed in ESC in the absence of PrP^{C} (Miranda et al., 2011; Peralta et al., 2011). Mesodermal and endodermal differentiations were, however, not affected. Finally, overexpressing PrP^{C} in differentiating ESC promotes proliferation and inhibits differentiation toward all the lineages (Lee and Baskakov, 2013).

Further supporting PrP^{C} function in stem cell biology, PrP^{C} is expressed in hematopoietic stem cells and promotes their self-renewal (Zhang et al., 2006). PrP^{C} is therefore involved in the self-renewal and differentiation of stem cells although its role and requirement seem to vary along the differentiation process and the cell lineage considered. Identifying the molecular mechanisms at play is a particularly challenging task since self-renewal, differentiation and proliferation are highly intertwined processes impacting each other.

THE ROLE OF PrP^C IN DEVELOPING AND RENEWING TISSUES OUTSIDE THE NERVOUS SYSTEM

Looking outside the central nervous system in newborn to full-grown animals allowed pinpointing subtle phenotypes in the absence of PrP^C (see **Table 1**), in perfect consistency with PrP^C widespread expression. PrP^C is expressed in developing human and murine teeth and study of dental cell cultures and teeth from *Prnp*$^{-/-}$ animals revealed multiple alterations during tooth formation (Schneider et al., 2007; Zhang et al., 2011). *In vitro*, embryonic dental mesenchymal cells from *Prnp*$^{-/-}$ embryos proliferate more rapidly and differentiate earlier and coherently, *Prnp*$^{-/-}$ newborn mice exhibit earlier formation of mesenchymally- and epithelially-derived dentin and enamel, respectively (Zhang et al., 2011).

PrP^C is also abundantly expressed in muscles and, although no abnormalities could be observed in a physiological context, it is involved in the regeneration of skeletal muscles in adult mice. Regeneration of locally damaged muscle fibers occurs at slower pace in *Prnp*$^{-/-}$ animals. Yet total recovery is finally achieved (Stella et al., 2010). This is associated with a longer phase of proliferation and a delayed maturation of muscle precursors, likely due to a reduced level of released myogenic factors (Stella et al., 2010). However, the identity of the releasing cells is difficult to establish, showing again the limits of the *in vivo* paradigm.

Another extra-neural tissue where PrP^C depletion induces alterations is the intestinal epithelium, a constantly renewing tissue. PrP^C is expressed by human enterocytes (Morel et al., 2004) and *Prnp*$^{-/-}$ animals exhibit shorter intestinal villi associated to an increased number of mitotic cells (Morel et al., 2008).

Collectively, these observations support a role of PrP^C in the regulation of cell cycle and cell differentiation extraneurally. But how can these functions be exerted?

PrP^C REGULATES CELLULAR ADHESION, EXTRA-CELLULAR MATRIX INTERACTIONS, AND CYTOSKELETON REMODELING

PrP-1 KD-induced gastrulation arrest in zebrafish embryos is due to impaired morphogenetic cell movements. This defect is, at least partially, caused by the disruption of cellular adhesion (Malaga-Trillo et al., 2009). The study revealed that PrP-1 normally accumulates at cell-cell contacts at early embryonic stages and regulates cell adhesion, including E-cadherin processing and/or storage, in a cell-autonomous way. Such functions may be conserved in mammals as injection of mouse mRNA *Prnp* partially rescued the *PrP-1* KD-induced phenotype (Malaga-Trillo et al., 2009). Consistent with this idea, a number of experimental evidence have linked PrP^C to cell adhesion and intimately associated processes such as cytoskeleton remodeling and interactions with extra-cellular matrix (Petit et al., 2013): (i) the expression pattern of PrP^C. PrP^C is expressed at the cell-to-cell contacts, growth cone, and other extending process tips in cultured neural progenitors and neurons (Santuccione et al., 2005; Devanathan et al., 2010; Miyazawa et al., 2010), cell-to-cell junctions in human intestinal epithelium villi (Morel et al., 2004), cell-to-cell contacts in primary cultured endothelial cells (Viegas et al., 2006), focal adhesions in HeLa cell lines (Schrock et al., 2009) and the apical face of ameloblasts of developing teeth (Zhang et al., 2011); (ii) relevant PrP^C partners and ligands, such as cell junction proteins including integrins, the cytoskeleton protein actin and components of the extra-cellular matrix (Graner et al., 2000; Schmitt-Ulms et al., 2001; Morel et al., 2004, 2008; Nieznanski et al., 2005; Hajj et al., 2007; Watts et al., 2009); (iii) alterations in cell adhesion, cytoskeleton dynamics, and extra-cellular matrix interactions upon ectopic expression or deletion of PrP^C in different cell systems. *Prnp* KD leads to alterations in the actin cytoskeleton, remodeling of focal adhesions and over-production of fibronectin in neuronal progenitors (Loubet et al., 2012), to disruption of adherens junctions in carcinoma cells (Solis et al., 2012) and to miss-localization of cell junction-related proteins in enterocytes (Morel et al., 2008). As for *Prnp* overexpression, it promotes aggregation and filopodia formation as well as alterations of the focal adhesion dynamics in neuroblastoma N2a cells (Mange et al., 2002; Schrock et al., 2009). Subtle changes were even noticed *in vivo* such as a decrease in the length of desmosomes in the intestinal epithelium of *Prnp*$^{-/-}$ mice (Morel et al., 2008).

Interestingly, the alterations induced by the deletion of PrP^C can be partially rescued by neutralizing antibodies to specific integrins: use of β1 and β5 integrins was found to restore neuritogenesis in differentiating PrP-deficient neuronal progenitors (Loubet et al., 2012) and *Nanog* expression in differentiating PrP-deficient ESC (Miranda et al., 2011), respectively. Moreover, modulation of integrin activity was suspected to be a spontaneous compensatory mechanism occurring in the absence of PrP^C (Hajj et al., 2007). This last observation supports an involvement of PrP^C in extra-cellular matrix interactions and suggests a possible adaptive mechanism to the absence of PrP^C, which may account for the lack of drastic phenotypes in *Prnp*$^{-/-}$ animals.

Taken together, these observations demonstrate a clear involvement of PrP^C in the regulation of cell adhesion, the interactions with the extra-cellular matrix and the cytoskeleton dynamics. The links between these functions and the role of PrP^C in the regulation of cell proliferation and differentiation have been the focus of few studies (Miranda et al., 2011; Loubet et al., 2012). Are they under-estimated? And are these functions still ongoing in adult organism?

STUDYING PrP^C FUNCTIONS DURING AGING

Little attention has been devoted to the potential functions of PrP^C during aging (see **Table 1** and also Gasperini and Legname, 2014). It is commonly accepted that the cellular machinery deteriorates or, at least, evolves with aging. As the potential development of adaptive mechanisms to the absence of PrP^C is assumed to complicate the study of PrP^C functions in *Prnp*$^{-/-}$ mice, the study of aged animals could be a way to bypass such difficulty: because of the age-related alterations accumulation, the tolerance to PrP^C depletion could partially break. In line with this scenario, behavioral differences between WT and *Prnp*$^{-/-}$ mice seem more pronounced when performed on aged animals (Rial et al., 2009; Massimino et al., 2013). Age-related physiological traits can also be found amplified in absence of PrP^C. Myelin abnormalities in the peripheral nervous system accumulate with aging (Verdu et al., 2000) and sciatic nerves from moderately aged mice devoid of PrP^C (in selected or all cell populations) showed highly precocious and increased demyelination in the absence of neuronal PrP^C (Bremer et al., 2010). If the axons-glia interactions are known to play a key role in myelination during development (Sherman and Brophy, 2005), their role in myelin maintenance in adulthood is less well-understood. However, myelin maintenance is clearly established as an active process (Bremer et al., 2011). The early demyelination observed in the absence of neuronal PrP^C suggests axonal PrP^C participates in myelin maintenance and is likely required, directly or indirectly, in axons-Schwann cells communication.

Thus, aging may perhaps be the ultimate stage to unravel the mystery of PrP^C function(s). However, it has to be considered that aging could also impact on PrP^C and/or its function(s). For instance, the age-related change in the cholesterol/sphingolipids ratio in lipid rafts is suspected to alter PrP^C compartmentalization (Agostini et al., 2013), which could ultimately impact on its functions. A functional study of PrP^C during aging could therefore be difficult to analyse due to the numerous changes occurring at the same time.

CONCLUDING REMARKS

A major function of PrP^C, consistent with its cell surface expression consists in the modulation of signaling pathways in response to various cues. The panel of putative PrP^C ligands is quite large, including soluble ligands, extra-cellular matrix components, and adhesion molecules. Their modes of interaction with PrP^C are variable, including cooperation (Santos et al., 2013) and both cis- and trans-interactions for a same ligand (Santuccione et al., 2005). The cellular context can strongly influence the composition of the PrP^C-multiprotein complexes and the sub-cellular localization of PrP^C is highly dynamic: it can cycle rapidly from the plasma membrane to internal membranes (Griffiths et al., 2007) possibly in response to a given stimulus (Lee et al., 2001; Brown and Harris, 2003; Caetano et al., 2008). All these elements taken together may allow a specific response to a given signal with regard to the cell type and differentiation state. This may explain why PrP^C appears to be involved in a multitude of functions. Regarding development, this view perfectly fits the requirement of different stem and progenitor cells by eliciting appropriate cell responses to adapt efficiently to different microenvironments. But how PrP^C precisely impacts on stemness maintenance and differentiation processes remains unresolved. The potential implication of PrP^C in cell polarization and cytoskeleton remodeling in differentiating stem cells is a very attractive hypothesis and should obviously deserve further investigation. Cell polarity and cytoskeleton dynamics are life-long important cell processes, notably during cell fate determination. They have been studied more particularly during drosophilia neurogenesis and in mammalian neuroepithelial cells (Fietz and Huttner, 2011), but they are fundamental processes for all sort of stem and progenitor cells such as hematopoietic progenitor cells (Bullock et al., 2007). Our unpublished data (Halliez et al., in preparation) reveal a polarized expression of PrP^C in neural and cardiovascular progenitor cells of the developing mouse embryo and therefore are compatible with such idea.

If PrP^C is involved in so many cell lineages, understanding the lack of drastic phenotype in PrP^C-depleted animals remains an ongoing issue. Yet, a modulation of the cell maturation phase (generally delayed but also shortened) is consistently observed in different models (*Prnp* KO but also KD and overexpression) and may therefore be the general price to pay for the loss of PrP^C (see **Table 1**). This may have a more selective impact on free-range than on laboratory animals (for example acquisition of perfect motor control) and justify the high degree of PrP conservation through evolution. Moreover, building an organism is a highly complex task requiring spatial and temporal fine-tuning of a broad number of biological processes via multiple regulatory networks. Cells, tissues and organs do not develop separately along each other but rely on multiple interactions to develop in a concerted, interconnected, and time-scheduled manner. Partial redundancies into and between signaling pathways ensure robustness to the system. That two or more proteins may exert overlapping functions during development and therefore allow tolerance to the loss of one of them is not unprecedented (Yoon et al., 2005; Santamaria and Ortega, 2006; Nicolae et al., 2007). Moreover, it is not unheard-of, notably in plants where environment adaptation is largely studied, that redundant genes in a regular context allow survival under environmental modifications and stress (Wu et al., 2002). Interestingly, *PrP* KO animals seem to react differentially than WT animals to stressful situations (Nico et al., 2005) and to alcohol (Rial et al., 2014) and show increased susceptibility to convulsant agents (Walz et al., 1999; Fleisch et al., 2013) (see **Table 1**). The general function of PrP^C could be to facilitate responses to external stimulus/factors at the cell- and at the organism-scale. Yet, laboratory animals are raised in standardized and protective environment and are often homogenous genetically. In such conditions, the time window to apprehend the cell specific functions of PrP^C may be restricted to the developmental and perinatal phases, with so much information provided and so much transformation occurring at the same time.

REFERENCES

Agostini, F., Dotti, C. G., Perez-Canamas, A., Ledesma, M. D., Benetti, F., and Legname, G. (2013). Prion protein accumulation in lipid rafts of mouse aging brain. *PLoS ONE* 8:e74244. doi: 10.1371/journal.pone.0074244

Arantes, C., Nomizo, R., Lopes, M. H., Hajj, G. N., Lima, F. R., and Martins, V. R. (2009). Prion protein and its ligand stress inducible protein 1 regulate astrocyte development. *Glia* 57, 1439–1449. doi: 10.1002/glia.20861

Bendheim, P. E., Brown, H. R., Rudelli, R. D., Scala, L. J., Goller, N. L., Wen, G. Y., et al. (1992). Nearly ubiquitous tissue distribution of the scrapie agent precursor protein. *Neurology* 42, 149–156. doi: 10.1212/WNL.42.1.149

Benestad, S. L., Austbo, L., Tranulis, M. A., Espenes, A., and Olsaker, I. (2012). Healthy goats naturally devoid of prion protein. *Vet. Res.* 43, 87. doi: 10.1186/1297-9716-43-87

Borchelt, D. R., Rogers, M., Stahl, N., Telling, G., and Prusiner, S. B. (1993). Release of the cellular prion protein from cultured cells after loss of its glycoinositol phospholipid anchor. *Glycobiology* 3, 319–329. doi: 10.1093/glycob/3.4.319

Bremer, J., Baumann, F., Tiberi, C., Wessig, C., Fischer, H., Schwarz, P., et al. (2010). Axonal prion protein is required for peripheral myelin maintenance. *Nat. Neurosci.* 13, 310–318. doi: 10.1038/nn.2483

Bremer, M., Frob, F., Kichko, T., Reeh, P., Tamm, E. R., Suter, U., et al. (2011). Sox10 is required for Schwann-cell homeostasis and myelin maintenance in the adult peripheral nerve. *Glia* 59, 1022–1032. doi: 10.1002/glia.21173

Bribian, A., Fontana, X., Llorens, F., Gavin, R., Reina, M., Garcia-Verdugo, J. M., et al. (2012). Role of the cellular prion protein in oligodendrocyte precursor cell proliferation and differentiation in the developing and adult mouse CNS. *PLoS ONE* 7:e33872. doi: 10.1371/journal.pone.0033872

Brown, L. R., and Harris, D. A. (2003). Copper and zinc cause delivery of the prion protein from the plasma membrane to a subset of early endosomes and the Golgi. *J. Neurochem.* 87, 353–363. doi: 10.1046/j.1471-4159.2003.01996.x

Bueler, H., Aguzzi, A., Sailer, A., Greiner, R. A., Autenried, P., Aguet, M., et al. (1993). Mice devoid of PrP are resistant to scrapie. *Cell* 73, 1339–1347. doi: 10.1016/0092-8674(93)90360-3

Bueler, H., Fischer, M., Lang, Y., Bluethmann, H., Lipp, H. P., DeArmond, S. J., et al. (1992). Normal development and behaviour of mice lacking the neuronal cell-surface PrP protein. *Nature* 356, 577–582. doi: 10.1038/356577a0

Bullock, T. E., Wen, B., Marley, S. B., and Gordon, M. Y. (2007). Potential of CD34 in the regulation of symmetrical and asymmetrical divisions by hematopoietic progenitor cells. *Stem Cells* 25, 844–851. doi: 10.1634/stemcells.2006-0346

Caetano, F. A., Lopes, M. H., Hajj, G. N., Machado, C. F., Pinto Arantes, C., Magalhaes, A. C., et al. (2008). Endocytosis of prion protein is required for ERK1/2 signaling induced by stress-inducible protein 1. *J. Neurosci.* 28, 6691–6702. doi: 10.1523/JNEUROSCI.1701-08.2008

Caughey, B., and Baron, G. S. (2006). Prions and their partners in crime. *Nature* 443, 803–810. doi: 10.1038/nature05294

Chadi, S., Young, R., Le Guillou, S., Tilly, G., Bitton, F., Martin-Magniette, M. L., et al. (2010). Brain transcriptional stability upon prion protein-encoding gene invalidation in zygotic or adult mouse. *BMC Genomics* 11:448. doi: 10.1186/1471-2164-11-448

Collinge, J., Whittington, M. A., Sidle, K. C., Smith, C. J., Palmer, M. S., Clarke, A. R., et al. (1994). Prion protein is necessary for normal synaptic function. *Nature* 370, 295–297. doi: 10.1038/370295a0

Cotto, E., Andre, M., Forgue, J., Fleury, H. J., and Babin, P. J. (2005). Molecular characterization, phylogenetic relationships, and developmental expression patterns of prion genes in zebrafish (*Danio rerio*). *FEBS J.* 272, 500–513. doi: 10.1111/j.1742-4658.2004.04492.x

Crecelius, A. C., Helmstetter, D., Strangmann, J., Mitteregger, G., Frohlich, T., Arnold, G. J., et al. (2008). The brain proteome profile is highly conserved between Prnp-/- and Prnp+/+ mice. *Neuroreport* 19, 1027–1031. doi: 10.1097/WNR.0b013e3283046157

Daude, N., Wohlgemuth, S., Brown, R., Pitstick, R., Gapeshina, H., Yang, J., et al. (2012). Knockout of the prion protein (PrP)-like Sprn gene does not produce embryonic lethality in combination with PrP(C)-deficiency. *Proc. Natl. Acad. Sci. U.S.A.* 109, 9035–9040. doi: 10.1073/pnas.1202130109

Devanathan, V., Jakovcevski, I., Santuccione, A., Li, S., Lee, H. J., Peles, E., et al. (2010). Cellular form of prion protein inhibits Reelin-mediated shedding of Caspr from the neuronal cell surface to potentiate Caspr-mediated inhibition of neurite outgrowth. *J. Neurosci.* 30, 9292–9305. doi: 10.1523/JNEUROSCI.5657-09.2010

Fietz, S. A., and Huttner, W. B. (2011). Cortical progenitor expansion, self-renewal and neurogenesis-a polarized perspective. *Curr. Opin. Neurobiol.* 21, 23–35. doi: 10.1016/j.conb.2010.10.002

Fleisch, V. C., Leighton, P. L., Wang, H., Pillay, L. M., Ritzel, R. G., Bhinder, G., et al. (2013). Targeted mutation of the gene encoding prion protein in zebrafish reveals a conserved role in neuron excitability. *Neurobiol. Dis.* 55, 11–25. doi: 10.1016/j.nbd.2013.03.007

Gadotti, V. M., Bonfield, S. P., and Zamponi, G. W. (2012). Depressive-like behaviour of mice lacking cellular prion protein. *Behav. Brain Res.* 227, 319–323. doi: 10.1016/j.bbr.2011.03.012

Gasperini, L., and Legname, G. (2014). Prion protein and aging. *Front. Cell Dev. Biol.* 2:44. doi: 10.3389/fcell.2014.00044

Graner, E., Mercadante, A. F., Zanata, S. M., Forlenza, O. V., Cabral, A. L., Veiga, S. S., et al. (2000). Cellular prion protein binds laminin and mediates neuritogenesis. *Brain Res. Mol. Brain Res.* 76, 85–92. doi: 10.1016/S0169-328X(99)00334-4

Griffiths, R. E., Heesom, K. J., and Anstee, D. J. (2007). Normal prion protein trafficking in cultured human erythroblasts. *Blood* 110, 4518–4525. doi: 10.1182/blood-2007-04-085183

Hajj, G. N., Lopes, M. H., Mercadante, A. F., Veiga, S. S., da Silveira, R. B., Santos, T. G., et al. (2007). Cellular prion protein interaction with vitronectin supports axonal growth and is compensated by integrins. *J. Cell Sci.* 120, 1915–1926. doi: 10.1242/jcs.03459

Harris, D. A., Huber, M. T., van Dijken, P., Shyng, S. L., Chait, B. T., and Wang, R. (1993). Processing of a cellular prion protein: identification of N- and C-terminal cleavage sites. *Biochemistry* 32, 1009–1016. doi: 10.1021/bi00055a003

Hartmann, C. A., Martins, V. R., and Lima, F. R. (2013). High levels of cellular prion protein improve astrocyte development. *FEBS Lett.* 587, 238–244. doi: 10.1016/j.febslet.2012.11.032

Hidaka, K., Shirai, M., Lee, J. K., Wakayama, T., Kodama, I., Schneider, M. D., et al. (2010). The cellular prion protein identifies bipotential cardiomyogenic progenitors. *Circ. Res.* 106, 111–119. doi: 10.1161/CIRCRESAHA.109.209478

Hirsch, T. Z., Hernandez-Rapp, J., Martin-Lannerée, S., Launay, J. M., and Mouillet-Richard, S. (2014). PrP signalling in neurons: from basics to clinical challenges. *Biochimie.* 104, 2–11. doi: 10.1016/j.biochi.2014.06.009

Khalife, M., Young, R., Passet, B., Halliez, S., Vilotte, M., Jaffrezic, F., et al. (2011). Transcriptomic analysis brings new insight into the biological role of the prion protein during mouse embryogenesis. *PLoS ONE* 6:e23253. doi: 10.1371/journal.pone.0023253

Klyubin, I., Nicoll, A. J., Khalili-Shirazi, A., Farmer, M., Canning, S., Mably, A., et al. (2014). Peripheral administration of a humanized anti-PrP antibody blocks Alzheimer's disease abeta synaptotoxicity. *J. Neurosci.* 34, 6140–6145. doi: 10.1523/JNEUROSCI.3526-13.2014

Lauren, J., Gimbel, D. A., Nygaard, H. B., Gilbert, J. W., and Strittmatter, S. M. (2009). Cellular prion protein mediates impairment of synaptic plasticity by amyloid-beta oligomers. *Nature* 457, 1128–1132. doi: 10.1038/nature07761

Lee, K. S., Magalhaes, A. C., Zanata, S. M., Brentani, R. R., Martins, V. R., and Prado, M. A. (2001). Internalization of mammalian fluorescent cellular prion protein and N-terminal deletion mutants in living cells. *J. Neurochem.* 79, 79–87. doi: 10.1046/j.1471-4159.2001.00529.x

Lee, Y. J., and Baskakov, I. V. (2010). Treatment with normal prion protein delays differentiation and helps to maintain high proliferation activity in human embryonic stem cells. *J. Neurochem.* 114, 362–373. doi: 10.1111/j.1471-4159.2010.06601.x

Lee, Y. J., and Baskakov, I. V. (2013). The cellular form of the prion protein is involved in controlling cell cycle dynamics, self-renewal, and the fate of human embryonic stem cell differentiation. *J. Neurochem.* 124, 310–322. doi: 10.1111/j.1471-4159.2012.07913.x

Le Pichon, C. E., Valley, M. T., Polymenidou, M., Chesler, A. T., Sagdullaev, B. T., Aguzzi, A., et al. (2009). Olfactory behavior and physiology are disrupted in prion protein knockout mice. *Nat. Neurosci.* 12, 60–69. doi: 10.1038/nn.2238

Lima, F. R., Arantes, C. P., Muras, A. G., Nomizo, R., Brentani, R. R., and Martins, V. R. (2007). Cellular prion protein expression in astrocytes modulates neuronal survival and differentiation. *J. Neurochem.* 103, 2164–2176. doi: 10.1111/j.1471-4159.2007.04904.x

Loubet, D., Dakowski, C., Pietri, M., Pradines, E., Bernard, S., Callebert, J., et al. (2012). Neuritogenesis: the prion protein controls beta1 integrin signaling activity. *FASEB J.* 26, 678–690. doi: 10.1096/fj.11-185579

Makhzami, S., Passet, B., Halliez, S., Castille, J., Moazami-Goudarzi, K., Duchesne, A., et al. (2014). The prion protein family: a view from the placenta. *Front. Cell Dev. Biol.* 2:35. doi: 10.3389/fcell.2014.00035

Malaga-Trillo, E., Solis, G. P., Schrock, Y., Geiss, C., Luncz, L., Thomanetz, V., et al. (2009). Regulation of embryonic cell adhesion by the prion protein. *PLoS Biol.* 7:e55. doi: 10.1371/journal.pbio.1000055

Mallucci, G., Dickinson, A., Linehan, J., Klohn, P. C., Brandner, S., and Collinge, J. (2003). Depleting neuronal PrP in prion infection prevents disease and reverses spongiosis. *Science* 302, 871–874. doi: 10.1126/science.1090187

Mange, A., Milhavet, O., Umlauf, D., Harris, D., and Lehmann, S. (2002). PrP-dependent cell adhesion in N2a neuroblastoma cells. *FEBS Lett.* 514, 159–162. doi: 10.1016/S0014-5793(02)02338-4

Manson, J. C., Clarke, A. R., Hooper, M. L., Aitchison, L., McConnell, I., and Hope, J. (1994a). 129/Ola mice carrying a null mutation in PrP that abolishes mRNA production are developmentally normal. *Mol. Neurobiol.* 8, 121–127. doi: 10.1007/BF02780662

Manson, J. C., Clarke, A. R., McBride, P. A., McConnell, I., and Hope, J. (1994b). PrP gene dosage determines the timing but not the final intensity or distribution of lesions in scrapie pathology. *Neurodegeneration* 3, 331–340.

Manson, J., West, J. D., Thomson, V., McBride, P., Kaufman, M. H., and Hope, J. (1992). The prion protein gene: a role in mouse embryogenesis? *Development* 115, 117–122.

Martin-Lannerée, S., Hirsch, T. Z., Hernandez-Rapp, J., Halliez, S., Vilotte, J.-L., Launay, J.-M., et al. (2014). PrP^C from stem cells to cancer. *Front. Cell Dev. Biol.* 2:55. doi: 10.3389/fcell.2014.00055

Martins, V. R., Beraldo, F. H., Hajj, G. N., Lopes, M. H., Lee, K. S., Prado, M. A., et al. (2010). Prion protein: orchestrating neurotrophic activities. *Curr. Issues Mol. Biol.* 12, 63–86.

Massimino, M. L., Redaelli, M., Bertoli, A., Sorgato, M. C., and Mucignat-Caretta, C. (2013). Altered behavioral aspects of aged mice lacking the cellular prion protein. *Physiol. Behav.* 119, 86–91. doi: 10.1016/j.physbeh.2013.06.006

Meotti, F. C., Carqueja, C. L., Gadotti Vde, M., Tasca, C. I., Walz, R., and Santos, A. R. (2007). Involvement of cellular prion protein in the nociceptive response in mice. *Brain Res.* 1151, 84–90. doi: 10.1016/j.brainres.2007.03.024

Miranda, A., Pericuesta, E., Ramirez, M. A., and Gutierrez-Adan, A. (2011). Prion protein expression regulates embryonic stem cell pluripotency and differentiation. *PLoS ONE* 6:e18422. doi: 10.1371/journal.pone.0018422

Miyazawa, K., Emmerling, K., and Manuelidis, L. (2010). Proliferative arrest of neural cells induces prion protein synthesis, nanotube formation, and cell-to-cell contacts. *J. Cell. Biochem.* 111, 239–247. doi: 10.1002/jcb.22723

Morel, E., Fouquet, S., Chateau, D., Yvernault, L., Frobert, Y., Pincon-Raymond, M., et al. (2004). The cellular prion protein PrPc is expressed in human enterocytes in cell-cell junctional domains. *J. Biol. Chem.* 279, 1499–1505. doi: 10.1074/jbc.M308578200

Morel, E., Fouquet, S., Strup-Perrot, C., Pichol Thievend, C., Petit, C., Loew, D., et al. (2008). The cellular prion protein PrP(c) is involved in the proliferation of epithelial cells and in the distribution of junction-associated proteins. *PLoS ONE* 3:e3000. doi: 10.1371/journal.pone.0003000

Nico, P. B., de-Paris, F., Vinade, E. R., Amaral, O. B., Rockenbach, I., Soares, B. L., et al. (2005). Altered behavioural response to acute stress in mice lacking cellular prion protein. *Behav. Brain Res.* 162, 173–181. doi: 10.1016/j.bbr.2005.02.003

Nicolae, C., Ko, Y. P., Miosge, N., Niehoff, A., Studer, D., Enggist, L., et al. (2007). Abnormal collagen fibrils in cartilage of matrilin-1/matrilin-3-deficient mice. *J. Biol. Chem.* 282, 22163–22175. doi: 10.1074/jbc.M610994200

Nicolas, O., Gavin, R., and del Rio, J. A. (2009). New insights into cellular prion protein (PrPc) functions: the "ying and yang" of a relevant protein. *Brain Res. Rev.* 61, 170–184. doi: 10.1016/j.brainresrev.2009.06.002

Nieznanski, K., Nieznanska, H., Skowronek, K. J., Osiecka, K. M., and Stepkowski, D. (2005). Direct interaction between prion protein and tubulin. *Biochem. Biophys. Res. Commun.* 334, 403–411. doi: 10.1016/j.bbrc.2005.06.092

Nourizadeh-Lillabadi, R., Seilo Torgersen, J., Vestrheim, O., Konig, M., Alestrom, P., and Syed, M. (2010). Early embryonic gene expression profiling of zebrafish prion protein (Prp2) morphants. *PLoS ONE* 5:e13573. doi: 10.1371/journal.pone.0013573

Nuvolone, M., Kana, V., Hutter, G., Sakata, D., Mortin-Toth, S. M., Russo, G., et al. (2013). SIRPalpha polymorphisms, but not the prion protein, control phagocytosis of apoptotic cells. *J. Exp. Med.* 210, 2539–2552. doi: 10.1084/jem.20131274

Passet, B., Young, R., Makhzami, S., Vilotte, M., Jaffrezic, F., Halliez, S., et al. (2012). Prion protein and Shadoo are involved in overlapping embryonic pathways and trophoblastic development. *PLoS ONE* 7:e41959. doi: 10.1371/journal.pone.0041959

Peralta, O. A., Huckle, W. R., and Eyestone, W. H. (2011). Expression and knockdown of cellular prion protein (PrPC) in differentiating mouse embryonic stem cells. *Differentiation* 81, 68–77. doi: 10.1016/j.diff.2010.09.181

Petit, C. S., Besnier, L., Morel, E., Rousset, M., and Thenet, S. (2013). Roles of the cellular prion protein in the regulation of cell-cell junctions and barrier function. *Tissue Barriers* 1:e24377. doi: 10.4161/tisb.24377

Prestori, F., Rossi, P., Bearzatto, B., Laine, J., Necchi, D., Diwakar, S., et al. (2008). Altered neuron excitability and synaptic plasticity in the cerebellar granular layer of juvenile prion protein knock-out mice with impaired motor control. *J. Neurosci.* 28, 7091–7103. doi: 10.1523/JNEUROSCI.0409-08.2008

Prodromidou, K., Papastefanaki, F., Sklaviadis, T., and Matsas, R. (2014). Functional cross-talk between the cellular prion protein and the neural cell adhesion molecule is critical for neuronal differentiation of neural stem/precursor cells. *Stem Cells* 32, 1674–1687. doi: 10.1002/stem.1663

Prusiner, S. B., Groth, D., Serban, A., Koehler, R., Foster, D., Torchia, M., et al. (1993). Ablation of the prion protein (PrP) gene in mice prevents scrapie and facilitates production of anti-PrP antibodies. *Proc. Natl. Acad. Sci. U.S.A.* 90, 10608–10612. doi: 10.1073/pnas.90.22.10608

Rial, D., Duarte, F. S., Xikota, J. C., Schmitz, A. E., Dafre, A. L., Figueiredo, C. P., et al. (2009). Cellular prion protein modulates age-related behavioral and neurochemical alterations in mice. *Neuroscience* 164, 896–907. doi: 10.1016/j.neuroscience.2009.09.005

Rial, D., Pandolfo, P., Bitencourt, R. M., Pamplona, F. A., Moreira, K. M., Hipolide, D., et al. (2014). Cellular prion protein (PrP(C)) modulates ethanol-induced behavioral adaptive changes in mice. *Behav. Brain Res.* 271, 325–332. doi: 10.1016/j.bbr.2014.05.067

Richt, J. A., Kasinathan, P., Hamir, A. N., Castilla, J., Sathiyaseelan, T., Vargas, F., et al. (2007). Production of cattle lacking prion protein. *Nat. Biotechnol.* 25, 132–138. doi: 10.1038/nbt1271

Rivera-Milla, E., Oidtmann, B., Panagiotidis, C. H., Baier, M., Sklaviadis, T., Hoffmann, R., et al. (2006). Disparate evolution of prion protein domains and the distinct origin of Doppel- and prion-related loci revealed by fish-to-mammal comparisons. *FASEB J.* 20, 317–319. doi: 10.1096/fj.05-4279fje

Roffe, M., Beraldo, F. H., Bester, R., Nunziante, M., Bach, C., Mancini, G., et al. (2010). Prion protein interaction with stress-inducible protein 1 enhances neuronal protein synthesis via mTOR. *Proc. Natl. Acad. Sci. U.S.A.* 107, 13147–13152. doi: 10.1073/pnas.1000784107

Roucou, X., Gains, M., and LeBlanc, A. C. (2004). Neuroprotective functions of prion protein. *J. Neurosci. Res.* 75, 153–161. doi: 10.1002/jnr.10864

Sanchez-Alavez, M., Criado, J. R., Klein, I., Moroncini, G., and Conti, B. (2008). Hypothalamic-pituitary-adrenal axis disregulation in PrPC-null mice. *Neuroreport* 19, 1473–1477. doi: 10.1097/WNR.0b013e32830f1e90

Santamaria, D., and Ortega, S. (2006). Cyclins and CDKS in development and cancer: lessons from genetically modified mice. *Front. Biosci.* 11, 1164–1188. doi: 10.2741/1871

Santos, T. G., Beraldo, F. H., Hajj, G. N., Lopes, M. H., Roffe, M., Lupinacci, F. C., et al. (2013). Laminin-gamma1 chain and stress inducible protein 1 synergistically mediate PrPC-dependent axonal growth via Ca2+ mobilization in dorsal root ganglia neurons. *J. Neurochem.* 124, 210–223. doi: 10.1111/jnc.12091

Santos, T. G., Silva, I. R., Costa-Silva, B., Lepique, A. P., Martins, V. R., and Lopes, M. H. (2011). Enhanced neural progenitor/stem cells self-renewal via the interaction of stress-inducible protein 1 with the prion protein. *Stem Cells* 29, 1126–1136. doi: 10.1002/stem.664

Santuccione, A., Sytnyk, V., Leshchyns'ka, I., and Schachner, M. (2005). Prion protein recruits its neuronal receptor NCAM to lipid rafts to activate p59fyn and to enhance neurite outgrowth. *J. Cell Biol.* 169, 341–354. doi: 10.1083/jcb.200409127

Schmitt-Ulms, G., Legname, G., Baldwin, M. A., Ball, H. L., Bradon, N., Bosque, P. J., et al. (2001). Binding of neural cell adhesion molecules (N-CAMs) to the cellular prion protein. *J. Mol. Biol.* 314, 1209–1225. doi: 10.1006/jmbi.2000.5183

Schneider, K., Korkmaz, Y., Addicks, K., Lang, H., and Raab, W. H. (2007). Prion protein (PrP) in human teeth: an unprecedented pointer to PrP's function. *J. Endod.* 33, 110–113. doi: 10.1016/j.joen.2006.11.010

Schrock, Y., Solis, G. P., and Stuermer, C. A. (2009). Regulation of focal adhesion formation and filopodia extension by the cellular prion protein (PrPC). *FEBS Lett.* 583, 389–393. doi: 10.1016/j.febslet.2008.12.038

Sherman, D. L., and Brophy, P. J. (2005). Mechanisms of axon ensheathment and myelin growth. *Nat. Rev. Neurosci.* 6, 683–690. doi: 10.1038/nrn1743

Solis, G. P., Schrock, Y., Hulsbusch, N., Wiechers, M., Plattner, H., and Stuermer, C. A. (2012). Reggies/flotillins regulate E-cadherin-mediated cell contact formation by affecting EGFR trafficking. *Mol. Biol. Cell* 23, 1812–1825. doi: 10.1091/mbc.E11-12-1006

Steele, A. D., Emsley, J. G., Ozdinler, P. H., Lindquist, S., and Macklis, J. D. (2006). Prion protein (PrPc) positively regulates neural precursor proliferation during developmental and adult mammalian neurogenesis. *Proc. Natl. Acad. Sci. U.S.A.* 103, 3416–3421. doi: 10.1073/pnas.0511290103

Steele, A. D., Lindquist, S., and Aguzzi, A. (2007). The prion protein knockout mouse: a phenotype under challenge. *Prion* 1, 83–93. doi: 10.4161/pri.1.2.4346

Stella, R., Massimino, M. L., Sandri, M., Sorgato, M. C., and Bertoli, A. (2010). Cellular prion protein promotes regeneration of adult muscle tissue. *Mol. Cell. Biol.* 30, 4864–4876. doi: 10.1128/MCB.01040-09

Stuermer, C. A. (2011). Reggie/flotillin and the targeted delivery of cargo. *J. Neurochem.* 116, 708–713. doi: 10.1111/j.1471-4159.2010.07007.x

Taylor, D. R., Whitehouse, I. J., and Hooper, N. M. (2009). Glypican-1 mediates both prion protein lipid raft association and disease isoform formation. *PLoS Pathog.* 5:e1000666. doi: 10.1371/journal.ppat.1000666

Tobler, I., Gaus, S. E., Deboer, T., Achermann, P., Fischer, M., Rulicke, T., et al. (1996). Altered circadian activity rhythms and sleep in mice devoid of prion protein. *Nature* 380, 639–642. doi: 10.1038/380639a0

Tremblay, P., Bouzamondo-Bernstein, E., Heinrich, C., Prusiner, S. B., and DeArmond, S. J. (2007). Developmental expression of PrP in the post-implantation embryo. *Brain Res.* 1139, 60–67. doi: 10.1016/j.brainres.2006.12.055

van Rosmalen, J. W., Born, J. M., and Martens, G. J. (2006). Prion protein mRNA expression in *Xenopus laevis*: no induction during melanotrope cell activation. *Brain Res.* 1075, 20–25. doi: 10.1016/j.brainres.2005.12.105

Verdu, E., Ceballos, D., Vilches, J. J., and Navarro, X. (2000). Influence of aging on peripheral nerve function and regeneration. *J. Peripher. Nerv. Syst.* 5, 191–208. doi: 10.1046/j.1529-8027.2000.00026.x

Viegas, P., Chaverot, N., Enslen, H., Perriere, N., Couraud, P. O., and Cazaubon, S. (2006). Junctional expression of the prion protein PrPC by brain endothelial cells: a role in trans-endothelial migration of human monocytes. *J. Cell Sci.* 119, 4634–4643. doi: 10.1242/jcs.03222

Walz, R., Amaral, O. B., Rockenbach, I. C., Roesler, R., Izquierdo, I., Cavalheiro, E. A., et al. (1999). Increased sensitivity to seizures in mice lacking cellular prion protein. *Epilepsia* 40, 1679–1682. doi: 10.1111/j.1528-1157.1999.tb01583.x

Watts, J. C., Huo, H., Bai, Y., Ehsani, S., Jeon, A. H., Shi, T., et al. (2009). Interactome analyses identify ties of PrP and its mammalian paralogs to oligomannosidic N-glycans and endoplasmic reticulum-derived chaperones. *PLoS Pathog.* 5:e1000608. doi: 10.1371/annotation/9eb11869-6acb-49b0-978e-abedc3cc545a

Wong, B. S., Liu, T., Li, R., Pan, T., Petersen, R. B., Smith, M. A., et al. (2001). Increased levels of oxidative stress markers detected in the brains of mice devoid of prion protein. *J. Neurochem.* 76, 565–572. doi: 10.1046/j.1471-4159.2001.00028.x

Wopfner, F., Weidenhofer, G., Schneider, R., von Brunn, A., Gilch, S., Schwarz, T. F., et al. (1999). Analysis of 27 mammalian and 9 avian PrPs reveals high conservation of flexible regions of the prion protein. *J. Mol. Biol.* 289, 1163–1178. doi: 10.1006/jmbi.1999.2831

Wu, Z., Liang, F., Hong, B., Young, J. C., Sussman, M. R., Harper, J. F., et al. (2002). An endoplasmic reticulum-bound Ca(2+)/Mn(2+) pump, ECA1, supports plant growth and confers tolerance to Mn(2+) stress. *Plant Physiol.* 130, 128–137. doi: 10.1104/pp.004440

Yoon, B. S., Ovchinnikov, D. A., Yoshii, I., Mishina, Y., Behringer, R. R., and Lyons, K. M. (2005). Bmpr1a and Bmpr1b have overlapping functions and are essential for chondrogenesis *in vivo*. *Proc. Natl. Acad. Sci. U.S.A.* 102, 5062–5067. doi: 10.1073/pnas.0500031102

Young, R., Passet, B., Vilotte, M., Cribiu, E. P., Béringue, V., Le Provost, F., et al. (2009). The prion or the related Shadoo protein is required for early mouse embryogenesis. *FEBS Lett.* 583, 3296–3300. doi: 10.1016/j.febslet.2009.09.027

Yu, G., Chen, J., Xu, Y., Zhu, C., Yu, H., Liu, S., et al. (2009). Generation of goats lacking prion protein. *Mol. Reprod. Dev.* 76, 3. doi: 10.1002/mrd.20960

Zhang, C. C., Steele, A. D., Lindquist, S., and Lodish, H. F. (2006). Prion protein is expressed on long-term repopulating hematopoietic stem cells and is important for their self-renewal. *Proc. Natl. Acad. Sci. U.S.A.* 103, 2184–2189. doi: 10.1073/pnas.0510577103

Zhang, Y., Kim, S. O., Opsahl-Vital, S., Ho, S. P., Souron, J. B., Kim, C., et al. (2011). Multiple effects of the cellular prion protein on tooth development. *Int. J. Dev. Biol.* 55, 953–960. doi: 10.1387/ijdb.113348yz

Conflict of Interest Statement: The authors declare that the research was conducted in the absence of any commercial or financial relationships that could be construed as a potential conflict of interest.

11

Induced pluripotent stem cells: applications in regenerative medicine, disease modeling, and drug discovery

Vimal K. Singh[1]*, Manisha Kalsan[2], Neeraj Kumar[2], Abhishek Saini[2] and Ramesh Chandra[3]

[1] *INSPIRE Faculty, Stem Cell Research Laboratory, Department of Biotechnology, Delhi Technological University, Delhi, India*
[2] *Stem Cell Research Laboratory, Department of Biotechnology, Delhi Technological University, Delhi, India*
[3] *B. R. Ambedkar Centre for Biomedical Research, University of Delhi, Delhi, India*

Edited by:
Thimios Mitsiadis, University of Zurich, Switzerland

Reviewed by:
Eumorphia Remboutsika, Biomedical Sciences Research Center "Alexander Fleming," Greece
Marco Emilio Bianchi, Università Vita Salute San Raffaele, Italy
Nicola Baldini, University of Bologna, Italy

Correspondence:
Vimal K. Singh, Stem Cell Research Laboratory, Department of Biotechnology, Delhi Technological University, Shahbad Daulatpur, Bawana Road, Delhi-110042, India
e-mail: vim_kissor@yahoo.co.in; vimalkishorsingh@gmail.com; vimalksingh@dce.ac.in

Recent progresses in the field of Induced Pluripotent Stem Cells (iPSCs) have opened up many gateways for the research in therapeutics. iPSCs are the cells which are reprogrammed from somatic cells using different transcription factors. iPSCs possess unique properties of self renewal and differentiation to many types of cell lineage. Hence could replace the use of embryonic stem cells (ESC), and may overcome the various ethical issues regarding the use of embryos in research and clinics. Overwhelming responses prompted worldwide by a large number of researchers about the use of iPSCs evoked a large number of peple to establish more authentic methods for iPSC generation. This would require understanding the underlying mechanism in a detailed manner. There have been a large number of reports showing potential role of different molecules as putative regulators of iPSC generating methods. The molecular mechanisms that play role in reprogramming to generate iPSCs from different types of somatic cell sources involves a plethora of molecules including miRNAs, DNA modifying agents (viz. DNA methyl transferases), NANOG, etc. While promising a number of important roles in various clinical/research studies, iPSCs could also be of great use in studying molecular mechanism of many diseases. There are various diseases that have been modeled by uing iPSCs for better understanding of their etiology which maybe further utilized for developing putative treatments for these diseases. In addition, iPSCs are used for the production of patient-specific cells which can be transplanted to the site of injury or the site of tissue degeneration due to various disease conditions. The use of iPSCs may eliminate the chances of immune rejection as patient specific cells may be used for transplantation in various engraftment processes. Moreover, iPSC technology has been employed in various diseases for disease modeling and gene therapy. The technique offers benefits over other similar techniques such as animal models. Many toxic compounds (different chemical compounds, pharmaceutical drugs, other hazardous chemicals, or environmental conditions) which are encountered by humans and newly designed drugs may be evaluated for toxicity and effects by using iPSCs. Thus, the applications of iPSCs in regenerative medicine, disease modeling, and drug discovery are enormous and should be explored in a more comprehensive manner.

Keywords: iPSC, reprogramming, pluripotency, differentiation, disease modeling, drug discovery, gene therapy

INTRODUCTION

Discovery of self-renewal by any living cell was one of the major breakthrough reported by Till and McCulloch (1961) who while subjecting the mice with lethal doses of radiation followed by injection of bone marrow cells found that these cells formed clumps due to cells cloned from them which was the main reason of survival of the mice (Till and McCulloch, 1961). Later studies defined their potential for differentiation into different cell types and self-renewal without senescence, and termed as Stem Cells. Stem cells can be defined on the basis of their origin and potency into Adult Stem Cells and Embryonic Stem Cells (ESC), similarly, considering their potency as the base of classification, stem cells can be classified into unipotent, multipotent, oligopotent, pluripotent, and totipotent (Lan et al., 2012). Totipotent are those which can differentiate into embryonic as well as extra-embryonic tissues such as the placenta. Pluripotent stem cells are those which can differentiate into other cells of the adult body. This property exists for only a specific time period of pre-implantation development in the cells forming Inner Cell Mass (ICM). As the cells differentiate into other cell lineages, their self renewing potential decreases due to various epigenetic changes which leads to the loss of pluripotency. Further research conducted on human stem cells (HSCs) made burgeoning use of human ESCs for which embryo needed to be isolated on regular basis that evoked several ethical issues among socio-research communities. There have been efforts by different research groups to bypass these ethical/technical problems. Subsequent studies demonstrated many advancements in the related techniques such as cloning in frog (Gurdon, 1962), generation of mouse Stem Cells (SCs) (Evans and Kaufman, 1981), cloning in sheep (Wilmut et al., 1997), generation of HSCs (Thomson et al., 1998), and development of Embryonic stem cell fusion technique (Tada et al., 2001). These findings contributed greatly to the development of cells that could eliminate ethical issues (**Figure 1**). However, the major breakthrough came in 2006 when Takahashi and Yamanaka introduced the concept of induced pluripotent stem cells (iPSCs) by generating stem cells that were having properties relating to ESCs. iPSCs were generated by using a combination of 4 reprogramming factors, including Oct4 (Octamer binding transcription factor-4), Sox2 (Sex determining region Y)-box 2, Klf4 (Kruppel Like Factor-4), and c-Myc and were demonstrated both self-renewing and differentiating like ESCs, and thus, could be used as an alternative for hESCs in various clinics/research. Since then, a number of different reprogramming factors/methods have been established. iPSCs generation may employ combination of different reprogramming factors, viz., a cocktail of various reprogramming factors, direct use of proteins, miRNA, peptide etc. Factors (LIN28 + Nanog, Esrrb, Pax5 shRNA, C/EBPa, p53 siRNA, UTF1, DNMT shRNA, Wnt3a, SV40 LT(T), hTERT) or chemicals (BIX-01294, BayK8644, RG108, AZA, dexamethasone, VPA, TSA, SAHA, PD025901 + CHIR99021(2i), A-83-01) have been found which are able to replace one or the other reprogramming factors from basal reprogramming factors or to enhance the efficiency of reprogramming (**Table 1**) (Yu et al., 2007; Dimos et al., 2008; Hanna et al., 2008; Huangfu et al., 2008a,b; Mali et al., 2008; Marson et al., 2008; Mikkelsen et al., 2008; Park et al., 2008a; Shi et al., 2008a,b; Silva et al., 2008; Zhao et al., 2008; Zhou et al., 2008; Feng et al., 2009; Li et al., 2009; Fong et al., 2011).

Similarly, there have been significant advancements in methods/techniques employed to deliver these factors such as Nuclear Transfer, Transcription Factors (Oct4, Sox2, Klf4 and c-Myc) and different reprogramming technologies, including Integration viral vectors and Integration free vectors (Plasmid DNA and transferase, Viral vectors, and Excision of integrated transgenes) (Hochedlinger and Jaenisch, 2002; Takahashi and Yamanaka, 2006a; Takahashi et al., 2006b; Yu et al., 2007; Stadtfeld et al., 2008a,b,a,b; Fusaki et al., 2009; Gonzalez et al., 2009; Yusa et al., 2009) (**Table 2**). The various sources, delivery methods and their cognate approaches with different combinations of transcription factors differ significantly in their efficiency (**Tables 2, 3**).

Since their discovery, iPSCs have been used in many research and clinical studies including *disease modeling, regenerative medicine, and drug discovery/drug cytotoxicity studies*. For disease modeling, somatic cells possessing the characteristics of diseased cells from patient are used for the generation of iPSCs and are further utilized for studying the diseases. Similarly, iPSCs are widely utilized for the regeneration of tissue-specific cells for the transplantation to patients of various injuries or degenerative diseases. Studies like drug discovery which implies screening of small molecules, testing of toxicity for assessment of safety, have successfully exploited iPSCs based technique. A large number of reports have accumulated indicating a need for discussing various apects of these techniques and their future perspectives in clinical/research studies. This article focuses the importance of these cells for therapeutics and research benefits in different biological sciences.

REGULATORY FACTORS OF iPSCs GENERATION: AN OVERVIEW

GENERATION OF iPSCs

Theoretically, iPSCs can be generated by using any somatic cell by employing appropriate reprogramming factors and most convenient method for their introduction to somatic cells. iPSC

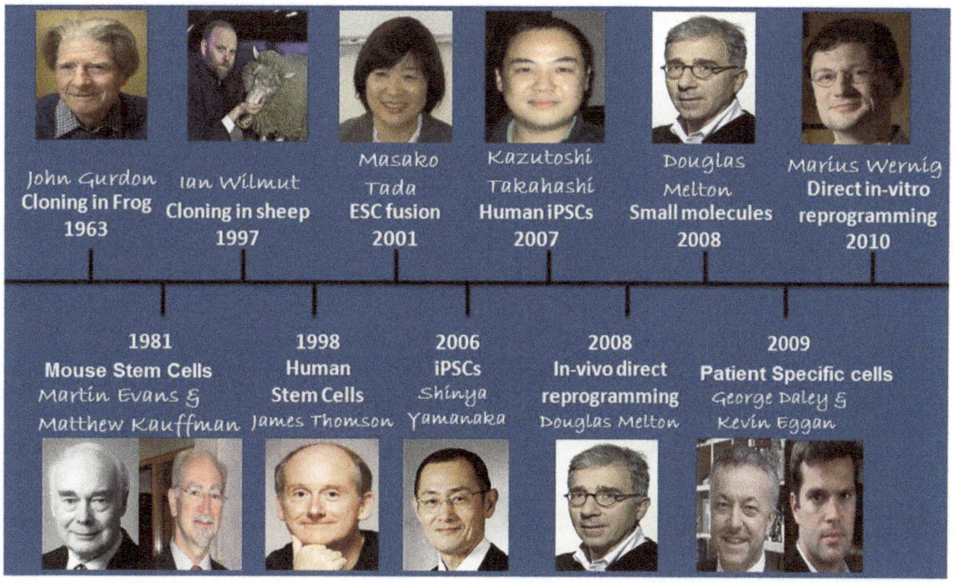

FIGURE 1 | Historical timeline showing events that led to the development of iPSCs and the recent advances that have occurred in the field.

Table 1 | Many factors and chemicals are able to replace the basal transcriptions factors used for reprogramming (O- Oct4; S- Sox2; K- Klf4; M- c-Myc; M*- N-Myc).

Factor/chemical	Function	Able to replace	References
Nanog	ESC-specific transcription factor	Together with Lin28, able of replacing K and M	Yu et al., 2007
Lin28	ESC-specific RNA-binding protein	Together with Nanog, able of replacing K and M	Yu et al., 2007
Esrrb	Orphan nuclear receptor	K	Feng et al., 2009
SV40 LT (T)	SV40 large T antigen used for cell transformation	K; M* and Lin28, Nanog	Mali et al., 2008
BIX-01294	Inhibitor of G9a histone methyltransferase	S, O	Shi et al., 2008a,b
VPA	Inhibitor of histone deacetylase	K and M	Huangfu et al., 2008a

Table 2 | Different delivery methods for transfer of different combinations of transcription factors have different efficiencies of reprogramming (O- Oct4; S- Sox2; K- Klf4; M- c-Myc).

	Reprogramming	Factors	Cell type	Efficiency %	References
Integrating methods	Retroviral transduction	OSKM	Mouse fibroblast	0.001–1	Takahashi and Yamanaka, 2006a
		OSK + VPA	Neonatal	1	Huangfu et al., 2008a
	Lentiviral	OSKM	Human fibroblast	0.1–1	Yu et al., 2007
		OK + parnate + CHIR99021	Neonatal	0.02	Li et al., 2009
	Inducible lentiviral	OSKM	Human fibroblast	0.1–2	Maherali et al., 2008
Non-integrating methods	Sendai virus	OSKM	Human fibroblast	~0.1	Fusaki et al., 2009
	Adeno viral transduction	OSKM	Mouse fibroblast	~0.001	Stadtfeld et al., 2008a
	Plasmid DNA transfer	OSK	Fibroblast	0.00	Okita et al., 2008
	lox p lentivirus	OSKM	Fibroblast	0.1–1	Somers et al., 2010
	PiggyBAC	OSKM	Fibroblast	0.01	Woltjen et al., 2009
	Polyarginine tagged polypeptide	OSKM	Neonatal fibroblast	0.00	Kim et al., 2009b
	RNA modified synthetic mRNA	OSKM	Human fibroblast	4.40	Warren et al., 2010

generation is reported by using cells from different sources such as fibroblasts, cord blood, peripheral blood (**Table 1**). There are large numbers of reports showing the whole process in a detailed manner that can be summarized in three major steps—(i) establishment of initial cell culture, (ii) induction of iPSCs and, (iii) characterization and expansion of iPSCs (**Figure 2**). Firstly, the source cells are isolated and cultured. The reprogramming factors are then introduced into those cultured cells. There are two types of methods for the introduction of reprogramming factors- Integrating Systems (retroviral, lentiviral, inducible lentiviral) and Non-Integrating Methods [non-intergrating viral vectors (sendai virus, adeno virus), plasmid DNA transfer, transgenes, recombinant proteins, synthetic mRNA] (**Table 2**). These transfected cells are incubated on feeder layers (a number of cells have been demonstrated as feeder cells, example., fibroblasts, keratinocytes) under appropriate media conditions and after the expression of these reprogramming factors, generation of iPSCs occurs. The cultured iPSC colonies can be characterized by different morphological and physiochemical methods. The morphological examination of iPSCs can be done on the basis of round shape, large nucleolus, and scant cytoplasm of iPSCs as similar to ESCs (Thomson et al., 1998). Reprogrammed colonies are tightly packed, sharp edged, flat, mitotically very active due to the possession of property of self renewal. Though it is hard to define iPSCs solely on the basis of their morphology, these characteristics give a quick idea about their states. Further, iPSCs may be defined on the basis of expression of different cell surface proteins (SSEA-4, alkaline phosphatase) and transcription factors Oct4, Sox2, Nanog which can also be used for the characterization (Adewumi et al., 2007). **Figure 3** describes the necessary requirements (Yamanaka's cocktail, alternative factors or chemicals for the replacement of one or more basal reprogramming factors enhancement of efficiency of generation) for the generation of iPSCs.

Table 3 | Different cell sources and different combinations of reprogramming factors have been used by different groups for reprogramming to iPSCs (O- Oct4; S- Sox2; K- Klf4; M- c-Myc; N-Nanog, L- Lin28).

Type of cells	Reprogramming factors	References
Fibroblast	OSKM	Takahashi and Yamanaka, 2006a
	OSLN	Yu et al., 2007
Keratinocytes	OSKM	Aasen et al., 2008
Cord blood endothelial cells	OSLN	Haase et al., 2009
Cord blood stem cells	OSKM	Ye et al., 2009
Neural stem cells	O	Kim et al., 2009a
Melanocytes	OSKM	Utikal et al., 2009
Amniotic cells	OSKM	Li et al., 2009
Adipose derived stem cells	OSKM	Sugii et al., 2010
Hepatocytes	OSKM	Liu et al., 2010
Circulating T cells	OSKM	Seki et al., 2010
Astrocytes	OSKM	Ruiz et al., 2010
Peripheral blood	OSKM	Kunisato et al., 2011
Kidney mesangial cells	OSKM	Song et al., 2011
Urine cells	OS	Zhou et al., 2012

DIFFERENT FACTORS THAT PLAY ROLE IN THE GENERATION OF iPSCs

As discussed above, the generation of iPSCs from somatic cells requires the introduction of reprogramming factors into the somatic cells. This can be done by two types of methods—Integrating Viral Vector Systems and Non-Integrating Systems. Integrating methods are those in which the viral vector gets integrated into the host cell genome. The use of retrovirus and lentivirus comes under this category (Takahashi et al., 2006b; Brambrink et al., 2008; Stadtfeld et al., 2008a,b). This method has a high efficiency but it poses the risks of cancer formation and hence, different approaches have also been worked on. Non-Integrating methods are those in which no integration in host cell genome occurs. Different approaches such as viral vectors (Adeno virus and Sendai virus), Plasmid DNA, and the use of RNA and proteins come under this category. Adeno virus (dsDNA) and Sendai virus (RNA) are non-integrating viral vectors which have been used in iPSC generation (Takahashi et al., 2006b; Stadtfeld et al., 2008a; Fusaki et al., 2009). Plasmids such as oriP/EBNA1 (derived from Epstein-bar virus), containing reprogramming factors has been used for reprogramming, but the efficiency was found to be low (Yu et al., 2009). Direct delivery of reprogramming proteins has also been carried out by fusing them with a cell penetrating peptide (Kim et al., 2009b). Yashioka et al. used a different approach using a single self replicating RNA replicon which expressed high levels of Yamanaka's factors for transfection into fibroblasts to be reprogrammed into iPSCs and found the iPSCs to be having all properties of pluripotent stem cells (Yoshioka et al., 2013).

Once the reprogramming factors are delivered to the somatic cells, they come into play for transforming those somatic cells into iPSCs. Not only they have their individual roles, but they interact with each other complementarily.

APPLICATIONS OF iPSCs

The treatment of many diseases is difficult because of the lack of information about the mechanisms that play a role in the disease progression. For this reason, diseases need to be modeled so that treatment could be developed aiming the main cause of the disease. There are a large number of disease testing models which have developed during previous era. Some of them are capable of mimicking human cellular microenvironment and metabolism to some extent. Many animal models such as rat, mice, dogs monkey, dog, and primates have been used for disease modeling. However, the use of animals as disease models is limited due to existing variability in the genetic make-up of them that is highly responsible for the biological functions and hence differences are exhibited while compared with human individuals. Secondly, the problem further gets complicated when the individuals are of two different species. Different species have different genetic makeup and hence different proteins. And thus, none of the animal model is able to fully mimick the human cell microenvironment. So, a different approach which can provide the same environment as in human cells is required and iPSCs pose to be a good alternative with some advantages as well. In case of iPSCs, there is no need of proliferation again and again, and, their derivatives are functional *in-vitro* as well as *in-vivo* after transplantation. Induced pluripotent stem cells are widely used in therapeutics for disease modeling, regenerative medicine, and drug discovery (**Figure 4**).

DISEASE MODELING

The use of iPSCs for disease modeling is based on the fact that these cells are capable of self renewing and that these cells can differentiate into all types of cells of the human body which can be utilized for the preparation of different disease models to study those diseases. Moreover, a patient specific iPSC could be of enormous use as far as development of specific therapeutics regimen/drug is concerned. By combining 3D culture with extracellular matrix proteins, *in-vivo* microenvironment can be mimicked. Lee et al. used iPSCs for the modeling of pathogenesis in Familial Dysautonomia (Lee et al., 2009). Since then, there have been many cases in which iPSCs have helped out in studying various mechanisms that play role in different diseases, a few have been described below.

Moad et al. used human prostate and urinary tract cells for the formation of iPSCs and further for studying the mechanisms that regulate the differentiation of prostate and urinary tract cells. With their study, they reported the first successful reprogramming of bladder, prostate and ureter stromal fibroblasts into a pluripotent state and concluded that iPSCs generated from prostate and urinary tract had better efficiency of differentiation to cells of prostate and urinary tract as compared to iPSCs derived from skin fibroblasts which showed that organ of origin plays an important role in terms of efficiency of differentiation (Moad et al., 2013).

Various types of diseases which are caused by some deficiency have been studied by using iPSCs. Park et al. used

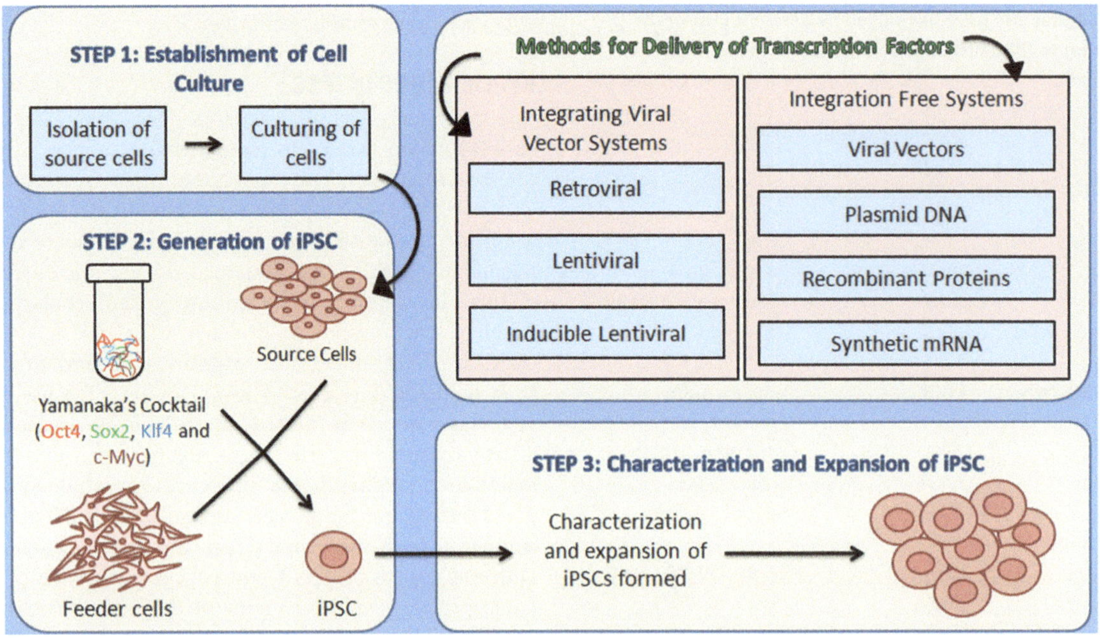

FIGURE 2 | An overview of the methodology for the generation of iPSCs. (1) Establishment of culture: the source cells are cultured for further use as host cells for the delivery of reprogramming proteins. (2) The cultured source cells are then transfected with the four factors from Yamanaka's cocktail and incubated on feeder layers that provide the nourishment to host cells and are responsible for the formation of extra cellular matrix, under suitable conditions of media. Two types of methods for the delivery of reprogramming factors into the somatic cells can be used- Integrating viral vector systems and Non-Integrating methods. (3) After the formation of iPSCs, they are characterized by different morphological and physicochemical analyses, which is followed by the expansion of iPSCs.

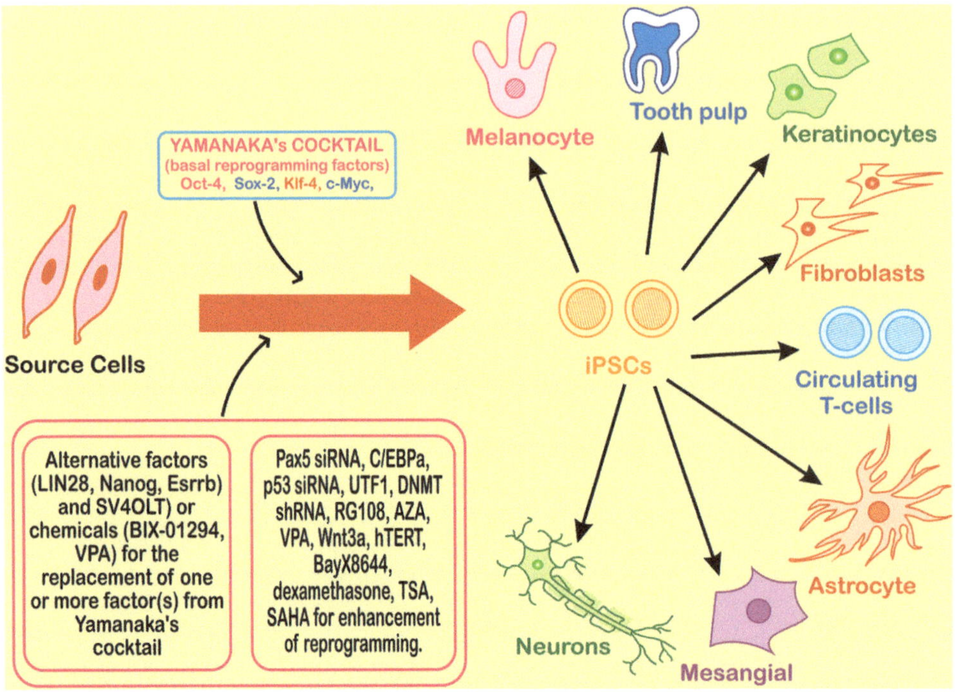

FIGURE 3 | Introduction of the four transcription factors (Oct-4, Sox-2, Klf-4, and c-Myc) leads to reprogramming of a somatic cell to an Induced Pluripotent Stem Cell (iPSC) which can further differentiate into different types of cells. Many factors or chemicals are able to replace one of the factors from the basic four factors required for reprogramming, and for the enhancement many other small molecule chemicals or factor are also used.

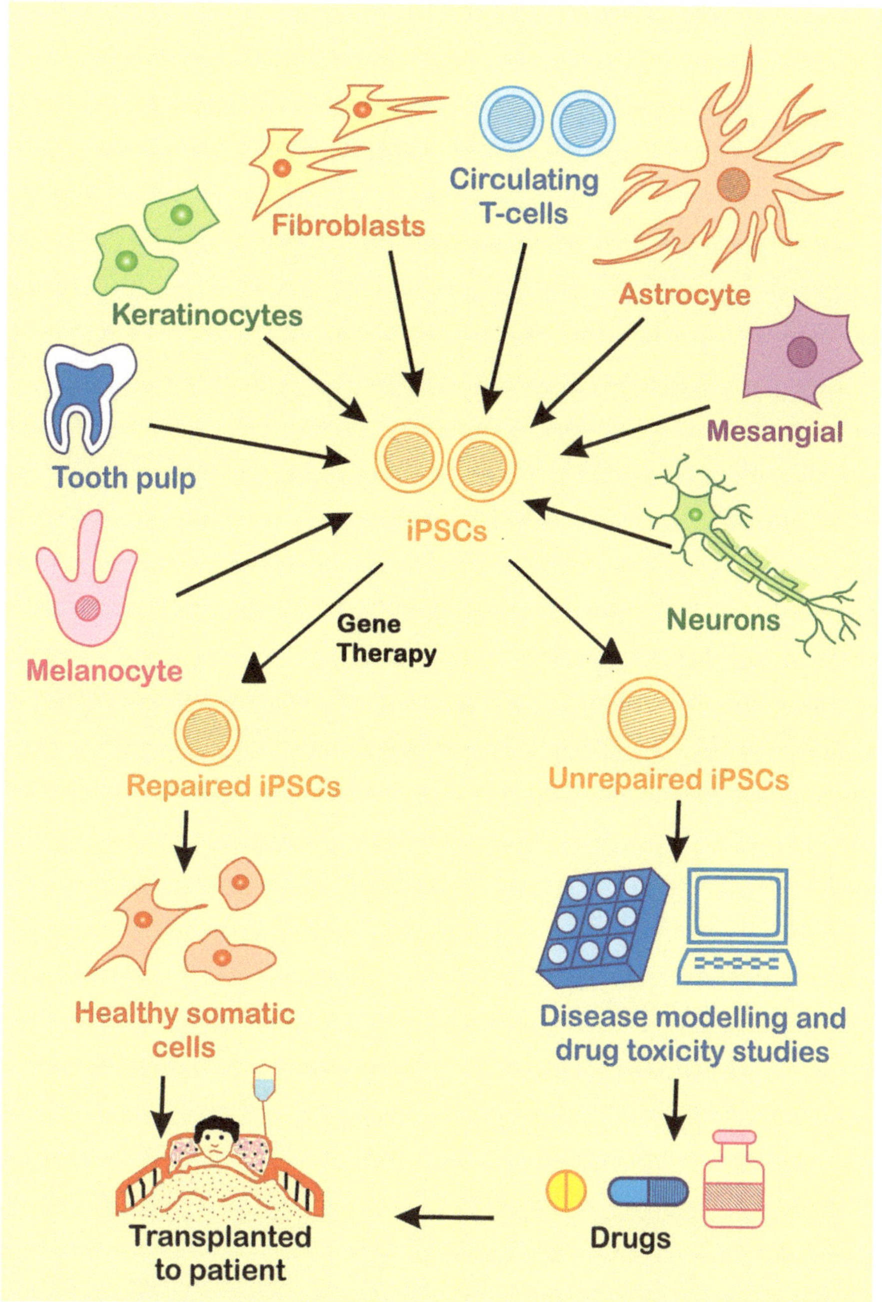

FIGURE 4 | There are many applications of iPSCs in the fields of gene therapy, disease modeling and drug discovery. Somatic cells from the patient are used for the generation of diseased iPSCs. These diseased iPSCs may be repaired by Gene Therapy and further used for the generation of healthy somatic cells to be transplanted to the patient, or they may be used to produce unrepaired somatic cells for disease modeling or drug screening.

iPSCs from patients of various diseases like *Adenosine Deaminase Deficiency-related Severe Combined ImmunoDeficiency (ADA-SCID), Shwachman-Bodian-Diamond syndrome (SBDS), Gaucher disease (GD) type III* for the study of disease models and drug discovery. They used dermal fibroblasts or bone-marrow derived mesenchymal stem cells for the generation of human iPSCs by the transduction of all four or three (excluding c-Myc) transcription factors. It was found from their study that ADA-SCID, SBDS, and Gaucher's disease type III are inherited in a classical Mendelian Inheritance manner like congenital disorders which are autosomal recessive. These diseases were shown to be caused by point mutations in those genes which were vital for normal hematopoiesis and immunological function. They also reported the point mutations in ADA-SCID, SBDS, and GD type III. They also studied *Duchenne (DMD) and Becker muscular dystrophy (BMD), Parkinson disease (PD), Huntington disease (HD), juvenile-onset, type 1 diabetes mellitus (JDM), Down syndrome (DS) /trisomy 21 and the carrier state of Lesch-Nyhan syndrome* (Park et al., 2008b).

There are many syndromes which are caused by the existence of one or more extra copies of a chromosome. Downs syndrome is one such example. Briggs et al. used iPSCs for the identification of molecular networks that drive the different aspects related to pathogenesis in Down's Syndrome. iPSCs in combination with microarray and RNA sequencing technology, can be used to generate phenotype-genotype maps of complex diseases by linking various defects with phenotypes, like in Down's Syndrome using Chromosome engineering of Down Syndrome-iPSCs (Briggs et al., 2013).

Neurogenerative diseases have also been studied using iPSCs as disease models. Parkinson's Disease (PD) is a very common neurodegenerative disease, in which, dopaminergic neurons of substantia nigra (a structure in mid brain) get lost and formation of Lewy's bodies (inclusions in the cytoplasm of neurons all over the body) occurs. Treatment of this disease had not been possible due to the reason that by the time, PD gets clinically manifested, the neurons have already lost, which makes it very difficult to be able to study the underlying mechanisms of PD so as to develop a treatment of it. In such a situation, iPSCs can be used and experiments have also been carried out in this aspect. Nguyen et al. studied G2019S mutation in LRRK2 (Leucine Rich Repeat Kinase2) gene. This mutation has been reported in cases of sporadic and familial PD. Fibroblasts from a 60-year old female homologous for G2019S mutation were used as a source for the generation of iPSCs. Retrovirus was used for the transduction of Oct4, Sox2, and Klf4 into the fibroblasts, which after 2 months of transfection developed 3 colonies of human ESC, which were analyzed and found out to be similar to hESCs and wild type iPSCs. For differentiation of G2019S iPSCs to dopaminergic neurons, a combination of mib-brain selective patterning factors, sonic hedgehog (SHH) (induces the differentiation of dopaminergic neurons in substantia nigra), and Fibroblast growth factor 8 beta (FGF8b) [induces the development of cerebellum by the activation of Ras-extracellular signal-regulated kinase (ERK) pathway] and small molecules was used. The neurons generated by this method were functionally active and were found to be able to form active neural networks. The findings suggested LRRK2 to be having an important role in survival of neurons and disruption of which can cause PD. (Hurtado-Lorenzo et al., 2004; Sato et al., 2004; Nguyen et al., 2011). Similarly, a different group has also worked for the generation of iPSCs in PD. Devine et al. developed iPSCs from fibroblasts taken from a PD affected person possessing triplication of Synuclein gene by the transduction of four basic transcription factors. These iPSCs were then directed to differentiate into dopaminergic neurons *in vitro* for the study of PD (Devine et al., 2011).

These findings from modeling different diseases (mutations which are causing them, the pathological phenotypes caused by defects in various networks, genes playing an important role) help out in knowing the molecular mechanisms underlying the disease better, which ultimately carries the work forward to knowing the disease better for the development of a treatment. This is evident from burgeoning reports on use of IPSc for the the purpose of disease modeling (**Table 4**). However, the technique suffers from various inherited risks and limitations as discussed ahead.

REGENERATIVE MEDICINE

In regenerative medicine the injured or degenerated tissues are repaired by the generation of those tissues with the help of iPSCs in labs and then transplanting them to the site of injury or degeneration. Important issues associated with gene therapy are availability of tissue or organ and immunorejection. It is quite often that patients die because of the non-availability of donors which is a result of the fact that the population is non-aware of the scarcity of donors and the ever inceasing need of organs due to accidents and degenerative diseases. Also, a patient can only be transplated with the cells, tissues, or organs from the person who does not have any disease and whose physiological profile matches with the patient. Keeping these risks in mind, various tests are conducted before transplating tissues or organs into the patient's body. The use of iPSCs offers a good approach for these treatments as the cells that will be transplanted to patient's body will be differentiated from the repaired iPSCs generated from the somatic cells from patient's own body. iPSCs have been used in treating a number of injuries and degenerative diseases. A comprehensive detail of all those would be beyond the scope of this article, however, major arguments are discussed below. Different injuries as a result of human activities, accidents or natural calamities can also be treated by the gene therapy utilizing iPSCs. The various conditions that can be treated are Hematopoietic disorders, Musculoskeletal injury, Spinal cord injury, liver damage by the generation of specific cells with the help of iPSCs (Liu et al., 2011b; Nori et al., 2011; Tan et al., 2012; Suzuki et al., 2013).

Other than accidental injuries, diseases can also be treated with the help of iPSCs. Kazuki et al. corrected the genetic deficiency in iPSCs from a human Duchhene Muscular Dystrophy (DMD) patient. They used Human Artificial Chromosome (HAC) for the expression of complete sequence of Dystrophin (DYS). They used fibroblasts from the DMD patient for the generation of iPSCs. HAC was used for the correction in the deletion or mutation in the Dystrophin gene present in iPSCs by transfer of DYS-HAC (complete genomic sequence of dystrophin in HAC) using Microcell-mediated chromosome transfer (MMCT) (Kazuki et al., 2009). MMCT is a technique in which micro cells containing few chromosomes are fused with whole cells (Doherty and Fisher, 2003).

iPSCS can also be used in cases of hepatocytes where loss of function of hepatocytes in culture and limited organ availability act as obstacles when fetal or adult progenitors are used for the development of hepatocytes. This lead to making use of iPScs for the generation of hepatocytes for various liver problems.

Recently, there are accumulating data on the use of iPSC for *ex-vivo* blood expansion of various blood components. They can be used for the generation of Red Blood Cells (RBCs) which could be utilized for the generation of blood that is required over the globe for the purpose of treatment of various damages or diseases.

Table 4 | List of diseases where iPSCs have been used for gene therapy and disease modeling.

Disease/syndrome	Cause(s)	Features	Cells used for iPSC generation	System used	References
Parkinson's Disease (PD)	Familial forms caused by α-synuclein, ubiquitin carboxy terminal hydroxylase L1, parkin, DJ-1, putative serine threonine kinase 1 and leucine rich repeat kinase 2	Loss in nigrostriated dopaminergic neurons in substantia nigra; presence of Lewy bodies	Dermal fibroblasts of patient with idiopathic PD	Lentiviral	Soldner et al., 2009
Huntington disease (HD)	CAG repeats (36 or more) in the first exon of *htt* gene gets expanded abnormally	Degeneration in striatum and cerebral cortex	Fibroblasts	Lentiviral	Park et al., 2008b; Kaye and Finkbeiner, 2013
ALS or Lou Gehrig's disease	Autosomal dominant mutation in superoxide demutase (SOD1)	Death of motor neurons of the motor cortex, brain stem and spinal cord	Fibroblasts	Lentiviral	Rosen et al., 1993; Chestkov et al., 2014
Friedreich's ataxia (FRDA)	GAA trinucleotide repeat in the first exon of the frataxin gene gets expanded	Accumulation of mitochondrial iron, specific enzymes in mitochondria become defective, sensitivity to oxidative stress increases, cell death mediated by free radicals	Fibroblasts	Lentiviral	Campuzano et al., 1996; Ku et al., 2010
Lesch-Nhyan syndrome (carrier state)	Deficiency of hypoxanthine guanine phospho ribosyl transferase (HPRT)	Over-production of uric acid, low or medium level of mental retardation, megaloblastic anemia is frequent	Dermal fibroblasts	Lentiviral	Park et al., 2008b
Shwachman-Bodian-Diamond syndrome (SBDS)	Mutations in the Shwachman-Bodian-Diamond syndrome (SBDS) gene	Exocrine pancreatic insufficiency, predisposition to leukemia, hematopoietic dysfunction	Fibroblasts	Lentiviral	Tulpule et al., 2013
Gaucher's type III	Deficiency of acid hydrolase, β-glucocerebrosidase, or glucosylceramidase	Myoclonal epilepsy, nerve deafness	Fibroblasts	Lentiviral	Park et al., 2008b
Becker type muscular dystrophy (BMD)	Mutation in dystrophin gene	Loss of walking ability, but progression slower than DMD	Fibroblasts	Lentiviral	Park et al., 2008b
Downs syndrome/trisomy 21	Trisomy of chromosome 21	Cardiac and cognitive defects, premature Alzheimers disease and aging, dysmorphic facial features	Fibroblasts	Lentiviral	Park et al., 2008b; Briggs et al., 2013

(Continued)

Table 4 | Continued

Disease/syndrome	Cause(s)	Features	Cells used for iPSC generation	System used	References
Familial dysautonomia (FD) or Riley-Day syndrome	Autosomal recessive disorder caused by a single mutation in exon 20 in I-K-B kinase complex associated protein (IKBKAP) gene	Dysfunction of small fiber sensory neurons	Fibroblasts	Lentiviral	Lee and Studer, 2011
Childhood cerebral adreno leuko dystrophy (CCALD)	Mutation in *ABCD1* gene	Adrenal cortex, nervous system and testes get affected, leading to rapid cerebral demyelination and adrenocortical atrophy.	Skin fibroblasts	Retroviral	Wang et al., 2012
Rett's syndrome	Classic form caused by loss-of-function mutation in Methyl-CpG-binding protein 2 (MECP2) gene on the X - chromosome, variants caused by mutations in FOXG1 or CDKL1 on chromosome 14 and X-chromosome, respectively	Neurocognitive regression and autistic behavior	Fibroblasts	Retroviral	Amenduni et al., 2011; Farra et al., 2012
Duchenne type muscular dystrophy (DMD)	Biochemical and genetic defects in Dystrophin-glycoprotein complex	Loss of walking ability	Tail tip fibroblasts (mouse)	Retroviral	Gangopadhyay et al., 1992; Filareto et al., 2013
Generation of human prostate and urinary tract cells	NA	NA	Human prostate and urinary tract cells	Lentiviral	Moad et al., 2013
Adenosine deaminase deficiency-related severe combined immunodeficiency (ADA-SCID)	Defects in Adenosine deaminase (AD) gene	Impaired development and functioning of T, B, and NK cells; complete absence of humoral and cellular immunity; recurrence of infections.	Bone Marrow derived mesenchymal cells	Lentiviral	Park et al., 2008b; Sauer et al., 2012
Type 1 diabetes mellitus (DM)	Progressive β-cell destruction	Long term micro and macro-vascular complications.	Fibroblasts	Lentiviral	Park et al., 2008b; Soejitno and Prayudi, 2011
Hemophilia A	Deficiency of factor VIII	Decreased protein production, inefficient clotting of blood	Fibroblasts	Retroviral	Xu et al., 2009
Glycogen storage disease 1a	Mutation in Glucose-6-Phosphate gene	Increased accumulation of lipids and glycogen	Dermal fibroblasts	Retroviral	Lei et al., 1995; Rashid et al., 2010
Familial hypercholestrolaemia	Mutation in low density lipoprotein receptor (LDLR) gene	Deficiency of LDL-receptor mediated uptake of cholesterol	Dermal fibroblasts	Retroviral	Rashid et al., 2010

(Continued)

Table 4 | Continued

Disease/syndrome	Cause(s)	Features	Cells used for iPSC generation	System used	References
Spinal muscular atrophy	Mutation in survival of motor neuron 1 (SMN1) gene	Paralysis, muscle weakness and often death	Fibroblasts	Lentiviral	Ebert et al., 2009
Hutchinson-Gilford progeria syndrome	Point mutations in lamin A	Premature atherosclerosis, vascular smooth muscles gets degraded	Fibroblasts	Retroviral	Liu et al., 2011a
Alzheimer disease	Duplication of amyloid β precursor protein (APP)	Presence of neurofibrillary tangles and amyloid plaques in the brain	Fibroblasts	Retroviral	Israel et al., 2012
LEOPARD syndrome	Mutation in protein tyrosine phosphatase non-receptor type 11 (PTPN11) gene	Cardiac abnormalities, ocular hypertelorism, and growth retardation.	Fibroblasts	Retroviral	Legius et al., 2002; Carvajal-Vergara et al., 2010.
Timothy's syndrome	CACNA1C	Webbed fingers and toes, autism, immune deficiency	Fibroblasts	Retroviral	Yazawa et al., 2011
Dyskeratosis congentia	Mutation in dyskerin (DKC1) gene	Increased failure of bone marrow, pulmonary fibrosis and cancer, oral leykoplakia, abnormal skin pigmentation and nail dystrophy	Fibroblasts	Retroviral	Batista et al., 2011
α1-antitrypsin deficiency	Mutation in α1-antitrypsin (A1AT) gene	misfolded α1-antitrypsin gets aggregated in the endoplasmic reticulum	Dermal fibroblasts	Retroviral	Rashid et al., 2010

There are various techniques by which we can use ESCs/iPSCs for the production of RBCs (Lim et al., 2013).

iPSCs can also be used for the generation of various cells which can help in the repairmen of many tissues, for example, cardiovascular cells for the repairment of heart valves, vessels and ischemic tissues, but are limitations like safe delivery, post treatment adverse effects and standardization of protocols to generate large amounts of pure good quality cells. These obstacles once overcome, offer great opportunities for the applications of iPSCs for generating cardiovascular cells and studying corresponding diseases (Laflamme et al., 2007; Cao et al., 2008; Levenberg et al., 2010).

Various degenerative diseases in which cells death causes many symptoms have been treated by gene therapy using iPSCs. One such case is Retinitis pigmentosa (RP) where eye's retina degeneration causes impaired vision. For the treatment of RP, iPSCs were generated from the patient suffering with the disease which were then showed to differentiate into rod photoreceptor cells (Yoshida et al., 2014). The differentiation of iPSCs to Retinal pigmented epithelium is proved to be beneficial for patients of Retinal Pigmentosa and Age-related Macular Degeneration (AMD) (Li et al., 2012).

Gene therapy using iPSCs has also been used to treat various primary immunodeficiencies like *Chronic Granulomatous disorder (CGD),* an autosomal recessive or X-linked disorder which affects the function of neutrophils, and *Wiskott–Aldrich syndrome (WAS),* X-linked deficiency caused by mutation in WAS gene encoding WAS protein (WASp) which is an actin cytoskeletol regulator in hematopoietic lineage (Klein et al., 2003; Jiang et al., 2012).

The technology of genome editing or genome engineering, which includes the use of different genome editing technologies like Zinc Finger Nucleases (ZFNs), Transcription Activators like effector nucleases (TALENs), CRISPR/Cas systems, Meganucleases, Adeno-associated viruses, and adenoviruses, offers an opportunity for the efficient introduction of various insertions or deletions that range from single nucleotides to insertion/deletion of whole genes which has a capability of taking disease modeling to new heights (Suzuki et al., 2008; Grizot et al., 2009; Hockemeyer et al., 2009, 2011; Khan et al., 2010). Choi et al. reported the gene correction in alpha 1 anti trypsin (AAT) deficiency by the use of TALENs. A pair of TALEN expression vectors was constructed. This construct recognized the sequences flanking the Z mutation of the AAT gene. After co-nucleofection of targeting vector and TALEN expression vector was carried out, PCR and Southern blot was done. The results showed a highly efficient and simultaneous targeting of both alleles (Choi et al., 2013).

Once the specific cells are formed they can be transplanted to the specific site, in case of degenerative diseases, for the cure of the disease. And, in case where diseased cells which possess some mutation, the mutation is first corrected to form normal iPSCs and then these iPSCs are differentiated into specific cell types by providing the specific conditions required for the development of those cells. These repaired cells can then be transplanted into the body of the organism from which the cells for the generation of iPSCs were isolated.

DRUG DISCOVERY AND CYTOTOXICITY STUDIES

Another important role offered by the use of iPSCs based methodology is their complementarity to drug discovery or prediction of toxicity via animal models. Animals or *in-vitro* animal derived cells are used as testing systems butare limited by their inability to to replicate the "exact" human physiological conditions and related phenotypic attributions. Sometimes, the benefits demonstrated in the animal models do not come out to be beneficial in humans, for example, the SOD (superoxidase dimutase) gene associated with ALS (Amyotrophic Lateral Sclerosis) allowed the identification of Vitamin E and Creatine to be relievers of the diseased phenotype which failed to cause any improvements in humans (Desnuelle et al., 2001; Groeneveld et al., 2003; Shefner et al., 2004). And also, animal models are not good testing models for drug toxicity as a chemical may be toxic to an animal but may not be toxic to a different animal. For carcinogenic agents as well, different agents pose different levels of carcinogenicity in different animals, for example, formaldehyde is more carcinogenic to rats as compared to mice (Kerns et al., 1983). Finally, a newly discovered drug or therapy must be tested on human cells or human test models itself. These reasons make it more important to be able to use the systems closer to humans. Moreover, these studies need to be done in a system where the results could be directly extrapolated to humans. These studies include steps such as prediction/identification of a potential drug molecule followed by its synthesis, generation of iPSCs, their differentiation to specific somatic cells, and testing for toxic or non-toxic effects of the synthesized drug on the somatic cells. For toxicity studies, iPSCs from normal and diseased cells are used to generate neurons, hepatocytes, cardiomyocytes etc. Toxicity and potential side-effects are often most common cause to rule out most of the therapeutic molecules.

There is a long pipeline between the time when a hit is identified and the time when it reaches the market. Before using drugs on humans, their toxic effects must be properly evaluated for a safe administration of those drugs which is very costly. Only 10% of the drugs that enter clinical trials are able to reach market approval stage. The cost of developing a drug is increasing with the estimated cost of whole process being US $1.2–1.7 billion per drug compound (Kaitin, 2008; Sollano et al., 2008; Gunaseeli et al., 2010). The development of 30% of the medicines was abandoned because of lack of efficacy and 30% due to concerns associated with safety (cardiotoxicity, hepatotoxicity) (Laustriat et al., 2010). The blocking of human ether-a-go-go related gene 1 (*hERG1*) channel by drug is associated with longer duration of ventricular repolarization which causes long QT syndrome. The response of iPSC-derived cardiomyocytes have been reported in response of cardiac and non-cardiac drugs in terms of electrophysiological capacity (Tanaka et al., 2009; Yokoo et al., 2009; Asai et al., 2010; Yoshida and Yamanaka, 2010). The generation of functional hepatocytes using iPSCs has also been reported (Sullivan et al., 2010). Due to lack in early detection of drug toxicity in human tissues, the process becomes more costly and extensively time consuming. Development of toxicity models that predict more accurately before clinical trials may help to reduce costs by demonstrating cardiotoxicity or hepatotoxicity caused by the drugs much before they reach clinical trials; reducing the time taken for clinical trials of the drugs which will fail due to cytotoxicity in the later stages of trials. iPSCs have been explored by many research groups in order to identify compounds which are toxic or pose side-effects. The use of iPSCs offer better alternative to conventional tests of toxicology and drug research and offer better chemical safety assessment as they provide a more similar environment to human physiological conditions than is provided by conventional testing systems which make use of animals. For example, iPSCs have been used to establish test systems for cytotoxicity–cardiotoxicity, hepatotoxicity, and embryotoxicity testing (Kimmel and Young, 1983; Seiler et al., 2004).

iPSCs have the potential of providing native cells for splicing and post translational modifications that does not come with the small molecule screening systems using purified protein or recombinant cell lines used for the over-expression of the protein of interest which has aided in the process of drug screening. The use of iPSCs in drug screening has also been reported. Stem Cells enable the identification of molecules therapeutically useful and able to modulate the behavior of tumorigenic and normal stem cells. But as the use of ESC is limited because of the concerned ethical issues, iPSCs (normal and tumorigenic) can be used as an alternative.

McNeisch et al. used neurons that were expressing a-amino- 3-hydroxyl-5-methyl-4-isoxazoleproprionate (AMPA) subtype glutamate receptors using mouse iPSCs and used them for the identification of small molecule AMPA potentiators (McNeish et al., 2010). Choi et al. used hepatic cells derived from iPSCs generated from patients with alpha-1 anti-trypsin deficiency for the screening of 3131 compounds of Johns Hopkins Drug Library. After immunofluorescence, total AAT fluorescence intensity was measured for each well. Out of 262 compounds which inhibited the accumulation of AAT by more than 50%, 43 compounds after literature survey were found to be FDA approved, clinically used internationally, and exerting no major side effect. These compounds were further tested by using four different lines of AAT- deficiency iPSCs and final 5 hits, which gave consistent results in all the four iPSC lines were selected (Choi et al., 2013). iPSCs also help in studying ADME (Absorption, Distribution, Metabolism, Excretion) properties of drugs which has highly speeded up drug discovery and development (Bahadduri et al., 2010).

ADVANTAGES OF iPSC TECHNOLOGY

iPSCs offer various advantages as compared with gene therapy by the use of nuclear transfer or Embryonic Cell mediated gene therapy (Kazuki et al., 2009).

THE ETHICAL ISSUES ARE ELIMITED BY THE USE OF iPSCs

The use of ESCs in research is laden with ethical issues regarding personhood, justice toward humankind and human dignity that are associated with the use of human life in its earliest form, the embryo. Additional concerns that surround the use of ESCs are informed consent, improper incentives, and health and safety concerns of the women donating eggs for the generation of embryos by *in-vitro* fertilization. iPSCs have solved the controversy over the destruction of embryos associated with the use of ESCs in research.

REDUCED CHANCES OF IMMUNOREJECTION

iPSCs are generated from the somatic cells of one's own body and hence there is no risk of immunorejection of these autologous cells. Guha et al. differentiated mouse iPSCs into embryoid bodies and different tissue specific cells. After the transplantation of these iPSC derived EBs or tissue specific cells, no evidence of increased Tcell proliferation or an antigen- specific secondary immune response was found (Guha et al., 2013).

THROUGHPUT SCREENING FOR PREDICTING TOXICITY/THERAPEUTIC RESPONSES OF NEWLY DEVELOPED DRUGS

The concept of using iPSCs to predict toxicology and therapeutic responses of drugs in based on the property of iPSCs to continuously self renew which make it possible to generate libraries and their ability to give rise to all types of body cells make them suitable to be used for prediction of toxicity and possible side effects of newly developed drugs in different body cells (Wobus and Loser, 2011).

LOWERING THE OVERALL COST AND RISK OF CLINICAL TRIALS

For every new drug to reach market, around 5000–10,000 compounds need to be tested during preclinical trials. Therefore, any strategy which standardizes the prediction of toxicity would impact the cost. Many drugs fail after Phase III, for example BMS-094 (hepatitis C drug) failed after death of a patient. The cost used for the clinical trials could be reduced by using iPSCs to provide the toxicity details of the drug by different cytotoxicity assays. Much of the cost for preclinical testing is due to the requirement of animal models for estimation of bioavailability of new drug. As animal models do not mimic the microenvironment of the human cells fully, the use of iPSCs for these tests may cut the cost associated with providing animal models which will ultimately decrease the overall costs of the clinical trials.

DEVELOPMENT OF A PERSONALIZED APPROACH FOR ADMINISTRATION OF DRUGS

As, iPSCs are derived from individual patients, these offer scientists an opportunity for modeling diseases on a patient-by-patient basis. This enables screening the genomic differences between individuals that may help in the progression of disease, and the screening of pharmacological agents to find the ideal one for each individual (Chun et al., 2011).

GENE TARGETING AND CORRECTION TECHNOLOGIES (GENE THERAPY)

Reprogramming of somatic cells from any genetic background to iPSCs has allowed the generation of cell lines possessing disease-causing mutations. The ability of modifying the specific sites in a genome for altering the genes of interest becomes highly important here. To realize the full potential of iPScs, methods for efficient genetic modifications are required.

FOR DISEASE MODELING, THE PHENOTYPES NEED TO BE CONSISTENT EVERY TIME, WHICH IS A POSSIBILITY IN CASE OF iPSCs

Many research groups have modeled various diseases using iPSCs and the cell lines developed were found to be consistent phenotypically. Fong et al. developed Tauopathy derived iPSCs carrying a TAU-A152T mutation. The phenotypes observed in the cells from iPSCs were consistent with those in patients with the mutation (Fong et al., 2013).

LIMITATIONS AND/OR RISKS ASSOCIATED WITH THE USE OF iPSCs IC CLINICS

There are some limitations/disadvantages associated with iPSCs as well. Generation of iPSCs make use of retroviral or lentiviral systems, so, it needs to be concerned if the viral systems get incorporated with the host genome. The genetic material inserted via retroviral vectors may randomly integrate into the genome of the host which can cause genetic aberration and teratoma formation (Howe et al., 2008). The transcription of the transgenes inserted via vector systems may resume in the cells that have been formed by the differentiation of iPSCs (Okita et al., 2007). The assessment also becomes complex as it depends on number of variables, like small molecules for the enhancement of iPSCs production are used or not, whether the factors used for reprogramming are contained in single vectors or multiple vectors, and which source has been used for the generation of iPSCs, contribute to the differences in the epigenome, genotype, and phenotype of the iPSCs produced which make it difficult to determine whether the problems occurred due to the technology or the reprogramming method(s) (Howe et al., 2008; Li et al., 2011). It is possible to use iPSCs for modeling of the diseases which have a phenotype and are monogenic but in case of complex diseases the role of genetic and epigenetic factors is uncertain which makes modeling of disease a tough job. It is not possible for iPSCs to model every disease successfully, hence, it should be noted if the iPSCs derivatives are getting properly matured to the cells of interest or not in order to know if they are suitable to be used for the purpose of our interest or not. The altered expressions of the basal reprogramming factors has also been reported to cause diseases. The overexpression of Oct4 may lead to epithelial cell dysplasia (Hochedlinger et al., 2005). The aberrant expression of Sox2 has been reported to cause mucinous colon carcinoma (Park et al., 2008c). Klf4 has role in the formation of breast tumors (Ghaleb et al., 2005). c-Myc plays an important role in the formation of around 70% of human cancers (Kuttler and Mai, 2006). For this reason, retroviral introduction of c-Myc is also advised to be avoided for use in clinical applications (Okita et al., 2007). Zhang et al. reported that out of 593 genes that express in iPSCs, 209 genes were found to express in cancer tissues and tumor cells. Five oncogenes were found to overexpress in iPSCs whereas the oncogene RAB25 was found to express in cells derived from iPSCs (Zhang et al., 2012). RAB25 (member of RAS protein family) plays an important role in survival of cancer cells (Cheng et al., 2013). The different pros and cons of the use of iPSCs in different applications has been mentioned in **Table 5**.

QUALITY CONTROL TESTING OF iPSCs

There are many things which need to be considered before using iPSCs clinically for regenerative medicine. The manufacture of clinical grade iPSCs demand quality control at different levels of the manufacturing process. Firstly, it is important to decide whether the therapy would be autologous or allogeneic. Autologous therapy should be such that it is able to manage the differences in the patient in terms of age, health and gender which could affect the quality, consistency and safety of the final product. It requires long term study of the increased risk of tumor formation as the patient would not be expected to have any immune response. The alternative, allogeneic therapy offer well-characterized cell banks for therapeutics to be established. Donors need to meet strict eligibility requirements and the donor must be aware of the use and testing of his sample and that the

Table 5 | The pros and cons associated with the use of iPSCs.

	Pros	Cons
Due to characteristics of iPSCs	Eliminates ethical issues	Premature aging
	Reduced chances of immunorejection (Guha et al., 2013)	High rate of apoptosis
	Reduced risks of clinical trials	Low level DNA damage repair (Zhang et al., 2012)
	Consistent phenotypes for disease modeling (Fong et al., 2013)	Sensitive to ionizing radiation (Zhang et al., 2013)
	Differentiation to any cell type	Low rate of reprogramming
Due to technology of development	Continuous cell supply	Insertional mutagenesis (Okita et al., 2007; Howe et al., 2008)
	Possible preservation	Tumourogenesis (Okita et al., 2007).
	Availability and accessability of source cells	Chances of development of diseases due to factors used (Ghaleb et al., 2005; Hochedlinger et al., 2005; Kuttler and Mai, 2006; Park et al., 2008c)
	Personalization of treatment (Chun et al., 2011)	Suboptimal standardization (Pappas and Yang, 2008)
Applications	High-throughput screening of drugs and toxicity prediction (Wobus and Loser, 2011; Choi et al., 2013)	Complex assessment
	Reduced cost	Complex diseases become difficult to be modeled
	Gene correction therapies add to the benefits from iPSCs (Choi et al., 2013)	Immature cells cause problems during cell line development

collection of sample would be carried out in an ethical way. After the collection and culture of the source cell, a suitable reprogramming method is to be decided. The use of retroviral vectors has risks of insertional mutagenesis and hence their use is not ideal for clinical applications. Non-viral methods address this question and the method is chosen according to pros and cons associated. Once the iPSCs have generated, proper characterization is vitally required. For the verification that the iPSCs generated are really pluripotent, the expression of different pluripotency markers is assessed. It must be checked if the iPSCs generated can differentiate to all the three germ layers. To further check the ability of the iPSCs to differentiate, the efficiency of generation of the cells which are required for the therapy is determined. The cells generated must be tested to be functional, so the physiological properties of cells generated from iPSCs were compared with that of the cells generated from ESCs. The final product has variability which depends on the quality of starting material used as cell source, difference in other raw materials used for cell culture, and the protocol followed. Identifying and controlling these is important to standardize any manufacturing process to demonstrate reproducibility and consistent production meeting the specifications. Research based methods that have manual nature may not prove to be ideal for the large production of cells. Quality control testing for final product and preparation procedures must also be designed. For administration of fresh cells, there is need of a rapid method for detection of contamination. Plans must be there in case if the sample fails the sterility test, if the patient has already been administered with the product. Once the quality testing has been done, it should be ensured that the final cell therapeutic is meeting all the required specifications. For allogeneic therapies, a cell bank must be established which would contain the starting cell material and it should be tested to check the consistency of that material. Establishment of a large cell bank is not important for autologous therapeutics; a small bank would serve the purpose of identification and minimization of variability that may result from initial reprogramming. The colonies which meet the specifications of karyotype, marker expression, mycoplasma/bacterial/fungal contaminate and growth characteristics, are expanded to build a bank. Additional knowledge about the post release handling of the cell product is also required. It is important to ensure that the final cell product is representative of therapeutic that will be subjected to human clinical trials. Preclinical animal studies should be able to address the safety of the cell product which includes the affects of the contaminants present in the cell population from donor, chances of teratoma formation and generation of karyotypically unstable cells. It should be ensured that the changes in process associated with change from animal studies to production for clinical use would not cause any significant problems.

FUTURE PERSPECTIVES

Since the emergence of the field of iPSCs, it has highly developed and successfully expanded to different fields of regenerative medicine. Their application in clinical research will surely benefit the patients in future. At present there are few, but important applications of iPSCs but, if this field keeps on growing at the present pace, it wouldn't take long to expand the applications of iPSCs to more biological fields to aid research and treatments. iPSCs are important in disease modeling as they do not have ethical issues concerned as in case of ESCs and provide with the physiological conditions more close to humans which is not an advantage with the animal disease models. iPSCs also have a possibility of giving consistent phenotypes every time, which is an important issue for disease modeling. More and more types of diseased cells are being tried to be modeled, but, it still has many limitations that are encountered while working with iPSCs, which need to be overcome. Also, it is not possible to be able to model every disease, so, further exploration of modeling diseases with iPSCs is required. As iPSCs eliminate the chances of immunorejection, they hold great importance in Gene therapy. But the generation of iPSCS itself has many issues associated it like incorporation of vectors into host genomes, and the dependence on the various factors like usage of small molecules, single or multiple vectors, source of cells, which make the generation of iPSCs a risky task and hence, more about iPSCs need to be explored to make it easier to generate them and to be able to apply them to various other important and attractive areas of medical sciences for their proper utilization to achieve the advantages that use of iPSCs can have. And, different advances in the field, including the solution of many limitations mentioned above would prove to be beneficial for increasing therapeutic potential. The present situation of the applications of iPSCs deals with the laboratory scale production and testing assay. A lot has to be done to realize the applications of iPSCs at a higher level, for the use in hospitals, to be beneficial for the patients.

CONCLUSION

The iPSC therapy is in infancy but a constant progress in reprogramming technologies has occurred; new and improved methods are gathering. As iPSCs can differentiate to different cell types, we can generate the required cells for the study or treatment of numerous diseases and determination of drug toxicity. The therapy using HSCs is being used since a long time. The change from allogeneic to autologous iPSCs has been a major breakthrough. Autologous therapy also offers the advantage of personalized medicine.

ACKNOWLEDGMENTS

We thank the Honourable Chairman and the Honourable Vice-Chancellor of Delhi Technological University, Shahbad Daulatpur, Bawana Road, Delhi-42, for the support. Dr. Vimal K. Singh particularly thanks the Department of Science and Technology (DST) and Indian National Science Academy (INSA), India, for the research grant.

REFERENCES

Aasen, T., Raya, A., Barrero, M. J., Garreta, E., Consiglio, A., Gonzalez, F., et al. (2008). Efficient and rapid generation of induced pluripotent stem cells from human keratinocytes. *Nat. Biotechnol.* 26, 1276–1284. doi: 10.1038/nbt.1503

Adewumi, O., Aflatoonian, B., Ahrlund-Richter, L., Amit, M., Andrews, P. W., Beighton, G., et al. (2007). Characterization of human embryonic stem cell lines by the International Stem Cell Initiative. *Nat. Biotechnol.* 25, 803–816. doi: 10.1038/nbt1318

Amenduni, M., De Filippis, R., Cheung, A. Y. L., Disciglio, V., Epistolato, M. C., Ariani, F., et al. (2011). iPS cells to model CDKL5-related disorders. *Eur. J. Hum. Genet.* 19, 1246–1255. doi: 10.1038/ejhg.2011

Asai, Y., Tada, M., Otsuji, T. G., and Nakatsuji, N. (2010). Combination of functional cardiomyocytes derived from human stem cells and a highly-efficient microelectrode array system: an ideal hybrid model assay for drug development. *Curr. Stem Cell Res. Ther.* 5, 227–232. doi: 10.2174/157488810791824502

Bahadduri, P. M., Polli, J. E., Swaan, P. W., and Ekins, S. (2010). Targeting drug transporters - combining *in silico* and *in vitro* approaches to predict *in vivo*. *Methods Mol. Biol.* 637, 65–103. doi: 10.1007/978-1-60761-700-6_4

Batista, L. F., Pech, M. F., Zhong, M. F., Nguyen, H. N., Xie, K. T., Zaug, A. J., et al. (2011). Telomere shortening and loss of self-renewal in dyskeratosis congenita induced pluripotent stem cells. *Nature* 474, 399–402. doi: 10.1038/nature10084

Brambrink, T., Foreman, R., Welstead, G. G., Lengner, C. J., Wernig, M., Suh, H., et al. (2008). Sequential expression of pluripotency markers during direct reprogramming of mouse somatic cells. *Cell Stem Cell* 2, 151–159. doi: 10.1016/j.stem.2008.01.004

Briggs, J. A., Mason, E. A., Ovchinnikov, D. A., Wells, C. A., and Wolvetang, E. J. (2013). Concise review: new paradigms for down syndrome research using induced pluripotent stem cells: tackling complex human genetic disease. *Stem Cells Transl. Med.* 2, 175–184. doi: 10.5966/sctm.2012-0117

Campuzano, V., Montermini, L., Molto, M. D., Pianese, L., Cossee, M., Cavalcanti, F., et al. (1996). Friedreich's ataxia: autosomal recessive disease caused by an intronic GAA triplet repeat expansion. *Science* 271, 1423–1427. doi: 10.1126/science.271.5254.1423

Cao, F., Wagner, R. A., Wilson, K. D., Xie, X., Fu, J. D., Drukker, M., et al. (2008). Transcriptional and functional profiling of human embryonic stem cell-derived cardiomyocytes. *PLoS ONE* 3:e3474. doi: 10.1371/journal.pone.0003474

Carvajal-Vergara, X., Sevilla, A., D'Souza, S. L., Ang, Y. S., Schaniel, C., Lee, D. F., et al. (2010). Patient-specifi c induced pluripotent stem-cell derived models of LEOPARD syndrome. *Nature* 465, 808–812. doi: 10.1038/nature09005

Cheng, K. W., Agarwal, R., Mitra, S., and Mills, G. B. (2013). Rab25 small GTPase mediates secretion of tumor necrosis factor receptor superfamily member 11b (osteoprotegerin) protecting cancer cells from effects of TRAIL. *J. Genet. Syndr. Gene Ther.* 4:153. doi: 10.4172/2157-7412.1000153

Chestkov, I. V., Vasilieva, E. A., Illarioshkin, S. N., Lagarkova, M. A., and Kiselev, S. L. (2014). Patient-specific induced pluripotent stem cells for SOD1-associated amyotrophic lateral sclerosis pathogenesis studies. *Acta Naturae* 6, 54–60.

Choi, M. S., Kim, Y., Shim, J. S., Park, J. T., Wang, R., Leach, S. D., et al. (2013). Efficient drug screening and gene correction for treating liver disease using patient-specific stem cells. *Hepatology* 57, 2458–2468. doi: 10.1002/hep.26237

Chun, Y. S., Byun, K., and Lee, B. (2011). Induced pluripotent stem cells and personalized medicine: current progress and future perspectives. *Anat. Cell Biol.* 44, 245–255. doi: 10.5115/acb.2011.44.4.245

Desnuelle, C., Dib, M., Garrel, C., and Favier, A. (2001). A double-blind, placebo-controlled randomized clinical trial of alpha-tocopherol (vitamin E) in the treatment of amyotrophic lateral sclerosis. ALS riluzole-tocopherol Study Group. *Amyotroph. Lateral Scler. Other Motor Neuron Disord.* 2, 9–18. doi: 10.1080/146608201300079364

Devine, M. J., Ryten, M., Vodicka, P., Thomson, A. J., Burdon, T., Houlden, H., et al. (2011). Parkinson's disease induced pluripotent stem cells with triplication of the α-synuclein locus. *Nat. Commun.* 2, 440. doi: 10.1038/ncomms1453

Dimos, J. T., Rodolfa, K. T., Niakan, K. K., Weisenthal, L. M., Mitsumoto, H., Chung, W., et al. (2008). Induced pluripotent stem cells generated from patients with ALS can be differentiated into motor neurons. *Science* 321, 1218–1221. doi: 10.1126/science.1158799

Doherty, A. M., and Fisher, E. M. (2003). Microcell-mediated chromosome transfer (MMCT): small cells with huge potential. *Mamm. Genome* 14, 583–592. doi: 10.1007/s00335-003-4002-0

Ebert, A. D., Yu, J., Rose, F. F. Jr., Mattis, V. B., Lorson, C. L., Thomson, J. A., et al. (2009). Induced pluripotent stem cells from a spinal muscular atrophy patient. *Nature* 457, 277–280. doi: 10.1038/nature07677

Evans, M. J., and Kaufman, M. H. (1981). Establishment in culture of pluripotential cells from mouse embryos. *Nature* 292, 154–156. doi: 10.1038/292154a0

Farra, N., Zhang, W. B., Pasceri, P., Eubanks, J. H., Salter, M. W., and Ellis, J. (2012). Rett syndrome induced pluripotent stem cell-derived neurons reveal novel neurophysiological alterations. *Mol. Psychiatry* 17, 1261–1271. doi: 10.1038/mp.2011.180

Feng, B., Jiang, J., Kraus, P., Ng, J. H., Heng, J. C., Chan, Y. S., et al. (2009). Reprogramming of fibroblasts into induced pluripotent stem cells with orphan nuclear receptor Esrrb. *Nat. Cell Biol.* 11, 197–203. doi: 10.1038/ncb1827

Filareto, A., Parker, S., Darabi, R., Borges, L., Iacovino, M., Schaaf, T., et al. (2013). An *ex vivo* gene therapy approach to treat muscular dystrophy using inducible pluripotent stem cells. *Nat. Commun.* 4, 1549. doi: 10.1038/ncomms2550

Fong, H., Wang, C., Knoferle, J., Walker, D., Balestra, M. E., and Tong, L. M. (2013). Genetic correction of tauopathy phenotypes in neurons derived from human induced pluripotent stem cells. *Stem Cell Reports* 1, 226–234. doi: 10.1016/j.stemcr.2013.08.001

Fong, Y. W., Inouye, C., Yamaguchi, T., Cattoglio, C., Grubisic, I., and Tjian, R. (2011). A DNA repair complex functions as an Oct4/Sox2 coactivator in embryonic stem cells. *Cell* 147, 120–131. doi: 10.1016/j.cell.2011.08.038

Fusaki, N., Ban, H., Nishiyama, A., Saeki, K., and Hasegawa, M. (2009). Efficient induction of transgene-free human pluripotent stem cells using a vector based on Sendai virus, an RNA virus that does not integrate into the host genome. *Proc. Jpn. Acad. Ser. B Phys. Biol. Sci.* 85, 348–362. doi: 10.2183/pjab.85.348

Gangopadhyay, S. B., Sherratt, T. G., Heckmatt, J. Z., Dubowitz, V., Miller, G., Shokeir, M., et al. (1992). Dystrophin in frameshift deletion patients with Becker muscular dystrophy. *Am. J. Hum. Genet.* 51, 562–570.

Ghaleb, A. M., Nandan, M. O., Chanchevalap, S., Dalton, W. B., Hisamuddin, I. M., and Yang, V. W. (2005). Krüppel-like factors 4 and 5: the yin and yang regulators of cellular proliferation. *Cell Res.* 15, 92–96. doi: 10.1038/sj.cr.7290271

Gonzalez, F., Barragan Monasterio, M., Tiscornia, G., Montserrat Pulido, N., Vassena, R., Batlle Morera, L., et al. (2009). Generation of mouse-induced pluripotent stem cells by transient expression of a single nonviral polycistronic vector. *Proc. Natl. Acad. Sci. U.S.A.* 106, 8918–8922. doi: 10.1073/pnas.0901471106

Grizot, S., Smith, J., Daboussi, F., Prieto, J., Redondo, P., Merino, N., et al. (2009). Efficient targeting of a SCID gene by an engineered single-chain homing endonuclease. *Nucleic Acids Res.* 37, 5405–5419. doi: 10.1093/nar/gkp548

Groeneveld, G. J., Veldink, J. H., van der Tweel, I., Kalmijn, S., Beijer, C., de Visser, M., et al. (2003). A randomized sequential trial of creatine in amyotrophic lateral sclerosis. *Ann. Neurol.* 53, 437–445. doi: 10.1002/ana.10554

Guha, P., Morgan, J. W., Mostoslavsky, G., Rodrigues, N. P., and Boyd, A. S. (2013). Lack of immune response to differentiated cells derived from syngeneic induced pluripotent stem cells. *Cell Stem Cell* 12, 407–412. doi: 10.1016/j.stem.2013.01.006

Gunaseeli, I., Doss, M. X., Antzelevitch, C., Hescheler, J., and Sachinidis, A. (2010). Induced pluripotent stem cells as a model for accelerated patient- and disease-specific drug discovery. *Curr. Med. Chem.* 17, 759–766. doi: 10.2174/092986710790514480

Gurdon, J. B. (1962). The developmental capacity of nuclei taken from intestinal epithelium cells of feeding tadpoles. *J. Embryol. Exp. Morphol.* 10, 622–640.

Haase, A., Olmer, R., Schwanke, K., Wunderlich, S., Merkert, S., Hess, C., et al. (2009). Generation of induced pluripotent stem cells from human cord blood. *Cell Stem Cell* 5, 434–441. doi: 10.1016/j.stem.2009.08.021

Hanna, J., Markoulaki, S., Schorderet, P., Carey, B. W., Beard, C., Wernig, M., et al. (2008). Direct reprogramming of terminally differentiated mature B lymphocytes to pluripotency. *Cell* 133, 250–264. doi: 10.1016/j.cell.2008.03.028

Hochedlinger, K., and Jaenisch, R. (2002). Nuclear transplantation: lessons from frogs and mice. *Curr. Opin. Cell Biol.* 14, 741–748. doi: 10.1016/S0955-0674(02)00380-0

Hochedlinger, K., Yamada, Y., Beard, C., and Jaenisch, R. (2005). Ectopic expression of Oct-4 blocks progenitor-cell differentiation and causes dysplasia in epithelial tissues. *Cell* 121, 465–477. doi: 10.1016/j.cell.2005.02.018

Hockemeyer, D., Soldner, F., Beard, C., Gao, Q., Mitalipova, M., DeKelver, R. C., et al. (2009). Efficient targeting of expressed and silent genes in human ESCs and iPSCs using zinc-finger nucleases. *Nat. Biotechnol.* 27, 851–857. doi: 10.1038/nbt.1562

Hockemeyer, D., Wang, H., Kiani, S., Lai, C. S., Gao, Q., Cassady, J. P., et al. (2011). Genetic engineering of human pluripotent cells using TALE nucleases. *Nat. Biotechnol.* 29, 731–734. doi: 10.1038/nbt.1927

Howe, S. J., Mansour, M. R., Schwarzwaelder, K., Bartholomae, C., Hubank, M., Kempski, H., et al. (2008). Insertional mutagenesis combined with acquired somatic mutations causes leukemogenesis following gene therapy of SCID-X1 patients. *J. Clin. Invest.* 118, 3143–3150. doi: 10.1172/JCI35798

Huangfu, D., Maehr, R., Guo, W., Eijkelenboom, A., Snitow, M., Chen, A. E., et al. (2008a). Induction of pluripotent stem cells by defined factors is greatly improved by small-molecule compounds. *Nat. Biotechnol.* 26, 795–797. doi: 10.1038/nbt1418

Huangfu, D., Osafune, K., Maehr, R., Guo, W., Eijkelenboom, A., Chen, S., et al. (2008b). Induction of pluripotent stem cells from primary human fibroblasts with only Oct4 and Sox2. *Nat. Biotechnol.* 26, 1269–1275. doi: 10.1038/nbt.1502

Hurtado-Lorenzo, A., Millan, E., Gonzalez-Nicolini, V., Suwelack, D., Castro, M. G., and Lowenstein, P. R. (2004). Differentiation and transcription factor gene therapy in experimental Parkinson's disease: sonic hedgehog and Gli-1, but Not Nurr-1, protect nigrostriatal cell bodies from 6-OHDA-induced neurodegeneration. *Mol. Ther.* 10, 507–524. doi: 10.1016/j.ymthe.2004.05.021

Israel, M. A., Yuan, S. H., Bardy, C., Reyna, S. M., Mu, Y., Herrera, C., et al. (2012). Probing sporadic and familial Alzheimer's disease using induced pluripotent stem cells. *Nature* 482, 216–220. doi: 10.1038/nature10821

Jiang, Y., Cowley, S. A., Siler, U., Melguizo, D., Tilgner, K., Browne, C., et al. (2012). Derivation and functional analysis of patient-specific induced pluripotent stem cells as an *in vitro* model of chronic granulomatous disease. *Stem Cells* 30, 599–611. doi: 10.1002/stem.1053

Kaitin, K. I. (2008). Obstacles and opportunities in new drug development. *Clin. Pharmacol. Ther.* 83, 210–212. doi: 10.1038/sj.clpt.6100462

Kaye, J. A., and Finkbeiner, S. (2013). Modeling Huntington's disease with induced pluripotent stem cells. *Mol. Cell. Neurosci.* 56, 50–64. doi: 10.1016/j.mcn.2013.02.005

Kazuki, Y., Hiratsuka, M., Takiguchi, M., Osaki, M., Kajitani, M., Hoshiya, H., et al. (2009). Complete genetic correction of iPS cells from Duchenne muscular dystrophy. *Am. Soc. Gene Cell Ther.* 18, 386–393. doi: 10.1038/mt.2009.274

Kerns, W. D., Pavkov, K. L., Donofrio, D. J., Gralla, E. J., and Swenberg, J. A. (1983). Carcinogenicity of formaldehyde in rats and mice after long-term inhalation exposure. *Cancer Res.* 43, 4382–4392.

Khan, I. F., Hirata, R. K., Wang, P. R., Li, Y., Kho, J., Nelson, A., et al. (2010). Engineering of human pluripotent stem cells by AAV-mediated gene targeting. *Mol. Ther.* 18, 1192–1199. doi: 10.1038/mt.2010.55

Kim, D., Kim, C. H., Moon, J. I., Chung, Y. G., Chang, M. Y., Han, B. S., et al. (2009b). Generation of human induced pluripotent stem cells by direct delivery of reprogramming proteins. *Cell Stem Cell* 4, 472–476. doi: 10.1016/j.stem.2009.05.005

Kim, J. B., Greber, B., Araúzo-Bravo, M. J., Meyer, J., Park, K. I., Zaehres, H., et al. (2009a). Direct reprogramming of human neural stem cells by OCT4. *Nature* 461, 649–643. doi: 10.1038/nature08436

Kimmel, C. A., and Young, J. F. (1983). Correlating pharmacokinetics and teratogenic endpoints. *Fundam. Appl. Toxicol.* 3, 250–255. doi: 10.1016/S0272-0590(83)80136-5

Klein, C., Nguyen, D., Liu, C. H., Mizoguchi, A., Bhan, A. K., Miki, H., et al. (2003). Gene therapy for Wiskott-Aldrich syndrome: rescue of T-cell signaling and amelioration of colitis upon transplantation of retrovirally transduced hematopoietic stem cells in mice. *Blood* 101, 2159–2166. doi: 10.1182/blood-2002-05-1423

Ku, S., Soragni, E., Campau, E., Thomas, E. A., Altun, G., Laurent, L. C., et al. (2010). Friedreich's ataxia induced pluripotent stem cells model intergenerational GAA•TTC triplet repeat instability. *Cell Stem Cell* 7, 631–637. doi: 10.1016/j.stem.2010.09.014

Kunisato, A., Wakatsuki, M., Shinba, H., Ota, T., Ishida, I., Nagao, K., et al. (2011). Direct generation of induced pluripotent stem cells from human nonmobilized blood. *Stem Cells Dev.* 20, 159–168. doi: 10.1089/scd.2010.0063

Kuttler, F., and Mai, S. (2006). c-Myc, genomic instability and disease. *Genome Dyn.* 1, 171–190. doi: 10.1159/000092507

Laflamme, M. A., Chen, K. Y., Naumova, A. V., Muskheli, V., Fugate, J. A., Dupras, S. K., et al. (2007). Cardiomyocytes derived from human embryonic stem cells in pro-survival factors enhance function of infarcted rat hearts. *Nat. Biotechnol.* 25, 1015–1024. doi: 10.1038/nbt1327

Lan, M. L., Acharya, M. M., Tran, K. K., Bahari-Kashani, J., Patel, N. H., Strnadel, J. et al. (2012). Characterizing the radioresponse of pluripotent and multipotent human stem cells. *PLoS ONE* 7:e50048. doi: 10.1371/journal.pone.0050048

Laustriat, D., Gide, J., and Peschanski, M. (2010). Human pluripotent stem cells in drug discovery and predictive toxicology. *Biochem. Soc. Trans.* 38, 1051–1057. doi: 10.1042/BST0381051

Lee, G., Papapetrou, E. P., Kim, H., Chambers, S. M., Tomishima, M. J., Fasano, C. A., et al. (2009). Modelling pathogenesis and treatment of familial dysautonomia using patient-specific iPSCs. *Nature* 461, 402–406. doi: 10.1038/nature 08320

Lee, G., and Studer, L. (2011). Modelling familial dysautonomia in human induced pluripotent stem cells. *Phil. Trans. R. Soc. B.* 366, 2286–2296. doi: 10.1098/rstb.2011.0026

Legius, E., Schrander-Stumpel, C., Schollen, E., Pulles-Heintzberger, C., Gewillig, M., and Fryns, J. P. (2002). PTPN11 mutations in LEOPARD syndrome. *J. Med. Genet.* 39, 571–574. doi: 10.1136/jmg.39.8.571

Lei, K., Chen, Y., Chen, H., Wong, L. C., Liu, J., McConkie-Rosell, A., et al. (1995). Genetic basis of glycogen storage disease type 1a: prevalent mutations at the glucose-6-phosphatase locus. *Am. J. Hum. Genet.* 57, 766–771.

Levenberg, S., Ferreira, L. S., Chen-Konak, L., Kraehenbuehl, T. P., and Langer, R. (2010). Isolation, differentiation and characterization of vascular cells derived from human embryonic stem cells. *Nat. Protoc.* 5, 1115–1126. doi: 10.1038/nprot.2010.31

Lim, W. F., Inoue-Yokoo, T., Tan, K. S., Lai I. M., and Sugiyama, D. (2013). Hematopoietic cell differentiation from embryonic and induced pluripotent stem cells. *Stem Cell Res. Ther.* 4:71. doi: 10.1186/scrt222

Li, W., Wei, W., Zhu, S., Zhu, J., Shi, Y., Lin, T., et al. (2009). Generation of rat and human induced pluripotent stem cells by combining genetic reprogramming and chemical inhibitors. *Cell Stem Cell* 4, 16–19. doi: 10.1016/j.stem.2008.11.014

Li, Y., Tsai, Y., Hsu, C., Erol, D., Yang, J., Wu, W., et al. (2012). Long-term safety and efficacy of human-induced pluripotent stem cell (iPS) grafts in a preclinical model of retinitis pigmentosa. *Mol. Med.* 18:1312–1319. doi: 10.2119/molmed.2012.00242

Li, Y., Zhang, Q., Yin, X., Yang, W., Du, Y., Hou, P., et al. (2011). Generation of iPSCs from mouse fibroblasts with a single gene, Oct4, and small molecules. *Cell Res.* 21, 196–204. doi: 10.1038/cr.2010.142

Liu, G. H., Barkho, B. Z., Ruiz, S., Diep, D., Qu, J., Yang, S. L., et al. (2011a). Recapitulation of premature ageing with iPSCs from Hutchinson–Gilford progeria syndrome. *Nature* 472, 221–225. doi: 10.1038/nature09879

Liu, H., Kim, Y., Sharkis, S., Marchionni, L., and Jang, Y. (2011b). *In vivo* liver regeneration potential of human induced pluripotent stem cells from diverse origins. *Sci. Transl. Med.* 3, 82ra39. doi: 10.1126/scitranslmed.3002376

Liu, H., Ye, Z., Kim, Y. H., Sharkis, S., and Jang, Y. Y. (2010). Generation of endoderm derived human ips cells from primary hepatocytes. *Hepatology* 51, 1810–1819. doi: 10.1002/hep.23626

Maherali, N., Ahfeldt, T., Rigamonti, A., Utikal, J., Cowan, C., and Hochedlinger, K. (2008). A high-efficiency system for the generation and study of human induced pluripotent stem cells. *Cell Stem Cell* 3, 340–345. doi: 10.1016/j.stem.2008.08.003

Mali, P., Ye, Z., Hommond, H. H., Yu, X., Lin, J., Chen, G., et al. (2008). Improved efficiency and pace of generating induced pluripotent stem cells from human adult and fetal fibroblasts. *Stem Cells* 26, 1998–2005. doi: 10.1634/stemcells.2008-0346

Marson, A., Foreman, R., Chevalier, B., Bilodeau, S., Kahn, M., Young, R. A., et al. (2008). Wnt signaling promotes reprogramming of somatic cells to pluripotency. *Cell Stem Cell* 3, 132–135. doi: 10.1016/j.stem.2008.06.019

McNeish, J., Roach, M., Hambor, J., Mather, R. J., Weibley, L., Lazzaro, J., et al. (2010). High throughput screening in embryonic stem cell derived neurons identifies potentiators of a-amino-3-hydroxyl-5-methyl-4-isoxazolepropionate-type glutamate receptors. *J. Biol. Chem.* 285, 17209–17217. doi: 10.1074/jbc.M109.098814

Mikkelsen, T. S., Hanna, J., Zhang, X., Ku, M., Wernig, M., Schorderet, P., et al. (2008). Dissecting direct reprogramming through integrative genomic analysis. *Nature* 454, 49–55. doi: 10.1038/nature07056

Moad, M., Pal, D., Hepburn, A. C., Williamson, S. C., Wilson, L., Lako, M., et al. (2013). A novel model of urinary tract differentiation, tissue regeneration, and disease: reprogramming human prostate and bladder cells into induced pluripotent stem cells. *Eur. Urol.* 64, 753–761. doi: 10.1016/j.eururo.2013.03.054

Nguyen, H. N., Byers, B., Cord, B., Shcheglovitov, A., Byrne, J., Gujar, P., et al. (2011). LRRK2 mutant iPSC-derived DA neurons demonstrate increased susceptibility to oxidative stress. *Cell Stem Cell.* 8, 267–280. doi: 10.1016/j.stem.2011.01.013

Nori, S., Okada, Y., Yasuda, K., Tsuji, O., Takahashi, Y., Kobayashi, Y., et al. (2011). Human-induced pluripotent stem-cell–derived neurospheres promote motor functional recovery after spinal cord injury in mice. *Proc. Natl. Acad. Sci. U.S.A.* 108, 16825–16830. doi: 10.1073/pnas.1108077108

Okita, K., Ichisaka, T., and Yamanaka, S. (2007). Generation of germline-competent induced pluripotent stem cells. *Nature* 448, 313–317. doi: 10.1038/nature05934

Okita, K., Nakagawa, M., Hyenjong, H., Ichisaka, T., and Yamanaka, S. (2008). Generation of mouse induced pluripotent stem cells without viral vectors. *Science* 322, 949–953. doi: 10.1126/science.1164270

Pappas, J. J., and Yang, P. C. (2008). Human ESC vs. iPSC- Pros and Cons. *J. Cardiovasc. Trans. Res.* 1, 96–99. doi: 10.1007/s12265-008-9032-2

Park, E. T., Gum, J. R., Kakar, S., Kwon, S. W., Deng, G., and Kim, Y. S. (2008c). Aberrant expression of SOX2 upregulates MUC5AC gastric foveolar mucin in mucinous cancers of the colorectum and related lesions. *Int. J. Cancer* 122, 1253–1260. doi: 10.1002/ijc.23225

Park, I. H., Arora, N., Huo, H., Maherali, N., Ahfeldt, T., Shimamura, A., et al. (2008b). Disease-specific induced pluripotent stem cells. *Cell* 134, 877–886. doi: 10.1016/j.cell.2008.07.041

Park, I. H., Zhao, R., West, J. A., Yabuuchi, A., Huo, H., Ince, T. A., et al. (2008a). Reprogramming of human somatic cells to pluripotency with defined factors. *Nature* 451, 141–146. doi: 10.1038/nature06534

Rashid, S. T., Corbineau, S., Hannan, N., Marciniak, S. J., Miranda, E., Alexander, G., et al. (2010). Modeling inherited metabolic disorders of the liver using human induced pluripotent stem cells. *J. Clin. Invest.* 120, 3127–3136. doi: 10.1172/JCI43122

Rosen, D. R., Siddique, T., Patterson, D., Figlewicz, D. A., Sapp, P., Hentati, A., et al. (1993). Mutations in Cu/Zn superoxide dismutase gene are associated with familial amyotrophic lateral sclerosis. *Nature* 362, 59–62. doi: 10.1038/362059a0

Ruiz, S., Brennand, K., Panopoulos, A. D., Herrerias, A., Gage, F. H., and Izpisua-Belmonte, J. C. (2010). High-efficient generation of induced pluripotent stem cells from human astrocytes. *PLoS ONE* 5:e15526. doi: 10.1371/journal.pone.0015526

Sato, T., Joyner, A. L., and Nakamura, H. (2004). How does Fgf signaling from the isthmic organizer induce midbrain and cerebellum development? *Dev. Growth Differ.* 46, 487–494. doi: 10.1111/j.1440-169x.2004.00769.x

Sauer, A. V., Morbach, H., Brigida, I., Ng, Y. S., Aiuti, A., and Meffre, E. (2012). Defective B cell tolerance in adenosine deaminase deficiency is corrected by gene therapy. *J. Clin. Invest.* 122, 2141–2152. doi: 10.1172/JCI61788

Seiler, A., Visan, A., Buesen, R., Genschow, E., and Spielmann, H. (2004). Improvement of an *in vitro* stem cell assay for developmental toxicity: the use of molecular endpoints in the embryonic stem cell test. *Reprod. Toxicol.* 18, 231–240. doi: 10.1016/j.reprotox.2003.10.015

Seki, T., Yuasa, S., Oda, M., Egashira, T., Yae, K., Kusumoto, D., et al. (2010). Generation of induced pluripotent stem cells from human terminally differentiated circulating T cells. *Cell Stem Cell* 7, 11–14. doi: 10.1016/j.stem.2010.06.003

Shefner, J. M., Cudkowicz, M. E., Schoenfeld, D., Conrad, T., Taft, J., Chilton, M., et al. (2004). NEALS Consortium. A clinical trial of creatine in ALS. *Neurology* 63, 1656–1661. doi: 10.1212/01.WNL.0000142992.81995.F0

Shi, Y., Desponts, C., Do, J. T., Hahm, H. S., Scholer, H. R., and Ding, S. (2008a). Induction of pluripotent stem cells from mouse embryonic fibroblasts by Oct4 and Klf4 with small-molecule compounds. *Cell Stem Cell* 3, 568–574. doi: 10.1016/j.stem.2008.10.004

Shi, Y., Do, J. T., Desponts, C., Hahm, H. S., Scholer, H. R., and Ding, S. (2008b). A combined chemical and genetic approach for the generation of induced pluripotent stem cells. *Cell Stem Cell* 2, 525–528. doi: 10.1016/j.stem.2008.05.011

Silva, J., Barrandon, O., Nichols, J., Kawaguchi, J., Theunissen, T. W., and Smith, A. (2008). Promotion of reprogramming to ground state pluripotency by signal inhibition. *PLoS Biol.* 6:e253. doi: 10.1371/journal.pbio.0060253

Soejitno, A., and Prayudi, P. K. A. (2011). The prospect of induced pluripotent stem cells for diabetes mellitus treatment. *Ther. Adv. Endocrinol. Metab.* 2, 197–210. doi: 10.1177/2042018811420198

Soldner, F., Hockemeyer, D., Beard, C., Gao, Q., Bell, G. W., Cook, E. G., et al. (2009). Parkinson's disease patient derived induced pluripotent stem cells free of viral reprogramming factors. *Cell* 136, 964–977. doi: 10.1016/j.cell.2009.02.013

Sollano, J. A., Kirsch, J. M., Bala, M. V., Chambers, M. G., and Harpole, L. H. (2008). The economics of drug discovery and the ultimate valuation of pharmacotherapies in the marketplace. *Clin. Pharmacol. Ther.* 84, 263–266. doi: 10.1038/clpt.2008.117

Somers, A., Jean, J. C., Sommer, C. A., Omari, A., Ford, C. C., Mills, J. A., et al. (2010). Generation of transgene-free lung disease-specific human induced pluripotent stem cells using a single excisable lentiviral stem cell cassette. *Stem Cells* 28, 1728–1740. doi: 10.1002/stem.495

Song, B., Niclis, J. V., Alikhan, M. A., Sakkal, S., Sylvain, A., Kerr, P. G., et al. (2011). Generation of induced pluripotent stem cells from human kidney mesangial cells. *J. Am. Soc. Nephrol.* 22, 1213–1220. doi: 10.1681/ASN.2010101022

Stadtfeld, M., Maherali, N., Breault, D. T., and Hochedlinger, K. (2008a). Defining molecular cornerstones during fibroblast to iPS cell reprogramming in mouse. *Cell Stem Cell* 2, 230–240. doi: 10.1016/j.stem.2008.02.001

Stadtfeld, M., Nagaya, M., Utikal, J., Weir, G., and Hochedlinger, K. (2008b). Induced pluripotent stem cells generated without viral integration. *Science* 322, 945–949. doi: 10.1126/science.1162494

Sugii, S., Kida, Y., Kawamura, T., Suzuki, J., Vassena, R., Yin, Y. Q., et al. (2010). Human and mouse adipose-derived cells support feeder-independent induction of pluripotent stem cells. *Proc. Natl. Acad. Sci. U.S.A.* 107, 3558–3563. doi: 10.1073/pnas.0910172106

Sullivan, G. J., Hay, D. C., Park, I. H., Fletcher, J., Hannoun, Z., Payne, C. M., et al. (2010). Generation of functional human hepatic endoderm from human induced pluripotent stem cells. *Hepatology* 51, 329–335. doi: 10.1002/hep.23335

Suzuki, K., Mitsui, K., Aizawa, E., Hasegawa, K., Kawase, E., Yamagishi, T., et al. (2008). Highly efficient transient gene expression and gene targeting in primate embryonic stem cells with helper-dependent adenoviral vectors. *Proc. Natl. Acad. Sci. U.S.A.* 105, 13781–13786. doi: 10.1073/pnas.0806976105

Suzuki, N., Yamazaki, S., Yamaguchi, T., Okabe, M., Masaki, H., Takaki, S., et al. (2013). Generation of engraftable hematopoietic stem cells from induced pluripotent stem cells by way of teratoma formation. *Mol. Ther.* 21, 1424–1431. doi: 10.1038/mt.2013.71

Tada, M., Takahama, Y., Abe, K., Nakatsuji, N., and Tada, T. (2001). Nuclear reprogramming of somatic cells by *in vitro* hybridization with ES cells. *Curr. Biol.* 11, 1553–1558. doi: 10.1016/S0960-9822(01)00459-6

Takahashi, K., Ichisaka, T., and Yamanaka, S. (2006b). Identification of genes involved in tumor-like properties of embryonic stem cells. *Methods Mol. Biol.* 329, 449–458. doi: 10.1385/1-59745-037-5:449

Takahashi, K., and Yamanaka, S. (2006a). Induction of pluripotent stem cells from mouse embryonic and adult fibroblast cultures by defined factors. *Cell* 126, 663–676. doi: 10.1016/j.cell.2006.07.024

Tan, Q., Lui, P. P. Y., Rui, Y. F., and Wong, Y. M. (2012). Comparison of potentials of stem cells isolated from tendon and bone marrow for musculoskeletal tissue engineering. *Tissue Eng. Part A* 18, 840–851. doi: 10.1089/ten.tea.2011.0362

Tanaka, T., Tohyama, S., Murata, M., Nomura, F., Kaneko, T., Chen, H., et al. (2009). *In vitro* pharmacologic testing using human induced pluripotent stem cell-derived cardiomyocytes. *Biochem. Biophys. Res. Commun.* 385, 497–502. doi: 10.1016/j.bbrc.2009.05.073

Thomson, J. A., Itskovitz-Eldor, J., Shapiro, S. S., Waknitz, M. A., Swiergiel, J. J., Marshall, V. S., et al. (1998). Embryonic stem cell lines derived from human blastocysts. *Science* 282, 1145–1147. doi: 10.1126/science.282.5391.1145

Till, J. E., and McCulloch, E. A. (1961). A direct measurement of the radiation sensitivity of normal mouse bone marrow cells. *Radiat. Res.* 14, 213–222. doi: 10.2307/3570892

Tulpule, A., Kelley, J. M., Lensch, M. W., McPherson, J., Park, I. H., Hartung, O., et al. (2013). Pluripotent stem cell models of Shwachman-Diamond syndrome reveal a common mechanism for pancreatic and hematopoietic dysfunction. *Cell Stem Cell* 12, 727–736. doi: 10.1016/j.stem.2013.04.002

Utikal, J., Maherali, N., Kulalert, W., and Hochedlinger, K. (2009). Sox2 is dispensable for the reprogramming of melanocytes and melanoma cells into induced pluripotent stem cells. *J. Cell Sci.* 122, 3502–3510. doi: 10.1242/jcs.054783

Wang, X. M., Yik, W. Y., Zhang, P., Lu, W., Dranchak, P. K., Shibata, D., et al. (2012). The gene expression profiles of induced pluripotent stem cells from individuals with childhood cerebral adrenoleukodystrophy are consistent with proposed mechanisms of pathogenesis. *Stem Cell Res. Ther.* 3, 39. doi: 10.1186/scrt130

Warren, L., Manos, P. D., Ahfeldt, T., Loh, Y. H., Li, H., Lau, F., et al. (2010). Highly efficient reprogramming to pluripotency and directed differentiation of human cells with synthetic modified mRNA. *Cell Stem Cell* 7, 618–630. doi: 10.1016/j.stem.2010.08.012

Wilmut, I., Schnieke, A. E., McWhir, J., Kind, A. J., and Campbell, K. H. (1997). Viable offspring derived from fetal and adult mammalian cells. *Nature* 385, 810–813. doi: 10.1038/385810a0

Wobus, A. M., and Loser, P. (2011). Present state and future perspectives of using pluripotent stem cells in toxicology research. *Arch. Toxicol.* 85, 79–117 doi: 10.1007/s00204-010-0641-6

Woltjen, K., Michael, I. P., Mohseni, P., Desai, R., Mileikovsky, M., Hämäläinen, R., et al. (2009). piggyBac transposition reprograms fibroblasts to induced pluripotent stem cells. *Nature* 458, 766–70. doi: 10.1038/nature07863

Xu, D., Alipio, Z., Fink, L. M., Adcock, D. M., Yang, J., and Ward, D. C., (2009). Phenotypic correction of murine hemophilia A using an iPS cell-based therapy. *Proc. Natl. Acad. Sci. U.S.A.* 106, 808–813. doi: 10.1073/pnas.0812090106

Yazawa, M., Hsueh, B., Jia, X., Pasca, A. M., Bernstein, J. A., Hallmayer, J., et al. (2011). Using induced pluripotent stem cells to investigate cardiac phenotypes in Timothy syndrome. *Nature* 471, 230–234. doi: 10.1038/nature09855

Ye, Z., Zhan, H., Mali, P., Dowey, S., Williams, D. M., Jang, Y. Y., et al. (2009). Human-induced pluripotent stem cells from blood cells of healthy donors and patients with acquired blood disorders. *Blood* 114, 5473–5480. doi: 10.1182/blood-2009-04-217406

Yokoo, N., Baba, S., Kaichi, S., Niwa, A., Mima, T., Doi, H., et al. (2009). The effects of cardioactive drugs on cardiomyocytes derived from human induced pluripotent stem cells. *Biochem. Biophys. Res. Commun.* 387, 482–488. doi: 10.1016/j.bbrc.2009.07.052

Yoshida, T., Ozawa, Y., Suzuki, K., Yuki, K., Ohyama, M., Akamatsu, W., et al. (2014). The use of induced pluripotent stem cells to reveal pathogenic gene mutations and explore treatments for retinitis pigmentosa. *Mol. Brain.* 7:45. doi: 10.1186/1756-6606-7-45

Yoshida, Y., and Yamanaka, S. (2010). Recent stem cell advances: induced pluripotent stem cells for disease modeling and stem cell-based regeneration. *Circulation* 122, 80–87. doi: 10.1161/CIRCULATIONAHA.109.881433

Yoshioka, N., Gros, E., Li, H., Kumar, S., Deacon, D. C., and Maron, C. (2013). Efficient generation of human iPSCs by a synthetic self-replicative RNA. *Cell Stem Cell* 13, 246–254. doi: 10.1016/j.stem.2013.06.001

Yu, J., Hu, K., Smuga-Otto, K., Tian, S., Stewart, R., Slukvin, I. I., et al. (2009). Human induced pluripotent stem cells free of vector and transgene sequences. *Science* 324, 797–801. doi: 10.1126/science.1172482

Yu, J., Vodyanik, M. A., Smuga-Otto, K., Antosiewicz-Bourget, J., Frane, J. L., Tian, S., et al. (2007). Induced pluripotent stem cell lines derived from human somatic cells. *Science* 318, 1917–1920. doi: 10.1126/science.1151526

Yusa, K., Rad, R., Takeda, J., and Bradley, A. (2009). Generation of transgene-free induced pluripotent mouse stem cells by the piggyBac transposon. *Nat. Methods* 6, 363–369. doi: 10.1038/nmeth.1323

Zhang, G., Shang, B., Yang, P., Cao, Z., Pan, Y., and Zhou, Q. (2012). Induced pluripotent stem cell consensus genes: implication for the risk of tumorigenesis and cancers in induced pluripotent stem cell therapy. *Stem Cells Dev.* 21, 955–964. doi: 10.1089/scd.2011.0649

Zhang, M., Yang, C., Liu, H., and Sun, Y. (2013). Induced pluripotent stem cells are sensitive to DNA damage. *Genomics Proteomics Bioinformatics* 11, 320–326. doi: 10.1016/j.gpb.2013.09.006

Zhao, Y., Yin, X., Qin, H., Zhu, F., Liu, H., Yang, W., et al. (2008). Two supporting factors greatly improve the efficiency of human iPSC generation. *Cell Stem Cell* 3, 475–479. doi: 10.1016/j.stem.2008.10.002

Zhou, Q., Brown, J., Kanarek, A., Rajagopal, J., and Melton, D. A. (2008). *In vivo* reprogramming of adult pancreatic exocrine cells to β-cells. *Nature* 455, 627–632. doi: 10.1038/nature07314

Zhou, T., Benda, C., Dunzinger, S., Huang, Y., Ho, J. C., Yang, J., et al. (2012). Generation of human induced pluripotent stem cells from urine samples. *Nat. Protoc.* 7, 2080–2089. doi: 10.1038/nprot.2012.115

Conflict of Interest Statement: The authors declare that the research was conducted in the absence of any commercial or financial relationships that could be construed as a potential conflict of interest.

12

Prion protein (PrP) gene-knockout cell lines: insight into functions of the PrP

Akikazu Sakudo[1]* and Takashi Onodera[2]

[1] Laboratory of Biometabolic Chemistry, Faculty of Medicine, School of Health Sciences, University of the Ryukyus, Nishihara, Japan
[2] Research Center for Food Safety, School of Agricultural and Life Sciences, University of Tokyo, Tokyo, Japan

Edited by:
Sophie Mouillet-Richard, Institut National de la Santé et de la Recherche Médicale, France

Reviewed by:
Sophie Mouillet-Richard, Institut National de la Santé et de la Recherche Médicale, France
Yong-Sun Kim, Hallym University, South Korea
Marilene Hohmuth Lopes, University of Sao Paulo, Brazil

***Correspondence:**
Akikazu Sakudo, Laboratory of Biometabolic Chemistry, Faculty of Medicine, School of Health Sciences, University of the Ryukyus, 207 Uehara, Nishihara, Okinawa 903-0215, Japan
e-mail: sakudo@med.u-ryukyu.ac.jp

Elucidation of prion protein (PrP) functions is crucial to fully understand prion diseases. A major approach to studying PrP functions is the use of PrP gene-knockout (*Prnp*$^{-/-}$) mice. So far, six types of *Prnp*$^{-/-}$ mice have been generated, demonstrating the promiscuous functions of PrP. Recently, other PrP family members, such as Doppel and Shadoo, have been found. However, information obtained from comparative studies of structural and functional analyses of these PrP family proteins do not fully reveal PrP functions. Recently, varieties of *Prnp*$^{-/-}$ cell lines established from *Prnp*$^{-/-}$ mice have contributed to the analysis of PrP functions. In this mini-review, we focus on *Prnp*$^{-/-}$ cell lines and summarize currently available *Prnp*$^{-/-}$ cell lines and their characterizations. In addition, we introduce the recent advances in the methodology of cell line generation with knockout or knockdown of the PrP gene. We also discuss how these cell lines have provided valuable insights into PrP functions and show future perspectives.

Keywords: prion protein, *Prnp*$^{-/-}$ cell line, Doppel, Shadoo, knockout mouse

INTRODUCTION

Prion diseases, also called transmissible spongiform encephalopathies (TSE), are a group of neurodegenerative disorders that affect both humans and animals (Prusiner, 1998). Examples of these disorders in humans include Creutzfeldt-Jakob disease (CJD), Gerstmann-Sträussler-Scheinker syndrome (GSS), kuru and fatal familial insomnia (FFI). TSEs found in animals include scrapie, bovine spongiform encephalopathy (BSE) and chronic wasting disease (CWD). The representative characteristics of prion diseases are spongiform changes, which are associated with neuronal loss, vacuolation, astrocytosis, formation of amyloid plaques and a failure to induce an inflammatory response (Sakudo and Ikuta, 2009a,b). A common characteristic of prion diseases is rapid symptom progression after a prolonged incubation period. The damaged region of the brain differs among each prion disease and determines the clinical signs and symptoms of the specific disease. Although the incidence of prion diseases is rare, it is fatal because of a lack of effective treatment to either cure or delay disease progression. Furthermore, the risk of prion transmission is a serious threat to public health due to the high resistance of prion agents to standard sterilization techniques (Sakudo et al., 2011a).

Prion protein (PrP) is the most important factor for prion infection. This finding has been verified by experiments using PrP gene (*Prnp*)-knockout (*Prnp*$^{-/-}$) mice and cells derived from these mice (Sakudo et al., 2006, 2007a,b). The first important finding was that *Prnp*$^{-/-}$ mice are resistant to infection with prion agents (Bueler et al., 1993; Prusiner et al., 1993; Manson et al., 1994a,b). Secondly, *Prnp*$^{-/-}$ primary culture neurons are not killed by the toxicity of prion agents or PrP(106-126) (Brandner et al., 1996a; Giese et al., 1998). Thirdly, prions cannot proliferate in the brain of *Prnp*$^{-/-}$ mice (Brandner et al., 1996b). Therefore, it is apparent that PrP plays an important role in the mechanisms of infection and contributes to the pathogenesis of prion diseases. This is also consistent with the notion that the conversion of cellular PrP (PrP^C) into abnormal PrP (PrP^{Sc}) constitutes a fundamental feature of prion diseases (Prusiner, 1998). It seems that PrP^C acts as a cellular receptor of PrP^{Sc}. Because PrP^C is thought to function as a dimeric form (Meyer et al., 2000; Kaimann et al., 2008), PrP^C-PrP^{Sc} interaction might alternate between PrP^C-PrP^{Sc} and PrP^{Sc}-PrP^{Sc} after prion infection. As shown above, *Prnp*$^{-/-}$ mice and the derived cells have confirmed the importance of PrP in the pathogenicity of prion diseases and have greatly contributed to the understanding of these disorders.

However, despite the numerous studies on *Prnp*$^{-/-}$ mice, the biological functions of PrP^C, which are crucial for understanding prion diseases, remain elusive. Cell lines are useful for detailed analysis of gene function. Therefore, to investigate PrP^C functions in detail, *Prnp*$^{-/-}$ cell lines derived from *Prnp*$^{-/-}$ mice have been established. The present review will summarize currently available *Prnp*$^{-/-}$ cell lines. In addition, recent advances in strategies for producing cell lines in which *Prnp* is knocked out or knocked down will be presented. Valuable insights into PrP^C functions made possible by the availability of these cell lines will be introduced as well.

THE ANALYSIS OF *Prnp*$^{-/-}$ MICE FOR UNDERSTANDING PrP^C FUNCTION

One approach to studying protein function is to analyze the effect of knocking out the corresponding gene. In the case of PrP^C,

Prnp$^{-/-}$ mice have been widely used for this purpose. Therefore, studies using *Prnp*$^{-/-}$ mice should be mentioned before we discuss *Prnp*$^{-/-}$ cell lines. *Prnp*$^{-/-}$ mice have been used for elucidating the functions of PrP^{C} *via* analysis of the phenotype of *Prnp*$^{-/-}$ mice (Weissmann and Flechsig, 2003). Six lines of *Prnp*$^{-/-}$ mice, designated Zrch I (Bueler et al., 1992), Zrch II (Rossi et al., 2001), Npu (Manson et al., 1994a), Ngsk (Sakaguchi et al., 1996), Rcm0 (Moore et al., 1999) and Rikn (Yokoyama et al., 2001), have been generated (**Figure 1**). However, there were some discrepancies among the phenotypes of the knockout mice. The first and second knockout lines, Zrch I and Npu, were generated by disrupting the PrP coding region located in exon 3 of the *Prnp* (Bueler et al., 1992; Manson et al., 1994a). Studies on these lines did not show any severe abnormality. Subsequently, Ngsk, Rikn, Rcm0, and Zrch II mouse lines were generated in which the entire coding region and part of intron 2 was deleted (Sakaguchi et al., 1996; Moore et al., 1999; Rossi et al., 2001; Yokoyama et al., 2001). Because of the structure of the targeted *Prnp* allele, intergenic splicing between *Prnp* and the surrounding gene led to ectopic expression of the surrounding gene in the brains of these mice. This prompted the discovery of the gene *Prnd* located 16 kbp downstream of *Prnp*, encoding the prion-related protein Doppel (Dpl) (Moore et al., 1999), which shares ~25% identity with two-thirds of the C-terminal region of PrP. Ectopic expression of Dpl leads to the development of late-onset ataxia in *Prnp*$^{-/-}$ mice. Therefore, the ataxic phenotype of some lines of *Prnp*$^{-/-}$ mice (Ngsk, Rikn, Rcm0, ZrchII) is due to ectopic expression of Dpl (derived from *Prnp/Prnd* chimeric mRNAs through intergenic splicing) as a result of the disruption of the splicing acceptor of *Prnp* exon 3 (Moore et al., 1999; Li et al., 2000a; Rossi et al., 2001). In this review article, to discriminate between *Prnp*$^{-/-}$ mice with and without ectopic expression of Dpl, we term the former ataxic *Prnp*$^{-/-}$ mice (Ngsk, Rikn, Rcm0, Zrch II) as type 2, while the latter non-ataxic *Prnp*$^{-/-}$ mice (Zrch I, Npu) as type-1 (**Figure 1**).

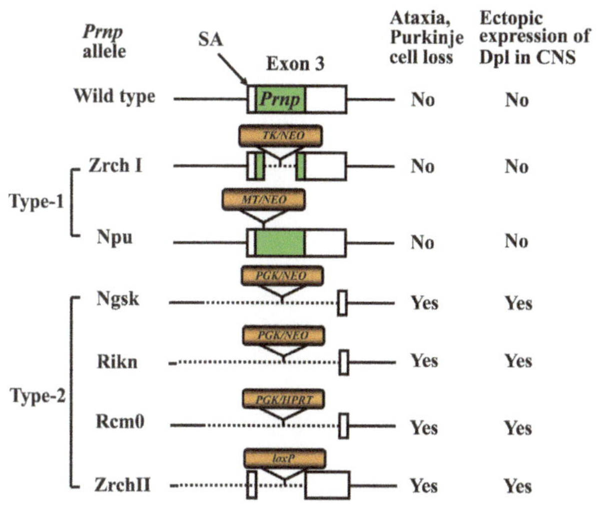

FIGURE 1 | Knockout constructs of type-1 and type-2 of prion protein gene (*Prnp*)-knockout (*Prnp*$^{-/-}$) mice. Structures of the constructs used to produce six lines of *Prnp*$^{-/-}$ mice. The *Prnp*$^{-/-}$ mice are divided into type-1 and type-2 *Prnp*$^{-/-}$ mice depending on whether the splicing acceptor of exon 3 is disrupted. The disruption of exon 3 is correlated with the development of late-onset ataxia and Purkinje cell loss, which is induced by ectopic expression of Doppel (Dpl) in the central nervous system (CNS). The structure of wild-type (WT) *Prnp* exon 3 and prion protein (PrP) coding region (green box) is shown at the top. The selection markers are indicated by orange boxes. The presence and absence of the exon 3 splicing acceptor (SA) is correlated with the development of late-onset ataxia. The selection markers were PGK, mouse phosphoglycerate kinase promoter; NEO, neomycin phosphotransferase; HPRT, mouse hypoxanthine phosphoribosyltransferase; TK, human herpes simplex virus type 1 thymidine kinase promoter; MT, mouse metallothionein promoter; loxP, a 34-bp recombination site from phage P1.

The type-1 *Prnp*$^{-/-}$ mice display no major abnormalities. Indeed, only small abnormalities could be found in type-1 *Prnp*-knockout mice. These include deficits related to neuronal and other cellular functions such as abnormal learning and sleep patterns, olfactory deficits, susceptibility to seizures, abnormal neuromuscular function and depressive disorder as well as abnormal glucose tolerance and deficits related to protection against pathophysiological stress (e.g., abnormal inflammatory responses, decreased protection against viral bacterial infection, accelerated symptoms of experimental autoimmune encephalomyelitis (EAE) and susceptibility to ischemic brain injury (Onodera et al., 2014). In addition, the most consistent phenotype of *Prnp*$^{-/-}$ mice is myelin degeneration (Nishida et al., 1999; Baumann et al., 2007; Bremer et al., 2010). Some of these abnormalities may be due to the genetic background (Steele et al., 2007). Therefore, experiments using backcrossed mice should be performed to confirm whether the observed phenotypes are attributable to PrP. There are also possibilities of synergistic phenotypes in a combination of mouse strain genetic backgrounds and knockout constructs. In addition, the breeding condition of the mice may affect emergence of some phenotypes. For knockout mouse studies, confirmation of findings using mice with distinct knockout constructs and genetic backgrounds in an independent laboratory will be required to firmly establish conclusions regarding PrP function. Indeed, we have recently reviewed the topic of different *Prnp*$^{-/-}$ mouse phenotypes to assess the physiological function of PrP^{C} (Onodera et al., 2014).

PrP FAMILY PROTEINS AND THEIR KNOCKOUT MICE

Recently, new members of the PrP family of proteins have been identified; namely, Dpl, encoded by *Prnd*, and Shadoo (Sho), encoded by *Sprn*. These PrP family proteins are all subject to endoproteolytic processing (Mays et al., 2014) and localize to similar membrane environments (Li et al., 2013). Dpl and Sho pose interesting and important questions concerning PrP function. Given that the function of PrP remains elusive, information obtained from comparative structural and functional studies of these PrP family proteins are of great interest. Specifically, complementary, overlapping or interference of function among PrP family members might be found from these analyses.

There are reports indicating functional interaction between PrP and Dpl. It has been shown that the ataxic phenotype observed in type-2 *Prnp*$^{-/-}$ mice can be neutralized by crossing the mice with those overexpressing wild-type mouse PrP

(Nishida et al., 1999; Rossi et al., 2001). These findings suggest that PrP has a neuroprotective role and functionally interacts with Dpl. Another report showed that the toxicity of Dpl overexpression in mice and cells is inhibited by PrP expression. In a separate report, using immortalized *Prnp*$^{-/-}$ neuronal cells, a decrease of cellular copper levels brought about by serum deprivation was also inhibited by PrP. However, the inhibitory effect of PrP on decreased copper levels was prevented by Dpl overexpression (Sakudo et al., 2004). A further study demonstrated that endoproteolysis of PrP was affected by Dpl overexpression in *Prnp*$^{-/-}$ cells (Sakudo and Onodera, 2011b). Therefore, PrP-Dpl interaction functions both *in vitro* and *in vivo*.

Dpl is localized in the brain for only a limited duration in the developmental process (Li et al., 2000b), whereas analysis of *Prnd*$^{-/-}$ mice showed that Dpl deficiency did not interfere with embryonic and postnatal development (Behrens et al., 2002; Genoud et al., 2003). Dpl was produced at a late stage of spermatogenesis. Spermatids of *Prnd*$^{-/-}$ mice were reduced in numbers, immobile, malformed and unable to fertilize oocytes, resulting in male sterility. PrP is also expressed in testis, but unlike other cells the PrP is expressed as N-terminally truncated isoforms (Peoc'h et al., 2002) or a C-terminally truncated isoform (Shaked et al., 1999). The C-terminal region of PrP resembles Dpl, suggesting that these proteins share a common function in testis in the case of N-terminally truncated PrP (Peoc'h et al., 2002) or show a complementary function to each other in the case of C-terminally truncated PrP (Shaked et al., 1999). Currently, there is no evidence that Dpl is associated with the pathogenesis of prion diseases. For example, the level of Dpl expression does not correlate with the onset of prion disease (Weissmann and Aguzzi, 1999; Tuzi et al., 2002).

By contrast, there is evidence that Sho is implicated in prion diseases. For example, the expression of Sho decreases with the accumulation of PrP^{Sc} after prion infection (Watts et al., 2007a, 2011). Sho protein levels were decreased in the brains of *Prnp*^{a} and *Prnp*^{b} mice (Westaway et al., 2011), hamsters, meadow, voles and sheep infected with natural and experimental prion strains (Watts et al., 2011). In addition, time course experiments showed that the levels of PrP^{Sc} vs. Sho protein were inversely proportional (Watts et al., 2011). Membrane anchoring and the N-terminal domain of PrP both influenced the inverse relationship between PrP^{Sc} and Sho (Watts et al., 2011). By contrast, increased expression of Sho did not influence prion replication (Watts et al., 2011), suggesting Sho merely acts as a marker for prion disease. Therefore, it remains unclear how Sho contributes to the pathogenesis of prion diseases. Indeed, depletion of Sho appears to be unimportant in terms of triggering prion diseases and in the processing and degradation of PrP^{Sc}.

In terms of physiological action, Sho can exhibit neuroprotective properties similar to PrP^{C} (Watts and Westaway, 2007b; Watts et al., 2007a) and share a number of binding partners common with PrP^{C} (Watts et al., 2009). In addition, as Sho has a similar structure to the N-terminal region of PrP (Watts and Westaway, 2007b), the relationships between the structural similarity and functional similarity of PrP and Sho are interesting. *Sprn*$^{-/-}$ mice underwent a subtle alteration of body weight (Passet et al., 2013), which was not noted in the case of double-knockout (*Sprn* and *Prnp*) mice. In addition, *Prnp/Sprn* knockout mice survived to over 600 days of age without any severe abnormality, suggesting the existence of a discrete signaling pathway of *Prnp* and *Sprn* to maintain neuronal survival. Sho was also found to be expressed in the trophoblast cells of the placenta (Passet et al., 2012). Comparative transcriptomic analyses performed between E6.5 and E7.5 *Sprn*-knockdown (using RNAi) embryo and their wild-type counterparts suggested that Sho has functions complementary, not necessary overlapping, with those of PrP, associated with cellular movement and hematological system development and differentiation (Passet et al., 2012). Interestingly, the expression profile of *Sprn* in testis and ovary resemble that of *Prnp*. Thus, interaction among three PrP family members may play important roles in reproductive tissues. Although the contribution of PrP family members to embryogenesis is suggested, *Prnp/Sprn* knockout mice are healthy and fertile (Daude and Westaway, 2012a; Daude et al., 2012b). Therefore, further studies on reproductive tissues are required to resolve the apparent discrepancy in the data. The topic of Sho is also discussed in detail in a review article in this research topic (Makzhami et al., 2014). As mentioned above, analysis of the phenotypes of knockout mice and comparison of PrP family members does not fully elucidate the functions of PrP. Therefore, other approaches to analyze PrP function are required. Next, we discuss the use of *Prnp*$^{-/-}$ cell lines to study the role of PrP.

CHALLENGES TO INVESTIGATING PrP^{C} FUNCTIONS *IN VITRO*

Cells have been classified into approximately 200 different types (Obinata, 2007). Through development and maintenance during adult life, cells are differentiated and obtain characteristics to elicit specialized functions. Differentiation can be chemically induced by various stimuli such as growth factors, cytokines, and hormones (Alberts et al., 2008). Histological location of the cells also influences cell differentiation. In order to study each cell type possessing a specialized function, the cells need to be separated from each other. Examination of large, pure populations of specific cell types is extremely valuable in advancing our understanding of the cells. Primary cultures are composed of a heterogeneous cell population and can be maintained only for a limited period of time. Although PrP^{C} is highly expressed in neurons, primary neuron cultures can be maintained for just a few weeks. Moreover, cells belonging to a rare cell type are often difficult to isolate and culture. For detailed functional gene analysis, an efficient transfection procedure is required. However, the transfection efficiency of primary cultures is generally low, especially when using primary neurons. If gene transfer to a primary culture of neurons or other cell types with low efficiency is necessary, a viral vector is required for the introduction of inducible genes. In this respect, cell lines are useful for detailed analysis of gene function because the transfection efficiency is usually high compared to that of primary cultures. In addition, cell lines can be maintained almost indefinitely.

However, studying pure cell lines has the disadvantage that it can only reflect a cell-autonomous condition, whereas biological processes in an organism involve a heterogeneous cell population.

Co-cultures of multiple cell populations may help to mimic the heterogeneous condition. Additionally, because there are a lot of cell types in an organism, it is very difficult to match cell types between cell lines when a comparison is required. Thus, comparisons between a wild-type and a knockout cell line are suboptimal for analyzing a target gene. Alternatively, a cloned knockout cell line should be compared in the absence and presence of the target gene.

PrP^C EXPRESSION AND PrP^C-INTERACTING PROTEINS

PrP^C is expressed in a variety of tissues. Northern blot analysis has shown that the levels of PrP mRNA vary among tissues, with the highest levels found in the brain and placenta; moderate levels in the testis, heart, and lung; and lower levels in the spleen and kidney (Horiuchi et al., 1995; Saeki et al., 1996). In sheep, 3–5 μg PrP^C/g of tissue is present in the brain, whereas the level of PrP^C per gram of tissue is 100 ng in the heart and skeletal muscle, 200 ng in the lung, 40 ng in the spleen, and 3 ng in the liver (Moudjou et al., 2001). Thus, the quantity of PrP^C in the brain is 20- to 50-fold higher than in the other tissues. This observation suggests that PrP^C plays an important role in several tissues, particularly in the brain.

Recently, it has been shown that PrP^C binds a variety of partner molecules. Using radioisotope or enzyme-labeled PrP, interaction of PrP with glial fibrillary acidic protein (GFAP), a NF-E2 related factor 2 (Nrf2), amyloid precursor protein 1 (Aplp1), F-box protein-6, neural F-box protein 42 kDa (NFB42), postsynaptic density 95 kDa (PSD-95)/ SAP-90 associated protein, protein tyrosine phosphatase non-receptor type-21, and predicted protein KIAA0443 was reported (Yehiely et al., 1997). A two hybrid system showed that laminin (Graner et al., 2000), 37-kDa/67-kDa laminin receptor (Gauczynski et al., 2001), 37-kDa laminin receptor precursor (Rieger et al., 1997), heat shock protein 60 kDa (Hsp60) (Edenhofer et al., 1996) and Bcl-2 (Kurschner and Morgan, 1995) all bind PrP. Co-immunoprecipitation experiments were used to identify PrP-interacting molecules, which included prion interactor 1 (Pint1) (Spielhaupter and Schatzl, 2001), synapsin Ib (Spielhaupter and Schatzl, 2001), growth factor receptor-bound protein 2 (Grb2) (Spielhaupter and Schatzl, 2001), and neurotrophin receptor interacting MAGE (melanoma-associated antigens) homolog (NRAGE) (Bragason and Palsdottir, 2005), dystroglycan (Keshet et al., 2000), the neuronal isoform of nitric oxide synthase (nNOS) (Keshet et al., 2000), glucose regulated protein 94 (Grp94) (Capellari et al., 1999), protein disulphide isomerase (Capellari et al., 1999), calnexin (Capellari et al., 1999), calreticulin (Capellari et al., 1999) and zeta-associated protein-70 (ZAP-70) (Mattei et al., 2004). A cross-linking method demonstrated interaction of PrP with tubulin (Nieznanski et al., 2005) and neural adhesion molecule (N-CAM) (Schmitt-Ulms et al., 2001). Complementary hydropathy and pull-down assays were able to show the interaction of PrP with stress inducible protein 1 (STI1) (Martins et al., 1997) and Hsp60 of *Brucella abortus* (Watarai et al., 2003). Intriguingly, PrP interacts with caveolin-1 (Toni et al., 2006), while cross-linking of cell-surface PrP stimulated caveolin-1-dependent interaction with Fyn tyrosine kinase (Mouillet-Richard et al., 2000), resulting in neurite outgrowth and differentiation of neuronal cells (Mouillet-Richard et al., 2000; Pantera et al., 2009). Thus, PrP contributes to the control of the cellular redox state and homeostasis of neuronal cells (Mouillet-Richard et al., 2007). Because Fyn is involved in various signaling pathways, the interaction implies that PrP^C has diverse functions. Most interestingly, a wealth of recent studies has established that PrP interacts with Amyloid β protein (Aβ), which is generated by the abnormal processing of the amyloid precursor protein (APP) by β-secretase, β-site APP cleaving enzyme (BACE1) and involved in the pathogenesis of Alzheimer's disease (Larson et al., 2012; Um et al., 2012; Um and Strittmatter, 2013; Dohler et al., 2014). In addition, several reports have shown that PrP^C interacts with APP (Yehiely et al., 1997; Kaiser et al., 2012). Several reports have further demonstrated an involvement of PrP in the toxicity of Aβ, although the use of different in *vitro* or transgenic models has yielded contrasting results (Schwarze-Eicker et al., 2005; Laurén et al., 2009; Balducci et al., 2010; Calella et al., 2010; Chung et al., 2010; Kessels et al., 2010; Morales et al., 2010; Ordóñez-Gutiérrez et al., 2013; Gasperini and Legname, 2014). Some groups have also reported that Fyn kinase mediates signal transduction downstream of the PrP^C-Aβ complex (Larson et al., 2012; Um et al., 2012; Um and Strittmatter, 2013). Because PrP^C inhibits BACE1 either by direct interaction (Griffiths et al., 2011) or indirectly without interaction (Parkin et al., 2007; McHugh et al., 2012), reduction of the PrP^C level may increase Aβ. Therefore, PrP^C may be involved in the pathogenesis of Alzheimer's disease not only by transducing Aβ toxic signals but also *via* regulation of neurotoxic Aβ production. Taken together, most of the interacting proteins are important factors involved in survival, proliferation, differentiation, development, and stress response. However, it should be mentioned that this interaction may depend on the specific cell type and/or the surrounding tissue environment.

Currently, *Prnp*$^{-/-}$ cell lines have predominantly been established from brain as well as fibroblast and macrophage cell lines (**Figure 2**). Next, we will introduce the *Prnp*$^{-/-}$ cell lines established so far.

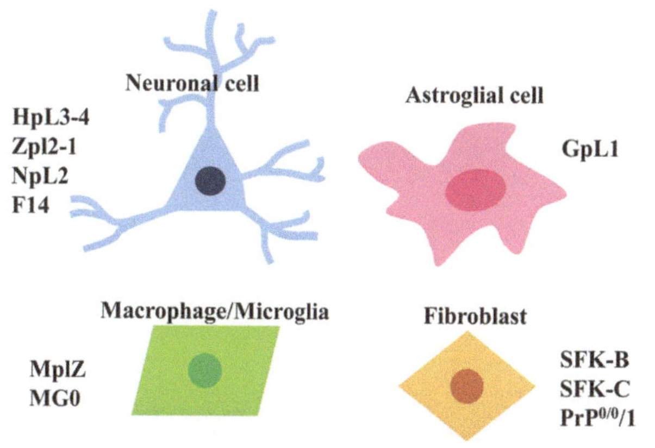

FIGURE 2 | Types of *Prnp*$^{-/-}$ cell lines that are available. Currently available *Prnp*$^{-/-}$ cell lines can be divided into four types i.e., neuronal cells (HpL3-4, Zpl2-1, NpL2, F14), astroglial cells (GpL1), macrophage/microglia (MpLZ, MG0), and fibroblasts (SFK-B, SFK-C, PrP$^{0/0}$/1).

HpL, *Prnp*$^{-/-}$ NEURONAL CELL LINES

To gain further insights into PrP functions, cell lines (HpL and HW) were established using the gene transfer of oncogenes by our group (Kuwahara et al., 1999). HpL was the first *Prnp*$^{-/-}$ cell line, while HW is the wild-type counterpart. These cell lines were from the hippocampal area of Rikn *Prnp*$^{-/-}$ mice and *Prnp*$^{+/+}$ mice using a target for infection with a recombinant retrovirus vector containing the simian virus 40 (SV40) T-antigen gene (Ryder et al., 1990). The virus vector was used for transfection of ψ2 packaging cells, which are stable cells containing all of the structural retroviral protein genes (*gag*, *pol*, and *env*). The retrovirus particles produced from the transfected ψ2 packaging cells were used for immortalization by introduction of the SV40 large T antigen gene. The hippocampal primary cultures derived from *Prnp*$^{-/-}$ and *Prnp*$^{+/+}$ mice were infected with the above retrovirus, then incubated, selected, expanded and finally subjected to limiting dilution to obtain cloned cell lines. Three *Prnp*$^{-/-}$ cell lines, HpL2-1, HpL-3-2 and HpL3-4, and 3 *Prnp*$^{+/+}$ cell lines HW8, HW9, HW19, were thereby obtained (Kuwahara et al., 1999).

The HpL and HW cells were exposed to dibutyl cyclic AMP (dcAMP) at 0.4–6 mM, retinoic acid (RA) at 0.1–10 μM, and phorbol 12-myristate 13-acetate (PMA) at 160–1600 nM, for up to 10 days of culture. These compounds all induced morphological changes, i.e., elongation of neurite-like processes, in both HpL and HW cells. Among the tested compounds, the morphological change was best observed by stimulating the HpL and HW cells in medium with dcAMP or PMA without fetal calf serum (FCS). To further characterize the cell types of the cell lines, reverse transcription polymerase chain reaction (RT-PCR) was applied to monitor the expression of neurofilament genes. Treatment with PMA resulted in the amplification of the predicted size products corresponding to neurofilament (NF)-68K and NF-200K genes. The NF-68K gene was transcribed in all 3 HpL and 3 HW cell lines with or without PMA treatment. After treatment with PMA, the NF-200K gene was transcribed in all three HW cell lines and in only one of the HpL cell lines (HpL2-1). The transcription of GFAP, a glial cell marker, was not detected in HpL or HW cells. Taken together, the results suggest that all six of these cell lines belong to the neuronal precursor cell lineage.

Next, the characteristics of HpL cells were compared to those of HW cells. The most remarkable feature of HpL cells by comparison to HW cells was observed when the serum was removed from the culture medium (Kuwahara et al., 1999). Serum deprivation caused comparable morphological changes within 1 h in all cell lines whether or not they were treated with dcAMP. However, within 4 days the HpL cells had died in the serum-free culture. The process started by the rounding up of the cells, cytoplasmic condensation and neurite retraction. By contrast, all three HW cell lines maintained their structural integrity under the same culture conditions. DNA fragmentation, a characteristic of apoptosis, was firstly observed at 12 h and peaked at 24 h after serum deprivation in the HpL cells, whether or not they were treated with PMA or cAMP. However, no such DNA fragmentation was detected in any of the HW cells after serum deprivation. Cultured cells in serum-free medium for 6, 12, or 24 h were ethanol-fixed and stained with propidium iodide before analysis by flow cytometry (Sakudo et al., 2003a). A significant peak in the fluorescence low-intensity area (hypodiploid cells), which corresponds to the fraction of apoptotic cells, was detected only in the samples from HpL cells at 6, 12, and 24 h after serum withdrawal. Because HpL3-4 cells started to generate superoxide anion at 6 h after serum deprivation, superoxide anion generation seems to induce apoptosis of the HpL3-4 cells (Sakudo et al., 2003b). To investigate whether the abnormality observed in HpL cells was due to the absence of PrP, the HpL cell lines (HpL2-1 and HpL3-4) were transfected with PrP expression vector. Cells transfected with PrP-expressing vector, designated HpL2-1TR and HpL3-4TR, survived under serum-free conditions (Kuwahara et al., 1999). In contrast, HpL2-1 and HpL3-4 cells transfected with empty vector showed apoptotic features similar to those of HpL2-1 and HpL3-4 cells under serum-free conditions. These results suggest that HpL cells are susceptible to serum deprivation and that apoptotic cell death is suppressed by reintroduction of *Prnp*. Moreover, levels of superoxide anion generated from HpL3-4 cells after serum deprivation were suppressed by reintroduction of *Prnp* (Sakudo et al., 2003b). Therefore, these results suggest that PrP functions by displaying anti-oxidative and anti-apoptotic activity. Recent studies have shown that the anti-apoptotic activity of PrP is species specific, as indicated by evidence that apoptosis was suppressed by mouse PrP but not by bovine or hamster PrP (Wu et al., 2008). Moreover, as serum-free supernatants of HW cell cultures could not rescue the HpL cells from apoptosis, molecules secreted from HW cells do not appear to be involved in the neuroprotective mechanism of PrP (Kuwahara et al., 1999). In addition to the cell death, HpL cell lines show significantly shorter neurite extension in comparison with HW cell lines after treatment with PMA, suggesting that PrP is involved not only in survival but also in differentiation (Kuwahara et al., 1999).

Recently, in addition to our group, several other groups have independently confirmed the neuroprotective function of PrP using the above-mentioned PrP-deficient cell lines or other cell lines. Kim et al. reproduced our findings using HpL3-4 cells, and detected significantly greater changes in the calcium ion level, transmembrane potential and cytochrome c level in the mitochondria of HpL3-4 cells over those observed in *Prnp*-reintroduced HpL3-4 cells after serum deprivation (Kim et al., 2004). Mange et al. also confirmed that cell viability was increased by the reintroduction of *Prnp* into HpL3-4 cells (Mange et al., 2004). By contrast, Christensen et al. confirmed a moderate but significant cytoprotective effect of PrP in serum-deprived HpL3-4 cells, but concluded the degree of neuroprotection offered by PrP was either not physiologically relevant or that the cell culture systems cannot adequately model the situation *in vivo* (Christensen and Harris, 2008). However, we believe the HpL3-4 cells used in these experiments were not appropriate for this analysis because they did not express neuronal markers (i.e., neuronal nuclear antigen (NeuN), microtubule-associated protein 2 (MAP2) or NF-L) (Christensen and Harris, 2008). This opinion is also supported by the evidence that *Prnp-Prnd* chimeric mRNA transcribed from the *Prnp* promoter was not detected by RT-PCR and immunofluorescence analysis in the cells used by Christensen et al. By contrast, our results were confirmed by

Prnp-Prnd chimeric mRNA expression in HpL3-4 cells (Sakudo et al., 2005b).

HpL3-4 cells are susceptible to various forms of stress. For example, Vassallo et al. showed HpL3-4 cells are prone to cell death induced by 3-morpholinosydnonimine (SIN-1) (Vassallo et al., 2005). Vassallo et al. further investigated the signal cascade of apoptosis induced by serum deprivation and SIN-1, and showed that recruitment of phosphatidylinositol 3-kinase by PrP contributes to cellular survival under conditions of oxidative stress imposed by SIN-1 or serum deprivation in HpL3-4 cells. *Prnp*-transfected HpL3-4 cells show autophagy after amyloid β treatment. Amyloid β_{1-42} ($A\beta_{42}$) enhanced autophagic flux, such as the formation of autophagic vacuoles, was observed in HpL3-4 cells in which *Prnp* was reintroduced. However, autophagy was impaired in HpL3-4 cells transfected with empty vector (Nah et al., 2013). In addition, PrP-dependent recruitment of Beclin1 (BECN1) into lipid rafts was observed in the HpL3-4 cell line. Because BECN1 is associated with phosphatidylinositol 3-kinase catalytic subunit type 3 (PIK3C3), there will be recruitment of BECN1-PIK3C3 complex into the lipid raft, which is an essential step for autophagy. The observed autophagy could be due to interaction of PrP and BECN1. Therefore, studies using HpL3-4 cells revealed that PrP is involved in $A\beta_{42}$ induced autophagy.

HpL3-4 cells were also susceptible to a monomeric, highly α-helical form of PrP known as toxic PrP (Zhou et al., 2012). Toxic PrP was identified as a recombinant form of PrP that was subjected to denaturation, fractionation and dilution refolding. However, there is no information on the comparative effect of toxic PrP on PrP-expressing and non-expressing HpL3-4 cells. Similarly, in another report, although treatment with $A\beta_{42}$ showed a cytotoxic effect and increased levels of reactive oxygen species (ROS) in HpL3-4 cells, there was no comparison with the HpL3-4 counterpart in which *Prnp* was reintroduced (Hyeon et al., 2012).

It has also been reported that treatment of HpL3-4 cells with hydrogen peroxide increases intracellular calcium levels in a PrP expression dependent manner, suggesting PrP functions as a sensor of hydrogen peroxide (Krebs et al., 2007). Therefore, it appears that PrP^C plays an important role in anti-oxidative stress. As such, the signaling cascade and oxidative metabolism were investigated in HpL3-4 cells after serum deprivation (Sakudo et al., 2003a,b). The studies revealed decreased expression of anti-apoptotic proteins Bcl-2 and Bcl-x_L in HpL3-4 cells during apoptosis induced by serum deprivation. Suppression of apoptosis by overexpression of Bcl-2 and Bcl-x_L, suggested that cell death in this model system is apoptotic, and Bcl-2 and Bcl-x_L play important roles in the apoptosis of HpL3-4 cells (Sakudo et al., 2003a). Furthermore, HpL3-4 cells subjected to serum deprivation showed a more marked decrease in intracellular copper concentration, greater increases of intracellular superoxide anion and caspase-3/9 activation, and a smaller increase in intracellular hydrogen peroxide levels by comparison to HpL3-4 cells in which *Prnp* was reintroduced (Sakudo et al., 2003b). HpL3-4 cells also showed a decrease of cellular superoxide dismutase (SOD) activity and cellular copper content compared to HpL3-4 cells expressing PrP (Sakudo et al., 2003b). By contrast, levels of *Prnp-Prnd* chimeric mRNA and *Prnd* mRNA were unchanged in HpL3-4 cells after serum deprivation (Sakudo et al., 2005c). Higher coxsackievirus B3 and poliovirus production in HpL3-4 cells showed that PrP may be involved not only in the inhibition of virus replication but also anti-apoptotic functions against virus-induced apoptosis (Nakamura et al., 2003b; Baj et al., 2005). Taken together, our studies of HpL cell lines have revealed that PrP plays important roles in differentiation, proliferation and cell survival (i.e., roles in anti-apoptosis, anti-oxidative and anti-viral effect *via* regulating a variety of cell signal cascades). The results obtained from the study of HpL cell lines are consistent with those derived from other studies using primary cultures and other cell lines, which also showed the involvement of PrP in neuroprotection and differentiation (Milhavet and Lehmann, 2002; Roucou et al., 2004).

Studies using HpL cells have contributed to identifying PrP-interacting molecules or PrP-regulated genes because HpL does not express PrP. Moreover, *Prnp*-transfected HpL cells are an ideal counterpart for PrP analysis. Zafar et al. used HpL3-4 cells in combination with STrEP-Tactin chromatography and quadrupole time-of-flight tandem mass spectrometry (Q-TOF MS/MS) analysis to search for PrP-interacting proteins and identified 28 new hits (Zafar et al., 2011). Among them, Rab7a, a regulator of vesicular transport located in a specific intracellular compartment (early to late endosome) and involved in vesicle biogenesis and vesicle fusion to lysosomes, was found to interact with PrP and regulate PrP trafficking in HpL3-4 cells. Furthermore, 23 PrP-interacting proteins were identified by using deletion mutants of PrP [PrP(Δ23-230)], which lack the N-terminal signal peptide sequence and C-terminal glycosylphoshatidylinositol (GPI) anchor site (Zafar et al., 2014). Among them, one novel protein (pyruvate kinase isozymes M1/M2) (PKM2) was found to bind PrP(Δ23-230). In addition, PrP(Δ23-230) exhibits reduced anti-apoptotic activity against staurosporine-induced cell stress with higher expression of PKM2 than full length PrP in HpL3-4 cells, suggesting that PrP(Δ23-230) regulates PKM2 and renders the cells susceptible to stress.

Taking advantage of the absence of *Prnp*, HpL3-4 cells have been used for the analysis of exogenous PrP. Hachiya et al. reported an N-terminal PrP fragment and C-terminal fragment showed distinct intracellular distribution using transfection of double-labeled fluorescent PrP into HpL3-4 cells (Hachiya et al., 2004). The N-terminal PrP fragment was associated with microtubules and had an intracellular localization, but did not co-localize with any intracellular organelle markers.

OTHER *Prnp*$^{-/-}$ NEURONAL CELL LINES

As described earlier, PrP exhibits a neuroprotective effect against apoptosis induced by serum deprivation in HpL3-4 cells, which are derived from type-2 *Prnp*$^{-/-}$ mice (Rikn) with ectopic Dpl expression. To investigate whether PrP inhibits apoptotic neuronal cell death in the absence of Dpl, neuronal cell lines were established from the brain of type-1 *Prnp*$^{-/-}$ mice (Zrch I), which did not show ectopic Dpl expression (Nishimura et al., 2007). The results showed that reintroduction of *Prnp* potently inhibited serum-withdrawal apoptotic cell death in a Zrch I neuronal *Prnp*$^{-/-}$ cell line (NpL2). Moreover, PrP expression was found to upregulate cellular SOD activity in NpL2 cells. Therefore, we concluded that ectopic Dpl production did not affect the anti-apoptotic and anti-oxidative functions of PrP. Thus, it was confirmed that the presence of PrP is directly correlated with protection against oxidative stress.

Kim et al. established neuronal cell lines from hippocampal neurons of Zrch I *Prnp*$^{-/-}$ mice by lipofection of SV40 large T antigen-expressing vector and found higher proliferation of neuronal cell lines from *Prnp*$^{-/-}$ mice (Zpl) than those derived from *Prnp*$^{+/+}$ mice (ZW) (Kim et al., 2005). Zpl cells do not express Dpl because the cells are derived from type 1 *Prnp*$^{-/-}$ mice (Zrch I). Moreover, Zpl cells showed higher levels of apoptosis and autophagy compared to ZW cells after serum deprivation (Oh et al., 2008). The neuroprotective activity of PrP in Zpl cells was confirmed by reintroduction of *Prnp* into Zpl cells. Cleaved caspase-3 levels for apoptotic index and LC3-II levels for autophagy index were correlated with the expression of PrP. Furthermore, PrP(Δ53-94), octapeptide repeat (OR)-deficient PrP, did not display a protective activity in Zpl cells, indicating the importance of the OR region in the neuroprotective activity of PrP. These results are consistent with the HpL3-4 studies using deletion PrP mutants as described in the following section.

The F14 neuronal cell line is another *Prnp*$^{-/-}$ derived cell line, obtained by fusion of *Prnp*$^{-/-}$ cerebellar cells and mouse neuroblastoma cells (Holme et al., 2003). The cells were used for investigating the effect of PrP mutations on PrP membrane orientation, although the characteristics of PrP-expressing cells and PrP-deficient cells have not been compared.

Although not *Prnp*-knockout cells, Loubet et al. used the 1C11 cell line, a neuroepithelial progenitor for shRNA (short hairpin RNA)-mediated knockdown of *Prnp*, to investigate the gene silencing effect of *Prnp* (Loubet et al., 2012). 1C11 cells lack neuron-associated functions and acquire, upon differentiation, the functions of serotogenic or noradrenergic neuronal cells associated with cytoskeleton remodeling along neuritogenesis. The results showed that *Prnp* knockdown caused impaired neuronal polarization with inhibition of the initial sprouting of neurite. Further analysis of the signaling pathway suggests that PrP^C contributes to neurite polarization by modulating integrin interactions with the extracellular matrix.

ASTROGLIAL *Prnp*$^{-/-}$ CELL LINES

PrP^C is expressed not only in neuronal cells but also in non-neuronal cells such as glial cells. Therefore, studies on PrP^C functions using non-neuronal cell lines are necessary. To address this issue, the *Prnp*$^{-/-}$ astroglial cell line GpL1 from hippocampal cells of Zrch I *Prnp*$^{-/-}$ mice were established by retrovirus-mediated SV40 large T antigen-immortalization using a retrovirus vector employing methods described above (Nishimura et al., 2008). GpL1 cells were confirmed to express GFAP but not MAP-2, indicating that GpL1 is an astroglial cell line. Transfection of *Prnp* suppressed cell death in GpL1 cells under serum-free conditions. This finding is similar to the results of *Prnp*$^{-/-}$ neuronal cell lines HpL3-4 and NpL2. The *Prnp*-transfected GpL1 cells (GpL1-PrP) showed increased SOD activity compared to control GpL1 cells transfected with empty vector (GpL1-EM). Thus, this study extended the notion that PrP^C prevents apoptosis by its anti-oxidative function in not only a *Prnp*$^{-/-}$ neuronal cell line but also a *Prnp*$^{-/-}$ astroglial cell line after serum withdrawal.

In addition to PrP^C function in independent cells, we have investigated whether PrP expression on the cell surface of astroglia influences neuronal survival because neurons and astroglia are closely coupled in their metabolic activities (Verkhratsky et al., 2014). Thus, the effect of PrP expression on cell viability of NpL2 cells under co-culture with GpL1-EM or GpL1-PrP cells in the absence of serum was investigated (Onodera and Sakudo, unpublished results). For co-culture experiments, GpL1-EM and GpL1-PrP cells were prepared and replaced with a serum-free DMEM. NpL2-EM and NpL2-PrP cells were plated onto coverslips, which were then transferred to a dish containing GpL1-EM or GpL1-PrP cells. After 4 days of co-culture, the coverslips were retracted and cell viability measured. The results showed that the NpL2 cell viability was significantly increased under co-culture with GpL1 cells transfected with *Prnp* (GpL1-PrP) compared to GpL1 cells transfected with empty vector (GpL1-EM). Similar results were observed when we used both NpL2 cells transfected with *Prnp* (NpL2-PrP) and NpL2 cells transfected with empty vector (NpL2-EM). These findings suggest that the neuronal cell viability depended on the PrP expression in feeder astroglia in the co-culture system. Taken together, astroglial cell lines are useful for revealing the mechanism of PrP^C not only in independent astroglial cells but also in the function of the neuron-glia relationship.

MACROPHAGE/MICROGLIA *Prnp*$^{-/-}$ CELL LINES

Recent studies have shown that cells of the immune system, such as macrophages, dendritic cells (DCs) or lymphocytes, can act as a replication site or as a reservoir for prions (Aucouturier and Carnaud, 2002). Follicular dendritic cells (FDCs) in the germinal centers of lymphoid organs are reported to be sites of PrP^{Sc} accumulation (Mcbride et al., 1992; Hill et al., 1999; Aucouturier and Carnaud, 2002). Current data suggest that the prion agent might be acquired by migratory DCs and macrophages (Maignien et al., 2005). However, the function of PrP^C in macrophages remains unclear.

Macrophages express PrP^C at a very low level. To study PrP^C functions in macrophage, the mouse bone marrow-derived macrophages (BMM) from FVB/N *Prnp*$^{+/+}$ and Zrch I *Prnp*$^{-/-}$ mice were transformed with a replication-defective retrovirus encoding SV40 large T antigen by a similar method previously used for the establishment of HpL, NpL, and GpL cell lines (Uraki et al., 2010). The cells were then selected and cloned. Among the obtained clones, we used MWF3-3 (from FVB/N *Prnp*$^{+/+}$ mice) and MplZ4-3 (from Zrch I *Prnp*$^{-/-}$ mice) for comparison, because these cells expressed macrophage specific proteins (F4/80 and MOMA-2) and displayed phagocytotic properties. Because it was previously reported that *Prnp*$^{-/-}$ neuronal HpL and NpL cells and astroglial GpL cells are sensitive to oxidative stress in serum-free conditions, MWF3-3 and MplZ4-3 cells were subjected to serum deprivation. The MplZ4-3 cells died after withdrawal of serum, whereas most MWF3-3 cells survived under the same culture conditions. These findings suggest that PrP^C increases the survival rate of macrophages.

With regard to the characteristics of the *Prnp*$^{-/-}$ macrophage cell line, we have analyzed their cell morphology and phagocytotic activity. The *Prnp*$^{-/-}$ macrophage cell line (MplZ4-3) showed shorter pseudopodium extension and less phagocytotic activity of latex beads than a *Prnp*$^{+/+}$ macrophage cell line (MWF3-3). These findings are consistent with the results using peritoneal macrophages from *Prnp*$^{-/-}$ and *Prnp*$^{+/+}$ mice using latex beads (Nitta et al., 2009). However, the findings are inconsistent with a previous study showing that primary cultures of Zrch I *Prnp*$^{-/-}$ macrophages have increased rates of phagocytosis of zymosan particles, suggesting PrP negatively regulates this process (De Almeida et al., 2005). One possible explanation for this discrepancy may be the different mouse strains used in these experiments. Whereas the previous report examined primary cells derived from C57BL/6, we used a macrophage cell line derived from FVB/N. Similarly, *Prnp*$^{-/-}$ mice are resistant to infection by *Brucella abortus*, which is supported in the report by De Almeida et al. implying that the deletion of PrP contributes to host defense against bacterial infection (Watarai et al., 2003). However, this was not confirmed in the case of *Brucella suis* in another laboratory (Fontes et al., 2005). In addition, a recent report by Nuvolone et al. suggests that the increased phagocytosis of apoptotic cells, which was reported as attributable to the absence of *Prnp* in *Prnp*$^{-/-}$ mice, has been shown instead to be caused by differences in a linked locus encoding signal regulatory protein-α (*Sirpa*) (Nuvolone et al., 2013). Thus, further studies would be required for investigating whether *Sirpa*-polymorphism influences the responses to zymosan or phagocytosis of latex beads.

Microglia are the primary immune cells of the central nervous system (CNS), and are highly similar to peripheral macrophages. These cells play important roles as the major inflammatory cell type in the brain, and respond to pathogens and injury and contribute to destroying pathogens as well as removing damaged cells. Several studies have indicated that microglia are implicated in the pathogenic events of prion diseases (Williams et al., 1994; Brown, 2001b). However, the pathophysiological function of PrP^C in microglia remains unclear. Iwamaru and Kitani et al. obtained the MG0 *Prnp*$^{-/-}$ microglial cell line, which is derived from the brains of Rikn *Prnp*$^{-/-}$ mice (Iwamaru et al., 2007). To immortalize these microglial cells for establishing the MG0 cell line, *c-myc*-containing retroviral vector was used. The MG0 cells were confirmed to be positive for the microglial markers Mac-1 and F4/80 and negative for the astroglial marker GFAP and neuronal marker MAP2. Moreover, these cells were shown to have phagocytotic activity and to produce inflammatory cytokines, such as tumor necrosis factor α, interleukin-1α and interleukin-6, when stimulated with lipopolysaccharide. However, there are no reports on the comparison between MG0 and the wild-type counterpart MG6 cell line or PrP-overexpressed microglial cell line MG20. Persistent infection of scrapie and BSE in these cells has been reported. Further analysis of PrP function in microglia using the MG0 cell line is keenly awaited.

FIBROBLAST *Prnp*$^{-/-}$ CELL LINES

Satoh et al. established skin fibroblast *Prnp*$^{-/-}$ cell lines (SFK) derived from continuous cultures of abdominal skin explants of Ngsk *Prnp*$^{-/-}$ mice. SFK cells showed decreased expression of Ras- and Rac-related proteins compared to *Prnp*$^{+/+}$ skin fibroblast cell lines (SFH) (Satoh et al., 1998, 2000), while heat shock proteins were expressed at the same level in SFK and SFH cells in the presence and absence of heat stress (Satoh et al., 1998). Other groups established *Prnp*$^{-/-}$ fibroblasts ($PrP^{0/0}$/1 cell line) from Ngsk *Prnp*$^{-/-}$ mice by immortalization with the chemical mutagen 3-methylcholanthrene (Prcina et al., 2010). The proliferation rate of $PrP^{0/0}$/1 cells expressing human PrP was lower those of non-transfected $PrP^{0/0}$/1 cells, and was negatively correlated with the PrP expression level. However, the authors also stated that these observations may be due to the antibiotic selection marker on the vector.

MECHANISMS BY WHICH PrP PREVENTS APOPTOSIS IN THE HpL CELL LINE

One way to identify the important regions of PrP is to generate partially deleted or mutated forms of PrP and transfect them into the *Prnp*$^{-/-}$ cell line. Endogenous PrP expression in the *Prnp*$^{+/+}$ cell line may interfere with these experiments. Hence, the *Prnp*$^{-/-}$ cell line is particularly useful for this approach. Here, we introduce our studies using HpL3-4 cells for investigating the important regions for PrP function (**Figure 3**).

PrP contains disulfide links (S–S), Asn-linked glycosylation sites (CHO), a signal peptide sequence (SP) and a GPI anchor attached to its C-terminus (Brown, 2001a). In addition, PrP^C contains two domains evolutionarily conserved among mammals; viz., the OR (amino acid residue 51–90 in mouse PrP) and the hydrophobic region (HR) (amino acid residue 112–145 in mouse PrP) domains.

OR, which is composed of five octapeptide repeats, P(Q/H)GGG(G/-)WGQ, is highly conserved among mammals and can selectively bind copper (Hornshaw et al., 1995; Brown et al., 1997; Miura et al., 1999; Viles et al., 1999; Kramer et al., 2001). Zinc, manganese, and nickel also bind to PrP but with lower affinity (Pan et al., 1992; Brown et al., 2000; Jackson et al., 2001). The binding of metals occurs *via* histidine residues located in the OR domain. The binding of copper to PrP is thought to be indispensable for the SOD activity of PrP (Brown et al., 1999; Cui et al., 2003). In addition, as metals such as copper are essential for myelin formation and maintenance (Skripuletz et al., 2008; Benetti et al., 2010), regulation of copper by PrP^C may be involved in demyelination of *Prnp*$^{-/-}$ mice (Nishida et al., 1999; Baumann et al., 2007; Bremer et al., 2010).

HR, which contains the highly conserved sequence motif AGAAAAGA (Gasset et al., 1992; Schatzl et al., 1995), has been shown to be critical for PrP topology in the endoplasmic reticulum membrane (Hegde et al., 1998), STI1 binding (Zanata et al., 2002) as well as normal metabolic cleavage (Chen et al., 1995; Mange et al., 2004) such as α-cleavage and β-cleavage.

To investigate the roles of OR and HR in the anti-apoptotic function of PrP, several deletions within these domains of mouse PrP or fusions with mouse PrP and Dpl have been

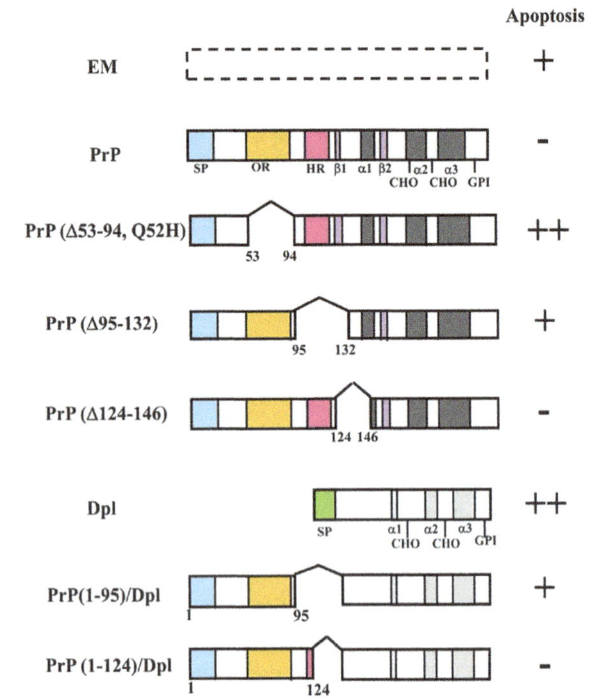

FIGURE 3 | Schematic representations of PrP deletion mutants or PrP-Dpl fusion proteins and *Prnp*$^{-/-}$ cells (HpL3-4) expressing these mutants that undergo apoptosis under serum deprivation. Schematic representations of deletion mutants of mouse PrP and PrP-Dpl fusion protein are shown. Mutants of mouse PrP [PrP: wild-type PrP; Δ#1, PrP(Δ53-94, Q52H); Δ#2, PrP(Δ95-132); Δ#3, PrP(Δ124-146)] were prepared using restriction digestion/ligation. Schematic locations of the deletions as compared with the wild-type protein are shown by a space within the bar next to the indicated protein. Dpl lacks sequences homologous to the octapeptide repeat (OR) and hydrophobic regions (HR) of PrP. ORDpl [PrP(1-95)-Dpl] is Dpl fused with amino acid residues 1-95 of PrP containing the OR. ORHRDpl [PrP(1-124)-Dpl] is Dpl fused with amino acid residues 1-124 of PrP containing the OR and N-terminal half of HR. The numbers refer to the amino acid residues in the mouse PrP sequence or Dpl sequence. The Asn-linked glycosylation sites (CHO), signal peptide sequence (SP), octapeptide repeat region (OR) and hydrophobic region (HR) are shown. The regions forming an α-helix secondary structure are shown as α1-α3, whereas those forming a β-sheet are shown as β1 and β2. Both PrP and Dpl have a glycosylphosphatidylinositol (GPI) anchor attached to its C-terminus. HpL3-4 cells expressing wild-type PrP (PrP), PrP(Δ53-94, Q52H) (Δ#1), PrP(Δ95-132) (Δ#2), PrP(Δ124-146) (Δ#3), Dpl (Dpl), PrP(1-95)-Dpl (ORDpl), PrP(1-124)-Dpl (ORHRDpl) or the empty vector alone (EM) were serum-deprived. The rate of apoptosis of the cells after serum deprivation for 24 h is shown on the right. The number of pluses (+) indicates the degree of apoptosis. Minus (−) indicates a lesser degree of apoptosis. Updated from Figure 4 in Sakudo et al. (2006) and Figure 3 in Sakudo et al. (2007b) with permission from Bentham Science Publishers, Ltd. and John Wiley & Sons, Inc., respectively.

made (**Figure 3**). The anti-apoptotic function of three PrP deletion constructs and PrP-Dpl fusion constructs as well as the control constructs was tested following stable transfection into *Prnp*$^{-/-}$ neuronal cells HpL3-4 [HpL3-4 cells expressing wild-type PrP (PrP: HpL3-4-PrP), PrP(Δ53-94, Q52H) (Δ#1: HpL3-4-Δ#1), PrP(Δ95-132) (Δ#2: HpL3-4-Δ#2), PrP(Δ124-146) (Δ#3: HpL3-4-Δ#3), PrP(1-95)-Dpl (ORDpl: HpL3-4-ORDpl), PrP(1-124)-Dpl (ORHRDpl: HpL3-4-ORHRDpl) or empty vector alone (EM: HpL3-4-EM)]. Western blotting confirmed that the cell clones expressed the deletion mutants and fusions at similar levels to those of clones expressing the full-length protein. In the absence of serum, HpL3-4 cells undergo cell death with features of apoptosis, whereas *Prnp*$^{-/-}$ cells transfected with *Prnp* are resistant to serum deprivation (Kuwahara et al., 1999; Sakudo et al., 2003b, 2005a). Serum deprivation showed that the control transfectants (HpL3-4-EM) underwent rapid cell death, while the full-length PrP-transfected cells were significantly protected. In contrast, the cells expressing OR-deficient PrP (HpL3-4-Δ#1) showed higher levels of cell death compared with HpL3-4-EM cells. Interestingly, the cells expressing the N-terminal half of HR-deficient PrP (HpL3-4-Δ#2) underwent cell death at equivalent levels to HpL3-4-EM cells. In contrast, the cells expressing the C-terminal half of HR-deficient PrP (HpL3-4-Δ#3) underwent cell death at equivalent levels to HpL3-4-PrP cells. The cells expressing Dpl fused to OR (HpL3-4-ORDpl) died at slightly higher levels compared to HpL3-4-EM cells, while the cells expressing Dpl fused to OR and N-terminal half of HR (HpL3-4-ORHRDpl) displayed elevated levels of survival over HpL3-4-EM cells (Lee et al., 2006).

Therefore, the data suggest that the anti-apoptotic function of PrP^C can be regulated by not only the OR domain but also the N-terminal half of HR. In addition, a fusion protein containing OR of PrP did not protect the cells against stress, whereas extending the fusion to the N-terminal half of HR did provide some protection. These findings support the notion that both the OR and HR are important domains for PrP function. Moreover, measurement of cellular SOD activity showed that OR and the N-terminal half of HR are necessary for the activation of cellular SOD in HpL3-4 cells (Sakudo et al., 2005c). Therefore, regulation of the anti-oxidative defense systems, such as SOD activity, by PrP suggests this domain contributes to the anti-oxidative and anti-apoptotic activity during serum deprivation in HpL3-4 cells.

Mouse PrP is cleaved at the β site located in OR and near His96 as well as the α site located at position 111/112 (Mange et al., 2004). A PrP fragment was immunocaptured from HpL3-4 cells expressing wild-type PrP using an antibody recognizing the N-terminal region of PrP. This fragment mainly includes the N-terminal portion of the protein generated after cleavage at the β site. Interestingly, β cleavage was observed after oxidative stress treatment *in vitro* (Mcmahon et al., 2001). Furthermore, HpL3-4 cells expressing PrP(Δ110-119) lacking the α cleavage site were more protected against stress than those expressing an equivalent level of wild-type PrP (Mange et al., 2004). These observations suggest that PrP(Δ110-119) is the dominant positive form of PrP. Because the absence of α cleavage was found to enhance cell viability, α cleavage appears to suppress the protective activity of PrP. By contrast, β cleavage may contribute to protection. Further studies are required to analyze the effect of β cleavage on the neuroprotection mechanism of PrP.

Regarding OR and HR, several potential interactors of PrP have been reported. Copper specifically binds the OR of PrP (Hornshaw et al., 1995; Brown et al., 1997; Viles et al., 1999; Kramer et al., 2001). STI1, binds with amino acid residues

113-128 located in the N-terminal half of the HR of PrP (Zanata et al., 2002). Indeed, immunoprecipitation studies suggest that STI1 interacts with PrP in HpL3-4-PrP cells (Sakudo et al., 2005a). Therefore, we performed additional studies using HpL3-4 cells in order to determine how STI1 and copper might contribute to PrP-dependent anti-oxidative signaling. To investigate whether the STI1 is important for the biological activities displayed by PrP, the effect of the inhibitory peptides against PrP-STI1 on HpL3-4-EM cells was compared to that on HpL3-4-PrP cells under serum-free conditions (Sakudo et al., 2005a). The inhibitory peptides are toxic to HpL3-4-PrP cells due to inhibiting the SOD activity, although this is not the case for HpL3-4-EM cells (Sakudo et al., 2005a). Regarding copper, the cellular copper concentration was decreased in HpL3-4-EM cells, but not in HpL3-4-PrP cells under serum deprivation (Sakudo et al., 2004). Therefore, it is proposed that STI1 and copper might be involved in PrP-dependent SOD activation that can inhibit apoptosis *via* the OR and N-terminal half of PrP.

By considering the data reported using HpL cell lines, it is further proposed that the susceptibility of HpL3-4 cells to cell death is probably due to reduced SOD activity, induced, at least in part, by the absence of PrP-STI1 signals and the abnormality of copper homeostasis as well as the lack of β cleavage. Taken together, these findings suggest that PrP plays an anti-oxidative role and functions only under extreme stress such as oxidative conditions.

CONCLUSIONS AND FUTURE PERSPECTIVES

The list of abnormalities caused by PrP-deficiency from the results of phenotype analysis of $Prnp^{-/-}$ mice and from comparison between PrP-non-expressing cells and PrP-expressing cells continues to grow. Nonetheless, there has been little consensus from these studies regarding the overall main cellular function of PrP^C. To enable us to analyse PrP^C in detail, $Prnp^{-/-}$ cell lines have been established. Recently, the number of available $Prnp^{-/-}$ cell lines has increased, including neuronal cells, astroglial cells, microglia, macrophages, and fibroblasts (**Table 1**). Rescue of *Prnp* by reintroducing into $Prnp^{-/-}$ cell lines will provide convincing data. If further evidence is required, genetic complementation using the PrP transgene should be performed using a vector driven *Prnp* promoter. Indeed, PrP^C function may differ depending on its expression level. As a consequence, levels of PrP^C expression that are either too high or too low may mask the true biological function of the protein. Indeed, overexpression may cause the emergence of different functions from the real physiological function. Thus, if possible, analysis should be performed using physiological expression levels of PrP^C in each cell type. However, in the case of low PrP^C expressing cells, it becomes very difficult to detect the PrP^C expression and analyse PrP^C function (Raeber et al., 1999; Prinz et al., 2004).

The use of the $Prnp^{-/-}$ cell lines has facilitated the identification of abnormalities caused by PrP-deficiency at the cellular level (**Table 2**). Establishment of more $Prnp^{-/-}$ cell lines will further contribute to our understanding of PrP^C function. For example, PrP family members are known to be expressed in reproductive tissues, such as testis, ovary, and placenta (Bonnet and Pailhoux, 2014; Makzhami et al., 2014). Thus, cells derived from reproductive tissues should also be used for the establishment of $Prnp^{-/-}$ cell lines. The importance of PrP^C in leukocyte function has also been suggested because $Prnp^{-/-}$ lymphocytes show a reduction of mitogen response, cytokine production and proliferation (Kubosaki et al., 2003; Bainbridge and Walker, 2005). Recent studies have also demonstrated that PrP^C plays a role in T cell and dendritic cell interactions (Ballerini et al., 2006). Therefore, the establishment of a $Prnp^{-/-}$ leukocyte cell line would be interesting.

As knockout animals lack all PrP expression, it is unclear whether the alteration and phenotypes in $Prnp^{-/-}$ mice are attributable to defects in cell function or development. Function of PrP^C in the maturation process is feasible (Martin-Lannerée et al., 2014). Recent developments in regenerative medicine enable us to use various stem cells, such as embryonic stem (ES) cells and induced pluripotent stem (iPS) cells (Yamanaka, 2007). Establishment of $Prnp^{-/-}$ ES or iPS cells would open the opportunity for detailed analysis of PrP^C function in reprogramming, expansion, isolation and differentiation.

The most studied phenotype of $Prnp^{-/-}$ mice is myelin degeneration. This phenotype is seen in both type 1 and type 2 $Prnp^{-/-}$ mice (Nishida et al., 1999; Baumann et al., 2007; Bremer et al., 2010). The myelin degeneration phenotype is caused by a deficiency of PrP^C, suggesting that myelin maintenance may be a representative physiological function of PrP^C. The process of demyelination is usually related to an abnormality of the immune system (Ramesh et al., 2013). The abnormal activity of immune cells may cause demyelination. Indeed, defects of immune cells in $Prnp^{-/-}$ mice have been reported in several papers. Myelin sheath is composed of Schwann cells and oligodendrocytes (Alberts et al., 2008). As Schwann cell and oligodendroglia is one of glial cells, the study using glial $Prnp^{-/-}$ cell line GpL1might reveal some characteristics of PrP^C related to myelin maintenance.

If knockout of *Prnp* is not required, suppression of *Prnp* is relatively easily achieved by gene silencing technology using siRNA (small interference RNA), miRNA (micro-RNA), bi-shRNA (bifunctional short hairpin RNA), or shRNA (Kubowicz et al., 2013). Indeed, shRNA of *Prnp* has been used for producing PrP^C-depleted 1C11 neuroectodermal cells (Loubet et al., 2012). If a knockout of *Prnp* is required, the relevant $Prnp^{-/-}$ cell line should be generated (**Figure 4**). Recent development of immortalization techniques enables cells derived from $Prnp^{-/-}$ mice to be immortalized using oncogene transfection, radiation, fusion with cancer cells or treatment with chemical mutagens. In other cases, cells derived from transgenic mice with an oncogene (e.g., temperature sensitive SV40 largeT or p53) that is overexpressed, mutated or deleted can proliferate and be maintained in culture medium, resulting in the establishment of a cell line (Obinata, 2007). Alternatively, gene editing techniques such as CRISPR (Clustered Regularly Interspaced Short Palindromic Repeat) (Hsu et al., 2014), ZFN (Zinc Finger Nuclease) (Swarthout et al., 2011) or TALEN (Transcription activator-like effector nuclease) (Wright et al., 2014) can be used to knockout *Prnp*, resulting in the production of $Prnp^{+/-}$ or $Prnp^{-/-}$ cell lines.

$Prnp^{-/-}$ cell lines are mainly used for two experimental purposes; analysis of PrP^C function and the study of prions.

Table 1 | *Prnp*$^{-/-}$ cell lines and their characteristics.

Names	Method of production of cell lines	Cell type	*Prnp*$^{-/-}$ mouse origin	Culture medium	Main characterization compared to PrP expressing cells	References
HpL2-1, HpL3-2, HpL4-3	Retrovirus-mediated method by SV40 large T antigen expressing vector	Neuronal-precursor cells (expessing NF-68K)	Rikn	10% FCS-DMEM	Susceptible to serum deprivation	Kuwahara et al., 1999
Zpl2-1, Zpl2-4, Zpl3-4	Lipofection of SV40 large T antigen expressing vector	Neuronal cells (expressing NeuN)	Zrchl	DMEM	Higher prolifertion rate	Kim et al., 2005
SFK-B, SFK-C	Continuous culture of abdominal skin explants	Skin fibroblast cells	Ngsk	10% FCS-DMEM	Decreased expression of Ras and Rac related proteins	Satoh et al., 1998, 2000
F14	Fusion of PrP-knockout cerebellar cells and mouse neuroblastoma cells	Neuronal cells (expressing NeuN, MAP-2, and synaptophysin)	Npu	Serum containing DMEM	Similar distribution of GFP-PrP fusion protein	Holme et al., 2003
NpL2	Retrovirus-mediated method by SV40 large T antigen expressing vector	Neuronal cells (expressing NF-L, NF-M, NF-H, and MAP-2)	Zrchl	10% FCS-NB/B27 medium	Susceptible to serum deprivation	Nishimura et al., 2007
MG0	Retrovirus-mediated method by c-myc expressing vector	Microglial cell (expressing Mac-1 and F4/80)	Rikn	10% FCS-DMEM supplemented with 100 μM β-mercaptoethanol, 10 μg/ml insulin	Comparison has not been performed	Iwamaru et al., 2007
MpLZ4-3	Retrovirus-mediated method by SV40 large T antigen expressing vector	Macrophage (expressing MOMA-2 and F4/80)	Zrchl	10% FCS-DMEM	Shorter pseudopodium extension and less phagocytotic activity	Uraki et al., 2010
PrP$^{0/0}$/1	Treatment with chemical mutagen (3-methylcholanthrene)	Embryonic fibroblast (expressing fibronectin)	Ngsk	10% FCS-MEMD supplemented with 2 mM-glutamine	Higher prolifertion rate (probably artificial)	Prcina et al., 2010
GpL1	Retrovirus-mediated method by SV40 large T antigen expressing vector	Glial cells (expressing GFAP)	Zrchl	10% FCS-DMEM	Susceptible to serum deprivation	Nishimura et al., 2008

DMEM, Dulbecco's modified Eagle's medium.

MCA, Methylcholanthrene, which is a highly carcinogenic polycyclic aromatic hydrocarbon produced by burning organic compounds at very high temperatures.

NB/B27 medium, neurobasal medium (NB) (Gibco BRL, Gaithersburg, MD, USA) supplemented with B27 supplement (B27) (Gibco) and glutamine.

Updated from Table 3 in Sakudo et al. (2006) with permission from Bentham Science Publishers, Ltd.

For example, there are several reports of the successful production of monoclonal and polyclonal antibodies against PrP by immunizing *Prnp*$^{-/-}$ mice with recombinant murine PrP (Zanusso et al., 1998), human recombinant PrP folded into α or β (Beringue et al., 2003), purified PrP^{C} and PrP^{Sc} (Prusiner et al., 1993; Tayebi et al., 2011), scrapie-infected mouse neuroblastoma cells (Nakamura et al., 2003a), DNA plasmids encoding *Prnp* (Krasemann et al., 1996a,b), PrP-displaying retrovirus particles (Nikles et al., 2005), and PrP-coated microbeads (Tayebi et al., 2011). Thus, by immunizing *Prnp*$^{-/-}$ mice with the derived *Prnp*$^{-/-}$ cell lines transfected with *Prnp* or mutated *Prnp*, specific antibodies to wild-type or mutated PrP could be readily obtained. Furthermore, to obtain specific antibodies to PrP^{Sc}, immunization of *Prnp*$^{-/-}$ mice with prion-infected *Prnp*$^{-/-}$ cells transfected with *Prnp* could be a promising approach.

There is a species barrier in prion susceptibility, which is determined by the difference between the species of the host expressing PrP and species of prion agent (Beringue et al., 2008).

Table 2 | Abnormality of PrP gene-deficient cell lines.

Names	Main characterization compared to PrP expressing cells	References
HpL2-1, HpL3-2, HpL4-3	Higher rate of apoptosis by serum deprivation	Kuwahara et al., 1999; Sakudo et al., 2003a, 2005a,b,c; Kim et al., 2004; Vassallo et al., 2005; Wu et al., 2008
HpL3-4	Decrease of intracellular copper concentration after serum deprivation	Sakudo et al., 2004
HpL3-4	Decrease of cellular SOD activity	Sakudo et al., 2003a
HpL3-4	Bigger increase of intracellular superoxide anion after serum deprivation	Sakudo et al., 2003a
HpL3-4	Reduced increase of intracellular hydrogen peroxide after serum deprivation	Sakudo et al., 2003a
HpL3-4	Bigger increase of caspase-3/9 activation after serum deprivation	Sakudo et al., 2003a
HpL3-4	Increased sensitivity of poliovirus infection	Baj et al., 2005
HpL3-4	Increased coxsackievirus B3 production and apoptotic cell death	Nakamura et al., 2003a
HpL3-4	Alteration of Ca^{2+} concentration in mitochondria after serum deprivation	Kim et al., 2004
HpL3-4	Alteration of transmembrane potentials in mitochondria after serum deprivation	Kim et al., 2004
HpL3-4	Alteration of cytochrome c level in mitochondria after serum deprivation	Kim et al., 2004
HpL3-4	Higher rate of apoptosis by SIN-1	Vassallo et al., 2005
HpL3-4	Reduced PI 3-kinase activity	Vassallo et al., 2005
HpL3-4	Decrease of cellular viability	Mange et al., 2004; Christensen and Harris, 2008
HpL3-4	Shorter neurite extension after differentiation	Kuwahara et al., 1999
HpL3-4	Suceptible to amyloid beta toxicity and amyloid beta-inducing autophagy	Nah et al., 2013
HpL3-4	Decreased rise in intracellular calcium following hydrogen peroxide treatment	Krebs et al., 2007
HpL3-4	Neutralization of Dpl toxicity	Sakudo et al., 2005c
NpL2	Decrease of SOD activity	Nishimura et al., 2007
NpL2	Higher rate of apoptosis by serum deprivation	Nishimura et al., 2007
Zpl2-1, Zpl2-4, Zpl3-4	Higher prolifertion rate	Kim et al., 2005
Zpl3-4	Increased autophagy induced by serum deprivation	Oh et al., 2008
Zpl3-4	Higher rate of apoptosis and autophagy by serum deprivation	Oh et al., 2008
GpL1	Decrease of SOD activity	Nishimura et al., 2008
GpL1	Higher rate of apoptosis by serum deprivation	Nishimura et al., 2008
GpL1	Decrease of cellular viability of co-cultured neuronal cells	Onodera and Sakudo, unpublished results
MplZ3-4	Shorter pseudopedium extension and reduced phagocytotic activity	Uraki et al., 2010
$PrP^{0/0}/1$	Higher proliferation rate (may be non-specific)	Prcina et al., 2010
SFK-B, SFK-C	Decreased expression of Ras and Rac related proteins	Satoh et al., 1998, 2000

PI3, phosphatidylinositol 3.

SIN-1, 3-morpholinosydnonimine.

SOD, superoxide dismutase.

Updated from Table 4 in Sakudo et al. (2006) with permission from Bentham Science Publishers, Ltd.

Prnp$^{-/-}$ cell lines are useful for producing susceptible cells to target species of prion agent after *Prnp* transfection of the target species. Importantly, endogenous PrP is not present in the *Prnp*$^{-/-}$ cell line. Thus, interference by endogenous PrP does not occur in a *Prnp*$^{-/-}$ cell line. The *Prnp*$^{-/-}$ cell line transfected with deletion mutants of PrP can be used for investigating important regions of PrP for prion infection. Previous results using the HpL3-4 cell line and chandler scrapie strain showed that OR is an essential region for the production of PrP^{Sc} during the early stage of infection (Sakudo et al., 2008). This was supported by other reports using the HpL3-4 cell line and 22L scrapie strain, indicating that substitution of amino acid residues of mouse PrP at position 96, located in the region between OR and HR, decreased PrP^{Sc} formation (Maas et al., 2007). The report also showed that the C-terminal half of PrP impacts on conversion efficiency. The C-terminal half of PrP includes regions forming a secondary structure of α-helix, termed α1-α3, while that of β-sheet is termed β1 and β2. Substitution of amino acid residues of mouse PrP at position 132 (located between β1 and α1), 150 (α1 region), 167 (located between β2 and α2), 189 (α2 region) and 204 (α3 region)

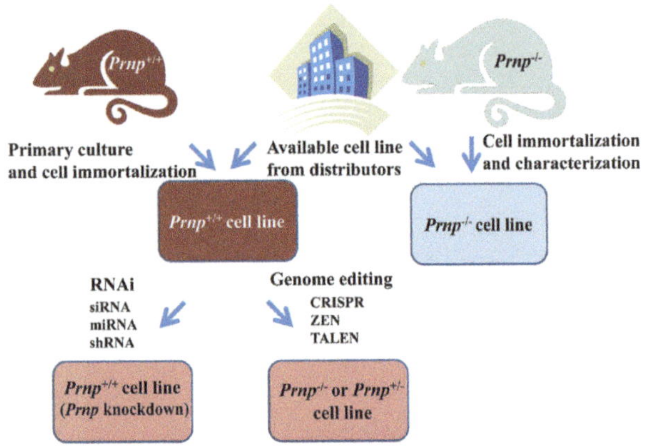

FIGURE 4 | Strategies for obtaining cell lines in which *Prnp* is suppressed or knocked out. Established *Prnp*$^{-/-}$ cell lines can be obtained from a number of distributors. If suppression of *Prnp* is required, rather than a knockout of *Prnp*, gene silencing technology using siRNA (small interference RNA), miRNA (micro-RNA) bi-shRNA (bifunctional short hairpin RNA) or shRNA (short hairpin RNA) is applicable. In other cases, when a new *Prnp*$^{-/-}$ cell line is generated, immortalization of cells derived from *Prnp*$^{-/-}$ mice is usually performed. Alternatively, gene editing techniques such as CRISPR (Clustered Regularly Interspaced Short Palindromic Repeat), ZFN (Zinc Finger Nuclease), or TALEN (Transcription activator-like effector nuclease) can be used to knockout *Prnp*, resulting in the production of *Prnp*$^{+/-}$ or *Prnp*$^{-/-}$ cell lines. Cells can be immortalized using oncogene transfection, radiation, fusion with cancer cells, or treatment with chemical mutagens. In other cases, cells derived from transgenic mice with an oncogene (e.g., temperature sensitive SV40 (Simian virus 40) large T, p53 etc.) that is overexpressed, mutated or deleted can proliferate and be maintained in culture medium. Continuous fibroblast culture can also be used. A cell line may be established after cloning and selection of these cells.

abolished the efficient conversion of PrP^{Sc}. Taken together, these results suggest OR, α-helices and the surrounding region contributes to the susceptibility of prion infection. Newly generated PrP^{C} and PrP^{Sc} antibodies together with cell lines that are susceptible or non-susceptible to target species of prion agents will be extremely useful for exploring prion biology.

Finally, we would like to emphasize that development of "OMICS" such as genomics, epigenomics, proteomics, metabolomics and transcriptomics will enable us to obtain a global view of the network of processes that are regulated by PrP^{C}. By exploiting the combination of *Prnp*$^{-/-}$ cell lines, comparison of PrP-expressing and PrP-non-expressing cells by OMICS will provide invaluable information on the effect of PrP-deficiency and fundamental differences caused by PrP-deficiency. The accumulated knowledge using these systems will be of great help in understanding PrP^{C} functions, which is also important in terms of clarifying the etiology of prion diseases.

AUTHOR CONTRIBUTIONS

All authors contributed to the conception of this review and approved the submitted version.

ACKNOWLEDGMENTS

This work was supported by Grant-in-Aids for Scientific Research from the Ministry of Education, Science, Culture and Technology of Japan (25450447, 23780299, 20780219, 17780228), Grants-in-Aid from the Research Committee of Prion disease and Slow Virus Infection, the Ministry of Health, Labour and Welfare of Japan.

REFERENCES

Alberts, B., Johnson, A., Lewis, J., Raff, M., Roberts, K., and Walter, P. (2008). *Molecular Biology of the Cell, 5th Edn.* New York, NY: Garland Science.

Aucouturier, P., and Carnaud, C. (2002). The immune system and prion diseases: a relationship of complicity and blindness. *J. Leukoc. Biol.* 72, 1075–1083.

Bainbridge, J., and Walker, K. B. (2005). The normal cellular form of prion protein modulates T cell responses. *Immunol. Lett.* 96, 147–150. doi: 10.1016/j.imlet.2004.08.006

Baj, A., Bettaccini, A., Nishimura, T., Onodera, T., and Toniolo, A. (2005). Poliovirus type 1 infection of murine PRNP-knockout neuronal cells. *J. Neurovirol.* 11, 237–246. doi: 10.1080/13550280590952745

Balducci, C., Beeg, M., Stravalaci, M., Bastone, A., Sclip, A., Biasini, E., et al. (2010). Synthetic amyloid-beta oligomers impair long-term memory independently of cellular prion protein. *Proc. Natl. Acad. Sci. U.S.A.* 107, 2295–2300. doi: 10.1073/pnas.0911829107

Ballerini, C., Gourdain, P., Bachy, V., Blanchard, N., Levavasseur, E., Gregoire, S., et al. (2006). Functional implication of cellular prion protein in antigen-driven interactions between T cells and dendritic cells. *J. Immunol.* 176, 7254–7262. doi: 10.4049/jimmunol.176.12.7254

Baumann, F., Tolnay, M., Brabeck, C., Pahnke, J., Kloz, U., Niemann, H. H., et al. (2007). Lethal recessive myelin toxicity of prion protein lacking its central domain. *EMBO J.* 26, 538–547. doi: 10.1038/sj.emboj.7601510

Behrens, A., Genoud, N., Naumann, H., Rulicke, T., Janett, F., Heppner, F. L., et al. (2002). Absence of the prion protein homologue Doppel causes male sterility. *EMBO J.* 21, 3652–3658. doi: 10.1093/emboj/cdf386

Benetti, F., Ventura, M., Salmini, B., Ceola, S., Carbonera, D., Mammi, S., et al. (2010). Cuprizone neurotoxicity, copper deficiency and neurodegeneration. *Neurotoxicology* 31, 509–517. doi: 10.1016/j.neuro.2010.05.008

Beringue, V., Mallinson, G., Kaisar, M., Tayebi, M., Sattar, Z., Jackson, G., et al. (2003). Regional heterogeneity of cellular prion protein isoforms in the mouse brain. *Brain* 126, 2065–2073. doi: 10.1093/brain/awg205

Beringue, V., Vilotte, J. L., and Laude, H. (2008). Prion agent diversity and species barrier. *Vet. Res.* 39, 47. doi: 10.1051/vetres:2008024

Bonnet, A., and Pailhoux, E. (2014). Role of the prion protein family in the gonads. *Front. Cell Dev. Biol.* 2:56. doi: 10.3389/fcell.2014.00056

Bragason, B. T., and Palsdottir, A. (2005). Interaction of PrP with NRAGE, a protein involved in neuronal apoptosis. *Mol. Cell. Neurosci.* 29, 232–244. doi: 10.1016/j.mcn.2005.02.013

Brandner, S., Isenmann, S., Raeber, A., Fischer, M., Sailer, A., Kobayashi, Y., et al. (1996a). Normal host prion protein necessary for scrapie-induced neurotoxicity. *Nature* 379, 339–343. doi: 10.1038/379339a0

Brandner, S., Raeber, A., Sailer, A., Blattler, T., Fischer, M., Weissmann, C., et al. (1996b). Normal host prion protein (PrP^{C}) is required for scrapie spread within the central nervous system. *Proc. Natl. Acad. Sci. U.S.A.* 93, 13148–13151. doi: 10.1073/pnas.93.23.13148

Bremer, J., Baumann, F., Tiberi, C., Wessig, C., Fischer, H., Schwarz, P., et al. (2010). Axonal prion protein is required for peripheral myelin maintenance. *Nat. Neurosci.* 13, 310–318. doi: 10.1038/nn.2483

Brown, D. R. (2001a). Copper and prion disease. *Brain Res. Bull.* 55, 165–173. doi: 10.1016/S0361-9230(01)00453-1

Brown, D. R. (2001b). Microglia and prion disease. *Microsc. Res. Tech.* 54, 71–80. doi: 10.1002/jemt.1122

Brown, D. R., Hafiz, F., Glasssmith, L. L., Wong, B. S., Jones, I. M., Clive, C., et al. (2000). Consequences of manganese replacement of copper for prion protein function and proteinase resistance. *EMBO J.* 19, 1180–1186. doi: 10.1093/emboj/19.6.1180

Brown, D. R., Qin, K., Herms, J. W., Madlung, A., Manson, J., Strome, R., et al. (1997). The cellular prion protein binds copper *in vivo*. *Nature* 390, 684–687. doi: 10.1038/37733

Brown, D. R., Wong, B. S., Hafiz, F., Clive, C., Haswell, S. J., and Jones, I. M. (1999). Normal prion protein has an activity like that of superoxide dismutase. *Biochem. J.* 344(Pt 1), 1–5. doi: 10.1042/0264-6021:3440001

Bueler, H., Aguzzi, A., Sailer, A., Greiner, R. A., Autenried, P., Aguet, M., et al. (1993). Mice devoid of PrP are resistant to scrapie. *Cell* 73, 1339–1347. doi: 10.1016/0092-8674(93)90360-3

Bueler, H., Fischer, M., Lang, Y., Bluethmann, H., Lipp, H. P., Dearmond, S. J., et al. (1992). Normal development and behaviour of mice lacking the neuronal cell-surface PrP protein. *Nature* 356, 577–582. doi: 10.1038/356577a0

Calella, A. M., Farinelli, M., Nuvolone, M., Mirante, O., Moos, R., Falsig, J., et al. (2010). Prion protein and Abeta-related synaptic toxicity impairment. *EMBO Mol. Med.* 2, 306–314. doi: 10.1002/emmm.201000082

Capellari, S., Zaidi, S. I., Urig, C. B., Perry, G., Smith, M. A., and Petersen, R. B. (1999). Prion protein glycosylation is sensitive to redox change. *J. Biol. Chem.* 274, 34846–34850. doi: 10.1074/jbc.274.49.34846

Chen, S. G., Teplow, D. B., Parchi, P., Teller, J. K., Gambetti, P., and Autilio-Gambetti, L. (1995). Truncated forms of the human prion protein in normal brain and in prion diseases. *J. Biol. Chem.* 270, 19173–19180. doi: 10.1074/jbc.270.32.19173

Christensen, H. M., and Harris, D. A. (2008). Prion protein lacks robust cytoprotective activity in cultured cells. *Mol. Neurodegener.* 3:11. doi: 10.1186/1750-1326-3-11

Chung, E., Ji, Y., Sun, Y., Kascsak, R. J., Kascsak, R. B., Mehta, P. D., et al. (2010). Anti-PrP^C monoclonal antibody infusion as a novel treatment for cognitive deficits in an Alzheimer's disease model mouse. *BMC Neurosci.* 11:130. doi: 10.1186/1471-2202-11-130

Cui, T., Daniels, M., Wong, B. S., Li, R., Sy, M. S., Sassoon, J., et al. (2003). Mapping the functional domain of the prion protein. *Eur. J. Biochem.* 270, 3368–3376. doi: 10.1046/j.1432-1033.2003.03717.x

Daude, N., and Westaway, D. (2012a). Shadoo/PrP (Sprn(0/0) /Prnp(0/0)) double knockout mice: more than zeroes. *Prion* 6, 420–424. doi: 10.4161/pri.21867

Daude, N., Wohlgemuth, S., Brown, R., Pitstick, R., Gapeshina, H., Yang, J., et al. (2012b). Knockout of the prion protein (PrP)-like Sprn gene does not produce embryonic lethality in combination with PrP(C)-deficiency. *Proc. Natl. Acad. Sci. U.S.A.* 109, 9035–9040. doi: 10.1073/pnas.1202130109

De Almeida, C. J., Chiarini, L. B., Da Silva, J. P., Pm, E. S., Martins, M. A., and Linden, R. (2005). The cellular prion protein modulates phagocytosis and inflammatory response. *J. Leukoc. Biol.* 77, 238–246. doi: 10.1189/jlb.1103531

Dohler, F., Sepulveda-Falla, D., Krasemann, S., Altmeppen, H., Schlüter, H., Hildebrand, D., et al. (2014). High molecular mass assemblies of amyloid-β oligomers bind prion protein in patients with Alzheimer's disease. *Brain* 137, 873–886. doi: 10.1093/brain/awt375

Edenhofer, F., Rieger, R., Famulok, M., Wendler, W., Weiss, S., and Winnacker, E. L. (1996). Prion protein PrP^C interacts with molecular chaperones of the Hsp60 family. *J. Virol.* 70, 4724–4728.

Fontes, P., Alvarez-Martinez, M. T., Gross, A., Carnaud, C., Kohler, S., and Liautard, J. P. (2005). Absence of evidence for the participation of the macrophage cellular prion protein in infection with Brucella suis. *Infect. Immun.* 73, 6229–6236. doi: 10.1128/IAI.73.10.6229-6236.2005

Gasperini, L., and Legname, G. (2014). Prion protein and aging. *Front. Cell Dev. Biol.* 2:44. doi: 10.3389/fcell.2014.00044

Gasset, M., Baldwin, M. A., Lloyd, D. H., Gabriel, J. M., Holtzman, D. M., Cohen, F., et al. (1992). Predicted alpha-helical regions of the prion protein when synthesized as peptides form amyloid. *Proc. Natl. Acad. Sci. U.S.A.* 89, 10940–10944. doi: 10.1073/pnas.89.22.10940

Gauczynski, S., Peyrin, J. M., Haik, S., Leucht, C., Hundt, C., Rieger, R., et al. (2001). The 37-kDa/67-kDa laminin receptor acts as the cell-surface receptor for the cellular prion protein. *EMBO J.* 20, 5863–5875. doi: 10.1093/emboj/20.21.5863

Genoud, N., Behrens, A., Arrighi, I., and Aguzzi, A. (2003). Prion proteins and infertility: insight from mouse models. *Cytogenet. Genome Res.* 103, 285–289. doi: 10.1159/000076814

Giese, A., Brown, D. R., Groschup, M. H., Feldmann, C., Haist, I., and Kretzschmar, H. A. (1998). Role of microglia in neuronal cell death in prion disease. *Brain Pathol.* 8, 449–457. doi: 10.1111/j.1750-3639.1998.tb00167.x

Graner, E., Mercadante, A. F., Zanata, S. M., Forlenza, O. V., Cabral, A. L., Veiga, S. S., et al. (2000). Cellular prion protein binds laminin and mediates neuritogenesis. *Brain Res. Mol. Brain Res.* 76, 85–92. doi: 10.1016/S0169-328X(99)00334-4

Griffiths, H. H., Whitehouse, I. J., Baybutt, H., Brown, D., Kellett, K. A., Jackson, C. D., et al. (2011). Prion protein interacts with BACE1 protein and differentially regulates its activity toward wild type and Swedish mutant amyloid precursor protein. *J. Biol. Chem.* 286, 33489–33500. doi: 10.1074/jbc.M111.278556

Hachiya, N. S., Watanabe, K., Sakasegawa, Y., and Kaneko, K. (2004). Microtubules-associated intracellular localization of the NH2-terminal cellular prion protein fragment. *Biochem. Biophys. Res. Commun.* 313, 818–823. doi: 10.1016/j.bbrc.2003.11.167

Hegde, R. S., Mastrianni, J. A., Scott, M. R., Defea, K. A., Tremblay, P., Torchia, M., et al. (1998). A transmembrane form of the prion protein in neurodegenerative disease. *Science* 279, 827–834. doi: 10.1126/science.279.5352.827

Hill, A. F., Butterworth, R. J., Joiner, S., Jackson, G., Rossor, M. N., Thomas, D. J., et al. (1999). Investigation of variant Creutzfeldt-Jakob disease and other human prion diseases with tonsil biopsy samples. *Lancet* 353, 183–189. doi: 10.1016/S0140-6736(98)12075-5

Holme, A., Daniels, M., Sassoon, J., and Brown, D. R. (2003). A novel method of generating neuronal cell lines from gene-knockout mice to study prion protein membrane orientation. *Eur. J. Neurosci.* 18, 571–579. doi: 10.1046/j.1460-9568.2003.02780.x

Horiuchi, M., Yamazaki, N., Ikeda, T., Ishiguro, N., and Shinagawa, M. (1995). A cellular form of prion protein (PrP^C) exists in many non-neuronal tissues of sheep. *J. Gen. Virol.* 76 (Pt 10), 2583–2587. doi: 10.1099/0022-1317-76-10-2583

Hornshaw, M. P., Mcdermott, J. R., and Candy, J. M. (1995). Copper binding to the N-terminal tandem repeat regions of mammalian and avian prion protein. *Biochem. Biophys. Res. Commun.* 207, 621–629. doi: 10.1006/bbrc.1995.1233

Hsu, P. D., Lander, E. S., and Zhang, F. (2014). Development and applications of CRISPR-Cas9 for genome engineering. *Cell* 157, 1262–1278. doi: 10.1016/j.cell.2014.05.010

Hyeon, J. W., Kim, S. Y., Park, J. S., Choi, B. Y., Lee, S. M., Ju, Y. R., et al. (2012). The association between prion proteins and Abeta(1)(-)(4)(2) oligomers in cytotoxicity and apoptosis. *Biochem. Biophys. Res. Commun.* 424, 214–220. doi: 10.1016/j.bbrc.2012.06.056

Iwamaru, Y., Takenouchi, T., Ogihara, K., Hoshino, M., Takata, M., Imamura, M., et al. (2007). Microglial cell line established from prion protein-overexpressing mice is susceptible to various murine prion strains. *J. Virol.* 81, 1524–1527. doi: 10.1128/JVI.01379-06

Jackson, G. S., Murray, I., Hosszu, L. L., Gibbs, N., Waltho, J. P., Clarke, A. R., et al. (2001). Location and properties of metal-binding sites on the human prion protein. *Proc. Natl. Acad. Sci. U.S.A.* 98, 8531–8535. doi: 10.1073/pnas.151038498

Kaimann, T., Metzger, S., Kuhlmann, K., Brandt, B., Birkmann, E., Holtje, H. D., et al. (2008). Molecular model of an alpha-helical prion protein dimer and its monomeric subunits as derived from chemical cross-linking and molecular modeling calculations. *J. Mol. Biol.* 376, 582–596. doi: 10.1016/j.jmb.2007.11.035

Kaiser, D. M., Acharya, M., Leighton, P. L., Wang, H., Daude, N., Wohlgemuth, S., et al. (2012). Amyloid beta precursor protein and prion protein have a conserved interaction affecting cell adhesion and CNS development. *PLoS ONE* 7:e51305. doi: 10.1371/journal.pone.0051305

Keshet, G. I., Bar-Peled, O., Yaffe, D., Nudel, U., and Gabizon, R. (2000). The cellular prion protein colocalizes with the dystroglycan complex in the brain. *J. Neurochem.* 75, 1889–1897. doi: 10.1046/j.1471-4159.2000.0751889.x

Kessels, H. W., Nguyen, L. N., Nabavi, S., and Malinow, R. (2010). The prion protein as a receptor for amyloid-beta. *Nature* 466, E3–E4. doi: 10.1038/nature09217

Kim, B. H., Kim, J. I., Choi, E. K., Carp, R. I., and Kim, Y. S. (2005). A neuronal cell line that does not express either prion or doppel proteins. *Neuroreport* 16, 425–429. doi: 10.1097/00001756-200504040-00002

Kim, B. H., Lee, H. G., Choi, J. K., Kim, J. I., Choi, E. K., Carp, R. I., et al. (2004). The cellular prion protein (PrP^C) prevents apoptotic neuronal cell death and mitochondrial dysfunction induced by serum deprivation. *Brain Res. Mol. Brain Res.* 124, 40–50. doi: 10.1016/j.molbrainres.2004.02.005

Kramer, M. L., Kratzin, H. D., Schmidt, B., Romer, A., Windl, O., Liemann, S., et al. (2001). Prion protein binds copper within the physiological concentration range. *J. Biol. Chem.* 276, 16711–16719. doi: 10.1074/jbc.M006554200

Krasemann, S., Groschup, M. H., Harmeyer, S., Hunsmann, G., and Bodemer, W. (1996b). Generation of monoclonal antibodies against human prion proteins in $PrP^{0/0}$ mice. *Mol. Med.* 2, 725–734.

Krasemann, S., Groschup, M., Hunsmann, G., and Bodemer, W. (1996a). Induction of antibodies against human prion proteins (PrP) by DNA-mediated immunization of $PrP^{0/0}$ mice. *J. Immunol. Methods* 199, 109–118. doi: 10.1016/S0022-1759(96)00165-2

Krebs, B., Wiebelitz, A., Balitzki-Korte, B., Vassallo, N., Paluch, S., Mitteregger, G., et al. (2007). Cellular prion protein modulates the intracellular calcium response to hydrogen peroxide. *J. Neurochem.* 100, 358–367. doi: 10.1111/j.1471-4159.2006.04256.x

Kubosaki, A., Nishimura-Nasu, Y., Nishimura, T., Yusa, S., Sakudo, A., Saeki, K., et al. (2003). Expression of normal cellular prion protein (PrP(c)) on T lymphocytes and the effect of copper ion: analysis by wild-type and prion protein gene-deficient mice. *Biochem. Biophys. Res. Commun.* 307, 810–813. doi: 10.1016/S0006-291X(03)01263-4

Kubowicz, P., Zelaszczyk, D., and Pekala, E. (2013). RNAi in clinical studies. *Curr. Med. Chem.* 20, 1801–1816. doi: 10.2174/09298673113209990118

Kurschner, C., and Morgan, J. I. (1995). The cellular prion protein (PrP) selectively binds to Bcl-2 in the yeast two-hybrid system. *Brain Res. Mol. Brain Res.* 30, 165–168. doi: 10.1016/0169-328X(95)00013-I

Kuwahara, C., Takeuchi, A. M., Nishimura, T., Haraguchi, K., Kubosaki, A., Matsumoto, Y., et al. (1999). Prions prevent neuronal cell-line death. *Nature* 400, 225–226. doi: 10.1038/22241

Larson, M., Sherman, M. A., Amar, F., Nuvolone, M., Schneider, J. A., Bennett, D. A., et al. (2012). The complex PrP(c)-Fyn couples human oligomeric Aβ with pathological tau changes in Alzheimer's disease. *J. Neurosci.* 32, 16857–16871. doi: 10.1523/JNEUROSCI.1858-12.2012

Laurén, J., Gimbel, D. A., Nygaard, H. B., Gilbert, J. W., and Strittmatter, S. M. (2009). Cellular prion protein mediates impairment of synaptic plasticity by amyloid-beta oligomers. *Nature* 457, 1128–1132. doi: 10.1038/nature 07761

Lee, D. C., Sakudo, A., Kim, C. K., Nishimura, T., Saeki, K., Matsumoto, Y., et al. (2006). Fusion of Doppel to octapeptide repeat and N-terminal half of hydrophobic region of prion protein confers resistance to serum deprivation. *Microbiol. Immunol.* 50, 203–209. doi: 10.1111/j.1348-0421.2006. tb03787.x

Li, A., Sakaguchi, S., Atarashi, R., Roy, B. C., Nakaoke, R., Arima, K., et al. (2000a). Identification of a novel gene encoding a PrP-like protein expressed as chimeric transcripts fused to PrP exon 1/2 in ataxic mouse line with a disrupted PrP gene. *Cell. Mol. Neurobiol.* 20, 553–567. doi: 10.1023/A:10070598 27541

Li, A., Sakaguchi, S., Shigematsu, K., Atarashi, R., Roy, B. C., Nakaoke, R., et al. (2000b). Physiological expression of the gene for PrP-like protein, PrPLP/Dpl, by brain endothelial cells and its ectopic expression in neurons of PrP-deficient mice ataxic due to Purkinje cell degeneration. *Am. J. Pathol.* 157, 1447–1452. doi: 10.1016/S0002-9440(10)64782-7

Li, Q., Chevalier, C., Henry, C., Richard, C. A., Moudjou, M., and Vidic, J. (2013). Shadoo binds lipid membranes and undergoes aggregation and fibrillization. *Biochem. Biophys. Res. Commun.* 438, 519–525. doi: 10.1016/j.bbrc.2013. 07.104

Loubet, D., Dakowski, C., Pietri, M., Pradines, E., Bernard, S., Callebert, J., et al. (2012). Neuritogenesis: the prion protein controls β1 integrin signaling activity. *FASEB J.* 26, 678–690. doi: 10.1096/fj.11-185579

Maas, E., Geissen, M., Groschup, M. H., Rost, R., Onodera, T., Schatzl, H., et al. (2007). Scrapie infection of prion protein-deficient cell line upon ectopic expression of mutant prion proteins. *J. Biol. Chem.* 282, 18702–18710. doi: 10.1074/jbc.M701309200

Maignien, T., Shakweh, M., Calvo, P., Marce, D., Sales, N., Fattal, E., et al. (2005). Role of gut macrophages in mice orally contaminated with scrapie or BSE. *Int. J. Pharm.* 298, 293–304. doi: 10.1016/j.ijpharm.2005.02.042

Makzhami, S., Passet, B., Halliez, S., Castille, J., Moazami-Goudarzi, K., Duchesne, A., et al. (2014). The prion protein family: a view from the placenta. *Front. Cell Dev. Biol.* 2:35. doi: 10.3389/fcell.2014.00035

Mange, A., Beranger, F., Peoc'h, K., Onodera, T., Frobert, Y., and Lehmann, S. (2004). Alpha- and beta- cleavages of the amino-terminus of the cellular prion protein. *Biol. Cell* 96, 125–132. doi: 10.1016/j.biolcel.2003.11.007

Manson, J. C., Clarke, A. R., Hooper, M. L., Aitchison, L., Mcconnell, I., and Hope, J. (1994a). 129/Ola mice carrying a null mutation in PrP that abolishes mRNA production are developmentally normal. *Mol. Neurobiol.* 8, 121–127. doi: 10.1007/BF02780662

Manson, J. C., Clarke, A. R., Mcbride, P. A., Mcconnell, I., and Hope, J. (1994b). PrP gene dosage determines the timing but not the final intensity or distribution of lesions in scrapie pathology. *Neurodegeneration* 3, 331–340.

Martin-Lannerée, S., Hirsch, T. Z., Hernandez-Rapp, J., Halliez, S., Vilotte, J., Launay, J., et al. (2014). PrP^C from stem cells to cancer. *Front. Cell Dev. Biol.* 2:55. doi: 10.3389/fcell.2014.00055

Martins, V. R., Graner, E., Garcia-Abreu, J., De Souza, S. J., Mercadante, A. F., Veiga, S. S., et al. (1997). Complementary hydropathy identifies a cellular prion protein receptor. *Nat. Med.* 3, 1376–1382. doi: 10.1038/nm1297-1376

Mattei, V., Garofalo, T., Misasi, R., Circella, A., Manganelli, V., Lucania, G., et al. (2004). Prion protein is a component of the multimolecular signaling complex involved in T cell activation. *FEBS Lett.* 560, 14–18. doi: 10.1016/S0014-5793(04)00029-8

Mays, C. E., Coomaraswamy, J., Watts, J. C., Yang, J., Ko, K. W., Strome, B., et al. (2014). Endoproteolytic processing of the mammalian prion glycoprotein family. *FEBS J.* 281, 862–876. doi: 10.1111/febs.12654

Mcbride, P. A., Eikelenboom, P., Kraal, G., Fraser, H., and Bruce, M. E. (1992). PrP protein is associated with follicular dendritic cells of spleens and lymph nodes in uninfected and scrapie-infected mice. *J. Pathol.* 168, 413–418. doi: 10.1002/path.1711680412

McHugh, P. C., Wright, J. A., Williams, R. J., and Brown, D. R. (2012). Prion protein expression alters APP cleavage without interaction with BACE-1. *Neurochem. Int.* 61, 672–680. doi: 10.1016/j.neuint.2012.07.002

Mcmahon, H. E., Mange, A., Nishida, N., Creminon, C., Casanova, D., and Lehmann, S. (2001). Cleavage of the amino terminus of the prion protein by reactive oxygen species. *J. Biol. Chem.* 276, 2286–2291. doi: 10.1074/jbc.M007243200

Meyer, R. K., Lustig, A., Oesch, B., Fatzer, R., Zurbriggen, A., and Vandevelde, M. (2000). A monomer-dimer equilibrium of a cellular prion protein (PrP^C) not observed with recombinant PrP. *J. Biol. Chem.* 275, 38081–38087. doi: 10.1074/jbc.M007114200

Milhavet, O., and Lehmann, S. (2002). Oxidative stress and the prion protein in transmissible spongiform encephalopathies. *Brain Res. Brain Res. Rev.* 38, 328–339. doi: 10.1016/S0165-0173(01)00150-3

Miura, T., Hori-i, A., Mototani, H., and Takeuchi, H. (1999). Raman spectroscopic study on the copper(II) binding mode of prion octapeptide and its pH dependence. *Biochemistry* 38, 11560–11569. doi: 10.1021/bi99 09389

Moore, R. C., Lee, I. Y., Silverman, G. L., Harrison, P. M., Strome, R., Heinrich, C., et al. (1999). Ataxia in prion protein (PrP)-deficient mice is associated with upregulation of the novel PrP-like protein doppel. *J. Mol. Biol.* 292, 797–817. doi: 10.1006/jmbi.1999.3108

Morales, R., Estrada, L. D., Diaz-Espinoza, R., Morales-Scheihing, D., Jara, M. C., Castilla, J., et al. (2010). Molecular cross talk between misfolded proteins in animal models of Alzheimer's and prion diseases. *J. Neurosci.* 30, 4528–4535. doi: 10.1523/JNEUROSCI.5924-09.2010

Moudjou, M., Frobert, Y., Grassi, J., and La Bonnardiere, C. (2001). Cellular prion protein status in sheep: tissue-specific biochemical signatures. *J. Gen. Virol.* 82, 2017–2024.

Mouillet-Richard, S., Ermonval, M., Chebassier, C., Laplanche, J. L., Lehmann, S., Launay, J. M., et al. (2000). Signal transduction through prion protein. *Science* 289, 1925–1928. doi: 10.1126/science.289.5486.1925

Mouillet-Richard, S., Schneider, B., Pradines, E., Pietri, M., Ermonval, M., Grassi, J., et al. (2007). Cellular prion protein signaling in serotonergic neuronal cells. *Ann. N. Y. Acad. Sci.* 1096, 106–119. doi: 10.1196/annals.1397.076

Nah, J., Pyo, J. O., Jung, S., Yoo, S. M., Kam, T. I., Chang, J., et al. (2013). BECN1/Beclin 1 is recruited into lipid rafts by prion to activate autophagy in response to amyloid beta 42. *Autophagy* 9, 2009–2021. doi: 10.4161/auto. 26118

Nakamura, N., Miyamoto, K., Shimokawa, M., Nishida, N., Mohri, S., Kitamoto, T., et al. (2003a). Generation of antibodies against prion protein by scrapie-infected cell immunization of PrP(0/0) mice. *Hybrid. Hybridomics* 22, 263–266. doi: 10.1089/153685903322328992

Nakamura, Y., Sakudo, A., Saeki, K., Kaneko, T., Matsumoto, Y., Toniolo, A., et al. (2003b). Transfection of prion protein gene suppresses coxsackievirus B3 replication in prion protein gene-deficient cells. *J. Gen. Virol.* 84, 3495–3502. doi: 10.1099/vir.0.19222-0

Nieznanski, K., Nieznanska, H., Skowronek, K. J., Osiecka, K. M., and Stepkowski, D. (2005). Direct interaction between prion protein and tubulin. *Biochem. Biophys. Res. Commun.* 334, 403–411. doi: 10.1016/j.bbrc.2005.06.092

Nikles, D., Bach, P., Boller, K., Merten, C. A., Montrasio, F., Heppner, F. L., et al. (2005). Circumventing tolerance to the prion protein (PrP): vaccination with PrP-displaying retrovirus particles induces humoral immune responses against the native form of cellular PrP. *J. Virol.* 79, 4033–4042. doi: 10.1128/JVI.79.7.4033-4042.2005

Nishida, N., Tremblay, P., Sugimoto, T., Shigematsu, K., Shirabe, S., Petromilli, C., et al. (1999). A mouse prion protein transgene rescues mice deficient for the prion protein gene from purkinje cell degeneration and demyelination. *Lab. Invest.* 79, 689–697.

Nishimura, T., Sakudo, A., Hashiyama, Y., Yachi, A., Saeki, K., Matsumoto, Y., et al. (2007). Serum withdrawal-induced apoptosis in ZrchI prion protein (PrP) gene-deficient neuronal cell line is suppressed by PrP, independent of Doppel. *Microbiol. Immunol.* 51, 457–466. doi: 10.1111/j.1348-0421.2007.tb03920.x

Nishimura, T., Sakudo, A., Xue, G., Ikuta, K., Yukawa, M., Sugiura, K., et al. (2008). Establishment of a new glial cell line from hippocampus of prion protein gene-deficient mice. *Biochem. Biophys. Res. Commun.* 377, 1047–1050. doi: 10.1016/j.bbrc.2008.10.087

Nitta, K., Sakudo, A., Masuyama, J., Xue, G., Sugiura, K., and Onodera, T. (2009). Role of cellular prion proteins in the function of macrophages and dendritic cells. *Protein Pept. Lett.* 16, 239–246. doi: 10.2174/092986609787601705

Nuvolone, M., Kana, V., Hutter, G., Sakata, D., Mortin-Toth, S. M., Russo, G., et al. (2013). SIRPα polymorphisms, but not the prion protein, control phagocytosis of apoptotic cells. *J. Exp. Med.* 210, 2539–2552. doi: 10.1084/jem.20131274

Obinata, M. (2007). The immortalized cell lines with differentiation potentials: their establishment and possible application. *Cancer Sci.* 98, 275–283. doi: 10.1111/j.1349-7006.2007.00399.x

Oh, J. M., Shin, H. Y., Park, S. J., Kim, B. H., Choi, J. K., Choi, E. K., et al. (2008). The involvement of cellular prion protein in the autophagy pathway in neuronal cells. *Mol. Cell. Neurosci.* 39, 238–247. doi: 10.1016/j.mcn.2008.07.003

Onodera, T., Sakudo, A., Tsubone, H., and Itohara, S. (2014). Review of studies that have used knockout mice to assess normal function of prion protein under immunological or pathophysiological stress. *Microbiol. Immunol.* 58, 361–374. doi: 10.1111/1348-0421.12162

Ordóñez-Gutiérrez, L., Torres, J. M., Gavín, R., Antón, M., Arroba-Espinosa, A. I., Espinosa, J. C., et al. (2013). Cellular prion protein modulates β-amyloid deposition in aged APP/PS1 transgenic mice. *Neurobiol. Aging* 34, 2793–2804. doi: 10.1016/j.neurobiolaging.2013.05.019

Pan, K. M., Stahl, N., and Prusiner, S. B. (1992). Purification and properties of the cellular prion protein from Syrian hamster brain. *Protein Sci.* 1, 1343–1352. doi: 10.1002/pro.5560011014

Pantera, B., Bini, C., Cirri, P., Paoli, P., Camici, G., Manao, G., et al. (2009). PrP^C activation induces neurite outgrowth and differentiation in PC12 cells: role for caveolin-1 in the signal transduction pathway. *J. Neurochem.* 110, 194–207. doi: 10.1111/j.1471-4159.2009.06123.x

Parkin, E. T., Watt, N. T., Hussain, I., Eckman, E. A., Eckman, C. B., Manson, J. C., et al. (2007). Cellular prion protein regulates beta-secretase cleavage of the Alzheimer's amyloid precursor protein. *Proc. Natl. Acad. Sci. U.S.A.* 104, 11062–11067. doi: 10.1073/pnas.0609621104

Passet, B., Halliez, S., Beringue, V., Laude, H., and Vilotte, J. L. (2013). The prion protein family: looking outside the central nervous system. *Prion* 7, 127–130. doi: 10.4161/pri.22851

Passet, B., Young, R., Makhzami, S., Vilotte, M., Jaffrezic, F., Halliez, S., et al. (2012). Prion protein and Shadoo are involved in overlapping embryonic pathways and trophoblastic development. *PLoS ONE* 7:e41959. doi: 10.1371/journal.pone.0041959

Peoc'h, K., Serres, C., Frobert, Y., Martin, C., Lehmann, S., Chasseigneaux, S., et al. (2002). The human "prion-like" protein Doppel is expressed in both Sertoli cells and spermatozoa. *J. Biol. Chem.* 277, 43071–43078. doi: 10.1074/jbc.M206357200

Prcina, M., Filipcik, P., and Kontsekova, E. (2010). Establishment of the cell line expressing human prion protein on $PrP^{0/0}$ background. *Acta Virol.* 54, 297–302. doi: 10.4149/av_2010_04_297

Prinz, M., Montrasio, F., Furukawa, H., Van Der Haar, M. E., Schwarz, P., Rulicke, T., et al. (2004). Intrinsic resistance of oligodendrocytes to prion infection. *J. Neurosci.* 24, 5974–5981. doi: 10.1523/JNEUROSCI.0122-04.2004

Prusiner, S. B. (1998). Prions. *Proc. Natl. Acad. Sci. U.S.A.* 95, 13363–13383. doi: 10.1073/pnas.95.23.13363

Prusiner, S. B., Groth, D., Serban, A., Koehler, R., Foster, D., Torchia, M., et al. (1993). Ablation of the prion protein (PrP) gene in mice prevents scrapie and facilitates production of anti-PrP antibodies. *Proc. Natl. Acad. Sci. U.S.A.* 90, 10608–10612. doi: 10.1073/pnas.90.22.10608

Raeber, A. J., Sailer, A., Hegyi, I., Klein, M. A., Rulicke, T., Fischer, M., et al. (1999). Ectopic expression of prion protein (PrP) in T lymphocytes or hepatocytes of PrP knockout mice is insufficient to sustain prion replication. *Proc. Natl. Acad. Sci. U.S.A.* 96, 3987–3992. doi: 10.1073/pnas.96.7.3987

Ramesh, G., Maclean, A. G., and Philipp, M. T. (2013). Cytokines and chemokines at the crossroads of neuroinflammation, neurodegeneration, and neuropathic pain. *Mediators Inflamm.* 2013:480739. doi: 10.1155/2013/480739

Rieger, R., Edenhofer, F., Lasmezas, C. I., and Weiss, S. (1997). The human 37-kDa laminin receptor precursor interacts with the prion protein in eukaryotic cells. *Nat. Med.* 3, 1383–1388. doi: 10.1038/nm1297-1383

Rossi, D., Cozzio, A., Flechsig, E., Klein, M. A., Rulicke, T., Aguzzi, A., et al. (2001). Onset of ataxia and Purkinje cell loss in PrP null mice inversely correlated with Dpl level in brain. *EMBO J.* 20, 694–702. doi: 10.1093/emboj/20.4.694

Roucou, X., Gains, M., and Leblanc, A. C. (2004). Neuroprotective functions of prion protein. *J. Neurosci. Res.* 75, 153–161. doi: 10.1002/jnr.10864

Ryder, E. F., Snyder, E. Y., and Cepko, C. L. (1990). Establishment and characterization of multipotent neural cell lines using retrovirus vector-mediated oncogene transfer. *J. Neurobiol.* 21, 356–375. doi: 10.1002/neu.480210209

Saeki, K., Matsumoto, Y., Hirota, Y., and Onodera, T. (1996). Three-exon structure of the gene encoding the rat prion protein and its expression in tissues. *Virus Genes* 12, 15–20. doi: 10.1007/BF00369996

Sakaguchi, S., Katamine, S., Nishida, N., Moriuchi, R., Shigematsu, K., Sugimoto, T., et al. (1996). Loss of cerebellar Purkinje cells in aged mice homozygous for a disrupted PrP gene. *Nature* 380, 528–531. doi: 10.1038/380528a0

Sakudo, A., Ano, Y., Onodera, T., Nitta, K., Shintani, H., Ikuta, K., et al. (2011a). Fundamentals of prions and their inactivation (review). *Int. J. Mol. Med.* 27, 483–489. doi: 10.3892/ijmm.2011.605

Sakudo, A., and Ikuta, K. (2009a). Fundamentals of prion diseases and their involvement in the loss of function of cellular prion protein. *Protein Pept. Lett.* 16, 217–229. doi: 10.2174/092986609787601732

Sakudo, A., and Ikuta, K. (2009b). Prion protein functions and dysfunction in prion diseases. *Curr. Med. Chem.* 16, 380–389. doi: 10.2174/092986709787002673

Sakudo, A., Lee, D. C., Li, S., Nakamura, T., Matsumoto, Y., Saeki, K., et al. (2005a). PrP cooperates with STI1 to regulate SOD activity in PrP-deficient neuronal cell line. *Biochem. Biophys. Res. Commun.* 328, 14–19. doi: 10.1016/j.bbrc.2004.12.132

Sakudo, A., Lee, D. C., Nakamura, I., Taniuchi, Y., Saeki, K., Matsumoto, Y., et al. (2005b). Cell-autonomous PrP-Doppel interaction regulates apoptosis in PrP gene-deficient neuronal cells. *Biochem. Biophys. Res. Commun.* 333, 448–454. doi: 10.1016/j.bbrc.2005.05.128

Sakudo, A., Lee, D. C., Nishimura, T., Li, S., Tsuji, S., Nakamura, T., et al. (2005c). Octapeptide repeat region and N-terminal half of hydrophobic region of prion protein (PrP) mediate PrP-dependent activation of superoxide dismutase. *Biochem. Biophys. Res. Commun.* 326, 600–606. doi: 10.1016/j.bbrc.2004.11.092

Sakudo, A., Lee, D. C., Saeki, K., Matsumoto, Y., Itohara, S., and Onodera, T. (2003a). Tumor necrosis factor attenuates prion protein-deficient neuronal cell death by increases in anti-apoptotic Bcl-2 family proteins. *Biochem. Biophys. Res. Commun.* 310, 725–729. doi: 10.1016/j.bbrc.2003.09.068

Sakudo, A., Lee, D. C., Saeki, K., Nakamura, Y., Inoue, K., Matsumoto, Y., et al. (2003b). Impairment of superoxide dismutase activation by N-terminally truncated prion protein (PrP) in PrP-deficient neuronal cell line. *Biochem. Biophys. Res. Commun.* 308, 660–667. doi: 10.1016/S0006-291X(03)01459-1

Sakudo, A., Lee, D. C., Yoshimura, E., Nagasaka, S., Nitta, K., Saeki, K., et al. (2004). Prion protein suppresses perturbation of cellular copper homeostasis under oxidative conditions. *Biochem. Biophys. Res. Commun.* 313, 850–855. doi: 10.1016/j.bbrc.2003.12.020

Sakudo, A., Nakamura, I., Ikuta, K., and Onodera, T. (2007a). Recent developments in prion disease research: diagnostic tools and *in vitro* cell culture models. *J. Vet. Med. Sci.* 69, 329–337. doi: 10.1292/jvms.69.329

Sakudo, A., and Onodera, T. (2011b). Tissue- and cell type-specific modification of prion protein (PrP)-like protein Doppel, which affects PrP endoproteolysis. *Biochem. Biophys. Res. Commun.* 404, 523–527. doi: 10.1016/j.bbrc.2010.12.016

Sakudo, A., Onodera, T., and Ikuta, K. (2007b). Prion protein gene-deficient cell lines: powerful tools for prion biology. *Microbiol. Immunol.* 51, 1–13. doi: 10.1111/j.1348-0421.2007.tb03877.x

Sakudo, A., Onodera, T., Suganuma, Y., Kobayashi, T., Saeki, K., and Ikuta, K. (2006). Recent advances in clarifying prion protein functions using knockout mice and derived cell lines. *Mini Rev. Med. Chem.* 6, 589–601. doi: 10.2174/138955706776876159

Sakudo, A., Wu, G., Onodera, T., and Ikuta, K. (2008). Octapeptide repeat region of prion protein (PrP) is required at an early stage for production of abnormal prion protein in PrP-deficient neuronal cell line. *Biochem. Biophys. Res. Commun.* 365, 164–169. doi: 10.1016/j.bbrc.2007.10.158

Satoh, J., Kuroda, Y., and Katamine, S. (2000). Gene expression profile in prion protein-deficient fibroblasts in culture. *Am. J. Pathol.* 157, 59–68. doi: 10.1016/S0002-9440(10)64517-8

Satoh, J., Yukitake, M., Kurohara, K., Nishida, N., Katamine, S., Miyamoto, T., et al. (1998). Cultured skin fibroblasts isolated from mice devoid of the prion protein gene express major heat shock proteins in response to heat stress. *Exp. Neurol.* 151, 105–115. doi: 10.1006/exnr.1998.6796

Schatzl, H. M., Da Costa, M., Taylor, L., Cohen, F. E., and Prusiner, S. B. (1995). Prion protein gene variation among primates. *J. Mol. Biol.* 245, 362–374. doi: 10.1006/jmbi.1994.0030

Schmitt-Ulms, G., Legname, G., Baldwin, M. A., Ball, H. L., Bradon, N., Bosque, P. J., et al. (2001). Binding of neural cell adhesion molecules (N-CAMs) to the cellular prion protein. *J. Mol. Biol.* 314, 1209–1225. doi: 10.1006/jmbi.2000.5183

Schwarze-Eicker, K., Keyvani, K., Görtz, N., Westaway, D., Sachser, N., and Paulus, W. (2005). Prion protein (PrP^C) promotes beta-amyloid plaque formation. *Neurobiol. Aging* 26, 1177–1182. doi: 10.1016/j.neurobiolaging.2004.10.004

Shaked, Y., Rosenmann, H., Talmor, G., and Gabizon, R. (1999). A C-terminal-truncated PrP isoform is present in mature sperm. *J. Biol. Chem.* 274, 32153–32158. doi: 10.1074/jbc.274.45.32153

Skripuletz, T., Lindner, M., Kotsiari, A., Garde, N., Fokuhl, J., Linsmeier, F., et al. (2008). Cortical demyelination is prominent in the murine cuprizone model and is strain-dependent. *Am. J. Pathol.* 172, 1053–1061. doi: 10.2353/ajpath.2008.070850

Spielhaupter, C., and Schatzl, H. M. (2001). PrP^C directly interacts with proteins involved in signaling pathways. *J. Biol. Chem.* 276, 44604–44612. doi: 10.1074/jbc.M103289200

Steele, A. D., Lindquist, S., and Aguzzi, A. (2007). The prion protein knockout mouse: a phenotype under challenge. *Prion* 1, 83–93. doi: 10.4161/pri.1.2.4346

Swarthout, J. T. P. D., Raisinghani, M., and Cui, X. (2011). Zinc Finger nucleases: a new era for transgenic animals. *Ann. Neurosci.* 18, 25–28. doi: 10.5214/ans.0972.7531.1118109

Tayebi, M., Jones, D. R., Taylor, W. A., Stileman, B. F., Chapman, C., Zhao, D., et al. (2011). PrP(Sc)-specific antibodies with the ability to immunodetect prion oligomers. *PLoS ONE* 6:e19998. doi: 10.1371/journal.pone.0019998

Toni, M., Spisni, E., Griffoni, C., Santi, S., Riccio, M., Lenaz, P., et al. (2006). Cellular prion protein and caveolin-1 interaction in a neuronal cell line precedes Fyn/Erk 1/2 signal transduction. *J. Biomed. Biotechnol.* 2006:69469. doi: 10.1155/JBB/2006/69469

Tuzi, N. L., Gall, E., Melton, D., and Manson, J. C. (2002). Expression of doppel in the CNS of mice does not modulate transmissible spongiform encephalopathy disease. *J. Gen. Virol.* 83, 705–711.

Um, J. W., Nygaard, H. B., Heiss, J. K., Kostylev, M. A., Stagi, M., Vortmeyer, A., et al. (2012). Alzheimer amyloid-β oligomer bound to postsynaptic prion protein activates Fyn to impair neurons. *Nat. Neurosci.* 15, 1227–1235. doi: 10.1038/nn.3178

Um, J. W., and Strittmatter, S. M. (2013). Amyloid-β induced signaling by cellular prion protein and Fyn kinase in Alzheimer disease. *Prion* 7, 37–41. doi: 10.4161/pri.22212

Uraki, R., Sakudo, A., Ando, S., Kitani, H., and Onodera, T. (2010). Enhancement of phagocytotic activity by prion protein in PrP-deficient macrophage cells. *Int. J. Mol. Med.* 26, 527–532. doi: 10.3892/ijmm_00000495

Vassallo, N., Herms, J., Behrens, C., Krebs, B., Saeki, K., Onodera, T., et al. (2005). Activation of phosphatidylinositol 3-kinase by cellular prion protein and its role in cell survival. *Biochem. Biophys. Res. Commun.* 332, 75–82. doi: 10.1016/j.bbrc.2005.04.099

Verkhratsky, A., Parpura, V., Pekna, M., Pekny, M., and Sofroniew, M. (2014). Glia in the pathogenesis of neurodegenerative diseases. *Biochem. Soc. Trans.* 42, 1291–1301. doi: 10.1042/BST20140107

Viles, J. H., Cohen, F. E., Prusiner, S. B., Goodin, D. B., Wright, P. E., and Dyson, H. J. (1999). Copper binding to the prion protein: structural implications of four identical cooperative binding sites. *Proc. Natl. Acad. Sci. U.S.A.* 96, 2042–2047. doi: 10.1073/pnas.96.5.2042

Watarai, M., Kim, S., Erdenebaatar, J., Makino, S., Horiuchi, M., Shirahata, T., et al. (2003). Cellular prion protein promotes Brucella infection into macrophages. *J. Exp. Med.* 198, 5–17. doi: 10.1084/jem.20021980

Watts, J. C., Drisaldi, B., Ng, V., Yang, J., Strome, B., Horne, P., et al. (2007a). The CNS glycoprotein Shadoo has PrP(C)-like protective properties and displays reduced levels in prion infections. *EMBO J.* 26, 4038–4050. doi: 10.1038/sj.emboj.7601830

Watts, J. C., Huo, H., Bai, Y., Ehsani, S., Jeon, A. H., Shi, T., et al. (2009). Interactome analyses identify ties of PrP and its mammalian paralogs to oligomannosidic N-glycans and endoplasmic reticulum-derived chaperones. *PLoS Pathog.* 5:e1000608. doi: 10.1371/annotation/9eb11869-6acb-49b0-978e-abedc3cc545a

Watts, J. C., Stohr, J., Bhardwaj, S., Wille, H., Oehler, A., Dearmond, S. J., et al. (2011). Protease-resistant prions selectively decrease Shadoo protein. *PLoS Pathog.* 7:e1002382. doi: 10.1371/journal.ppat.1002382

Watts, J. C., and Westaway, D. (2007b). The prion protein family: diversity, rivalry, and dysfunction. *Biochim. Biophys. Acta* 1772, 654–672. doi: 10.1016/j.bbadis.2007.05.001

Weissmann, C., and Aguzzi, A. (1999). Perspectives: neurobiology. PrP's double causes trouble. *Science* 286, 914–915. doi: 10.1126/science.286.5441.914

Weissmann, C., and Flechsig, E. (2003). PrP knock-out and PrP transgenic mice in prion research. *Br. Med. Bull.* 66, 43–60. doi: 10.1093/bmb/66.1.43

Westaway, D., Genovesi, S., Daude, N., Brown, R., Lau, A., Lee, I., et al. (2011). Down-regulation of Shadoo in prion infections traces a pre-clinical event inversely related to PrP(Sc) accumulation. *PLoS Pathog.* 7:e1002391. doi: 10.1371/journal.ppat.1002391

Williams, A. E., Lawson, L. J., Perry, V. H., and Fraser, H. (1994). Characterization of the microglial response in murine scrapie. *Neuropathol. Appl. Neurobiol.* 20, 47–55. doi: 10.1111/j.1365-2990.1994.tb00956.x

Wright, D. A., Li, T., Yang, B., and Spalding, M. H. (2014). TALEN-mediated genome editing: prospects and perspectives. *Biochem. J.* 462, 15–24. doi: 10.1042/BJ20140295

Wu, G., Nakajima, K., Takeyama, N., Yukawa, M., Taniuchi, Y., Sakudo, A., et al. (2008). Species-specific anti-apoptotic activity of cellular prion protein in a mouse PrP-deficient neuronal cell line transfected with mouse, hamster, and bovine *Prnp. Neurosci. Lett.* 446, 11–15. doi: 10.1016/j.neulet.2008.09.020

Yamanaka, S. (2007). Strategies and new developments in the generation of patient-specific pluripotent stem cells. *Cell Stem Cell* 1, 39–49. doi: 10.1016/j.stem.2007.05.012

Yehiely, F., Bamborough, P., Da Costa, M., Perry, B. J., Thinakaran, G., Cohen, F. E., et al. (1997). Identification of candidate proteins binding to prion protein. *Neurobiol. Dis.* 3, 339–355. doi: 10.1006/nbdi.1997.0130

Yokoyama, T., Kimura, K. M., Ushiki, Y., Yamada, S., Morooka, A., Nakashiba, T., et al. (2001). *In vivo* conversion of cellular prion protein to pathogenic isoforms, as monitored by conformation-specific antibodies. *J. Biol. Chem.* 276, 11265–11271. doi: 10.1074/jbc.M008734200

Zafar, S., Asif, A. R., Ramljak, S., Tahir, W., Schmitz, M., and Zerr, I. (2014). Anchorless 23-230 PrP^C interactomics for elucidation of PrP^C protective role. *Mol. Neurobiol.* 49, 1385–1399. doi: 10.1007/s12035-013-8616-2

Zafar, S., Von Ahsen, N., Oellerich, M., Zerr, I., Schulz-Schaeffer, W. J., Armstrong, V. W., et al. (2011). Proteomics approach to identify the interacting partners of cellular prion protein and characterization of Rab7a interaction in neuronal cells. *J. Proteome Res.* 10, 3123–3135. doi: 10.1021/pr2001989

Zanata, S. M., Lopes, M. H., Mercadante, A. F., Hajj, G. N., Chiarini, L. B., Nomizo, R., et al. (2002). Stress-inducible protein 1 is a cell surface ligand for cellular prion that triggers neuroprotection. *EMBO J.* 21, 3307–3316. doi: 10.1093/emboj/cdf325

Zanusso, G., Liu, D., Ferrari, S., Hegyi, I., Yin, X., Aguzzi, A., et al. (1998). Prion protein expression in different species: analysis with a panel of new mAbs. *Proc. Natl. Acad. Sci. U.S.A.* 95, 8812–8816. doi: 10.1073/pnas.95.15.8812

Zhou, M., Ottenberg, G., Sferrazza, G. F., and Lasmezas, C. I. (2012). Highly neurotoxic monomeric alpha-helical prion protein. *Proc. Natl. Acad. Sci. U.S.A.* 109, 3113–3118. doi: 10.1073/pnas.1118090109

Conflict of Interest Statement: The authors declare that the research was conducted in the absence of any commercial or financial relationships that could be construed as a potential conflict of interest.

13

Polyethylenimine-mediated expression of transgenes in the acinar cells of rats salivary glands *in vivo*

Monika Sramkova†, Laura Parente†, Timothy Wigand, Myo-Pale' Aye, Akiko Shitara and Roberto Weigert*

Intracellular Membrane Trafficking Unit, Oral and Pharyngeal Cancer Branch, National Institute of Dental and Craniofacial Research, National Institutes of Health, Bethesda, MD, USA

Edited by:
Ken William Dunn, Indiana University School of Medicine, USA

Reviewed by:
Heike Folsch, Northwestern University Feinberg School of Medicine, USA
Robert Bacallao, Indiana University School of Medicine, USA

****Correspondence:***
Roberto Weigert, Intracellular Membrane Trafficking Unit, Oral and Pharyngeal Cancer Branch, National Institute of Dental and Craniofacial Research, National Institutes of Health, 30 Convent Dr. 303A, Bethesda, MD 20892-4340, USA
e-mail: weigertr@mail.nih.gov

†These authors have contributed equally to this work.

Non viral-mediated transfection of plasmid DNA provides a fast and reliable way to express various transgenes in selected cell populations in live animals. Here, we show an improvement of a previously published method that is based on injecting plasmid DNA into the ductal system of the salivary glands in live rats. Specifically, using complexes between plasmid DNA and polyethyleneimine (PEI) we show that the expression of the transgenes is directed selectively to the salivary acinar cells. PEI does not affect the ability of cells to undergo regulated exocytosis, which was one of the main drawbacks of the previous methods. Moreover PEI does not affect the proper localization and targeting of transfected proteins, as shown for the apical plasma membrane water channel aquaporin 5 (AQP5). Overall, this approach, coupled with the use of intravital microscopy, permits to conduct localization and functional studies under physiological conditions, in a rapid, reliable, and affordable fashion.

Keywords: intravital microscopy, *in vivo* transfection, polyethyleneimine, aquaporin 5, non-viral gene delivery, salivary glands, acinar cells

INTRODUCTION

Intravital microscopy (IVM) has enabled researchers to image several biological processes in live multicellular organisms and in real time (Pittet and Weissleder, 2011; Ritsma et al., 2012; Weigert et al., 2013). In the last decade, this approach has permitted to image subcellular organelles in live rodents with temporal and spatial resolution almost identical to that achieved in cell cultures (Dunn et al., 2002; Weigert et al., 2013; Weigert, 2014). This has been possible by the development of procedures aimed at minimizing the motion artifacts due to the heartbeat and respiration, which otherwise would make imaging at a submicron scale an impossible task. In particular, we have developed a rodent salivary gland model that has enabled (1) tracking individual structures for extended periods of time, and (2) extracting quantitative data in a rigorous fashion. This approach has been used to study various aspects of the dynamics of endocytosis and regulated exocytosis in the salivary epithelium of both rats and mice (Masedunskas and Weigert, 2008; Masedunskas et al., 2011a, 2012), and recently, it has revealed a novel modality of mitochondrial metabolism that cannot be observed in *in vitro* models (Porat-Shliom et al., 2014).

So far, pharmacological tools have been the main approach to study the molecular mechanisms regulating the dynamics of subcellular events in live animals (Sandoval et al., 2004; Masedunskas et al., 2011a). One of the current challenges is to generate molecular tools that can be rapidly expressed *in vivo*, and interfere with the process of interest. An elegant approach is to develop transgenic mice engineered to express fluorescently tagged molecules, which act as reporters for the pathway of interest, and, at the same time, lack or express the specific molecule under investigation. Although the application of gene editing techniques to mice (e.g., zinc finger or CRISPR is rapidly growing (Sung et al., 2012; Yang et al., 2013), these methods are still expensive and time consuming. Viable alternatives are *in vivo* transient transfection of transgenes and/or shRNAs, which can be accomplished by either viral or non-viral mediated delivery (Boulaiz et al., 2005; Kawakami et al., 2008; Wang et al., 2013). Viral-mediated gene transfer is the most widely used approach since it allows to transfect a high number of cells and to efficiently deliver and express the transgenes in non-dividing or slowly dividing cells (Boulaiz et al., 2005). This approach provides with the opportunity to perform biochemical analysis on the transfected tissues. In small animals, the main viral vectors currently used are: recombinant adenovirus serotype 5 (rAD5), adeno-associated virus serotype 2 (AAV2), and lenti viruses (Snyder and Francis, 2005; Adriaansen et al., 2010; Wilson et al., 2013). However, viral-mediated transfections have some limitations, which include adverse reactions due to the host immune response (Sun et al., 2003). Moreover, the generation of viruses is time consuming and subjected to safety restrictions. To overcome these issues, non viral-mediated techniques have been attempted in various organs since the early 90's, and in particular, direct injections of plasmid DNA in rodent skeletal muscle have proven to be quite successful (Wolff et al., 1992). Following up on early studies in SGs (Goldfine et al., 1997), we have developed and characterized a transfection method that can be used with commercially available plasmids, and thus suitable for rapid screenings and mutation analysis. We have shown that plasmid DNA can be transfected in the SGs of live rats using three different modalities of administration, which ensure to direct the expression of the transgene to selected cell populations (Sramkova et al., 2009, 2012). Although the yield of non viral-based transfections is an order of magnitude lower than viral-based methods, this approach is robust and enables performing single cell imaging *in vivo*. Here we show a variation of one of these transfection methods that is based on the use of polyethyleneimine (PEI) a molecule previously used as a vehicle to compact and deliver siRNA *in vivo* (Kircheis et al., 2001; Nimesh, 2012). This approach permits the selective transfection of acinar cells in the SG epithelium in a rapid and reliable fashion without affecting the properties and the localization of the transgenes.

RESULTS AND DISCUSSION

PEI FACILITATES THE TARGETING OF THE PLASMID DNA TO THE ACINAR CELLS OF THE SALIVARY GLANDS *IN VIVO*

We have previously shown that plasmid DNA can be delivered to the submandibular SGs in live rats by retro-injection trough the salivary duct (or Wharton's duct). Plasmid DNA is loaded into the ductal system and the acinar canaliculi, thus gaining access to the salivary epithelium from the apical plasma membrane (APM) (**Figure 1A**). Specifically, we have shown that although plasmid DNA is uptaken by all the cells in the epithelium, only few of the proliferating cells of the intercalated ducts express the transgenes (Sramkova et al., 2012) (**Figures 1B,F**). On the other hand, plasmid DNA mixed with empty rAD5 particles elicited the expression of the transgenes primarily in large ducts (Sramkova et al., 2009). Finally, we have shown that plasmid DNA is targeted exclusively to the acinar cells during β adrenergic-mediated stimulation of regulated exocytosis (Sramkova et al., 2009). Indeed, during secretion, plasmid DNA is internalized into the acinar cells by virtue of compensatory endocytosis that is elicited to maintain the homeostasis of the APM (Sramkova et al., 2009; Masedunskas et al., 2011b) (**Figures 1B,F**). However, one of the main drawbacks of this approach is the fact that β-adrenergic stimulation induces an extensive degranulation of the acinar cells (Masedunskas et al., 2011a) (**Figure 1C**, central panel). Since the biogenesis of the secretory granules is a slow process, their full complement is reconstituted in 24–36 h, which is beyond the optimal expression of the transgene (16–24 from the administration of the plasmid). Therefore, this method cannot not be used to study processes such as regulated exocytosis, and trafficking from the TGN, where the granules are originated. To overcome this issue we sought to establish a method that could enhance the low level of endocytosis that occurs at the APM of the acinar cells without inducing exocytosis (Sramkova et al., 2012). Specifically, we tested PEI, which has been extensively used to deliver siRNAs both *in vitro* and *in vivo*. Although its precise mechanisms of action is not fully understood, PEI has been suggested to: (1) facilitate the packing of nucleotide strands, (2) increase the endocytosis of the complex PEI-siRNA, and (3) promote endosomal rupture, thus releasing the complex into the cytoplasm (Neu et al., 2005; Creusat et al., 2010). PEI was mixed with plasmid DNA encoding for the monomeric YFP (see

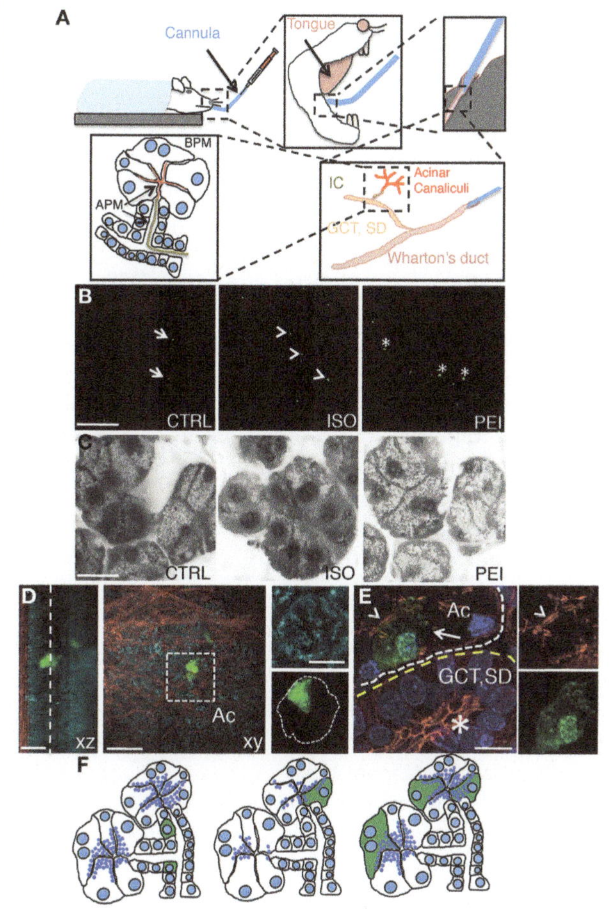

FIGURE 1 | Non viral-mediated expression of transgenes in rat salivary glands. (A) Diagram of the transfection procedures in anesthetized rats. A thin polyethylene cannula is inserted into the Wharton's duct, as described in the Methods sections. The plasmid DNA is gently injected and diffuses into the large ducts (granular convoluted tubules or GCT, and the striated ducts or SD), then into the intercalated ducts (IC), and finally into the acinar canaliculi. Plasmid DNA accesses the epithelium through the apical plasma membrane (APM, arrows). **(B)** *In vivo* transfection with plasmid DNA -12 μg of Plasmid DNA encoding for pVenus were injected alone (CTRL, left panel), after SC injection of isoproterenol (ISO, center panel), or mixed with PEI as described in the Methods (PEI, right panel). After 16 h the submandibular SGs were excised and immediately imaged by two-photon microscopy using a 10X objective (excitation 930 nm) to reveal the cells expressing pVenus (arrows). Scale bar, 1 mm. **(C)** Effect of PEI on regulated exocytosis. Mice expressing cytoplasmic GFP were treated as described in B and the SGs imaged by confocal microscopy (excitation 488 nm) to estimate exocytosis of the secretory granules, as previously reported (Milberg et al., 2013). In the inset the apical plasma membrane (arrows) and the secretory granules (arrowheads) are highlighted. Scale bar, 20 μm. **(D–E)** PEI-mediated transfection in SGs drives expression into acinar cells. **(D)** Excised glands transfected with pVenus/PEI were imaged by two-photon microscopy using a 60x water immersion objective **(D)** or processed for immunofluorescence and imaged by confocal microscopy **(E)**. **(D)** Two Z-stacks from the same area of the sample were acquired twice by using two different excitation wavelengths, respectively: 740 nm to reveal the epithelium of the gland (cyan) and 930 nm to reveal pVenus (green) and collagen fibers (red). The Z-stacks were combined and shown as xz (left) or xy (center) projection. The cell expressing pVenus is part of an acinus (inset, right). Scale bar, 20 μm. **(E)** The SGs were labeled for phallodin (red) and the nuclear staining Hoechst (blue). The cell expressing pVenus (green) is *(Continued)*

FIGURE 1 | Continued
part of an acinus (white broken line) as identified by the actin-labeled acinar canaliculi (arrowhead). Large ducts (yellow broken line), as identified by the characteristic actin pattern (asterisk) did not express any pVenus. Scale bar, 10 μm. **(F)** Diagram summarizing the pattern of expression of plasmid DNA in SGs. pVenus was expressed in cells of the IC under control conditions (left) and in acinar cells either upon stimulation with ISO (center) or administration with PEI (right).

Materials and Methods) and injected through the Wharton's duct of anesthetized rats. After 16 h the animals were anesthetized and the SGs were exposed and imaged by intravital two-photon microscopy (**Figure 1B**). Using a low magnification objective we sampled large areas of the tissue and observed that $\approx$10 cells/mm^2 expressed the transgene, a level comparable to that observed in β adrenergic- stimulated rats (**Figure 2B**). To determine whether PEI stimulated regulated exocytosis we used mice expressing cytoplasmic GFP that enable the visualization of the secretory granules (Masedunskas et al., 2011a; Milberg et al., 2013). Notably, SC injection of isoproterenol elicited robust exocytosis whereas PEI did not (**Figure 1C** and insets). The yield of transfection was maximal between 16–24 h from the administration (not shown). Moreover, we found that the transgene was expressed in acinar cells, as shown by performing either two-photon microscopy in freshly excised glands (**Figure 1D**) or immunocytochemistry and confocal microscopy (**Figure 1E**).

PEI facilitates the expression of molecules other than YFP, as shown by administering plasmid DNAs encoding for various molecules, such as cytoplasmic (e.g., actin, myosin Vc), nuclear (e.g., H2B), membrane-associated (e.g., Rab5), single and multiple spanning transmembrane proteins (e.g., TGN38, TfnR, VAMP8) (**Figures 2A,B**). The efficiency of transgene expression did not correlate with any characteristics of the expressed proteins or the size of the plasmid, as shown by expressing a transgene generated by fusing one, two or three copies of YFP (**Figure 2C**).

Our data show that PEI is an effective agent in facilitating the delivery and expression of transgenes into the acinar cells of rat salivary glands *in vivo*. Moreover, PEI does not cause any obvious cellular toxicity, as shown by the fact that regulated exocytosis and the dynamics of assembly of the actin cytoskeleton were not affected (**Figure 2D**). Finally, we did not observe any general signs of toxicity in the whole animal, at least within 24–48 h from the administration of PEI (not shown).

PEI DOES NOT AFFECT PROTEIN TARGETING TO SPECIALIZED DOMAINS OF THE PLASMA MEMBRANE

Next, we checked whether PEI affected the proper targeting of plasma membrane proteins to different domains of the acinar cells in the live animal. To this end, we investigated the expression and the localization of aquaporin 5 (AQP5), a water channel that regulates fluid secretion in mammalian SGs and is targeted to the APM via a sorting signal encoded in its C-terminus (Raina et al., 1995; Wellner et al., 2005; Delporte and Steinfeld, 2006; Horsefield et al., 2008).

Trafficking, localization, and targeting mechanism of AQP5 have been primarily investigated in cell cultures grown on solid substrates (Tada et al., 1999), in polarized MDCK cells (Nejsum and Nelson, 2007; Karabasil et al., 2009), and only to a small extent in native tissue in rodents (Gresz et al., 2004; Wellner et al., 2005; Matsuzaki et al., 2006). For these studies AQP5 has been tagged with fluorescent proteins (e.g., GFP or RFP) and the behavior of these constructs has been shown to be dependent on the experimental system and the position of the tag. We began by characterizing various constructs tagged either at the C- or the N-terminus (both human and rat AQP5) in cultured salivary glands-derived HSG cells (Royce et al., 1993). We used the monomeric form of the yellow fluorescent proteins (YFP) that has better quantum efficiency than the GFP when excited by 2-photon microscopy (Nagai et al., 2002). In HSG cells, we observed that AQP5-YFP was expressed but did not get exported out of the endoplasmic reticulum (**Figure 3A**, upper panel, arrows). On the other hand, YFP-AQP5 was properly targeted to the PM as previously reported (Wellner et al., 2005), and no intracellular vesicles were detected (**Figure 3A**, lower panel, arrowhead). Next, we generated stable HSG cells expressing YFP-AQP5 grown either as a confluent monolayer or in matrigel (Hoffman et al., 1996, 1998). In matrigel, HSG cells form acinar-like structures, which exhibit a partial polarization, as shown by phalloidin labeling (**Figure 3B**) and expression of amylase, a marker for acinar differentiation (**Figure 3C**) (Hoffman et al., 1996). YFP-AQP5 was properly targeted to the plasma membrane both in monolayers and in matrigel (**Figure 3D**). However, in acinar-like structures AQP5 did not exhibit any polarized localization, suggesting that other components may be required for the proper sorting, such as the μ1-b subunit of the AP1 adaptor complex (**Figure 3D**) (Ohno et al., 1999). Similar results were obtained expressing rat YFP-AQP5 (not shown).

Next, we injected the plasmid encoding for rat YFP-AQP5 into the salivary ducts of anesthetized rats by using PEI as vehicle. As expected, YFP-AQP5 was expressed in the SGs (**Figure 4A**, arrows), and primarily in acinar cells that were identified by endogenous fluorescence emissions (excitation 740 nm), which highlight the cellular architecture of the salivary epithelium (Masedunskas and Weigert, 2008) (**Figure 4A**, lower panel). YFP-AQP5 was properly targeted to the APM as shown by two-photon microscopy (**Figure 4B**) and indirect immunofluorescence (**Figure 4C**, arrows). Interestingly, we observed that YFP-AQP5 was localized onto a series of intracellular vesicles localized in the sub-apical area of the acinar cells (**Figures 4B,C**, arrows) differently from what observed in cultured cells. These vesicles were not due to the fixation and the processing of the glands since they could be observed in the live animal (**Figure 4B**) and they were not an artifact of the ectopic expression of YFP-AQP5, since endogenous AQP5 was also localized both at the APM and in vesicles (**Figure 4C**). Notably, a sub population of the YFP-AQP5 vesicles was found in close proximity of the TGN (**Figures 4B,E**). Although we could not determine whether these vesicles are endocytic or exocytic in nature, a time-lapse analysis revealed that they were stationary under basal conditions, possibly indicating that AQP5 is stored in an intracellular reservoir that may be mobilized during stimulation of water or protein secretion (**Figure 4E**), as shown for AQP2 in the kidney (Fushimi et al., 1997).

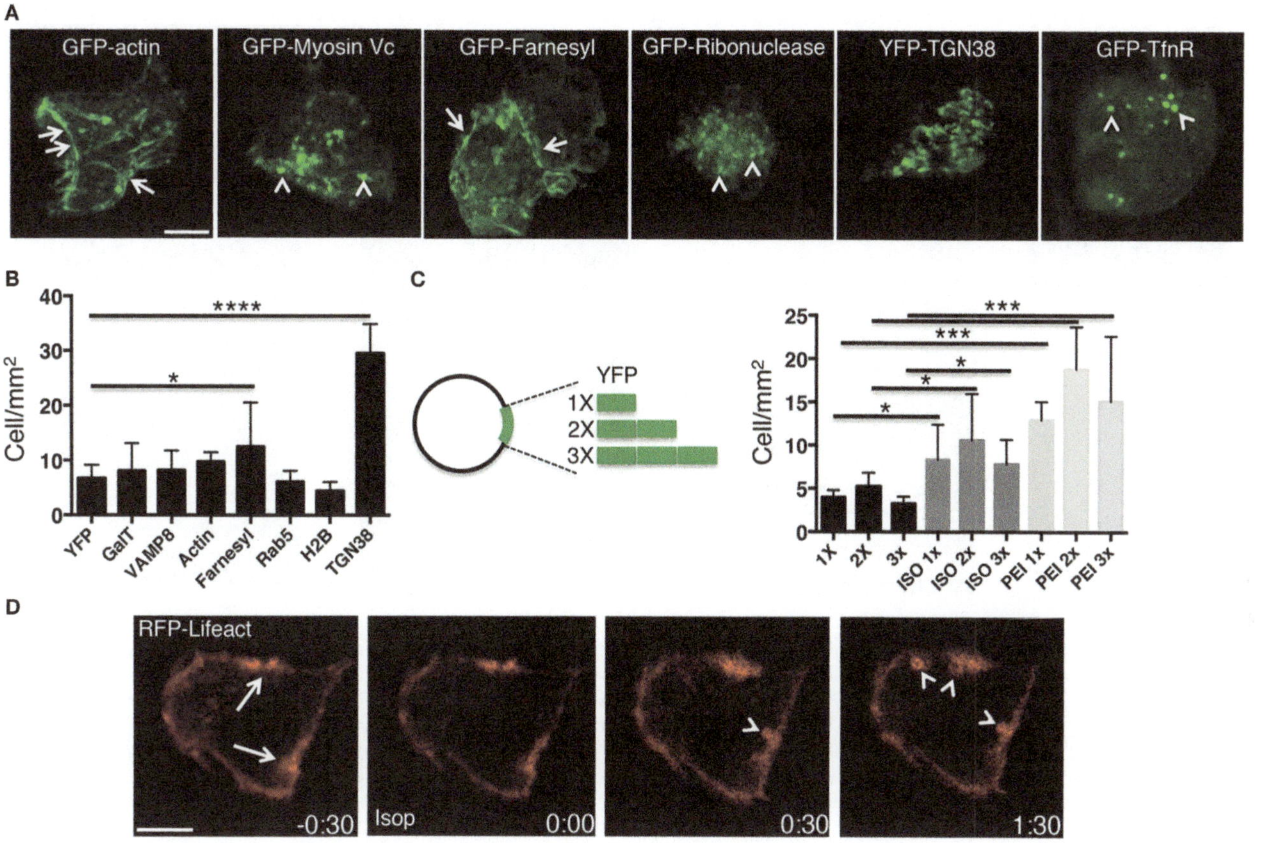

FIGURE 2 | PEI enables the expression of several transgenes in the acinar cells of the SGs. (A–C) Plasmid DNAs encoding either for selected fluorescently tagged proteins **(A,B)** or for multiple copies of pVenus **(C)** were injected into the SGs of anesthetized rats under different conditions. **(A)** High magnification images acquired by two-photon microscopy (excitation 930 nm, 60x water-immersion objective) to reveal the intracellular localization of selected transfected transgenes. **(B–C)** The efficiency of expression of the transgene was determined as described in the Methods and expressed as cells/mm2. Values represent average ±S.D. ($N = 3$). Statistical significance was calculated by using a *t*-test $^{*}p < 0.1$ $^{***}p < 0.001$ $^{****}P < 0.0001$. **(B)** Plasmids encoding for various cellular proteins and mixed with PEI are shown. **(C)** Plasmids containing 1 (1X), 2 (2X) or 3 (3X) copies of pVenus fused together where injected into the salivary duct alone, after stimulation with ISO, or mixed with PEI. **(D)** Plasmid DNA encoding for the F-actin reporter RFP-lifeact was mixed with PEI and injected into the salivary duct of anesthetized rats. After 24 h the SGs were exposed and imaged by confocal microscopy (excitation 561 nm). In acinar cells, RFP-life act was loclaized at the plasma membrane and in particular at the APM (arrows). SC injection of isoproterenol (time 0:00) elicited regulated exocytosis. As previously shown (Masedunskas et al., 2011a), secretory granules fused with the APM and recruited a F-actin coat (arrowheads). Scale bar, 10 μm.

In summary PEI-mediated *in vivo* transfection is a powerful tool to rapidly and reliably express transgenes in rat salivary glands *in vivo*. Although this method is suitable to perform single cell imaging and immunocytochemistry, its main limitation is the low yield of transfection (1–2% of the cells of the parenchyma). Therefore, PEI is not suitable to carry out biochemical analysis or for gene therapy. Nonetheless, this approach provides the opportunity to perform localization and functional studies under physiological conditions using commercially available plasmids and in a relatively short time.

MATERIALS AND METHODS

PLASMIDS AND CELL LINES

Human AQP5 and rat AQP5 were obtained from Dr. Bruce Baum (NIDCR, NIH) and subcloned in the following vectors pEGFP-C1, pEGFP-N1, pEYFP-C1, pEYFP-N1 (Clontech). TGN38-mCherry and YFP-TGN38 were a generous gift from Dr. Sarah Hamm-Alvarez (UCSC). GFP-actin, GFP-myosin Vc and GFP-TfnR were obtained from Dr. Julie Donaldson (NHLBI, NIH). GFP-Ribonuclease was a generous gift from Dr. David Yule (Rochester University). The 2X-EYFP vector was generated by inserting a copy of the YFP into a pEYFP-C and used to generate the 3X-EYFP using the same procedure. HSG-cell lines stably expressing YFP-AQP5 were prepared by lentiviral expression, as previously reported (Amornphimoltham et al., 2013).

CELL CULTURES

HSG cells were cultured in Dulbecco's modified Eagle's medium/Ham's F-12 (1:1) (Invitrogen, CA) in the presence of 5% fetal bovine serum, 100 U/mL penicillin, and 100 mg/mL streptomycin (Sigma-Aldrich, MO) at 37°C in a 5% CO_2 humidified atmosphere.

Growth factor-reduced matrigel (Trevigen, MD) was diluted in complete medium at the appropriate concentration

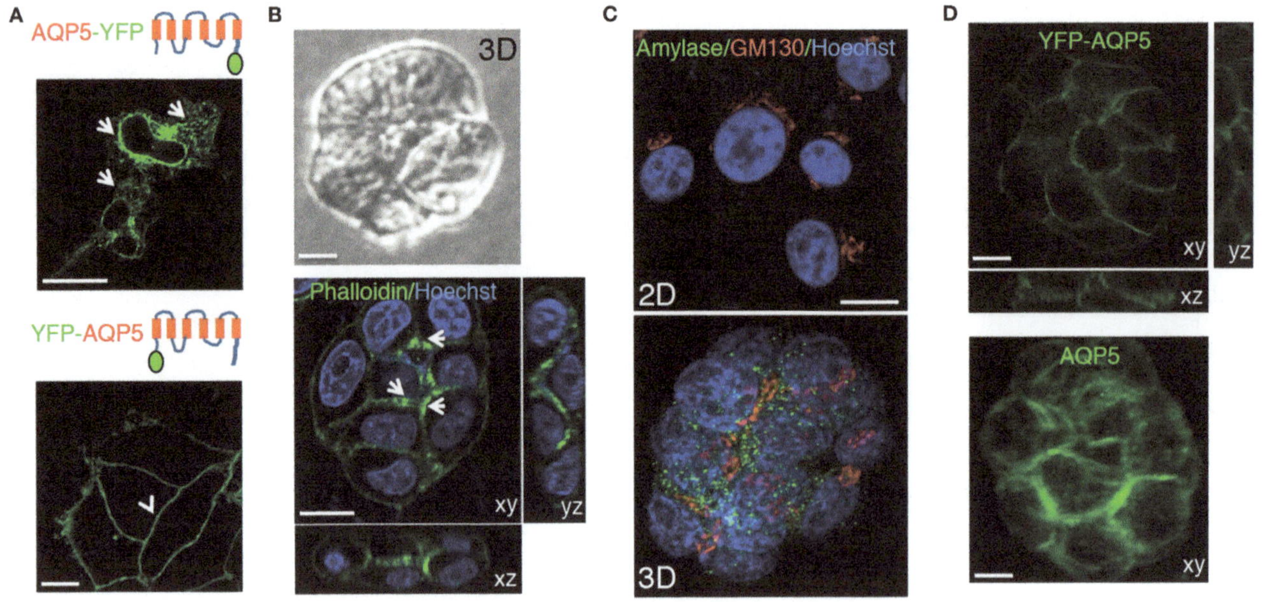

FIGURE 3 | Characterization of the fluorescently tagged AQP5. (A) Human AQP5 was tagged either at the **(C)** or the N-terminus and transiently expressed in HSG cells grown on glass slides. AQP5-YFP is localized in the ER and the nuclear envelope (arrows), whereas YFP-AQP5 is localized at the APM (arrowheads). **(B–D)** HSG cells (**B**, **C**, and **D** lower panel) or HSG cells stably expressing YFP-AQP5 (**D**, upper panel) were grown for 96 h in matrigel to form acinar-like structures or on glass coverslips (**C**, upper panel) and imaged by either phase contrast (**B**, upper panel) or confocal microscopy (**B**, **C**, lower panel and **C**) in order to acquire Z-stacks. Cells were fixed and processed to reveal F-actin and the nuclei (**B**, lower panel), YFP-AQP5 (**C**, upper panel), endogenous AQP5 (**C**, lower panel), amylase **(C)** and the Golgi apparatus (GM130, **C**). Maximal projections of the xy view and side views (xz, yz) show that F-actin is partially polarized in the acinar-like cells (arrows) whereas both endogenous and ectopically expressed AQP5 did not. Scale bars, 5 μm.

(2.0–6.0 mg/mL), added to a 12-well tissue culture dish, and incubated for 1 h at 37°C before culturing of the cells. 50 cells/mm^2 were added onto the pre-solidified matrigel and grown for different times. Acinar cells were recovered from the matrigel by incubating in cell recovery solution (Trevigen, MD) for 30 min on ice. Acinar cells released from the matrigel were spread on glass coverslips and dehydrated at 37°C for 15 min. Cells were then fixed in 2% formaldehyde/PBS, incubated first in blocking solution (10% FBS, 0.02% sodium azide in PBS) for 15 min, and later with various primary antibodies for 1 h at RT. Cells were then washed and incubated with the appropriate secondary antibodies for 1 h at RT. After 3 washes with blocking solution, nuclei were labeled by incubating the cells with Hoechst 33342 before mounting.

ANIMAL PROCEDURES

All the experiments were approved by the National Institute of Dental and Craniofacial Research (NIDCR, National Institute of Health, Bethesda, MD, USA) Animal Care and Use Committee. Sprague–Dawley male rats weighing 150–250 g were obtained from Harlan Laboratories Inc. (Frederick, MD) whereas the mice expressing cytoplasmic GFP were from Jackson laboratory (Masedunskas et al., 2011a). The animals were acclimated for 1 week before used for the procedures. Water and food were provided *ad libitum*. The animals were anesthetized by an IM injection of a mixture of Ketamine and Xylazine (100 mg/Kg an 20 mg/Kg respectively) with additional injections as needed.

PEI-PLASMID DNA COMPLEX

Polyethyleneimine (PEI) was purchased from Polyplus Transfection (New York, NY) as *in vivo*-jet PEI. Efficient transfection is achieved by using 12–24 μg of plasmid DNA/gland. Mix 50 μl of 10% glucose with the plasmid DNA and adjust the volume to 100 μl (solution 1). Mix 50 μl of 10% glucose with 7.5 μl of Jet PEI and adjust the volume to 100 μl (solution 2). Mix solution 1 and solution 2 and incubate them for 30 min at room temperature. Aspirate the transfection mixture with a syringe (30-gage needle) and make sure that no air bubbles are released when injecting the fluid.

IN VIVO TRANSFECTION

Anesthetized rats were secured into a previously described stereotactic device (Masedunskas et al., 2013) with the mandibles wide open and the cheeks extended to the sides. The tongue was folded toward the back of the mouth to expose the ductal orifices without obstructing the airways. The stereotactic device was at about 45 degrees and positioned under a stereo-microscope (SZX7, Olympus America, Center Valley, PA) to visualize the area below the tongue and locate the two orifices of the Wharton's ducts. A 30–40 cm PE-5 cannula (Strategic Applications, Libertyville, IL) was gently pushed into the orifice and introduced into the duct using bent sharp tweezers. To seal the cannula a small drop of histoacryl tissue glue was applied to the orifice. The cannula was connected to the syringe containing the transfection mixture that was injected e gradually over a 5 min time period by applying gentle pressure on the plunger. The cannula was removed from the

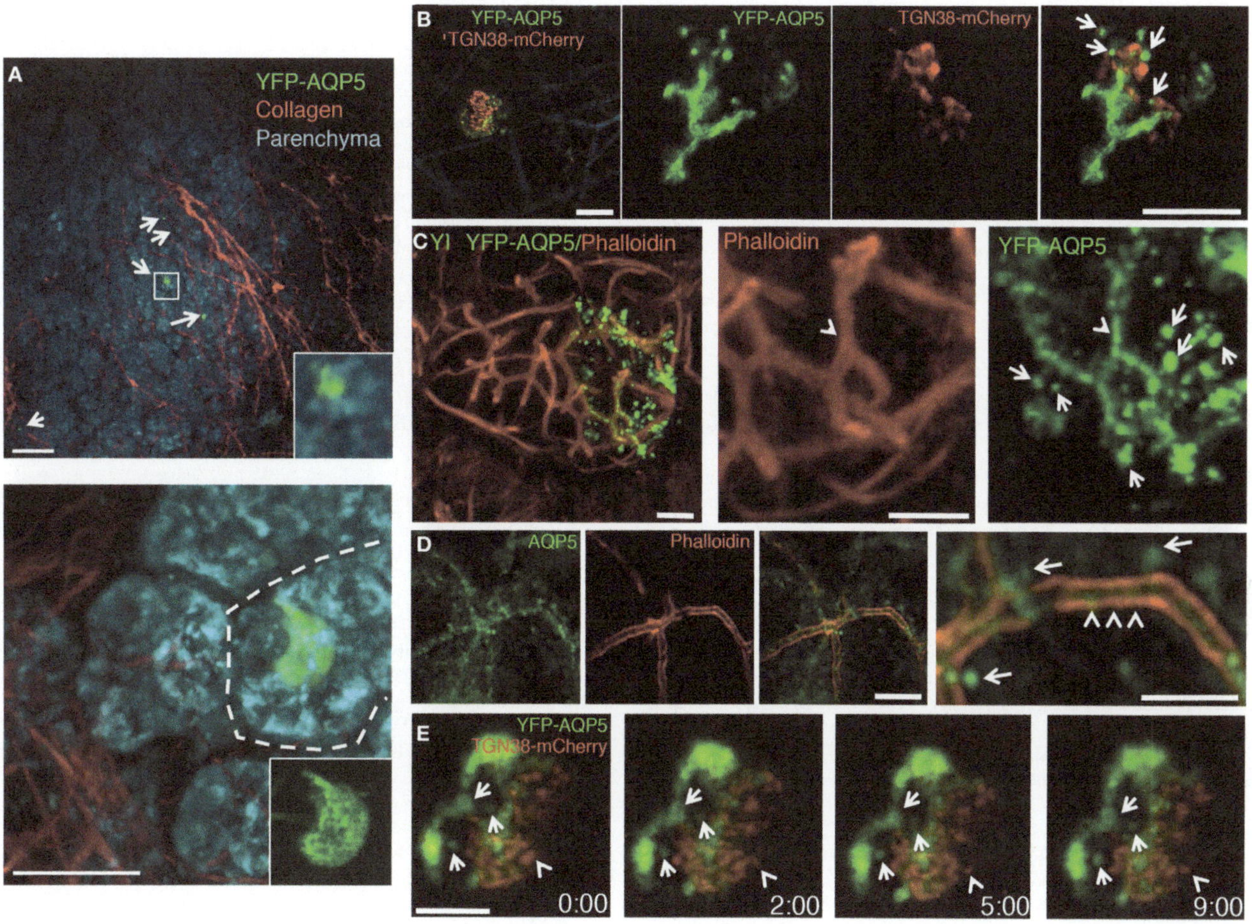

FIGURE 4 | Targeting of ectopically expressed AQP5 in the SG *in vivo*. The SGs of anesthetized rats were left untreated **(D)** or injected with a plasmid encoding for rat YFP-AQP5 either alone **(A,B)** or in combination with TGN38-mcherry, a marker for the Trans-Golgi Network **(C,E)**. The salivary glands were exposed and imaged by two-photon microscopy **(A,B,E)** or fixed, processed for immunofluorescence and imaged by confocal microscopy **(C,D)**. **(A)** Two Z-stacks from the same area of the sample were acquired twice by using two different excitation wavelengths, respectively: 740 nm to reveal the epithelium of the gland (cyan) and 930 nm to reveal the YFP (green) and collagen fibers (red). The Z-stacks were combined as described in the Methods and shown as maximal projections. Upper panel shows a low magnification of the parenchyma (20X objective) with 5 cells expressing YFP-AQP5 (arrows). Lower panel show an acinus (broken line) with a transfected cell (60X objective). Scale bars, 40 μm. **(B)** Maximal projection of an individual acinar cells (Left panel) and high magnification of a single optical slice (60X objective) showing that YFP-AQP5 (green) is localized a at the APM and in small vesicles (arrows). Scale bars, 5 μm. **(C,D)** Cryosections from SGs were labeled with Rhodamine-Phalloidin (red) alone **(C)**, or with an antibody directed against rat AQP5 (**D**, green). Both ectopically expressed and endogenous AQP5 are localized at the APM (arrowheads) and in small vesicles (arrows). Scale bars, 5 μm. **(E)** Time lapse imaging of an individual acinar cells show that YFP-AQP5-containing vesicles are stationary over a long period of time. Note a TGN38-containng vesicle that pinches off the TGN (arrowhead). Scale bar 3 μm.

mouth and the animal was allowed to recover. A warm environment was provided to facilitate the recovery and the animal was monitored for at least 2 h. The optimal yield of transfection was after 16–24 from the injection.

TWO-PHOTON AND CONFOCAL MICROSCOPY

Two-photon and confocal microscopy were performed on an IX81 inverted confocal microscope (Olympus, Melville, NY) modified to perform two-photon microscopy, as described previously (Masedunskas and Weigert, 2008). Intact and excised glands, and cultured cells, were imaged in the inverted setting by using the following objectives: UPLSAPO x10 NA 0.4, XLUMPFL20XW, 20X N.A. 0.95, and UPLSAPO 60X NA 1.2 (Olympus America Inc., Center Valley, PA). Intravital time-lapse imaging (**Figure 4E**) was performed as described in Masedunskas et al. (2011a) (acquisition speed 1 frame/sec).

IMMUNOFLUORESCENCE

The excised SGs were placed in optimum cutting temperature compound (Sakura Finetek USA Inc., Torrance, CA) and snap frozen in 2-methylbutane and liquid nitrogen. The glands were cut on silanated glass slides (10 μm thickness) and cryosections were incubated with blocking solution for 45 min followed by incubation with the appropriate primary antibody or with Texas-Red-Phalloidin (Invitrogen). Antibody against aquaporin 5 was purchased from Alomone labs.

ACKNOWLEDGMENTS

This research was supported by the Intramural Research Program of the NIH, National Institute of Dental and Craniofacial Research. We thank Dr. Natalie Porat-Shliom for the critical reading of the manuscript.

REFERENCES

Adriaansen, J., Zheng, C., Perez, P., and Baum, B. J. (2010). Production and sorting of transgenic, modified human parathyroid hormone *in vivo* in rat salivary glands. *Biochem. Biophys. Res. Commun.* 391, 768–772. doi: 10.1016/j.bbrc.2009.11.135

Amornphimoltham, P., Rechache, K., Thompson, J., Masedunskas, A., Leelahavanichkul, K., Patel, V., et al. (2013). Rab25 regulates invasion and metastasis in head and neck cancer. *Clin. Cancer Res.* 19, 1375–1388. doi: 10.1158/1078-0432.CCR-12-2858

Boulaiz, H., Marchal, J. A., Prados, J., Melguizo, C., and Aranega, A. (2005). Non-viral and viral vectors for gene therapy. *Cell. Mol. Biol. (Noisy-le-grand).* 51, 3–22.

Creusat, G., Rinaldi, A. S., Weiss, E., Elbaghdadi, R., Remy, J. S., Mulherkar, R., et al. (2010). Proton sponge trick for pH-sensitive disassembly of polyethylenimine-based siRNA delivery systems. *Bioconjug. Chem.* 21, 994–1002. doi: 10.1021/bc100010k

Delporte, C., and Steinfeld, S. (2006). Distribution and roles of aquaporins in salivary glands. *Biochim. Biophys. Acta* 1758, 1061–1070. doi: 10.1016/j.bbamem.2006.01.022

Dunn, K. W., Sandoval, R. M., Kelly, K. J., Dagher, P. C., Tanner, G. A., Atkinson, S. J., et al. (2002). Functional studies of the kidney of living animals using multicolor two-photon microscopy. *Am. J. Physiol. Cell Physiol.* 283, C905–C916. doi: 10.1152/ajpcell.00159.2002

Fushimi, K., Sasaki, S., and Marumo, F. (1997). Phosphorylation of serine 256 is required for cAMP-dependent regulatory exocytosis of the aquaporin-2 water channel. *J. Biol. Chem.* 272, 14800–14804. doi: 10.1074/jbc.272.23.14800

Goldfine, I. D., German, M. S., Tseng, H. C., Wang, J., Bolaffi, J. L., Chen, J. W., et al. (1997). The endocrine secretion of human insulin and growth hormone by exocrine glands of the gastrointestinal tract. *Nat. Biotechnol.* 15, 1378–1382. doi: 10.1038/nbt1297-1378

Gresz, V., Kwon, T. H., Gong, H., Agre, P., Steward, M. C., King, L. S., et al. (2004). Immunolocalization of AQP-5 in rat parotid and submandibular salivary glands after stimulation or inhibition of secretion *in vivo*. *Am. J. Physiol. Gastrointest. Liver Physiol.* 287, G151–G161. doi: 10.1152/ajpgi.00480.2003

Hoffman, M. P., Kibbey, M. C., Letterio, J. J., and Kleinman, H. K. (1996). Role of laminin-1 and TGF-beta 3 in acinar differentiation of a human submandibular gland cell line (HSG). *J. Cell Sci.* 109(Pt 8), 2013–2021.

Hoffman, M. P., Nomizu, M., Roque, E., Lee, S., Jung, D. W., Yamada, Y., et al. (1998). Laminin-1 and laminin-2 G-domain synthetic peptides bind syndecan-1 and are involved in acinar formation of a human submandibular gland cell line. *J. Biol. Chem.* 273, 28633–28641. doi: 10.1074/jbc.273.44.28633

Horsefield, R., Norden, K., Fellert, M., Backmark, A., Tornroth-Horsefield, S., Terwisscha van Scheltinga, A. C., et al. (2008). High-resolution x-ray structure of human aquaporin 5. *Proc. Natl. Acad. Sci. U.S.A.* 105, 13327–13332. doi: 10.1073/pnas.0801466105

Karabasil, M. R., Hasegawa, T., Azlina, A., Purwanti, N., Purevjav, J., Yao, C., et al. (2009). Trafficking of GFP-AQP5 chimeric proteins conferred with unphosphorylated amino acids at their PKA-target motif ((152)SRRTS) in MDCK-II cells. *J. Med. Invest.* 56, 55–63. doi: 10.2152/jmi.56.55

Kawakami, S., Higuchi, Y., and Hashida, M. (2008). Nonviral approaches for targeted delivery of plasmid DNA and oligonucleotide. *J. Pharm. Sci.* 97, 726–745. doi: 10.1002/jps.21024

Kircheis, R., Wightman, L., and Wagner, E. (2001). Design and gene delivery activity of modified polyethylenimines. *Adv. Drug Deliv. Rev.* 53, 341–358. doi: 10.1016/S0169-409X(01)00202-2

Masedunskas, A., Porat-Shliom, N., Rechache, K., Aye, M. P., and Weigert, R. (2012). Intravital microscopy reveals differences in the kinetics of endocytic pathways between cell cultures and live animals. *Cells* 1, 1121–1132. doi: 10.3390/cells1041121

Masedunskas, A., Sramkova, M., Parente, L., Sales, K. U., Amornphimoltham, P., Bugge, T. H., et al. (2011a). Role for the actomyosin complex in regulated exocytosis revealed by intravital microscopy. *Proc. Natl. Acad. Sci. U.S.A.* 108, 13552–13557. doi: 10.1073/pnas.1016778108

Masedunskas, A., Sramkova, M., Parente, L., and Weigert, R. (2013). Intravital microscopy to image membrane trafficking in live rats. *Methods Mol. Biol.* 931, 153–167. doi: 10.1007/978-1-62703-056-4_9

Masedunskas, A., Sramkova, M., and Weigert, R. (2011b). Homeostasis of the apical plasma membrane during regulated exocytosis in the salivary glands of live rodents. *Bioarchitecture* 1, 225–229. doi: 10.4161/bioa.18405

Masedunskas, A., and Weigert, R. (2008). Intravital two-photon microscopy for studying the uptake and trafficking of fluorescently conjugated molecules in live rodents. *Traffic* 9, 1801–1810. doi: 10.1111/j.1600-0854.2008.00798.x

Matsuzaki, T., Ablimit, A., Suzuki, T., Aoki, T., Hagiwara, H., and Takata, K. (2006). Changes of aquaporin 5-distribution during release and reaccumulation of secretory granules in isoproterenol-treated mouse parotid gland. *J. Electron Microsc. (Tokyo)* 55, 183–189. doi: 10.1093/jmicro/dfl023

Milberg, O., Porat-Shliom, N., Tora, M., Parente, L., Masedunskas, A., and Weigert, R. (2013). "Intravital microscopy and its application to study regulated exocytosis in the exocrine glands of live rodents," in *Methods in the Study of Exocytosis*, ed P. Thorn (Springer).

Nagai, T., Ibata, K., Park, E. S., Kubota, M., Mikoshiba, K., and Miyawaki, A. (2002). A variant of yellow fluorescent protein with fast and efficient maturation for cell-biological applications. *Nat. Biotechnol.* 20, 87–90. doi: 10.1038/nbt0102-87

Nejsum, L. N., and Nelson, W. J. (2007). A molecular mechanism directly linking E-cadherin adhesion to initiation of epithelial cell surface polarity. *J. Cell Biol.* 178, 323–335. doi: 10.1083/jcb.200705094

Neu, M., Fischer, D., and Kissel, T. (2005). Recent advances in rational gene transfer vector design based on poly(ethylene imine) and its derivatives. *J. Gene Med.* 7, 992–1009. doi: 10.1002/jgm.773

Nimesh, S. (2012). Polyethylenimine as a promising vector for targeted siRNA delivery. *Curr. Clin. Pharmacol.* 7, 121–130. doi: 10.2174/157488412800228857

Ohno, H., Tomemori, T., Nakatsu, F., Okazaki, Y., Aguilar, R. C., Foelsch, H., et al. (1999). Mu1B, a novel adaptor medium chain expressed in polarized epithelial cells. *FEBS Lett.* 449, 215–220. doi: 10.1016/S0014-5793(99)00432-9

Pittet, M. J., and Weissleder, R. (2011). Intravital imaging. *Cell* 147, 983–991. doi: 10.1016/j.cell.2011.11.004

Porat-Shliom, N., Chen, Y., Tora, M., Shitara, A., Masedunskas, A., and Weigert, R. (2014). *In vivo* tissue-wide synchronization of mitochondrial metabolic oscillations. *Cell Rep.* doi: 10.1016/j.celrep.2014.09.022

Raina, S., Preston, G. M., Guggino, W. B., and Agre, P. (1995). Molecular cloning and characterization of an aquaporin cDNA from salivary, lacrimal, and respiratory tissues. *J. Biol. Chem.* 270, 1908–1912. doi: 10.1074/jbc.270.4.1908

Ritsma, L., Ponsioen, B., and van Rheenen, J. (2012). Intravital imaging of cell signaling in mice. *Intravital* 1, 2–10. doi: 10.4161/intv.20802

Royce, L. S., Kibbey, M. C., Mertz, P., Kleinman, H. K., and Baum, B. J. (1993). Human neoplastic submandibular intercalated duct cells express an acinar phenotype when cultured on a basement membrane matrix. *Differentiation* 52, 247–255. doi: 10.1111/j.1432-0436.1993.tb00637.x

Sandoval, R. M., Kennedy, M. D., Low, P. S., and Molitoris, B. A. (2004). Uptake and trafficking of fluorescent conjugates of folic acid in intact kidney determined using intravital two-photon microscopy. *Am. J. Physiol. Cell Physiol.* 287, C517–C526. doi: 10.1152/ajpcell.00006.2004

Snyder, R. O., and Francis, J. (2005). Adeno-associated viral vectors for clinical gene transfer studies. *Curr. Gene Ther.* 5, 311–321. doi: 10.2174/1566523054065066

Sramkova, M., Masedunskas, A., Parente, L., Molinolo, A., and Weigert, R. (2009). Expression of plasmid DNA in the salivary gland epithelium: novel approaches to study dynamic cellular processes in live animals. *Am. J. Physiol. Cell Physiol.* 297, C1347–C1357. doi: 10.1152/ajpcell.00262.2009

Sramkova, M., Masedunskas, A., and Weigert, R. (2012). Plasmid DNA is internalized from the apical plasma membrane of the salivary gland epithelium in live animals. *Histochem. Cell Biol.* 138, 201–213. doi: 10.1007/s00418-012-0959-7

Sun, J. Y., Anand-Jawa, V., Chatterjee, S., and Wong, K. K. (2003). Immune responses to adeno-associated virus and its recombinant vectors. *Gene Ther.* 10, 964–976. doi: 10.1038/sj.gt.3302039

Sung, Y. H., Baek, I. J., Seong, J. K., Kim, J. S., and Lee, H. W. (2012). Mouse genetics: catalogue and scissors. *BMB Rep.* 45, 686–692. doi: 10.5483/BMBRep.2012.45.12.242

Tada, J., Sawa, T., Yamanaka, N., Shono, M., Akamatsu, T., Tsumura, K., et al. (1999). Involvement of vesicle-cytoskeleton interaction in AQP5 trafficking in AQP5-gene-transfected HSG cells. *Biochem. Biophys. Res. Commun.* 266, 443–447. doi: 10.1006/bbrc.1999.1828

Wang, W., Li, W., Ma, N., and Steinhoff, G. (2013). Non-viral gene delivery methods. *Curr. Pharm. Biotechnol.* 14, 46–60. doi: 10.2174/138920113804805278

Weigert, R. (2014). Imaging the dynamics of endocytosis in live mammalian tissues. *Cold Spring Harb. Perspect. Biol.* 6:a017012. doi: 10.1101/cshperspect.a017012

Weigert, R., Porat-Shliom, N., and Amornphimoltham, P. (2013). Imaging cell biology in live animals: ready for prime time. *J. Cell Biol.* 201, 969–979. doi: 10.1083/jcb.201212130

Wellner, R. B., Hong, S., Cotrim, A. P., Swaim, W. D., and Baum, B. J. (2005). Modifying the NH2 and COOH termini of aquaporin-5: effects on localization in polarized epithelial cells. *Tissue Eng.* 11, 1449–1458. doi: 10.1089/ten.2005.11.1449

Wilson, A. A., Kwok, L. W., Porter, E. L., Payne, J. G., McElroy, G. S., Ohle, S. J., et al. (2013). Lentiviral delivery of RNAi for *in vivo* lineage-specific modulation of gene expression in mouse lung macrophages. *Mol. Ther.* 21, 825–833. doi: 10.1038/mt.2013.19

Wolff, J. A., Dowty, M. E., Jiao, S., Repetto, G., Berg, R. K., Ludtke, J. J., et al. (1992). Expression of naked plasmids by cultured myotubes and entry of plasmids into T tubules and caveolae of mammalian skeletal muscle. *J Cell Sci.* 103(Pt 4), 1249–1259.

Yang, H., Wang, H., Shivalila, C. S., Cheng, A. W., Shi, L., and Jaenisch, R. (2013). One-step generation of mice carrying reporter and conditional alleles by CRISPR/Cas-mediated genome engineering. *Cell* 154, 1370–1379. doi: 10.1016/j.cell.2013.08.022

Conflict of Interest Statement: The authors declare that the research was conducted in the absence of any commercial or financial relationships that could be construed as a potential conflict of interest.

14

The neural crest lineage as a driver of disease heterogeneity in Tuberous Sclerosis Complex and Lymphangioleiomyomatosis

Sean P. Delaney[1,2,3†], Lisa M. Julian[1,2†] and William L. Stanford[1,2,3,4]*

[1] Sprott Centre for Stem Cell Research, Regenerative Medicine Program, Ottawa Hospital Research Institute, Ottawa, ON, Canada
[2] Faculty of Graduate and Postdoctoral Studies, University of Ottawa, Ottawa, ON, Canada
[3] Department of Cellular and Molecular Medicine, University of Ottawa, Ottawa, ON, Canada
[4] Department of Biochemistry, Microbiology, and Immunology, University of Ottawa, Ottawa, ON, Canada

Edited by:
Thimios Mitsiadis, University of Zurich, Switzerland

Reviewed by:
Eumorphia Remboutsika, Biomedical Research Centre "Alexander Fleming", Greece
Amancio Carnero, Instituto de Biomedicina de Sevilla, Spain
Takahiko Hara, Tokyo Metropolitan Institute of Medical Science, Japan
Shuibin Lin, Harvard Medical School, USA

***Correspondence:**
William L. Stanford, Sprott Centre for Stem Cell Research, Ottawa Hospital Research Institute, 501 Smyth Road, Box 511, Ottawa, ON K1H 8L6, Canada
e-mail: wstanford@ohri.ca

† These authors have contributed equally to this work.

Lymphangioleiomyomatosis (LAM) is a rare neoplastic disease, best characterized by the formation of proliferative nodules that express smooth muscle and melanocytic antigens within the lung parenchyma, leading to progressive destruction of lung tissue and function. The pathological basis of LAM is associated with Tuberous Sclerosis Complex (TSC), a multi-system disorder marked by low-grade tumors in the brain, kidneys, heart, eyes, lung and skin, arising from inherited or spontaneous germ-line mutations in either of the *TSC1* or *TSC2* genes. LAM can develop either in a patient with TSC (TSC-LAM) or spontaneously (S-LAM), and it is clear that the majority of LAM lesions of both forms are characterized by an inactivating mutation in either *TSC1* or *TSC2*, as in TSC. Despite this genetic commonality, there is considerable heterogeneity in the tumor spectrum of TSC and LAM patients, the basis for which is currently unknown. There is extensive clinical evidence to suggest that the cell of origin for LAM, as well as many of the TSC-associated tumors, is a neural crest cell, a highly migratory cell type with extensive multi-lineage potential. Here we explore the hypothesis that the types of tumors that develop and the tissues that are affected in TSC and LAM are dictated by the developmental timing of *TSC* gene mutations, which determines the identities of the affected cell types and the size of downstream populations that acquire a mutation. We further discuss the evidence to support a neural crest origin for LAM and TSC tumors, and propose approaches for generating humanized models of TSC and LAM that will allow cell of origin theories to be experimentally tested. Identifying the cell of origin and developing appropriate humanized models is necessary to truly understand LAM and TSC pathology and to establish effective and long-lasting therapeutic approaches for these patients.

Keywords: Lymphangioleiomyomatosis, Tuberous Sclerosis, neural crest, cell of origin, disease modeling

INTRODUCTION

CLINICAL FEATURES IN TSC AND LAM REVEAL A HETEROGENEOUS DISEASE SPECTRUM

Tuberous sclerosis complex (TSC) is a multisystem disorder that arises as a consequence of inherited or spontaneously acquired mutations in either the *TSC1* or *TSC2* gene. Along with Neurofibromatosis type 1, TSC is among the most common neurocutaneous diseases, occurring in an estimated 1 in 6000 births (Kandt, 2003; Kristof and Moss, 2011). TSC affects both children and adults, often with clinical manifestations initiating during embryonic development, accompanied by the gradual presentation of additional symptoms throughout childhood and into adulthood. Clinical features of TSC (**Figure 1**) include the appearance of low-grade tumors and malformations in the brain, heart, lungs, kidneys, eyes, skin, and bone, and loss of heterozygosity or second-hit mutations of the wild-type *TSC1* or *TSC2* allele are thought to be responsible for the formation of most of these lesions (Henske et al., 1996; Carsillo et al., 2000; Crino et al., 2010; Qin et al., 2010a, 2011; Tyburczy et al., 2014). Cortical tubers are prevalent in TSC and account for the majority of the debilitative neurological symptoms, including epilepsy, mental retardation and autism (Webb et al., 1996; Goh et al., 2005; Crino et al., 2006; Wong, 2006; Crino and Tsai, 2012). Tubers, along with cardiac rhabdomyomas, are formed during embryogenesis and can be detected prenatally (Park et al., 1997), whereas other tumors and lesions form during childhood and into adulthood. These include retinal astrocytomas, subependymal giant cell astrocytomas (SEGAs), angiofibromas and hypomelanotic macules in the skin, sub-ependymal nodules/giant-cell tumors, and renal angiomyolipomas (AMLs).

TSC is also associated with pulmonary and lymphatic manifestations in the form of Lymphangioleiomyomatosis (LAM), a progressive neoplasm of the lung that occurs in at least 30% of women with TSC (TSC-LAM), with an average age of diagnosis

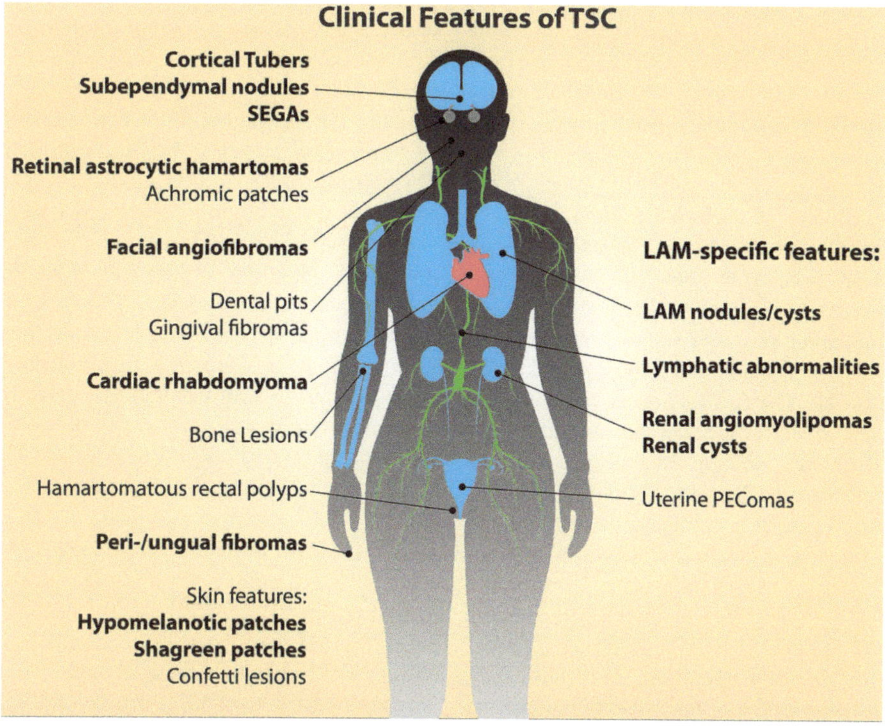

FIGURE 1 | The clinical manifestations of TSC and LAM are diverse and affect multiple organs and tissues. The major diagnostic features of TSC are indicated in bold type (Northrup et al., 2013). Cortical tubers and cardiac rhabdomyomas occur during fetal development. Facial angiofibromas, hypomelanic macules, and retinal astrocytic hamartomas can be detected in infancy, while the other features continue to present themselves throughout development into adulthood.

of about 35 years (Henske and McCormack, 2012). LAM is characterized by the presence of multiple neoplastic nodules within the lung interstitium, composed of proliferating smooth muscle-like cells and abnormally large epithelial cells that express melanocytic markers. The proliferative LAM nodules form cystic lesions within the lungs, which lead to the destruction of the parenchyma, resulting in progressive shortness of breath, chylous pleural effusions, pneumothorax, and eventual respiratory failure (Kitaichi et al., 1995; Chu et al., 1999; Urban et al., 1999).

LAM can also occur in a sporadic form (S-LAM), where lung lesions and associated renal and lymphatic manifestations are similar to those observed in TSC patients, but other TSC-associated tumors are absent (Costello et al., 2000; Moss et al., 2001). In addition to pulmonary manifestations, other common features of LAM include lymphatic abnormalities, such as lymphadenopathy (Chu et al., 1999; Urban et al., 1999), renal AMLs, and uterine PEComas (Perivascular Epitheloid Cell tumors). AMLs are benign growths composed of adipocytes and, much like their pulmonary counterparts, smooth muscle cells, and are asymptomatic in most cases (Bissler and Kingswood, 2004). Uterine PEComas are, curiously, also characterized by cells of an epithelial morphology that express smooth muscle and melanocytic markers, similar to the cells that comprise LAM lung nodules (Martignoni et al., 2007; Hayashi et al., 2011; Henske and McCormack, 2012). Thus, TSC is associated with the development of a broad spectrum of pathological lesions, affecting a wide diversity of tissues and cell types.

GENETIC FEATURES IN LAM AND TSC

TSC, TSC-LAM, and S-LAM patients typically possess inactivating mutations in *TSC1* or, much more commonly, *TSC2* (Smolarek et al., 1998; Carsillo et al., 2000; Sato et al., 2002). Loss of heterozygosity in either of these genes from second-hit mutations is thought to be important in driving tumor development (Green et al., 1994; Sepp et al., 1996; Niida et al., 2001; Han et al., 2004; Henske and McCormack, 2012). According to the Leiden Open Variation Database (Fokkema et al., 2011), as of 2014 there are over 680 and 1500 reported unique pathogenic DNA variants in *TSC1* and *TSC2*, respectively. The wide array of mutations spans all exons of both *TSC1* and *TSC2*, with no specificity for any particular coding region or mutation type, and includes a number of mutation types [such as genomic deletions, missense and non-sense mutations and protein truncations (Cheadle et al., 2000; Crino et al., 2006; Au et al., 2007)]. Although mutations in either of these genes lead to the development of disease features, extensive genotype-phenotype analysis has been performed and found that mutations in *TSC2* are associated with more severe phenotypes than are mutations in TSC1 (Niida et al., 1999; van Slegtenhorst et al., 1999; Au et al., 2007).

Intriguingly, the genetic features in S-LAM patients mirror those of TSC patients, but pathological manifestations in S-LAM are limited to those in the lung, kidneys, lymphatics, liver and uterus (Henske and McCormack, 2012), and *TSC* gene mutations in S-LAM patients are not wide-spread or found within the germline (Astrinidis et al., 2000). Currently, the conditions that underlie the heterogeneity in the disease spectrum between TSC and LAM patients are unknown. Clinical observations, and some experimental evidence, offer clues, however, that the cell of origin in concert with the developmental timing of *TSC1* or *TSC2* mutation are likely key factors in determining which disease features ultimately manifest.

MOLECULAR CHARACTERISTICS OF TSC AND LAM LESIONS

Aberrant mTOR signaling broadly characterizes TSC lesions

TSC and LAM lesions are believed to share the common feature of elevated mTOR pathway signaling, ultimately as a consequence of inactivating mutations in *TSC1* or *TSC2* (Green et al., 1994; Niida et al., 2001; Han et al., 2004; Henske and McCormack, 2012). The *TSC1* and *TSC2* genes encode Hamartin and Tuberin, respectively, which form heterodimers and inhibit the GTPase activity of the Ras homolog Rheb, a direct activator of the Mammalian Target of Rapamycin (mTOR). mTOR signals through two macromolecular complexes, mTORC1 and mTORC2, to integrate the external environment (growth factor stimulation, cellular stress, oxygen availability, energy and nutrient levels) with diverse cellular functions, including proliferation, survival, growth and size, death and autophagy, protein translation, angiogenesis, movement, and migration. *TSC1* and *TSC2* function as critical gatekeepers of this multi-functional cellular signaling network.

The therapeutic efficacy of Rapamycin, a molecular inhibitor of mTOR, in treating many of the manifestations in TSC and LAM (Franz et al., 2006; Bissler et al., 2008; Krueger et al., 2010; Micozkadioglu et al., 2010; Casanova et al., 2011; Dabora et al., 2011; DeKlotz et al., 2011; McCormack et al., 2011; Tiberio et al., 2011; Canpolat et al., 2014; Park et al., 2014) has established elevated mTOR signaling as a key feature driving these tumors. The cells that comprise LAM and TSC tumors reflect many features of hyper-active mTOR signaling: abnormally large or "giant" cells, altered morphology, aberrant migration, abnormal proliferation, increased cell survival and reduced autophagy are common features of many lesions (Crino, 2004; Crino and Tsai, 2012; Henske and McCormack, 2012). Increased angiogenesis is also a feature of many TSC lesions, including LAM (Nguyen-Vu et al., 2001; Arbiser et al., 2002; Kumasaka et al., 2004; Li et al., 2013).

Unique molecular characteristics distinguish TSC-associated lesions

Despite the common element of hyperactive mTOR signaling, the types of lesions that develop in TSC and LAM are diverse, in terms of both the tissues that are affected and the molecular characteristics of the aberrant cells. Furthermore, different types of TSC-associated tumors exhibit additional behavioral and signaling characteristics that are distinct from one another. For example, LAM cells are uniquely responsive to estrogen and prolactin signaling, and express these receptors on their surface (Terasaki et al., 2010; Gao et al., 2014)(reviewed in Yu and Henske, 2010). They also express markers of smooth muscle cells and melanocytes, cell types not normally found throughout the lung interstitium.

Alternatively, the aberrant cells that comprise cortical tubers and SEGAs in TSC patients express markers of early neuronal and glial cell differentiation (Lopes et al., 1996; Ess et al., 2005; Zhou et al., 2011) and recent mouse models have clarified that these tumors can result from mTOR-dependent promotion of aberrant neural progenitor cell (NPC) expansion and premature differentiation, accompanied by defective maturation and abnormal migration (Magri et al., 2011; Magri and Galli, 2013). Thus, in the TSC lesions confined to the CNS, mTOR hyper-activation leads to a direct reprogramming of the self-renewal and differentiation capacity of the cells endogenous to the tissue in which the lesions are found, while the etiology of mTOR hyper-activation to the appearance of LAM-associated lesions in the lung, kidneys, and uterus is much less clear. TSC and LAM are further differentiated by the time-point of disease presentation, where most TSC features typically present in infancy or during childhood and LAM presents in adulthood. These differences suggest that, despite their common genetic characteristic, distinct TSC and LAM lesions exhibit unique pathological mechanisms.

POTENTIAL CELLS OF ORIGIN FOR TSC AND LAM

The cell(s) of origin for TSC and LAM lesions are largely unknown, however clinical observations and limited experimental evidence have offered some key clues. The extensive multi-system involvement in these diseases is remarkable, and while there is evidence for both common and independent second-hit *TSC1* or *TSC2* mutations across multiple tumors within a given individual (Henske et al., 1996; Smolarek et al., 1998; Carsillo et al., 2000; Karbowniczek et al., 2003; Tyburczy et al., 2014), a single first-hit mutation appears to characterize all of a patients' tumors. The lesions that commonly arise in TSC are generally comprised of immature cells that exhibit molecular and behavioral features of either neuronal (central nervous system (CNS) lesions) or mesenchymal differentiation. While embryonic neural stem cells (NSCs) appear to function as a cell of origin for the CNS manifestations in TSC (discussed further below), the mesenchymal neural crest lineage has been postulated to broadly give rise to other types of lesions in TSC and LAM.

With the potential involvement of both NSCs and the neural crest lineage in disease heterogeneity, the influence of the developmental time-point at which *TSC1* or *TSC2* mutations occur on disease severity has also been called into question. However, these theories have not been experimentally confirmed, and no animal or humanized model currently exists to directly test these suppositions. Here, we explore the evidence supporting the cell of origin hypotheses that have emerged from the TSC and LAM field, and, based on these theories, we further propose approaches for generating relevant humanized models of TSC and LAM that will enable the identification of improved therapeutic avenues and open the door for patient-centered approaches to treatment.

EXPERIMENTAL EVIDENCE THAT NEURAL STEM CELLS DRIVE CNS PATHOLOGY IN TSC

Mouse models of TSC gene deficiency in the CNS

Experimental approaches to identify the cell of origin in TSC have focused to date almost exclusively on validating the hypothesis that the CNS tumors and behavioral phenotypes are driven by *TSC1* or *TSC2* mutations in the NSC lineage. Since the CNS manifestations cause significant morbidity and are apparent at such a young age, this focus is not surprising. As mouse models of both TSC1- and TSC2-deficiency using gene targeting lead to early embryonic lethality, and mice heterozygous for either *TSC* gene do not develop TSC-related CNS pathologies (likely due to the absence of a second-hit mutation during embryonic development) (Kobayashi et al., 1999, 2001; Onda et al., 1999; Kwiatkowski et al., 2002), more sophisticated models have recently emerged to assess the effects of homozygous TSC gene mutation in select populations of NPCs and differentiating or differentiated neuronal cells (Uhlmann et al., 2002; Meikle et al., 2007; Ehninger et al., 2008; Way et al., 2009; Goto et al., 2011; Magri et al., 2011; Zeng et al., 2011; Zhou et al., 2011; Carson et al., 2012; Feliciano et al., 2012, 2011; Magri and Galli, 2013; Normand et al., 2013; Prabhakar et al., 2013). These models have included a variety of drug inducible and non-inducible conditional knockouts, as well as adenoviral-mediated approaches or *in utero* electroporation to induce gene knock-out stochastically. These approaches have allowed for targeted induction of TSC1- or TSC2-deficiency in distinct neural cell populations, limited by regional location, the size of the affected cell population, or the developmental time-point of induction.

Arising from these efforts, mouse models of TSC1- or TSC2-deficiency now exist that closely model the CNS manifestations of TSC that occur in human patients, including neuronal network dysfunction and seizure activity, sub-ependymal nodules that lead to the development of SEGAs and elements of cortical tubers, accompanied by abnormal neuronal migration, enhanced astrogliosis, enlarged cells and cortical lamination defects (Goto et al., 2011; Magri et al., 2011; Carson et al., 2012; Magri and Galli, 2013; Normand et al., 2013; Prabhakar et al., 2013). Importantly, hyper-activation of mTORC1 signaling has indeed been confirmed as causative in the majority of TSC-like phenotypes in these models, confirmed by extensive phenotypic rescue following post-natal Rapamycin treatment in many of these studies. Interestingly, these models have revealed that while abnormal, enlarged cells and nodule-like formations can be induced to some degree by *TSC* gene deletion in virtually all neuronal populations (progenitors and post-mitotic cells) and at different developmental stages (throughout embryogenesis and postnatally), neurological features such as seizure activity are only clearly reproduced by targeting NPCs and neurons prenatally, during peak stages of neurogenesis. Furthermore, the overall severity of CNS manifestations following *TSC* gene deletion at different stages occurs on a gradient scale, with the most severe and extensive phenotypes associated with mutations in NPC populations at the earliest stages of neurogenesis (in neuroepithelial cells and radial glia at E9.5 and E12.5, respectively). As appearance of the full spectrum of TSC-associated CNS lesions seems to require *TSC* gene mutation in early NPCs, these studies together provide solid evidence to support the hypothesis that a neural stem or progenitor cell is the cell of origin for the CNS manifestations in TSC.

The proportion of TSC1/ TSC2-deficient cells affects phenotypic severity in mouse models

Not only have the CNS mouse models revealed that *TSC* gene deficiency at different stages of neural development can lead to different types of manifestations, but also that the degree of mosaicism of TSC1- or TSC2-deficient cells correlates with phenotypic severity. The first key element in generating mouse models that closely recapitulate the TSC condition was induction of *TSC* gene deficiency in specified neuronal sub-populations, as opposed to homozygous mutations in ESCs. This demonstrated that while wide-spread homozygous TSC1- or TSC2-deficiency within the germline results in overly severe, lethal phenotypes, restricted *TSC* gene mutations within the neuronal lineage can be better tolerated but lead to TSC-like phenotypes in the brain. Again, mutation at late vs. early developmental stages, in which fewer downstream cells would be directly affected, leads to less severe phenotypes.

Some recent mouse models have directly demonstrated experimentally, within distinct cell populations, that the degree of mosaicism for *TSC* gene deficiency within neural cells also leads to a spectrum of phenotypes. Specifically, mutation of *TSC1* in a subset of thalamic progenitor cells caused substantially greater neurological dysfunction, indicative of TSC-like pathology, if the mutation was made at mid-gestation (E12.5) vs. late gestation (E18.5) in an embryonic mouse model (Normand et al., 2013), and experiments in which subsets of NPCs were knocked-out for *TSC1* in variable numbers using adenoviral-mediated Cre injection into the cerebral ventricles showed that the severity of resulting TSC-like pathology was dependent on the dose and serotype of the injected virus (Prabhakar et al., 2013). These models have provided strong evidence that the CNS lesions and neuropathologies associated with TSC can be driven by TSC1- or TSC2-deficiency and consequent mTOR deregulation in NPCs, and together demonstrate that the severity of the TSC-associated CNS phenotypes exist on a spectrum that is strongly influenced by the developmental timing and degree of mosaicism for *TSC1* or *TSC2*.

HETEROGENEITY IN DISEASE SPECTRUM SUGGESTS DISTINCT CELLS OF ORIGIN AND POTENTIAL MOSAICISM IN TSC

A clear distinction between individuals with *TSC* gene mutations who acquire Tuberous Sclerosis vs. S-LAM is that the somatic mutations in *TSC1* or *TSC2* are more widespread in those with TSC, and are typically also present in the germline (Dabora et al., 2001; Strizheva et al., 2001). This was a substantial discovery toward our understanding of how S-LAM may develop in the absence of other TSC manifestations, and pushed forth the concept that TSC and LAM lesions may develop from different founding cell types, and thus that TSC gene mutations are restricted to a smaller population of cells in S-LAM patients. The fact that there is considerable heterogeneity in the clinical presentation and severity of the disease spectrum among both LAM and TSC patients (Henske and McCormack, 2012; Curatolo and Maria, 2013) and evidence for gonadal and/or somatic mosaicism in some TSC patients that can correlate with phenotypic variability (Rose et al., 1999; Verhoef et al., 1999; Qin et al., 2010b; Boronat et al., 2014a,b) lend further support to the concept that disease spectrum is significantly affected by the proportion of cells, and likely the identity and fate of those cells, that possess a *TSC* gene mutation.

It is also possible that distinct *TSC1* or *TSC2* mutations present in different patients may also contribute to disease heterogeneity by encoding TSC proteins with varying degrees of function. Indeed, hypomorphic mutations have been observed in some TSC patients, and experimental and clinical evidence suggest that these mutations result in reduced phenotypic severity (Khare et al., 2001; O'Connor et al., 2003; Nellist et al., 2005, 2008; Jansen et al., 2006; Pollizzi et al., 2009; Yuan et al., 2012; Fu and Ess, 2013). However, these cases appear to represent a minority of known TSC patients. Nevertheless, in cases where a hypomorphic *TSC* gene mutation is present, it is likely that this feature acts in conjunction with the identity and proportion of cells that carry the mutation to influence phenotypic severity.

The findings from the TSC CNS-related mouse models discussed above provide strong experimental support for the concept presented here. Although these studies were limited to the CNS manifestations, it is likely that this theme persists throughout the full spectrum of the disease and that in general a more restricted mutational burden will lead to the development of fewer types of lesions and fewer numbers of lesions. With this in mind, one could postulate that while *TSC* gene mutations in NSCs lead to the CNS lesions, mutations within the neural crest lineage are responsible for the development of most other TSC lesions, including those associated with LAM, and that variable mutational burden within the neural crest (NC) lineage leads to the greatest heterogeneity in TSC disease phenotypes.

EVIDENCE SUPPORTING A NEURAL CREST CELL OF ORIGIN FOR LAM AND OTHER TSC-ASSOCIATED LESIONS

Neural crest cell activity during embryogenesis

The neural crest is a transient population of migratory progenitor cells that emerge at the interface between the dorsal region of the neural plate and the non-neural ectoderm in embryonic development during neurulation. Although the migratory patterns of the NC can differ slightly from species to species, studies performed in model organisms, such as *Xenopus*, mouse, quail and chick embryos, have offered great insight into the contributions of the NC population to various tissues in the developing embryo. Although exact migratory patterning in humans is unknown, lineage tracing has revealed enough commonality between these model species to establish a paradigm for neural crest cell (NCC) migration and differentiation during human development. Through contact mediated interactions, NCCs emerge, undergo epithelial-to-mesenchymal transition (EMT), and migrate in specific succession and order in an anterior to posterior sequence (Theveneau and Mayor, 2012). Of note, both the neural plate and the surface ectoderm contribute cells to the NC population (Moury and Jacobson, 1990; Selleck and Bronner-Fraser, 1995). Once NCCs have emerged and delaminated, they follow distinct migratory patterns depending on their eventual fate, the majority of which are summarized in **Table 1**.

NCCs can be generally categorized into four major subgroups based on their migratory cohort: cranial, cardiac, vagal and trunk. Cranial NCCs contribute to the tissues of the craniofacial region, differentiating into the widest array of cell types of the neural crest subgroups, including vascular smooth muscle cells and pericytes, chondrocytes, adipocytes, osteocytes, odontoblasts, melanocytes, connective tissue, and sensory and parasympathetic ganglia

Table 1 | Neural crest cells can be categorized into four major subgroups, based on their respective migratory pattern and the tissues to which they contribute.

NC Subtype	Tissues	Cell types	
Cranial	Craniofacial skeleton Skin Cornea Connective tissue	Fibroblasts Melanocytes Adipocytes Osteocytes Odontoblasts Chondrocytes	Neurons Glia Mesenchymal cells Myocytes Pericytes
Cardiac	Branchial arches Cardiac septum Parasympathetic cardiac ganglia	Myocytes Pericytes Neurons Glia	
Vagal	Enteric nervous system	Neurons Glia	
Trunk	Peripheral nervous system Skin Adrenal medulla	Neurons Glia Melanocytes Chromafinn cells	

It should be noted that individual NCC multipotency varies between subgroups as well as within subgroups themselves.

(D'Amico-Martel and Noden, 1983; Köntges and Lumsden, 1996; Etchevers et al., 2001; Billon et al., 2007; Grenier et al., 2009). Cardiac NCCs differentiate into parasympathetic cardiac ganglia, as well as smooth muscle of the branchial arches and the cardiac outflow tract, and form the aorticopulmonary septum (Kirby et al., 1983; Kirby and Stewart, 1983). Vagal NCCs primarily give rise to the neurons and glia of the enteric nervous system (Yntema and Hammond, 1954; Le Douarin and Teillet, 1973). Finally, trunk NCCs give rise to melanocytes and the neurons and glia of the peripheral nervous system (Raible and Ungos, 2006).

Potential role for neural crest subtypes in TSC disease heterogeneity

NCCs are able to differentiate into a wide array of cell types that contribute to many of the tissues associated with TSC tumors. Theoretically, distinct NC subtypes could be specifically associated with some of the multi-systemic tumors observed in TSC. For instance, the cranial NC may be implicated in the development of facial angiofibromas, dental pits, and retinal tumors. Indeed, cases of retinal lesions have been reported in connection with neurofibromatosis, a disease also thought to be of neural crest origin (Ishiko et al., 2006; Sachdeva et al., 2010). Cortical tubers could also potentially be attributed to the cranial neural crest, a theory that has previously been postulated (Hindman et al., 1997). Extending this theory, TSC1- or TSC2-deficient cardiac NC may be the cell of origin for cardiac rhabdomyomas and even pulmonary LAM lesions (due to the involvement in aorticopulmonary septum development), and dysfunctional trunk NC may be responsible for "confetti" skin lesions and hypomelanotic macules. Within this frame of thought, the determining factor in which tumor types ultimately form is perhaps both spatial and temporal. If the first-hit mutation occurs early on during embryogenesis within a cell of the developing neural plate or surface ectoderm before neurulation, the potential number of affected progeny capable of acquiring a second-hit mutation would be much greater and could potentially affect all NC subgroups, leading to a more severe TSC phenotype affecting multiple tissues. If the first-hit mutation occurs later in development, perhaps in an emerging NC subtype, the affected cell may only have influence on the specific subtype emerging from its respective anterior/posterior position along the developing neural tube. Such a scenario could explain the extreme differences in TSC disease spectrum that are observed in highly affected TSC patients vs. individuals with S-LAM.

Neural crest stem cells persist in adult tissues associated with TSC lesions

While tissue innervation by migrating NCCs is a fundamental process during embryonic development, NCCs with stem cell-like properties are now known to persist in a number of adult tissues. Just as tissues that are innervated by embryonic NCCs during development theoretically have the potential to give rise to LAM-like cells or cells that comprise other TSC mesenchymal lesions, these adult neural crest stem cells (NCSCs) are a particularly promising population as a potential LAM and TSC cell of origin. The tissues in which NCSCs reside in the adult are diverse, and include the gut, bone marrow, heart, dental mesenchyme, dorsal root ganglia, and cornea (reviewed in Dupin and Sommer, 2012). Intriguingly, a number of distinct multipotent stem cells with neural and/or NC differentiation potential have also been identified in the skin (Dupin and Sommer, 2012). As NCCs possess inherent migratory and transdifferentiation potential (discussed further below), and exhibit extensive interactions with their surrounding microenvironments that can instruct migratory, differentiation and invasiveness behaviors, many of these cell types could potentially serve as the cell of origin for TSC lesions (Takahashi et al., 2013).

The susceptibility of the skin to ultraviolet (UV) radiation implicates resident skin NCSCs as promising cell of origin candidates, as UV-induced damage could induce second-hit mutations. For example, skin-derived precursors (SKPs), multipotent NCSCs found in the dermal sheath and papillae of hair follicles in both rodents and humans (Toma et al., 2001, 2005; Fernandes et al., 2004; Biernaskie et al., 2009; Jinno et al., 2010), have been heavily implicated as the cell of origin for the dermal tumors associated with the neurocristopathic disorder Neurofibromatosis type I (Le et al., 2009). A recent study has also provided compelling evidence that precursor cells found within the skin may serve as the cell of origin for facial angiofibromas in TSC patients, following the acquisition of UV-induced second-hit *TSC2* mutations (Tyburczy et al., 2014). Given this potential, it is a distinct possibility that SKPs may similarly serve as the cell of origin for LAM and the cells that comprise other TSC-associated lesions, providing that SKPs possessing *TSC* gene mutations are able to invade the lymphatic and/or cardiovascular circulation.

Clinical evidence suggests a neural crest origin for TSC and LAM lesions

Sequencing of tissue from LAM nodules or disseminated LAM cells from body fluids has indicated that germline *TSC1* or *TSC2* mutations are not prevalent in S-LAM patients (Astrinidis et al., 2000; Carsillo et al., 2000). Thus, the first-hit mutation that drives LAM must typically arise *de novo* in a somatic cell. The lack of CNS involvement in the pathology of S-LAM provides strong clinical evidence that the neural stem cell lineage is not affected in S-LAM patients. The molecular markers present in LAM cells within its classical lesions [alpha-SMA, HMB-45, TRP-1, MART-1, GD3 (Carbone, 2009; Gilbert et al., 2013)], as well as the tissue types affected in the majority of TSC-associated tumors, do however present significant clues to the potential cell(s) of origin. The tissues in which lesions/abnormalities are observed in TSC patients are diverse, but all of them share the common feature that they are wholly or in part comprised of neural crest progeny (**Figure 1**, **Table 1**) (Dupin and Sommer, 2012). These tissues include the heart (cardiac rhabdomyomas), the skin (angiofibromas, most predominant in the facial region), dental enamel, eyes (retinal astrocytic hamartomas), and the innervation of the kidneys (AMLs), uterus and lungs (Curatolo and Maria, 2013). Additionally, the abnormal cells observed in LAM lesions express molecular markers of cells that are generated by the neural crest; these include smooth muscle, melanocytes, and adipocytes.

In addition to key molecular markers, there is also compelling evidence that TSC and LAM cells are able to migrate between different tissue sites (Smolarek et al., 1998; Karbowniczek et al., 2003), reminiscent of the highly migratory behavior of neural crest cells. Furthermore, LAM lung lesions contain a heterogeneous population of cells, with both highly proliferative spindle shaped cells and large epithelial cells making up LAM clusters. Moreover, both of these cell types express molecular markers of mesenchymal neural crest cells *in vivo* (e.g., smooth muscle actin, and HMB45 or gp100 in the epithelial cells). This suggests the intriguing possibility that, if indeed LAM cells originate from the neural crest, they may possess the ability to undergo EMT or even mesenchymal-to-epithelial transition (MET) within the lesion itself, accounting for the presence of these distinct cell types within single lesions.

LAM cells exhibit extensive migration and EMT

Just as neural crest cells represent a highly migratory cell population, a striking characteristic of LAM cells is their ability to travel from one tissue site to another, leading to multi-tissue tumor formation. For example, mutation analysis has revealed that the same mutations are present in LAM cells of both pulmonary LAM nodules and AMLs, indicating that the tumors originated from a common progenitor cell that was able to migrate or metastasize to and/or from either of these tissues (Sato et al., 2002). Although identical *TSC1* or *TSC2* mutations among tumor cells are not always observed in TSC-LAM and S-LAM patients, it is clear that LAM cells are mobile; markedly, LAM cells are able to infiltrate and repopulate healthy donor lungs following transplantation (Bittmann et al., 2003; Karbowniczek et al., 2003). They are also are able to metastasize between the lungs, kidneys, lymphatics, and uterus using mechanisms that do not fit the classic cancer metastasis paradigm. LAM cells can be found throughout the lymphatics in LAM patients, and their expression of VEGF-C and VEGF-D along with VEGF receptor-3 indicates their lymphangiogenic potential (Glasgow et al., 2012), indicating that the extensive network of lymphatics may serve as the primary niche for LAM cells and facilitates their spread.

LAM cells do not easily succumb to anoikis and can be found in the chylous fluid, urine, and circulating in the blood of LAM patients, a phenomenon that may be directly related to estrogen signaling (Crooks et al., 2004; Yu et al., 2009; Gu et al., 2013). Although these cells do not present themselves as malignant or highly proliferative compared to most cancer cells, their apparent robustness and infiltrative ability is certainly not benign.

A complete mechanistic model for LAM cell "metastasis" has yet to be confirmed; however, numerous studies have shown that LAM cells display increased motility and invasive properties *in vitro*. This invasive phenotype can be directly linked to loss-of-function mutations in either *TSC1* or *TSC2*, leading to pro-migratory cytoskeletal rearrangement and changes in focal adhesions. Primary LAM cells with TSC2 loss-of-function mutations display increased invasive and migratory properties as a result of RhoA GTPase-mediated cytoskeletal rearrangement and modulation of focal adhesions. RhoA GTPase is normally regulated by TSC1 when complexed with TSC2, and this invasive phenotype is mitigated upon TSC2 re-expression or TSC1 knockdown (Lamb et al., 2000; Goncharova et al., 2006). Perhaps contributing to the robustness of migratory LAM cells, dysfunctional TSC2 leads to a loss in membrane bound E-cadherin via a Rapamycin sensitive pathway. This, in turn, leads to reduced cell-cell adhesion, EMT, cell detachment, and resistance to anoikis (Barnes et al., 2010).

Molecular features and behavior of LAM cells parallel the neural crest

Clearly, dysfunction of either TSC1 or TSC2 will result in a cell that is primed to undergo EMT and migrate, a scenario that is reminiscent of neural crest delamination and migration during embryonic development. RhoA is critical in regulating the directionality of migrating neural crest cells, greatly affecting spatial-temporal patterning upon its loss (Rupp and Kulesa, 2007; Matthews et al., 2008). Cadherins play a crucial role during neurulation and help define the non-neural ectoderm and the neural plate (Taneyhill, 2008; Dady et al., 2012), the tissues whose interface from which neural crest cells delaminate and from which they migrate. Specifically, the transition from E-cadherin expression to N-cadherin plays a critical role in this process, and interestingly, a similar phenomenon is observed in LAM cells. As mentioned above, TSC2-deficiency results in internalization of E-cadherin leading to EMT; moreover, when LAM patient-derived cells are exposed to estrogen, E-cadherin expression is decreased and N-cadherin is upregulated, similar to the segregation of the germ layers during neurulation (Gu et al., 2013). Pairing well with the decrease in E-cadherin expression in TSC2-deficient cells, there is an upregulation in the expression of Snail, an important transcription factor involved in initiating EMT, migration, and neural crest specification (Cano et al., 2000; Aybar et al., 2003; Barnes et al., 2010). A second-hit mutation in *TSC2* therefore causes a change in the expression profile in LAM cells that resembles a neural crest progenitor undergoing migration and fate specification.

Adding fuel to the fire, LAM cells have been shown to excrete matrix metalloproteinases (MMPs), particularly MMP-2 and MMP-9, that likely contribute to their invasiveness and lead to the slow destruction of the lung parenchyma (Hayashi et al., 1997; Lee et al., 2010a). MMP2 and MMP9 are required in neural crest cells for their initial delamination and migration, and inhibition of these MMPs during development results in the complete halt of NC migration and differentiation (Monsonego-Ornan et al., 2012).

Extending this comparative train of thought, changes in TGF-ß signaling in LAM cells also relates directly to the NC and its lineages. Histopathological analysis of LAM cells, along with TSC2-null cell lines, show increased levels of TGF-ß and smooth muscle actin (SMA) (Zhe and Schuger, 2004; Barnes et al., 2010; Lee et al., 2010a). Although expression of TGF-ß and SMA is common to the EMT process, TGF-ß signaling is involved in NC-derived smooth muscle specification *in vivo* and is a common element of smooth muscle differentiation protocols of neural crest cells *in vitro* (Shah et al., 1996; Curchoe et al., 2010; Xie et al., 2013). The initiation of signaling cascades resulting in a smooth muscle-like cell fate may help explain the presence of spindle-like SMA+ proliferative cells in LAM nodules and angiomyolipomas. A loss of TSC2 function could destabilize a NC-derived cell, causing it to dedifferentiate and undergo EMT. This cell could then potentially migrate to a favorable niche (e.g., lung, kidney, lymphatics) and, due to increased TGF-ß signaling, differentiate into the smooth muscle cells and form the characteristic tumors associated with LAM.

The molecular signaling events that occur upon a second-hit mutation in TSC1 or TSC2 may act in some capacity to recapitulate the developmental processes of neural crest migration and specification. Such an orchestration may be enough to perhaps destabilize a dormant NCSC or a fully differentiated NC cell, whose epigenetic memory is already prone to EMT, migration, and cell motility. Indeed, with the proper stimulus, fully differentiated NC progeny display the ability to dedifferentiate and migrate *in vivo*, as well as return to a multipotent state when isolated and cultured *in vitro* (Real et al., 2005, 2006; Nagoshi et al., 2011; Dupin and Sommer, 2012).

The NC has already been implicated in the reactivation of EMT in fully differentiated and mature cells in the form of the metastatic cancers melanoma and neuroblastoma (Grimmer and Weiss, 2006; Shakhova, 2014). Furthermore, dedifferentiation and redifferentiation (EMT-MET switch) is a proposed mechanism of metastatic cancer supported by experimental evidence (Ocaña et al., 2012; Tsai et al., 2012). Neural crest cells have the unique ability to transdifferentiate during embryonic development from a neural ectodermal lineage to specialized cells that are ultimately considered to be mesodermal. This phenotypic plasticity is a unique quality of the neural crest, which clearly requires precise signaling and environmental cues during development for precise specification. Similar to the notion that a machine with many small moving parts is more likely to malfunction, neural crest-derived cells may be more prone to undergo EMT, migrate, and differentiate based on their inherent plasticity if given the proper cues.

Current models to investigate neural crest contribution to TSC

Few studies have directly investigated the hypothesis that *TSC1* or *TSC2* mutations in neural crest stem cells drive the development of non-CNS tumors in TSC, including LAM. Mice containing a conditional *TSC1* knockout allele crossed with mice expressing Cre driven by the *WNT1* promoter die prenatally, for unknown reasons, with no indication of renal or lung pathology (Kwiatkowski, 2010). No additional mouse or cell culture-based models of TSC1 or TSC2-deficiency in the neural crest lineage have been reported to date. The lack of clarity provided by existing mouse models, and the paucity of such models, regarding the development and progression of LAM in humans, heavily underscores the need for improved models that more closely recapitulate human pathology.

It is quite possible that the biology of TSC1- or TSC2-deficiency is substantially different in mouse vs. human NCCs, and that establishing the same models (e.g., Wnt1-cre conditional models of TSC1/2-deficiency) in a humanized system will allow us to generate accurate disease models. Alternatively, the lethality of the Wnt1-cre *TSC1* knockout mouse model may also serve as an indicator that homozygous *TSC* gene deletions within the neural crest lineage are not compatible with embryonic development and that TSC-like phenotypes will only be observed under conditions where second-hit mutations can be induced in a controlled manner within sub-populations of cells. Speaking to this, it is possible that different tumor types that are prevalent in TSC patients, as well as those that define LAM disease, are derived from TSC1- or TSC2-deficient cells of different neural crest subtypes. Identification of the NCC subtype(s) responsible for the development of LAM tumors may importantly result in less severe phenotypes and prolonged survival in animal models that should enable longer-term experiments to be performed than have been possible with current models. Additionally, this will inevitably result in disease phenotypes more specific to the LAM condition, allowing us to generate the most appropriate models possible.

STRATEGIES FOR GREATER UNDERSTANDING AND IMPROVED TREATMENT OPTIONS IN TSC AND LAM

Theoretical model of neural crest contribution to disease heterogeneity

While there is extensive correlative evidence for a potential role of the NC lineage as the cell of origin for LAM and many TSC manifestations, this hypothesis remains to be adequately tested experimentally. We present a model (**Figure 2**) whereby the diverse, heterogeneous tumor burden in TSC patients is dependent on when first- and second-hit mutations in *TSC1 or TSC2* occur during development and into adulthood. In this model, mutations that are carried in the germline or that are acquired during early stages of embryogenesis lead to a greater disease burden than those that occur at later time points. For instance, the majority of TSC patients have germline mutations, and thus all cell lineages, including both neural and neural crest progenitors, are vulnerable to the acquisition of second-hit mutations. Second-hit mutations may occur independently within the neural and neural crest lineages, or perhaps in a cell very early in development (i.e., before neurulation) (**Figure 2A**). In any case, TSC patients that carry a germline *TSC1* or *TSC2* mutation are likely to acquire wide-spread and severe disease manifestations due to the diverse pool of progeny susceptible to acquiring second-hit mutations (i.e., both the NCC and NSC lineages).

Similarly, a first-hit mutation acquired at a time point as early as neurulation (**Figure 2A**) would give rise to sporadic TSC, explaining the genotypic mosaicism observed in some patients. In this scenario, the severity of the resulting disease spectrum depends on the distribution pattern of "first-hit" cells into downstream lineages as well as in which cell populations second-hit *TSC* gene mutation(s) are acquired. Here, both the NCC and NSC lineages are vulnerable to being infiltrated by *TSC* mutant cells, however mosaicism may lead to reduced disease severity by restricting the number of cells or the cell lineages that carry *TSC* mutations.

Following this model further, the more limited disease manifestations observed in S-LAM, and potentially milder forms of TSC, would originate from a primary *TSC* gene mutation in a much more restricted population limited to the NC lineage. This could occur either during embryogenesis within a common or sub-type restricted neural crest progenitor cell (**Figure 2B**) or in adulthood (**Figure 2C**), affecting NCSCs or NC-derived cells, such as melanocytes or SKPs. The former possibility could foreseeably occur either via mosaicism within the broader NC lineage, in which only certain NC subtypes carry enough of a mutational burden to drive the development of a lesion, or restriction of first- and/or second-hit mutations within a functionally defined sub-population of NC. A likely candidate is the cardiac NC lineage due to its involvement in development of the aorticopulmonary septum. In either scenario, only cells that were derived from NCCs during development or NCSCs that persist in adult tissues are vulnerable to a second-hit mutation later in life that may endow them with the ability to drive the formation of TSC and LAM lesions. This highlights the possibility that the heterogeneity within TSC and S-LAM patients is predominantly dictated by the cells that carry a first-hit *TSC* gene mutation and, consequently, the resulting progeny left vulnerable to acquiring a second-hit mutation.

As mosaicism within the NC lineage may potentially lead to the more restricted phenotypes associated with LAM, so too may gonadal or somatic mosaicism lead to varying degrees of phenotypic severity in TSC patients, as has been observed clinically (Rose et al., 1999; Verhoef et al., 1999; Qin et al., 2010b; Boronat et al., 2014a,b). This type of model is therefore likely to underlie phenotypic differences between TSC vs. LAM alone, as well as the heterogeneity in disease features observed in both TSC and LAM patient cohorts themselves. Thus, a *TSC1* or *TSC2* mutation at an earlier developmental stage has the potential to result in a more complete spectrum of TSC phenotypes, while a later mutation will lead to a more restricted pathological potential, resulting in limited TSC phenotypes or LAM only.

Current treatment options for TSC and LAM

While double lung transplant is currently the gold standard treatment option for LAM, this approach is not possible in all patients;

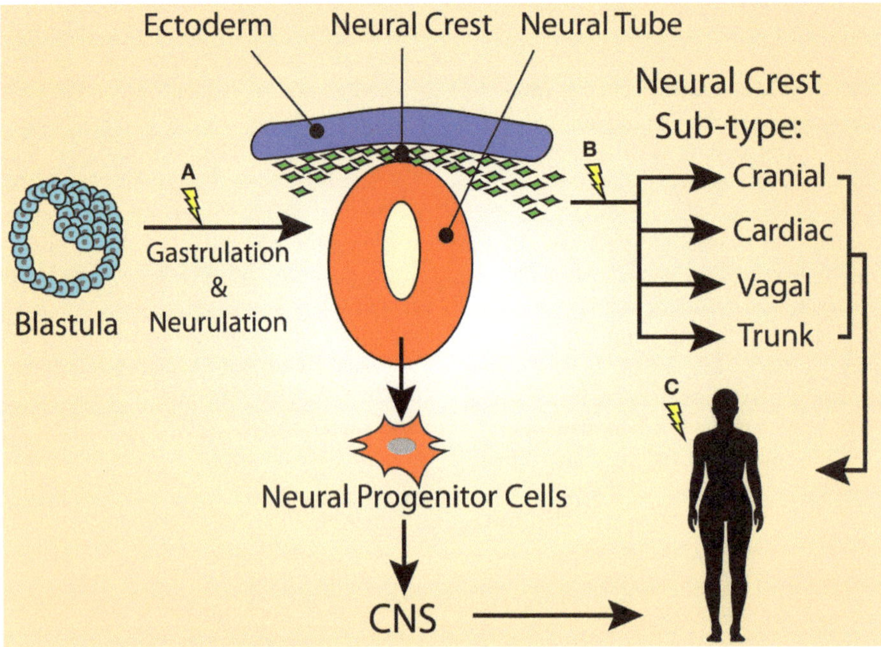

FIGURE 2 | Model illustrating that the variation of tumor heterogeneity in TSC may be related to the stage of development in which loss of function mutations in *TSC1* or *TSC2* occurs. (A) For TSC patients with germline mutations, second-hit mutations may be acquired during early stages of embryogenesis, such as during gastrulation or neurulation. This would result in a large population of cells that carry the mutation and would lead to a severe TSC phenotype, possibly displaying the full spectrum of TSC manifestations. First-hit mutations at this time point would result in mosaicism. The affected cells could contribute not only to the neural crest lineage, but also the neural progenitor population of the neural tube, resulting in the CNS manifestations of TSC. **(B)** Second-hit mutations occurring in the emerging neural crest population likely results in many of the non-CNS symptoms of TSC, potentially including LAM. First-hit mutations at this time-point may result in S-LAM, as only the neural crest lineage would be vulnerable to second-hit mutations. It is possible that a sub-population of NCCs, such as the cardiac NC, is responsible for the more restricted lesions associated with S-LAM. **(C)** Alternatively, first- and/or second-hit mutations acquired within an adult NCC could also be responsible for driving LAM and some TSC-associated lesions (e.g., skin lesions).

additionally, LAM lesions have been shown to recur in the donor lung in some patients (Bittmann et al., 2003; Karbowniczek et al., 2003). Thus, transplantation does not address the underlying mechanisms that drive the development of LAM cells and associated lesions. The mTOR-specific inhibitor Rapamycin is a promising molecular intervention currently under intensive investigation as a potential treatment for LAM and other manifestations in TSC. Rapamycin has proven clinically effective for the treatment of most TSC-associated lesions (Franz et al., 2006; Bissler et al., 2008; Krueger et al., 2010; Micozkadioglu et al., 2010; Casanova et al., 2011; Dabora et al., 2011; DeKlotz et al., 2011; Tiberio et al., 2011; Canpolat et al., 2014; Park et al., 2014), and in the context of LAM can lead to reduction in the size of AMLs, provide stabilization of lung function, and a marked improvement in quality of life measures (McCormack et al., 2011).

A major caveat with this approach, however, is that Rapamycin is tumoristatic rather than tumoricidal. Thus, disease progression resumes once Rapamycin is withdrawn. Therefore, patients must remain on the drug permanently to provide sustained treatment. Long-term Rapamycin treatment is also not without potential challenges, as there is concern that LAM tumors, as well as other TSC lesions, may become refractory to Rapamycin over an extended time period. Additionally, while Rapamycin treatment can reduce mTOR signaling, it may also have the undesirable effect of promoting the survival of TSC1 or TSC2-null cells (reviewed in Henske and McCormack, 2012). In this regard, one of the key downstream processes controlled by mTORC1 signaling is autophagy (Laplante and Sabatini, 2012), a key cellular survival mechanism, and inhibition of autophagy is a typical consequence of TSC1- or TSC2-deficiency (Crino et al., 2006; Parkhitko et al., 2011; Crino and Tsai, 2012; Henske and McCormack, 2012). Thus, while Rapamycin appears to generally inhibit the growth of TSC1- or TSC2-deficient cells, it may concurrently promote their survival by inducing autophagy. Rapamycin has also been shown to induce expression of pro-survival micro-RNAs (Trindade et al., 2013). Thus, Rapamycin treatment may therefore result in a precarious situation whereby the growth and proliferation of TSC1- or TSC2-deficient cells is halted, but the cells remain alive *in situ*, allowing them to re-expand once treatment is stopped or an alternative mTOR signaling pathway is triggered in the extensive mTOR signaling network to bypass mTORC1 inhibition.

As long-term Rapamycin treatment carries a significant risk of acquired treatment resistance and adverse side-effects (McCormack et al., 2011), additional therapeutic options are needed to provide life-long treatment options to TSC and LAM patients. Importantly, a number of proteins thought to be important in LAM pathogenesis are unaffected by Rapamycin treatment in LAM cells, including MMP2 (Lee et al., 2010b). Due to this fact and observations that alternative pathways are activated by Rapamycin treatment that promote cell survival, there is a strong indication for the development of alternative or combination therapies. Thus, the development of new therapeutic approaches is currently of the highest priority within the LAM and TSC fields. In concert with this, the development of effective models to aid in the identification and testing of potentially effective therapeutic interventions must also be at the top of the list.

Identification of the cell of origin for LAM and other TSC-associated tumors is imperative for the development of more effective, long-lasting treatment strategies. Only by inducing first- and second-hit mutations in the proper cell of origin, and in the proper proportion of that population, will we be able to fully understand and therapeutically target the molecular mechanisms that drive the initiation and progression of distinct TSC lesions. mTOR activation can lead to different primary phenotypes in distinct cell types, such as precocious differentiation in the CNS (Hartman et al., 2013), increased cycling of hematopoietic stem cells (Gan et al., 2008), and exit of satellite cells from quiescence upon injury (Rodgers et al., 2014). Therefore, different TSC-associated lesions may require different targeted therapeutic approaches. Furthermore, the downstream targets of mTOR are diverse and affect multiple cellular signaling pathways, and mTOR participates in a number of feed-back signaling mechanisms (Laplante and Sabatini, 2012). Due to this complexity, truly effective treatment approaches are likely to require combination therapies in which multiple signaling nodes are targeted, and the most promising approaches may well be different for distinct tumor types.

A need for humanized disease models

The TSC and LAM fields have benefited significantly from cell culture-based and rodent models of TSC1- or TSC2-deficiency; however, these models carry a number of caveats. Rodent models of spontaneous (e.g., the Eker rat) or genetically engineered *TSC* gene deficiency, and cells cultured from these animals (e.g., ELT3 cells, isolated fibroblasts), have offered important insights into the basic molecular and cellular phenotypes caused by *TSC* ablation and have provided avenues for experimental drug testing. However, these models do not fully recapitulate the tumor spectrum or phenotype of many TSC lesions or their derivative cells, including LAM. While these animal models exhibit a high penetrance of renal cystadenomas and liver hemangiomas with age, and a partial penetrance (about 30%) of lung adenomas, these tumors do not recapitulate the phenotypes of those observed in human TSC patients (Kobayashi et al., 1999, 2001; Onda et al., 1999; Kwiatkowski et al., 2002; Kwiatkowski, 2010). Therefore, they cannot be used to understand the pathophysiology of the most clinically significant manifestations of TSC, nor as accurate pre-clinical models for drug testing to treat or eliminate these lesions.

Presently, there is a lack of humanized-models of TSC and LAM. Primary TSC1- or TSC2-deficient cells from TSC or LAM patients unfortunately cannot be efficiently propagated in culture without viral transformation, and even then some elements of LAM cell identity observed *in vivo* are not recapitulated in culture. Additionally, patient-derived cells lack proper control cell lines, and induction of TSC1- or TSC2-deficiency (disruption of both alleles or a high-degree mRNA knockdown), as is the case for most of the existing models, bypasses mechanisms that lead to second-hit mutations. As this important feature of TSC and LAM tumor pathogenesis is absent, these systems cannot be used to understand the processes leading to tumor induction and subsequent disease progression. Thus, to accurately model TSC and LAM it will be essential to generate or obtain human cell lines in which first- and second-hit *TSC* gene mutations can be induced in a controlled manner, and in which cellular phenotypes can be clearly assessed following differentiation into putative cell of origin lineages. Identification of the cell types of origin for TSC and LAM lesions is an essential step toward the development of accurate pre-clinical humanized disease models. Such models will not only allow for a more relevant understanding of the mechanisms that drive disease, but will also greatly improve therapeutic options for patients, making personalized therapeutics a potentially viable option for TSC.

Approaches for generating humanized disease models

While it is clear from mouse models that it is at least possible to induce TSC-like brain lesions by manipulating mTOR signaling in the NSC lineage, no humanized models exist to test whether the same events are indeed responsible for driving CNS pathologies in humans. Furthermore, while a NC origin for non-CNS lesions in TSC seems intuitive, direct experimental evidence testing this hypothesis in human cells is needed. To perform these pivotal experiments, we need better disease models. First, we need to generate human cell lines that will recapitulate phenotypes observed in distinct TSC and LAM lesions, and can potentially give rise to these lesions *in vitro* and in *in vivo* mouse models, following controlled induction of TSC1- or TSC2-deficiency. Second, the behavior of the cells that comprise TSC lesions appears to be highly dependent on interactions with their tissue microenvironment. Thus, proper disease models should incorporate *in vivo* and *in vitro* culture-based approaches that closely recapitulate the natural, *in situ* environment in which TSC and LAM lesions normally develop.

One promising approach for the generation of humanized models in TSC is to couple genetic engineering to induce TSC1/2-deficiency with embryonic stem cell (ESC) or patient-derived induced pluripotent stem cell (iPSC) technology. ESCs and iPSCs are pluripotent stem cell populations that have the capacity to differentiate into any cell type found in the body, including candidate cell types of origin for TSC and LAM lesions. Furthermore, iPSCs importantly offer an avenue for patient-specific modeling, as it allows fully differentiated cells from TSC and LAM patients harboring *TSC* gene mutations to be reprogrammed to an ESC-like state, enabling subsequent directed differentiation into putative cell of origin populations.

Directed differentiation of human ESCs or iPSCs into neural and neural crest lineages *in vitro* and *in vivo*, coupled with the ability to inducibly knock-out or knock-down TSC1 or TSC2 at different stages of differentiation, will directly allow cell of origin theories to be tested. This will also allow for the establishment of human cell lines that reflect TSC and LAM cell phenotypes and, following *in vivo* injection of TSC1/2-deficient cell lines, the generation of humanized animal models of LAM and TSC lesions. Subsequently, growth and differentiation of these cell lines in not only 2-dimensional, but also 3-dimensional culture conditions, as well as injection into immune-compromised mice, will enable the behavior of these cells to be studied in environments that more closely recapitulate the *in vivo* environment of distinct TSC and LAM lesions.

A second, highly promising approach that would offer opportunities for patient-centered discovery and treatment is to use endogenous sources of NC-like stem cells resident within adult tissues, such as SKPs, to generate TSC1- or TSC2-deficient lines. SKPs reside in a location that is highly accessible for tissue extraction and subsequent NC cell isolation, thus they may be a particularly useful cell population to allow for the development of patient-specific models. Development of models in which a patient's own cells can be used as the source material to follow disease initiation and progression will importantly allow us to understand not only the basic mechanisms of TSC and LAM tumor formation, but also what drives the heterogeneous phenotypes that are often observed between different patients.

CONCLUSION

The TSC and LAM research communities have moved at an impressive pace in the last two decades. Since discovering the *TSC2* gene in 1993 (European Chromosome 16 Tuberous Sclerosis Consortium, 1993), investigative research efforts have quickly produced a plethora of experimental data leading to the development of the first generation of therapeutics for TSC and LAM. However, due to the pitfalls associated with the largely tumoristatic action of Rapamycin and closely associated "rapalogs," it is apparent that a wider array of advanced treatment options is necessary to combat these multisystem disorders. Current disease models of TSC and LAM have been invaluable in providing insight into molecular disease pathology, provided a basis for clinical trials of rapalogs and other potential therapeutics, and has supported the investigation of neural stem cell of origin theories for TSC CNS manifestations. However, these same disease models have their limitations and do not fully recapitulate the nature or progression of TSC and LAM observed in humans.

There is sufficient evidence to suggest that the NC lineage is integrally involved in TSC and LAM tumor heterogeneity; however, this notion has largely been limited to discussion due to a lack of an effective means to study such theories. Currently, there are no humanized models of TSC and LAM that provide a framework for investigating this theory, nor to investigate the influence of *TSC* gene mutations acquired during early development on disease heterogeneity. Yet, with the current advances in tissue culture methods, stem cell-based disease modeling, and the advent of iPSC technology, this will likely change in quick order. Using humanized models to identify the cell of origin for LAM and other TSC tumors is crucial in identifying and targeting the specific molecular mechanisms responsible for distinct TSC and LAM lesions, and will help pave the way for the next generation of TSC and LAM therapeutics.

ACKNOWLEDGMENTS

Our TSC and LAM research is being funded by a Special Accelerated Discovery grant by the McEwen Centre for Regenerative Medicine, supported by Green Eggs and LAM to William L. Stanford and a grant (W81XWH-14-1-0434) awarded to WLS from the United States Department of Defense via the Tuberous Sclerosis Complex Research Program of the Congressionally Directed Medical Research Program. William L. Stanford is supported by a Tier 1 Canada Research Chair in Integrated Stem Cell Biology. Lisa M. Julian is supported by a Canadian Institutes of Health Research Banting Fellowship and an Ontario Stem Cell Initiative Postdoctoral Fellowship.

REFERENCES

Arbiser, J. L., Brat, D., Hunter, S., D'Armiento, J., Henske, E. P., Arbiser, Z. K., et al. (2002). Tuberous sclerosis-associated lesions of the kidney, brain, and skin are angiogenic neoplasms. *J. Am. Acad. Dermatol.* 46, 376–380. doi: 10.1067/mjd.2002.120530

Astrinidis, A., Khare, L., Carsillo, T., Smolarek, T., Au, K. S., Northrup, H., et al. (2000). Mutational analysis of the tuberous sclerosis gene TSC2 in patients with pulmonary lymphangioleiomyomatosis. *J. Med. Genet.* 37, 55–57. doi: 10.1136/jmg.37.1.55

Au, K. S., Williams, A. T., Roach, E. S., Batchelor, L., Sparagana, S. P., Delgado, M. R., et al. (2007). Genotype/phenotype correlation in 325 individuals referred for a diagnosis of tuberous sclerosis complex in the United States. *Genet. Med.* 9, 88–100. doi: 10.1097/GIM.0b013e31803068c7

Aybar, M. J., Nieto, M. A., and Mayor, R. (2003). Snail precedes slug in the genetic cascade required for the specification and migration of the Xenopus neural crest. *Development* 130, 483–494. doi: 10.1242/dev.00238

Barnes, E. A., Kenerson, H. L., Jiang, X., and Yeung, R. S. (2010). Tuberin regulates E-cadherin localization: implications in epithelial-mesenchymal transition. *Am. J. Pathol.* 177, 1765–1778. doi: 10.2353/ajpath.2010.090233

Biernaskie, J., Paris, M., Morozova, O., Fagan, B. M., Marra, M., Pevny, L., et al. (2009). SKPs derive from hair follicle precursors and exhibit properties of adult dermal stem cells. *Cell Stem Cell* 5, 610–623. doi: 10.1016/j.stem.2009.10.019

Billon, N., Iannarelli, P., Monteiro, M. C., Glavieux-Pardanaud, C., Richardson, W. D., Kessaris, N., et al. (2007). The generation of adipocytes by the neural crest. *Development* 134, 2283–2292. doi: 10.1242/dev.002642

Bissler, J. J., and Kingswood, J. C. (2004). Renal angiomyolipomata. *Kidney Int.* 66, 924–934. doi: 10.1111/j.1523-1755.2004.00838.x

Bissler, J. J., McCormack, F. X., Young, L. R., Elwing, J. M., Chuck, G., Leonard, J. M., et al. (2008). Sirolimus for angiomyolipoma in tuberous sclerosis complex or lymphangioleiomyomatosis. *N. Engl. J. Med.* 358, 140–151. doi: 10.1056/NEJMoa063564

Bittmann, I., Rolf, B., Amann, G., and Löhrs, U. (2003). Recurrence of lymphangioleiomyomatosis after single lung transplantation: new insights into pathogenesis. *Hum. Pathol.* 34, 95–98. doi: 10.1053/hupa.2003.50

Boronat, S., Caruso, P., and Thiele, E. A. (2014a). Absence of subependymal nodules in patients with tubers suggests possible neuroectodermal mosaicism in tuberous sclerosis complex. *Dev. Med. Child Neurol.* 56, 1207–1211. doi: 10.1111/dmcn.12523

Boronat, S., Shaaya, E. A., Doherty, C. M., Caruso, P., and Thiele, E. A. (2014b). Tuberous sclerosis complex without tubers and subependymal nodules: a phenotype-genotype study. *Clin. Genet.* 86, 149–154. doi: 10.1111/cge.12245

Cano, A., Pérez-Moreno, M. A., Rodrigo, I., Locascio, A., Blanco, M. J., del Barrio, M. G., et al. (2000). The transcription factor snail controls epithelial-mesenchymal transitions by repressing E-cadherin expression. *Nat. Cell Biol.* 2, 76–83. doi: 10.1038/35000025

Canpolat, M., Per, H., Gumus, H., Yikilmaz, A., Unal, E., Patiroglu, T., et al. (2014). Rapamycin has a beneficial effect on controlling epilepsy in children with tuberous sclerosis complex: results of 7 children from a cohort of 86. *Childs Nerv. Syst.* 30, 227–240. doi: 10.1007/s00381-013-2185-6

Carbone, M. (2009). Feasability of immunotherapy for lymphangioleiomyomatosis. *Am. J. Pathol.* 175, 2252–2254. doi: 10.2353/ajpath.2009.090900

Carsillo, T., Astrinidis, A., and Henske, E. P. (2000). Mutations in the tuberous sclerosis complex gene TSC2 are a cause of sporadic pulmonary lymphangioleiomyomatosis. *Proc. Natl. Acad. Sci. U.S.A.* 97, 6085–6090. doi: 10.1073/pnas.97.11.6085

Carson, R. P., Van Nielen, D. L., Winzenburger, P. A., and Ess, K. C. (2012). Neuronal and glia abnormalities in Tsc1-deficient forebrain and partial rescue by rapamycin. *Neurobiol. Dis.* 45, 369–380. doi: 10.1016/j.nbd.2011.08.024

Casanova, A., María Girón, R., Acosta, O., Barrón, M., Valenzuela, C., and Ancochea, J. (2011). Lymphangioleiomyomatosis treatment with sirolimus. *Arch. Bronconeumol.* 47, 470–472. doi: 10.1016/j.arbres.2011.01.014

Cheadle, J. P., Reeve, M. P., Sampson, J. R., and Kwiatkowski, D. J. (2000). Molecular genetic advances in tuberous sclerosis. *Hum. Genet.* 107, 97–114. doi: 10.1007/s004390000348

Chu, S. C., Horiba, K., Usuki, J., Avila, N. A., Chen, C. C., Travis, W. D., et al. (1999). Comprehensive evaluation of 35 patients with lymphangioleiomyomatosis. *Chest* 115, 1041–1052. doi: 10.1378/chest.115.4.1041

Costello, L. C., Hartman, T. E., and Ryu, J. H. (2000). High frequency of pulmonary lymphangioleiomyomatosis in women with tuberous sclerosis complex. *Mayo Clin. Proc.* 75, 591–594. doi: 10.4065/75.6.591

Crino, P., and Tsai, V. (2012). Tuberous sclerosis complex: genetic basis and management strategies. *Adv. Genomics Genet.* 2, 19–31. doi: 10.2147/AGG.S19837

Crino, P. B. (2004). Molecular pathogenesis of tuber formation in tuberous sclerosis complex. *J. Child Neurol.* 19, 716–725. doi: 10.1177/08830738040190091301

Crino, P. B., Aronica, E., Baltuch, G., and Nathanson, K. L. (2010). Biallelic TSC gene inactivation in tuberous sclerosis complex. *Neurology* 74, 1716–1723. doi: 10.1212/WNL.0b013e3181e04325

Crino, P. B., Nathanson, K. L., and Henske, E. P. (2006). The tuberous sclerosis complex. *N. Engl. J. Med.* 355, 1345–1356. doi: 10.1056/NEJMra055323

Crooks, D. M., Pacheco-Rodriguez, G., DeCastro, R. M., McCoy, J. P., Wang, J.-A., Kumaki, F., et al. (2004). Molecular and genetic analysis of disseminated neoplastic cells in lymphangioleiomyomatosis. *Proc. Natl. Acad. Sci. U.S.A.* 101, 17462–17467. doi: 10.1073/pnas.0407971101

Curatolo, P., and Maria, B. L. (2013). Tuberous sclerosis. *Handb. Clin. Neurol.* 111, 323–331. doi: 10.1016/B978-0-444-52891-9.00038-5

Curchoe, C. L., Maurer, J., McKeown, S. J., Cattarossi, G., Cimadamore, F., Nilbratt, M., et al. (2010). Early acquisition of neural crest competence during hESCs neuralization. *PLoS ONE* 5:e13890. doi: 10.1371/journal.pone.0013890

Dabora, S. L., Franz, D. N., Ashwal, S., Sagalowsky, A., DiMario, F. J., Miles, D., et al. (2011). Multicenter phase 2 trial of sirolimus for tuberous sclerosis: kidney angiomyolipomas and other tumors regress and VEGF- D levels decrease. *PLoS ONE* 6:e23379. doi: 10.1371/journal.pone.0023379

Dabora, S. L., Jozwiak, S., Franz, D. N., Roberts, P. S., Nieto, A., Chung, J., et al. (2001). Mutational analysis in a cohort of 224 tuberous sclerosis patients indicates increased severity of TSC2, compared with TSC1, disease in multiple organs. *Am. J. Hum. Genet.* 68, 64–80. doi: 10.1086/316951

Dady, A., Blavet, C., and Duband, J.-L. (2012). Timing and kinetics of E- to N-cadherin switch during neurulation in the avian embryo. *Dev. Dyn.* 241, 1333–1349. doi: 10.1002/dvdy.23813

D'Amico-Martel, A., and Noden, D. M. (1983). Contributions of placodal and neural crest cells to avian cranial peripheral ganglia. *Am. J. Anat.* 166, 445–468. doi: 10.1002/aja.1001660406

DeKlotz, C. M. C., Ogram, A. E., Singh, S., Dronavalli, S., and MacGregor, J. L. (2011). Dramatic improvement of facial angiofibromas in tuberous sclerosis with topical rapamycin: optimizing a treatment protocol. *Arch. Dermatol.* 147, 1116–1117. doi: 10.1001/archdermatol.2011.254

Dupin, E., and Sommer, L. (2012). Neural crest progenitors and stem cells: from early development to adulthood. *Dev. Biol.* 366, 83–95. doi: 10.1016/j.ydbio.2012.02.035

Ehninger, D., Han, S., Shilyansky, C., Zhou, Y., Li, W., Kwiatkowski, D. J., et al. (2008). Reversal of learning deficits in a Tsc2+/- mouse model of tuberous sclerosis. *Nat. Med.* 14, 843–848. doi: 10.1038/nm1788

Ess, K. C., Kamp, C. A., Tu, B. P., and Gutmann, D. H. (2005). Developmental origin of subependymal giant cell astrocytoma in tuberous sclerosis complex. *Neurology* 64, 1446–1449. doi: 10.1212/01.WNL.0000158653.81008.49

Etchevers, H. C., Vincent, C., Le Douarin, N. M., and Couly, G. F. (2001). The cephalic neural crest provides pericytes and smooth muscle cells to all blood vessels of the face and forebrain. *Development* 128, 1059–1068.

Feliciano, D. M., Quon, J. L., Su, T., Taylor, M. M., and Bordey, A. (2012). Postnatal neurogenesis generates heterotopias, olfactory micronodules and cortical infiltration following single-cell Tsc1 deletion. *Hum. Mol. Genet.* 21, 799–810. doi: 10.1093/hmg/ddr511

Feliciano, D. M., Su, T., Lopez, J., Platel, J.-C., and Bordey, A. (2011). Single-cell Tsc1 knockout during corticogenesis generates tuber-like lesions and reduces seizure threshold in mice. *J. Clin. Invest.* 121, 1596–1607. doi: 10.1172/JCI44909

Fernandes, K. J. L., McKenzie, I. A., Mill, P., Smith, K. M., Akhavan, M., Barnabé-Heider, F., et al. (2004). A dermal niche for multipotent adult skin-derived precursor cells. *Nat. Cell Biol.* 6, 1082–1093. doi: 10.1038/ncb1181

Fokkema, I. F., Taschner, P. E., Schaafsma, G. C., Celli, J., Laros, J. F., and den Dunnen, J. T. (2011). LOVD v.2.0: the next generation in gene variant databases. *Hum. Mutat.* 32, 557–563. doi: 10.1002/humu.21438

Franz, D. N., Leonard, J., Tudor, C., Chuck, G., Care, M., Sethuraman, G., et al. (2006). Rapamycin causes regression of astrocytomas in tuberous sclerosis complex. *Ann. Neurol.* 59, 490–498. doi: 10.1002/ana.20784

Fu, C., and Ess, K. C. (2013). Conditional and domain-specific inactivation of the TSC2 gene in neural progenitor cells. *Genesis.* 51, 284–292. doi: 10.1002/dvg.22377

Gan, B., Sahin, E., Jiang, S., Sanchez-Aguilera, A., Scott, K. L., Chin, L., et al. (2008). mTORC1-dependent and -independent regulation of stem cell renewal, differentiation, and mobilization. *Proc. Natl. Acad. Sci.* 105, 19384–19389. doi: 10.1073/pnas.0810584105

Gao, L., Yue, M. M., Davis, J., Hyjek, E., and Schuger, L. (2014). In pulmonary lymphangioleiomyomatosis expression of progesterone receptor is frequently higher than that of estrogen receptor. *Virchows Arch.* 464, 495–503. doi: 10.1007/s00428-014-1559-9

Gilbert, E. R., Eby, J. M., Hammer, A. M., Klarquist, J., Christensen, D. G., Barfuss, A. J., et al. (2013). Positioning Ganglioside D3 as an immunotherapeutic target in lymphangioleiomyomatosis. *Am. J. Pathol.* 183, 226–234. doi: 10.1016/j.ajpath.2013.04.002

Glasgow, C. G., El-Chemaly, S., and Moss, J. (2012). Lymphatics in lymphangioleiomyomatosis and idiopathic pulmonary fibrosis. *Eur. Respir. Rev.* 21, 196–206. doi: 10.1183/09059180.00009311

Goh, S., Kwiatkowski, D. J., Dorer, D. J., and Thiele, E. A. (2005). Infantile spasms and intellectual outcomes in children with tuberous sclerosis complex. *Neurology* 65, 235–238. doi: 10.1212/01.wnl.0000168908.78118.99

Goncharova, E. A., Goncharov, D. A., Lim, P. N., Noonan, D., and Krymskaya, V. P. (2006). Modulation of cell migration and invasiveness by tumor suppressor TSC2 in lymphangioleiomyomatosis. *Am. J. Respir. Cell Mol. Biol.* 34, 473–480. doi: 10.1165/rcmb.2005-0374OC

Goto, J., Talos, D. M., Klein, P., Qin, W., Chekaluk, Y. I., Anderl, S., et al. (2011). Regulable neural progenitor-specific Tsc1 loss yields giant cells with organellar dysfunction in a model of tuberous sclerosis complex. *Proc. Natl. Acad. Sci.* 108, E1070–E1079. doi: 10.1073/pnas.1106454108

Green, A. J., Smith, M., and Yates, J. R. (1994). Loss of heterozygosity on chromosome 16p13.3 in hamartomas from tuberous sclerosis patients. *Nat. Genet.* 6, 193–196. doi: 10.1038/ng0294-193

Grenier, J., Teillet, M.-A., Grifone, R., Kelly, R. G., and Duprez, D. (2009). Relationship between neural crest cells and cranial mesoderm during head muscle development. *PLoS ONE* 4:e4381. doi: 10.1371/journal.pone.0004381

Grimmer, M. R., and Weiss, W. A. (2006). Childhood tumors of the nervous system as disorders of normal development. *Curr. Opin. Pediatr.* 18, 634–638. doi: 10.1097/MOP.0b013e32801080fe

Gu, X., Yu, J. J., Ilter, D., Blenis, N., Henske, E. P., and Blenis, J. (2013). Integration of mTOR and estrogen-ERK2 signaling in lymphangioleiomyomatosis pathogenesis. *Proc. Natl. Acad. Sci.* 110, 14960–14965. doi: 10.1073/pnas.1309110110

Han, S., Santos, T. M., Puga, A., Roy, J., Thiele, E. A., McCollin, M., et al. (2004). Phosphorylation of tuberin as a novel mechanism for somatic inactivation of the tuberous sclerosis complex proteins in brain lesions. *Cancer Res.* 64, 812–816. doi: 10.1158/0008-5472.CAN-03-3277

Hartman, N. W., Lin, T. V., Zhang, L., Paquelet, G. E., Feliciano, D. M., and Bordey, A. (2013). mTORC1 targets the translational repressor 4E-BP2, but not S6 kinase 1/2, to regulate neural stem cell self-renewal *in vivo*. *Cell Rep.* 5, 433–444. doi: 10.1016/j.celrep.2013.09.017

Hayashi, T., Fleming, M. V., Stetler-Stevenson, W. G., Liotta, L. A., Moss, J., Ferrans, V. J., et al. (1997). Immunohistochemical study of matrix metalloproteinases (MMPs) and their tissue inhibitors (TIMPs) in pulmonary lymphangioleiomyomatosis (LAM). *Hum. Pathol.* 28, 1071–1078. doi: 10.1016/S0046-8177(97)90061-7

Hayashi, T., Kumasaka, T., Mitani, K., Terao, Y., Watanabe, M., Oide, T., et al. (2011). Prevalence of uterine and adnexal involvement in pulmonary lymphangioleiomyomatosis: a clinicopathologic study of 10 patients. *Am. J. Surg. Pathol.* 35, 1776–1785. doi: 10.1097/PAS.0b013e318235edbd

Henske, E. P., and McCormack, F. X. (2012). Lymphangioleiomyomatosis - a wolf in sheep's clothing. *J. Clin. Invest.* 122, 3807–3816. doi: 10.1172/JCI58709

Henske, E. P., Scheithauer, B. W., Short, M. P., Wollmann, R., Nahmias, J., Hornigold, N., et al. (1996). Allelic loss is frequent in tuberous sclerosis kidney lesions but rare in brain lesions. *Am. J. Hum. Genet.* 59, 400–406.

Hindman, B. W., Gill, H. K., and Zuppan, C. W. (1997). Primitive neuroectodermal tumor in a child with tuberous sclerosis. *Skeletal Radiol.* 26, 184–187. doi: 10.1007/s002560050217

Ishiko, S., Yoshida, A., Kato, Y., and Kagokawa, H. (2006). Occult retinal and choroidal lesions in neurofibromatosis type 1. *Br. J. Ophthalmol.* 90, 1067–1068. doi: 10.1136/bjo.2006.092718

Jansen, A. C., Sancak, O., D'Agostino, M. D., Badhwar, A., Roberts, P., Gobbi, G., et al. (2006). Unusually mild tuberous sclerosis phenotype is associated with *TSC2* R905Q mutation. *Ann. Neurol.* 60, 528–239. doi: 10.1002/ana.21037

Jinno, H., Morozova, O., Jones, K. L., Biernaskie, J. A., Paris, M., Hosokawa, R., et al. (2010). Convergent genesis of an adult neural crest-like dermal stem cell from distinct developmental origins. *Stem Cells* 28, 2027–2040. doi: 10.1002/stem.525

Kandt, R. S. (2003). Tuberous sclerosis complex and neurofibromatosis type 1: the two most common neurocutaneous diseases. *Neurol. Clin.* 21, 983–1004. doi: 10.1016/S0733-8619(03)00004-5

Karbowniczek, M., Astrinidis, A., Balsara, B. R., Testa, J. R., Lium, J. H., Colby, T. V., et al. (2003). Recurrent lymphangiomyomatosis after transplantation: genetic analyses reveal a metastatic mechanism. *Am. J. Respir. Crit. Care Med.* 167, 976–982. doi: 10.1164/rccm.200208-969OC

Khare, L., Strizheva, G. D., Bailey, J. N., Au, K. S., Northrup, H., Smith, M., et al. (2001). A novel missense mutation in the GTPase activating protein homology region of TSC2 in two large families with tuberous sclerosis complex. *J. Med. Genet.* 38, 347–349. doi: 10.1136/jmg.38.5.347

Kirby, M. L., Gale, T. F., and Stewart, D. E. (1983). Neural crest cells contribute to normal aorticopulmonary septation. *Science* 220, 1059–1061. doi: 10.1126/science.6844926

Kirby, M. L., and Stewart, D. E. (1983). Neural crest origin of cardiac ganglion cells in the chick embryo: identification and extirpation. *Dev. Biol.* 97, 433–443. doi: 10.1016/0012-1606(83)90100-8

Kitaichi, M., Nishimura, K., Itoh, H., and Izumi, T. (1995). Pulmonary lymphangioleiomyomatosis: a report of 46 patients including a clinicopathologic study of prognostic factors. *Am. J. Respir. Crit. Care Med.* 151, 527–533. doi: 10.1164/ajrccm.151.2.7842216

Kobayashi, T., Minowa, O., Kuno, J., Mitani, H., Hino, O., and Noda, T. (1999). Renal carcinogenesis, hepatic hemangiomatosis, and embryonic lethality caused by a germ-line Tsc2 mutation in mice. *Cancer Res.* 59, 1206–1211.

Kobayashi, T., Minowa, O., Sugitani, Y., Takai, S., Mitani, H., Kobayashi, E., et al. (2001). A germ-line Tsc1 mutation causes tumor development and embryonic lethality that are similar, but not identical to, those caused by Tsc2 mutation in mice. *Proc. Natl. Acad. Sci. U.S.A.* 98, 8762–8767. doi: 10.1073/pnas.151033798

Köntges, G., and Lumsden, A. (1996). Rhombencephalic neural crest segmentation is preserved throughout craniofacial ontogeny. *Development* 122, 3229–3242.

Kristof, A. S., and Moss, J. (2011). "Lymphangioleiomyomatosis," in *Interstitial Lung Disease*, eds M. I. Schwarz and T. E. King (Shelton, CT: People's Medical Publishing House), 1061–1077.

Krueger, D. A., Care, M. M., Holland, K., Agricola, K., Tudor, C., Mangeshkar, P., et al. (2010). Everolimus for subependymal giant-cell astrocytomas in tuberous sclerosis. *N. Engl. J. Med.* 363, 1801–1811. doi: 10.1056/NEJMoa1001671

Kumasaka, T., Seyama, K., Mitani, K., Sato, T., Souma, S., Kondo, T., et al. (2004). Lymphangiogenesis in lymphangioleiomyomatosis: its implication in the progression of lymphangioleiomyomatosis. *Am. J. Surg. Pathol.* 28, 1007–1016. doi: 10.1097/01.pas.0000126859.70814.6d

Kwiatkowski, D. J. (2010). Animal models of lymphangioleiomyomatosis (LAM) and tuberous sclerosis complex (TSC). *Lymphat. Res. Biol.* 8, 51–57. doi: 10.1089/lrb.2009.0013

Kwiatkowski, D. J., Zhang, H., Bandura, J. L., Heiberger, K. M., Glogauer, M., el-Hashemite, N., et al. (2002). A mouse model of TSC1 reveals sex-dependent lethality from liver hemangiomas, and up-regulation of p70S6 kinase activity in Tsc1 null cells. *Hum. Mol. Genet.* 11, 525–534. doi: 10.1093/hmg/11.5.525

Lamb, R. F., Roy, C., Diefenbach, T. J., Vinters, H. V., Johnson, M. W., Jay, D. G., et al. (2000). The TSC1 tumour suppressor hamartin regulates cell adhesion through ERM proteins and the GTPase Rho. *Nat. Cell Biol.* 2, 281–287. doi: 10.1038/35010550

Laplante, M., and Sabatini, D. M. (2012). mTOR signaling in growth control and disease. *Cell* 149, 274–293. doi: 10.1016/j.cell.2012.03.017

Le, L. Q., Shipman, T., Burns, D. K., and Parada, L. F. (2009). Cell of origin and microenvironment contribution for NF1-associated dermal neurofibromas. *Cell Stem Cell* 4, 453–463. doi: 10.1016/j.stem.2009.03.017

Le Douarin, N. M., and Teillet, M. A. (1973). The migration of neural crest cells to the wall of the digestive tract in avian embryo. *J. Embryol. Exp. Morphol.* 30, 31–48.

Lee, C.-H., Hong, C.-H., Yu, H.-S., Chen, G.-S., and Yang, K.-C. (2010a). Transforming growth factor-β enhances matrix metalloproteinase-2 expression and activity through AKT in fibroblasts derived from angiofibromas in patients with tuberous sclerosis complex. *Br. J. Dermatol.* 163, 1238–1244. doi: 10.1111/j.1365-2133.2010.09971.x

Lee, P.-S., Tsang, S. W., Moses, M. A., Trayes-Gibson, Z., Hsiao, L.-L., Jensen, R., et al. (2010b). Rapamycin-insensitive up-regulation of MMP2and other genes in tuberous sclerosis complex 2–deficient lymphangioleiomyomatosis-like cells. *Am. J. Respir. Cell Mol. Biol.* 42, 227–234. doi: 10.1165/rcmb.2009-0050OC

Li, S., Moss, J., and Darling, T. N. (2013). Molecular mechanisms and therapeutics of skin hamartomas in tuberous sclerosis complex. *J. Dermatol. Clin. Res.* 2, 1007–1009.

Lopes, M. B., Altermatt, H. J., Scheithauer, B. W., Shepherd, C. W., and VandenBerg, S. R. (1996). Immunohistochemical characterization of subependymal giant cell astrocytomas. *Acta Neuropathol.* 91, 368–375. doi: 10.1007/s004010050438

Magri, L., Cambiaghi, M., Cominelli, M., Alfaro-Cervello, C., Cursi, M., Pala, M., et al. (2011). Sustained activation of mTOR pathway in embryonic neural stem cells leads to development of tuberous sclerosis complex-associated lesions. *Cell Stem Cell* 9, 447–462. doi: 10.1016/j.stem.2011.09.008

Magri, L., and Galli, R. (2013). mTOR signaling in neural stem cells: from basic biology to disease. *Cell. Mol. Life Sci.* 70, 2887–2898. doi: 10.1007/s00018-012-1196-x

Martignoni, G., Pea, M., Reghellin, D., Zamboni, G., and Bonetti, F. (2007). PEComas: the past, the present and the future. *Virchows Arch.* 452, 119–132. doi: 10.1007/s00428-007-0509-1

Matthews, H. K., Marchant, L., Carmona Fontaine, C., Kuriyama, S., Larraín, J., Holt, M. R., et al. (2008). Directional migration of neural crest cells *in vivo* is regulated by Syndecan-4/Rac1 and non-canonical Wnt signaling/RhoA. *Development* 135, 1771–1780. doi: 10.1242/dev.017350

McCormack, F. X., Inoue, Y., Moss, J., Singer, L. G., Strange, C., Nakata, K., et al. (2011). Efficacy and safety of sirolimus in lymphangioleiomyomatosis. *N. Engl. J. Med.* 364, 1595–1606. doi: 10.1056/NEJMoa1100391

Meikle, L., Talos, D. M., Onda, H., Pollizzi, K., Rotenberg, A., Sahin, M., et al. (2007). A mouse model of tuberous sclerosis: neuronal loss of Tsc1 causes dysplastic and ectopic neurons, reduced myelination, seizure activity, and limited survival. *J. Neurosci.* 27, 5546–5558. doi: 10.1523/JNEUROSCI.5540-06.2007

Micozkadioglu, H., Koc, Z., Ozelsancak, R., and Yildiz, I. (2010). Rapamycin therapy for renal, brain, and skin lesions in a tuberous sclerosis patient. *Ren. Fail.* 32, 1233–1236. doi: 10.3109/0886022X.2010.517345

Monsonego-Ornan, E., Kosonovsky, J., Bar, A., Roth, L., Fraggi-Rankis, V., Simsa, S., et al. (2012). Matrix metalloproteinase 9/gelatinase B is required for neural crest cell migration. *Dev. Biol.* 364, 162–177. doi: 10.1016/j.ydbio.2012.01.028

Moss, J., Avila, N. A., Barnes, P. M., Litzenberger, R. A., Bechtle, J., Brooks, P. G., et al. (2001). Prevalence and clinical characteristics of lymphangioleiomyomatosis (LAM) in patients with tuberous sclerosis complex. *Am. J. Respir. Crit. Care Med.* 164, 669–671. doi: 10.1164/ajrccm.164.4.2101154

Moury, J. D., and Jacobson, A. G. (1990). The origins of neural crest cells in the axolotl. *Dev. Biol.* 141, 243–253. doi: 10.1016/0012-1606(90)90380-2

Nagoshi, N., Shibata, S., Hamanoue, M., Mabuchi, Y., Matsuzaki, Y., Toyama, Y., et al. (2011). Schwann cell plasticity after spinal cord injury shown by neural crest lineage tracing. *Glia* 59, 771–784. doi: 10.1002/glia.21150

Nellist, M., Sancak, O., Goedbloed, M., Adriaans, A., Wessels, M., Maat-Kievit, A., et al. (2008). Functional characterisation of the TSC1-TSC2 complex to assess multiple *TSC2* variants identified in single families affected by tuberous sclerosis complex. *BMC Med. Genet.* 9:10. doi: 10.1186/1471-2350-9-10

Nellist, M., Sancak, O., Goedbloed, M., Rohe, C., van Netten, D., Mayer, K., et al. (2005). Distinct effects of single amino-acid changes to tuberin on the function of the tuberin-hamartin complex. *Eur. J. Hum. Genet.* 13, 59–68. doi: 10.1038/sj.ejhg.5201276

Nguyen-Vu, P. A., Fackler, I., Rust, A., DeClue, J. E., Sander, C. A., Volkenandt, M., et al. (2001). Loss of tuberin, the tuberous-sclerosis-complex-2 gene product is associated with angiogenesis. *J. Cutan. Pathol.* 28, 470–475. doi: 10.1034/j.1600-0560.2001.028009470.x

Niida, Y., Lawrence-Smith, N., Banwell, A., Hammer, E., Lewis, J., Beauchamp, R., et al. (1999). Analysis of both *TSC1* and *TSC2* for germline mutations in 126 unrelated patients with tuberous sclerosis. *Hum. Mutat.* 14, 412–422.

Niida, Y., Stemmer-Rachamimov, A. O., Logrip, M., Tapon, D., Perez, R., Kwiatkowski, D. J., et al. (2001). Survey of somatic mutations in tuberous sclerosis complex (TSC) hamartomas suggests different genetic mechanisms for pathogenesis of TSC lesions. *Am. J. Hum. Genet.* 69, 493–503. doi: 10.1086/321972

Normand, E. A., Crandall, S. R., Thorn, C. A., Murphy, E. M., Voelcker, B., Browning, C., et al. (2013). Temporal and mosaic tsc1 deletion in the developing thalamus disrupts thalamocortical circuitry, neural function, and behavior. *Neuron* 78, 895–909. doi: 10.1016/j.neuron.2013.03.030

Northrup, H., Krueger, D. A., and International Tuberous Sclerosis Complex Consensus Group (2013). Tuberous sclerosis complex diagnostic criteria update: recommendations of the 2012 iinternational tuberous sclerosis complex consensus conference. *Pediatr. Neurol.* 49, 243–254. doi: 10.1016/j.pediatrneurol.2013.08.001

Ocaña, O. H., Córcoles, R., Fabra, A., Moreno-Bueno, G., Acloque, H., Vega, S., et al. (2012). Metastatic colonization requires the repression of the epithelial-mesenchymal transition inducer Prrx1. *Cancer Cell* 22, 709–724. doi: 10.1016/j.ccr.2012.10.012

O'Connor, S. E., Kwiatkowski, D. J., Roberts, P. S., Wollmann, R. L., and Huttenlocher, P. R. (2003). A family with seizures and minor features of tuberous sclerosis and a novel TSC2 mutation. *Neurology* 61, 409–412. doi: 10.1212/01.WNL.0000073272.47681.BB

Onda, H., Lueck, A., Marks, P. W., Warren, H. B., and Kwiatkowski, D. J. (1999). Tsc2+/− mice develop tumors in multiple sites that express gelsolin and are influenced by genetic background. *J. Clin. Invest.* 104, 687–695. doi: 10.1172/JCI7319

Park, J., Yun, S.-K., Cho, Y.-S., Song, K.-H., and Kim, H.-U. (2014). Treatment of angiofibromas in tuberous sclerosis complex: the effect of topical rapamycin and concomitant laser therapy. *Dermatology (Basel)* 228, 37–41. doi: 10.1159/000357033

Park, S. H., Pepkowitz, S. H., Kerfoot, C., De Rosa, M. J., Poukens, V., Wienecke, R., et al. (1997). Tuberous sclerosis in a 20-week gestation fetus: immunohistochemical study. *Acta Neuropathol.* 94, 180–186. doi: 10.1007/s004010050691

Parkhitko, A., Myachina, F., Morrison, T. A., Hindi, K. M., Auricchio, N., Karbowniczek, M., et al. (2011). Tumorigenesis in tuberous sclerosis complex is autophagy and p62/sequestosome 1 (SQSTM1)-dependent. *Proc. Natl. Acad. Sci. U.S.A.* 108, 12455–12460. doi: 10.1073/pnas.1104361108

Pollizzi, K., Malinowska-Kolodziej, I., Doughty, C., Betz, C., Ma, J., Goto, J., et al. (2009). A hypomorphic allele of TSC2 highlights the role of TSC1-TSC2 in signaling to AKT and models mild human TSC2 alleles. *Hum. Mol. Genet.* 18, 2378–2387. doi: 10.1093/hmg/ddp176

Prabhakar, S., Goto, J., Zuang, X., Sena-Esteves, M., Bronson, R., Brockmann, J., et al. (2013). Stochastic model of Tsc1 lesions in mouse brain. *PLoS ONE* 8:e64224. doi: 10.1371/journal.pone.0064224

Qin, W., Bajaj, V., Malinowska, I., Lu, X., MacConaill, L., Wu, C.-L., et al. (2011). Angiomyolipoma have common mutations in TSC2 but No other common genetic events. *PLoS ONE* 6:e24919. doi: 10.1371/journal.pone.0024919

Qin, W., Chan, J. A., Vinters, H. V., Mathern, G. W., Franz, D. N., Taillon, B. E., et al. (2010a). Analysis of TSC cortical tubers by deep sequencing of TSC1, TSC2 and KRAS demonstrates that small second-hit mutations in these genes are rare events. *Brain Pathol.* 20, 1096–1105. doi: 10.1111/j.1750-3639.2010.00416.x

Qin, W., Kozlowski, P., Taillon, B. E., Bouffard, P., Holmes, A. J., Janne, P., et al. (2010b). Ultra deep sequencing detects a low rate of mosaic mutations in tuberous sclerosis complex. *Hum. Genet.* 127, 573–582. doi: 10.1007/s00439-010-0801-z

Raible, D. W., and Ungos, J. M. (2006). Specification of sensory neuron cell fate from the neural crest. *Adv. Exp. Med. Biol.* 589, 170–180. doi: 10.1007/978-0-387-46954-6_10

Real, C., Glavieux-Pardanaud, C., Le Douarin, N. M., and Dupin, E. (2006). Clonally cultured differentiated pigment cells can dedifferentiate and generate multipotent progenitors with self-renewing potential. *Dev. Biol.* 300, 656–669. doi: 10.1016/j.ydbio.2006.09.032

Real, C., Glavieux-Pardanaud, C., Vaigot, P., Le-Douarin, N., and Dupin, E. (2005). The instability of the neural crest phenotypes: schwann cells can differentiate into myofibroblasts. *Int. J. Dev. Biol.* 49, 151–159. doi: 10.1387/ijdb.041940cr

Rodgers, J. T., King, K. Y., Brett, J. O., Cromie, M. J., Charville, G. W., Maguire, K. K., et al. (2014). mTORC1 controls the adaptive transition of quiescent stem cells from G0 to G(Alert). *Nature* 510, 393–396. doi: 10.1038/nature13255

Rose, V. M., Au, K. S., Pollom, G., Roach, E. S., Prashner, H. R., and Northrup, H. (1999). Germ-line mosaicism in tuberous sclerosis: how common? *Am. J. Hum. Genet.* 64, 986–992. doi: 10.1086/302322

Rupp, P. A., and Kulesa, P. M. (2007). A role for RhoA in the two-phase migratory pattern of post-otic neural crest cells. *Dev. Biol.* 311, 159–171. doi: 10.1016/j.ydbio.2007.08.027

Sachdeva, R., Rothner, D. A., Traboulsi, E. I., Hayden, B. C., and Rychwalski, P. J. (2010). Astrocytic hamartoma of the optic disc and multiple café-au-lait macules in a child with neurofibromatosis type 2. *Ophthalmic Genet.* 31, 209–214. doi: 10.3109/13816810.2010.512356

Sato, T., Seyama, K., Fujii, H., Maruyama, H., Setoguchi, Y., Iwakami, S.-I., et al. (2002). Mutation analysis of the TSC1 and TSC2 genes in Japanese patients with pulmonary lymphangioleiomyomatosis. *J. Hum. Genet.* 47, 20–28. doi: 10.1007/s10038-002-8651-8

European Chromosome 16 Tuberous Sclerosis Consortium (1993). Identification and characterization of the tuberous sclerosis gene on chromosome 16. *Cell* 75, 1305–1315.

Selleck, M. A., and Bronner-Fraser, M. (1995). Origins of the avian neural crest: the role of neural plate-epidermal interactions. *Development* 121, 525–538.

Sepp, T., Yates, J. R., and Green, A. J. (1996). Loss of heterozygosity in tuberous sclerosis hamartomas. *J. Med. Genet.* 33, 962–964. doi: 10.1136/jmg.33.11.962

Shah, N. M., Groves, A. K., and Anderson, D. J. (1996). Alternative neural crest cell fates are instructively promoted by TGFbeta superfamily members. *Cell* 85, 331–343. doi: 10.1016/S0092-8674(00)81112-5

Shakhova, O. (2014). Neural crest stem cells in melanoma development. *Curr. Opin. Oncol.* 26, 215–221. doi: 10.1097/CCO.0000000000000046

Smolarek, T. A., Wessner, L. L., McCormack, F. X., Mylet, J. C., Menon, A. G., and Henske, E. P. (1998). Evidence that lymphangiomyomatosis is caused by TSC2 mutations: chromosome 16p13 loss of heterozygosity in angiomyolipomas and lymph nodes from women with lymphangiomyomatosis. *Am. J. Hum. Genet.* 62, 810–815. doi: 10.1086/301804

Strizheva, G. D., Carsillo, T., Kruger, W. D., Sullivan, E. J., Ryu, J. H., and Henske, E. P. (2001). The spectrum of mutations in TSC1 and TSC2 in women with tuberous sclerosis and lymphangiomyomatosis. *Am. J. Respir. Crit. Care Med.* 163, 253–258. doi: 10.1164/ajrccm.163.1.2005004

Takahashi, Y., Sipp, D., and Enomoto, H. (2013). Tissue interactions in neural crest cell development and disease. *Science* 341, 860–863. doi: 10.1126/science.1230717

Taneyhill, L. A. (2008). To adhere or not to adhere: the role of Cadherins in neural crest development. *Cell Adh. Migr.* 2, 223–230. doi: 10.4161/cam.2.4.6835

Terasaki, Y., Yahiro, K., Pacheco-Rodriguez, G., Steagall, W. K., Stylianou, M. P., Evans, J. F., et al. (2010). Effects of prolactin on TSC2-null eker rat cells and in pulmonary lymphangioleiomyomatosis. *Am. J. Respir. Crit. Care Med.* 182, 531–539. doi: 10.1164/rccm.200911-1737OC

Theveneau, E., and Mayor, R. (2012). Neural crest delamination and migration: from epithelium-to-mesenchyme transition to collective cell migration. *Dev. Biol.* 366, 34–54. doi: 10.1016/j.ydbio.2011.12.041

Tiberio, D., Franz, D. N., and Phillips, J. R. (2011). Regression of a cardiac rhabdomyoma in a patient receiving everolimus. *Pediatrics* 127, e1335–e1337. doi: 10.1542/peds.2010-2910

Toma, J. G., Akhavan, M., Fernandes, K. J., Barnabé-Heider, F., Sadikot, A., Kaplan, D. R., et al. (2001). Isolation of multipotent adult stem cells from the dermis of mammalian skin. *Nat. Cell Biol.* 3, 778–784. doi: 10.1038/ncb0901-778

Toma, J. G., McKenzie, I. A., Bagli, D., and Miller, F. D. (2005). Isolation and characterization of multipotent skin-derived precursors from human skin. *Stem Cells* 23, 727–737. doi: 10.1634/stemcells.2004-0134

Trindade, A. J., Medvetz, D. A., Neuman, N. A., Myachina, F., Yu, J., Priolo, C., et al. (2013). MicroRNA-21 is induced by rapamycin in a model of tuberous sclerosis (TSC) and lymphangioleiomyomatosis (LAM). *PLoS ONE* 8:e60014. doi: 10.1371/journal.pone.0060014

Tsai, J. H., Donaher, J. L., Murphy, D. A., Chau, S., and Yang, J. (2012). Spatiotemporal regulation of epithelial-mesenchymal transition is essential for squamous cell carcinoma metastasis. *Cancer Cell* 22, 725–736. doi: 10.1016/j.ccr.2012.09.022

Tyburczy, M. E., Wang, J. A., Li, S., Thangapazham, R., Chekaluk, Y., Moss, J., et al. (2014). Sun exposure causes somatic second-hit mutations and angiofibroma development in tuberous sclerosis complex. *Hum. Mol. Genet.* 23, 2023–2029. doi: 10.1093/hmg/ddt597

Uhlmann, E. J., Apicelli, A. J., Baldwin, R. L., Burke, S. P., Bajenaru, M. L., Onda, H., et al. (2002). Heterozygosity for the tuberous sclerosis complex (TSC) gene products results in increased astrocyte numbers and decreased p27-Kip1 expression in TSC2+/− cells. *Oncogene* 21, 4050–4059. doi: 10.1038/sj.onc.1205435

Urban, T., Lazor, R., Lacronique, J., Murris, M., Labrune, S., Valeyre, D., et al. (1999). Pulmonary lymphangioleiomyomatosis. A study of 69 patients. Groupe d'Etudes et de Recherche sur les Maladies "Orphelines" Pulmonaires (GERM'O'P). *Medicine (Baltimore)* 78, 321–337. doi: 10.1097/00005792-199909000-00004

van Slegtenhorst, M., Verhoef, S., Tempelaars, A., Bakker, L., Wang, Q., Wessels, M., et al. (1999). Mutational spectrum of the TSC1 gene in a cohort of 225 tuberous sclerosis complex patients: no evidence for genotype-phenotype correlation. *J. Med. Genet.* 36, 285–289.

Verhoef, S., Bakker, L., Tempelaars, A. M., Hesseling-Janssen, A. L., Mazurczak, T., Jozwiak, S., et al. (1999). High rate of mosaicism in tuberous sclerosis complex. *Am. J. Hum. Genet.* 64, 1632–1637. doi: 10.1086/302412

Way, S. W., McKenna, J., Mietzsch, U., Reith, R. M., Wu, H. C.-J., and Gambello, M. J. (2009). Loss of Tsc2 in radial glia models the brain pathology of tuberous sclerosis complex in the mouse. *Hum. Mol. Genet.* 18, 1252–1265. doi: 10.1093/hmg/ddp025

Webb, D. W., Fryer, A. E., and Osborne, J. P. (1996). Morbidity associated with tuberous sclerosis: a population study. *Dev. Med. Child Neurol.* 38, 146–155. doi: 10.1111/j.1469-8749.1996.tb12086.x

Wong, V. (2006). Study of the relationship between tuberous sclerosis complex and autistic disorder. *J. Child Neurol.* 21, 199–204. doi: 10.2310/7010.2006.00046

Xie, W.-B., Li, Z., Shi, N., Guo, X., Tang, J., Ju, W., et al. (2013). Smad2 and myocardin-related transcription factor B cooperatively regulate vascular smooth muscle differentiation from neural crest cells. *Circ. Res.* 113, e76–e86. doi: 10.1161/CIRCRESAHA.113.301921

Yntema, C. L., and Hammond, W. S. (1954). The origin of intrinsic ganglia of trunk viscera from vagal neural crest in the chick embryo. *J. Comp. Neurol.* 101, 515–541. doi: 10.1002/cne.901010212

Yu, J., and Henske, E. P. (2010). mTOR activation, lymphangiogenesis, and estrogen-mediated cell survival: the "perfect storm" of pro-metastatic factors in LAM pathogenesis. *Lymphat. Res. Biol.* 8, 43–49. doi: 10.1089/lrb.2009.0020

Yu, J. J., Robb, V. A., Morrison, T. A., Ariazi, E. A., Karbowniczek, M., Astrinidis, A., et al. (2009). Estrogen promotes the survival and pulmonary metastasis of tuberin-null cells. *Proc. Natl. Acad. Sci.* 106, 2635–2640. doi: 10.1073/pnas.0810790106

Yuan, E., Tsai, P. T., Greene-Colozzi, E., Sahin, M., Kwiatkowski, D. J., and Malinowska, I. A. (2012). Graded loss of tuberin in an allelic series of brain models of TSC correlates with survival, and biochemical, histological and behavioral features. *Hum. Mol. Genet.* 21, 4286–4300. doi: 10.1093/hmg/dds262

Zeng, L.-H., Rensing, N. R., Zhang, B., Gutmann, D. H., Gambello, M. J., and Wong, M. (2011). Tsc2 gene inactivation causes a more severe epilepsy phenotype than Tsc1 inactivation in a mouse model of tuberous sclerosis complex. *Hum. Mol. Genet.* 20, 445–454. doi: 10.1093/hmg/ddq491

Zhe, X., and Schuger, L. (2004). Combined smooth muscle and melanocytic differentiation in lymphangioleiomyomatosis. *J. Histochem. Cytochem.* 52, 1537–1542. doi: 10.1369/jhc.4A6438.2004

Zhou, J., Shrikhande, G., Xu, J., McKay, R. M., Burns, D. K., Johnson, J. E., et al. (2011). Tsc1 mutant neural stem/progenitor cells exhibit migration deficits and give rise to subependymal lesions in the lateral ventricle. *Genes. Dev.* 25, 1595–1600. doi: 10.1101/gad.16750211

Conflict of Interest Statement: The authors declare that the research was conducted in the absence of any commercial or financial relationships that could be construed as a potential conflict of interest.

15

Cellular reprogramming for understanding and treating human disease

Riya R. Kanherkar[1], Naina Bhatia-Dey[1], Evgeny Makarev[2] and Antonei B. Csoka[1]*

[1] *Epigenetics Laboratory, Department of Anatomy, Howard University, Washington, DC, USA*
[2] *InSilico Medicine, Emerging Technology Center, Johns Hopkins University Eastern, Baltimore, MD, USA*

Edited by:
Dan S. Kaufman, University of Minnesota, USA

Reviewed by:
In-Hyun Park, Yale University, USA
Shree Ram Singh, National Cancer Institute, USA
George Scaria, University of Minnesota, USA

****Correspondence:***
Antonei B. Csoka, Epigenetics Laboratory, Department of Anatomy, Howard University, 520 W St. NW, Mudd 431, Washington, DC 20059, USA
e-mail: antonei.csoka@howard.edu

In the last two decades we have witnessed a paradigm shift in our understanding of cells so radical that it has rewritten the rules of biology. The study of cellular reprogramming has gone from little more than a hypothesis, to applied bioengineering, with the creation of a variety of important cell types. By way of metaphor, we can compare the discovery of reprogramming with the archeological discovery of the Rosetta stone. This stone slab made possible the initial decipherment of Egyptian hieroglyphics because it allowed us to see this language in a way that was previously impossible. We propose that cellular reprogramming will have an equally profound impact on understanding and curing human disease, because it allows us to perceive and study molecular biological processes such as differentiation, epigenetics, and chromatin in ways that were likewise previously impossible. Stem cells could be called "cellular Rosetta stones" because they allow also us to perceive the connections between development, disease, cancer, aging, and regeneration in novel ways. Here we present a comprehensive historical review of stem cells and cellular reprogramming, and illustrate the developing synergy between many previously unconnected fields. We show how stem cells can be used to create *in vitro* models of human disease and provide examples of how reprogramming is being used to study and treat such diverse diseases as cancer, aging, and accelerated aging syndromes, infectious diseases such as AIDS, and epigenetic diseases such as polycystic ovary syndrome. While the technology of reprogramming is being developed and refined there have also been significant ongoing developments in other complementary technologies such as gene editing, progenitor cell production, and tissue engineering. These technologies are the foundations of what is becoming a fully-functional field of regenerative medicine and are converging to a point that will allow us to treat almost any disease.

Keywords: reprogramming, stem cells, aging, disease, epigenetics

INTRODUCTION

PARADIGM SHIFTS AND SEEING THINGS IN NEW WAYS

Stem cell biology is sometimes thought of as a field that offers promises that can't be kept; for example the potential to one day treat heart attacks or repair spinal cord injuries. But focusing on regenerative medicine's so-far small impact on patient care, misses a shift in our understanding of cells so radical that it has rewritten the rules of biology in less than two decades.

How much have things changed? By way of analogy, the Rosetta stone was a stone slab found in 1799 that bore parallel inscriptions in Greek, Demotic characters, and Egyptian hieroglyphics that made possible an unprecedented decipherment of the latter. We propose that the discovery of the mechanisms for producing and modulating stem cells, in particular induced pluripotent stem cells, will have an equally profound impact in understanding human health and disease (**Figure 1**). Stem cells can be conceptualized as "cellular Rosetta stones" because they are enabling us to create and compare cells with diseases or cellular phenomena that are poorly understood (non-deciphered Hieroglyphics), with isogenic cells that are completely normal (well-understood, Ancient Greek or Demotic). Because of this, reprogrammed stem cells will be the quintessential tool for studying epigenetics, aging, cancer, and regeneration. This new possibility will open up opportunities to study cells that go awry in disease and to one day use patients' own cells to heal them.

Stem cells are emerging as the ultimate tool for understanding biological processes at the molecular and cellular level, as well as treating debilitating diseases. By using reprogramming as the means to make "cellular Rosetta stones," it will be possible to form an accurate understanding of currently unknown biological phenomena, and perhaps eventually develop a universal synthesis of cell biology (**Figure 1**).

In this review we present a comprehensive summary of the history of cellular reprogramming from its initial conception to the present day. We also describe how these techniques are being used to better understand basic biology and human disease, and lay

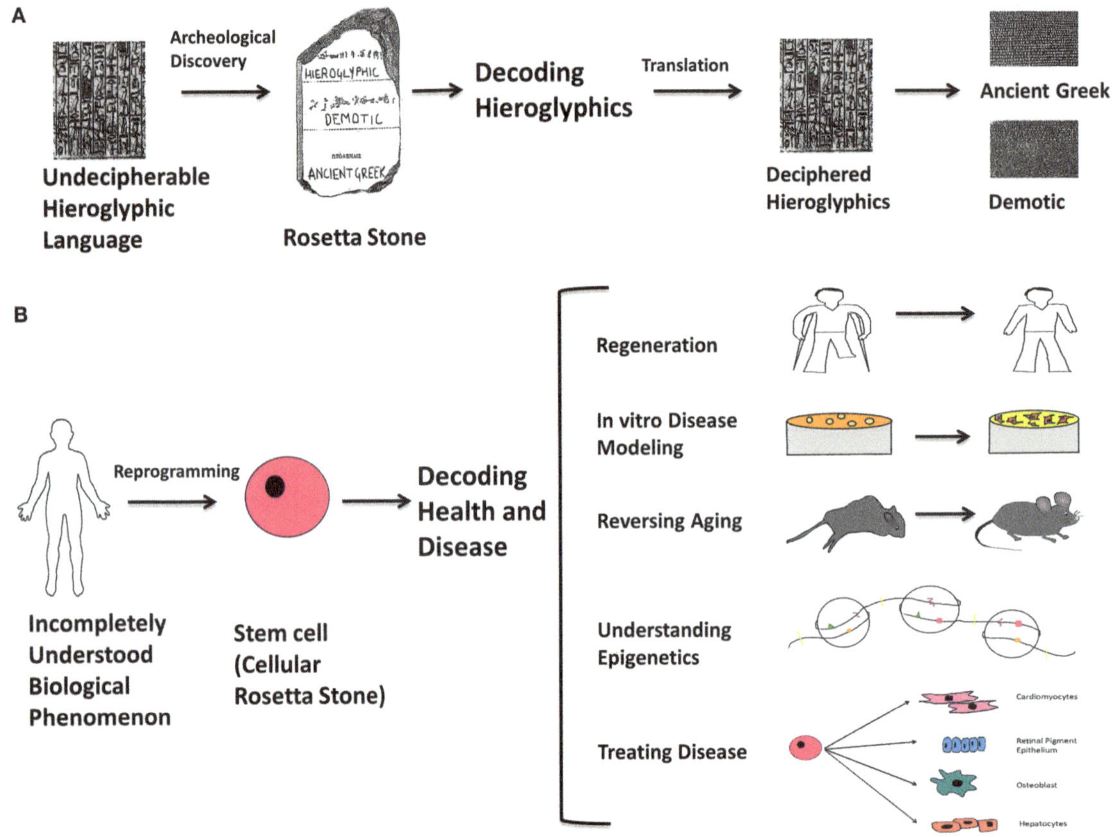

FIGURE 1 | The Rosetta stone analogy. The Rosetta stone is an archeological slab with the same text in three different languages; Hieroglyphics, Demotic, and Ancient Greek, and its discovery proved to be a turning point in understanding the Hieroglyphic language. By direct comparison of the three languages, it was possible to decipher previously unintelligible Hieroglyphics from the other two known languages. An analogy can be made with respect to stem cells: they can be thought of as "cellular Rosetta stones" because they are key to understanding the multi-faceted mysteries underlying human health and disease. Stem cells can be conceptualized as "cellular Rosetta stones" because they are enabling us to create and compare cells with diseases or cellular phenomena that are poorly understood (non-deciphered Hieroglyphics), with isogenic cells that are completely normal (well-understood, Ancient Greek or Demotic). This is the contribution of stem cells toward modeling and treating diseases and generating functional cell types. In addition, they are becoming quintessential for studying epigenetics, aging, cancer and regeneration. It won't be long before they emerge as the ultimate practical tool and make their mark on the conceptualization of biological science in terms of understanding molecular and cellular events and treating debilitating diseases. By using reprogramming as the means to make cellular Rosetta stones, it will be possible to form a universal understanding of currently unknown biological phenomena and develop an accurate philosophy for cellular processes, disease and therapy. **(A)** Illustrates the rationale behind the way the Rosetta stone was used to decode hieroglyphics. **(B)** Illustrates the analogously unexplained biological phenomena to which stem cells can be made with various characteristics on an isogenic background and used to understand them; for example (1) limb-regeneration (2) disease modeling at the cellular level (3) treating organismal aging (4) understanding epigenetic mechanisms underlying diseases (5) generating lineage specific cell types to treat degenerative and chronic diseases, or acute injuries.

the foundations for a fully functional discipline of regenerative medicine.

Therefore, to begin, what exactly are stem cells?

STEM CELLS

Stem cells are undifferentiated cells that possess two unique characteristics:

(1) The ability to self-renew through numerous cycles of cell division while maintaining an undifferentiated state, and
(2) Pluripotent potential with the ability to generate progenitors of multiple lineages (International Stem Cell et al., 2007; Mitalipov and Wolf, 2009). These qualities confer unique regenerative abilities upon stem cells, and make them a desirable commodity in the endeavor to replenish, regenerate and repair human tissues (Gurtner et al., 2007; Wu et al., 2007). There are essentially two types of stem cells in mammals: somatic stem cells (SSCs, also sometimes called progenitor cells), and embryonic stem cells (ESCs) (Niwa et al., 2000).

SSCs are found in various adult tissues, such as bone marrow, adipose tissue and blood (including umbilical cord blood) (Jiang et al., 2002; Terai et al., 2006; Gimble et al., 2007; Luna-Zurita and Bruneau, 2013). They are partially differentiated cells found throughout the body that possess the capacity to divide in order to replenish damaged tissue. They are able to differentiate into more than one cell type, but unlike ESCs, they are restricted to a specific cellular lineage. The advantage of SSCs is that their production doesn't require the destruction of embryos, or reprogramming (see below) (Gardner, 2002).

ESCs are produced from eggs derived from the inner cell mass of fertilized embryos. Contrasted with SSCs, ESCs are pluripotent, and can differentiate into *all* of the three primary germ layers (ectoderm, endoderm, and mesoderm) and their derivatives. ESCs are characterized by long-term self-renewal, and can be grown in cell culture as an undifferentiated, pluripotent population. Regulation of pluripotency networks is important for maintaining the undifferentiated state of such cells in culture, or during differentiation to obtain desired cell types. The transcription factor (TF), Oct 3/4 is the master regulator of pluripotency, and its precise levels during development are responsible for the differentiation of ESCs into specific lineages, whereas repression of Oct 3/4 results in loss of pluripotency and formation of trophoectoderm (Niwa et al., 2000).

ESCs can be directed to differentiate into a particular cell type through alteration of culture conditions and/or the supplementation of differentiation signals. Understanding the differentiation process has provided insights into de-differentiation and trans-differentiation strategies as well. Dedifferentiation is the formation of pluripotent or multipotent stem cells from terminally differentiated somatic cells, i.e., reverting to a state of increased developmental plasticity, and becoming ready to accept a new identity (Halley-Stott et al., 2013). Transdifferentiation is the process in which a particular somatic cell is switched from one lineage-specific identity to a completely different identity (Graf, 2011; Vierbuchen and Wernig, 2012); in other words, the direct conversion of one type of somatic cell into another type, bypassing the intermediate step of dedifferentiation.

The discovery of ESCs (Evans and Kaufman, 1981; Martin, 1981) eventually prompted the search for discovering artificial dedifferentiation techniques to confer the properties of ESCs onto somatic cells by altering epigenomic activity, such that the derived cells are pluripotent and capable of giving rise to embryonic-like stem cells. These techniques are collectively referred to as cellular reprogramming. But before we describe these various techniques, we will provide some background on the history of how we arrived at today's reprogramming technology.

HISTORY AND DEVELOPMENT OF CELLULAR REPROGRAMMING

In 1909, Ethel Browne Harvey, who was known for her work on sea urchins, was the first to show that cell transplants could induce a secondary axis of polarity in the host. Harvey's experiments were the basis for the discovery of Spemann's organizer (Lenhoff, 1991). In 1928, Hans Spemann and Hilde Mangold, in a quest to discover the factors responsible for embryonic determination and cell differentiation, performed classical embryology experiments with salamanders and demonstrated cell-to-cell induction, in which a group of cells or organizing centers signal differentiation in neighboring cells and hence regulate their fate in the embryo (De Robertis, 2006). The cells responsible for this kind of phenomenon came to be known as the Spemann organizer, which over subsequent decades led to many experiments in molecular embryology aimed at finding inducing factors responsible for early embryonic determination and cell fate (Grunz, 2001). Further, Spemann had proposed an experiment to determine whether differentiated cells could be restored to an embryonic state, or if the cells continued to remain specialized (Subramanyam, 2013). Spemann reasoned that if a nucleus from a differentiated cell implanted in a previously enucleated egg developed into a normal embryo, this would prove that the transplanted nucleus retained a genome fully capable of directing all types of differentiation. In other words, a differentiated nucleus could still be totipotent.

SOMATIC CELL NUCLEAR TRANSFER

In 1938, Spemann published an account of his experiments with a prototypical nuclear transfer technique (Spemann, 1938). Using a piece of hair wrapped around a newly-fertilized salamander egg, he separated the egg's nucleus on one side, with the cytoplasm on the other. After the nucleated side divided four times, creating a 16-cell embryo, he removed the hair and allowed a nucleus to slide back into the separated cytoplasm. Cell division now began on this side as well, and by putting the hair loop back again tightly, he broke the two embryos apart. The result was a twin set of salamanders.

This important work showed that the nucleus remained totipotent after undergoing four divisions, but Spemann wondered whether nuclei from much older embryos, or even adult animals, had similar potential. He wrote that transplanting an older nucleus into an egg would be a "fantastical experiment." What he was proposing is what eventually became known as somatic cell nuclear transfer (SCNT). However, for the next 14 years, scientists struggled with making it work.

But in 1952 Briggs and King performed the first successful SCNT experiment showing the reversal of cellular identity. They showed that the transfer of a frog nucleus from a blastula cell to an enucleated egg gave rise to a cleaved blastula (Briggs and King, 1952). This was followed by the pioneering experiments of John Gurdon in 1962, involving transfer of nuclei from terminally differentiated frog intestinal epithelium cells into enucleated eggs to produce normal tadpoles, which proved that the nucleus contains all the genetic information required to give rise to all the differentiated cells in the organism (Gurdon, 1962). After incremental improvements in SCNT techniques by various laboratories over the intervening three decades, the cloning of "Dolly" in 1996 was a major breakthrough, and the first ever successful mammalian cloning experiment, which involved transplanting quiescent nuclei from cultured adult sheep mammary gland cells into enucleated sheep eggs (Campbell et al., 1996). The creation of SCNT-derived human ESCs required further improvements in technology and almost another two decades, but in 2013, after a long and challenging pursuit, Mitalipov and colleagues finally succeeded in creating SCNT-derived human embryonic stem cell (hESC) lines (Tachibana et al., 2013).

SCNT has, over the decades, evolved as a technique, which can be used for reproductive cloning as well as to produce lines of hESCs. The attraction of using an egg for nuclear reprogramming is that, in nature, it has near 100% efficiency in reprogramming sperm nuclei (Gurdon and Melton, 2008; Teperek and Miyamoto, 2013). The egg uses molecular chaperones and enzymes to erase the epigenetic signatures during synkaryon formation and cell division, resulting in extremely efficient reprogramming (Kikyo and Wolffe, 2000). Variations in the technique used to produce offspring through SCNT such as improvement of processes like oocyte maturation, enucleation, donor nucleus transfer, activation and culture, along with prior epigenetic modification of the donor nucleus, can improve the efficiency of the process (Campbell et al., 2007).

INDUCED PLURIPOTENCY

Once SCNT in mammals was achieved, researchers wondered if the same result could be achieved without using eggs. Thanks to the pioneering experiments of Yamanaka, it is now possible to create close equivalents of ESCs without using embryos by engineering their creation *in vitro*. In 2006 and 2007, Yamanaka and colleagues identified conditions that allowed somatic cells to be genetically reprogrammed into induced pluripotent stem cells (iPSCs) that are almost as potent as ESCs. This was a significant breakthrough and was achieved by transfection of mouse and human fibroblasts with the transcription factors (TFs) Oct4, Sox2, Klf4, and c-Myc, (OSKM) using retroviral vectors to induce a "forced" expression of specific genes that results in iPSC formation (Takahashi and Yamanaka, 2006; Okita et al., 2007). Subsequent manipulation using different reprogramming factors such as Nanog rendered the reprogramming process more efficient (Wernig et al., 2007). There are now many different such pathways of inducing pluripotency, because different combinations of reprogramming factors can all achieve complete reprogramming (Takahashi, 2012). In addition to such reprogramming factors, introduction of improvements such as p53 knockdown and telomerase overexpression, along with small molecules that exert epigenetic effects (see Section Small Molecules below) are now being used routinely (Batista, 2014; Campos-Sanchez and Cobaleda, 2014). Besides using different combinations of TFs, iPSCs have also been induced using methods other than retroviral vectors, such as DNA and RNA (integrating plasmids, episomal plasmids and transposons), cell penetrating peptides, and small molecules (see Small Molecules below) (Li et al., 2014).

At present, a primary concern is the ability to achieve pluripotency in iPSCs while maintaining the functionality of ESC's—irrespective of the type of somatic cells used at the outset. For example, it is possible that less differentiated cells of the hematopoietic lineage reprogram more efficiently than differentiated cells, suggesting that the use of SSCs/progenitor cells as starting material could maximize the efficiency of reprogramming (Eminli et al., 2009). This may be because less differentiated progenitors possess less condensed chromatin in specific regions, which is more accessible to reprogramming factors. Also progenitors in general may be more susceptible to the disruption of their transcriptional networks, or possess a transcriptome that has higher resemblance to the transcriptome of ESCs (Papp and Plath, 2011).

A very recent refinement to induced pluripotency has been achieved which allows the *in vitro* reprogramming of cells back to a ground state of pluripotency that is even closer to ESCs. Novel human naïve stem cells have been derived by adding a unique combination of cytokines and small molecule inhibitors to the standard reprogramming cocktail (Gafni et al., 2013). These ground state stem cells closely resemble mouse iPSCs and ESCs (relatively more pluripotent than human ESCs or iPSCs), exhibiting hallmarks of naïve pluripotency that include driving Oct 4 transcription by its distal enhancer, retaining a pre-inactivation X chromosome state and a global reduction in DNA methylation and an H3K27me3 repressive chromatin mark deposition on developmental regulatory gene promoters (Gafni et al., 2013). The epigenetic changes induced by naïve human stem cell medium (NSHM) conditions indicated that naïve conditions are likely to resolve previously described technical phenotypes of epigenetic memory, lineage differentiation biases and aberrant reprogramming in human iPSCs and ESCs, thus providing a stable source for treating and modeling diseases (Gafni et al., 2013).

Also, recent studies in mice have shown that the *in vivo* transitory induction of OSKM generated teratomas, indicating that *in situ* reprogramming can give rise to iPSCs from a variety of cell types, and these *in vivo* iPSCs more closely resembled ESCs than *in vitro* generated iPSCs, with a more primitive and plastic state (Abad et al., 2013) (**Figure 3**). These *in vivo* iPSCs can further generate desired cell types in the presence of appropriate signals for re-differentiation (shown in **Figure 3**).

The most attractive application of induced pluripotency using the newest techniques is of course the production of patient-specific iPSCs for replacement of damaged, aged, or non-functional tissue, which will be discussed in later sections. While both ESCs and iPSCs are desirable for reasons of potency, the availability of ESCs is hindered by practical (egg availability) and ethical concerns, while iPSC production still remains relatively inefficient.

CELL FUSION

Another possible alternative to SCNT is cell fusion, in which two cells fuse together in the presence of agents like Sendai virus, polyethylene glycol (PEG), chimeric hemagglutinins, or an electric pulse to give rise to homokaryons from the same type of cells or heterokaryons from different types of cell (Soza-Ried and Fisher, 2012). In the case of heterokaryon formation, the dominant cell is the larger and actively dividing partner that imposes its own pattern of gene expression on the other. But cells fused in this way do not usually proliferate well, limiting their therapeutic value (Gurdon and Melton, 2008).

CELL-EXTRACT TREATMENT

In recent years, cell extracts (usually extracted from embryonic-like stem cells) have been used to derive pluripotent cells that can give rise to diverse cell lineages following differentiation (Patel and Yang, 2010). The use of extracts derived from pluripotent cells such as ESCs can also trigger the formation of ESC-like colonies with the upregulation of pluripotency genes and the downregulation of somatic genes such as Lamin A (*LMNA*) (Alberio et al., 2006). Notably, remodeling of the nuclear lamina has been used as a marker for reprogramming events in experiments involving the incubation of *Xenopus laevis* somatic cells in egg-extract in which reprogramming events could be tracked by the resulting nuclear configuration which acted as an indicator of increased plasticity

(Alberio et al., 2006). *LMNA* is also a marker of differentiation of human ESC's into somatic cells (Constantinescu et al., 2006).

SMALL MOLECULES

Another recent addition to the different types of reprogramming techniques is the use of small molecules, with or without a combination of TFs. It is thought that small molecules might lead to a better clinical approach because they involve fewer genetic manipulations than TF-based reprogramming (Pandian et al., 2014). A recent breakthrough study showed that a combination of just seven small-molecules was enough to chemically reprogram mouse embryonic fibroblasts (MEFs) such that Oct4 was dispensable for reprogramming (Hou et al., 2013). Such chemically induced pluripotent stem cells (CiPSCs) were similar to ESCs in terms of gene expression profile and epigenetic state, and did not require any exogenous expression of master pluripotency genes at all (Hou et al., 2013).

TRANSDIFFERENTIATION/DIRECT REPROGRAMMING

Transdifferentiation techniques might make it possible to produce patient-specific cells in a faster and more efficient manner by avoiding the intermediate stage of iPSC formation, and thus preventing the risk of tumorigenesis and immunogenicity associated with iPSCs (Ma et al., 2013) (**Figure 3**). For example, transdifferentiation studies *in vivo* have already demonstrated the conversion of adult murine pancreatic exocrine cells to β-cell-like populations using the TFs Ngn3, Pdx1, and Mafa (Zhou et al., 2008).

iPSC TF-based transdifferentiation is a new method of transdifferentiation that uses the transient overexpression of iPSC related TFs and cell-type-specific signals (growth factors, cytokines, and small molecules) to reprogram somatic cells to diverse lineage-specific cells or multipotent progenitors without transitioning through the iPSC stage (Ma et al., 2013). An example of such inter-lineage transdifferentiation is the generation of expandable neural progenitors from fibroblasts by the transient induction of OSKM along with appropriate signaling inputs (Kim et al., 2011). Another example is the conversion of human dermal fibroblasts to multipotent hematopoetic progenitors of myeloid, erythroid and megakaryocytic lineages via induction of OCT4 and its specific binding to regulatory regions of hematopoetic-specific genes along with cytokine supplementation (Szabo et al., 2010).

REGENERATION

Finally, the process of regeneration involves local dedifferentiation of somatic cells into a blastema composed of SSCs, which are multipotent and capable of proliferating, and re-differentiating into different lineages, and which can repopulate the damaged or degenerated tissue with functional cells. In human adults somatic cells cannot normally dedifferentiate and the process of wound healing involves scar tissue formation caused by the deposition of collagen, which is the most abundant part of the extracellular matrix secreted during wound healing. Thus, as compared to lower vertebrates like salamanders that are capable of complete regeneration with very little or no scar tissue, human adults have limited regenerative capacity. Regenerative capacity in humans seems to have diminished during the course of evolution compared to our lower vertebrate cousins as a mechanism to fight cancer.

However, regeneration in humans is not completely non-existent. The developing fetus is capable of scarless wound healing and this capacity is lost in gestation. For example, in the case of cardiac regeneration in neonates, the same extracellular matrix that contributes to scar formation might be involved in signaling an increase in proliferative capacity (Porrello and Olson, 2014). Thus, the much-coveted mammalian fetal regeneration capacity including wound healing without scar-tissue formation is a result of the extracellular matrix modulators, growth factors as well as cytokines (Colwell et al., 2005).

In spite of having wide-ranging information about the regenerative capacity of invertebrates, absolute knowledge of the limits of mammalian regenerative capacity is obscure because our understanding of mammalian stem cell biology usually comes from isolated stem cells, rather than from actually studying them *in vivo*. Given the fact that cells are dynamic entities and their properties change according to the environment they are in, the actual information on the number and functionality of stem cells *in vivo* is not accurately known (Sanchez Alvarado and Yamanaka, 2014).

Apart from harvesting the intrinsic regenerative potential of stem cells, mammalian regeneration can be induced through the administration of TFs known to locally dedifferentiate cells (transient) followed by re-differentiation. In a landmark study, it was shown that the upregulation of a single transcription factor, FOXN1, resulted in regeneration of the thymus in aged mice to an extent that it restored thymopoesis, a ground breaking example of *in vivo* mammalian regeneration (Bredenkamp et al., 2014). Such recent efforts in regeneration research have helped to advance our knowledge to the point where restoration of regenerative potential in mammals is perhaps within reach.

All of the above cellular reprogramming methods remodel chromatin and cause epigenetic changes in the target cells. But before we describe the application of the above methods to understanding and treating disease, we will first provide an explanation of what exactly is meant by "epigenetics," with a brief summary of this rapidly developing field.

EPIGENETICS

OVERVIEW OF EPIGENETICS

Epigenetics is the study of heritable changes in gene activity exclusive of direct modification of the DNA sequence, and includes DNA methylation and post-translational histone modification (Berger et al., 2009). The epigenome is a collection of the DNA methylation states and covalent modifications of histone proteins along the genome, and it differs in each cell type. Epigenetic mechanisms play an important role in the control of gene expression by organizing the nuclear architecture of chromosomes, restricting or facilitating TF access to DNA, and preserving a memory of past transcriptional activities (Rivera and Ren, 2013).

Epigenetic theory explains how the genome and environment work in tandem, involving mechanisms that affect DNA by regulating gene expression (Weinhold, 2006), and this interaction can operate across the entire human lifespan (Kanherkar et al., 2014).

Epigenetic modifications are of course crucially important for driving a cell toward an appropriate function during differentiation. The complexity of signaling during differentiation causes a cell's DNA to acquire specific epigenetic marks that restrict the expression of specific genes, and this inactivation survives cell division. DNA methylation is a signaling tool that occurs naturally on cytosine bases at CpG island promoter sequences and inactivates genes (Phillips, 2008). This type of epigenetic modification is associated with regulation of gene transcription, X-chromosome inactivation, and regulation of cellular development and differentiation (Bird, 2007). Another method of gene regulation is through the remodeling of chromatin. Remodeling occurs by post-translational modification of the amino acids that make up histone proteins via acetylation, methylation, phosphorylation and ubiquitination (Lunyak and Rosenfeld, 2008). Such epigenetic signatures are maintained by histone modifying enzymes such as histone acetyltransferases and histone methyltransferases (known as the "writers") and histone demethylases and histone deacetylases (known as the "erasers") which act as co-activators or co-repressors of OSKM respectively at different stages of reprogramming, and thereby influence iPSC formation (Apostolou and Hochedlinger, 2013) (**Figure 2**).

EPIGENETICS DURING CELLULAR REPROGRAMMING

Cellular reprogramming affects epigenetics at two different levels: firstly, through histone modification at the level of chromatin, and secondly through regulation of epigenetic marks on pluripotency genes and lineage-specifying genes that promote differentiation of various progenitors cells (Festuccia et al., 2013; Van Oevelen et al., 2013; Watanabe et al., 2013) (**Figure 2**). A major mechanism affecting epigenetic plasticity is genome-wide methylation and its regulation (Watanabe et al., 2013); nevertheless, it has to be accompanied by other epigenetic modulators that affect the state of pluripotency, totipotency, and their erasure (Loh and Lim, 2012). These modulators include histone acetylases and deacetylases such as Brgl and Baf155 (**Figure 2**). While initial efforts to improve the reprogramming efficiency targeted the roles of transcriptional regulators, current studies suggest the involvement of signaling pathways in this process, and that the upregulation and downregulation of major signaling pathways can help in improving pluripotency reprogramming, lineage reprogramming and/or cell differentiation (Fritz et al., 2014).

The role of epigenetics during cellular reprogramming can be shown figuratively as an "epigenetic landscape." An epigenetic landscape graphically represents the process of cell fate decision during development. Such a landscape is in reality a product of complex gene networks that are epigenetically regulated but can be represented figuratively in the form of a "mountain" with numerous "valleys." Cells (represented by balls) reside in the valleys (epigenetically stable networks) and a cell can go down any valley depending upon which stimuli it receives to attain a differentiated state (**Figure 3**). Such epigenetic landscapes and the related complex gene regulatory networks that regulate cell fate can be used to predict the TFs associated with particular cell fates (Lang et al., 2014).

Reprogramming cell fate works through the manipulation of networks governing an epigenetic state. Different types of reprogramming can result in producing undifferentiated cells with varying degree of "stemness" by moving the cell back to the top of the mountain. Four cellular reprogramming methods, namely SCNT, and induced pluripotency (*in vivo, in vitro, and in vitro* using a special media) are noteworthy and are described below.

The totipotent cell resides at the very top of the landscape, and currently SCNT is the only method capable of reprogramming a somatic cell to a totipotent cell capable of forming a complete organism (**Figure 3A**). Below this would be a near-totipotent stem cell that has been generated through one of the following two methods: *in vivo* reprogramming using OSKM that generates iPSCs that are additionally capable of contributing to the trophoectoderm lineage and express embryonic and extra-embryonic markers (Abad et al., 2013) (**Figure 3B**), or *in vitro* reprogramming to ground state naïve pluripotency using NHSM media (see Section Induced Pluripotency above) that generates iPSCs more similar to mouse naïve ESCs and demonstrates the potential to overcome problems related to epigenetic memory and lineage differentiation biases (Gafni et al., 2013) (**Figure 3C**). The conventional *in vitro* method of TF-based reprogramming generates iPSCs that are similar to ESCs but often have problems related to epigenetic memory and lack totipotent cell-like features (Papp and Plath, 2011; De Los Angeles and Daley, 2013) (**Figure 3D**).

Other processes that also involve manipulation of the epigenetic regulatory networks include transdifferentiation (**Figure 3E**) and regeneration (**Figure 3F**). In transdifferentiation one differentiated cell type is converted to another through direct reprogramming (Takahashi, 2012), by avoiding going all the way back up the mountain to an intermediate pluripotent state. Regeneration is similar in that it results in derivation of SSCs or progenitor-like cells from a differentiated cell during wound healing, for example the formation of a blastema (an undifferentiated cell mass) in amphibians (Gurtner et al., 2008) that involves the cell going half-way back up the mountain.

A new framework has been developed in order to elucidate the role of epigenetic landscapes in reprogramming that combines the techniques of whole genome expression profiling and spin glass physics (Lang et al., 2014). This model has verified that partially reprogrammed cells co-express TFs associated with multiple cell fates, has reproduced known reprogramming protocols, and has the potential to generate new reprogramming protocols to create novel cell types (Lang et al., 2014). Defining epigenetic landscapes can not only help to gain better insights into reprogramming processes, but also to design highly efficient protocols. This can be done by targeting epigenetic modifiers involved in the differentiation process at specific positions in the landscape, as well as identifying easy access points in the hierarchy at which reprogramming can be induced for high efficiency.

Now that reprogramming technology has been described in detail we will move on to practical applications.

MODELING HUMAN DISEASE *IN VITRO*

We can group human diseases into three broad categories: genetic, epigenetic and acute environmental (Cherry and Daley, 2012). Modeling of all three types is possible *in vitro* using stem cells, and an excellent way to study the intricate mechanisms and

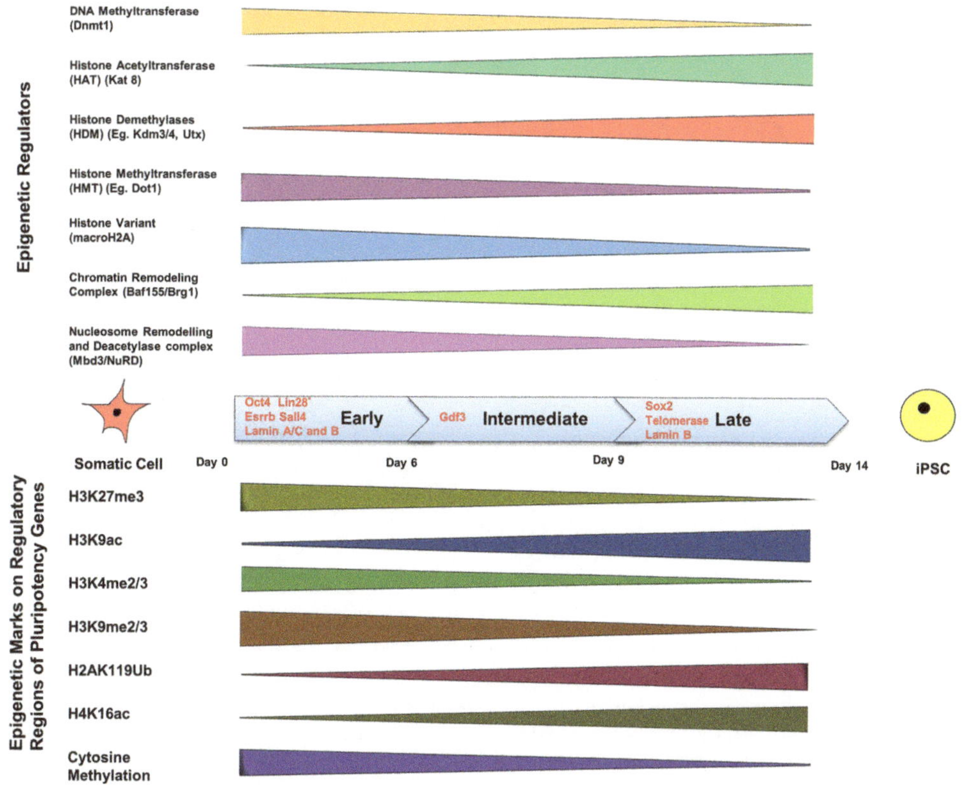

FIGURE 2 | The epigenetics of induced pluripotency. The process of induced pluripotency using OSKM involves turning on pluripotency genes and turning off genes responsible for the maintenance of a differentiated somatic cell state. The timeline of formation of an iPSC can be divided into three stages; early, intermediate, and late, and involves the presence of different transcription factors during each of these phases. Transcription factors Oct4, Esrrb, and Sall4, are expressed during the early stage. Lin 28, suggested to be expressed in the early phase, is a controversial marker of early stages of reprogramming. Gdf3 is expressed during the intermediate stage, and Sox2 during the late stage. Lamins A/C and B are expressed in somatic cells but not in iPSCs, whereas telomerase expression is upregulated in iPSCs, but not in somatic cells. The color bars tapering toward either side of the timeline indicate a decline in expression or activity of the epigenetic regulator or epigenetic mark. The top panel shows the range of activity of some well-defined epigenetic regulators of chromatin in a somatic cell during the three phases. Methyl group-adding enzymes are shown on the very top, next is histone acetyltranferase that adds acetyl groups, followed by demethylases that remove methylation marks; the presence of histone variant: macroH2A, and finally chromatin remodeling complexes are depicted in the last two rows. The bottom panel includes the list of known epigenetic tags present on regulatory regions of pluripotency genes and their presence during the early, late and intermediate phases of induction in a somatic cell to pluripotency using reprogramming factors. DNA methylation on pluripotency genes decreases in the course of reprogramming because of a decline in the activity of Dnmt (DNA Methyltransferase), which is responsible for methylation. Histone acetylation increases during reprogramming due to the increased activity of histone acetyltransferase. Insufficient histone acetylation and hypermethylated DNA are the "epigenetic barriers," which need to be overcome during reprogramming. Reprogramming leads to acquisition of active histone marks (e.g., H3K9ac) and loss of repressive histone marks (H3K4me2) on pluripotency genes, which facilitates the opening up of a compact chromatin structure and thereby allowing exposure of pluripotency gene promoters and binding of pluripotency factors like Oct4. (Buganim et al., 2012; Hansson et al., 2012; Loh and Lim, 2012; Apostolou and Hochedlinger, 2013; Liang and Zhang, 2013; Luna-Zurita and Bruneau, 2013; Papp and Plath, 2013).

pathways underlying the etiology and pathophysiology of disease (**Figure 4**). Stem cells in general are ideal for creating "disease-in-a-dish" models because of their capacity for self-renewal and differentiation, their potential for recapitulating disease pathogenesis, and also their amenability for developing and testing therapeutics (Sterneckert et al., 2014).

The first and simplest types of diseases are monogenic disorders. For example Fanconi anemia (FA) has been studied using patient-derived, integration-free iPSCs with and without *in situ* gene correction (Liu et al., 2014). These FA-iPSCs are also useful for drug screening and shortlisting compounds that can trigger their hematopoietic differentiation potential (Liu et al., 2014) (**Figure 4A**). Likewise, iPSCs have been derived from cells obtained from patients suffering from a wide and rapidly-growing number of diseases such as Hutchinson-Gilford Progeria Syndrome (HGPS, see below), cystic fibrosis, Parkinson's disease, familial amyotrophic lateral sclerosis (ALS), dyskeratosis congenita, epilepsy, autism, spinal muscular atrophy, Huntington's disease, adenosine deaminase severe combined immunodeficiency, sickle cell anemia, long QT syndrome type I, glycogen storage disease type 1a, Alzheimer's disease, diabetes, and Down's syndrome and can be differentiated toward disease-specific lineages in order to study the genotype-phenotype interactions of the disease, as well as to screen new drugs that could be capable of reversing cellular pathology (Liu et al., 2011a; Unternaehrer and Daley, 2011; Cherry and Daley, 2012; Ess, 2013; Miller et al., 2013;

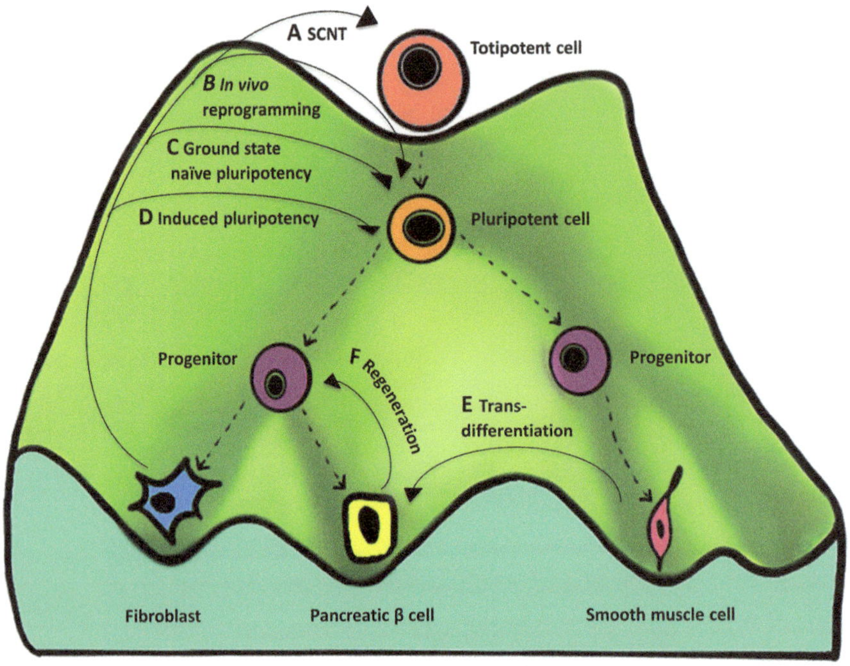

FIGURE 3 | Importance of the "epigenetic landscape" in cellular reprogramming. An epigenetic landscape represents the process of cell fate decisions during development and is a graphical rendering of complex regulatory networks. The figure represents such an epigenetic landscape denoted by a "mountain" with its various "valleys." The top of the mountain displays a totipotent cell that is completely undifferentiated and represents ultimate "stemness," whereas at the bottom of the mountain are differentiated cells resulting from differentiation of their predecessors. Proceeding from the top of the landscape to the bottom, the cell changes from undifferentiated (totipotent and pluripotent), to partially differentiated (progenitor), to terminally differentiated (eg., fibroblast). **(A–F)** represent the different methods and their course on an epigenetic landscape, showing reversal of cellular identity to a more primitive state. There are four types of methods that are currently known for reprogramming to **(A)** totipotency, **(B,C)** near totipotency, and **(D)** pluripotency. (For simplicity, we have excluded reprogramming methods using only small molecules, cell-extract treatment, and cell fusion that can also result in ESCs-like cell formation). **(A)** SCNT gives rise to a totipotent cell capable of forming a complete organism. **(B)** *In vivo* reprogramming using OSKM generates iPSCs that are additionally capable of contributing to the trophoectoderm lineage and express embryonic and extra-embryonic markers. **(C)** *In vitro* reprogramming to ground state naïve pluripotency using NHSM media generates iPSCs more similar to mouse naïve ESCs. **(D)** Induced pluripotency generates iPSCs similar to but less pluripotent than **(A–C)**. Other processes that also involve manipulation of the epigenetic regulatory networks include **(E)** transdifferentiation—one differentiated cell type (smooth muscle cell) is converted to another (pancreatic beta cell) and **(F)** regeneration—derivation of a progenitor-like cell from a differentiated cell during wound healing.

Abdelalim et al., 2014; Acimovic et al., 2014; Hibaoui et al., 2014; Mohamet et al., 2014) (**Figure 4A**).

Monogenic diseases are easy to model in culture, but it is becoming increasingly clear that most diseases are a manifestation of altered genetics and/or epigenetics, with factors like age and environment complicating things further. It is necessary to consider such interlinked effects on different cell types in the body, especially in the case of late-onset disease, and for such purposes iPSCs can serve as an excellent model (Csobonyeiova et al., 2014). Thus, in addition to modeling monogenic diseases, cellular diseased phenotypes derived from iPSCs can be useful in studying complex diseases having genetic and epigenetic components, or having unknown sporadic genetic or acute environmental etiologies (Grskovic et al., 2011; Zhu et al., 2011). Such cellular phenotypes are also useful for screening drugs and understanding the pathophysiology of complex diseases (Grskovic et al., 2011).

Epigenetic diseases can be subtly contrasted with environmental diseases based on the type and extent of damage caused by the epigenetic or environmental factor. Specifically, when iPSCs are exposed to epigenetic factors, the iPSCs and redifferentiated cells may gain temporary or permanent epigenetic marks and manifest the diseased phenotype. Modeling epigenetic diseases in such a way can elucidate the trans-generational potential of the epigenetic factors i.e., whether the epigenetic mark is strong enough to pass on through the germ line to the next generation. Compared to epigenetic diseases, acute environmental insults do not cause heritable genetic or epigenetic effects, but rather the effect is seen at the somatic DNA sequence or protein level (Cherry and Daley, 2012). For example, an acute environmental disease like melanoma that is frequently caused by excessive sunburn (ultraviolet ray damage) can lead to direct DNA damage in the form of pyrimidine dimers (Anna et al., 2007). Such sporadic environmental insults, including spontaneous deaminations, change the DNA sequence as well as alter the cellular biochemistry through upregulation of DNA repair processes, alter expression of cell cycle regulators, or possibly cause apoptosis (Fulda et al., 2010).

Using patient-specific iPSCs we can recapitulate the suspected effects of the environmental or epigenetic component known

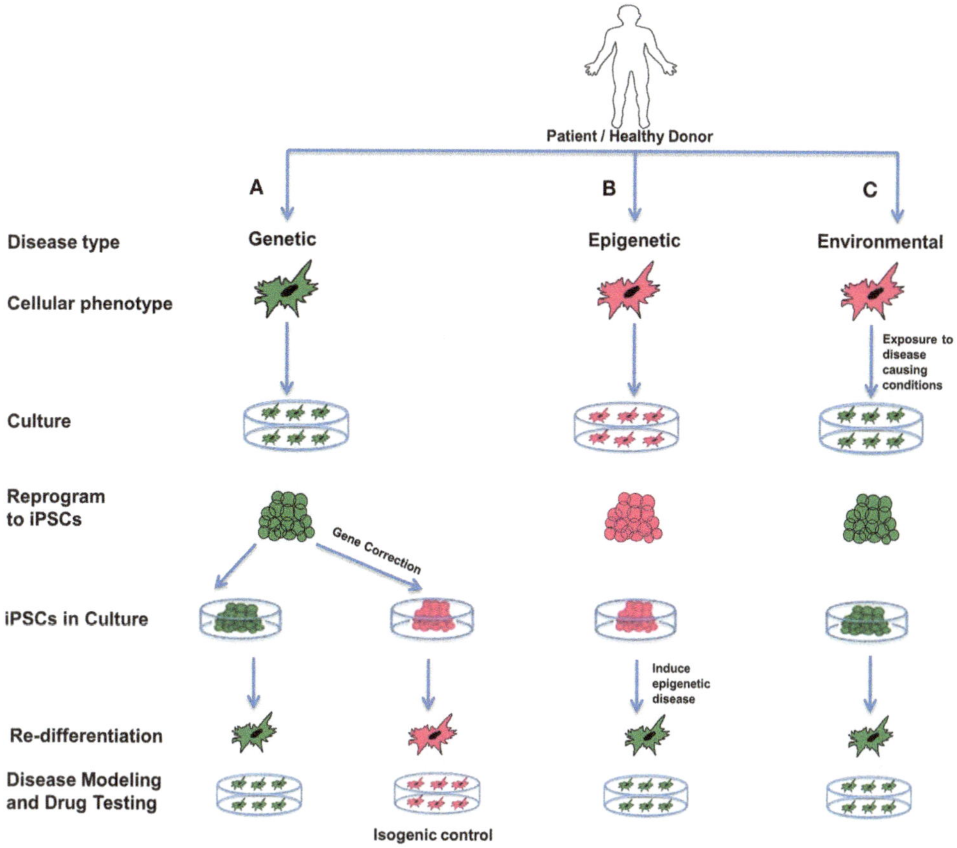

FIGURE 4 | Disease modeling. iPSCs are an excellent source for modeling genetic, epigenetic, and environmental diseases. Such cellular models representing diseased phenotypes can be used for understanding the interplay between the genetics, epigenetics and environment involved in the disease, and can expose unknown details about disease pathophysiology, and can be used for screening drugs. In the figure all green cells represent diseased cells, and all pink cells represent healthy cells. **(A)** Genetic diseases can be modeled by reprogramming diseased cells to iPSCs and then re-differentiating them to produce a diseased phenotype. Additionally, these iPSCs can be corrected for the genetic mutation involved in the disease using gene-editing technology. On re-differentiation, corrected iPSCs produce healthy cells that can be used as isogenic controls. **(B)** Epigenetic diseases can be modeled using healthy cells that are reprogrammed to iPSCs and then induced toward an epigenetic disease state by recapitulating an environment containing the epigenetic factor(s) contributing to the disease. If iPSCs retain an epigenetic mark when in culture, or after being redifferentiated to the desired cell type, it indicates that the epigenetic mark is permanent and is likely to be passed on to offspring or carried by germ-line cells. It can also mean that the particular cell type is predisposed to retaining that epigenetic mark. Patient-specific models can be used as special models, as they can involve known epigenetic factors contributing to the disease. **(C)** Acute environmental diseases can be modeled using healthy cells by exposing them to a disease-causing environment that results in genetic damage or instability in the cells. For disease modeling, such cells can be reprogrammed to iPSCs and redifferentiated to diseased phenotypes. All of the above models can help us gain better insight into the diverse factors affecting a complex disease in terms of susceptibility, prognosis as well outcomes.

to have contributed to the patient's disease in order to understand its severity or mechanism of action before, during, or after reprogramming (**Figures 4B,C**). Cells from any source can be epigenetically manipulated to mimic epigenetic (**Figure 4B**) or acute environmental (**Figure 4C**) alterations contributing to disease. Such models can help us gain better insight into the environmental or epigenetic factors affecting a complex disease in terms of susceptibility, prognosis as well as outcomes. Patient-specific models can also be used as special models, as they can involve known epigenetic changes contributing to the disease.

iPSCs can also be used for modeling disease at the organ level as well as understanding systemic diseases. For example human iPSCs can be used for *in vitro* development of complex structures like the retina by spatiotemporally recapitulating the *in vivo* steps of retina formation and making it possible to test future therapies *in vitro* (Zhong et al., 2014).

TREATING HUMAN DISEASE

Beyond modeling disease, reprogramming can also be used to treat disease. While the technology of cellular reprogramming has been developed and refined there have also been significant ongoing developments in other complementary technologies such as gene editing, progenitor cell production, and tissue engineering that we will describe below. These are converging to a point that will allow us to treat almost any disease. Personalized medicine may soon become a reality, with cellular harvest and reprogramming, genetic engineering and tissue engineering becoming a routine procedure. These technologies are the foundations of what is becoming a fully-functional field of regenerative medicine.

IN VITRO REPROGRAMMING AND COMPLEMENTARY TECHNOLOGIES

Induced pluripotent stem cells

The greatest advantage of using *in vitro* reprogramming is that we can generate patient-specific iPSCs (autologous) and eliminate the threat of rejection, or have well-characterized iPSC cell-lines (allogenic). Reprogramming *in vitro* is often more useful when used with complementary technologies like gene editing to treat genetic diseases by fixing the causative mutation and then repopulating damaged or non-functional tissue with the desired healthy cell type (**Figure 5A**).

Generation of iPSCs enables *in vitro* coordination of differentiation into cells of all three germ layers (endoderm, mesoderm, and ectoderm). Where the aim is simply to repopulate the non-functional or damaged tissue with functional cells, transplantation with iPSC-derived cells could cure the patient. For example, mouse embryonic fibroblasts or mouse pancreas-derived epithelial cells reprogrammed to iPSCs and then differentiated to functional pancreatic beta cells were capable of treating diabetes in mice (Jeon et al., 2012). These iPSC-derived beta-like cells released insulin and on transplantation into non-obese mice, normalized the blood-sugar level (Jeon et al., 2012). If treating a disease involved fixing a genetic mutation permanently, then gene-editing technologies (which are explained in detail in section Gene editing using CRISPR/Cas9, ZFN, TALENs etc) can be used as an additional step before differentiating the iPSCs into the desired cell type. For example in mice, the correction of a mutation in iPSCs by gene therapy using homologous recombination, followed by autologous transplantation back into the mouse, has led to the replacement of diseased cells and rejuvenation of functional tissue in single-gene defects like sickle cell anemia. Such procedures have overcome problems associated with immune rejection and validated the role of iPSCs in regenerative medicine (Hanna et al., 2007; Takahashi and Yamanaka, 2013). However, autologous iPSCs will incur a comparatively higher medical cost compared to allogenic ones, and their creation could take a longer time compared to allogenic iPSCs. Also they might not be available immediately when required for the treatment of acute injuries like spinal cord injury, implying the need for HLA-matched donor allogenic iPSCs (Takahashi and Yamanaka, 2013), unless they can be prepared beforehand.

One of the most promising cell types derived from iPSCs are neural cells (Svendsen, 2013). Neurons derived from iPSCs demonstrate great promise in revitalizing and repopulating the central nervous system, and have promise for use in the development of neurologically based disease models such as Huntington's and Parkinson's diseases (Kriks et al., 2011; Liu et al., 2013b).

Potential risks associated with clinical iPSC applications include the possibility that viral or other agents used for reprogramming could trigger harmful immune or inflammatory responses (Zhao et al., 2011; Guha et al., 2013). The prevention of vector or transgene integration into the host genome or residual transgene expression once reprogramming has been achieved is important in order to avoid such unwanted responses. This problem has now been resolved with the use of non-integrating episomal vectors to create integration-free iPSCs (Hu et al., 2011; Su et al., 2013). The use of non-viral vectors such as the transposon-based systems like Sleeping Beauty (SB) and piggyBac (PB) offer a cost-effective and safe alternative (Talluri et al., 2014). Several experiments to eliminate the use of more than one or two reprogramming factors during induced pluripotency have also been conducted (Radzisheuskaya and Silva, 2014). For example easily accessible dermal papilla cells (specialized mesenchymal cells endogenously expressing Sox2, Klf4, Myc) from the hair follicles of skin have been reprogrammed with Oct4 alone and without any other chemical compounds or small molecules (Tsai et al., 2011).

Gene editing using CRISPR/Cas9, ZFN, TALENs etc

As stated, the use of the combined techniques of reprogramming and genetic engineering followed by cellular therapy can potentially correct a genetic defect and repopulate a degenerated or damaged tissue with the required cell type and cure a genetic disease. Such technology can also be used to manipulate and study model organisms as well (Gaj et al., 2013).

Gene editing technologies are methods used to fix an endogenous mutation, or to make a gene inactive for the purpose of repairing or deleting a mutation. Recent progress in gene editing technology has made it possible to induce a site-specific DNA cleavage followed by repair. This results in high precision genome editing in cultured cells and whole organisms; useful in designing therapeutic treatments and creating models *in vitro* and *in vivo* (Kim and Kim, 2014).

Gene editing technology is based on the use of RNA sequences or proteins that bind to specific DNA sequences and recruit nucleases to introduce double strand breaks in the target sequence. Once DNA damage occurs, homologous or non-homologous end joining processes repair the DNA sequence. With homologous recombination, the wrong nucleotides/s can be replaced with the correct ones to repair a genetic mutation. Current precise gene editing methods include the following three types depending on the type of repair and nucleases involved: 1. Zinc finger nucleases (ZFN), 2. Transcription activator-like effector nuclease (TALEN), and 3. Clustered regularly interspaced short palindromic repeats (CRISPR).

1. ZFN- This is the most commonly used type at present, and is comprised of a eukaryotic zinc finger binding protein domain coupled with a nuclease.
2. TALEN- This type similarly uses a transcription activator-like effector (TALE) domain coupled with a nuclease.
   ZNFs and TALENs include programmable, sequence specific DNA binding modules linked to a non-specific DNA cleavage domain and are capable of introducing a variety of genetic modifications (Gaj et al., 2013).
3. CRISPR- This is the newest type of gene editing technology based on RNA interference and uses CRISPR and CRISPR associated genes (Cas) from bacterial and archeal immune systems. The CRISPR loci have short palindromic repeats

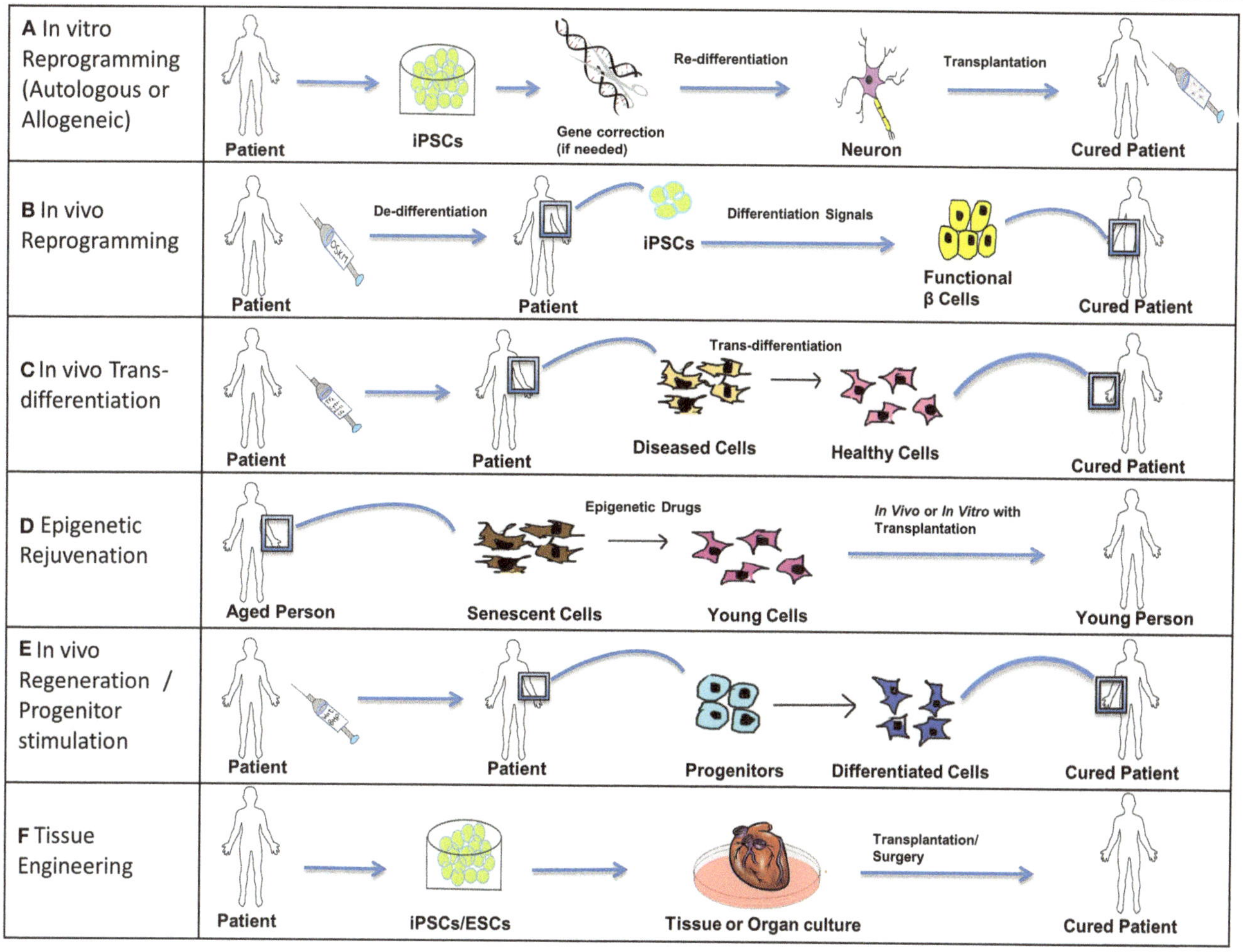

FIGURE 5 | Treating disease. Stem cells (iPSCs, ESCs, and SSCs/Progenitors) can be used in the following ways to treat acute, chronic, and degenerative diseases. **(A)** *In vitro* reprogramming. iPSCs reprogrammed *in vitro* using different combinations of transcription factors can be further differentiated to specific lineages and these cells can be transplanted into the patient suffering from a degenerative disease or injury. For example, autologous iPSCs generated from fibroblasts of a patient suffering from Parkinson's disease can be differentiated into functional dopamine secreting neurons and can be transplanted into the patient's brain through surgery. In cases where solely transplanting functional cells is inadequate to completely cure a genetic disease, a permanent fix of the genetic mutation involved is required. Thus, an additional step of gene correction can be performed on the patient-specific iPSCs using gene-editing technologies like CRISPR/Cas. Allogeneic iPSCs might be useful to treat traumatic acute injuries like spinal cord injury, as they would be available immediately when required. **(B)** *In vivo* reprogramming. iPSCs can be generated *in vivo* at the site of non-functional or damaged tissue (by administration of reprogramming factors like OSKM) and can then be differentiated toward a specific lineage (by the administration of differentiation signals). For example, iPSCs generated *in vivo* from pancreatic cells of a patient suffering from diabetes can be differentiated into functional β cells secreting insulin and thus restore the lost function by repopulating the pancreatic tissue with functional β cells. **(C)** *In vivo* trans-differentiation. A somatic cell can be trans-differentiated *in vivo* into a different type of somatic cell using transcription factors, growth factors, etc. such that the new cell type helps to restore the lost function of the diseased or damaged tissue. This process is noteworthy as it avoids the intermediate step of reprogramming somatic cells to pluripotency; however this process is not yet well-established. **(D)** Epigenetic rejuvenation. The idea behind epigenetic rejuvenation is that epigenetic drugs can be used to convert senescent cells to young cells such that aging is reversed without affecting the differentiation or specialized function of the cell. This can be done *in vivo* directly or *in vitro* followed by transplantation of the rejuvenated cells into the patient. **(E)** *In vivo* regeneration/progenitor cell stimulation. Progenitors can be induced to differentiate into their successors by injecting differentiation signals like Wnt and Bmp proteins in order to achieve regeneration *in vivo*, which will restore lost function of damaged or diseased tissue. This process avoids procedures involving induction of pluripotency as well as transplantation into the patient. **(F)** Tissue Engineering. Patient-specific iPSCs (with optional gene correction step) or ESCs can be used to culture tissues and organs *in vitro* using specialized scaffolds and organ molds or using bioprinting. Such cultured organs can then be transplanted into a patient through advanced surgical procedures to restore the function of an entire organ or organs, in the case of chronic or systemic diseases, as well as to treat acute illnesses like myocardial infarction with cardiac tissue transplants.

(24–37 base pairs) separated by unique spacer sequences of equal length that can integrate foreign DNA, [which in the source bacteria/archea can be used as a type of memory in the case of a future attack by the same foreign DNA (Richter et al., 2013)]. When a CRISPR sequence is transcribed, the foreign DNA is transcribed along with it. The processed RNA in the form of crRNA and tracrRNA is responsible for sequence specific silencing by recognizing the target foreign

DNA sequence and degrading it with Cas proteins that have nuclease activity (Gaj et al., 2013).

This RNA-guided genome editing tool could be used to correct disease- associated genes in zygotes and human cell lines (High et al., 2014). The CRISPR/Cas system can effectively replace the traditional method of producing mutant mice by sequential recombination since this technology allows a highly efficient single-step approach for production of ES cells with mutations in multiple genes, as well as generation of mutant mice by direct embryo manipulation (Wang et al., 2013).

CRISPR- based genome editing is also useful in disease modeling because it is very efficient at generating allelic isogenic models to study diseases *in vitro*. For example the combination of CRISPR technology and an antibody-based method to screen recombination events has been used to generate genetically modified human cell lines for the study of Huntington's disease (An et al., 2014). The Type II CRISPR/Cas9 system is gaining special recognition for its ability to accelerate the production of cell lines, especially in combination with reprogramming, and the production of animal models with a desired genetic makeup for drug screening and pre-clinical evaluation. Also it seems to be a desirable candidate for gene therapy alone (Yingze Zhao, 2014).

IN VIVO REPROGRAMMING

Other potential ways of using iPSCs for treating diseases would be by generating them *in vivo*. These *in vivo* iPSCs can further generate desired cell types in the presence of appropriate signals for re-differentiation (**Figure 5B**). As described in section Induced Pluripotency and Epigenetics during cellular reprogramming, *in vivo* transitory induction of OSKM in mice led to *in situ* reprogramming and gave rise to iPSCs from a variety of cell types (Abad et al., 2013). These iPSCs represented totipotent-like features and they resembled ESCs more closely compared to *in vitro* generated iPSCs (Abad et al., 2013). Another *in vivo* study showed the generation of induced adult neuroblasts (iANBs) from resident astrocytes using a single TF Sox2 and these iANBs formed mature neurons when targeted for differentiation using noggin, BDNF or histone deacetylases inhibitor (Niu et al., 2013).

IN VIVO TRANS-DIFFERENTIATION

Transdifferentiation is the process in which a particular somatic cell is switched from one lineage-specific identity to a completely different identity (Graf, 2011; Vierbuchen and Wernig, 2012); in other words, the direct conversion of one type of somatic cell into another type, bypassing the intermediate step of dedifferentiation (**Figure 5C**). As described in Section Transdifferentiation/direct reprogramming transdifferentiation *in vivo* converted adult murine pancreatic exocrine cells to β-cell like population using the transcription factors Ngn3, Pdx1, and Mafa (Zhou et al., 2008). Another example is the *in vivo* transdifferentiation of human peripheral blood $CD34^+$ cells into non-hematopoietic lineages such as cardiomyocytes (Yeh et al., 2003). However, it is important to note that such processes occur only under the conditions of severe tissue damage (Yeh et al., 2003).

EPIGENETIC REJUVENATION

Epigenetic rejuvenation is the concept of erasing an "aged" epigenome to obtain an epigenome that corresponds to the one seen in a "young" healthy cell along with maintaining its differentiated state and rejuvenating its function (Manukyan and Singh, 2012). The idea behind epigenetic rejuvenation is to treat aged cells from patients *in vitro* and transplant them back into the tissue and replace aged or unhealthy, damaged cells (Manukyan and Singh, 2012) (**Figure 5D**). (Please refer to Section Hutchinson-Gilford Progeria Syndrome for additional details.)

IN VIVO REGENERATION/PROGENITOR CELL STIMULATION

In vivo regeneration involves the repair of tissue damage by stimulation of endogenous stem cells (SSCs/progenitors) by growth factors, cytokines, transcription factors, and second messengers (**Figure 5E**). A promising strategy to treat degenerative and chronic disorders resulting from loss of function of a particular type of cell, is through harnessing the intrinsic regenerative potential of the progenitor cells which are responsible for creating the specific type of damaged cell and restoring their lost function. Studies involving the use of regenerative factors like developmental signals that play a role in the regeneration of a damaged tissue, suggest that the latent/dormant intrinsic potential can be triggered to harness the *in vivo* regeneration capacity in a disease or trauma-struck individual (**Figure 5E**). Such changes during regeneration are accompanied by epigenetic modifications. A few examples of signaling pathways and differentiation factors, which are being studied and could potentially activate regeneration (thus should be considered as suitable targets) are described below.

1. Wnt, a ligand working in conjunction with a receptor, involved in the transcription of β- catenin promotes morphogenesis and has the capacity to initiate the regeneration of broken bone to treat bone deformities or injuries, as well as cause Muller progenitors to differentiate into retinal progenitors, thereby ameliorating retinal trauma, and possibly blindness (Minear et al., 2010; Liu et al., 2013a).
2. Platelet-derived growth factor (PDGF) and insulin-like growth factor (IGF) have been shown to stimulate the regenerative capacity of mammalian skeletal and neural tissue (Rutherford et al., 1992; Mason et al., 2000).
3. Bone morphogenic protein (BMP) is essential for ESC renewal whereas basic Fibroblast growth factor (bFGF) increases the differentiation potential of MSCs into chondrogenic, osteogenic and adipogenic lineages and epidermal growth factor receptor (EGFR) increases the proliferative potential and survival capacity of MSCs (Bhattacharyya et al., 2012).
4. Lin28 is expressed during embryogenesis as it plays an important role in pluripotency and development, also its overexpression promotes epidermal hair regrowth, and digit repair in mice by altering the bioenergetic state during tissue repair following enhanced translation of metabolic enzymes that increase glycolysis and oxidative phosphorylation (Shyh-Chang et al., 2013).
5. The transient inactivation of two tumor suppressors ARF (alternative reading frame protein) and RB (retinoblastoma

protein) in mammalian cells resulted in dedifferentiation restoring regenerative capacity and produced differentiated cell types when cultured in the presence of proper signaling factors suggesting that terminally differentiated mammalian cells can be induced for regeneration without neoplastic transformation (Pajcini et al., 2010).

6. Hox genes play an important role in embryonic pattern formation and can be programmed such that their previous epigenetic memory is erased resulting in plasticity, as observed in iPSCs, that can in turn be used to trigger regenerative capability of injured tissue (Wang et al., 2009).

TISSUE ENGINEERING

Tissue engineering is the field of engineering tissues, or organs grown from *in vitro* cultures of stem cell-derived differentiated cells, followed by their transplantation (**Figure 5F**).

Tissue engineering is the *de novo* production of tissues or organs using starting material like cells, extracellular matrix and scaffolds for supporting the three dimensional process of tissue formation (Fisher and Mauck, 2013). This technology has been translated into actual therapies for skin replacement and cartilage repair (Berthiaume et al., 2011). The first step in tissue engineering is to procure the desired cell types required for constructing the tissue or organ. iPSCs as well ESCs serve as a good source for producing the cells through directed differentiation. For example, iPSC-derived multipotent neural crest stem cells (NCSCs) have been used to engineer nerve conduits by seeding NCSCs into nanofibrous tubular scaffolds and such nerve conduits were used for regenerating sciatic nerve (Wang et al., 2011). Tissue engineering uses conventional methods for scaffold fabrication that are probably inadequate (Seol et al., 2014), but the latest bioprinting technology can construct 2-D and 3-D tissues and organs using additive manufacturing technology that allows precise placement of cells, biomaterial and biomolecules in an adequately fabricated scaffold (Seol et al., 2014).

CLINICAL TRIALS FOR STEM CELL THERAPY IN HUMANS

The U.S. Food and Drug Administration defines somatic cell therapy as the administration to humans of autologous, allogeneic or xenogeneic living non-germ line cells, other than transfusion blood products, which have been manipulated, processed, propagated or expanded *ex vivo*, or are drug-treated (Health et al., 2001; Parson, 2006; Bieback et al., 2011; George, 2011).

With the FDA approving the first ESCs Phase I clinical trials for acute spinal cord injury in 2011, the science behind stem cell biology and regenerative medicine made its way to practical applications in the real world (**Table 1**). Extensive efforts from the California Institute of Regenerative Medicine (CIRM), National institute of Health (NIH) and other organizations around the world has allowed stem cell therapies to progress into clinical trials for a variety of diseases, many of which have made significant progress. Clinical trials using mesenchymal stem cells (MSCs) for treating bone and cartilage diseases, neurodegenerative diseases, heart diseases, gastrointestinal diseases, autoimmune/immune rejection diseases, and cancer have shown a lot of progress with massive funding from CIRM, their collaborating partners and others (Trounson et al., 2011) (**Table 1**).

The most advanced clinical trials listed by EuroStemCell include testing of therapies for bone, skin, corneal disease and injury, the majority of which use mesenchymal stem cells. The NIH lists two active clinical trials for the eyes diseases Age related Macular Degeneration (AMD) and Stargardt's muscular dystrophy using cells derived from human ESCs, being produced by Advanced Cell Technology. Other sources like bone marrow stem cells for treating cancer and blood disorders, human central nervous system stem cells for AMD and human spinal cord stem cells for ALS are also in clinical trials. Human MSCs are being used in clinical trials by Osiris therapeutics for Type 1 diabetes, cardiomyopathies, Graft vs. host disease and Crohn's disease. More details about these ongoing clinical trials can be found through the links for online resources provided in **Table 1**.

CANDIDATE DISEASES FOR TREATMENT USING REPROGRAMMING

In this section, we provide specific examples of how reprogramming and complementary technologies can be used to treat and eventually cure human disease.

AGING AND AGE-RELATED DISEASES

A few decades ago, Richard Cutler proposed that aging was a result of cells drifting away from their proper state of differentiation due to changes in gene expression, through a process

Table 1 | Resources for tracking progress in the Field of Regenerative Medicine.

Sr. No.	Topic	Affiliation	Access data
1.	Current status of therapies undergoing clinical trials	CIRM	http://www.cirm.ca.gov/sites/default/files/files/about_stemcells/Portfolio%20table_10-13.pdf
2.	Publically and Privately supported Clinical studies around the world	NIH	http://clinicaltrials.gov/ct2/home
3.	Clinical trials and stem cell treatment	EuroStemCell	http://www.eurostemcell.org/clinical-trials
4.	Adult stem cell clinical trials	Stem cell research facts	http://www.stemcellresearchfacts.org
5.	Disease information	CIRM	http://www.cirm.ca.gov/our-progress/disease-information
6.	Current status of funding and research	CIRM	http://www.cirm.ca.gov/sites/default/files/files/funding_page/Portfolio%20summary%2006-03.pdf

he termed "dysdifferentiation" (Kator et al., 1985). This process could be reinterpreted today as epigenetic drift. These changes are so predictable that recently an "epigenetic clock" has been formulated that can very accurately predict the age of an individual based solely on the methylation of specific genes in genomic DNA (Horvath, 2013). Indeed, epigenetic changes represent a major aspect of aging, and epigenetic manipulation may allow reversal of these damaging effects (Campisi and Vijg, 2009; Munoz-Najar and Sedivy, 2011).

It has been proposed that if we can reverse the epigenetic changes associated with aging, we could actually reverse aging itself. This epigenetic approach to reversing aging has been termed "epigenetic rejuvenation" (see Section Epigenetic Rejuvenation and **Figure 5D**). Although the very existence of animals cloned by SCNT proves that epigenetic rejuvenation is possible in principle, only recently have *in vitro* experiments on human cells shown that this is possible. Indeed, even centenarian fibroblasts can be rejuvenated, and young fibroblasts cultured to senescence can be reprogrammed to a pluripotent state (Lapasset et al., 2011; Villeda et al., 2011; Loffredo et al., 2013). Recently, very detailed experiments have shown that age-related and tissue-specific DNA methylation patterns remain erased in mesenchymal stem cells produced by induced pluripotency (Frobel et al., 2014).

Epigenetic reversal also crucially includes reactivation of telomerase expression, which resets the chromosome "aging clock" back to a young state. Reprogramming "old" cells to "young" cells will eventually enable the repopulation of an aged and/or degenerated tissue with cells capable of extended proliferation. Perhaps this strategy could eventually be applied to the whole body.

An example of a type of epigenetic rejuvenation already practically applied is heterochronic parabiosis: experiments with two animals of a different age joined together surgically, showed that tissue-specific stem cells from an aged animal could be rejuvenated by exposing them to a "young" environment (Rando and Chang, 2012; Conboy et al., 2013). This "young" environment involves factors such as chemokine's, cytokines and rejuvenating Wnt and TGFβ signaling pathways (Conboy et al., 2013). On the other hand "young" stem cells adopt a more aged structure and function (Conboy et al., 2005). Unlike reprogramming, heterochronic parabiosis maintains the differentiated states of the cells and rejuvenates the regenerative potential of old cells in the milieu of a young environment, thus uncoupling dedifferentiation and rejuvenation (Conboy et al., 2013). This type of rejuvenation might have an epigenetic basis regulated by a switch that could possibly convert a cell from "old" to "young" or vice versa.

Aging also involves mitochondrial dysfunction as well as defective mitochondrial DNA. SCNT can be used to correct diseases caused by defective mitochondrial DNA (mtDNA) because in NT-ESCs the mtDNA exclusively comes from the oocyte, and only the nucleus from the donor/patient, thus making it possible to avoid defective mtDNA being carried over from the donor/patient (Tachibana et al., 2013). Surprisingly, induced pluripotency rejuvenates mitochondira, but in a different way to SCNT (Suhr et al., 2010).

Hutchinson-Gilford Progeria Syndrome

Hutchinson-Gilford Progeria Syndrome (HGPS) is a segmental premature aging disease characterized by symptoms of premature aging, with a life expectancy in the early teens or twenties. The single point mutation responsible for causing this disease occurs in position 1824 (cytosine is replaced with thymine) of the Lamin A (*LMNA*) gene on chromosome 1. Lamin A is an intermediate filament protein that stabilizes the inner membrane of the nuclear envelope. The mutation results in a defective Lamin A that produces an abnormal protein called Progerin. Ultimately, this results in abnormally shaped, dysfunctional nuclei that in turn disrupt cell division and other cellular functions (Eriksson et al., 2003; McClintock et al., 2007; Korf, 2008).

Somatic cells from Progeria patients can be reprogrammed to iPSCs and the LMNA mutation can be corrected using gene-editing technology (see above). The fact that ESCs and iPSCs do not express *LMNA* (Constantinescu et al., 2006), allows the efficient correction of the mutation in iPSCs, and any disease-related pathology can be avoided in iPSC-derived differentiated cells. Notably, HGPS iPSCs exhibit downregulation of lamin A/C and Progerin, causing a lack of progeroid features in the iPSC population, rendering these iPSCs normal with respect to nuclear morphology, pluripotency, and epigenetic profiles (Liu et al., 2011a). HGPS iPSCs corrected for the *LMNA* mutation can then be differentiated into normal somatic cells (Liu et al., 2011b). These iPSC-derived differentiated cells could eventually be transplanted into patients using cellular therapy to repopulate the aged tissue with healthy proliferative cells free of the original Progeria mutation.

Age-related macular degeneration

Age-related macular degeneration (AMD) is a complex multifactorial diseases caused by the degeneration of the photoreceptors and retinal pigment epithelium (RPE) of the eye (Yonekawa and Kim, 2014). It is the most common cause of visual impairment in the elderly (+60 years) and it is broadly classified into two clinical categories, namely, the wet form (neovascular or exudative) and dry form (Cook et al., 2008). AMD, which is primarily known to be an age-related disease, is also affected by other risk factors like genetics, patient history, smoking, trauma, etc (Chakravarthy et al., 2010). Data from pre-clinical studies suggests that transplantation with ESCs-derived RPE or patient-specific iPSC-derived RPE (with correction of the gene mutation) may prevent photoreceptor degeneration in animal models of RPE degeneration (Schwartz et al., 2012; Heller and Martin, 2014). With the advent of techniques in *in vivo* reprogramming, another promising approach to cure AMD is to target the intrinsic potential of neural progenitors in a patient and drive their differentiation into RPE through the administration of appropriate differentiation signals. It has already been suggested that retinal stem cells (RSCs) that are multipotent, rare cells in the pigmented ciliary epithelium of peripheral retina can be enriched an directed to differentiate into photoreceptors using factors influencing neural retinal development (Ballios et al., 2012). Specialized knowledge of such differentiation and proliferation signals with the development of advanced techniques toward delivery and transplantation of stem cells or RPE lead us one step closer to the development of clinical therapies for AMD.

EPIGENETIC DISEASES

Epigenetic diseases are caused by altered expression of genes caused by abnormal changes in the epigenetic profile of a cell. Reprogramming can be used to reset the epigenetic profile of any cell so that the cell becomes pluripotent. It can then be differentiated to obtain the cell type with a desired epigenetic profile. Such procedures followed by cellular therapy might be a potential treatment for epigenetic diseases caused by known epigenetic factors. Let us consider the example of polycystic ovary syndrome (PCOS) as an epigenetic disease.

PCOS is a heterogeneous endocrine disorder responsible for a large proportion of female subfertility (Azziz et al., 2004). The hallmarks of the disease are anovulation, excessive androgenic hormones, and insulin resistance. The PCOS phenotype is characterized by hypersecretion of both luteinizing hormone as a result of compromised hypothalamic sensitivity, and insulin in response to increased abdominal adiposity (Escobar-Morreale et al., 2005). Familial clustering of cases and the heritability of endocrine and metabolic aspects of PCOS indicate a dominant regulatory gene with incomplete penetrance and strongly indicate a genetic etiology (Diamanti-Kandarakis et al., 2006).

As described in Section Epigenetics above, environmental factors can induce epigenetic alterations, which potentially involve trans-generational inheritance of methylation patterns. The exposure of female rats to a mixture of pesticides showed that the first and third generation offspring were afflicted with PCOS (Manikkam et al., 2012). This demonstrated that PCOS might have an epigenetic basis caused by a differential methylation pattern in the diseased offspring. In humans aberrant gene methylation has been reported in patients with PCOS, suggesting a role for a bZIP TF, CEBPB, in insulin resistance and regulation of the interleukin-1 response element in the IL-6 gene and other genes important for the immune and inflammatory components of this disease (Natsuka et al., 1992; Harries et al., 2012; Shen et al., 2013). Inappropriate epigenetic reprogramming is a contributing factor in PCOS. Epigenetic alterations of peroxisome proliferator-activated receptor gamma 1 (PPARG1), nuclear co-repressor 1 (NCOR1), and histone deacetylase 3 (HDAC3) genes in granulosa cells are induced by hyperandrogenism (Qu et al., 2012).

Recently, PCOS-derived hESCs have been successfully isolated from the inner cell masses of blastocysts. These hESCs retain pluripotency, which allows the dissection of the *in vitro* pathogenesis of the disease. Eventually cell therapies will be developed for this disease.

COMBINED GENETIC AND EPIGENETIC DISEASES

Some diseases long known to have genetic causes are now thought to also have epigenetic bases. Just as with epigenetic diseases, such diseases can be treated using reprogramming to re-establish the normal epigenetic profile but with the addition of gene correction. One of the most well-known such diseases is cancer.

Cancer

Cancer remains the leading cause of death worldwide with over 7.6 million deaths (13%) in 2008 (Jemal et al., 2011). Cancer does of course have a partly genetic basis: most human cancers are caused by gene mutations in one or more components of the p53 and/or the pRB pathway (Campisi, 2001). And most tumor suppressor genes can be classified into two types; Caretaker tumor suppressors which prevent DNA damage and Gatekeeper tumor suppressors which eliminate potential neoplastic cells via apoptosis (Campisi, 2005).

But besides the genetic mutations, cancer is also an epigenetic disease. The dysdifferentiation hypothesis of aging and cancer emphasizes the importance of instability of the differentiated state of cells due to improper gene regulation in the development of cancer (Zs-Nagy et al., 1988). Cancer is comprised of the following eight hallmarks, some of which may have a wholly epigenetic etiology: sustained proliferative signaling, evading growth suppressors, enabling replicative immortality, inducing angiogenesis, resisting cell death, activating invasion and metastasis, reprogramming of energy metabolism and evading immune destruction. (Hanahan and Weinberg, 2011). Reprogramming of metabolic pathways contributes to the development of the disease, which is evidently seen in cancer with upregulation of oxidative phosphorylation and glycolysis-associated tarnscriptomes (Lisanti et al., 2014). For example, cancer-associated fibroblasts found in tumor microenvironments have a distinctive energy metabolism reprogramming phenotype responsible for proliferation, migration, invasion and epithelial-mesenchymal transition (Tang et al., 2014).

Cancer cells have altered DNA methylation profiles such that the overall methylated DNA is less than that seen in normal cells. A study testing epigenetics as a molecular marker system for cancer noted 12 genes to be associated with promoter hypermethylation causing silencing of their respective genes. These were functional in cell cycle regulation, tumor suppression, DNA repair, metastatic potential and apoptosis and have been linked to 15 major tumor types, making epigenetics a significant tool in cancer detection and typing (Esteller et al., 2001). Epigenetic changes in cancer are becoming well-noted and epigenetic drugs like DNA methylation inhibitors and histone deacetylase inhibitors have shown promising results in clinical trials and are less toxic than the conventional chemotherapy drugs (Chen et al., 2014). Epigenetic modulators may be used to inhibit self-renewal and survival of quiescent cancer stem cells (CSCs) by making them sensitive to targeted or conventional chemotherapy (Li and Bhatia, 2011).

Other diseases

Genetic factors contributing to epigenetic variation like DNA sequence variants, differential expressions of genes that regulate chromatin remodeling and DNA methylation, can contribute toward an epigenetic etiology of a disease (Bjornsson et al., 2004). Familial forms of neurodegenerative diseases like Alzheimer's disease, Huntington's disease, Parkinson's disease, and ALS are known to have a genetic basis but increasing evidence points toward epigenetic alterations like DNA methylation, histone tail modifications, and RNA mediated mechanisms in sporadic forms (Kleivi, 2006).

Some autoimmune diseases such as systemic lupus erythematous, rheumatoid arthritis, multiple sclerosis, and type 1 diabetes mellitus suggest the contribution of environment to a genetically pre-disposed trait (Hewagama and Richardson, 2009). For example, Immunodeficiency chromosomal-instability facial anomalies syndrome (ICF) is one such disease, which is caused by mutation in the gene for DNA methyl transferase (DNMT3B) that further leads to abnormal methylation causing varied gene expression (Ehrlich, 2003; Bjornsson et al., 2004; Rodenhiser and Mann, 2006).

Other examples include the downregulation of the Reelin gene (RELN) in schizophrenia which leads to the reduction of RELN protein by 50% compared to a normal brain and this has been linked to an epigenetic modulation of the promoter of the RELN gene through increased expression of DNMT1 (Costa et al., 2002; Sharma, 2005; Rodenhiser and Mann, 2006), and Rett syndrome, mostly caused by a mutation in the gene MeCP2 that produces methyl-CpG- binding protein, is an X linked dominant neurodevelopmental disorder (Bienvenu et al., 2000). The loss of the MeCP2 protein results in loss of transcriptional repression and hence inappropriate transcription of downstream genes during brain development (Van Den Veyver and Zoghbi, 2001; Rodenhiser and Mann, 2006).

INFECTIOUS DISEASES

One of the most promising strategies to treat an infectious disease is to replace a receptor on a host target cell, which is used by a pathogen to gain entry into the cell, with a mutated one. This makes the receptors on the target cell non-functional and blocks the entry of the pathogen into the cell, preventing its multiplication and the progression of the disease. One important disease that may eventually be universally treated and perhaps eradicated this way is acquired immunodeficiency syndrome (AIDS). AIDS is a disease of the human immune system caused by infection with human immunodeficiency virus (HIV). It is a major public health concern worldwide, with more than 30 million people infected (Kallings, 2008; Unaids, 2013). There is no cure or vaccine at present, but antiretroviral treatment (ART) can slow the course of the disease and result in near-normal life expectancy. ART reduces the risk of death and complications from the disease, but is expensive and associated with side-effects. We need to find better treatments.

Of note, one of the molecular mechanisms of HIV infection and resulting AIDS is depletion of $CD4^+$ T cells. All cells susceptible to HIV infection (CD4 T cells, macrophages, and dendritic cells) are derived from hematopoietic stem cells. Following infection, most $CD4^+$ T cell loss occurs in the intestinal mucosa which harbors the majority of the lymphocytes found in the body (Mehandru et al., 2004).

C-C chemokine receptor type 5, also known as CCR5, is a protein on the surface of white blood cells. It functions as a receptor for chemokines allowing T cells to hone in on target tissues and organs. HIV often uses CCR5 to enter and infect host cells. The majority of mucosal $CD4^+$ T cells express the CCR5 protein, which HIV uses as a co-receptor to gain access to the cells (Brenchley et al., 2004). HIV seeks out and destroys CCR5-expressing $CD4^+$ T cells during acute infection. However, certain individuals carry a mutation known as CCR5-$\Delta32$, which renders the receptor non-functional and prevents entry of HIV into the cells. In humans the CCR5 gene is located on chromosome 3. A $\Delta32$ mutation results in the genetic deletion of a portion of the CCR5 gene. Homozygous carriers of this mutation are resistant to certain strains of HIV-1 infection (Samson et al., 1996). This mutation, although having negative effects on T-cell function, can protect these individuals against HIV.

Future reprogramming technology may allow generation of CCR5-deficient HSCs derived from hESCs/iPSCs, perhaps by creating iPSCs from a CCR5-deficient individual (Ledran et al., 2008; Hutter et al., 2009), or even better from an existing patient using gene editing technology. Diseases such as HIV could potentially be cured by means of hematopoietic cell transplantation whereby HSCs or even fibroblasts are first dedifferentiated into continuously growing iPSC lines using induced reprogramming, followed by gene editing to introduce the $\Delta32$ mutation and then directed differentiation toward the desired hematopoietic lineage followed by bone marrow transplantation (Kambal et al., 2011). The use of engineered zinc finger nucleases (ZFNs) to disrupt the CCR5 gene in human HSCs *in vivo* suggested that ZFN mediated autologous HSC modification could provide a permanent supply of HIV-resistant $CD34^+$ HSC progeny which as well would be capable of immune reconstitution and long-term protection against viral replication (Holt et al., 2010).

In principle this basic idea could be applied to other infectious diseases that rely on a single or even multiple receptors to gain entry into human cells.

CONCLUSION

The last decade has seen a boom in the field of stem cell biology and regenerative medicine coupled with extensive studies on the epigenetics involved in cellular reprogramming and disease. Nuclear reprogramming experiments beginning in the early 1950's introduced the concept of generation of pluripotent cells from somatic cells through nuclear transfer. The discovery of hESCs in the late 1990's acquainted us with a source for designing therapies to treat a broad range of diseases. Following that came the invention of iPSC technology, bringing stem cell production and its use in medicine out of an ethical quandary. Today, stem cell therapies have advanced into clinical trials and we predict that it will not be long before regenerative medicine units become established at health care centers.

Similarly, epigenetic studies have revolutionized the classical genetic approach toward studying and treating hereditary diseases. As we embark on this journey to understand every aspect of the interlocking pathways and processes underlying aging, cancer, regeneration, and repair, we still need to completely decipher the epigenetic code in order to design and implement effective and safe stem-cell based therapies to treat age-related diseases, acute injuries and perhaps chronic disorders.

A full understanding of epigenetics, reprogramming, senescence, cancer, and regeneration will drive remarkable progress in achieving cellular and rejuvenation therapies as well as designing drugs for diseases with an epigenetic basis. The best approach is to embrace all of these fields in a unifying synthesis.

ACKNOWLEDGMENTS

The Authors would like to thank Dr. Donald Orlic for critical review of the manuscript.

REFERENCES

Abad, M., Mosteiro, L., Pantoja, C., Canamero, M., Rayon, T., Ors, I., et al. (2013). Reprogramming *in vivo* produces teratomas and iPS cells with totipotency features. *Nature* 502, 340–345. doi: 10.1038/nature12586

Abdelalim, E. M., Bonnefond, A., Bennaceur-Griscelli, A., and Froguel, P. (2014). Pluripotent stem cells as a potential tool for disease modeling and cell therapy in diabetes. *Stem Cell Rev.* 10, 327–337. doi: 10.1007/s12015-014-9503-6

Acimovic, I., Vilotic, A., Pesl, M., Lacampagne, A., Dvorak, P., Rotrekl, V., et al. (2014). Human pluripotent stem cell-derived cardiomyocytes as research and therapeutic tools. *Biomed. Res. Int.* 2014:512831. doi: 10.1155/2014/512831

Alberio, R., Campbell, K. H., and Johnson, A. D. (2006). Reprogramming somatic cells into stem cells. *Reproduction* 132, 709–720. doi: 10.1530/rep.1.01077

An, M. C., O'brien, R. N., Zhang, N., Patra, B. N., De La Cruz, M., Ray, A., et al. (2014). Polyglutamine disease modeling: epitope based screen for homologous recombination using CRISPR/Cas9 system. *PLoS Curr.* 6:ecurrents.hd.0242d2e7ad72225efa72f6964589369a. doi: 10.1371/currents.hd.0242d2e7ad72225efa72f6964589369a

Anna, B., Blazej, Z., Jacqueline, G., Andrew, C. J., Jeffrey, R., and Andrzej, S. (2007). Mechanism of UV-related carcinogenesis and its contribution to nevi/melanoma. *Expert Rev. Dermatol.* 2, 451–469. doi: 10.1586/17469872.2.4.451

Apostolou, E., and Hochedlinger, K. (2013). Chromatin dynamics during cellular reprogramming. *Nature* 502, 462–471. doi: 10.1038/nature12749

Azziz, R., Woods, K. S., Reyna, R., Key, T. J., Knochenhauer, E. S., and Yildiz, B. O. (2004). The prevalence and features of the polycystic ovary syndrome in an unselected population. *J. Clin. Endocrinol. Metab.* 89, 2745–2749. doi: 10.1210/jc.2003-032046

Ballios, B. G., Clarke, L., Coles, B. L., Shoichet, M. S., and Van Der Kooy, D. (2012). The adult retinal stem cell is a rare cell in the ciliary epithelium whose progeny can differentiate into photoreceptors. *Biol. Open* 1, 237–246. doi: 10.1242/bio.2012027

Batista, L. F. (2014). Telomere biology in stem cells and reprogramming. *Prog. Mol. Biol. Transl. Sci.* 125, 67–88. doi: 10.1016/B978-0-12-397898-1.00003-7

Berger, S. L., Kouzarides, T., Shiekhattar, R., and Shilatifard, A. (2009). An operational definition of epigenetics. *Genes Dev.* 23, 781–783. doi: 10.1101/gad.1787609

Berthiaume, F., Maguire, T. J., and Yarmush, M. L. (2011). Tissue engineering and regenerative medicine: history, progress, and challenges. *Annu. Rev. Chem. Biomol. Eng.* 2, 403–430. doi: 10.1146/annurev-chembioeng-061010-114257

Bhattacharyya, S., Kumar, A., and Lal Khanduja, K. (2012). The voyage of stem cell toward terminal differentiation: a brief overview. *Acta Biochim. Biophys. Sin. (Shanghai).* 44, 463–475. doi: 10.1093/abbs/gms027

Bieback, K., Kinzebach, S., and Karagianni, M. (2011). Translating research into clinical scale manufacturing of mesenchymal stromal cells. *Stem Cells Int.* 2010:193519. doi: 10.4061/2010/193519

Bienvenu, T., Carrie, A., De Roux, N., Vinet, M. C., Jonveaux, P., Couvert, P., et al. (2000). MECP2 mutations account for most cases of typical forms of Rett syndrome. *Hum. Mol. Genet.* 9, 1377–1384. doi: 10.1093/hmg/9.9.1377

Bird, A. (2007). Perceptions of epigenetics. *Nature* 447, 396–398. doi: 10.1038/nature05913

Bjornsson, H. T., Fallin, M. D., and Feinberg, A. P. (2004). An integrated epigenetic and genetic approach to common human disease. *Trends Genet.* 20, 350–358. doi: 10.1016/j.tig.2004.06.009

Bredenkamp, N., Nowell, C. S., and Blackburn, C. C. (2014). Regeneration of the aged thymus by a single transcription factor. *Development* 141, 1627–1637. doi: 10.1242/dev.103614

Brenchley, J. M., Schacker, T. W., Ruff, L. E., Price, D. A., Taylor, J. H., Beilman, G. J., et al. (2004). CD4+ T cell depletion during all stages of HIV disease occurs predominantly in the gastrointestinal tract. *J. Exp. Med.* 200, 749–759. doi: 10.1084/jem.20040874

Briggs, R., and King, T. J. (1952). Transplantation of living nuclei from blastula cells into enucleated Frogs' eggs. *Proc. Natl. Acad. Sci. U.S.A.* 38, 455–463.

Buganim, Y., Faddah, D. A., Cheng, A. W., Itskovich, E., Markoulaki, S., Ganz, K., et al. (2012). Single-cell expression analyses during cellular reprogramming reveal an early stochastic and a late hierarchic phase. *Cell* 150, 1209–1222. doi: 10.1016/j.cell.2012.08.023

Campbell, K. H., Fisher, P., Chen, W. C., Choi, I., Kelly, R. D., Lee, J. H., et al. (2007). Somatic cell nuclear transfer: past, present and future perspectives. *Theriogenology* 68(Suppl. 1), S214–S231. doi: 10.1016/j.theriogenology.2007.05.059

Campbell, K. H., McWhir, J., Ritchie, W. A., and Wilmut, I. (1996). Sheep cloned by nuclear transfer from a cultured cell line. *Nature* 380, 64–66. doi: 10.1038/380064a0

Campisi, J. (2001). Cellular senescence as a tumor-suppressor mechanism. *Trends Cell Biol.* 11, S27–S31. doi: 10.1016/S0962-8924(01)02151-1

Campisi, J. (2005). Senescent cells, tumor suppression, and organismal aging: good citizens, bad neighbors. *Cell* 120, 513–522. doi: 10.1016/j.cell.2005.02.003

Campisi, J., and Vijg, J. (2009). Dos damage to DNA and other macromolecules play a role in aging? If so, how? *J. Gerentol. A. Biol. Sci.* 64, 4. doi: 10.1093/gerona/gln065

Campos-Sanchez, E., and Cobaleda, C. (2014). Tumoral reprogramming: plasticity takes a walk on the wild side. *Biochim. Biophys. Acta.* doi: 10.1016/j.bbagrm.2014.07.003. [Epub ahead of print].

Chakravarthy, U., Wong, T. Y., Fletcher, A., Piault, E., Evans, C., Zlateva, G., et al. (2010). Clinical risk factors for age-related macular degeneration: a systematic review and meta-analysis. *BMC Ophthalmol.* 10:31. doi: 10.1186/1471-2415-10-31

Chen, Q. W., Zhu, X. Y., Li, Y. Y., and Meng, Z. Q. (2014). Epigenetic regulation and cancer (Review). *Oncol. Rep.* 31, 523–532. doi: 10.3892/or.2013.2913

Cherry, A. B., and Daley, G. Q. (2012). Reprogramming cellular identity for regenerative medicine. *Cell* 148, 1110–1122. doi: 10.1016/j.cell.2012.02.031

Colwell, A. S., Longaker, M. T., and Lorenz, H. P. (2005). Mammalian fetal organ regeneration. *Adv. Biochem. Eng. Biotechnol.* 93, 83–100. doi: 10.1007/b99972

Conboy, I. M., Conboy, M. J., Wagers, A. J., Girma, E. R., Weissman, I. L., and Rando, T. A. (2005). Rejuvenation of aged progenitor cells by exposure to a young systemic environment. *Nature* 433, 760–764. doi: 10.1038/nature03260

Conboy, M. J., Conboy, I. M., and Rando, T. A. (2013). Heterochronic parabiosis: historical perspective and methodological considerations for studies of aging and longevity. *Aging Cell* 12, 525–530. doi: 10.1111/acel.12065

Constantinescu, D., Gray, H. L., Sammak, P. J., Schatten, G. P., and Csoka, A. B. (2006). Lamin A/C expression is a marker of mouse and human embryonic stem cell differentiation. *Stem Cells* 24, 177–185. doi: 10.1634/stemcells.2004-0159

Cook, H. L., Patel, P. J., and Tufail, A. (2008). Age-related macular degeneration: diagnosis and management. *Br. Med. Bull.* 85, 127–149. doi: 10.1093/bmb/ldn012

Costa, E., Chen, Y., Davis, J., Dong, E., Noh, J. S., Tremolizzo, L., et al. (2002). REELIN and schizophrenia: a disease at the interface of the genome and the epigenome. *Mol. Interv.* 2, 47–57. doi: 10.1124/mi.2.1.47

Csobonyeiova, M., Polak, S., Koller, J., and Danisovic, L. (2014). Induced pluripotent stem cells and their implication for regenerative medicine. *Cell Tissue Bank.* doi: 10.1007/s10561-014-9462-9. [Epub ahead of print].

De Los Angeles, A., and Daley, G. Q. (2013). Stem cells: reprogramming *in situ.* *Nature* 502, 309–310. doi: 10.1038/nature12559

De Robertis, E. M. (2006). Spemann's organizer and self-regulation in amphibian embryos. *Nat. Rev. Mol. Cell Biol.* 7, 296–302. doi: 10.1038/nrm1855

Diamanti-Kandarakis, E., Kandarakis, H., and Legro, R. S. (2006). The role of genes and environment in the etiology of PCOS. *Endocrine* 30, 19–26. doi: 10.1385/ENDO:30:1:19

Ehrlich, M. (2003). The ICF syndrome, a DNA methyltransferase 3B deficiency and immunodeficiency disease. *Clin. Immunol.* 109, 17–28. doi: 10.1016/S1521-6616(03)00201-8

Eminli, S., Foudi, A., Stadtfeld, M., Maherali, N., Ahfeldt, T., Mostoslavsky, G., et al. (2009). Differentiation stage determines potential of hematopoietic cells for reprogramming into induced pluripotent stem cells. *Nat. Genet.* 41, 968–976. doi: 10.1038/ng.428

Eriksson, M., Brown, W. T., Gordon, L. B., Glynn, M. W., Singer, J., Scott, L., et al. (2003). Recurrent de novo point mutations in lamin A cause Hutchinson-Gilford progeria syndrome. *Nature* 423, 293–298. doi: 10.1038/nature01629

Escobar-Morreale, H. F., Luque-Ramirez, M., and San Millan, J. L. (2005). The molecular-genetic basis of functional hyperandrogenism and the polycystic ovary syndrome. *Endocr. Rev.* 26, 251–282. doi: 10.1210/er.2004-0004

Ess, K. C. (2013). Patient heal thyself: modeling and treating neurological disorders using patient-derived stem cells. *Exp. Biol. Med. (Maywood).* 238, 308–314. doi: 10.1177/1535370213480713

Esteller, M., Corn, P. G., Baylin, S. B., and Herman, J. G. (2001). A gene hypermethylation profile of human cancer. *Cancer Res.* 61, 3225–3229.

Evans, M. J., and Kaufman, M. H. (1981). Establishment in culture of pluripotential cells from mouse embryos. *Nature* 292, 154–156.

Festuccia, N., Osorno, R., Wilson, V., and Chambers, I. (2013). The role of pluripotency gene regulatory network components in mediating transitions between pluripotent cell states. *Curr. Opin. Genet. Dev.* 23, 504–511. doi: 10.1016/j.gde.2013.06.003

Fisher, M. B., and Mauck, R. L. (2013). Tissue engineering and regenerative medicine: recent innovations and the transition to translation. *Tissue Eng. Part B Rev.* 19, 1–13. doi: 10.1089/ten.TEB.2012.0723

Fritz, A. L., Mao, S. R., West, M. G., and Schaffer, D. V. (2014). A medium-throughput analysis of signaling pathways involved in early stages of stem cell reprogramming. *Biotechnol. Bioeng.* doi: 10.1002/bit.25336. [Epub ahead of print].

Frobel, J., Hemeda, H., Lenz, M., Abagnale, G., Joussen, S., Denecke, B., et al. (2014). Epigenetic rejuvenation of mesenchymal stromal cells derived from induced pluripotent stem cells. *Stem Cell Reports* 3, 414–422. doi: 10.1016/j.stemcr.2014.07.003

Fulda, S., Gorman, A. M., Hori, O., and Samali, A. (2010). Cellular stress responses: cell survival and cell death. *Int. J. Cell Biol.* 2010:214074. doi: 10.1155/2010/214074

Gafni, O., Weinberger, L., Mansour, A. A., Manor, Y. S., Chomsky, E., Ben-Yosef, D., et al. (2013). Derivation of novel human ground state naive pluripotent stem cells. *Nature* 504, 282–286. doi: 10.1038/nature12745

Gaj, T., Gersbach, C. A., and Barbas, C. F. 3rd. (2013). ZFN, TALEN, and CRISPR/Cas-based methods for genome engineering. *Trends Biotechnol.* 31, 397–405. doi: 10.1016/j.tibtech.2013.04.004

Gardner, R. L. (2002). Stem cells: potency, plasticity and public perception. *J. Anat.* 200, 277–282. doi: 10.1046/j.1469-7580.2002.00029.x

George, B. (2011). Regulations and guidelines governing stem cell based products: clinical considerations. *Perspect. Clin. Res.* 2, 94–99. doi: 10.4103/2229-3485.83228

Gimble, J. M., Katz, A. J., and Bunnell, B. A. (2007). Adipose-derived stem cells for regenerative medicine. *Circ. Res.* 100, 1249–1260. doi: 10.1161/01.RES.0000265074.83288.09

Graf, T. (2011). Historical origins of transdifferentiation and reprogramming. *Cell Stem Cell* 9, 504–516. doi: 10.1016/j.stem.2011.11.012

Grskovic, M., Javaherian, A., Strulovici, B., and Daley, G. Q. (2011). Induced pluripotent stem cells–opportunities for disease modeling and drug discovery. *Nat. Rev. Drug Discov.* 10, 915–929. doi: 10.1038/nrd3577

Grunz, H. (2001). Developmental biology of amphibians after Hans Spemann in Germany. *Int. J. Dev. Biol.* 45, 39–50.

Guha, P., Morgan, J. W., Mostoslavsky, G., Rodrigues, N. P., and Boyd, A. S. (2013). Lack of immune response to differentiated cells derived from syngeneic induced pluripotent stem cells. *Cell Stem Cell* 12, 407–412. doi: 10.1016/j.stem.2013.01.006

Gurdon, J. B. (1962). The developmental capacity of nuclei taken from intestinal epithelium cells of feeding tadpoles. *J. Embryol. Exp. Morphol.* 10, 622–640.

Gurdon, J. B., and Melton, D. A. (2008). Nuclear reprogramming in cells. *Science* 322, 1811–1815. doi: 10.1126/science.1160810

Gurtner, G. C., Callaghan, M. J., and Longaker, M. T. (2007). Progress and potential for regenerative medicine. *Annu. Rev. Med.* 58, 299–312. doi: 10.1146/annurev.med.58.082405.095329

Gurtner, G. C., Werner, S., Barrandon, Y., and Longaker, M. T. (2008). Wound repair and regeneration. *Nature* 453, 314–321. doi: 10.1038/nature 07039

Halley-Stott, R. P., Pasque, V., and Gurdon, J. B. (2013). Nuclear reprogramming. *Development* 140, 2468–2471. doi: 10.1242/dev.092049

Hanahan, D., and Weinberg, R. A. (2011). Hallmarks of cancer: the next generation. *Cell* 144, 646–674. doi: 10.1016/j.cell.2011.02.013

Hanna, J., Wernig, M., Markoulaki, S., Sun, C. W., Meissner, A., Cassady, J. P., et al. (2007). Treatment of sickle cell anemia mouse model with iPS cells generated from autologous skin. *Science* 318, 1920–1923. doi: 10.1126/science.1152092

Hansson, J., Rafiee, M. R., Reiland, S., Polo, J. M., Gehring, J., Okawa, S., et al. (2012). Highly coordinated proteome dynamics during reprogramming of somatic cells to pluripotency. *Cell Rep.* 2, 1579–1592. doi: 10.1016/j.celrep.2012.10.014

Harries, L. W., Pilling, L. C., Hernandez, L. D., Bradley-Smith, R., Henley, W., Singleton, A. B., et al. (2012). CCAAT-enhancer-binding protein-beta expression *in vivo* is associated with muscle strength. *Aging Cell* 11, 262–268. doi: 10.1111/j.1474-9726.2011.00782.x

Health, U. S. D. O., Human Services, F., Drug Administration, C. F. B. E. (2001). Guidance for human somatic cell therapy and gene therapy. *Hum. Gene Ther.* 12, 303–314. doi: 10.1089/10430340150218431

Heller, J. P., and Martin, K. R. (2014). Enhancing RPE cell-based therapy outcomes for AMD: the role of Bruch's membrane. *Transl. Vis. Sci. Technol.* 3:11. doi: 10.1167/tvst.3.3.11

Hewagama, A., and Richardson, B. (2009). The genetics and epigenetics of autoimmune diseases. *J. Autoimmun.* 33, 3–11. doi: 10.1016/j.jaut.2009.03.007

Hibaoui, Y., Grad, I., Letourneau, A., Sailani, M. R., Dahoun, S., Santoni, F. A., et al. (2014). Modeling and rescuing neurodevelopmental defect of down syndrome using induced pluripotent stem cells from monozygotic twins discordant for trisomy 21. *EMBO Mol. Med.* 6, 259–277. doi: 10.1002/emmm.201302848

High, K., Gregory, P. D., and Gersbach, C. (2014). CRISPR technology for gene therapy. *Nat. Med.* 20, 476–477. doi: 10.1038/nm.3566

Holt, N., Wang, J., Kim, K., Friedman, G., Wang, X., Taupin, V., et al. (2010). Human hematopoietic stem/progenitor cells modified by zinc-finger nucleases targeted to CCR5 control HIV-1 *in vivo*. *Nat. Biotechnol.* 28, 839–847. doi: 10.1038/nbt.1663

Horvath, S. (2013). DNA methylation age of human tissues and cell types. *Genome Biol.* 14:R115. doi: 10.1186/gb-2013-14-10-r115

Hou, P., Li, Y., Zhang, X., Liu, C., Guan, J., Li, H., et al. (2013). Pluripotent stem cells induced from mouse somatic cells by small-molecule compounds. *Science* 341, 651–654. doi: 10.1126/science.1239278

Hu, K., Yu, J., Suknuntha, K., Tian, S., Montgomery, K., Choi, K. D., et al. (2011). Efficient generation of transgene-free induced pluripotent stem cells from normal and neoplastic bone marrow and cord blood mononuclear cells. *Blood* 117, e109–e119. doi: 10.1182/blood-2010-07-298331

Hutter, G., Nowak, D., Mossner, M., Ganepola, S., Mussig, A., Allers, K., et al. (2009). Long-term control of HIV by CCR5 Delta32/Delta32 stem-cell transplantation. *N. Engl. J. Med.* 360, 692–698. doi: 10.1056/NEJMoa0802905

International Stem Cell, I., Adewumi, O., Aflatoonian, B., Ahrlund-Richter, L., Amit, M., Andrews, P. W., et al. (2007). Characterization of human embryonic stem cell lines by the International Stem Cell Initiative. *Nat. Biotechnol.* 25, 803–816. doi: 10.1038/nbt1318

Jemal, A., Bray, F., Center, M. M., Ferlay, J., Ward, E., and Forman, D. (2011). Global cancer statistics. *CA Cancer J. Clin.* 61, 69–90. doi: 10.3322/caac.20107

Jeon, K., Lim, H., Kim, J. H., Thuan, N. V., Park, S. H., Lim, Y. M., et al. (2012). Differentiation and transplantation of functional pancreatic beta cells generated from induced pluripotent stem cells derived from a type 1 diabetes mouse model. *Stem Cells Dev.* 21, 2642–2655. doi: 10.1089/scd.2011.0665

Jiang, Y., Jahagirdar, B. N., Reinhardt, R. L., Schwartz, R. E., Keene, C. D., Ortiz-Gonzalez, X. R., et al. (2002). Pluripotency of mesenchymal stem cells derived from adult marrow. *Nature* 418, 41–49. doi: 10.1038/nature00870

Kallings, L. O. (2008). The first postmodern pandemic: 25 years of HIV/AIDS. *J. Intern. Med.* 263, 218–243. doi: 10.1111/j.1365-2796.2007.01910.x

Kambal, A., Mitchell, G., Cary, W., Gruenloh, W., Jung, Y., Kalomoiris, S., et al. (2011). Generation of HIV-1 resistant and functional macrophages from hematopoietic stem cell-derived induced pluripotent stem cells. *Mol. Ther.* 19, 584–593. doi: 10.1038/mt.2010.269

Kanherkar, R. R., Bhatia-Dey, N., and Csoka, A. B. (2014). Epigenetics across the human lifespan. *Front. Cell Dev. Biol.* 2:49. doi: 10.3389/fcell.2014.00049

Kator, K., Cristofalo, V., Charpentier, R., and Cutler, R. G. (1985). Dysdifferentiative nature of aging: passage number dependency of globin gene expression in normal human diploid cells grown in tissue culture. *Gerontology* 31, 355–361.

Kikyo, N., and Wolffe, A. P. (2000). Reprogramming nuclei: insights from cloning, nuclear transfer and heterokaryons. *J. Cell Sci.* 113(Pt 1), 11–20.

Kim, H., and Kim, J. S. (2014). A guide to genome engineering with programmable nucleases. *Nat. Rev. Genet.* 15, 321–334. doi: 10.1038/nrg3686

Kim, J., Efe, J. A., Zhu, S., Talantova, M., Yuan, X., Wang, S., et al. (2011). Direct reprogramming of mouse fibroblasts to neural progenitors. *Proc. Natl. Acad. Sci. U.S.A.* 108, 7838–7843. doi: 10.1073/pnas.1103113108

Kleivi, K. (2006). Advances in the genetics and epigenetics of gene regulation and human disease. *Genome Biol.* 7:325. doi: 10.1186/gb-2006-7-8-325

Korf, B. (2008). Hutchinson-Gilford progeria syndrome, aging, and the nuclear lamina. *N. Engl. J. Med.* 358, 552–555. doi: 10.1056/NEJMp0800071

Kriks, S., Shim, J. W., Piao, J., Ganat, Y. M., Wakeman, D. R., Xie, Z., et al. (2011). Dopamine neurons derived from human ES cells efficiently engraft in animal models of Parkinson's disease. *Nature* 480, 547–551. doi: 10.1038/nature10648

Lang, A. H., Li, H., Collins, J. J., and Mehta, P. (2014). Epigenetic landscapes explain partially reprogrammed cells and identify key reprogramming genes. *PLoS Comput. Biol.* 10:e1003734. doi: 10.1371/journal.pcbi.1003734

Lapasset, L., Milhavet, O., Prieur, A., Besnard, E., Babled, A., Ait-Hamou, N., et al. (2011). Rejuvenating senescent and centenarian human cells by reprogramming through the pluripotent state. *Genes Dev.* 25, 2248–2253. doi: 10.1101/gad.173922.111

Ledran, M. H., Krassowska, A., Armstrong, L., Dimmick, I., Renstrom, J., Lang, R., et al. (2008). Efficient hematopoietic differentiation of human embryonic stem cells on stromal cells derived from hematopoietic niches. *Cell Stem Cell* 3, 85–98. doi: 10.1016/j.stem.2008.06.001

Lenhoff, H. M. (1991). Ethel Browne, Hans Spemann, and the discovery of the discovery of the oganiser phenomenon. *Biol. Bull.* 181, 72–80.

Li, J., Song, W., Pan, G., and Zhou, J. (2014). Advances in understanding the cell types and approaches used for generating induced pluripotent stem cells. *J. Hematol. Oncol.* 7, 50. doi: 10.1186/s13045-014-0050-z

Li, L., and Bhatia, R. (2011). Stem cell quiescence. *Clin. Cancer Res.* 17, 4936–4941. doi: 10.1158/1078-0432.CCR-10-1499

Liang, G., and Zhang, Y. (2013). Embryonic stem cell and induced pluripotent stem cell: an epigenetic perspective. *Cell Res.* 23, 49–69. doi: 10.1038/cr.2012.175

Lisanti, S., Tavecchio, M., Chae, Y. C., Liu, Q., Brice, A. K., Thakur, M. L., et al. (2014). Deletion of the mitochondrial chaperone TRAP-1 uncovers global reprogramming of metabolic networks. *Cell Rep.* 8, 671–677. doi: 10.1016/j.celrep.2014.06.061

Liu, B., Hunter, D. J., Rooker, S., Chan, A., Paulus, Y. M., Leucht, P., et al. (2013a). Wnt signaling promotes muller cell proliferation and survival after injury. *Invest. Ophthalmol. Vis. Sci.* 54, 444–453. doi: 10.1167/iovs.12-10774

Liu, G. H., Barkho, B. Z., Ruiz, S., Diep, D., Qu, J., Yang, S. L., et al. (2011a). Recapitulation of premature ageing with iPSCs from Hutchinson-Gilford progeria syndrome. *Nature* 472, 221–225. doi: 10.1038/nature09879

Liu, G. H., Suzuki, K., Li, M., Qu, J., Montserrat, N., Tarantino, C., et al. (2014). Modeling Fanconi anemia pathogenesis and therapeutics using integration-free patient-derived iPSCs. *Nat. Commun.* 5, 4330. doi: 10.1038/ncomms5330

Liu, G. H., Suzuki, K., Qu, J., Sancho-Martinez, I., Yi, F., Li, M., et al. (2011b). Targeted gene correction of laminopathy-associated LMNA mutations in patient-specific iPSCs. *Cell Stem Cell* 8, 688–694. doi: 10.1016/j.stem.2011.04.019

Liu, Y., Weick, J. P., Liu, H., Krencik, R., Zhang, X., Ma, L., et al. (2013b). Medial ganglionic eminence-like cells derived from human embryonic stem cells correct learning and memory deficits. *Nat. Biotechnol.* 31, 440–447. doi: 10.1038/nbt.2565

Loffredo, F. S., Steinhauser, M. L., Jay, S. M., Gannon, J., Pancoast, J. R., Yalamanchi, P., et al. (2013). Growth differentiation factor 11 is a circulating factor that reverses age-related cardiac hypertrophy. *Cell* 153, 828–839. doi: 10.1016/j.cell.2013.04.015

Loh, K. M., and Lim, B. (2012). Epigenetics: actors in the cell reprogramming drama. *Nature* 488, 599–600. doi: 10.1038/488599a

Luna-Zurita, L., and Bruneau, B. G. (2013). Chromatin modulators as facilitating factors in cellular reprogramming. *Curr. Opin. Genet. Dev.* 23, 556–561. doi: 10.1016/j.gde.2013.07.002

Lunyak, V. V., and Rosenfeld, M. G. (2008). Epigenetic regulation of stem cell fate. *Hum. Mol. Genet.* 17, R28–R36. doi: 10.1093/hmg/ddn149

Ma, T., Xie, M., Laurent, T., and Ding, S. (2013). Progress in the reprogramming of somatic cells. *Circ. Res.* 112, 562–574. doi: 10.1161/CIRCRESAHA.111.249235

Manikkam, M., Tracey, R., Guerrero-Bosagna, C., and Skinner, M. K. (2012). Pesticide and insect repellent mixture (permethrin and DEET) induces epigenetic transgenerational inheritance of disease and sperm epimutations. *Reprod. Toxicol.* 34, 708–719. doi: 10.1016/j.reprotox.2012.08.010

Manukyan, M., and Singh, P. B. (2012). Epigenetic rejuvenation. *Genes Cells* 17, 337–343. doi: 10.1111/j.1365-2443.2012.01595.x

Martin, G. R. (1981). Isolation of a pluripotent cell line from early mouse embryos cultured in medium conditioned by teratocarcinoma stem cells. *Proc. Natl. Acad. Sci. U.S.A.* 78, 7634–7638.

Mason, J. L., Jones, J. J., Taniike, M., Morell, P., Suzuki, K., and Matsushima, G. K. (2000). Mature oligodendrocyte apoptosis precedes IGF-1 production and oligodendrocyte progenitor accumulation and differentiation during demyelination/remyelination. *J. Neurosci. Res.* 61, 251–262. doi: 10.1002/1097-4547(20000801)61:3%3C251::AID-JNR3%3E3.0.CO;2-W

McClintock, D., Ratner, D., Lokuge, M., Owens, D. M., Gordon, L. B., Collins, F. S., et al. (2007). The mutant form of lamin A that causes Hutchinson-Gilford progeria is a biomarker of cellular aging in human skin. *PLoS ONE* 2:e1269. doi: 10.1371/journal.pone.0001269

Mehandru, S., Poles, M. A., Tenner-Racz, K., Horowitz, A., Hurley, A., Hogan, C., et al. (2004). Primary HIV-1 infection is associated with preferential depletion of CD4+ T lymphocytes from effector sites in the gastrointestinal tract. *J. Exp. Med.* 200, 761–770. doi: 10.1084/jem.20041196

Miller, J. D., Ganat, Y. M., Kishinevsky, S., Bowman, R. L., Liu, B., Tu, E. Y., et al. (2013). Human iPSC-based modeling of late-onset disease via progerin-induced aging. *Cell Stem Cell* 13, 691–705. doi: 10.1016/j.stem.2013.11.006

Minear, S., Leucht, P., Jiang, J., Liu, B., Zeng, A., Fuerer, C., et al. (2010). Wnt proteins promote bone regeneration. *Sci. Transl. Med.* 2, 29ra30. doi: 10.1126/scitranslmed.3000231

Mitalipov, S., and Wolf, D. (2009). Totipotency, pluripotency and nuclear reprogramming. *Adv. Biochem. Eng. Biotechnol.* 114, 185–199. doi: 10.1007/10_2008_45

Mohamet, L., Miazga, N. J., and Ward, C. M. (2014). Familial Alzheimer's disease modeling using induced pluripotent stem cell technology. *World J. Stem Cells* 6, 239–247. doi: 10.4252/wjsc.v6.i2.239

Munoz-Najar, U., and Sedivy, J. M. (2011). Epigenetic control of aging. *Antioxid. Redox Signal.* 14, 241–259. doi: 10.1089/ars.2010.3250

Natsuka, S., Akira, S., Nishio, Y., Hashimoto, S., Sugita, T., Isshiki, H., et al. (1992). Macrophage differentiation-specific expression of NF-IL6, a transcription factor for interleukin-6. *Blood* 79, 460–466.

Niu, W., Zang, T., Zou, Y., Fang, S., Smith, D. K., Bachoo, R., et al. (2013). *In vivo* reprogramming of astrocytes to neuroblasts in the adult brain. *Nat. Cell Biol.* 15, 1164–1175. doi: 10.1038/ncb2843

Niwa, H., Miyazaki, J., and Smith, A. G. (2000). Quantitative expression of Oct-3/4 defines differentiation, dedifferentiation or self-renewal of ES cells. *Nat. Genet.* 24, 372–376. doi: 10.1038/74199

Okita, K., Ichisaka, T., and Yamanaka, S. (2007). Generation of germline-competent induced pluripotent stem cells. *Nature* 448, 313–317. doi: 10.1038/nature05934

Pajcini, K. V., Corbel, S. Y., Sage, J., Pomerantz, J. H., and Blau, H. M. (2010). Transient inactivation of Rb and ARF yields regenerative cells from postmitotic mammalian muscle. *Cell Stem Cell* 7, 198–213. doi: 10.1016/j.stem.2010.05.022

Pandian, G. N., Taniguchi, J., and Sugiyama, H. (2014). Cellular reprogramming for pancreatic beta-cell regeneration: clinical potential of small molecule control. *Clin. Transl. Med.* 3:6. doi: 10.1186/2001-1326-3-6

Papp, B., and Plath, K. (2011). Reprogramming to pluripotency: stepwise resetting of the epigenetic landscape. *Cell Res.* 21, 486–501. doi: 10.1038/cr.2011.28

Papp, B., and Plath, K. (2013). Epigenetics of reprogramming to induced pluripotency. *Cell* 152, 1324–1343. doi: 10.1016/j.cell.2013.02.043

Parson, A. (2006). The long journey from stem cells to medical product. *Cell* 125, 9–11. doi: 10.1016/j.cell.2006.03.024

Patel, M., and Yang, S. (2010). Advances in reprogramming somatic cells to induced pluripotent stem cells. *Stem Cell Rev.* 6, 367–380. doi: 10.1007/s12015-010-9123-8

Phillips, T. (2008). The role of methylation in gene expression. *Nat. Educ.* 1:116.

Porrello, E. R., and Olson, E. N. (2014). A neonatal blueprint for cardiac regeneration. *Stem Cell Res.* doi: 10.1016/j.scr.2014.06.003. [Epub ahead of print].

Qu, F., Wang, F. F., Yin, R., Ding, G. L., El-Prince, M., Gao, Q., et al. (2012). A molecular mechanism underlying ovarian dysfunction of polycystic ovary syndrome: hyperandrogenism induces epigenetic alterations in the granulosa cells. *J. Mol. Med. (Berl.)* 90, 911–923. doi: 10.1007/s00109-012-0881-4

Radzisheuskaya, A., and Silva, J. C. (2014). Do all roads lead to Oct4? the emerging concepts of induced pluripotency. *Trends Cell Biol.* 24, 275–284. doi: 10.1016/j.tcb.2013.11.010

Rando, T. A., and Chang, H. Y. (2012). Aging, rejuvenation, and epigenetic reprogramming: resetting the aging clock. *Cell* 148, 46–57. doi: 10.1016/j.cell.2012.01.003

Richter, H., Randau, L., and Plagens, A. (2013). Exploiting CRISPR/Cas: interference mechanisms and applications. *Int. J. Mol. Sci.* 14, 14518–14531. doi: 10.3390/ijms140714518

Rivera, C. M., and Ren, B. (2013). Mapping human epigenomes. *Cell* 155, 39–55. doi: 10.1016/j.cell.2013.09.011

Rodenhiser, D., and Mann, M. (2006). Epigenetics and human disease: translating basic biology into clinical applications. *CMAJ* 174, 341–348. doi: 10.1503/cmaj.050774

Rutherford, R. B., Niekrash, C. E., Kennedy, J. E., and Charette, M. F. (1992). Platelet-derived and insulin-like growth factors stimulate regeneration of periodontal attachment in monkeys. *J. Periodont. Res.* 27, 285–290.

Samson, M., Libert, F., Doranz, B. J., Rucker, J., Liesnard, C., Farber, C. M., et al. (1996). Resistance to HIV-1 infection in caucasian individuals bearing mutant alleles of the CCR-5 chemokine receptor gene. *Nature* 382, 722–725. doi: 10.1038/382722a0

Sanchez Alvarado, A., and Yamanaka, S. (2014). Rethinking differentiation: stem cells, regeneration, and plasticity. *Cell* 157, 110–119. doi: 10.1016/j.cell.2014.02.041

Schwartz, S. D., Hubschman, J. P., Heilwell, G., Franco-Cardenas, V., Pan, C. K., Ostrick, R. M., et al. (2012). Embryonic stem cell trials for macular degeneration: a preliminary report. *Lancet* 379, 713–720. doi: 10.1016/S0140-6736(12)60028-2

Seol, Y. J., Kang, H. W., Lee, S. J., Atala, A., and Yoo, J. J. (2014). Bioprinting technology and its applications. *Eur. J. Cardiothorac. Surg.* 46, 342–348. doi: 10.1093/ejcts/ezu148

Sharma, R. P. (2005). Schizophrenia, epigenetics and ligand-activated nuclear receptors: a framework for chromatin therapeutics. *Schizophr. Res.* 72, 79–90. doi: 10.1016/j.schres.2004.03.001

Shen, H. R., Qiu, L. H., Zhang, Z. Q., Qin, Y. Y., Cao, C., and Di, W. (2013). Genome-wide methylated DNA immunoprecipitation analysis of patients with polycystic ovary syndrome. *PLoS ONE* 8:e64801. doi: 10.1371/journal.pone.0064801

Shyh-Chang, N., Zhu, H., Yvanka De Soysa, T., Shinoda, G., Seligson, M. T., Tsanov, K. M., et al. (2013). Lin28 enhances tissue repair by reprogramming cellular metabolism. *Cell* 155, 778–792. doi: 10.1016/j.cell.2013.09.059

Soza-Ried, J., and Fisher, A. G. (2012). Reprogramming somatic cells towards pluripotency by cellular fusion. *Curr. Opin. Genet. Dev.* 22, 459–465. doi: 10.1016/j.gde.2012.07.005

Spemann, H. (1938). *Embryonic Development and Induction.* New Haven, CT: Yale University Press.

Sterneckert, J. L., Reinhardt, P., and Scholer, H. R. (2014). Investigating human disease using stem cell models. *Nat. Rev. Genet.* 15, 625–639. doi: 10.1038/nrg3764

Su, R. J., Baylink, D. J., Neises, A., Kiroyan, J. B., Meng, X., Payne, K. J., et al. (2013). Efficient generation of integration-free ips cells from human adult peripheral blood using BCL-XL together with Yamanaka factors. *PLoS ONE* 8:e64496. doi: 10.1371/journal.pone.0064496

Subramanyam, D. (2013). Cellular reprogramming—Turning the clock back. *Reson.* 18, 514–521. doi: 10.1007/s12045-013-0069-4

Suhr, S. T., Chang, E. A., Tjong, J., Alcasid, N., Perkins, G. A., Goissis, M. D., et al. (2010). Mitochondrial rejuvenation after induced pluripotency. *PLoS ONE* 5:e14095. doi: 10.1371/journal.pone.0014095

Svendsen, C. N. (2013). Back to the future: how human induced pluripotent stem cells will transform regenerative medicine. *Hum. Mol. Genet.* 22, R32–R38. doi: 10.1093/hmg/ddt379

Szabo, E., Rampalli, S., Risueno, R. M., Schnerch, A., Mitchell, R., Fiebig-Comyn, A., et al. (2010). Direct conversion of human fibroblasts to multilineage blood progenitors. *Nature* 468, 521–526. doi: 10.1038/nature09591

Tachibana, M., Amato, P., Sparman, M., Gutierrez, N. M., Tippner-Hedges, R., Ma, H., et al. (2013). Human embryonic stem cells derived by somatic cell nuclear transfer. *Cell* 153, 1228–1238. doi: 10.1016/j.cell.2013.05.006

Takahashi, K. (2012). Cellular reprogramming–lowering gravity on Waddington's epigenetic landscape. *J. Cell Sci.* 125, 2553–2560. doi: 10.1242/jcs.084822

Takahashi, K., and Yamanaka, S. (2006). Induction of pluripotent stem cells from mouse embryonic and adult fibroblast cultures by defined factors. *Cell* 126, 663–676. doi: 10.1016/j.cell.2006.07.024

Takahashi, K., and Yamanaka, S. (2013). Induced pluripotent stem cells in medicine and biology. *Development* 140, 2457–2461. doi: 10.1242/dev.092551

Talluri, T. R., Kumar, D., Glage, S., Garrels, W., Ivics, Z., Debowski, K., et al. (2014). Non-viral reprogramming of fibroblasts into induced pluripotent stem cells by sleeping beauty and piggybac transposons. *Biochem. Biophys. Res. Commun.* 450, 581–587. doi: 10.1016/j.bbrc.2014.06.014

Tang, S., Yang, L., Tang, X., and Liu, M. (2014). The role of oxidized ATM in the regulation of oxidative stress-induced energy metabolism reprogramming of CAFs. *Cancer Lett.* 353, 133–144. doi: 10.1016/j.canlet.2014.07.028

Teperek, M., and Miyamoto, K. (2013). Nuclear reprogramming of sperm and somatic nuclei in eggs and oocytes. *Reprod. Med. Biol.* 12, 133–149. doi: 10.1007/s12522-013-0155-z

Terai, S., Ishikawa, T., Omori, K., Aoyama, K., Marumoto, Y., Urata, Y., et al. (2006). Improved liver function in patients with liver cirrhosis after autologous bone marrow cell infusion therapy. *Stem Cells* 24, 2292–2298. doi: 10.1634/stemcells.2005-0542

Trounson, A., Thakar, R. G., Lomax, G., and Gibbons, D. (2011). Clinical trials for stem cell therapies. *BMC Med.* 9:52. doi: 10.1186/1741-7015-9-52

Tsai, S. Y., Bouwman, B. A., Ang, Y. S., Kim, S. J., Lee, D. F., Lemischka, I. R., et al. (2011). Single transcription factor reprogramming of hair follicle dermal papilla cells to induced pluripotent stem cells. *Stem Cells* 29, 964–971. doi: 10.1002/stem.649

Unaids. (2013). *2013 Global Fact Sheet.* Geneva: UNAIDS.

Unternaehrer, J. J., and Daley, G. Q. (2011). Induced pluripotent stem cells for modeling human diseases. *Philos Trans. R. Soc. Lond. B Biol. Sci.* 366, 2274–2285. doi: 10.1098/rstb.2011.0017

Van Den Veyver, I. B., and Zoghbi, H. Y. (2001). Mutations in the gene encoding methyl-CpG-binding protein 2 cause Rett syndrome. *Brain Dev.* 23(Suppl. 1), S147–S151. doi: 10.1016/S0387-7604(01)00376-X

Van Oevelen, C., Kallin, E. M., and Graf, T. (2013). Transcription factor-induced enhancer modulations during cell fate conversions. *Curr. Opin. Genet. Dev.* 23, 562–567. doi: 10.1016/j.gde.2013.07.003

Vierbuchen, T., and Wernig, M. (2012). Molecular roadblocks for cellular reprogramming. *Mol. Cell* 47, 827–838. doi: 10.1016/j.molcel.2012.09.008

Villeda, S. A., Luo, J., Mosher, K. I., Zou, B., Britschgi, M., Bieri, G., et al. (2011). The ageing systemic milieu negatively regulates neurogenesis and cognitive function. *Nature* 477, 90–94. doi: 10.1038/nature10357

Wang, A., Tang, Z., Park, I. H., Zhu, Y., Patel, S., Daley, G. Q., et al. (2011). Induced pluripotent stem cells for neural tissue engineering. *Biomaterials* 32, 5023–5032. doi: 10.1016/j.biomaterials.2011.03.070

Wang, H., Yang, H., Shivalila, C. S., Dawlaty, M. M., Cheng, A. W., Zhang, F., et al. (2013). One-step generation of mice carrying mutations in multiple genes by CRISPR/Cas-mediated genome engineering. *Cell* 153, 910–918. doi: 10.1016/j.cell.2013.04.025

Wang, K. C., Helms, J. A., and Chang, H. Y. (2009). Regeneration, repair and remembering identity: the three Rs of hox gene expression. *Trends Cell Biol.* 19, 268–275. doi: 10.1016/j.tcb.2009.03.007

Watanabe, A., Yamada, Y., and Yamanaka, S. (2013). Epigenetic regulation in pluripotent stem cells: a key to breaking the epigenetic barrier. *Philos. Trans. R. Soc. Lond. B Biol. Sci.* 368:20120292. doi: 10.1098/rstb.2012.0292

Weinhold, B. (2006). Epigenetics: the science of change. *Environ. Health Perspect.* 114, A160–A167. doi: 10.1289/ehp.114-a160

Wernig, M., Meissner, A., Foreman, R., Brambrink, T., Ku, M., Hochedlinger, K., et al. (2007). *In vitro* reprogramming of fibroblasts into a pluripotent ES-cell-like state. *Nature* 448, 318–324. doi: 10.1038/nature05944

Wu, D. C., Boyd, A. S., and Wood, K. J. (2007). Embryonic stem cell transplantation: potential applicability in cell replacement therapy and regenerative medicine. *Front. Biosci.* 12, 4525–4535. doi: 10.2741/2407

Yeh, E. T., Zhang, S., Wu, H. D., Korbling, M., Willerson, J. T., and Estrov, Z. (2003). Transdifferentiation of human peripheral blood CD34+-enriched cell population into cardiomyocytes, endothelial cells, and smooth muscle cells *in vivo*. *Circulation* 108, 2070–2073. doi: 10.1161/01.CIR.0000099501.52718.70

Yingze Zhao, Y. Y. A. Y. W. (2014). Developing CRISPR/Cas9 technologies for research and medicine. *MOJ Cell Sci. Report* 1, 7.

Yonekawa, Y., and Kim, I. K. (2014). Clinical characteristics and current treatment of age-related macular degeneration. *Cold Spring Harb. Perspect. Med.* doi: 10.1101/cshperspect.a017178. [Epub ahead of print].

Zhao, T., Zhang, Z. N., Rong, Z., and Xu, Y. (2011). Immunogenicity of induced pluripotent stem cells. *Nature* 474, 212–215. doi: 10.1038/nature10135

Zhong, X., Gutierrez, C., Xue, T., Hampton, C., Vergara, M. N., Cao, L. H., et al. (2014). Generation of three-dimensional retinal tissue with functional photoreceptors from human iPSCs. *Nat. Commun.* 5, 4047. doi: 10.1038/ncomms5047

Zhou, Q., Brown, J., Kanarek, A., Rajagopal, J., and Melton, D. A. (2008). *In vivo* reprogramming of adult pancreatic exocrine cells to beta-cells. *Nature* 455, 627–632. doi: 10.1038/nature07314

Zhu, H., Lensch, M. W., Cahan, P., and Daley, G. Q. (2011). Investigating monogenic and complex diseases with pluripotent stem cells. *Nat. Rev. Genet.* 12, 266–275. doi: 10.1038/nrg2951

Zs-Nagy, I., Cutler, R. G., and Semsei, I. (1988). Dysdifferentiation hypothesis of aging and cancer: a comparison with the membrane hypothesis of aging. *Ann. N. Y. Acad. Sci.* 521, 215–225.

Conflict of Interest Statement: The authors declare that the research was conductedin the absence of any commercial or financial relationships that could be construed as a potential conflict of interest.

16

Moonlighting functions of the NRZ (mammalian Dsl1) complex

Mitsuo Tagaya *, Kohei Arasaki, Hiroki Inoue and Hana Kimura

Department of Molecular Life Sciences, School of Life Sciences, Tokyo University of Pharmacy and Life Sciences, Hachioji, Japan

Edited by:
Brian Storrie, University of Arkansass for Medical Sciences, USA

Reviewed by:
Vladimir Lupashin, University of Arkansas for Medical Sciences, USA
Daniel Ungar, University of York, UK
Frederick Hughson, Princeton University, USA

****Correspondence:***
Mitsuo Tagaya, Department of Molecular Life Sciences, School of Life Sciences, Tokyo University of Pharmacy and Life Sciences, 1432-1 Horinouchi, Hachioji, Tokyo 192-0392, Japan
e-mail: tagaya@toyaku.ac.jp

The yeast Dsl1 complex, which comprises Dsl1, Tip20, and Sec39/Dsl3, has been shown to participate, as a vesicle-tethering complex, in retrograde trafficking from the Golgi apparatus to the endoplasmic reticulum. Its metazoan counterpart NRZ complex, which comprises NAG, RINT1, and ZW10, is also involved in Golgi-to-ER retrograde transport, but each component of the complex has diverse cellular functions including endosome-to-Golgi transport, cytokinesis, cell cycle checkpoint, autophagy, and mRNA decay. In this review, we summarize the current knowledge of the metazoan NRZ complex and discuss the "moonlighting" functions and intercorrelation of their subunits.

Keywords: autophagy, CATCHR tether complex, cell cycle, endoplasmic reticulum, mRNA decay, NAG, RINT1, ZW10

INTRODUCTION

The unicellular organisms *Escherichia coli* and yeast *Saccharomyces cerevisiae* have roughly 4300 and 6600 genes, respectively, whereas the multicellular organism human contains 22,000–26,000 genes. One may feel that the number of the human genes is too small, given the complex architecture and function of human being. The limited number of genes in multicellular organisms may be due to a strong selective pressure; cells might have evolved not to increase the number of genes because more energy and time are necessary to carry extra genes and accurately transcribe and translate them. To prevent the increase in gene number, multicellular organisms might adopt a strategy to reuse certain proteins for very different cellular processes. Jeffery (1999) coined the term "moonlighting protein" to describe proteins with multiple roles. The list of moonlighting proteins continued to expand and now includes components implicated in membrane trafficking (reviewed by Copley, 2012; Royle, 2013).

The Dsl1 complex in yeast comprises Dsl1 (VanRheenen et al., 2001), Tip20 (Sweet and Pelham, 1993), and Sec39/Dsl3 (Mnaimneh et al., 2004; Kraynack et al., 2005). This complex is a member of the Complex Associated with Tethering Containing Helical Rods (CATCHR) family (Yu and Hughson, 2010) and has been implicated in tethering of Golgi-derived transport vesicles with the endoplasmic reticulum (ER) (Andag et al., 2001; Reilly et al., 2001; Andag and Schmitt, 2003; Zink et al., 2009; Diefenbacher et al., 2011). In 2004, using an immunoaffinity method we isolated from rat liver membranes a large complex containing syntaxin 18, an ER-associated soluble *N*-ethylmaleimide-sensitive factor attachment protein receptor (SNARE) implicated in membrane fusion (Hatsuzawa et al., 2000), and found that the complex includes ZW10 and RINT1 (Hirose et al., 2004). Several lines of evidence suggest that ZW10 and RINT1 are the mammalian counterparts of yeast Dsl1 and Tip20, respectively, although the amino acid sequence identities between the yeast and mammalian proteins are very low (~14%). Intriguingly, both ZW10 and RINT1 had been discovered as cell cycle checkpoint proteins (Williams et al., 1992; Xiao et al., 2001). Nowadays, the mammalian Dsl1 complex is known to participate not only in membrane trafficking and cell cycle but also in other cellular processes including autophagy. In this review, we will focus on the moonlighting functions of the mammalian Dsl1complex, alternatively called the NRZ complex. Regarding the detailed structural and functional features of the Dsl1 complex and other CATCHR family complexes, we refer readers to some excellent reviews (Bröker et al., 2010; Brown and Pfeffer, 2010; Schmitt, 2010; Yu and Hughson, 2010; Bonifacino and Hierro, 2011; Spang, 2012; Hong and Lev, 2014).

DSL1 COMPLEX: A TETHER AT THE ER

In eukaryotic cells, communication between organelles in the secretory and endocytic pathways is mediated by membrane-bound vesicles that transit between organelles. Transport vesicles are formed at the donor compartment, traffic to their destination, lose their coat, and fuse with the acceptor compartment. Docking and fusion of transport vesicles with the target membrane involve an initial contact mediated by Rab GTPases and tethering factors, followed by SNARE-catalyzed membrane fusion.

Tethering factors, which not only facilitate long-range interactions between transport vesicles and the acceptor membrane but also coordinate SNARE complex assembly, can be classified into two groups, large proteins with extended coiled-coils and multisubunit protein complexes. The CATCHR complexes are a subfamily of the multisubunit tethering complexes consisting of the Dsl1, COG, exocyst, and GARP complexes (Yu and Hughson,

2010). Despite subtle sequence homology among their subunits, the CATCHR family tethers have strong three-dimensional structural homology.

The yeast Dsl1 complex, which comprises three subunits (Dsl1, Tip20, and Sec39/Dsl3) (**Figure 1A**), has been implicated in tethering of Golgi-derived transport vesicles on the ER (Andag et al., 2001; Reilly et al., 2001; Andag and Schmitt, 2003; Zink et al., 2009; Diefenbacher et al., 2011). Although all these proteins are rich in α-helices, their origins may be different; Dsl1 and Tip20 have a common α-helical fold, thus originating from an ancestral CATCHR protein (Tripathi et al., 2009), whereas Sec39/Dsl3 lacks the shared fold and thus appears to be derived from a non-related protein (Ren et al., 2009). The Dsl1 complex resides on ER membranes and binds to and regulates the assembly of the ER SNAREs Ufe1, Sec20, Use1/Slt1, and Sec22 (**Figure 2A**) (Kraynack et al., 2005; Ren et al., 2009; Diefenbacher et al., 2011). The Dsl1 complex has been predicted to form a 20-nm-tall tower from the ER surface, which can allow an interaction with COPI-coated vesicles (Ren et al., 2009). The COPI (α- and δ-COPs)-binding sites in Dsl1 have been defined to its central acidic region (Andag and Schmitt, 2003), which is located in the tip of the tower (Ren et al., 2009). An additional role of the Dsl1 complex is to assist uncoating of COPI-coated vesicles tethered on ER membranes. In this role, Dsl1 may block domains in COPI that drive repolymerization and the formation of large COPI aggregates (Zink et al., 2009).

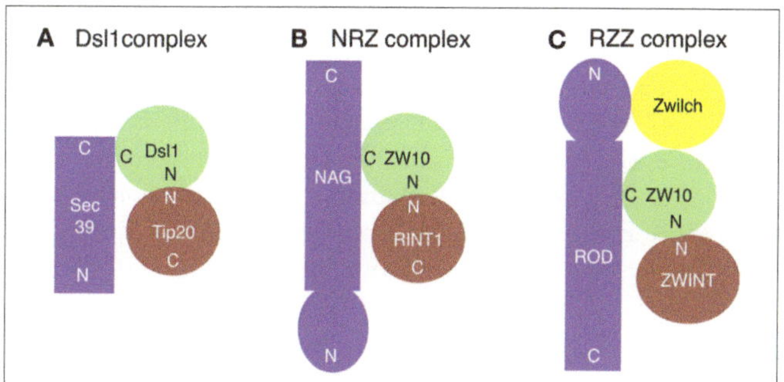

FIGURE 1 | Subunit compositions of the Dsl1 (A), NRZ (B), and RZZ (C) complexes. ZWINT is not included in the RZZ complex, but shown here because it may have a role corresponding to Tip20/RINT1 in the Dsl1/NRZ complexes. The ZW10-binding site on ROD has not been mapped. N and C indicate the N- and C-terminal regions, respectively.

While the roles of the yeast Dsl1 complex have been so far entirely limited to the tethering function on the ER, one study has suggested the involvement of the Dsl1 complex in peroxisome biogenesis. Previously, peroxisomes were recognized as an autonomous organelle, but recent studies have revealed a primary role of the ER in the *de novo* formation of peroxisomes (reviewed by Agrawal and Subramani, 2013). Searching for ER-associated proteins responsible for this role revealed that all Dsl1 complex components are required for peroxisome biogenesis (Perry et al., 2009). It was speculated that Dsl1 complex components may function as a tether for retrograde carriers from peroxisomes (Nagotu et al., 2010) or an anchor to recruit dynein for peroxisome biogenesis (Perry et al., 2009), but the mechanism by which Dsl1 complex subunits regulate peroxisome biogenesis should be elucidated in future studies.

The mammalian counterpart of the Dsl1 complex was identified by us (Hirose et al., 2004; Aoki et al., 2009) and was later called the NRZ complex for its subunit names, NAG (Sec39/Dsl3), RINT1 (Tip20), and ZW10 (Dsl1) (Civril et al., 2010) (**Figure 1B** and **Table 1**). Like the yeast Dsl1 complex (Sweet and Pelham, 1993; Kraynack et al., 2005; Ren et al., 2009; Tripathi et al., 2009; Diefenbacher et al., 2011), the NRZ complex is associated with the ER SNAREs syntaxin 18 (Ufe1), BNIP1 (Sec20), p31 (Use1/Slt1), and Sec22b (Sec22) (**Figure 2B**) (Hatsuzawa et al., 2000; Hirose et al., 2004; Nakajima et al., 2004; Aoki et al., 2009; Uemura et al.,

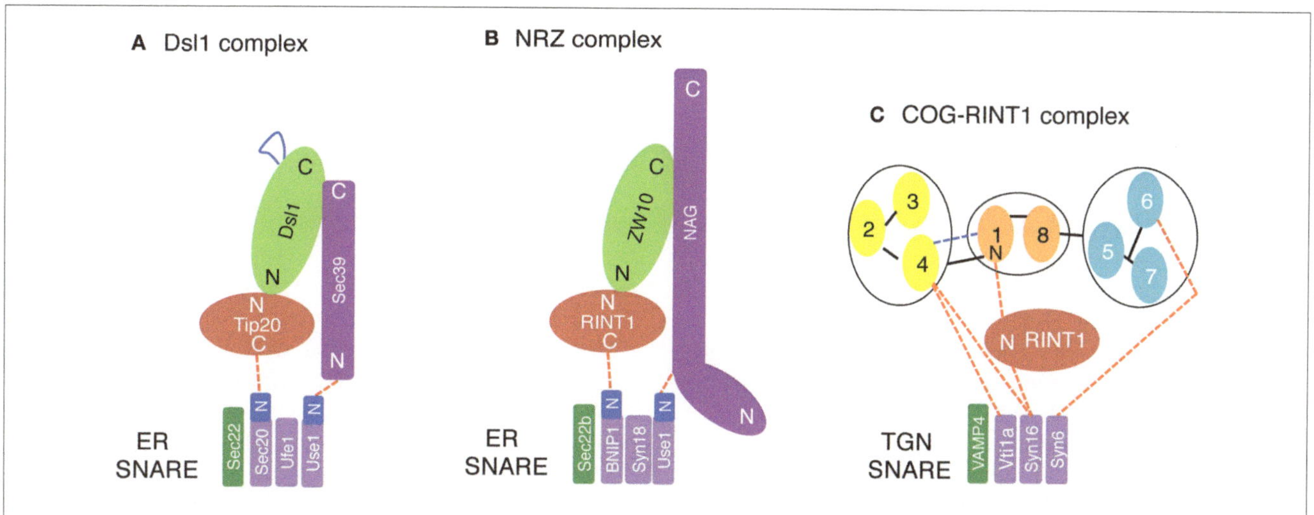

FIGURE 2 | Interactions between tethers and SNAREs. (A) On yeast ER membranes, Tip20 and Sec39 bind to the N-terminal regions (N) of Sec20 and Use1/Slt1, respectively. The region of Dsl1 shown in blue represents an acidic region that interacts with COPI components. **(B)** On the mammalian ER, RINT1 and NAG bind to the N-terminal regions of BNIP1 and Use1/p31, respectively. **(C)** On the mammalian TGN, SNAREs binds the COG complex directly or indirectly through RINT1. The COG complex model (Oka et al., 2005; Ungar et al., 2005) has been slightly modified. N and C indicate the N- and C-terminal regions, respectively.

Table 1 | Features of subunits of the human NRZ and RZZ complexes.

Proteins	Binding regions with partners	Structural features
ZW10 (779 aa*)	1–170 aa: RINT1, dynamitin 1–82 aa: ZWINT	Putative coiled-coil regions** (13–38 aa, 53–83 aa, 107–130 aa, 356–383 aa, 616–636 aa)
RINT1 (792 aa)	1–264 aa: ZW10, COG1 151–256 aa: UVRAG 358–440 aa: RBL2/p130 257–792 aa: Rad50	Putative coiled-coil regions (60–92 aa, 101–126 aa, 163–183 aa)
NAG (2371 aa)	1036–2371 aa: ZW10-RINT1	β-propeller (1–420 aa) Sec39-like (734–1355 aa)
ROD (2209 aa)	1–350 aa: Zwilch	β-propeller (1–390 aa) Sec39-like (557–1153 aa)
Zwilch (591 aa)	Not determined	Putative coiled-coil regions (69–89 aa, 290–311 aa)

*aa, amino acids.

**window size = 21.

2009). The yeast Dsl1 complex interacts with another SNARE, Ykt6 (Meiringer et al., 2011), but it is not clear whether or not the NRZ complex binds to the mammalian ortholog of this SNARE. The mechanism underlying SNARE complex assembly appears to be somewhat different between mammals and yeast. In mammals, the assembly of the ER SNAREs occurs in the absence of RINT1 (Arasaki et al., 2006), whereas yeast Tip20 plays a pivotal role in ER SNARE complex assembly (Kraynack et al., 2005; Diefenbacher et al., 2011). Moreover, the binding of Sec22b to syntaxin 18 in mammals creates high-affinity binding sites for BNIP1 and p31 (Aoki et al., 2008), whereas yeast Ufe1, Sec20, and Use1/Slt1 form a stable complex in the absence of Sec22 (Kraynack et al., 2005).

The Dsl1 complex is conserved in plants. Screening for *Arabidopsis* mutants that abnormally accumulate the precursors of storage proteins, 2S albumin and 12S globulin, in dry seeds identified a mutant, designated maigo2 (MAG2: At3g47700) (Li et al., 2006). MAG2 is the ortholog of RINT1 (Tip20). Affinity purification identified three MAG2-binding proteins, MIP1 (At2g32900), MIP2 (At5g24350), and MIP3 (At2g42700). MIP1 and MIP2 share sequence homology with ZW10 (Dsl1) and NAG (Sec39/Dsl3), respectively, (Li et al., 2013). MIP3 is a member of the Sec1 family domain-containing proteins, named SCFD2. SCFD2 is present in mammals, and may bind to ZW10 (Hutchins et al., 2010), although our original study failed to detect this protein in the syntaxin 18 immunoprecipitates (Hirose et al., 2004). Instead, we found Sly1/SCFD1 in the precipitates (Hirose et al., 2004), likely due to the direct binding of Sly1/SCFD1 to syntaxin 18 (Yamaguchi et al., 2002). The *Arabidopsis* Dsl1 complex is implicated in abscisic acid-mediated response to abiotic stresses. This response may be related to Dsl1 complex-mediated membrane trafficking between the ER and Golgi (Zhao et al., 2013). There is a homolog of MAG2, named MAG2-like (MAG2L: At1g08400) in *Arabidopsis* (Li et al., 2006), and MAG2L is partially redundant with MAG2 in response to environmental stresses (Zhao et al., 2013).

RZZ COMPLEX: ALTERNATE ZW10-CONTAINING COMPLEX

To preserve genetic information, the genomes of organisms must be accurately replicated and segregated before cell division. In eukaryotes, during mitosis and meiosis, sister-chromatid segregation occurs after all kinetochores form stable bipolar microtubule attachments. If not correctly attached to the spindle, kinetochores activate the spindle assembly checkpoint, leading to the block of cell cycle progression.

ZW10 (Zeste White 10) and ROD (Rough Deal) were discovered as proteins that are required for faithful chromosome segregation in *Drosophila* (Karess and Glover, 1989; Williams et al., 1992). Both proteins were found to be conserved in humans (Starr et al., 1998; Scaërou et al., 2001). At the onset of mitosis, these proteins redistribute from the cytoplasm to kinetochores (Williams et al., 1992; Scaërou et al., 1999) and recruit the core spindle checkpoint proteins Mad1-Mad2 to microtubule-unattached kinetochores (Buffin et al., 2005; Kops et al., 2005; Gassmann et al., 2010). Once all kinetochores become stably attached to the spindle, ZW10 and ROD are transported away from kinetochores by dynein (Howell et al., 2001; Wojcik et al., 2001; Basto et al., 2004). The removal of these proteins from kinetochores acts as a signal for the termination of the spindle checkpoint (Basto et al., 2000; Chan et al., 2000). ZW10 interacts with dynamitin, a subunit of the dynein-dynactin complex (Echeverri et al., 1996), thereby recruiting this motor to kinetochores (Starr et al., 1998). ZW10 and ROD form a large complex of 800 kDa, and immunoaffinity purification identified the third component, Zwilch (Williams et al., 2003), which encouraged Karess (2005) to call this complex RZZ for ROD-ZW10-Zwilch (**Figure 1C**). The binding of ROD and NAG to ZW10 is mutually exclusive, suggesting that the RZZ and NRZ complexes exist as distinct entities. Structural prediction revealed unexpected similarity between ROD and NAG. They share an N-terminal β propeller followed by an α solenoid, which is a characteristic structure of certain nucleoporins such as Nup133 and vesicle coat subunits such as clathrin heavy chain and α-COP (Civril et al., 2010). In Nup133, the C-terminal α solenoid structure is responsible for the localization at the nuclear pore complex in interphase and at kinetochores in mitotic cells, whereas the N-terminal β propeller domain interacts with CENPF, which can recruit dynein-dynactin via NudE or its homolog NudEL (Stehman et al., 2007; Bolhy et al., 2011). This protein–protein interaction chain (Nup133-CENPF-NudE or NudEL-dynein-dynactin) is reminiscent of the RZZ-mediated recruitment of dynein-dynactin to kinetochores. ROD binds to Zwilch through its N-terminal β propeller domain (Civril et al., 2010), as does Nup133 to CENPF.

Zwilch and RINT1, both of which are ZW10 binding partners, do not share any structural similarity, suggesting that Zwilch is not a descendant of the ancestral CATCHR protein (Civril et al., 2010). Moreover, in contrast to RINT1, Zwilch and ROD are not well conserved in eukaryotes (Schmitt, 2010), raising the possibility that the RZZ complex was generated after the occurrence of the

NRZ complex accompanied by the acquisition of an additional function by ZW10. This is consistent with the view that the RZZ complex is a fairly recent add-on to the core spindle checkpoint complex (Vleugel et al., 2012). ROD and Zwilch always co-occur in Opisthokonta, and their co-occurrence correlates well with the presence of flagella and centrioles in Opisthokonta (Schmitt, 2010). The co-appearance of ROD and Zwilch might allow microtubules and their motor dynein to function in the termination of spindle checkpoint.

In *Drosophila* there is no NAG ortholog, and ROD may function as a substituent for NAG in membrane trafficking (Wainman et al., 2012). In *Drosophila* spermatocytes, ROD and ZW10 accumulate at the Golgi apparatus, as well as the ER in the case of ZW10, and these proteins are required for Golgi stack integrity (Wainman et al., 2012). The presence of ZW10 on both ER and Golgi membranes has also been reported in COS-7 (Varma et al., 2006) and rodent cells (Arasaki et al., 2007). Depletion of *Drosophila* RINT1, as well as mutations in the *ZW10* and *ROD* genes, causes apparent alteration in Golgi morphology and reduces the number of Golgi stacks. In contrast, Zwilch is not associated with membranes and its mutation does not affect Golgi structure (Wainman et al., 2012). ZW10 and RINT1 are necessary for another membrane trafficking-related event, meiotic cytokinesis in spermatocytes, but not for mitotic one (Williams et al., 1992). In ZW10 or RINT1 mutant spermatocytes, regular central spindles and actomyosin rings can assemble, but furrow ingression halts prematurely due to a defective plasma membrane addition. On the other hand, neither ROD nor Zwilch is required for cytokinesis (Wainman et al., 2012).

MULTIPLE FUNCTIONS OF NRZ SUBUNITS

ZW10

During interphase, in addition to the retrograde transport from the Golgi to the ER, ZW10 may participate in dynein-mediated movement of endosomes and lysosomes (Varma et al., 2006). It is not clear whether ZW10 functions in the anterograde transport from the ER to the Golgi, which is mediated by dynein-dynactin that moves toward the minus-end of microtubules (Presley et al., 1997). One study showed that depletion of ZW10 causes the release of dynein from membranes (Varma et al., 2006). Moreover, ZW10 moves along microtubules toward their minus-end and accumulate at the centrosome in interphase and at kinetochores in mitosis, when the interaction of ZW10 with dynein-dynactin is stabilized (Inoue et al., 2008; Famulski et al., 2011). However, a recent study strongly suggests that ZW10 acts as a tether rather than a dynein anchor (Majeed et al., 2014). It should be clarified in future studies how ZW10 participates in tethering despite the lack of an acidic region responsible for the interaction with COPI-coated vesicles in yeast Dsl1 (Andag and Schmitt, 2003; Schmitt, 2010).

At the onset of mitosis, microtubules switch their role from the transport of membranous structures, such as Golgi, endosomes, lysosomes, and transport vesicles, to that of non-membranous chromosomes as well as of kinetochore proteins. ZW10, ROD, and perhaps many other kinetochore proteins are supposed to move along microtubules from the cytosol to the nucleus (Williams et al., 1992; Scaërou et al., 1999). The finding that Golgi disassembly is required for entry into mitosis (Sütterlin et al., 2002) might be related to this phenomenon. Presence of the Golgi structure might be an "obstacle" for massive protein transport along microtubules to the nucleus, and its loss may allow cytosolic and ER- and Golgi-associated kinetochore proteins to enter the nucleus just before and during nuclear envelope breakdown.

During mitosis, ZW10 interacts with several kinetochore-associated proteins, such as ZWINT (Starr et al., 2000), C19orf25 (Kops et al., 2005), and PIASy (Ryu and Azuma, 2010). ZWINT was first identified by yeast two-hybrid screen as a ZW10-binding protein (Starr et al., 2000). This 43-kDa protein, predicted to be largely coiled-coil, is recruited to kinetochores in early prophase, before the earliest detection of ZW10, and remains there until mid-anaphase (Starr et al., 2000; Wang et al., 2004). ZWINT and RINT1 may have equivalent functions in different ZW10-containing complexes. Like RINT1 (Inoue et al., 2008), ZWINT interacts with the N-terminal region of ZW10 and helps the recruitment of the ZW10-containing complex to the target architecture, kinetochores (Starr et al., 2000; Wang et al., 2004; Famulski et al., 2008).

ZWINT may also play a role in membrane trafficking in interphase cells. It interacts with membrane trafficking proteins such as Rab3c, SNAP25, SNAP29, and GM130 (Lee et al., 2002; van Vlijmen et al., 2008; Hutchins et al., 2010). Both Rab3c and SNAP25 bind to the same region of ZWINT (amino acids 79–179), which includes a putative coiled-coil region (van Vlijmen et al., 2008). Rat ZWINT, called SIP30 for SNAP-25 interacting protein of 30 kDa, is upregulated in the central nervous system in response to neuropathic pain (Zhang et al., 2009), and is required for neuropathic pain-evoked aversion (Han et al., 2014).

A very recent study has revealed an unexpected role for ZWINT. ZWINT has been shown to directly interact with Beclin1 whose loss results in a significant reduction of the outer kinetochore proteins including ZW10, leading to a defect in chromosome congression (Frémont et al., 2013). Beclin1 is the mammalian ortholog of yeast ATG6/VPS30 and a typical moonlighting protein participating not only in autophagy, but also in endocytosis, apoptosis, and inflammation (reviewed by Salminen et al., 2013).

RINT1

RINT1 was originally discovered as a Rad50-interacting protein required for G_2/M cell cycle checkpoint control (Xiao et al., 2001). Depletion of RINT1 causes not only partial Golgi fragmentation (Arasaki et al., 2006; Sun et al., 2007), but also defects in mitosis, including the formation of multiple spindle poles and frequent chromosome missegregation (Lin et al., 2007). RINT1 heterozygous mice quite often developed multiple tumors due to haploinsufficiency, suggesting that RINT1 serves as a tumor suppressor (Lin et al., 2007). However, a recent integrative functional genomics study has validated that *RINT1* is a glioblastoma oncogene that can confer tumorigenicity to primary nontransformed murine astrocytes *in vivo* (Quayle et al., 2012). An additional role for RINT1 is to interact with Rb-related 2 (RBL2/p130) and control telomere length (Kong et al., 2006). Yeast Tip20 may function in the nucleus because crystal-like structures are formed in the nucleus of Tip20 mutants (Spang, 2012).

Recent studies by us revealed an unanticipated involvement of RINT1 in endosome-to-*trans*-Golgi network (TGN) trafficking. Deletion of RINT1 causes a more severe dispersal of TGN proteins than that of *cis*-Golgi proteins due to a defect in endosome-to-TGN trafficking (Arasaki et al., 2013). In this context, RINT1 does not work with ZW10, but cooperates with the COG complex (**Figure 2C**), another member of the CATCHR family complexes (Yu and Hughson, 2010). RINT1 binds to COG1 through its N-terminal 264-amino acid region, which is also a ZW10-binding region (Hirose et al., 2004; Arasaki et al., 2006). This fact may explain why the COG-containing RINT1 complex is distinct from the NRZ complex. RINT1 binds to the SNARE (C-terminal coiled-coil) domain of syntaxin 16, a SNARE implicated in endosome-to-TGN trafficking (Mallard et al., 2002), and regulates SNARE complex assembly presumably at the TGN (Arasaki et al., 2013). The region of COG1 responsible for the association with RINT1 is the N-terminal 93-amino acid region containing a putative coiled-coil region, which has been identified as a VPS51-like domain in the Pfam database. VPS51 is a member of the GARP complex consisting of VPS51/Ang2, VPS52, VPS53, and VPS54 (Pérez-Victoria et al., 2010). Although RINT1 binds to the VPS51-like domain of COG1, it does not bind to Vps51 or other GARP subunits (Arasaki et al., 2013). The COG and GARP complexes are also involved in endosome-to-TGN trafficking through the interaction with TGN SNAREs (Pérez-Victoria and Bonifacino, 2009; Pérez-Victoria et al., 2010; Laufman et al., 2011, 2013; Willett et al., 2013). The requirement of many different tethers for endosome-to-TGN trafficking likely reflects the presence of multiple sources of vesicles that traffic to the TGN. Moreover, the participation of RINT1 in this pathway may reflect adaptation to the demand for more diverse transport routes from endosomes to the TGN in mammals compared with those in a unicellular organism, yeast.

Autophagy is a process to engulf, degrade, and recycle cytoplasmic contents, and is required for cell survival in response to starvation. Recent work by Liang and colleagues (He et al., 2013) revealed a connection between RINT1 and autophagy. UV-radiation resistance-associated gene (UVRAG) interacts with Beclin1 through its coiled-coil region, which in turn activates phosphatidylinositol 3-kinase for autophagy (Liang et al., 2006; Itakura et al., 2008). It also plays a role in endocytosis under basal conditions (Liang et al., 2008). He et al. (2013) discovered that, in fed cells, UVRAG is associated with RINT1 and participates in tethering of COPI vesicles in cooperation with the NRZ complex. Upon starvation, it dissociates from RINT1 and binds to Beclin1, which in turn promotes ATG9 translocation from the Golgi to the autophagosome biogenesis site (He et al., 2013). It should be noted that Beclin1 interacts with ZWINT (Frémont et al., 2013), suggesting a connection between membrane trafficking (RINT1), kinetochore function (ZWINT), and autophagy (Beclin1-UVRAG) (**Figure 3**).

NAG

The neuroblastoma-amplified gene (NAG), alternatively called neuroblastoma-amplified sequence (NBAS), was first identified as a gene co-amplified with the *MYCN* gene in neuroblastoma (Wimmer et al., 1999). This amplification occurred likely because a 2.8 Mb non-fragile region containing the *MYCN* and *NAG* genes flanks the *FRA2C* region, a common fragile site of chromosomes (Blumrich et al., 2011). The association of NAG with tumor progression and prognosis remains controversial (Scott et al., 2003; Weber et al., 2004; Kaneko et al., 2007). Although NAG is well conserved in eukaryotes, the molecular size is very different between fungi and other eukaryotes (Schmitt, 2010). In many fungi, the molecular size of the predicted NAG orthologs, such as yeast Sec39/Dsl3, is about 80–100 kDa, whereas in other eukaryotes, the size is almost 2 to 3-fold larger than fungal proteins because of the presence of a long N-terminal extension comprising a β propeller domain and a long C-terminal extension, in addition to the central Sec39-like domain (about 13% homology between human and yeast).

The larger size of metazoan NAGs may allow them to have additional functions. Indeed, NAG has been reported to participate in nonsense-mediated mRNA decay (NMD) (Longman et al., 2007, 2013; Anastasaki et al., 2011). The NMD pathway is an elaborate surveillance mechanism that triggers the degradation of mRNAs containing premature termination codons (PTCs) and also regulates ~10% of naturally occurring transcripts (reviewed by Chang et al., 2007). A genome-wide RNA interference screen in *Caenorhabditis elegans* identified NAG and a member of the DEAD/DEAH-box helicases, DHX34, as factors responsible for the NMD pathway independent of core NMD factors (SMG1-7) (Longman et al., 2007). This pathway is conserved in zebrafish and human (Anastasaki et al., 2011), and NAG and DHX34, like core NMD factors, co-regulate a significant proportion of genes (Longman et al., 2013). Given that DEAD/DEAH-box helicases are commonly involved in many aspects of RNA metabolism including transcription, pre-mRNA splicing, translation, and mRNA decay (reviewed by Fuller-Pace, 2006), the involvement of DHX34 in the NMD pathway is understandable. However, the mechanism underlying NAG-mediated NMD is totally unknown. Interestingly, a mutation in the NAG gene has been shown to be a cause for the hereditary short stature syndrome (SOPH syndrome) in Yakuts, who live in the far east of the Russian Federation (Maksimova et al., 2010). This autosomal recessive disorder is associated with optic nerve atrophy and Pelger-Huët anomaly, the latter of which is characterized by an abnormal nuclear shape in neutrophil granulocytes (Maksimova et al., 2010). These symptoms may be explained by the fact that NAG is involved in the NMD pathway and ER function. As the outer nuclear membrane is contiguous with the ER, disruption of the ER structure may affect the nuclear morphology.

Nicotiana benthamiana NAG comprises 2409 amino acids, thus can be classified to non-fungal NAGs, and its silencing causes ER stress and cell death (Lee et al., 2013). This phenotype may be relevant to that observed in MAG2 mutants (Zhao et al., 2013).

CONCLUSIONS AND PERSPECTIVES

Figure 3 shows the summary of protein–protein interactions between NRZ components. Many, but not all, interactions are mediated through putative coiled-coil regions. As described above, the NRZ complex, as well as the Dsl1complex, consists of two distinct structural units. ZW10 and RINT are derived from the CATCHR family ancestor, and NAG is from a different

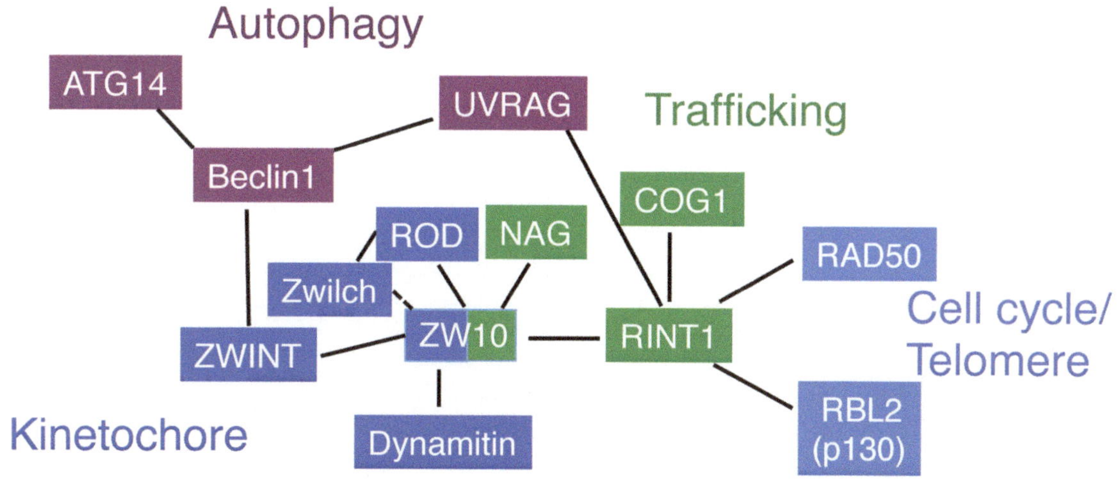

FIGURE 3 | Correlation diagram of NRZ complex subunits. ZW10 is the center for the interactions. Through its N-terminal, putative coiled-coil region, it interacts with RINT1 for membrane trafficking (Hirose et al., 2004; Inoue et al., 2008), dynamitin for recruiting the dynein-dynactin complex (Starr et al., 1998; Inoue et al., 2008), ZWINT for the association with kinetochores (Wang et al., 2004; Kops et al., 2005). The C-terminal region of ZW10 likely interacts with NAG and ROD (Kraynack et al., 2005; Aoki et al., 2009; Civril et al., 2010). The interaction of Zwilch with ZW10 is not tight in the absence of ROD (Civril et al., 2010). On the other hand, RINT1 interacts with COG1 through its N-terminal, putative coiled-coil region (Arasaki et al., 2013). This interaction mode may be a common for the interactions of CATCHR family members. The N-terminal region is also responsible for the interaction with UVRAG, but in this case, RINT1 keeps the binding to ZW10 (He et al., 2013). The RINT1 interacts with RAD50 and RBL2/p130. UVRAG interacts with RINT1 under basal conditions and, upon starvation, dissociates from RINT1 and interacts with Beclin1. Of note is that Beclin1 during mitosis interacts with ZWINT, a ZW10 partner.

one. Because a CATCHR family member RINT1 interacts with another CATCHR family member COG1 (Arasaki et al., 2013), it is tempting to speculate that RINT1 orthologs in other eukaryotes may also interact with CATCHR family members. Indeed, a comprehensive analysis of protein–protein interactions in yeast demonstrated that Tip20 can interact with COG4 (Uetz et al., 2000). Like RINT1, ZW10, and its orthologs may also interact with CATCHR family members. Given that SCFD1 and SCFD2 bind to the mammalian COG4 (Laufman et al., 2009) and plant MAG2 (Li et al., 2013), respectively, it is worth examining the interactions between CATCHR family members and Sly1 family (SCFD) proteins. The currently available data suggest that the role of RINT1 in autophagy is passive; it anchors UVRAG until autophagy induction. However, as the ER is a major source of autophagosome formation (reviewed by Lamb et al., 2013), RINT1 and other NRZ complex subunits may have more active roles in autophagy.

Schmitt (2010) has proposed that the ancestor of animals and fungi probably expressed both sets of ZW10-interacting proteins (NAG- RINT1 and ROD-Zwilch). Rod and Zwilch encoding genes have been assumed to be lost independently in different phylogenetic branches, and Sec39 likely lost the N-terminal β-propeller domain in parallel with the acquisition of the acidic COPI-interacting region by Dsl1. We would like to suggest another possibility. The ancestor of NAG/Sec39 acquired a β-propeller domain with a C-terminal extension for yet unrevealed functions, and the occurrence of Zwilch might facilitate the appearance of ROD by gene duplication. In this context, it is important to identify proteins that interact with the β-propeller domain and/or C-terminal extension of NAG. Moreover, elucidation of the molecular mechanism by which NAG mediates the NMD pathway may provide a clue to the reason for the association/cooperation of CATCHR proteins with NAG.

ACKNOWLEDGMENTS

This work was supported in part by Grants-in-Aid for Scientific Research, #23113726, #24657141, and #25291029 (to Mitsuo Tagaya), from the Ministry of Education, Culture, Sports, Science and Technology of Japan. We thank Dr. William J. Brown at Cornell University for reading the manuscript.

REFERENCES

Agrawal, G., and Subramani, S. (2013). Emerging role of the endoplasmic reticulum in peroxisome biogenesis. *Front. Physiol.* 4:286. doi: 10.3389/fphys.2013.00286

Anastasaki, C., Longman, D., Capper, A., Patton, E. E., and Cáceres, J. F. (2011). Dhx34 and Nbas function in the NMD pathway and are required for embryonic development in zebrafish. *Nucleic Acids Res.* 39, 3686–3694. doi: 10.1093/nar/gkq1319

Andag, U., Neumann, T., and Schmitt, H. D. (2001). The coatomer-interacting protein Dsl1p is required for Golgi-to-endoplasmic reticulum retrieval in yeast. *J. Biol. Chem.* 276, 39150–39160. doi: 10.1074/jbc.M105833200

Andag, U., and Schmitt, H. D. (2003). Dsl1p, an essential component of the Golgi-endoplasmic reticulum retrieval system in yeast, uses the same sequence motif to interact with different subunits of the COPI vesicle coat. *J. Biol. Chem.* 278, 51722–51734. doi: 10.1074/jbc.M308740200

Aoki, T., Ichimura, S., Itoh, A., Kuramoto, M., Shinkawa, T., Isobe, T., et al. (2009). Identification of the neuroblastoma-amplified gene product as a component of the syntaxin 18 complex implicated in Golgi-to-endoplasmic reticulum retrograde transport. *Mol. Biol. Cell* 20, 2639–2649. doi: 10.1091/mbc.E08-11-1104

Aoki, T., Kojima, M., Tani, K., and Tagaya, M. (2008). Sec22b-dependent assembly of endoplasmic reticulum Q-SNARE proteins. *Biochem. J.* 410, 93–100. doi: 10.1042/BJ20071304

Arasaki, K., Takagi, D., Furuno, A., Sohda, M., Misumi, Y., Wakana, Y., et al. (2013). A new role for RINT-1 in SNARE complex assembly at the *trans*-Golgi network in coordination with the COG complex. *Mol. Biol. Cell* 24, 2907–2917. doi: 10.1091/mbc.E13-01-0014

Arasaki, K., Taniguchi, M., Tani, K., and Tagaya, M. (2006). RINT-1 regulates the localization and entry of ZW10 to the syntaxin 18 complex. *Mol. Biol. Cell* 17, 2780–2788. doi: 10.1091/mbc.E05-10-0973

Arasaki, K., Uemura, T., Tani, K., and Tagaya, M. (2007). Correlation of Golgi localization of ZW10 and centrosomal accumulation of dynactin. *Biochem. Biophys. Res. Commun.* 359, 811–816. doi: 10.1016/j.bbrc.2007.05.188

Basto, R., Gomes, R., and Karess, R. E. (2000). Rough deal and Zw10 are required for the metaphase checkpoint in *Drosophila*. *Nat. Cell Biol.* 2, 939–943. doi: 10.1038/35046592

Basto, R., Scaerou, F., Mische, S., Wojcik, E., Lefebvre, C., Gomes, R., et al. (2004). *In vivo* dynamics of the rough deal checkpoint protein during *Drosophila* mitosis. *Curr. Biol.* 14, 56–61. doi: 10.1016/j.cub.2003.12.025

Blumrich, A., Zapatka, M., Brueckner, L. M., Zheglo, D., Schwab, M., and Savelyeva, L. (2011). The FRA2C common fragile site maps to the borders of MYCN amplicons in neuroblastoma and is associated with gross chromosomal rearrangements in different cancers. *Hum. Mol. Genet.* 20, 1488–1501. doi: 10.1093/hmg/ddr027

Bolhy, S., Bouhlel, I., Dultz, E., Nayak, T., Zuccolo, M., Gatti, X., et al. (2011). A Nup133-dependent NPC-anchored network tethers centrosomes to the nuclear envelope in prophase. *J. Cell Biol.* 192, 855–871. doi: 10.1083/jcb.201007118

Bonifacino, J. S., and Hierro, A. (2011). Transport according to GARP: receiving retrograde cargo at the *trans*-Golgi network. *Trends Cell Biol.* 21, 159–167. doi: 10.1016/j.tcb.2010.11.003

Bröker, C., Engelbrecht-Vandé, S., and Ungermann, C. (2010). Multisubunit tethering complexes and their role in membrane fusion. *Curr. Biol.* 20, R943–R952. doi: 10.1016/j.cub.2010.09.015

Brown, F. C., and Pfeffer, S. R. (2010). An update on transport vesicle tethering. *Mol. Membr. Biol.* 27, 457–461. doi: 10.3109/09687688.2010.501765

Buffin, E., Lefebvre, C., Huang, J., Gagou, M. E., and Karess, R. E. (2005). Recruitment of MAD2 to the kinetochore requires the Rod/Zw10 complex. *Curr. Biol.* 15, 856–861. doi: 10.1016/j.cub.2005.03.052

Chan, G. K. T., Jablonski, S. A., Starr, D. A., Goldberg, M. L., and Yen, T. J. (2000). Human Zw10 and ROD are mitotic checkpoint proteins that bind to kinetochores. *Nat. Cell Biol.* 2, 944–947. doi: 10.1038/35046598

Chang, Y. F., Imam, J. S., and Wilkinson, M. F. (2007). The nonsense-mediated decay RNA surveillance pathway. *Annu. Rev. Biochem.* 76, 51–74. doi: 10.1146/annurev.biochem.76.050106.093909

Civril, F., Wehenkel, A., Giorgi, F. M., Santaguida, S., Di Fonzo, A., Grigorean, G., et al. (2010). Structural analysis of the RZZ complex reveals common ancestry with multisubunit vesicle tethering machinery. *Structure* 18, 616–626. doi: 10.1016/j.str.2010.02.014

Copley, S. D. (2012). Moonlighting is mainstream: paradigm adjustment required. *Bioessays* 34, 578–588. doi: 10.1002/bies.201100191

Diefenbacher, M., Thorsteinsdottir, H., and Spang, A. (2011). The Dsl1 tethering complex actively participates in soluble NSF (N-ethylmaleimide-sensitive factor) attachment protein receptor (SNARE) complex assembly at the endoplasmic reticulum in *Saccharomyces cerevisiae*. *J. Biol. Chem.* 286, 25027–25038. doi: 10.1074/jbc.M110.215657

Echeverri, C. J., Paschal, B. M., Vaughan, K. T., and Vallee, R. B. (1996). Molecular characterization of the 50-kD subunit of dynactin reveals function for the complex in chromosome alignment and spindle organization during mitosis. *J. Cell Biol.* 132, 617–633. doi: 10.1083/jcb.132.4.617

Famulski, J. K., Vos, L., Sun, X., and Chan, G. (2008). Stable hZW10 kinetochore residency, mediated by hZwint-1 interaction, is essential for the mitotic checkpoint. *J. Cell Biol.* 180, 507–520. doi: 10.1083/jcb.200708021

Famulski, J. K., Vos, L. J., Rattner, J. B., and Chan, G. K. (2011). Dynein/Dynactin-mediated transport of kinetochore components off kinetochores and onto spindle poles induced by nordihydroguaiaretic acid. *PLoS ONE* 6:16494. doi: 10.1371/journal.pone.0016494

Frémont, S., Gérard, A., Galloux, M., Janvier, K., Karess, R. E., and Berlioz-Torrent, C. (2013). Beclin-1 is required for chromosome congression and proper outer kinetochore assembly. *EMBO Rep.* 14, 364–372. doi: 10.1038/embor.2013.23

Fuller-Pace, F. V. (2006). DExD/H box RNA helicases: multifunctional proteins with important roles in transcriptional regulation. *Nucleic Acids Res.* 34, 4206–4215. doi: 10.1093/nar/gkl460

Gassmann, R., Holland, A. J., Varma, D., Wan, X., Civril, F., Cleveland, D. W., et al. (2010). Removal of Spindly from microtubule-attached kinetochores controls spindle checkpoint silencing in human cells. *Genes Dev.* 24, 957–971. doi: 10.1101/gad.1886810

Han, M., Xiao, X., Yang, Y., Huang, R. Y., Cao, H., Zhao, Z. Q., et al. (2014). SIP30 is required for neuropathic pain-evoked aversion in rats. *J. Neurosci.* 34, 346–355. doi: 10.1523/JNEUROSCI.3160-13.2014

Hatsuzawa, K., Hirose, H., Tani, K., Yamamoto, A., Scheller, R. H., and Tagaya, M. (2000). Syntaxin 18, a SNAP receptor that functions in the endoplasmic reticulum, intermediate compartment, and *cis*-Golgi vesicle trafficking. *J. Biol. Chem.* 275, 13713–13720. doi: 10.1074/jbc.275.18.13713

He, S., Ni, D., Ma, B., Lee, J.-H., Zhang, T., Ghozalli, I., et al. (2013). PtdIns(3)P-bound UVRAG coordinates Golgi-ER retrograde and Atg9 transport by differential interactions with the ER tether and the beclin 1 complex. *Nat. Cell Biol.* 15, 1206–1219. doi: 10.1038/ncb2848

Hirose, H., Arasaki, K., Dohmae, N., Takio, K., Hatsuzawa, K., Nagahama, M., et al. (2004). Implication of ZW10 in membrane trafficking between the endoplasmic reticulum and Golgi. *EMBO J.* 23, 1267–1278. doi: 10.1038/sj.emboj.7600135

Hong, W., and Lev, S. (2014). Tetherig the assembly of SNARE complexes. *Trends Cell Biol.* 24, 35–43. doi: 10.1016/j.tcb.2013.09.006

Howell, B. J., McEwen, B. F., Canman, J. C., Hoffman, D. B., Farrar, E. M., Rieder, C. L., et al. (2001). Cytoplasmic dynein/dynactin drives kinetochore protein transport to the spindle poles and has a role in mitotic spindle checkpoint inactivation. *J. Cell Biol.* 155, 1159–1172. doi: 10.1083/jcb.200105093

Hutchins, J. R., Toyoda, Y., Hegemann, B., Poser, I., Hériché, J. K., Sykora, M. M., et al. (2010). Systematic analysis of human protein complexes identifies chromosome segregation proteins. *Science* 328, 593–599. doi: 10.1126/science.1181348

Inoue, M., Arasaki, K., Ueda, A., Aoki, T., and Tagaya, M. (2008). N-terminal region of ZW10 serves not only as a determinant for localization but also as a link with dynein function. *Genes Cells* 13, 905–914. doi: 10.1111/j.1365-2443.2008.01215.x

Itakura, E., Kishi, C., Inoue, K., and Mizushima, N. (2008). Beclin 1 forms two distinct phosphatidylinositol 3-kinase complexes with mammalian Atg14 and UVRAG. *Mol. Biol. Cell* 19, 5360–5372. doi: 10.1091/mbc.E08-01-0080

Jeffery, C. J. (1999). Moonlighting proteins. *Trends Biochem. Sci.* 24, 8–11. doi: 10.1016/S0968-0004(98)01335-8

Kaneko, S., Ohira, M., Nakamura, Y., Isogai, E., Nakagawara, A., and Kaneko, M. (2007). Relationship of DDX1 and NAG gene amplification/overexpression to the prognosis of patients with MYCN-amplified neuroblastoma. *J. Cancer Res. Clin. Oncol.* 133, 185–192. doi: 10.1007/s00432-006-0156-y

Karess, R. (2005). Rod-Zw10-Zwilch: a key player in the spindle checkpoint. *Trends Cell Biol.* 15, 386–392. doi: 10.1016/j.tcb.2005.05.003

Karess, R. E., and Glover, D. M. (1989). Rough deal: a gene required for proper mitotic segregation in *Drosophila*. *J. Cell Biol.* 1109, 2951–2961. doi: 10.1083/jcb.109.6.2951

Kong, L. J., Meloni, A. R., and Nevins, J. R. (2006). The Rb-related p130 protein controls telomere lengthening through an interaction with a Rad50-interacting protein, RINT-1. *Mol. Cell* 22, 63–71. doi: 10.1016/j.molcel.2006.02.016

Kops, G. J., Kim, Y., Weaver, B. A., Mao, Y., McLeod, I., Yates, J. R. 3rd., et al. (2005). ZW10 links mitotic checkpoint signaling to the structural kinetochore. *J. Cell Biol.* 169, 49–60. doi: 10.1083/jcb.200411118

Kraynack, B. A., Chan, A., Rosenthal, E., Essid, M., Umansky, B., Waters, M. G., et al. (2005). Dsl1p, Tip20p, and the novel Dsl3(Sec39) protein are required for the stability of the Q/t-SNARE complex at the endoplasmic reticulum in yeast. *Mol. Biol. Cell* 16, 3963–3977. doi: 10.1091/mbc.E05-01-0056

Lamb, C. A., Yoshimori, T., and Tooze, S. A. (2013). The autophagosome: origins unknown, biogenesis complex. *Nat. Rev. Mol. Cell Biol.* 14, 759–774. doi: 10.1038/nrm3696

Laufman, O., Hong, W., and Lev, S. (2011). The COG complex interacts directly with syntaxin 6 and positively regulates endosome-to-TGN retrograde transport. *J. Cell Biol.* 194, 459–472. doi: 10.1083/jcb.201102045

Laufman, O., Hong, W., and Lev, S. (2013). The COG complex interacts with multiple Golgi SNAREs and enhances fusogenic assembly of SNARE complexes. *J. Cell Sci.* 126, 1506–1516. doi: 10.1242/jcs.122101

Laufman, O., Kedan, A., Hong, W., and Lev, S. (2009). Direct interaction between the COG complex and the SM protein, Sly1, is required for Golgi SNARE pairing. *EMBO J.* 28, 2006–2017. doi: 10.1038/emboj.2009.168

Lee, H. K., Safieddine, S., Petralia, R. S., and Wenthold, R. J. (2002). Identification of a novel SNAP25 interacting protein (SIP30). *J. Neurochem.* 81, 1338–1347. doi: 10.1046/j.1471-4159.2002.00937.x

Lee, J. Y., Sarowar, S., Kim, H. S., Kim, H., Hwang, I., Kim, Y. J., et al. (2013). Silencing of *nicotiana benthamiana* neuroblastoma-amplified gene causes ER stress and cell death. *BMC Plant Biol.* 13:69. doi: 10.1186/1471-2229-13-69

Li, L., Shimada, T., Takahashi, H., Koumoto, Y., Shirakawa, M., Takagi, J., et al. (2013). MAG2 and three MAG2-INTERACTING PROTEINs form an ER-localized complex to facilitate storage protein transport in *Arabidopsis thaliana. Plant J.* 76, 781–791. doi: 10.1111/tpj.12347

Li, L., Shimada, T., Takahashi, H., Ueda, H., Fukao, Y., Kondo, M., et al. (2006). MAIGO2 is involved in exit of seed storage proteins from the endoplasmic reticulum in *Arabidopsis thaliana. Plant Cell* 18, 3535–3547. doi: 10.1105/tpc.106.046151

Liang, C., Feng, P., Ku, B., Dotan, I., Canaani, D., Oh, B. H., et al. (2006). Autophagic and tumour suppressor activity of a novel Beclin1-binding protein UVRAG. *Nat. Cell Biol.* 8, 688–699. doi: 10.1038/ncb1426

Liang, C., Lee, J. S., Inn, K. S., Gack, M. U., Li, Q., Roberts, E. A., et al. (2008). Beclin1-binding UVRAG targets the class C Vps complex to coordinate autophagosome maturation and endocytic trafficking. *Nat. Cell Biol.* 10, 776–787. doi: 10.1038/ncb1740

Lin, X., Liu, C. C., Gao, Q., Zhang, X., Wu, G., and Lee, W. H. (2007). RINT-1 serves as a tumor suppressor and maintains Golgi dynamics and centrosome integrity for cell survival. *Mol. Cell. Biol.* 27, 4905–4916. doi: 10.1128/MCB.02396-06

Longman, D., Hug, N., Keith, M., Anastasaki, C., Patton, E. E., Grimes, G., et al. (2013). DHX34 and NBAS form part of an autoregulatory NMD circuit that regulates endogenous RNA targets in human cells, zebrafish and *Caenorhabditis elegans. Nucleic Acids Res.* 41, 8319–8331. doi: 10.1093/nar/gkt585

Longman, D., Plasterk, R. H., Johnstone, I. L., and Cáceres, J. F. (2007). Mechanistic insights and identification of two novel factors in the *C. elegans* NMD pathway. *Genes Dev.* 21, 1075–1085. doi: 10.1101/gad.417707

Majeed, W., Liu, S., and Storrie, B. (2014). Distinct sets of Rab6 effectors contribute to ZW10- and COG-dependent Golgi homeostasis. *Traffic.* 15, 630–647. doi: 10.1111/tra.12167

Maksimova, N., Hara, K., Nikolaeva, I., Chun-Feng, T., Usui, T., Takagi, M., et al. (2010). Neuroblastoma amplified sequence gene is associated with a novel short stature syndrome characterised by optic nerve atrophy and Pelger-Huët anomaly. *J. Med. Genet.* 47, 538–548. doi: 10.1136/jmg.2009.074815

Mallard, F., Tang, B. L., Galli, T., Tenza, D., Saint-Pol, A., Yue, X., et al. (2002). Early/recycling endosomes-to-TGN transport involves two SNARE complexes and a Rab6 isoform. *J. Cell Biol.* 156, 653–664. doi: 10.1083/jcb.200110081

Meiringer, C. T., Rethmeier, R., Auffarth, K., Wilson, J., Perz, A., Barlowe, C., et al. (2011). The Dsl1 protein tethering complex is a resident endoplasmic reticulum complex, which interacts with five soluble NSF (N-ethylmaleimide-sensitive factor) attachment protein receptors (SNAREs): implications for fusion and fusion regulation. *J. Biol. Chem.* 286, 25039–25046. doi: 10.1074/jbc.M110.215327

Mnaimneh, S., Davierwala, A. P., Haynes, J., Moffat, J., Peng, W. T., Zhang, W., et al. (2004). Exploration of essential gene functions via titratable promoter alleles. *Cell* 118, 31–34. doi: 10.1016/j.cell.2004.06.013

Nagotu, S., Veenhuis, M., and van der Klei, I. J. (2010). Divide et impera: the dictum of peroxisomes. *Traffic* 11, 175–184. doi: 10.1111/j.1600-0854.2009.01019.x

Nakajima, K., Hirose, H., Taniguchi, M., Kurashina, H., Arasaki, K., Nagahama, M., et al. (2004). Involvement of BNIP1 in apoptosis and endoplasmic reticulum membrane fusion. *EMBO J.* 23, 3216–3226. doi: 10.1038/sj.emboj.7600333

Oka, T., Vasile, E., Penman, M., Novina, C. D., Dykxhoorn, D. M., Ungar, D., et al. (2005). Genetic analysis of the subunit organization and function of the conserved oligomeric golgi (COG) complex: studies of COG5- and COG7-deficient mammalian cells. *J. Biol. Chem.* 280, 32736–32745. doi: 10.1074/jbc.M505558200

Pérez-Victoria, F. J., and Bonifacino, J. S. (2009). Dual roles of the mammalian GARP complex in tethering and SNARE complex assembly at the *trans*-golgi network. *Mol. Cell. Biol.* 29, 5251–5263. doi: 10.1128/MCB.00495-09

Pérez-Victoria, F. J., Schindler, C., Magadán, J. G., Mardones, G. A., Delevoye, C., Romao, M., et al. (2010). Ang2/fat-free is a conserved subunit of the Golgi-associated retrograde protein complex. *Mol. Biol. Cell* 21, 3386–3395. doi: 10.1091/mbc.E10-05-0392

Perry, R. J., Mast, F. D., and Rachubinski, R. A. (2009). Endoplasmic reticulum-associated secretory proteins Sec20p, Sec39p, and Dsl1p are involved in peroxisome biogenesis. *Eukaryotic Cell* 8, 830–843. doi: 10.1128/EC.00024-09

Presley, J. F., Cole, N. B., Schroer, T. A., Hirschberg, K., Zaal, K. J., and Lippincott-Schwartz, J. (1997). ER-to-Golgi transport visualized in living cells. *Nature* 389, 81–85. doi: 10.1038/38891

Quayle, S. N., Chheda, M. G., Shukla, S. A., Wiedemeyer, R., Tamayo, P., Dewan, R. W., et al. (2012). Integrative functional genomics identifies RINT1 as a novel GBM oncogene. *Neuro Oncol.* 14, 1325–1331. doi: 10.1093/neuonc/nos246

Reilly, B. A., Kraynack, B. A., VanRheenen, S. M., and Waters, M. G. (2001). Golgi-to-endoplasmic reticulum (ER) retrograde traffic in yeast requires Dsl1p, a component of the ER target site that interacts with a COPI coat subunit. *Mol. Biol. Cell* 12, 3783–3796. doi: 10.1091/mbc.12.12.3783

Ren, Y., Yip, C. K., Tripathi, A., Huie, D., Jeffrey, P. D., Walz, T., et al. (2009). A structure-based mechanism for vesicle capture by the multisubunit tethering complex Dsl1. *Cell* 139, 1119–1129. doi: 10.1016/j.cell.2009.11.002

Royle, S. J. (2013). Protein adaptation: mitotic functions for membrane trafficking proteins. *Nat. Rev. Mol. Cell Biol.* 14, 592–599. doi: 10.1038/nrm3641

Ryu, H., and Azuma, Y. (2010). Rod/Zw10 complex is required for PIASy-dependent centromeric SUMOylation. *J. Biol. Chem.* 285, 32576–32585. doi: 10.1074/jbc.M110.153817

Salminen, A., Kaarniranta, K., and Kauppinen, A. (2013). Beclin 1 interactome controls the crosstalk between apoptosis, autophagy and inflammasome activation: impact on the aging process. *Ageing Res. Rev.* 12, 520–534. doi: 10.1016/j.arr.2012.11.004

Scaërou, F., Aguilera, I., Saunders, R., Kane, N., Blottiere, L., and Karess, R. (1999). The rough deal protein is a new kinetochore component required for accurate chromosome segregation in *Drosophila. J. Cell Sci.* 112, 3757–3768.

Scaërou, F., Starr, D. A., Piano, F., Papoulas, O., Karess, R. E., and Goldberg, M. L. (2001). The ZW10 and rough deal checkpoint proteins function together in a large, evolutionarily conserved complex targeted to the kinetochore. *J. Cell Sci.* 114, 3103–3114.

Schmitt, H. D. (2010). Dsl1p/Zw10: common mechanisms behind tethering vesicles and microtubules. *Trends Cell Biol.* 20, 257–268. doi: 10.1016/j.tcb.2010.02.001

Scott, D. K., Board, J. R., Lu, X., Pearson, A. D., Kenyon, R. M., and Lunec, J. (2003). The neuroblastoma amplified gene, NAG: genomic structure and characterisation of the 7.3 kb transcript predominantly expressed in neuroblastoma. *Gene* 307, 1–11. doi: 10.1016/S0378-1119(03)00459-1

Spang, A. (2012). The DSL1 complex: the smallest but not the least CATCHR. *Traffic* 13, 908–913. doi: 10.1111/j.1600-0854.2012.01362.x

Starr, D. A., Saffery, R., Li, Z., Simpson, A. E., Choo, K. H., Yen, T. J., et al. (2000). HZwint-1, a novel human kinetochore component that interacts with HZW10. *J. Cell Sci.* 113, 1939–1950.

Starr, D. A., Williams, B. C., Hays, T. S., and Goldberg, M. L. (1998). ZW10 helps recruit dynactin and dynein to the kinetochore. *J. Cell Biol.* 142, 763–774. doi: 10.1083/jcb.142.3.763

Stehman, S. A., Chen, Y., McKenney, R. J., and Vallee, R. B. (2007). NudE and NudEL are required for mitotic progression and are involved in dynein recruitment to kinetochores. *J. Cell Biol.* 178, 583–594. doi: 10.1083/jcb.200610112

Sun, Y., Shestakova, A., Hunt, L., Sehgal, S., Lupashin, V., and Storrie, B. (2007). Rab6 regulates both ZW10/RINT-1 and conserved oligomeric Golgi complex-dependent Golgi trafficking and homeostasis. *Mol. Biol. Cell* 18, 4129–4142. doi: 10.1091/mbc.E07-01-0080

Sütterlin, C., Hsu, P., Mallabiabarrena, A., and Malhotra, V. (2002). Fragmentation and dispersal of the pericentriolar Golgi complex is required for entry into mitosis in mammalian cells. *Cell* 109, 359–369. doi: 10.1016/S0092-8674(02)00720-1

Sweet, D. J., and Pelham, H. R. (1993). The TIP1 gene of *Saccharomyces cerevisiae* encodes an 80 kDa cytoplasmic protein that interacts with the cytoplasmic domain of Sec20p. *EMBO J.* 12, 2831–2840.

Tripathi, A., Ren, Y., Jeffrey, P. D., and Hughson, F. M. (2009). Structural characterization of Tip20p and Dsl1p, subunits of the Dsl1p vesicle tethering complex. *Nat. Struct. Mol. Biol.* 16, 114–123. doi: 10.1038/nsmb.1548

Uemura, T., Sato, T., Aoki, T., Yamamoto, A., Okada, T., Hirai, R., et al. (2009). p31 deficiency influences endoplasmic reticulum tubular morphology and cell survival. *Mol. Cell. Biol.* 29, 1869–1881. doi: 10.1128/MCB.01089-08

Uetz, P., Giot, L., Cagney, G., Mansfield, T. A., Judson, R. S., Knight, J. R., et al. (2000). A comprehensive analysis of protein-protein interactions in *Saccharomyces cerevisiae. Nature* 403, 623–627. doi: 10.1038/35001009

Ungar, D., Oka, T., Vasile, E., Krieger, M., and Hughson, F. M. (2005). Subunit map of the conserved oligomeric Golgi complex. *J. Biol. Chem.* 280, 32729–32735. doi: 10.1074/jbc.M504590200

VanRheenen, S. M., Reilly, B. A., Chamberlain, S. J., and Waters, M. G. (2001). Dsl1p, an essential protein required for membrane traffic at the endoplasmic reticulum/Golgi interface in yeast. *Traffic* 2, 212–231. doi: 10.1034/j.1600-0854.2001.020307.x van Vlijmen, T., Vleugel, M., Evers, M., Mohammed, S., Wulf, P. S., Heck, A. J., et al. (2008). A unique residue in rab3c determines the interaction with novel binding protein Zwint-1. *FEBS Lett.* 582, 2838–2842. doi: 10.1016/j.febslet.2008.07.012

Varma, D., Dujardin, D. L., Stehman, S. A., and Vallee, R. B. (2006). Role of the kinetochore/cell cycle checkpoint protein ZW10 in interphase cytoplasmic dynein function. *J. Cell Biol.* 172, 655–662. doi: 10.1083/jcb.200510120

Vleugel, M., Hoogendoorn, E., Snel, B., and Kops, G. J. (2012). Evolution and function of the mitotic checkpoint. *Dev. Cell* 23, 239–250. doi: 10.1016/j.devcel.2012.06.013

Wainman, A., Giansanti, M. G., Goldberg, M. L., and Gatti, M. (2012). The *Drosophila* RZZ complex—roles in membrane trafficking and cytokinesis. *J. Cell Sci.* 125, 4014–4025. doi: 10.1242/jcs.099820

Wang, H., Hu, X., Ding, X., Dou, Z., Yang, Z., Shaw, A. W., et al. (2004). Human Zwint-1 specifies localization of Zeste White 10 to kinetochores and is essential for mitotic checkpoint signaling. *J. Biol. Chem.* 279, 54590–54598. doi: 10.1074/jbc.M407588200

Weber, A., Imisch, P., Bergmann, E., and Christiansen, H. (2004). Coamplification of DDX1 correlates with an improved survival probability in children with MYCN-amplified human neuroblastoma. *J. Clin. Oncol.* 22, 2681–2690. doi: 10.1200/JCO.2004.07.192

Willett, R., Kudlyk, T., Pokrovskaya, I., Schönherr, R., Ungar, D., Duden, R., et al. (2013). COG complexes form spatial landmarks for distinct SNARE complexes. *Nat. Commun.* 4, 1553. doi: 10.1038/ncomms2535

Williams, B. C., Karr, T. L., Montgomery, J. M., and Goldberg, M. L. (1992). The *Drosophila* l(1) zw10 gene product, required for accurate mitotic chromosome segregation, is redistributed at anaphase onset. *J. Cell Biol.* 118, 759–773. doi: 10.1083/jcb.118.4.759

Williams, B. C., Li, Z., Liu, S., Williams, E. V., Leung, G., Yen, T. J., et al. (2003). Zwilch, a new component of the ZW10/ROD complex required for kinetochore functions. *Mol. Biol. Cell* 14, 1379–1391. doi: 10.1091/mbc.E02-09-0624

Wimmer, K., Zhu, X. X., Lamb, B. J., Kuick, R., Ambros, P. F., Kovar, H., et al. (1999). Co-amplification of a novel gene, NAG, with the N-myc gene in neuroblastoma. *Oncogene* 18, 233–238. doi: 10.1038/sj.onc.1202287

Wojcik, E., Basto, R., Serr, M., Scaërou, F., Karess, R., and Hays, T. (2001). Kinetochore dynein: its dynamics and role in the transport of the Rough deal checkpoint protein. *Nat. Cell Biol.* 3, 1001–1007. doi: 10.1038/ncb1101-1001

Xiao, J., Liu, C.-C., Chen, P.-L., and Lee, W.-H. (2001). RINT-1, a novel rad50-interacting protein, participates in radiation induced G_2/M checkpoint control. *J. Biol. Chem.* 276, 6105–6111. doi: 10.1074/jbc.M008893200

Yamaguchi, T., Dulubova, I., Min, S. W., Chen, X., Rizo, J., and Südhof, T. C. (2002). Sly1 binds to Golgi and ER syntaxins via a conserved N-terminal peptide motif. *Dev. Cell* 2, 295–305. doi: 10.1016/S1534-5807(02)00125-9

Yu, I. M., and Hughson, F. M. (2010). Tethering factors as organizers of intracellular vesicular traffic. *Annu. Rev. Cell Dev. Biol.* 26, 137–156. doi: 10.1146/annurev.cellbio.042308.113327

Zhang, Y. Q., Guo, N., Peng, G., Wang, X., Han, M., Raincrow, J., et al. (2009). Role of SIP30 in the development and maintenance of peripheral nerve injury-induced neuropathic pain. *Pain* 146, 130–140. doi: 10.1016/j.pain.2009.07.011

Zhao, P., Liu, F., Zhang, B., Liu, X., Wang, B., Gong, J., et al. (2013). MAIGO2 is involved in abscisic acid-mediated response to abiotic stresses and Golgi-to-ER retrograde transport. *Physiol. Plant* 148, 246–260. doi: 10.1111/j.1399-3054.2012.01704.x

Zink, S., Wenzel, D., Wurm, C. A., and Schmitt, H. D. (2009). A link between ER tethering and COP-I vesicle uncoating. *Dev. Cell* 17, 403–416. doi: 10.1016/j.devcel.2009.07.012

Conflict of Interest Statement: The authors declare that the research was conducted in the absence of any commercial or financial relationships that could be construed as a potential conflict of interest.

17

CK1δ in lymphoma: gene expression and mutation analyses and validation of CK1δ kinase activity for therapeutic application

B. Sophia Winkler[1†‡], Franziska Oltmer[2†‡], Julia Richter[2], Joachim Bischof[2], Pengfei Xu[2], Timo Burster[3], Frank Leithäuser[1] and Uwe Knippschild[2*]

[1] Department of Pathology, Ulm University Hospital, Ulm, Germany
[2] Department of General and Visceral Surgery, Surgery Center, Ulm University Hospital, Ulm, Germany
[3] Department of Neurosurgery, Ulm University Hospital, Ulm, Germany

Edited by:
You-Wen He, Duke University Medical Center, USA

Reviewed by:
Zhibin Chen, University of Miami Miller School of Medicine, USA
Ivan Lilyanov Dzhagalov, National Yang-Ming University, Taiwan

***Correspondence:**
Uwe Knippschild, Department of General and Visceral Surgery, Surgery Center, Ulm University Hospital, Albert-Einstein-Allee 23, 89081 Ulm, Germany
e-mail: uwe.knippschild@uniklinik-ulm.de

†Present address:
B. Sophia Winkler, Department of Obstetrics and Gynecology, University Hospital Aachen, Germany;
Franziska Oltmer, Department of Internal Medicine, University Hospital Oldenburg, Germany

‡These authors have contributed equally to this work.

The prognosis of lymphoid neoplasms has improved considerably during the last decades. However, treatment response for some lymphoid neoplasms is still poor, indicating the need for new therapeutic approaches. One promising new strategy is the inhibition of kinases regulating key signal transduction pathways, which are of central importance in tumorigenesis. Kinases of the CK1 family may represent an attractive drug target since CK1 expression and/or activity are associated with the pathogenesis of malignant diseases. Over the last years efforts were taken to develop highly potent and selective CK1-specific inhibitor compounds and their therapeutic potential has now to be proved in pre-clinical trials. Therefore, we analyzed expression and mutational status of CK1δ in several cell lines representing established lymphoma entities, and also measured the mRNA expression level in primary lymphoma tissue as well as in non-neoplastic blood cells. For a selection of lymphoma cell lines we furthermore determined CK1δ kinase activity and demonstrated therapeutic potential of CK1-specific inhibitors as a putative therapeutic option in the treatment of lymphoid neoplasms.

Keywords: CK1, lymphoma, therapy, inhibitor, mutation analysis, cell cycle

INTRODUCTION

Lymphoid neoplasms are heterogeneous malignancies of the hematopoietic and lymphoid tissues (Jaffe et al., 2001; Harris et al., 2008). In the United States, the annual incidence of all lymphoid malignancies is about 34 per 100,000.

According to the current WHO classification lymphoid neoplasms are divided into Hodgkin-(HL) and Non-Hodgkin-Lymphomas (NHL) (Jaffe et al., 2001; Harris et al., 2008). HL represents approximately 30% of lymphoid neoplasms and is dichotomized into classical Hodgkin Lymphoma and nodular lymphocyte predominant Hodgkin lymphoma. Although Hodgkin lymphoma is of B cell origin with only very few exceptions, its distinct histology, immunophenotype, and clinical presentation justify separating it from other mature B cell neoplasms (Jaffe et al., 2001; Schmitz et al., 2009). Mature NHL are broadly divided into B and T cell neoplasms and further sub-classified into a large number of distinct entities, according to phenotype, genotype and clinical properties. One of the major entities of mature B cell neoplasms is diffuse large B cell lymphoma, not otherwise specified (NOS), that contains various subtypes and morphologic variants, including immunoblastic lymphoma (Menon et al., 2012). Other B cell lymphoma entities are primary mediastinal B cell lymphoma and Burkitt lymphoma. Primary mediastinal B cell lymphoma is a diffuse large B cell lymphoma typically arising in the mediastinum (Stein et al., 2008). Burkitt lymphoma constitutes an aggressive, highly proliferative mature B cell lymphoma often located at extranodal sites (Leoncini et al., 2008). B and T lymphoblastic lymphomas are primarily pediatric, immature tumors belonging to the group of precursor lymphoid neoplasms (Borowitz and Chan, 2008).

Owing to increasingly elaborate therapy protocols, the prognosis of lymphoid neoplasms has continuously improved during the last decades. Nevertheless, for a significant number of lymphoid neoplasms the response to established treatment remains unsatisfactory, highlighting the need for new therapy strategies. Currently, one of the most promising approaches in oncology seems to be targeting of growth-promoting protein kinases by specific inhibitory substances. Such kinases have already been identified for a few lymphoid neoplasms, i.e., chronic lymphocytic leukemia, mantle cell lymphoma (Aalipour and Advani, 2014) and hairy cell leukemia (Dietrich et al., 2012). However, for the vast majority of malignant lymphomas suitable target molecules are still unknown. Therefore, large research interest is now being focused on the identification of new specific protein kinases to interrupt pivotal signaling pathways in lymphoma cells.

Casein kinase 1 (CK1) isoforms are ubiquitously expressed serine/threonine-specific kinases. In human, at least six different isoforms (α, γ1, γ2, γ3, δ, and ε) and a number of related splice variants have been identified (Green and Bennett, 1998; Fu et al., 2001; Burzio et al., 2002; Knippschild et al., 2014). While all isoforms and variants significantly differ in length and composition of their N- and C-terminal domains, the kinase domain is highly conserved within all isoforms. CK1δ and ε show the highest sequence homology of about 98%. CK1 family members are able to phosphorylate a broad range of substrates regulating important cellular processes like cytokinesis, DNA repair, cell cycle progression, differentiation processes, and apoptosis (Knippschild et al., 2005, 2014; Price, 2006; Cheong and Virshup, 2011). CK1 activity is also linked to pathways involving the central signal transduction molecules p53 or β-catenin (Cheong and Virshup, 2011; Cruciat, 2014; Knippschild et al., 2014). Accordingly, changes in cellular CK1 expression and/or activity contribute to the pathogenesis of various diseases including cancer. In recent years considerable research effort has addressed the development and characterization of CK1-specific inhibitors (for review see Knippschild et al., 2014). As a result, new inhibitor compounds specifically targeting CK1 isoforms are more and more proving their therapeutic potential for a variety of malignancies (Perez et al., 2010; Knippschild et al., 2014).

To investigate whether CK1δ could be a potential therapeutic target in lymphoid malignancies, we characterized the role of CK1δ in 18 human cell lines representing a number of established lymphoma entities, i.e., acute lymphoblastic B cell lymphoma, acute lymphoblastic T cell lymphoma, diffuse-large B cell lymphoma (NOS), primary mediastinal B cell lymphoma, Burkitt lymphoma, classical Hodgkin-lymphoma, and nodular lymphocyte-predominant Hodgkin Lymphoma. In the present study, we tested the effects of established and novel CK1-specific inhibitor compounds on selected lymphoma cell lines in order to evaluate the therapeutic potential of CK1 inhibition. Prior to functional testing, we determined expression, mutational status, and kinase activity of CK1δ in several established lymphoma cell lines to estimate the effects of CK1-specific kinase inhibition and to correlate kinase inhibition-mediated effects with CK1δ expression and/or activity levels or possible mutations in the CK1δ coding sequence.

MATERIALS AND METHODS

HUMAN LYMPHOMA TISSUE

In summary 12 patients suffering from diffuse large B cell lymphomas whose informed consent was obtained prior to surgery were included in the study. Lymphoma diagnosis was in accordance with the current World Health Organization classification. All samples were drawn from our archive of frozen tissues and pseudonymized to comply with the German law for correct usage of archival tissue for clinical research (Deutsches Ärzteblatt 2003; 100 A1632). Frozen tissue samples were used for gene expression analysis. The project was performed with the permission of the independent local ethics committee of the University of Ulm.

CELL LINES

Within the present study the following established lymphoma cell lines were used: DAUDI (Klein et al., 1968), DEV (Poppema et al., 1985), DOHH-2 (Kluin-Nelemans et al., 1991), HDLM2 (Drexler et al., 1986), JIYOYE (Pulvertaft, 1965), Jurkat (Schneider et al., 1977), KARPAS-1106P (Nacheva et al., 1994), KM-H2 (Kamesaki et al., 1986), L-1236 (Kanzler et al., 1996), L-428 (Diehl et al., 1981), L-540 (Diehl et al., 1981), MedB-1 (Moller et al., 2001), NALM-6 (Hurwitz et al., 1979), RAJI (Pulvertaft, 1964), RAMOS (Klein et al., 1975), SU-DHL-6 (Epstein et al., 1978), SU-DHL-8 (Epstein et al., 1978), U-H01 (Mader et al., 2007). Furthermore, the Epstein-Barr virus (EBV) transformed lymphoblastoid B cells from peripheral blood of healthy donors LCL1, LCL2, and BSM (Halder et al., 2000) were used. All cell lines were maintained in RPMI-1640 medium (Invitrogen, Karlsruhe, Germany) except for U-H01 (4:1 mixture of DMEM and RPMI-1640; Invitrogen, Karlsruhe, Germany) and MedB-1 (IMDM; Invitrogen, Karlsruhe, Germany). Media were supplemented with 2 mM glutamine, 100 units/ml penicillin, 100 μg/ml streptomycin and 10% heat-inactivated fetal calf serum (FCS) except for DEV, HDLM2, Jurkat, L-540, for which 20% FCS was used. Cells were kept in a humidified 5% carbon dioxide atmosphere at 37°C.

GENE EXPRESSION ANALYSIS

Total RNA was isolated from human lymphoma cells, blood cells, and tumors from lymphoid tissue using the RNeasy Mini kit (Qiagen, Hilden, Germany). Equal amounts of total RNA were reverse transcribed into complementary DNA using the AffinityScript cDNA Synthesis kit (Agilent Technologies, Waldbronn, Germany). Quantitative reverse-transcription PCR (qRT-PCR) data presented in **Figure 1** were generated using an iCycler real-time PCR system and iQ SYBR green supermix according to the manufacturer's instructions (both Biorad, Munich, Germany). Glycerinaldehyd-3-phosphat-dehydrogenase (GAPDH) was used as external standard, cryopreserved tissue of human tonsils as reference material. qPCR-conditions were denaturation at 95°C, annealing at 60°C (CK1δ-specific primers), and elongation at 72°C in 40 cycles. CK1δ was amplified using oligonucleotide primers AGCACATCCCCTATCGTGAG and CGTAGCCCAGAGACTCCAAG. SYBR green fluorescent signals were detected at the end of each annealing step. Relative gene expression of CK1δ in lymphoma cells was calculated using the ΔΔCt method. Real-time RT-PCR data shown in **Figure 7** were generated using the QuantiTect™ SYBR Green PCR kit (Qiagen, Hilden, Germany) and a LightCycler® 480 (Roche, Mannheim, Germany) real-time PCR instrument. QuantiTect Primer Assays for CK1δ and HPRT were purchased from Qiagen (Qiagen, Hilden, Germany) and used according to the manufacturer's instructions using the following cycling conditions: heat activation at 95°C for 15 min, 55 cycles of 15 s denaturation at 94°C, 20 s annealing at 55°C and 30 s synthesis at 72°C. Slope was always set at 20°C/s throughout amplification. The fluorescence data collection of the color signal was measured in real-time model

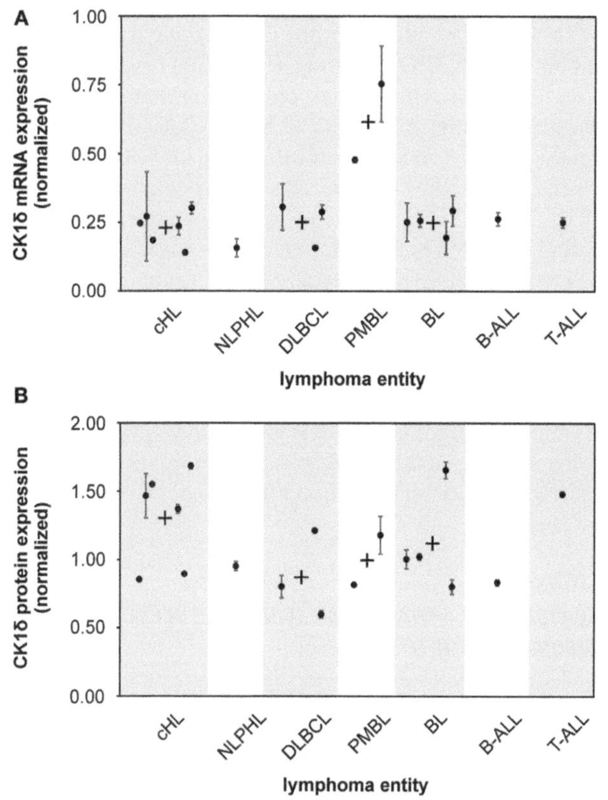

FIGURE 1 | mRNA and protein expression levels of CK1δ in various established malignant lymphoma cell lines. The expression levels of CK1δ in various established malignant lymphoma cell lines were analyzed by quantitative RT-PCR **(A)** and quantitative Western blot analyses **(B)**. Black dots represent gene or protein expression of one cell line relative to tonsil reference material. For clear presentation cell lines were assigned to their entity of lymphoma. Each black dot of gene expression analysis is representative for three replicates. Error bars indicate the fold-change range for the calculated ΔΔCt values. For quantitative Western blot analyses data from one representative analysis out of three are shown. In case more than one cell line per lymphoma entity was tested, mean values are indicated by a black cross. Abbreviations: *CK1δ*: Casein Kinase 1 isoform delta, c*HL*: classical Hodgkin lymphoma including HDLM2, KM-H2, L-1236, L-428, L-540, and U-H01 cells, *NLPHL:* nodular lymphocyte predominant Hodgkin lymphoma represented by DEV cells, *DLBCL*: diffuse large B cell lymphoma containing DOHH-2, SU-DHL-6, and SU-DHL-8 cells, *PMBL:* primary mediastinal large B cell lymphoma including KARPAS-1106P and MedB-1 cells, *BL:* Burkitt lymphoma including DAUDI, JIYOYE, RAJI, and RAMOS cells, B-*ALL:* B cell acute lymphoblastic lymphoma represented by NALM-6 cells, *T-ALL*: T cell acute lymphoblastic lymphoma represented by JURKAT cells.

at the end of each DNA synthesis cycle, and the crossing point was calculated. The specificity of DNA amlplification for CK1δ was examined analyzing the melting curve. mRNA contents were normalized to HPRT mRNA levels.

ISOLATION OF PRIMARY HUMAN IMMUNE CELLS

Human peripheral blood mononuclear cells (PBMCs) were freshly isolated from buffy coats of healthy blood donors by density gradient centrifugation. Conventional dendritic cells (cDC, CD1c$^+$), monocytes (CD14$^+$), T regulatory cells (CD4$^+$, CD25$^+$), cytotoxic T lymphocytes (CD8$^+$), natural killer cells (NK cells, CD56$^+$), and B cells (CD19$^+$) were positively selected using the appropriate magnetic cell separation kit (Miltenyi Biotec, Bergisch Gladbach, Germany) following the manufacturer's protocol. Use of patient material in this study was approved by the ethics committee at the University of Ulm.

SEQUENCE ANALYSIS

CK1δ was amplified from cDNA prepared as described above using PrimeSTAR HS DNA Polymerase. PCR was performed in a Primus 96 Plus Thermal Cycler (MWG Biotech, Ebersberg, Germany). To obtain adequate amounts of specific amplification product, a nested PCR was carried out applying the following conditions: denaturation at 98°C, annealing at 56°C, and elongation at 72°C in 35 cycles. For the first round of PCR the primers GCCCTTCACAGCAATAAGGA and CCAGAGTTCAGACCCAGGAA were used in 35 cycles. For the second round the primers GCACGACAGACTGAAGACCA and CCAGAGTTCAGACCCAGGAA were used in 20 cycles. Prior to sequencing, amplified CK1-products were separated on a 2% agarose gel and cleaned up using the peqGOLD MicroSpin gel elution kit (peqlab Biotechnologie GmbH, Erlangen, Germany). Sequencing primers for CK1δ were CCAGAGTTCAGACCCAG GAA, AGCACATCCCCTATCGTGAG, and GCACGACAGACT GAAGACCA. Sequencing of DNA was accomplished by LGC genomics (Berlin, Germany). Received CK1δ cDNA sequence data were evaluated using the database sequence for human CK1δ transcription variant 2 (accession ID: NM_139062) and the multiple sequence alignment tool ClustalW (Larkin et al., 2007; Goujon et al., 2010).

WESTERN BLOT

Cells were lysed in radio immunoprecipitation assay (RIPA) lysis buffer by incubation on ice for 30 min. Extracts were clarified by centrifugation and protein concentration was determined using the BCA protein assay (Thermo Fisher Scientific, Rockford, USA). Equal amounts of protein extracts were separated by SDS-PAGE and transferred to a PVDF membrane by semidry Western blotting. After blocking with 5% (w/v) milk the membrane was incubated with the specific antibody for CK1δ (C–18; Santa Cruz biotechnology, Heidelberg, Germany). Immunocomplexes were detected using HRP-conjugated anti-goat IgG (Santa Cruz biotechnology, Heidelberg, Germany), followed by chemiluminescence detection with SuperSignal West Dura (Thermo Fisher Scientific, Rockford, USA) and exposure to X-ray films. Densitometric quantification was performed using TINA 2.09 software. Equal loading of extracted proteins was determined by re-probing the membrane with a β-actin specific monoclonal antibody.

FRACTIONATION OF PROTEINS

Protein extracts for anion exchange chromatography were prepared in sucrose lysis buffer (20 mM Tris-HCl pH 7.0, 0.27 M sucrose, 1 mM EDTA, 1 mM EGTA, 1% (v/v) Triton X-100, 1 mM benzamidine, 25 μg/ml aprotinin, and 5 mM DTT) by incubation on ice for 30 min. Extracts were cleared by centrifugation and protein concentration was determined using the BCA protein assay (Thermo Fisher Scientific, Rockford, USA). Prior to column loading protein lysates were passed through a 0.4 μm filter. Fractionation of equal protein amounts was performed using an anion-exchange column (Resource Q; GE Healthcare, Chalfont St Giles, UK) attached to an ETTAN LC purifier (GE Healthcare, Chalfont St Giles, UK). Bound proteins were eluted with a linear ascending gradient between 0 and 1000 mM NaCl in fractionation buffer (50 mM Tris-HCl, pH 7.5, 1 mM EDTA, 5% (v/v) glycerol, 0.04% Brij, 1 mM benzamidine, 4 μg/ml leupeptin, and 0.1% (v/v) β-mercaptoethanol) and 250 μl of each protein fraction was collected.

EXPRESSION AND PURIFICATION OF GLUTATHIONE S-TRANSFERASE FUSION PROTEINS

Expression and purification of glutathione S-transferase (GST)-$p53^{1-64}$ fusion protein (FP267) (Milne et al., 2001) has previously been described in detail (Knippschild et al., 1997).

IN VITRO KINASE REACTIONS

In order to detect cellular CK1-specific kinase activity *in vitro* kinase assays were carried out using selected fractions of anion-exchange fractionated cellular protein extracts as source of kinase while the GST-$p53^{1-64}$ fusion protein (FP267) was used as substrate. Kinase reactions were performed in kinase buffer (25 mM Tris-HCl pH 7.5, 10 mM $MgCl_2$, 0.1 mM EDTA, 100 nM ATP) containing 2 μCi [γ-^{32}P]-ATP per reaction. Where indicated, given concentrations of CK1-specific inhibitor compounds [IC261 (Mashhoon et al., 2000), compound 1 (Richter et al., 2014a), and compound 17 (Peifer et al., 2009)] were added. Kinase reactions were incubated at 30°C for 30 min, stopped by the addition of 5 × SDS sample buffer [250 mM Tris-HCl, pH 6.8, 25% (v/v) β-mercaptoethanol, 50% (v/v) glycerol, 10% (w/v) SDS, 0.5% (w/v) bromphenol blue], and separated by SDS-PAGE. Radioactively labeled protein bands on dried gels were visualized by autoradiography. Phosphorylated protein bands were excised and phosphate incorporation was quantified by Cherenkov counting (LS6000IC, Beckman Coulter, USA). Subsequently *in vitro* kinase assays were carried out with the CK1δ peak activity fractions of RAMOS and KM-H2 cells in presence of CK1δ specific inhibitors. For each reaction 2 μl of the inhibitor diluted in DMSO was added. Following inhibitor concentrations were used: 3 μM of IC261, 200 nM of compound 1, and 60 nM of compound 17. DMSO controls were included.

CELL TREATMENT AND FACS ANALYSIS

For flow cytometry analysis 5 × 10^5/ml RAMOS, KM-H2, U-H01, and DOHH-2 cells were either grown in the presence of IC261 (0.4 μM and 1.6 μM), compound 1 (2 μM and 4 μM), or compound 17 (0.5 μM and 2 μM) for 24 h and 48 h, respectively. Untreated cells and cells treated with 0.01% DMSO served as controls. At the indicated time points cells were prepared for cell cycle analysis using "Cycle Test Plus" kit (BD, San Jose, USA). Cells were stained with propidium iodide and analyzed by flow cytometry using a FACScan flow cytometer (BD bioscience, San Jose, USA) and the CellQuest software (BD, bioscience, San Jose, USA).

INHIBITOR COMPOUNDS

In addition to the well-established CK1-specific inhibitor IC261 (Mashhoon et al., 2000; Cheong et al., 2011) two structurally different ATP-competitive small molecule inhibitors were used. Imidazole-derivative compound 17 has previously demonstrated increased potency and isoform selectivity for CK1δ as well as enhanced effects on cultured cells. Compound 17 is able to bind to the selectivity pocket of the CK1δ protein and therefore can be affected by certain mutations of the CK1δ gatekeeper amino acid residue (Peifer et al., 2009). Compound 1 represents a next generation CK1-specific inhibitor originating from a previously published set of benzimidazole-derived CK1-specific inhibitors (Bischof et al., 2012). By successful structure-activty relationship (SAR) based modification, a set of difluoro-dioxolo-benzoimidazole based inhibitors was developed with compound 1 showing improved inhibitory effects on CK1 isoforms δ and ε and the survival and viability of numerous tumor cell lines (Richter et al., 2014a).

RESULTS

ANALYSIS OF CK1δ mRNA AND PROTEIN LEVELS IN ESTABLISHED LYMPHOMA CELL LINES

Several studies indicate that deregulated expression and/or activity of CK1 is associated with tumorigenesis in a number of malignancies (Inuzuka et al., 2010; Elyada et al., 2011; Knippschild et al., 2014). However, for human malignant lymphoma the impact of CK1δ on tumor development or progression has not been systematically investigated so far. In order to determine CK1δ expression levels, we first conducted quantitative reverse-transcription PCR (qRT-PCR). CK1δ mRNA was found in all 18 cell lines investigated. Both PMBL (mediastinal large B cell lymphoma) cell lines, MedB-1 and KARPAS-1066P, showed about twofold higher amounts of CK1δ mRNA than the other lymphoma cell lines included in our study (**Figure 1A**). CK1δ protein expression was quantified by Western blotting analysis and could be detected in all tested cell lines of the various lymphoma entities. In contrast to the mRNA findings, PMBL cell lines did not show elevated CK1δ protein expression, whereas the cHL (classical Hodgkin lymphoma) group and the single tested T-ALL (T cell acute lymphoblastic lymphoma) cell line displayed slightly increased CK1δ protein levels in comparison to the other lymphoma entities (**Figure 1B**).

CK1δ-SPECIFIC KINASE ACTIVITY DIFFERS IN ESTABLISHED LYMPHOMA CELL LINES

CK1 activity is tightly regulated by various cellular mechanisms. Having demonstrated CK1δ expression, we therefore tested CK1δ-specific kinase activity in selected lymphoma cell lines: the two classical HL lines KM-H2 and U-H01, the diffuse large B cell lymhoma line DOHH-2, and the Burkitt lymphoma line RAMOS. Equal amounts of protein extracts were fractionated by ion-exhange chromatography in order to separate CK1 from other protein kinases as well as cellular modulators of kinase acitvity like activating/inhibiting binding partners and phosphatases. Resulting fractions were checked for CK1 kinase activity using the substrate GST-$p53^{1-64}$. Strong CK1-specific activity could be detected in extracts of KM-H2 and RAMOS cells while U-H01 and DOHH-2 cells only showed weak detectable activity (**Figure 2A**). In order to confirm a major contribution of CK1δ to the detected activity in RAMOS and KM-H2 cells, kinase activity in the peak fractions was assayed in the presence of different CK1-specific inhibitors, including IC261 (Mashhoon et al., 2000), compound 1 (Richter et al., 2014a), and compound 17 (Peifer et al., 2009). For both cell lines kinase activity in the peak fractions could be inhibited by all three substances. At least for compounds 1 and 17 strong CK1δ isoform-selectivity has been reported. Strongest activity-reduction of up to 70% was demonstrated with compound 1, confirming, that the detected activity peak indeed mainly was caused by CK1δ (**Figure 2B**).

SEQUENCE ANALYSIS OF CK1δ IN ESTABLISHED LYMPHOMA CELL LINES

Published data indicate that gene mutations may enhance the oncongenic potential of CK1δ (Tsai et al., 2007; Schittek and Sinnberg, 2014). Therefore, we screended for CK1δ coding region mutations by direct sequencing of RT-PCR-products. All 18 lymphoma cell lines were found to express wildtype CK1δ (**Figure 3**).

CK1-SPECIFIC INHIBITORS SHOW DIFFERENT EFFECTS ON CELL CYCLE DISTRIBUTION OF VARIOUS LYMPHOMA CELL LINES

Since expression of wildtype CK1δ could be detected in all analyzed cell lines and significant CK1δ-specific kinase activity could be detected at least in RAMOS and KM-H2 cells, the influence of CK1-specific inhibitors on cell cycle distribution was determined for selected lymphoma cell lines. Treatment with the CK1δ/ε-specific inhibitor IC261 showed obvious effects on RAMOS, KM-H2, and DOHH-2 cells by either increasing the amount of dead cells in case of RAMOS cells or by leading to cell cycle arrest in G2 phase for an increased number of KM-H2 and DOHH-2 cells. In contrast IC261 treatment only led to minor effects on U-H01 cells (**Figure 4**). While compound 17 showed no obvious effects on the tested cell lines treatment with only low concentrations of compound 1 resulted in an increase of dead cells already after 24 h (**Figures 5, 6**). Except for RAMOS these effects of compound 1 could not be potentiated but remained stable while treatment duration or inhibitor concentration was increased. All compounds showed most obvious effects on RAMOS cells after 24 h which could be enhanced by either increasing inhibitor concentration or treatment duration to 48 h.

CK1δ mRNA EXPRESSION LEVELS IN LYMPHOMA TISSUES CORRESPOND TO TUMOR CELL LINES BUT DIFFER FROM NON-NEOPLASTIC LYMPHOID CELLS

So far, our results demonstrate that CK1-specific inhibitors differently affect the growth of various lymphoma cell lines. Their use in novel therapeutic concepts could be considered, especially if alterations in the expression levels of CK1δ in tumor tissue are detectable. Therefore, we now analyzed CK1δ mRNA expression in 12 tissue samples of diffuse large B cell lymphoma, which is the most prevalent aggressive lymphoma type. We also measured the expression levels in EBV-immortalized lymphoid cells, and non-neoplastic lymphoid cell populations from healthy donors. In the tested tumor tissue samples the expression levels of CK1δ varied up to 5.9-fold (**Figure 7A, Table 1**). Similarly, the expression levels of CK1δ among blood cells and lymphocytes from healthy donors also varied up to 5.4-fold (**Figure 7B, Table 1**). However, CK1δ expression in non-neoplastic blood cells (i.e., B cells, Tregs, cytotoxic T cells, and monocytes) were markedly increased compared to expression levels in tumor samples, lymphoma cell lines, and EBV-transformed lymphoblastoid B cells (**Figures 7A,B**).

DISCUSSION

Protein kinases of the CK1 family are involved in essential cellular processes and CK1 dysregulation can be associated with the pathogenesis of certain diseases as well as tumorigenesis

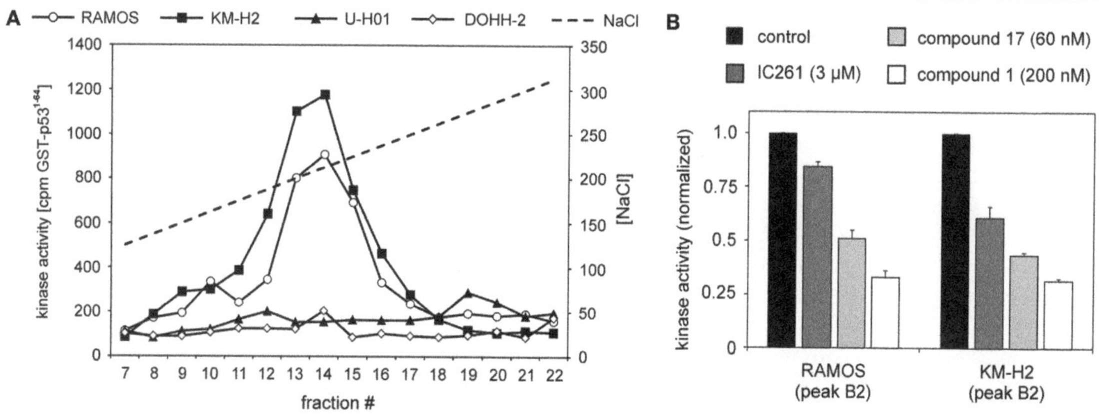

FIGURE 2 | Cell lines RAMOS, KM-H2, U-H01, and DOHH-2 show differences in CK1-specific kinase activity. (A) Protein extracts from RAMOS, KM-H2, U-H01, and DOHH-2 cells were ion-exchange fractionated and resulting fractions were used to phosphorylate GST-p53$^{1-64}$. Kinase reactions were separated in SDS-PAGE and phosphate incorporation into the substrate was quantified by Cherenkov counting. While cell lines RAMOS and KM-H2 show high CK1-specific activity no comparable kinase peak could be detected for U-H01 and DOHH-2 cells. **(B)** Peak kinase activity was assayed in presence of the CK1-specific inhibitors IC261 (Mashhoon et al., 2000), compound 1 (Richter et al., 2014a), and compound 17 (Peifer et al., 2009). Strong inhibition of peak kinase activity most of all by the CK1δ-specific compound 1 clearly confirms contribution of CK1δ. Experiments were performed in triplicate. Data are presented as mean values, error bars indicate standard error of the mean (SEM).

```
CK1delta   YRLGRKIGSGSFGDIYLGTDIAAGEEVAIKLECVKTKHPQLHIESKIYKMMQGGVGIPTI
RAMOS      YRLGRKIGSGSFGDIYLGTDIAAGEEVAIKLECVKTKHPQLHIESKIYKMMQGGVGIPTI
KM-H2      YRLGRKIGSGSFGDIYLGTDIAAGEEVAIKLECVKTKHPQLHIESKIYKMMQGGVGIPTI
U-H01      YRLGRKIGSGSFGDIYLGTDIAAGEEVAIKLECVKTKHPQLHIESKIYKMMQGGVGIPTI
DOHH-2     YRLGRKIGSGSFGDIYLGTDIAAGEEVAIKLECVKTKHPQLHIESKIYKMMQGGVGIPTI
           ************************************************************

CK1delta   RWCGAEGDYNVMVMELLGPSLEDLFNFCSRKFSLKTVLLLADQMISRIEYIHSKNFIHRD
RAMOS      RWCGAEGDYNVMVMELLGPSLEDLFNFCSRKFSLKTVLLLADQMISRIEYIHSKNFIHRD
KM-H2      RWCGAEGDYNVMVMELLGPSLEDLFNFCSRKFSLKTVLLLADQMISRIEYIHSKNFIHRD
U-H01      RWCGAEGDYNVMVMELLGPSLEDLFNFCSRKFSLKTVLLLADQMISRIEYIHSKNFIHRD
DOHH-2     RWCGAEGDYNVMVMELLGPSLEDLFNFCSRKFSLKTVLLLADQMISRIEYIHSKNFIHRD
           ************************************************************

CK1delta   VKPDNFLMGLGKKGNLVYIIDFGLAKKYRDARTHQHIPYRENKNLTGTARYASINTHLGI
RAMOS      VKPDNFLMGLGKKGNLVYIIDFGLAKKYRDARTHQHIPYRENKNLTGTARYASINTHLGI
KM-H2      VKPDNFLMGLGKKGNLVYIIDFGLAKKYRDARTHQHIPYRENKNLTGTARYASINTHLGI
U-H01      VKPDNFLMGLGKKGNLVYIIDFGLAKKYRDARTHQHIPYRENKNLTGTARYASINTHLGI
DOHH-2     VKPDNFLMGLGKKGNLVYIIDFGLAKKYRDARTHQHIPYRENKNLTGTARYASINTHLGI
           ************************************************************

CK1delta   EQSRRDDLESLGYVLMYFNLGSLPWQGLKAATKRQKYERISEKKMSTPIEVLCKGYPSEF
RAMOS      EQSRRDDLESLGYVLMYFNLGSLPWQGLKAATKRQKYERISEKKMSTPIEVLCKGYPSEF
KM-H2      EQSRRDDLESLGYVLMYFNLGSLPWQGLKAATKRQKYERISEKKMSTPIEVLCKGYPSEF
U-H01      EQSRRDDLESLGYVLMYFNLGSLPWQGLKAATKRQKYERISEKKMSTPIEVLCKGYPSEF
DOHH-2     EQSRRDDLESLGYVLMYFNLGSLPWQGLKAATKRQKYERISEKKMSTPIEVLCKGYPSEF
           ************************************************************

CK1delta   ATYLNFCRSLRFDDKPDYSYLRQLFRNLF
RAMOS      ATYLNFCRSLRFDDKPDYSYLRQLFRNLF
KM-H2      ATYLNFCRSLRFDDKPDYSYLRQLFRNLF
U-H01      ATYLNFCRSLRFDDKPDYSYLRQLFRNLF
DOHH-2     ATYLNFCRSLRFDDKPDYSYLRQLFRNLF
           *****************************
```

FIGURE 3 | Protein sequence alignment of CK1δ kinase domain amplified from cell lines RAMOS, KM-H2, U-H01, and DOHH-2. Messenger RNA from cell lines RAMOS, KM-H2, U-H01, and DOHH-2 was isolated, transcribed into cDNA, and the CK1δ coding sequence was amplified as described in Material and Methods. DNA sequences obtained from reactions using primers listed in Material and Methods were translated into protein sequences and the alignment was generated using ClustalW 2.1 (Larkin et al., 2007; Goujon et al., 2010).

(Knippschild et al., 2014). Recently, CK1α was suggested as therapeutic target for innovative strategies to treat malignancies arising from the hematopoietic system like multiple myeloma (Hu et al., 2014), acute myeloid leukemia (Jaras et al., 2014), or del(5q) myelodysplastic syndrome (Schneider et al., 2014). In the present study, we demonstrate the therapeutic potential of CK1-specific kinase inhibition by treating established lymphoma cell lines with CK1-specific small molecule inhibitors. In order to characterize the CK1-specific status and to estimate the expected treatment effects, we first analyzed CK1δ expression and mutation status in 18 cell lines covering a broad spectrum of lymphoid neoplasms. In all cell lines CK1δ mRNA and protein were found to be expressed strongly and at a similar level (except mRNA expression in PMBL cell lines) pointing to essential functions of CK1 in the regulation of apoptosis, NFκB, and Wnt-related signaling (Cruciat, 2014; Knippschild et al., 2014). Previous studies carried out in a murine model system already detected increased expression of CK1δ in hyperplastic B cell follicles and progressed B cell lymphomas in p53-deficient mice (Maritzen et al., 2003).

Mutations in the coding sequences of central signal transduction molecules play a pivotal role in tumorigenesis. In a number of genes, somatic mutations are quite common in malignant lymphoma and can even be characteristic for certain lymphoma entities (Kuppers, 2005). Mutations in the CK1δ protein may lead to altered binding and potency of certain inhibitor molecules (Peifer et al., 2009; Richter et al., 2014b). Therefore, mutation analysis is essential when determining the therapeutic potential of CK1-specific inhibitor compounds. Remarkably however, the analyzed human lymphoma cell lines without exception showed an unmutated state of the CK1δ coding sequence. This finding may point to an essential need for unmutated and functional CK1δ in signaling pathways required for intact lymphoid neoplasm cell function. Further analysis would be needed to validate this conclusion. An important mechanism leading to mutations

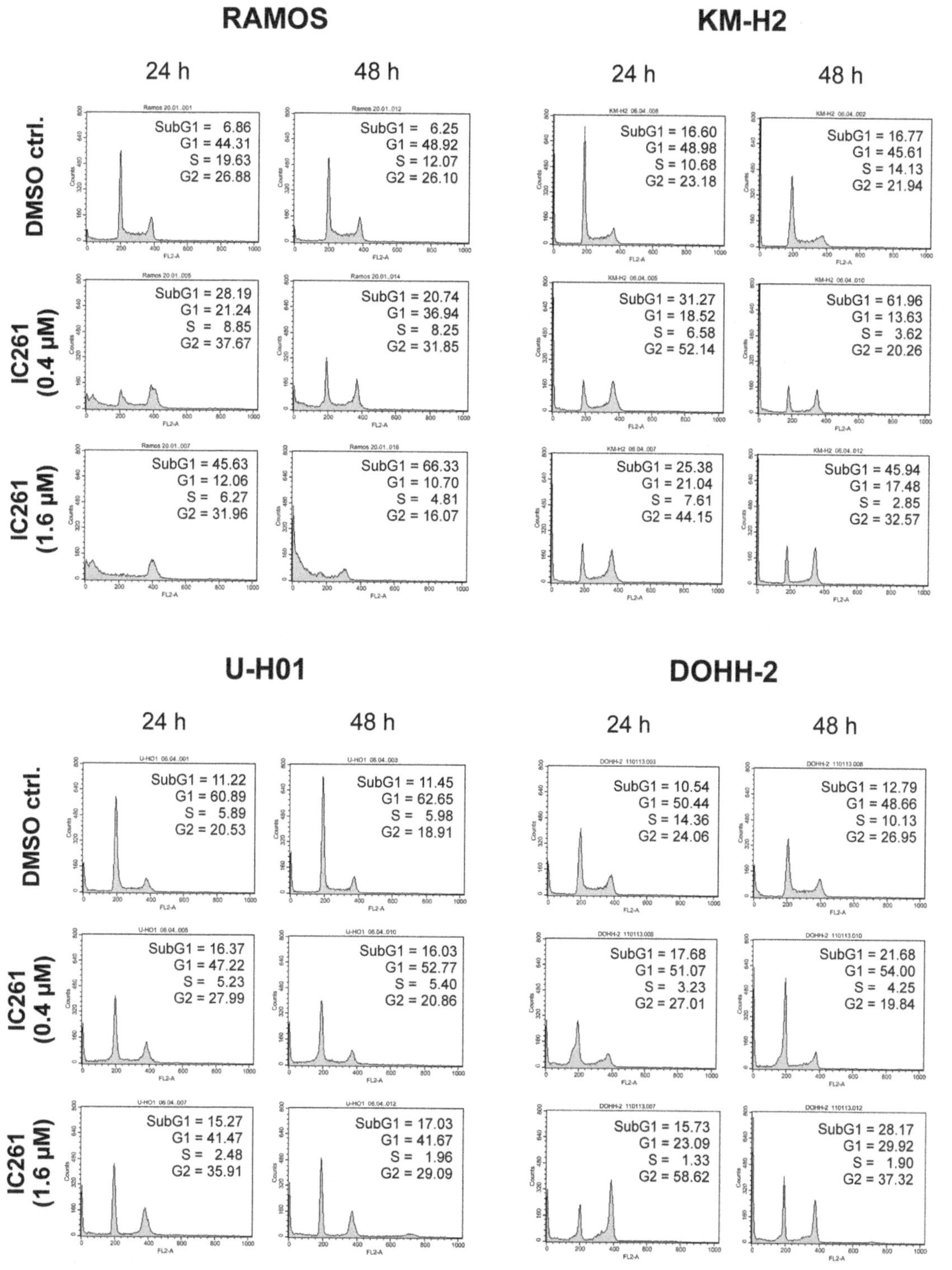

FIGURE 4 | Cell cycle analyses of RAMOS, KM-H2, U-H01, and DOHH-2 cells after treatment with the CK1δ/ε-specific inhibitor IC261. Cells were treated with different concentrations of IC261 for 24 h and 48 h and stained with propidium iodide in order to perform cell cycle analysis as described in Material and Methods. DMSO treated cells served as control. Each 20.000 cells in two replicates per condition were counted. One representative graph is shown. The determined fractions of cells in various phases of the cell cycle are given as percent values (%). Flow cytometry plots show the number of cells (y-axis) with each observed intensity of propidium iodide DNA staining (x-axis). DMSO treated cells showed a normal cell cycle distribution of asynchronously proliferating cells. Treatment with IC261 led to an increase of dead RAMOS cells, whereas in KM-H2, U-H01, and DOHH-2 cells an increase of G2 arrested cells could be observed.

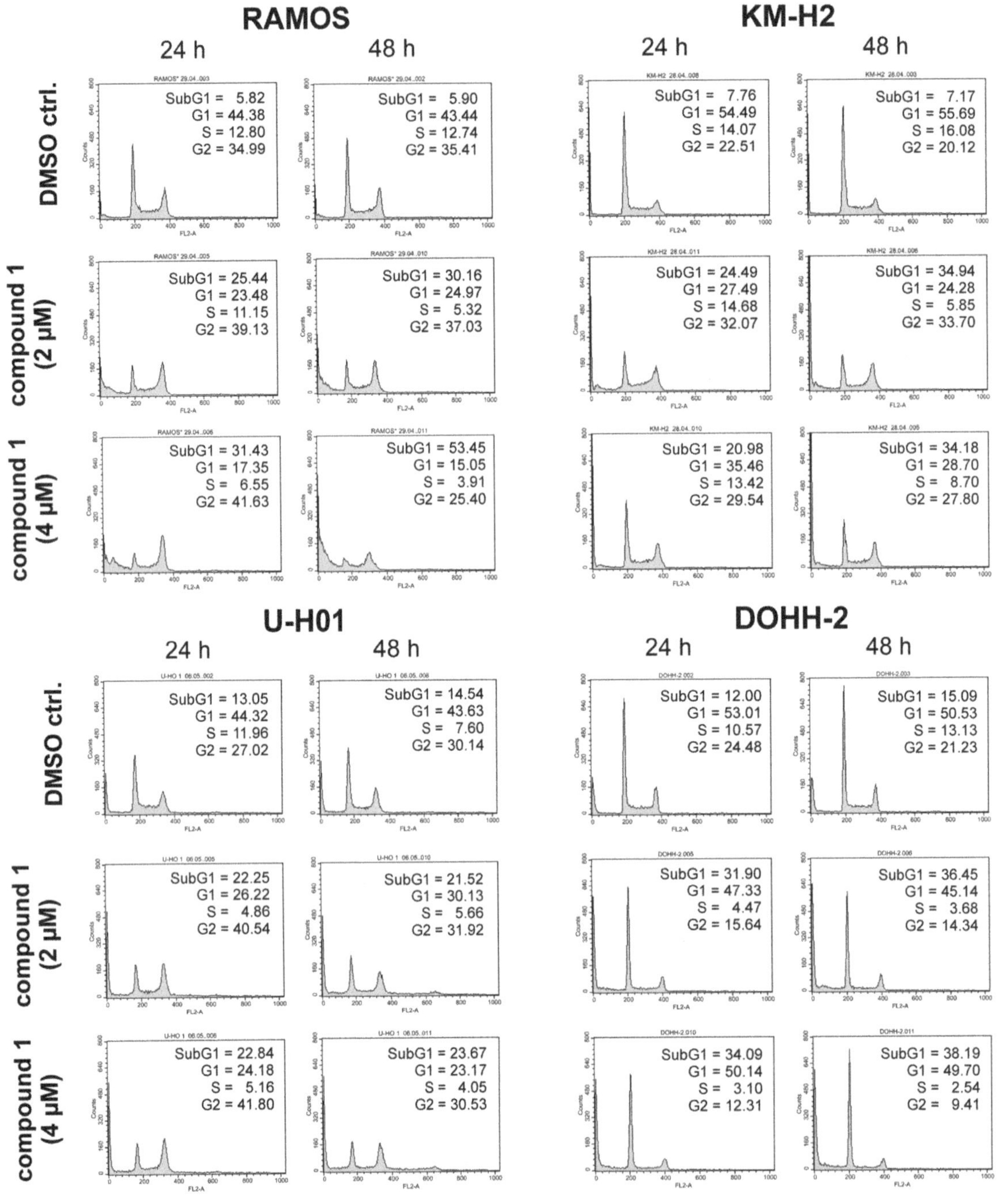

FIGURE 5 | Cell cycle analyses of RAMOS, KM-H2, U-H01, and DOHH-2 cells after treatment with the CK1δ-specific inhibitor compound 1. Cells were treated with different concentrations of compound 1 for 24 h and 48 h and stained with propidium iodide in order to perform cell cycle analysis as described in Material and Methods. DMSO treated cells served as control. Each 20.000 cells in two replicates per condition were counted. One representative graph is shown. The determined fractions of cells in various phases of the cell cycle are given as percent values (%). Flow cytometry plots show the number of cells (y-axis) with each observed intensity of propidium iodide DNA staining (x-axis). DMSO treated cells showed a normal cell cycle distribution of asynchronously proliferating cells. Treatment with compound 1 led to an increase of RAMOS, KM-H2, and U-H01 cells in G2 phase. Additionally, RAMOS cells accumulate in Sub G1 phase as well. No obvious effects could be observed in DOHH-2 cells.

in malignant lymphoma is cytidine deamination by activation induced cytidine deaminase (AID) and the subsequent incorrect repair of the affected base. Since chromatin-immunoprecipitation studies in murine B-lymphocytes revealed that AID is not able to substantially bind to CK1δ DNA sequence (Yamane et al., 2011), this could provide one explanation for the absence of CK1δ mutations while some of the analyzed cell lines carry mutations in multiple genes (Kuppers, 2005; Mader et al., 2007).

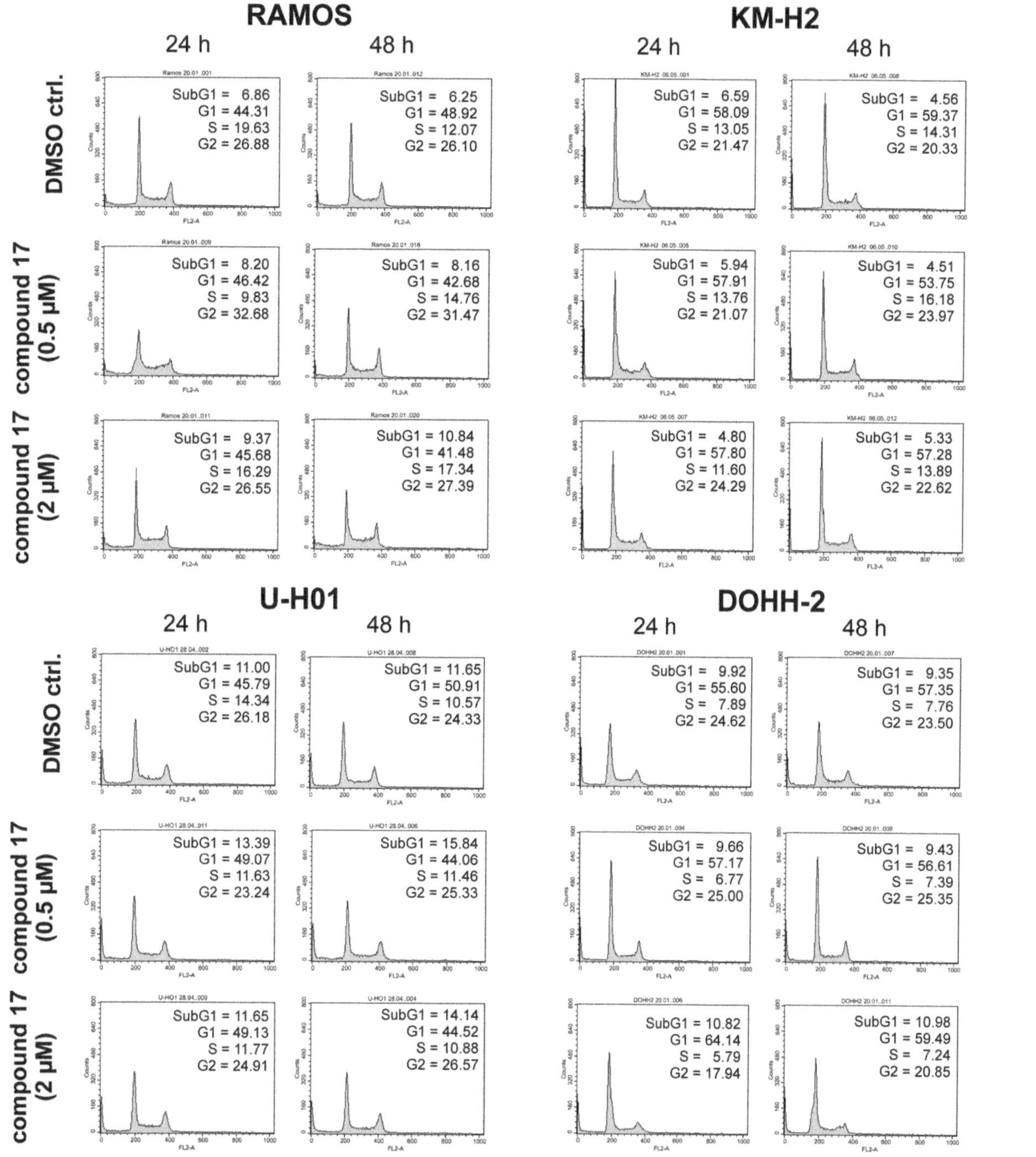

FIGURE 6 | Cell cycle analyses of RAMOS, KM-H2, U-H01, and DOHH-2 cells after treatment with the CK1δ-specific inhibitor compound 17. Cells were treated with different concentrations of compound 17 for 24 h and 48 h and stained with propidium iodide in order to perform cell cycle analysis as described in Material and Methods. DMSO treated cells served as control. Each 20.000 cells in two replicates per condition were counted. One representative graph is shown. The determined fractions of cells in various phases of the cell cycle are given as percent values (%). Flow cytometry plots show the number of cells (y-axis) with each observed intensity of propidium iodide DNA staining (x-axis). DMSO treated cells showed a normal cell cycle distribution of asynchronously proliferating cells. Treatment with compound 17 only led to minor effects on all tested cell lines.

For further analyses on the role of CK1δ in human lymphoma cell lines we concentrated on four cell lines, representing classical Hodgkin lymphoma (KM-H2 and U-H01), diffuse large B cell lymphoma (DOHH-2) and Burkitt lymphoma (RAMOS). Although all four cell lines share rather similar CK1δ expression levels, differences in the cellular kinase activity of CK1δ could be detected. Using the established CK1-specific small molecule inhibitor IC261 and two structurally independent CK1δ-specific inhibitors (Peifer et al., 2009; Richter et al., 2014a) the kinase activity detected using substrate GST-p53$^{1-64}$ (Knippschild et al.,

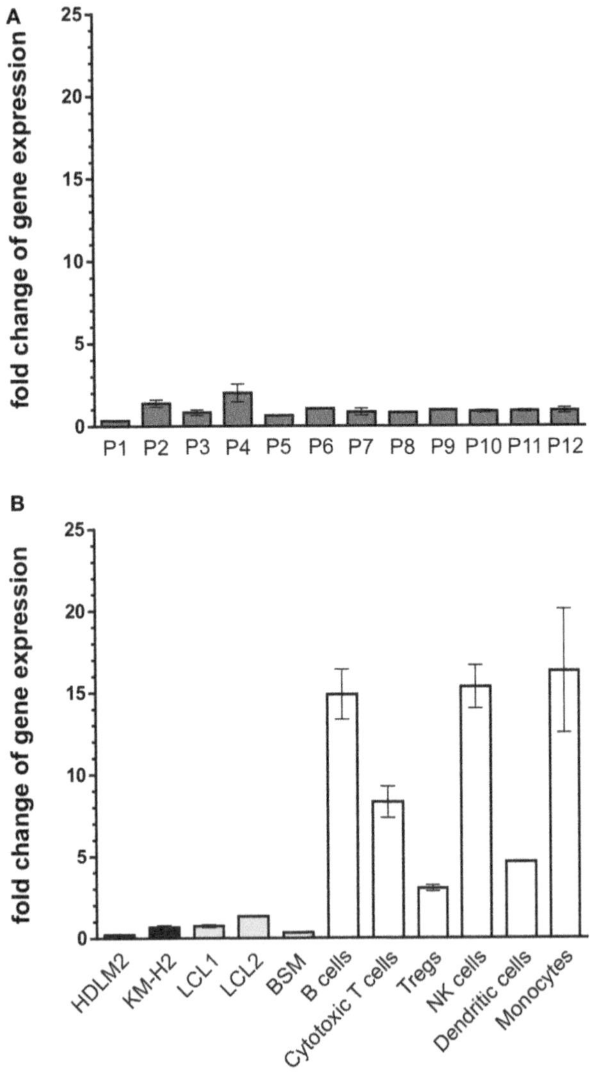

FIGURE 7 | mRNA expression levels of CK1δ in diffuse large B cell lymphoma tissues compared to lymphocytes from healthy donors. Comparison of CK1δ mRNA expression levels between **(A)** tumors of lymphoid tissues from 12 patients (P1–P12), **(B)** the established lymphoma cell lines HDLM2 and KM-H2 as well as the EBV-transformed lymphoblastoid B cells LCL1, LCL2, and BSM, and lymphocytes and other blood cells of healthy donors by quantitative real-time PCR as described in Materials and Methods. Results clearly demonstrate decreased mRNA expression levels of CK1δ in lymphoid tissues, lymphoma cell lines and EBV-transformed lymphoblastoid B cells compared to lymphocytes from healthy donors. Experiments were performed in duplicate. Data are presented as mean values of relative gene expression compared to HPRT mRNA levels, error bars indicate standard deviation (SD).

1997) could clearly be assigned to CK1δ. Cell line specific differences might be explained by posttranslational modifications of CK1δ and have already been reported in previous studies (Maritzen et al., 2003; Giamas et al., 2007). Mechanisms that regulate CK1δ activity are intramolecular autophosphorylation and phosphorylation through other cellular kinases (Rivers et al., 1998; Gietzen and Virshup, 1999; Giamas et al., 2007; Bischof et al., 2013).

Table 1 | CK1δ expression levels in tumors from lymphoid tissue, established lymphoma cell lines, EBV-immortalized lymphoblastoid B cells, and peripheral blood cell populations of healthy donors.

Sample	Fold change expression of CK1δ
P1	0.34 ± 0.06
P2	1.39 ± 0.30
P3	0.82 ± 0.23
P4	2.00 ± 0.76
P5	0.64 ± 0.01
P6	1.05 ± 0.01
P7	0.85 ± 0.30
P8	0.81 ± 0.03
P9	0.96 ± 0.05
P10	0.88 ± 0.09
P11	0.91 ± 0.07
P12	0.94 ± 0.23
HDLM2	0.22 ± 0.04
KM-H2	0.66 ± 0.16
LCL1	0.74 ± 0.12
LCL2	1.34 ± 0.05
BSM	0.33 ± 0.02
B cells	14.88 ± 2.17
Cytotoxic T cells	8.30 ± 1.35
Tregs	3.03 ± 0.26
NK cells	15.35 ± 1.87
Dendritic cells	4.64 ± 0.05
Monocytes	16.31 ± 5.34

Values represent fold change of relative CK1δ expression ± SD compared to HPRT mRNA levels.

In all four cell lines, FACS analyses demonstrated changes in cell cycle distribution after treatment with CK1δ-specific inhibitor compounds. Due to reduced stability and solubility of inhibitor compounds 1 and 17 much higher concentrations of these compounds have to be used for the treatment of cultured cells when being compared to *in vitro* application and to the use of IC261. The observed effects are dependent on inhibitor concentration, cell line, and duration of inhibitor exposure. Cell cycle arrest, as observed for some cell lines being treated with IC261 and compound 1, can be explained by CK1 effects on microtubule dynamics and chromosome segregation (Behrend et al., 2000; Stoter et al., 2014). However, Cheong and colleagues meanwhile demonstrated, that IC261 could have a similar effect on cellular microtubule dynamics as the spindle poison colchicine (Cheong et al., 2011). In order to exclude unspecific effects, we therefore performed all experiments with three structurally different CK1-inhibitors. Previous studies showed that the status of the tumor suppressor p53 plays a central role for the effects induced by inhibitor IC261. While treatment of wildtype p53 expressing cells with CK1-inhibitors leads to postmitotic arrest or apoptosis, the treatment of cells expressing mutated or no p53 at all is followed by endoreduplication (Behrend et al., 2000; Stoter et al., 2005). The present study is in partial accordance to these reports. Cell lines KM-H2 and DOHH-2 (wt p53, Drakos et al., 2007; Wang et al., 2011) showed increased apoptosis upon treatment with

IC261 and compound 1. Treatment with IC261, however, also arrested an increased number of KM-H2 and DOHH-2 cells in G2-phase of the cell cycle. Albeit expressing mutated p53 (Farrell et al., 1991), RAMOS cells showed a remarkable increase of apoptotic cells after treatment with the CK1-specific inhibitors. In DOHH-2 cells the chromosomal translocation t(14:18) results in overexpression of Bcl-2 which is associated with chemoresistance (Tsujimoto et al., 1984; Kluin-Nelemans et al., 1991; Reed, 2008). The anti-apoptotic effect of Bcl2 may explain the low level of apoptosis induction that could be achieved in DOHH-2 treated with CK1-specific inhibitors. Similarly, the reduced reactivity to inhibitor treatment of cell line U-H01 could be explained by mutation of the phosphatase PTPN1 which protects the cells from apoptosis and therefore provides an advantage for cell survival (Knecht et al., 2010). Generally, the low activity of CK1δ detected in U-H01 and DOHH-2 provides another reason why the observed effects are lower than in KM-H2 or RAMOS, two cell lines showing high cellular CK1δ kinase activity.

After having demonstrated significant effects of CK1-specific inhibitors on established lymphoma cell lines, the mRNA expression levels of CK1δ in primary lymphoid tumor tissue, EBV-immortalized lymphoid cells, and blood cells from healthy donors were compared. While the variation of CK1δ mRNA levels was nearly the same in primary tumor tissue and normal blood cells, CK1δ mRNA levels in general were remarkably low in all tested samples in comparison to blood cells from healthy donors. Cell lines HDLM2 and KM-H2, which were also included in this comparison, showed similarly low CK1δ mRNA levels as the analyzed primary tumor tissue samples. Therefore, these data provide further evidence that these cell lines may be used as a model for the validation of CK1-specific inhibitors in new lymphoma treatment strategies.

As a conclusion, its multiple functions may qualify CK1 as a new potential target molecule to treat lymphoid neoplasms. The results of the present study suggest for the first time that inhibition of CK1δ could have adjuvant effects in the therapy of malignant lymphoma. However, these effects are strongly dependent on the cellular background of the individual lymphoma cell population and for future application in novel concepts of personalized medicine still large efforts have to be made to increase solubility and selectivity of the inhibitor compounds as well as to improve their safe delivery to the target tissue to avoid severe side effects.

AUTHOR CONTRIBUTIONS

Project work was designed by UK and FL. Experimental work was performed by BW, FO, JR, JB, PX, and TB. Data analysis and interpretation was done by BW, FO, JR, JB, PX, TB, FL, and UK. All Authors (BW, FO, JR, JB, PX, TB, FL, and UK) were involved in writing passages of the present manuscript and participated in final approval and revision. Figures were created by BW, FO, JR, and JB.

ACKNOWLEDGMENTS

The authors would like to thank Nadine Süßner, Arnhild Grothey, Ulrike Kostezka, and Karola Dorsch for excellent technical assistance. Inhibitor compound 1 was kindly provided by 4SC Discovery GmbH (Planegg-Martinsried, Germany). Furthermore, we thank Christian Peifer (Department of Pharmaceutical Chemistry, Christian Albrechts University of Kiel, Germany) for providing inhibitor compound 17.

REFERENCES

Aalipour, A., and Advani, R. H. (2014). Bruton's tyrosine kinase inhibitors and their clinical potential in the treatment of B-cell malignancies: focus on ibrutinib. *Ther. Adv. Hematol.* 5, 121–133. doi: 10.1177/2040620714539906

Behrend, L., Milne, D. M., Stoter, M., Deppert, W., Campbell, L. E., Meek, D. W., et al. (2000). IC261, a specific inhibitor of the protein kinases casein kinase 1-delta and -epsilon, triggers the mitotic checkpoint and induces p53-dependent postmitotic effects. *Oncogene* 19, 5303–5313. doi: 10.1038/sj.onc.1203939

Bischof, J., Leban, J., Zaja, M., Grothey, A., Radunsky, B., Othersen, O., et al. (2012). 2-Benzamido-N-(1H-benzo[d]imidazol-2-yl)thiazole-4-carboxamide derivatives as potent inhibitors of CK1δ/ε. *Amino Acids* 43, 1577–1591. doi: 10.1007/s00726-012-1234-x

Bischof, J., Randoll, S. J., Sussner, N., Henne-Bruns, D., Pinna, L. A., and Knippschild, U. (2013). CK1delta kinase activity is modulated by Chk1-mediated phosphorylation. *PLoS ONE* 8:e68803. doi: 10.1371/journal.pone.0068803

Borowitz, M. J., and Chan, J. K. C. (2008). "B lymphoblastic leukaemia/lymphoma, not otherwise specified," in *WHO Classification of Tumours of Haematopoietic and Lymphoid Tissues,* eds S. H. Swerdlow, E. Campo, N. L. Harris, E. S. Jaffe, S. A. Pileri, H. Stein, J. Thiele and J. W. Vardiman. (Lyon: IARC), 168–170.

Burzio, V., Antonelli, M., Allende, C. C., and Allende, J. E. (2002). Biochemical and cellular characteristics of the four splice variants of protein kinase CK1alpha from zebrafish (Danio rerio). *J. Cell. Biochem.* 86, 805–814. doi: 10.1002/jcb.10263

Cheong, J. K., Hung, N. T., Wang, H., Tan, P., Voorhoeve, P. M., Lee, S. H., et al. (2011). IC261 induces cell cycle arrest and apoptosis of human cancer cells via CK1delta/varepsilon and Wnt/beta-catenin independent inhibition of mitotic spindle formation. *Oncogene* 30, 2558–2569. doi: 10.1038/onc.2010.627

Cheong, J. K., and Virshup, D. M. (2011). Casein kinase 1: complexity in the family. *Int. J. Biochem. Cell Biol.* 43, 465–469. doi: 10.1016/j.biocel.2010.12.004

Cruciat, C. M. (2014). Casein kinase 1 and Wnt/beta-catenin signaling. *Curr. Opin. Cell Biol.* 31C, 46–55. doi: 10.1016/j.ceb.2014.08.003

Diehl, V., Kirchner, H. H., Schaadt, M., Fonatsch, C., Stein, H., Gerdes, J., et al. (1981). Hodgkin's disease: establishment and characterization of four *in vitro* cell lies. *J. Cancer Res. Clin. Oncol.* 101, 111–124.

Dietrich, S., Glimm, H., Andrulis, M., Von Kalle, C., Ho, A. D., and Zenz, T. (2012). BRAF inhibition in refractory hairy-cell leukemia. *N. Engl. J. Med.* 366, 2038–2040. doi: 10.1056/NEJMc1202124

Drakos, E., Thomaides, A., Medeiros, L. J., Li, J., Leventaki, V., Konopleva, M., et al. (2007). Inhibition of p53-murine double minute 2 interaction by nutlin-3A stabilizes p53 and induces cell cycle arrest and apoptosis in Hodgkin lymphoma. *Clin. Cancer Res.* 13, 3380–3387. doi: 10.1158/1078-0432.CCR-06-2581

Drexler, H. G., Gaedicke, G., Lok, M. S., Diehl, V., and Minowada, J. (1986). Hodgkin's disease derived cell lines HDLM-2 and L-428: comparison of morphology, immunological and isoenzyme profiles. *Leuk. Res.* 10, 487–500.

Elyada, E., Pribluda, A., Goldstein, R. E., Morgenstern, Y., Brachya, G., Cojocaru, G., et al. (2011). CKIalpha ablation highlights a critical role for p53 in invasiveness control. *Nature* 470, 409–413. doi: 10.1038/nature09673

Epstein, A. L., Levy, R., Kim, H., Henle, W., Henle, G., and Kaplan, H. S. (1978). Biology of the human malignant lymphomas. IV. Functional characterization of ten diffuse histiocytic lymphoma cell lines. *Cancer* 42, 2379–2391.

Farrell, P. J., Allan, G. J., Shanahan, F., Vousden, K. H., and Crook, T. (1991). p53 is frequently mutated in Burkitt's lymphoma cell lines. *EMBO J.* 10, 2879–2887.

Fu, Z., Chakraborti, T., Morse, S., Bennett, G. S., and Shaw, G. (2001). Four casein kinase I isoforms are differentially partitioned between nucleus and cytoplasm. *Exp. Cell Res.* 269, 275–286. doi: 10.1006/excr.2001.5324

Giamas, G., Hirner, H., Shoshiashvili, L., Grothey, A., Gessert, S., Kuhl, M., et al. (2007). Phosphorylation of CK1delta: identification of Ser370 as the major phosphorylation site targeted by PKA *in vitro* and *in vivo*. *Biochem. J.* 406, 389–398. doi: 10.1042/BJ20070091

Gietzen, K. F., and Virshup, D. M. (1999). Identification of inhibitory autophosphorylation sites in casein kinase I epsilon. *J. Biol. Chem.* 274, 32063–32070.

Goujon, M., Mcwilliam, H., Li, W., Valentin, F., Squizzato, S., Paern, J., et al. (2010). A new bioinformatics analysis tools framework at EMBL-EBI. *Nucleic Acids Res.* 38, W695–W699. doi: 10.1093/nar/gkq313

Green, C. L., and Bennett, G. S. (1998). Identification of four alternatively spliced isoforms of chicken casein kinase I alpha that are all expressed in diverse cell types. *Gene* 216, 189–195. doi: 10.1016/S0378-1119(98)00291-1

Halder, T. M., Bluggel, M., Heinzel, S., Pawelec, G., Meyer, H. E., and Kalbacher, H. (2000). Defensins are dominant HLA-DR-associated self-peptides from CD34(-) peripheral blood mononuclear cells of different tumor patients (plasmacytoma, chronic myeloid leukemia). *Blood* 95, 2890–2896.

Harris, N. L., Campo, E., Jaffe, E. S., Pileri, S. A., Stein, H., Swerdlow, S. H., et al. (2008). *Introduction to the WHO Classification of Tumours of Haematopoietic and Lymphoid Tissues.* Lyon: IARC.

Hu, Y., Song, W., Cirstea, D., Lu, D., Munshi, N. C., and Anderson, K. C. (2014). CSNK1alpha1 mediates malignant plasma cell survival. *Leukemia* 29, 474–482. doi: 10.1038/leu.2014.202

Hurwitz, R., Hozier, J., Lebien, T., Minowada, J., Gajl-Peczalska, K., Kubonishi, I., et al. (1979). Characterization of a leukemic cell line of the pre-B phenotype. *Int. J. Cancer* 23, 174–180.

Inuzuka, H., Tseng, A., Gao, D., Zhai, B., Zhang, Q., Shaik, S., et al. (2010). Phosphorylation by casein kinase I promotes the turnover of the Mdm2 oncoprotein via the SCF(beta-TRCP) ubiquitin ligase. *Cancer Cell* 18, 147–159. doi: 10.1016/j.ccr.2010.06.015

Jaffe, E. S., Harris, N. L., Stein, H., and Vardiman, J. W. (2001). *World Health Organization Classification of Tumours: Pathology and Genetics of Tumours of Haematopoietic and Lymphoid Tissues.* Lyon: IARC Press.

Jaras, M., Miller, P. G., Chu, L. P., Puram, R. V., Fink, E. C., Schneider, R. K., et al. (2014). Csnk1a1 inhibition has p53-dependent therapeutic efficacy in acute myeloid leukemia. *J. Exp. Med.* 211, 605–612. doi: 10.1084/jem.20131033

Kamesaki, H., Fukuhara, S., Tatsumi, E., Uchino, H., Yamabe, H., Miwa, H., et al. (1986). Cytochemical, immunologic, chromosomal, and molecular genetic analysis of a novel cell line derived from Hodgkin's disease. *Blood* 68, 285–292.

Kanzler, H., Kuppers, R., Hansmann, M. L., and Rajewsky, K. (1996). Hodgkin and Reed-Sternberg cells in Hodgkin's disease represent the outgrowth of a dominant tumor clone derived from (crippled) germinal center B cells. *J. Exp. Med.* 184, 1495–1505.

Klein, E., Klein, G., Nadkarni, J. S., Nadkarni, J. J., Wigzell, H., and Clifford, P. (1968). Surface IgM-kappa specificity on a Burkitt lymphoma cell *in vivo* and in derived culture lines. *Cancer Res.* 28, 1300–1310.

Klein, G., Giovanella, B., Westman, A., Stehlin, J. S., and Mumford, D. (1975). An EBV-genome-negative cell line established from an American Burkitt lymphoma; receptor characteristics. EBV infectibility and permanent conversion into EBV-positive sublines by *in vitro* infection. *Intervirology* 5, 319–334.

Kluin-Nelemans, H. C., Limpens, J., Meerabux, J., Beverstock, G. C., Jansen, J. H., De Jong, D., et al. (1991). A new non-Hodgkin's B-cell line (DoHH2) with a chromosomal translocation t(14;18)(q32;q21). *Leukemia* 5, 221–224.

Knecht, H., Bruderlein, S., Wegener, S., Lichtensztejn, D., Lichtensztejn, Z., Lemieux, B., et al. (2010). 3D nuclear organization of telomeres in the Hodgkin cell lines U-HO1 and U-HO1-PTPN1: PTPN1 expression prevents the formation of very short telomeres including "t-stumps." *BMC Cell Biol.* 11:99. doi: 10.1186/1471-2121-11-99

Knippschild, U., Kruger, M., Richter, J., Xu, P., Garcia-Reyes, B., Peifer, C., et al. (2014). The CK1 family: contribution to cellular stress response and its role in carcinogenesis. *Front. Oncol.* 4:96. doi: 10.3389/fonc.2014.00096

Knippschild, U., Milne, D. M., Campbell, L. E., Demaggio, A. J., Christenson, E., Hoekstra, M. F., et al. (1997). p53 is phosphorylated *in vitro* and *in vivo* by the delta and epsilon isoforms of casein kinase 1 and enhances the level of casein kinase 1 delta in response to topoisomerase-directed drugs. *Oncogene* 15, 1727–1736.

Knippschild, U., Wolff, S., Giamas, G., Brockschmidt, C., Wittau, M., Wurl, P. U., et al. (2005). The role of the casein kinase 1 (CK1) family in different signaling pathways linked to cancer development. *Onkologie* 28, 508–514. doi: 10.1159/000087137

Kuppers, R. (2005). Mechanisms of B-cell lymphoma pathogenesis. *Nat. Rev. Cancer* 5, 251–262. doi: 10.1038/nrc1589

Larkin, M. A., Blackshields, G., Brown, N. P., Chenna, R., Mcgettigan, P. A., Mcwilliam, H., et al. (2007). Clustal W and Clustal X version 2.0. *Bioinformatics* 23, 2947–2948. doi: 10.1093/bioinformatics/btm404

Leoncini, L., Raphael, M., Stein, H., Harris, N. L., Jaffe, E. S., and Kluin, P. M. (2008). "Burkitt lymphoma," in *WHO Classification of Tumours of Haematopoietic and Lymphoid Tissues*, eds S. H. Swerdlow, E. Campo, N. L. Harris, E. S. Jaffe, S. A. Pileri, H. Stein, J. Thiele and J. W. Vardiman (Lyon: IARC), 262–264.

Mader, A., Bruderlein, S., Wegener, S., Melzner, I., Popov, S., Muller-Hermelink, H. K., et al. (2007). U-HO1, a new cell line derived from a primary refractory classical Hodgkin lymphoma. *Cytogenet. Genome Res.* 119, 204–210. doi: 10.1159/000112062

Maritzen, T., Lohler, J., Deppert, W., and Knippschild, U. (2003). Casein kinase I delta (CKIdelta) is involved in lymphocyte physiology. *Eur. J. Cell Biol.* 82, 369–378. doi: 10.1078/0171-9335-00323

Mashhoon, N., Demaggio, A. J., Tereshko, V., Bergmeier, S. C., Egli, M., Hoekstra, M. F., et al. (2000). Crystal structure of a conformation-selective casein kinase-1 inhibitor. *J. Biol. Chem.* 275, 20052–20060. doi: 10.1074/jbc.M001713200

Menon, M. P., Pittaluga, S., and Jaffe, E. S. (2012). The histological and biological spectrum of diffuse large B-cell lymphoma in the World Health Organization classification. *Cancer J.* 18, 411–420. doi: 10.1097/PPO.0b013e31826aee97

Milne, D. M., Looby, P., and Meek, D. W. (2001). Catalytic activity of protein kinase CK1 delta (casein kinase 1delta) is essential for its normal subcellular localization. *Exp. Cell Res.* 263, 43–54. doi: 10.1006/excr.2000.5100

Moller, P., Bruderlein, S., Strater, J., Leithauser, F., Hasel, C., Bataille, F., et al. (2001). MedB-1, a human tumor cell line derived from a primary mediastinal large B-cell lymphoma. *Int. J. Cancer* 92, 348–353. doi: 10.1002/ijc.1211

Nacheva, E., Dyer, M. J., Metivier, C., Jadayel, D., Stranks, G., Morilla, R., et al. (1994). B-cell non-Hodgkin's lymphoma cell line (Karpas 1106) with complex translocation involving 18q21.3 but lacking BCL2 rearrangement and expression. *Blood* 84, 3422–3428.

Peifer, C., Abadleh, M., Bischof, J., Hauser, D., Schattel, V., Hirner, H., et al. (2009). 3,4-Diaryl-isoxazoles and -imidazoles as potent dual inhibitors of p38alpha mitogen activated protein kinase and casein kinase 1delta. *J. Med. Chem.* 52, 7618–7630. doi: 10.1021/jm9005127

Perez, D. I., Gil, C., and Martinez, A. (2010). Protein kinases CK1 and CK2 as new targets for neurodegenerative diseases. *Med. Res. Rev.* 31, 924–954. doi: 10.1002/med.20207

Poppema, S., De Jong, B., Atmosoerodjo, J., Idenburg, V., Visser, L., and De Ley, L. (1985). Morphologic, immunologic, enzymehistochemical and chromosomal analysis of a cell line derived from Hodgkin's disease. Evidence for a B-cell origin of Sternberg-Reed cells. *Cancer* 55, 683–690.

Price, M. A. (2006). CKI, there's more than one: casein kinase I family members in Wnt and Hedgehog signaling. *Genes Dev.* 20, 399–410. doi: 10.1101/gad.1394306

Pulvertaft, J. V. (1964). Cytology of Burkitt's Tumour (African Lymphoma). *Lancet* 1, 238–240.

Pulvertaft, J. V. (1965). A study of malignant tumours in Nigeria by short-term tissue culture. *J. Clin. Pathol.* 18, 261–273.

Reed, J. C. (2008). Bcl-2-family proteins and hematologic malignancies: history and future prospects. *Blood* 111, 3322–3330. doi: 10.1182/blood-2007-09-078162

Richter, J., Bischof, J., Zaja, M., Kohlhof, H., Othersen, O., Vitt, D., et al. (2014a). Difluoro-dioxolo-benzoimidazol-benzamides as potent inhibitors of CK1delta and epsilon with Nanomolar inhibitory activity on cancer cell proliferation. *J. Med. Chem.* 57, 7933–7946. doi: 10.1021/jm500600b

Richter, J., Ullah, K., Xu, P., Alscher, V., Blatz, A., Peifer, C., et al. (2014b). Effects of altered expression and activity levels of CK1δ and ε on tumor growth and survival of colorectal cancer patients. *Int. J. Cancer.* doi: 10.1002/ijc.29346. [Epub ahead of print].

Rivers, A., Gietzen, K. F., Vielhaber, E., and Virshup, D. M. (1998). Regulation of casein kinase I epsilon and casein kinase I delta by an *in vivo* futile phosphorylation cycle. *J. Biol. Chem.* 273, 15980–15984.

Schittek, B., and Sinnberg, T. (2014). Biological functions of casein kinase 1 isoforms and putative roles in tumorigenesis. *Mol. Cancer* 13, 231. doi: 10.1186/1476-4598-13-231

Schmitz, R., Stanelle, J., Hansmann, M. L., and Kuppers, R. (2009). Pathogenesis of classical and lymphocyte-predominant Hodgkin lymphoma. *Annu. Rev. Pathol.* 4, 151–174. doi: 10.1146/annurev.pathol.4.110807.092209

Schneider, R. K., Adema, V., Heckl, D., Jaras, M., Mallo, M., Lord, A. M., et al. (2014). Role of casein kinase 1A1 in the biology and targeted therapy of del(5q) MDS. *Cancer Cell* 26, 509–520. doi: 10.1016/j.ccr.2014.08.001

Schneider, U., Schwenk, H. U., and Bornkamm, G. (1977). Characterization of EBV-genome negative "null" and "T" cell lines derived from children with acute lymphoblastic leukemia and leukemic transformed non-Hodgkin lymphoma. *Int. J. Cancer* 19, 621–626.

Stein, H., Warnke, R. A., Chan, W. C., Jaffe, E. S., Chan, J. K. C., Gatter, K. C., et al. (2008). "Diffuse large B-cell lymphoma, not otherwise specified," in *WHO Classification of Tumours of Haematopoietic and Lymphoid Tissues, 4th Edn.*, eds S. H. Swerdlow, E. Campo, N. L. Harris, E. S. Jaffe, S. A. Pileri, H. Stein, J. Thiele and J. W. Vardiman (Lyon: IARC), 233–237.

Stoter, M., Bamberger, A. M., Aslan, B., Kurth, M., Speidel, D., Loning, T., et al. (2005). Inhibition of casein kinase I delta alters mitotic spindle formation and induces apoptosis in trophoblast cells. *Oncogene* 24, 7964–7975. doi: 10.1038/sj.onc.1208941

Stoter, M., Kruger, M., Banting, G., Henne-Bruns, D., and Knippschild, U. (2014). Microtubules depolymerization caused by the CK1 inhibitor IC261 may be not mediated by CK1 blockage. *PLoS ONE* 9:e100090. doi: 10.1371/journal.pone.0100090

Tsai, I. C., Woolf, M., Neklason, D. W., Branford, W. W., Yost, H. J., Burt, R. W., et al. (2007). Disease-associated casein kinase I delta mutation may promote adenomatous polyps formation via a Wnt/beta-catenin independent mechanism. *Int. J. Cancer* 120, 1005–1012. doi: 10.1002/ijc.22368

Tsujimoto, Y., Finger, L. R., Yunis, J., Nowell, P. C., and Croce, C. M. (1984). Cloning of the chromosome breakpoint of neoplastic B cells with the t(14;18) chromosome translocation. *Science* 226, 1097–1099.

Wang, F., Liu, J., Robbins, D., Morris, K., Sit, A., Liu, Y. Y., et al. (2011). Mutant p53 exhibits trivial effects on mitochondrial functions which can be reactivated by ellipticine in lymphoma cells. *Apoptosis* 16, 301–310. doi: 10.1007/s10495-010-0559-8

Yamane, A., Resch, W., Kuo, N., Kuchen, S., Li, Z., Sun, H. W., et al. (2011). Deep-sequencing identification of the genomic targets of the cytidine deaminase AID and its cofactor RPA in B lymphocytes. *Nat. Immunol.* 12, 62–69. doi: 10.1038/ni.1964

Conflict of Interest Statement: The authors declare that the research was conducted in the absence of any commercial or financial relationships that could be construed as a potential conflict of interest.

Biochemical insight into the prion protein family

Danica Ciric and Human Rezaei*

Virologie et Immunologie Moléculaires, Institut National de la Recherche Agronomique, Jouy-en-Josas, France

Edited by:
Jean-Luc Vilotte, National Institute of Agronomical Research, France

Reviewed by:
Franca Fraternali, King's College London, UK
Jose A. Del Rio, Institute for Bioengineering of Catalonia, Spain

***Correspondence:**
Human Rezaei, Institut National de la Recherche Agronomique, Bat. Biotechnologies, Domaine de Vilvert, 78350 Jouy-en-Josas, France
e-mail: human.rezaei@jouy.inra.fr

Prion protein family comprises proteins, which share not only similarity in their primary structure, but also similarity in their fold. These two groups of similarity presume a parceling in their respective biological function through the common biochemical properties. In this review, biochemical and structural similarities of PrP and two other proteins, Doppel and Shadoo, are evocated. Some evidence demonstrating respectively similarity between PrP N-terminal and C-terminal domain with respectively Shadoo and Doppel is presented. We extended primary structure similarity analysis to the other PrP subdomain as 166-176 polyNQ domain and compare it to proteins using aggregation as a support for structural information transference and structural epigenetic. Finally, we questioned if prion protein family have conserved the PrP structural bistability, which should be at the origin of Prion phenomenon and if Prion pathology is not, ultimately, an exaptation of the physiological propensity of PrP to undergo a structural switch and polymerize.

Keywords: shadoo, doppel, PrP, prion protein, amyloid, evolution, dynamic systems

PRNP AND ITS PARALOGS

When prion gene was firstly identified it was named "sine," to indicate scrapie incubation period of the ME7 scrapie agent in mice (Dickinson et al., 1968), later it was shown that this gene corresponds to the murine prion gene, and is linked to the Prn-i gene, which determines incubation time in experimental scrapie (Oesch et al., 1985; Carlson et al., 1986). Human PRN gene locus contains three genes: PRNP, PRND, and a novel gene PRNT (Makrinou et al., 2002). Cellular prion protein (PrP^C) is encoded by PRNP gene as a single copy. This last has been mapped on the mouse chromosome 2, and on chromosome 20 in the human, where it was mapped to band 20p12-3pte (Sparkes et al., 1986). Human and hamster PRNP consists of two exons, with open reading frame (ORF) located in exon 2. In contrast PRNP of mouse, sheep, and rat contain three exons, with the entire ORF located in exon 3 (Lee et al., 1998).

First prion-related gene, Doppel (Dpl) or "downstream prion protein-like gene," was discovered during sequencing of cosmid clones, isolated from a $Prnp^{b/b}$ mouse (I/LnJ-4), due to the effort for characterization of the locus around PRNP (Moore et al., 1999). The gene encoding Dpl labeled PRND is located at the same locus with the PRNP gene. The PRND is a single copy gene and is located on the chromosome 2 in mouse 16 kb downstream from PRNP, and at chromosome 20 in human 27 kb downstream, and 52 kb downstream in ovine (Moore et al., 1999). PRND consists of two exons in human, or 4 and 5 exons in mouse depending on the different splicing. The two major transcripts are encoded by the part of exon 3 and exon 4 (Flicek et al., 2014). Current genomic evidence indicates that Dpl was present in the last common ancestor of tetrapods, but was lost in birds since there divergence from reptiles (Harrison et al., 2010). The third member of PRN locus, PRNT gene, was discovered 3 kb downstream from PRND. Even if these three gens are evolutionary related they show low primary structure homology which could suggest distinct biological function.

A new gene outside of PRN locus was discovered by Premzl in 2003 during an exploration for potential homolog of PrP in the in the NCBI non-redundant protein database. The gene coding for Shadoo protein was labeled SPRN and is located at the chromosome 7 in the mouse it has two exons, but the second exon has ORF. Unique transcript of 3374 bps is translated in to the protein product of 147 residues in mouse (Watts and Westaway, 2007). In the Enseml database 16 ortholog sequences of mouse SPRN gene was published in four classes of *Vertebtates* from bony fishes (*Osteichthyes*), reptiles (*Reptilia*), birds (*Aves*), and mammals (*Mammalia*) (Flicek et al., 2014).

Comparison of predicted amino acid sequences of Sho orthologs showed highly conserved signal peptide responsible for exportation and one Arg-rich repetitive region containing up to six tetra-repeats of consensus XXRG. Moreover, Sho has a hydrophobic region of 20 residues, with strong homology to PrP 106–126 poly Ala segment. Sho's C-terminal domain contains a conserved NXT glycosylation motif and signal peptide predicated for glycophosphotidylinositol (GPI)-anchor attachment (Premzl et al., 2003).

THE EVOLUTIONARY ORIGIN OF PRION GENES

Bioinformatics analyses of PRN loci revealed the evolutionary descent of prion genes from an ancestral ZIP metal ion transporter (Ehsani et al., 2011). During the emergence of metazoa, a cysteine-flanked core domain was inserted, or de novo arose, in a pre-existing ZIP ancestor gene to generate a prion-like ectodomain in a sub-branch of ZIP genes. Approximately a half-billion years later, a genomic insertion of a spliced transcript coding for such a prion-like ZIP ectodomain may have created the prion founder gene (Ehsani et al., 2011). Premzl and colleges were annotated the prion gene family (PrP-GF) in 42 complete eukaryotic genome assemblies, uncovering new genes and pseudo genes. According to this evidence it is likely that the Dpl gene was present in the last common ancestor of *Tetrapoda*, but it was lost in the bird lineage, since its divergence from reptiles. It has been suggested that PRNP and SPRN have evolved from the same ancestral gene into genes that may still share some functions, but may also have also gained new biological roles (Premzl et al., 2004). SPRN gene in mammals and fishes has conserved their genomic position. It is located close to the proximal adjacent gene, encoding a GTP-binding protein (GTP). This gene has tail-to-tail orientation relative to SPRN and it is conserved from fishes to mammals. The other most proximal gene encodes an amine oxidase (AO), is conserved between Fugu (*Arothron sp.)* and mammals, and it has also tail-to-tail orientation with SPRN. The block of three genes, with its conserved gene order (AO–GTP–SPRN) and orientation is an example of conserved contiguity between fishes and mammals, strongly indicates gene orthology. The genes distal to SPRN are not conserved between mouse and human indicating a chromosome rearrangement in either the mouse or human genome (Premzl et al., 2003). The proteins coded by PRNP and his two paralogs PRND and SPRN are grouped in the prion protein family (Prion, Dpl, and Shadoo protein respectively). In contrast, protein product of PRNT gene does not share any distinctive homology with any of proteins of PrP-GF excluding any functional relation (Harrison et al., 2010).

THE PrP PROTEIN

When PrP coding sequences were compared in 26 mammalian species it was found that part of sequence for glycosylation sites, positions of cysteines responsible for formation of disulfide bridge, and sequence for putative hydrophobic transmembrane region (Zhang et al., 1997) are perfectly conserved (Van Rheede et al., 2003). Human PrP precursor protein consists of 253 amino acids. It is processed in the ER and Golgi complex during the transport to the cell surface. In the ER its N-terminal signal sequence of 22 residues is cleaved, as well as 23 residues from C-terminal part, after addition of glycosyl phosphatidylinositol (GPI) anchor (Yusa et al., 2012). PrP is properly folded before transporting to the Golgi complex, where it can be differently glycosylated at N181 and N197 position. In cell surface, PrP^C can exist in unglycosylated, monoglycosylated, and diglycosylated form (Meyer et al., 1986). Mature PrP^C consists of 208 residues (human numbering). It is cell surface glycoprotein, attached to the membrane thought GPI anchor (**Figure 1**). The N-terminal domain of PrP consists of positively charged amino acid sequences and an octapeptide repeat sequence, which can bind copper ions. Middle region between residues 106–126 constitute conserved hydrophobic domain (HD) rich with alanine and valine. This last segment has been reported to be involved in several regulatory processes (Rezaei-Ghaleh et al., 2011; Béland and Roucou, 2012) and have been reported to be able to span membrane (James et al., 1997).

Tertiary structure of PrP globular domain has been resolved by NMR in 1996 (Riek et al., 1996) (**Figure 2A**). Since this first 3D structure, PrP tertiary structure of several other mammalian species have been resolved (Lysek et al., 2005). All of them, revealed similar and highly conserved fold. Moreover, the comparison between extractive PrP^C, purified from bovine brain, and bovine recPrP, produced in *E.coli*, revealed similar fold, meaning

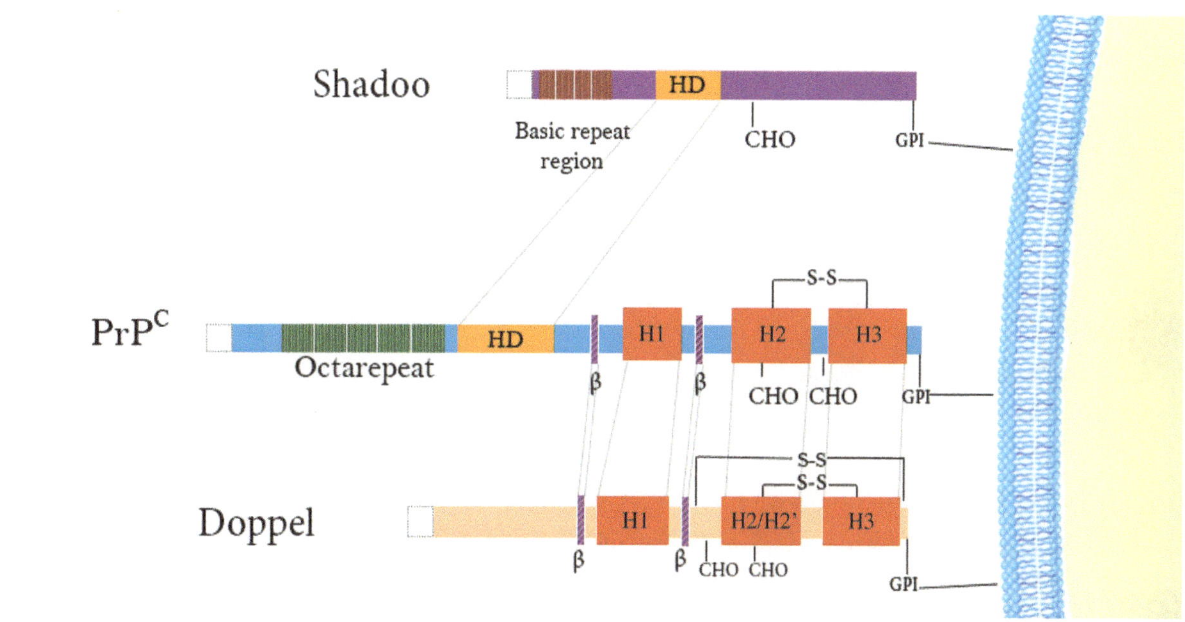

FIGURE 1 | Schematic representation of the domain architecture of PrP family members. Doppel and PrP^C have structured C-terminal domains consisting of 3 α-helices (in red) and 2 short β-strands (in purple) as well as positively-charged N-terminal regions. Disulfide bridges are indicated above the proteins (-S–S-) and N-glycosylation sites (CHO) are denoted below the proteins. Repetitive regions are found in both PrP^C (green) and Shadoo (dark red) with the former possessing octarepeats binding copper and the latter possessing tetrarepeats rich in arginine and glycine residues. A conserved hydrophobic domain is also illustrated in PrP^C and Shadoo (yellow) N-terminal signal sequences are cleaved in all of them (white).

that glycosylation and GPI anchor did not affect general PrP fold (Hornemann et al., 2004). The PrP 3D structure reveals a globular domain, which contains three α-helices comprising the residues 144-154 (H1), 173-194 (H2), and 200-228 (H3) and a short anti-parallel β-sheet comprising the residues 128–131 (β1) and 161–164 (β2). Within the globular domain there is three loops, between residues 167–171, at the end of H2 residues 187–194, and in the C-terminal part of H3 residues 219–228. PrP have a disulfide bridge between H2 and H3 helixes. The reduction of this S–S bond *in vitro* has been reported to be at origin of structural switch and formation of amyloid fibrils (Jackson et al., 1999). Moreover, it was demonstrated that the H2H3 segment constitutes an independent folding unit (Adrover et al., 2010; Xu et al., 2011) (**Figure 2B**).

DOPPEL PROTEIN

PRND gene coding Dpl a 179 residues protein, sharing 25% identity with PrP globular domain. As PrP, Dpl has a cell surface exportation amino acids signal sequence at its N-terminus (1–27 residues) and GPI anchoring signal (from 156 to 179), at its C-terminal domain. Dpl is processed in the ER and Golgi as PrP and has two glycosylation sites, one at the 111 residues in the form N-X-T occurring and second non-conserved N-V-T Asn-linked glycosylation site at residue 99 (**Figure 3A**). The GPI anchor is predict to be attached at Gly155. The Dpl, as PrP, is attached to outer cell surface trough GPI anchored (Silverman et al., 2000). Despite the fact that both Dpl and PrP^C are attached to the rafts, it was reported that they are attached to distinct microenvironments and not in the same raft domains. This observation

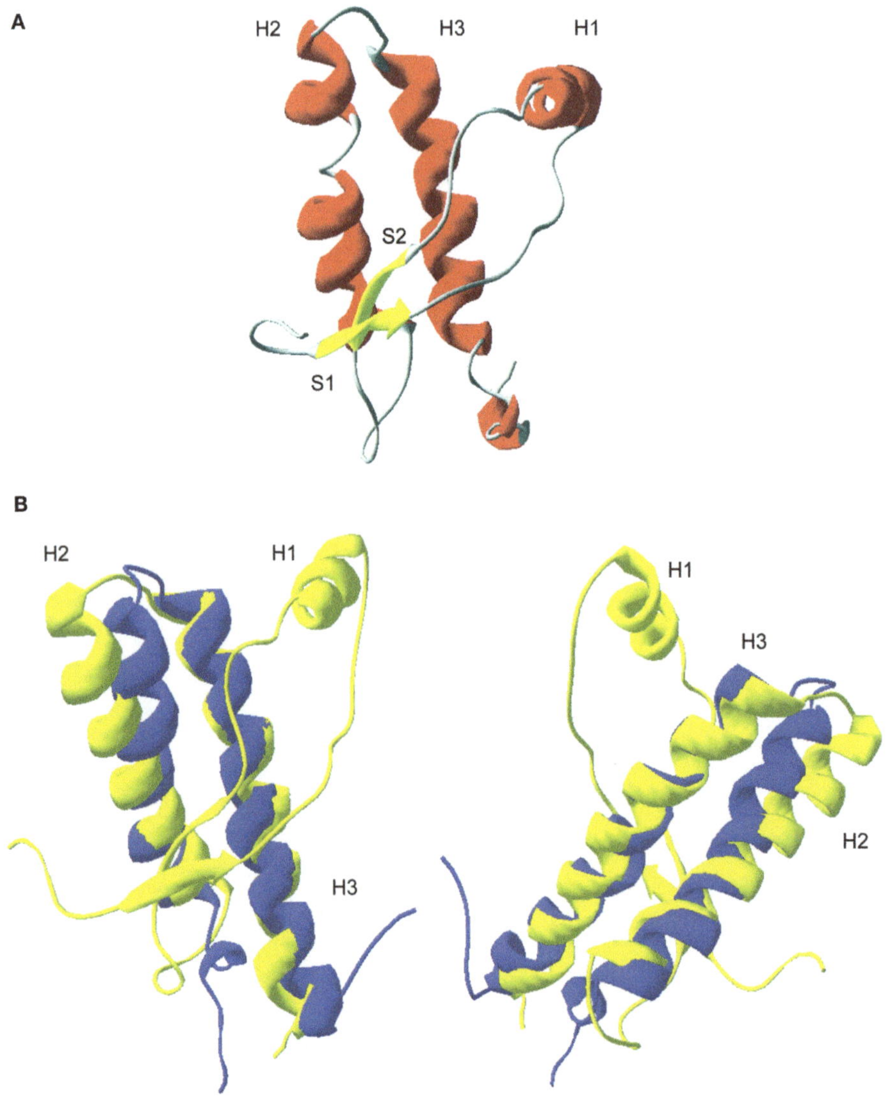

FIGURE 2 | 3D structure of residue 124-231 of recombinant PrP. (A) Ribbon representation of the C-terminal domain of murine PrP (PDB: 1AG2). H1, H2, and H3 represent respectively helix 1, helix 2, and helix 3. S1 and S2, β-sheet 1 and 2. **(B)** Superposition of ovine ARQ structure (in yellow, PDB: ITPX) with the structure of purified ovine H2H3 domain (in blue, residue 169–234, PDB: 2KTM). Left and right represent two different view of the superimposition. As shown, the H2H3 domain adopts quasi-similar fold the full-length globular domain indicating that segment S1-S2-H1 does not contribute the H2H3 folding.

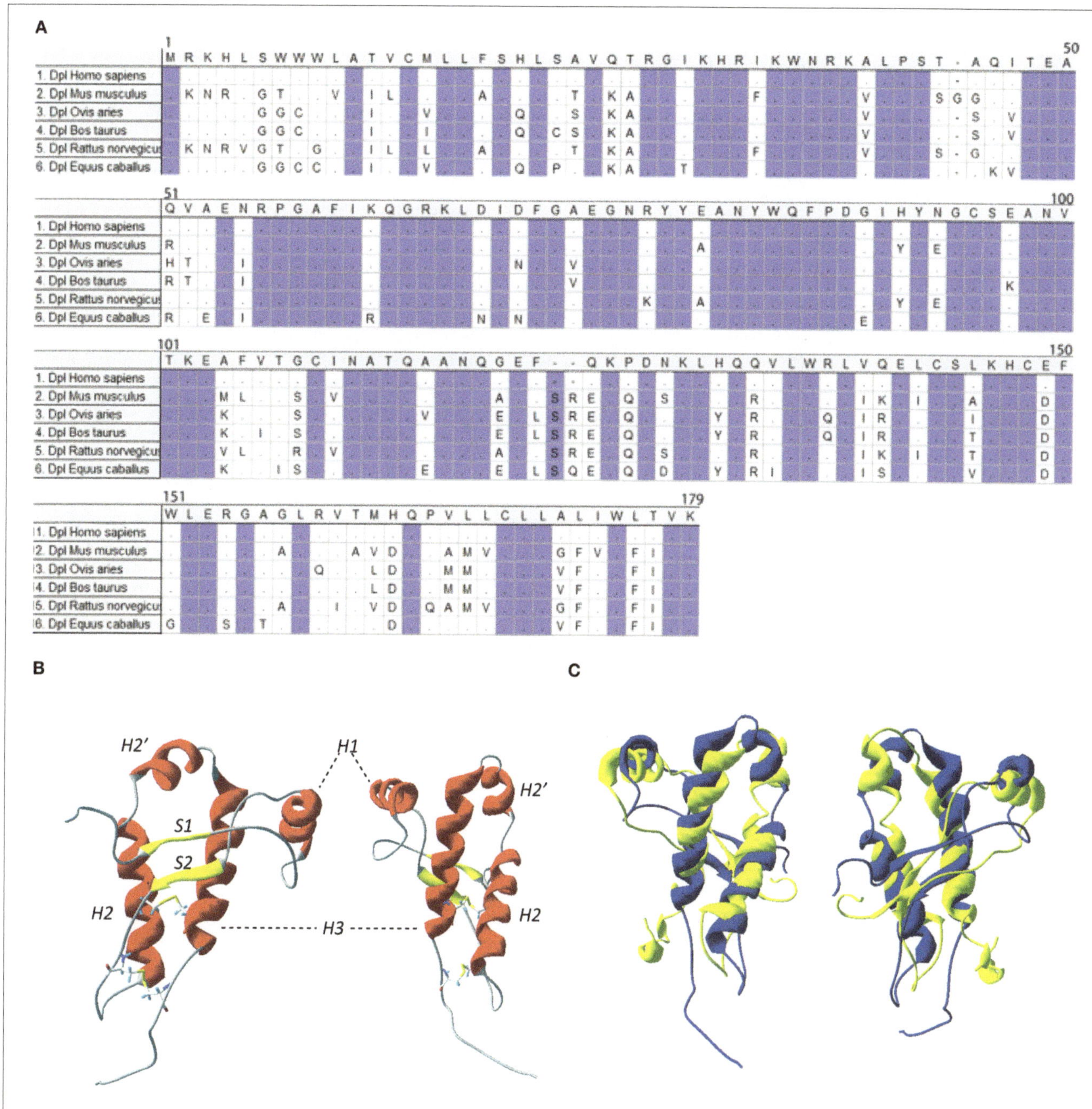

FIGURE 3 | (A) Protein sequence of Doppel. Multiple sequence alignment for Doppel protein sequences of different species of vertebrates. Highlighted conserved domains are in violet. Dots refers to identical residues, (-) represents gaps. **(B)** Two different view of mouse Dpl structure (PDB:1I17). **(C)** Two different views of the superimposition of the mouse PrP structure (in yellow, PDB: 1AG2) and mouse Dpl structure (in blue, PDB: 1I17).

makes the authors to propose that even though PrP and Dpl share some identities they might have distinct functions (Shaked et al., 2002).

NMR structure of the recombinant human Dpl protein shows a short flexibly disordered N-terminal domain comprising residues 24–51, and a globular domain extending from residues 52–149 (**Figure 3B**). The globular domain contains four α-helices comprising residues 72–80 (α1), 101–115 (α2a), 117–121 (α2b), and 127–141 (α3), and a short two-stranded anti-parallel β-sheet comprising residues 58–60 (β1) and 88–90 (β2). The C-terminal peptide segment 144–149 folds back onto the loop connecting β2 and α2 (Lührs et al., 2003). Dpl has two disulfide bonds (Cys95–Cys148 and Cys109–Cys143), in contrast to the PrP who has one. This additional disulfide bound could contribute to restriction of Dpl conformational dynamics, making it more rigid. Even if the PrP and Dpl present less than 25% of homology, their folds are similar (**Figure 3C**).

SHADOO PROTEIN

Nascent protein product encoded by SPRN ORF is subjected to the similar type of processing events as the other members of PrP family. The mature mouse Sho consists of 98 amino acids it is a GPI-anchored protein (Premzl et al., 2003; Watts et al., 2007). Sho also has functional ER targeting signal sequence, and it can be modified with complex glycans and targeted to the outer leaflet of the plasma membrane (Miesbauer et al., 2006). N-terminal segment, from residues 25-42 has very strictly conserved sequence (**Figure 4A**), with conserved methylation site at Arg27 (GGRGG). In human Sho the RGG repeats are consist of 15 residues: Arg residues are positively charged at physiological pH and 7 Gly residues, 6 of them grouped as dipeptides "GG" giving a high degree of flexibility to the backbone. The increased prevalence of Gly-Gly dipeptides, in the higher Eutherian mammals, could suggest evolutionary pressure for increased flexibility in this domain. In addition, proteins with an RGG-box motif can have RNA-binding function (Thandapani et al., 2013). It

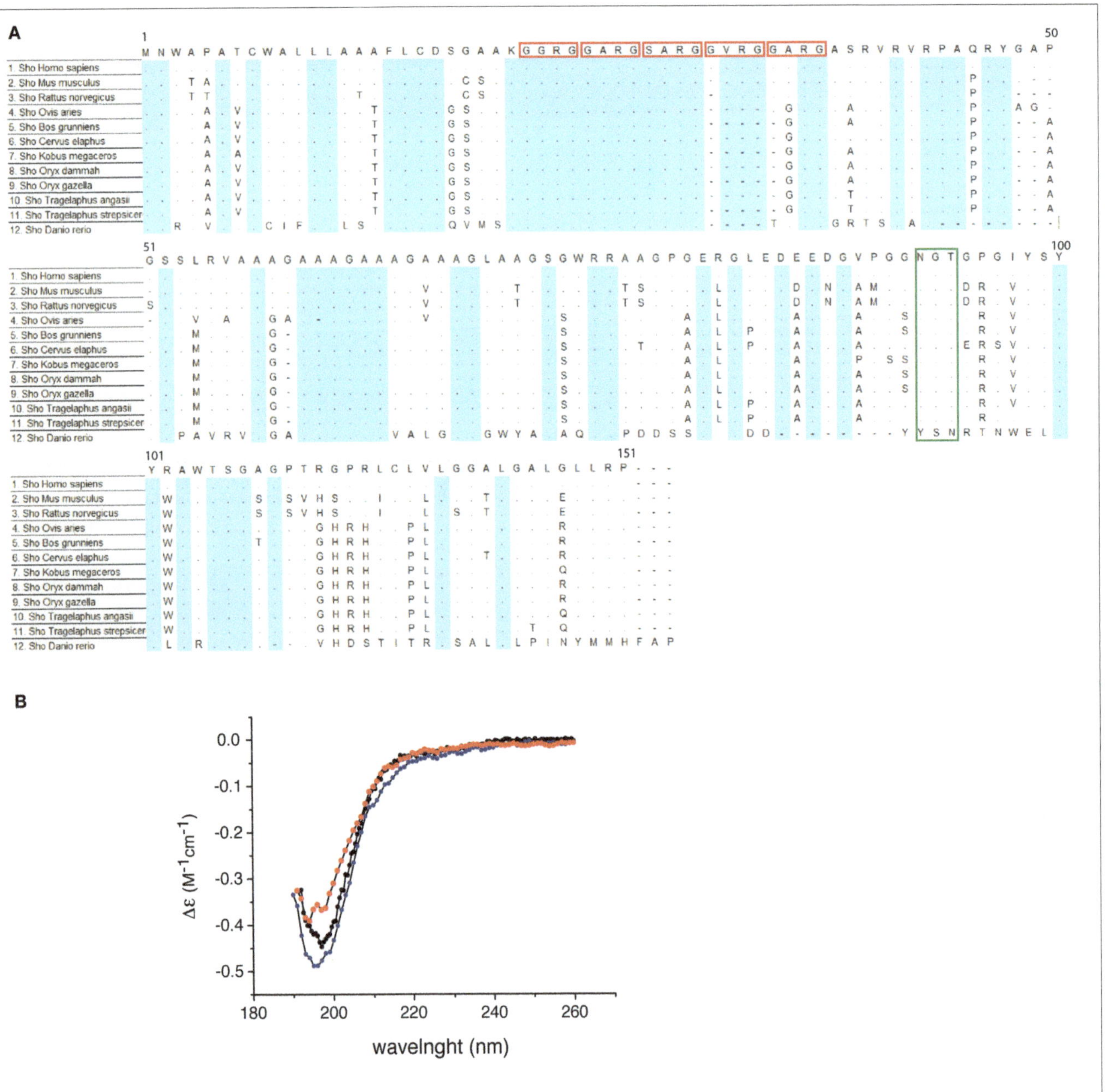

FIGURE 4 | (A) Conservation of Sho protein sequences. Alignment of Shadoo protein sequences in 12 species. Multiple alignments of protein referent sequences for Shadoo in vertebrates. In blue color are highlighted conserved domains. Red squares indicate the consensus XXRG repletion. The green box indicates the NXT N-glycosylation signal. The alignment was obtained and color coded by MEGA5 software, dots refers to identical residues, (-) represents gaps, according to MEGA5 convention. **(B)** Circular dichroism (CD) spectra of Sho at pH 3.8, 4.5, and 7.0. CD spectra indicate that Sho adopts a random coil configuration for 3.8 ≤ pH ≤ 7.0.

is suggested that Sho could bind mRNA directly (Corley and Gready, 2008) and thus play a role in neural plasticity as PrP, through his involvement in neural signaling pathways (Kanaani et al., 2005; Santuccione et al., 2005). HD in the middle of the protein is arranged in five tandem repeats (**Figure 4**), which consist of GxxxG motifs. These motifs can have a role in organization of transmembrane helixes and packaging of amyloidal fibers (Russ and Engelman, 2000). HD has a high degree of homology with PrP 106–126 domain. The primary structure analysis of Human Sho reveals the existence of a one putative N-glycosilation site (N111). Moreover, the different predication algorithms highly suggest that Sho could be devoid of secondary structure. The CD spectra of recombinant mouse Sho at different pH strongly suggest that Sho adopts a random coil structuration (**Figure 4B**). These observations strongly suggest that Sho belong to intrinsically disordered proteins (IDP) family.

STRUCTURAL DYNAMIC OF PRION PROTEIN FAMILY

In early 90th when conventional descriptions of prion pathology failed to describe prion disease, a new theory considering an infectious protein emerged (Griffith, 1967; Prusiner et al., 1982; Prusiner, 1998). The prion theory, now largely extended, stipulate an autonomous structural rearrangement of PrP^{C} into PrP^{Sc} conformer. The prion theory explicitly requires that PrP protein should exist, at least, in two conformations. However, the existence of several strains, for a given PrP primary structure, points out that PrP could exist not only in two conformations, but as plethora of conformations, each associated with a physiopathological state. The fact that PrP could adopt different conformations, in the quasi-similar environments, makes PrP a multi-stable protein, a hallmark of its plasticity. Therefore, one can question why PrP protein, in particular, kept all along the evolution such structural plasticity? The primary structure analysis of prion protein family, could highlight this point. Indeed, divergence of PRNP, PRND, and SPRN conduce to specific differences in PrP, Dpl, and Sho proteins. The existence of additional disulfide bridge in Dpl (**Figure 3**) highly reduces his plasticity, and therefore reduces Dpl's propensity to adopt several conformations. In terms of PrP-GF evolution, one can consider that additional disulfide bond has been selected to reduce Dpl conformational dynamic and hence, it's self-propensity to undergo a deleterious structural switch. A similar rational could be constructed for Sho. This last protein, during the selection process have discarded segments reported to be involved in the conformational switch of PrP, as it is the case of PrP globular part.

Another aspect of PRNP gene evolution, in relation with PrP structural landscape, should be also evocated. Compared to Sho and Dpl, the primary structure of PrP is highly conserved among the mammals (**Figure 5**). This high conservation could be either a hallmark of its folding, linked to its biological function, or linked to the fact that the amount of mutations and variation in the primary structure of PrP could lead to the appearance of deleterious

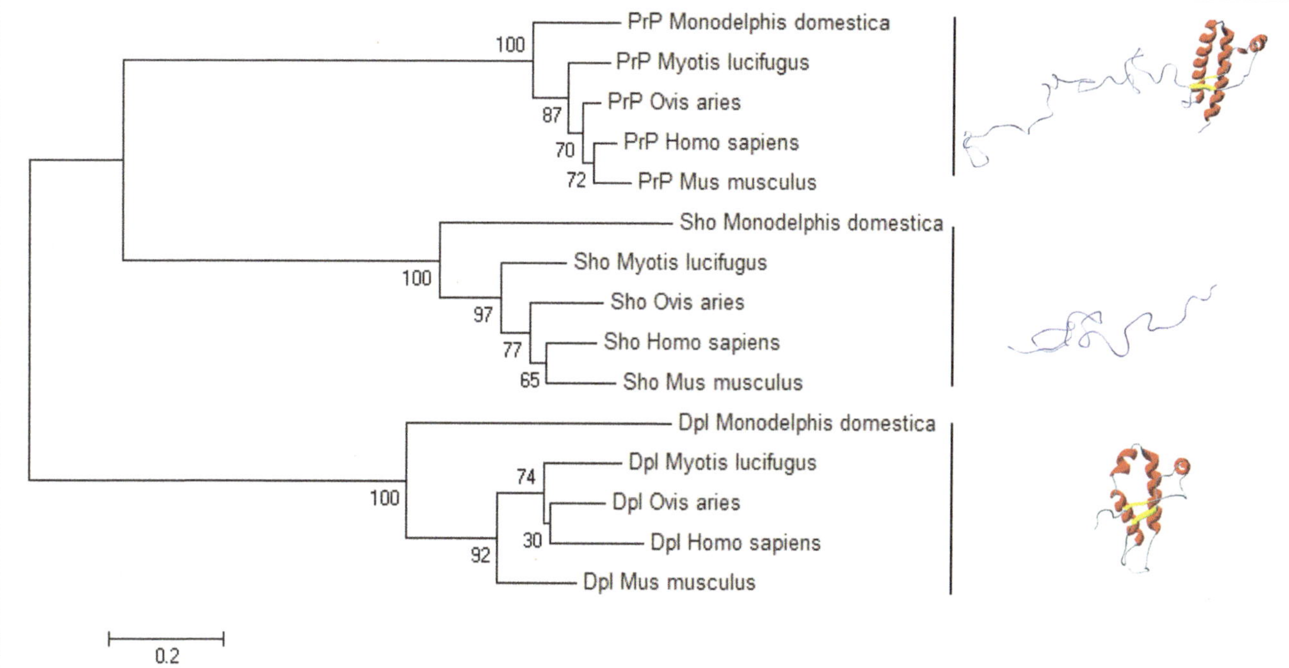

FIGURE 5 | The bootstrap consensus tree for Prion protein family. We used MEGA6 to infer the NJ phylogenetic tree in five species of mammals, using the pairwise deletion option for Shadoo, Doppel, and Prion protein sequences retrieved from NCBI Protein database: Mus musculus Dpl (NP_001265449.1), Mus musculus Sho (NP_898970.1), Mus musculus PrP (NP_001265185.1); Ovis aries Dpl (NP_001009261.1), Ovis aries Sho (NP_001156033.1), Ovis aries PrP (NP_001009481.1); Myotis lucifugus Dpl (XP_006096490.1), Myotis lucifugus Sho (XP_006099074.1), Myotis lucifugus PrP (XP_006096489.1); Monodelphis domestica Dpl (CAJ75789.1), Monodelphis domestica Sho (CAJ43800.1), Monodelphis domestica PrP (NP_001035117.1); Homo sapiens Dpl (NP_036541.2), Homo sapiens Sho (NP_001012526.2), Homo sapiens Prp (NP_001073592.1). The horizontal bar indicates 0.2 amino-acid substitution per site.

events, as it is the spontaneous conversion. This last phenomenon could be at the origin of PrP mutations responsible of the occurrence of GSS, FFI, and genetic CJD. To better understand this low variability of PrP primary structure we have to consider PrP primary structure in relation with its putative biological function. PrP protein could be segmented into two domains, the N-terminal domain, highly flexible, binding Cu^{2+} ions and the globular domain. So far, most of the PrP biological functions are reported to involve only N-terminal segment. Contribution of the globular domain in the PrP physiological function is not well understood (Béland and Roucou, 2012). The question may arise: why during the mammalian evolution, the propensity of the PrP globular domain to misfold, has not been suppressed, if PrP biological function is only restricted to the N-terminal domain? The answer to this question could be an entanglement between PrP folding and biological function. Indeed, if we consider that PrP biological function is intertwined with its structural bistability in a highly controlled process, evolution should manage with both PrP bistability and the homeostasis of this bistability. One of direct consequences of this hypothesis is that highly controlled PrP conversion could have a physiological role, and that prion pathology could emerge as a breakdown of homeostasis of PrP physiological conversion process, induced by appearance PrP^{Sc}.

REFERENCES

Adrover, M., Pauwels, K., Prigent, S., de Chiara, C., Xu, Z., Chapuis, C., et al. (2010). Prion fibrillization is mediated by a native structural element that comprises helices H2 and H3. *J. Biol. Chem.* 285, 21004–21012. doi: 10.1074/jbc.M110.111815

Béland, M., and Roucou, X. (2012). The prion protein unstructured N-terminal region is a broad-spectrum molecular sensor with diverse and contrasting potential functions. *J. Neurochem.* 120, 853–868. doi: 10.1111/j.1471-4159.2011.07613.x

Carlson, G. A., Kingsbury, D. T., Goodman, P. A., Coleman, S., Marshall, S. T., DeArmond, S., et al. (1986). Linkage of prion protein and scrapie incubation time genes. *Cell* 46, 503–511. doi: 10.1016/0092-8674(86)90875-5

Corley, S. M., and Gready, J. E. (2008). Identification of the RGG box motif in Shadoo: RNA-binding and signaling roles? *Bioinform. Biol. Insights* 2, 383–400.

Dickinson, A. G., Meikle, V. M. H., and Fraser, H. (1968). Identification of a gene which controls the incubation period of some strains of scrapie agent in mice. *J. Comp. Pathol.* 78, 293–299. doi: 10.1016/0021-9975(68)90005-4

Ehsani, S., Tao, R., Pocanschi, C. L., Ren, H., Harrison, P. M., and Schmitt-Ulms, G. (2011). Evidence for retrogene origins of the prion gene family. *PLoS ONE* 6:e26800. doi: 10.1371/journal.pone.0026800

Flicek, P., Amode, M. R., Barrell, D., Beal, K., Billis, K., Brent, S., et al. (2014). Ensembl 2014. *Nucleic Acids Res.* 42, D749–D755. doi: 10.1093/nar/gkt1196

Griffith, J. S. (1967). Self-replication and scrapie. *Nature* 215, 1043–1044. doi: 10.1038/2151043a0

Harrison, P. M., Khachane, A., and Kumar, M. (2010). Genomic assessment of the evolution of the prion protein gene family in vertebrates. *Genomics* 95, 268–277. doi: 10.1016/j.ygeno.2010.02.008

Hornemann, S., Schorn, C., and Wüthrich, K. (2004). NMR structure of the bovine prion protein isolated from healthy calf brains. *EMBO Rep.* 5, 1159–1164. doi: 10.1038/sj.embor.7400297

Jackson, G. S., Hosszu, L. L., Power, A., Hill, A. F., Kenney, J., Saibil, H., et al. (1999). Reversible conversion of monomeric human prion protein between native and fibrilogenic conformations. *Science* 283, 1935–1937. doi: 10.1126/science.283.5409.1935

James, T. L., Liu, H., Ulyanov, N. B., Farr-Jones, S., Zhang, H., Donne, D. G., et al. (1997). Solution structure of a 142-residue recombinant prion protein corresponding to the infectious fragment of the scrapie isoform. *Proc. Natl. Acad. Sci.* 94, 10086–10091. doi: 10.1073/pnas.94.19.10086

Kanaani, J., Prusiner, S. B., Diacovo, J., Baekkeskov, S., and Legname, G. (2005). Recombinant prion protein induces rapid polarization and development of synapses in embryonic rat hippocampal neurons *in vitro*. *J. Neurochem.* 95, 1373–1386. doi: 10.1111/j.1471-4159.2005.03469.x

Lee, I. Y., Westaway, D., Smit, A. F. A., Wang, K., Seto, J., Chen, L., et al. (1998). Complete genomic sequence and analysis of the prion protein gene region from three mammalian species. *Genome Res.* 8, 1022–1037.

Lührs, T., Riek, R., Güntert, P., and Wüthrich, K. (2003). NMR structure of the human doppel protein. *J. Mol. Biol.* 326, 1549–1557. doi: 10.1016/S0022-2836(02)01471-7

Lysek, D. A., Schorn, C., Nivon, L. G., Esteve-Moya, V., Christen, B., Calzolai, L., et al. (2005). Prion protein NMR structures of cats, dogs, pigs, and sheep. *Proc. Natl. Acad. Sci. U.S.A.* 102, 640–645. doi: 10.1073/pnas.0408937102

Makrinou, E., Collinge, J., and Antoniou, M. (2002). Genomic characterization of the human prion protein (PrP) gene locus. *Mamm. Genome* 13, 696–703. doi: 10.1007/s00335-002-3043-0

Meyer, R. K., McKinley, M. P., Bowman, K. A., Braunfeld, M. B., Barry, R. A., and Prusiner, S. B. (1986). Separation and properties of cellular and scrapie prion proteins. *Proc. Natl. Acad. Sci. U.S.A.* 83, 2310–2314. doi: 10.1073/pnas.83.8.2310

Miesbauer, M., Bamme, T., Riemer, C., Oidtmann, B., Winklhofer, K. F., Baier, M., et al. (2006). Prion protein-related proteins from zebrafish are complex glycosylated and contain a glycosylphosphatidylinositol anchor. *Biochem. Biophys. Res. Commun.* 341, 218–224. doi: 10.1016/j.bbrc.2005.12.168

Moore, R. C., Lee, I. Y., Silverman, G. L., Harrison, P. M., Strome, R., Heinrich, C., et al. (1999). Ataxia in prion protein (PrP)-deficient mice is associated with upregulation of the novel PrP-like protein doppel. *J. Mol. Biol.* 292, 797–817. doi: 10.1006/jmbi.1999.3108

Oesch, B., Westaway, D., Wälchli, M., McKinley, M. P., Kent, S. B., Aebersold, R., et al. (1985). A cellular gene encodes scrapie PrP 27-30 protein. *Cell* 40, 735–746. doi: 10.1016/0092-8674(85)90333-2

Premzl, M., Gready, J. E., Jermiin, L. S., Simonic, T., Graves, J. A. M., and Marshall Graves, J. A. (2004). Evolution of vertebrate genes related to prion and Shadoo proteins–clues from comparative genomic analysis. *Mol. Biol. Evol.* 21, 2210–2231. doi: 10.1093/molbev/msh245

Premzl, M., Sangiorgio, L., Strumbo, B., Marshall Graves, J. A., Simonic, T., and Gready, J. E. (2003). Shadoo, a new protein highly conserved from fish to mammals and with similarity to prion protein. *Gene* 314, 89–102. doi: 10.1016/S0378-1119(03)00707-8

Prusiner, S. B. (1998). Nobel lecture: prions. *Proc. Natl. Acad. Sci. U.S.A.* 95, 13363–13383. doi: 10.1073/pnas.95.23.13363

Prusiner, S. B., Bolton, D. C., Groth, D. F., Bowman, K. A., Cochran, S. P., and McKinley, M. P. (1982). Further purification and characterization of scrapie prions. *Biochemistry* 21, 6942–6950. doi: 10.1021/bi00269a050

Rezaei-Ghaleh, N., Andreetto, E., Yan, L. M., Kapurniotu, A., and Zweckstetter, M. (2011). Interaction between amyloid beta peptide and an aggregation blocker peptide mimicking islet amyloid polypeptide. *PLoS ONE* 6: e20289. doi: 10.1371/journal.pone.0020289

Riek, R., Hornemann, S., Wider, G., Billeter, M., Glockshuber, R., and Wüthrich, K. (1996). NMR structure of the mouse prion protein domain PrP(121-231). *Nature* 382, 180–182. doi: 10.1038/382180a0

Russ, W. P., and Engelman, D. M. (2000). The GxxxG motif: a framework for transmembrane helix-helix association. *J. Mol. Biol.* 296, 911–919. doi: 10.1006/jmbi.1999.3489

Santuccione, A., Sytnyk, V., Leshchyns'ka, I., and Schachner, M. (2005). Prion protein recruits its neuronal receptor NCAM to lipid rafts to activate p59fyn and to enhance neurite outgrowth. *J. Cell Biol.* 169, 341–354. doi: 10.1083/jcb.200409127

Shaked, Y., Hijazi, N., and Gabizon, R. (2002). Doppel and PrP(C) do not share the same membrane microenvironment. *FEBS Lett.* 530, 85–88. doi: 10.1016/S0014-5793(02)03430-0

Silverman, G. L., Qin, K., Moore, R. C., Yang, Y., Mastrangelo, P., Tremblay, P., et al. (2000). Doppel is an N-glycosylated, glycosylphosphatidylinositol-anchored protein. Expression in testis and ectopic production in the brains of Prnp(0/0) mice predisposed to Purkinje cell loss. *J. Biol. Chem.* 275, 26834–26841. doi: 10.1074/jbc.M003888200

Sparkes, R. S., Simon, M., Cohn, V. H., Fournier, R. E., Lem, J., Klisak, I., et al. (1986). Assignment of the human and mouse prion protein genes to homologous chromosomes. *Proc. Natl. Acad. Sci.* 83, 7358–7362. doi: 10.1073/pnas.83.19.7358

Thandapani, P., O'Connor, T. R., Bailey, T. L., and Richard, S. (2013). Defining the RGG/RG motif. *Mol. Cell* 50, 613–623. doi: 10.1016/j.molcel.2013.05.021

Van Rheede, T., Smolenaars, M. M. W., Madsen, O., and de Jong, W. W. (2003). Molecular evolution of the mammalian prion protein. *Mol. Biol. Evol.* 20, 111–121. doi: 10.1093/molbev/msg014

Watts, J. C., Drisaldi, B., Ng, V., Yang, J., Strome, B., Horne, P., et al. (2007). The CNS glycoprotein Shadoo has PrP(C)-like protective properties and displays reduced levels in prion infections. *EMBO J.* 26, 4038–4050. doi: 10.1038/sj.emboj.7601830

Watts, J. C., and Westaway, D. (2007). The prion protein family: diversity, rivalry, and dysfunction. *Biochim. Biophys. Acta* 1772, 654–672. doi: 10.1016/j.bbadis.2007.05.001

Xu, Z., Prigent, S., Deslys, J.-P., and Rezaei, H. (2011). Dual conformation of H2H3 domain of prion protein in mammalian cells. *J. Biol. Chem.* 286, 40060–40068. doi: 10.1074/jbc.M111.275255

Yusa, S., Oliveira-Martins, J. B., Sugita-Konishi, Y., and Kikuchi, Y. (2012). Cellular prion protein: from physiology to pathology. *Viruses* 4, 3109–3131. doi: 10.3390/v4113109

Zhang, H., Stockel, J., Mehlhorn, I., Groth, D., Baldwin, M. A., Prusiner, S. B., et al. (1997). Physical studies of conformational plasticity in a recombinant prion protein. *Biochemistry* 36, 3543–3553. doi: 10.1021/bi961965r

Conflict of Interest Statement: The Associate Editor Jean-Luc Vilotte declares that, despite being affiliated to the same institution as the author Human Rezaei, the review process was handled objectively and no conflict of interest exists. The authors declare that the research was conducted in the absence of any commercial or financial relationships that could be construed as a potential conflict of interest.

The regulation of MMP targeting to invadopodia during cancer metastasis

Abitha Jacob and Rytis Prekeris*

Department of Cell and Developmental Biology, School of Medicine, Anschutz Medical Campus, University of Colorado Denver, Aurora, CO, USA

Edited by:
Irina Kaverina, Vanderbilt University Medical Center, USA

Reviewed by:
Frederique Gaits-Iacovoni, Institut National de la Santé et de la Recherche Médicale U1048, France
Sébastien Roger, University of Tours, France

Correspondence:
Rytis Prekeris, Department of Cell and Developmental Biology, School of Medicine, Anschutz Medical Campus, University of Colorado Denver, 12801 East 17th Avenue, Bldg. RC-1, Room L18-12402, Aurora, CO, USA
e-mail: Rytis.Prekeris@ucdenver.edu

The dissemination of cancer cells from the primary tumor to a distant site, known as metastasis, is the main cause of mortality in cancer patients. Metastasis is a very complex cellular process that involves many steps, including the breaching of the basement membrane (BM) to allow the movement of cells through tissues. The BM breach occurs via highly regulated and localized remodeling of the extracellular matrix (ECM), which is mediated by formation of structures, known as invadopodia, and targeted secretion of matrix metalloproteinases (MMPs). Recently, invadopodia have emerged as key cellular structures that regulate the metastasis of many cancers. Furthermore, targeting of various cytoskeletal modulators and MMPs has been shown to play a major role in regulating invadopodia function. Here, we highlight recent findings regarding the regulation of protein targeting during invadopodia formation and function.

Keywords: MMP9, MMP2, Rab40B, membrane traffic, invadopodia, cancer, metastasis

INTRODUCTION

Although epithelial cancers are one of the leading causes of death, the mechanisms regulating the development and metastasis of carcinomas are not fully understood. Multiple studies suggest that the progression of tumors is dependent on the intrinsic properties of cancer cells, such as their ability to migrate and invade. Furthermore, many extrinsic factors, such as extracellular matrix (ECM) proteins, are also crucial for regulation of cancer metastasis. The ECM proteins that make up the specialized basement membrane (BM) serve as a barrier for cell invasion. However, the BM which is rich in laminin and collagen IV, also provides the substrate for adhesion of the migrating tumor cells. Furthermore, BM degradation results in the release/activation of various growth factors required for angiogenesis, tumor growth, and metastasis (Kalluri, 2003; Yurchenco, 2011). ECM degrading enzymes known as matrix metalloproteinases (MMPs) are known to play important roles in the degradation of the BM. Since several excellent reviews have already described the importance of MMPs in cancer cell growth and metastasis (Egeblad and Werb, 2002; Deryugina and Quigley, 2006; Fingleton, 2006; Gialeli et al., 2011), this review will focus on the mechanisms governing the targeting of MMPs to invadopodia.

THE ROLE OF THE FAMILY OF MATRIX METALLOPROTEINASES (MMPs) IN CANCER CELL METASTASIS

BM disruption involves a localized degradation of the ECM via the secretion of MMPs (Chamber and Matrisian, 1997). MMPs are a family of zinc endopeptidases that cleave ECM molecules and are subdivided into categories depending on their substrate specificity. The MMP family of enzymes includes not only the classical secreted and membrane bound MMPs, but also ADAMs (a disintegrin and metalloproteinase). ADAM metalloproteinases, also known as sheddases, are involved in cleaving various growth factors, cytokines, receptors, and adhesion molecules and are fundamental to development and homeostasis (Edwards et al., 2008). Like ADAMs, MMPs are also required for normal processes like tissue remodeling in embryonic development, wound healing, involution of mammary glands, angiogenesis, and ossification (Woessner, 1991). However, high levels of MMPs or aberrant MMP expression have often been correlated with pathological conditions like periodontitis, arthritis (Woessner, 1991) and have been implicated in multiple stages of cancer progression including invasion and metastasis (Egeblad and Werb, 2002). In this review, we will focus on the canonical MMPs, more specifically the ones that are targeted to the invadopodia and implicated in BM remodeling during metastasis of epithelial cancers. The association of MMPs with malignancies has been well documented with the majority of the evidence derived from mouse model studies and analysis of human patient samples. Based on substrate recognition, MMPs are categorized into interstitial collagenases, gelatinases, stromelysins, and membrane bound MMPs. Out of the 28 known MMPs, 14 have been implicated in cancer development and progression (Kohrmann et al., 2009). It has been shown that elevated expression of MMP1, 2, 3, 7, 9, 13, and 14 is positively correlated with tumor progression, metastasis, and poor overall prognosis (Lochter et al., 1998; Kerkela and Saarialho-Kere, 2003; Mook et al., 2004; Wagenaar-Miller et al., 2004; Ala-aho and Kahari, 2005; Bjorklund and Koivunen, 2005; Hofmann et al., 2005). Recently, it was shown that MMP9 drives tumor progression and metastasis of triple negative breast cancer (Mehner et al., 2014) and increased expression of MMP9 has been found in the early steps of melanoma (van den Oord et al., 1997). Expression of MMP9 has been associated both positively and negatively with survival rates in breast and colon cancer patients (Zeng et al., 1996; Takeha et al., 1997; Pacheco et al., 1998; Scorilas et al., 2001). Additionally, cancer cells have lesser capability to colonize the lungs of MMP2 or MMP9 deficient mice compared to wild type mice (Itoh et al., 1998, 1999) and cancer cell proliferation is decreased in tumors obtained from MMP9 knock-out mice (Bergers et al., 2000; Coussens et al., 2000). Overexpression of MMP3 and MMP14 (also known as MT1-MMP) has been shown to promote mammary carcinogenesis (Sternlicht et al., 1999; Ha et al., 2001). MMP12 expression in colon cancer has been correlated with increased survival (Yang et al., 2001). These studies indicate that several MMPs play a key role in cancer growth and metastasis. However, the expression levels and functions of individual MMPs are clearly dependent on the stage and type of cancer.

While MMP expression is increased in many cancers, the levels of activated rather than total MMPs appear to be a better indicator of tumor metastatic potential. There are two main mechanisms of post transcriptional regulation of MMP activity: activation of the latent precursor form (zymogen) and inhibition of the active enzyme by tissue inhibitors of MMPs or tissue inhibitors of metalloproteinases (TIMPs) (Polette et al., 2004). Most MMPs are secreted in an inactive pro-enzyme form and are activated extra-cellularly. An interesting property of MMPs is that they are capable of mutual activation. For example, MMP1 and MMP14 can activate MMP2 (Murphy and Crabbe, 1995; Strongin et al., 1995; Sang et al., 1996). The proteolytic activity of MMPs is also regulated by TIMPs, by binding to the zinc ion in their catalytic site (Gomis-Ruth et al., 1997; Fernandez-Catalan et al., 1998). There are four known TIMPs, of which TIMP1 and TIMP2 are the most promiscuous and inhibit the majority of MMPs. *In vivo* studies in mice have shown that overexpression of TIMP1 decreases metastasis to the brain and to the liver (Soloway et al., 1996; Kruger et al., 1997, 1998; Sternlicht and Werb, 2001). Overall, MMP activity is tightly regulated by different mechanisms and is involved in both normal and pathologic processes (Polette et al., 2004).

THE ROLE OF INVADOPODIA FORMATION DURING CANCER CELL INVASION

While the mechanisms mediating the movement of cells through the ECM remain to be fully characterized, it is now widely accepted that the formation of actin rich invasive protrusions is a key step during cancer cell invasion. These structures were identified in tissue culture cells and have been termed as podosomes or invadopodia (Tarone et al., 1985; Chen, 1989). While the functional differences between podosomes and invadopodia remain to be clearly defined, recent nomenclature has tried to distinguish podosomes as present in normal cells and invadopodia as present in cancer cells (Murphy and Courtneidge, 2011; Hoshino et al., 2013). Nevertheless, there are more similarities than differences between podosomes and invadopodia. Both these structures are actin rich and possess the ability to degrade the ECM (Linder and Kopp, 2005) However, they differ in their size, number, lifetime and location, which allows for differentiation between them during visualization (Linder and Kopp, 2005). Both podosomes and invadopodia are usually visualized with phalloidin, which stains filamentous actin and appears as punctate spots located below the nucleus. Podosomes have been observed in cells of monocytic lineage such as macrophages (Lehto et al., 1982; Linder et al., 1999) osteoclasts (Marchisio et al., 1984) and in induced smooth muscle cells (Gimona et al., 2003) as well as endothelial cells (Moreau et al., 2003; Osiak et al., 2005). In contrast, invadopodia are found in cells transformed with oncogenes (David-Pfeuty and Singer, 1980; Tarone et al., 1985) and are thought to protrude further into the matrix and invade more aggressively than podosomes (Weaver, 2008; Linder et al., 2011; Murphy and Courtneidge, 2011). A variety of actin regulators, scaffold proteins, small GTPases, and proteinases have been shown to play important roles in several steps of invadopodia formation. Several studies using animal xenografts and primary tumor cells from patients have also demonstrated the formation of invadopodia *in vivo*. Additionally, invadopodia has been observed in bladder cancer (Sutoh et al., 2010; Yamamoto et al., 2011), colorectal cancer (Schoumacher et al., 2010), breast cancer (Coopman et al., 1998; Yamaguchi et al., 2005a, 2011) squamous cell carcinoma (Takkunen et al., 2010) and glioblastoma (Stylli et al., 2008).

The ability of invadopodia to degrade ECM is attributed to the presence of matrix degrading enzymes such as MMPs. While the cellular machinery mediating the targeted release of MMPs from invadopodia remains to be defined, it is becoming clear that some MMPs are targeted and released from invadopodia to facilitate invasion (Nakahara et al., 1997). MMP14, MMP2, and MMP9 have all been shown to be important in cancer progression and enriched at the invadopodia (Monsky et al., 1993; Nakahara et al., 1997; Bourguignon et al., 1998; Artym et al., 2006; Clark and Weaver, 2008; Poincloux et al., 2009). MMP2 and MMP9 contain fibronectin repeats that help them recognize gelatin (denatured collagen) as a substrate (Polette et al., 2004) and Type IV collagen is the main constituent of the BM, one of the first barriers that cancer cells need to traverse to metastasize. In addition, MMP14 can recognize and cleave a broad spectrum of ECM substrates and also functions as an activator of MMP2 (Lebeau et al., 1999). Thus, the combined activity of MMP2, MMP9, and MMP14 is suggested to be an important step in initiating localized degradation of the BM during epithelial cancer metastasis (Nakahara et al., 1997; Chen and Wang, 1999). Even though this review focuses on the proteolytic aspect of MMP14, it is interesting to note that it can also function through a non-proteolytic mechanism. MMP14 can stimulate ATP production by activating Hypoxia- Inducible Factor-1(HIF-1) (Sakamoto and Seiki, 2010). The non-proteolytic activity of MMP14 also includes binding of its transmembrane domain to integrin $\beta1$, which leads to MAPK activation, thereby regulating branching in mammary epithelium (Mori et al., 2013).

Invadopodia formation and function are complex cellular events that involve substantial reorganization in cytoskeleton dynamics and membrane transport. Recent studies have attempted to define different stages of invadopodia formation and function by using various criteria, including the recruitment of actin, targeted release of MMPs and the localized degradation of the ECM. Based on these, invadopodia formation has been divided into three stages, namely, initiation, assembly, and maturation. The following sections describe the stages of invadopodia formation in detail.

INITIATION

In the initiation phase of invadopodia formation, invadopodial precursors, or "buds" form at the cell periphery which are usually marked by actin puncta (Yamaguchi et al., 2005b). The process of invadopodia formation is initiated by growth factors such as epidermal growth factor (EGF), platelet derived growth factor (PDGF) and transforming growth factor- β (TGFβ). Growth factor receptor signaling activates Phosphatidylinositide 3-Kinase (P13K) leading to Src activation, which in turn phosphorylates multiple proteins including Tks5 (Tyrosine kinase substrate). Since the PX domain (phospholipid binding domain) of Tks5 has been shown to bind to PI3P and $PI(3,4)P_2$ (Abram et al., 2003), it was suggested that Tks5 localizes to $PI(3,4)P_2$ enriched regions of the plasma membrane, thus targeting Tks5 to initiate the invadopodia "bud" (Courtneidge et al., 2005). Src phosphorylates synaptojanin 2 to activate its phosphatase activity, which dephosphorylates $PI(3,4,5)P_3$at the plasma membrane to form $PI(3,4)P_2$, thus forming the site for invadopodia formation (Chuang et al., 2012). Src mediated activation of the Abl-family kinase Arg also leads to the phosphorylation of cortactin, resulting in the recruitment of Nck1 to the invadopodia (Oser et al., 2009; Mader et al., 2011). Nck1 then recruits the Neural Wiskott-Aldrich syndrome protein (N-Wasp) complex to the invadopodia leading to Cdc42-dependent activation of N-Wasp. N-Wasp in turn induces actin polymerization through the Arp2/3 complex, resulting in formation of invadopodial precursors (Yamaguchi et al., 2005a). The co-localization of cortactin with Tks5 in invadopodial "buds," led to the hypothesis that Tks5 acts a scaffolding protein that recruits the other cellular components required for initiation of invadopodia formation. However, some recent evidence suggests that Tks5 might instead be involved in invadopodia maturation (Sharma et al., 2013).

ASSEMBLY

Invadopodia are highly dynamic and motile structures that have been divided into two types based on their motility and lifetimes, motile short-lived and stationary long-lived invadopodia (Yamaguchi et al., 2005a). The short-lived invadopodia are thought to be precursors of fully functional mature invadopodia and could also be equivalent to podosomes (Yamaguchi et al., 2005a). *In vitro*, newly formed or early invadopodia can move laterally within the plasma membrane that faces the ECM. These motile invadopodia are then anchored and stabilized by actin polymerization and extension of the invadopodia (Yamaguchi et al., 2005a). A plethora of proteins are recruited to the invadopodial "bud" converting it from motile to stationary invadopodia. The precise order of how proteins are recruited is still unknown. Since Tks5 has five tandem SH3 domains, it is thought that Tks5 can scaffold several proteins like Nck1, Nck2, Grb2 (Growth factor receptor bound protein 2) and N-Wasp. It has also been proposed that Tks5, along with cortactin, recruits various actin regulators leading to nucleation of branched actin filaments and the formation of a stable actin core in the invadopodia (Clark and Weaver, 2008; Oser et al., 2009). Consistent with this hypothesis, it has been shown that phosphorylation of cortactin leads to dissociation of the cortactin/cofilin complex, which is an essential step in invadopodia formation and elongation (Oser et al., 2009). The dissociation of the cortactin/cofilin complex and the polymerization of actin is also a pH-dependent process (Pope et al., 2004; Frantz et al., 2008).

MATURATION

Actin polymerization is crucial for formation and function of invadopodia. Inhibition of cofilin results in the formation of short-lived unstable invadopodia, which suggests that actin polymerization caused by cofilin is required for elongation and maturation of invadopodia (Yamaguchi et al., 2005a). Apart from the branched actin network, invadopodia also contain linear actin bundles (Li et al., 2010; Schoumacher et al., 2010). mDia2, a formin that induces the formation of linear actin networks, has been found to promote elongation and stability of invadopodia (Lizarraga et al., 2009). Fascin, another actin bundling protein has also been shown to promote elongation, stability and matrix degradation in invadopodia (Li et al., 2010; Schoumacher et al., 2010).

Src kinase is a major regulator of invadopodia formation and function. Interestingly, several other protein kinases including Abl kinases like Arg (Abl related gene) kinase have recently emerged as important players in invadopodia formation and maturation (Beaty et al., 2013). It was shown that β1 integrin interacts with Arg leading to stimulation of Arg mediated cortactin phosphorylation, a key switch in promoting invadopodial maturation (Beaty et al., 2013). Additionally, β1 integrin has been shown to localize to invadopodia and promote degradation of collagen type IV, the main constituent of the BM (Sameni et al., 2008), presumably by recruiting and docking proteases at the invadopodia. Separase, a gelatinolytic enzyme that has been shown to be enriched at the invadopodia, binds to α3β1 integrin resulting in the formation of functional invadopodia (Mueller et al., 1999). The ultrastructure of mature invadopodia has shown the presence of microtubules and many vesicles/endosomes indicative of active trafficking and a possible route for delivery of specific proteins like MMPs (Schoumacher et al., 2010). The activity of proteases docked at the invadopodia has been shown to be pH-dependent (Greco et al., 2014). Furthermore, it was demonstrated that the acidification of the peri-invadopodial space by the Na^+/H^+ exchanger (NHE1) promotes ECM proteolysis (Busco et al., 2010).

MECHANISMS MEDIATING MMP TARGETING TO INVADOPODIA

The final maturation stage of the invadopodia involves targeted delivery and exocytosis of MMP2, MMP9, and MMP14. The appearance of these MMPs is generally considered to be a mark of functional mature invadopodia. As the result, much effort has been invested in understanding the regulation of MMP targeting to invadopodia, leading to recent studies defining the machinery governing MMP transport during cancer cell invasion.

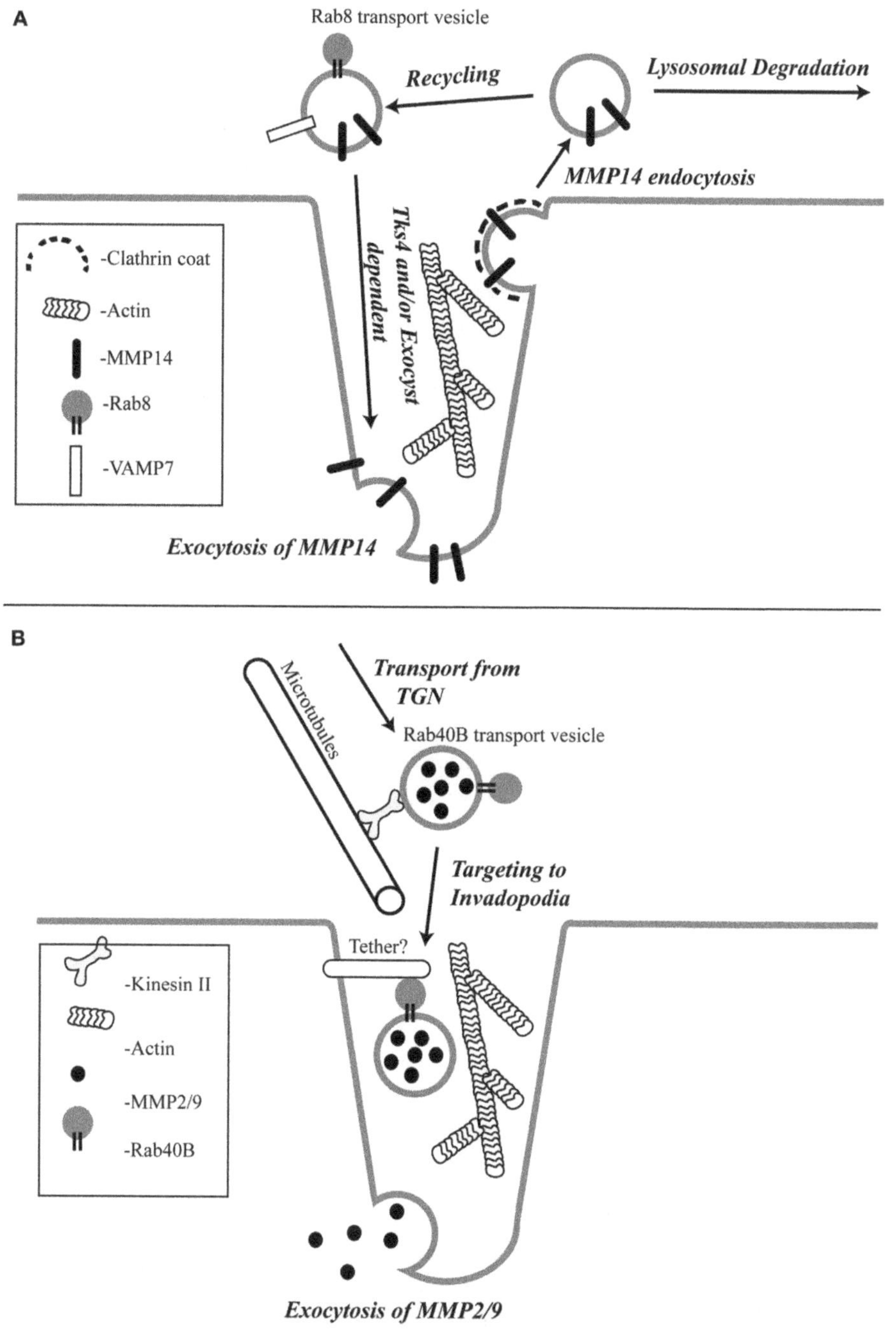

FIGURE 1 | The schematic representation of the pathways regulating MMP14 (A) and MMP2/9 (B) targeting to the invadopodia.

TARGETING OF MMP14

MMP14 is a membrane embedded MMP whose extracellular proteolytic activity is regulated by a balance between exocytosis and internalization via clathrin and/or caveolar mediated endocytosis (Remacle et al., 2003; **Figure 1A**). Once internalized, MMP14 is then either targeted to lysosomes for degradation (Jiang et al., 2001; Remacle et al., 2003) or shunted to endocytic recycling pathways, thus controlling the levels of active enzyme at the cell surface (Remacle et al., 2003; **Figure 1A**). However, invasive cancer cells have mechanisms to counteract the removal of the active enzyme from the plasma membrane. Consistently, enrichment of active MMP14 at the invadopodia associated plasma membrane of tumor cells has been demonstrated *in vitro* (Nakahara et al., 1997; Artym et al., 2006; Clark and Weaver, 2008; Steffen et al., 2008). Endocytic recycling (Poincloux et al., 2009) exocytosis (Monteiro et al., 2013), Rab 8 (Bravo-Cordero et al., 2007) and Tks4 (Buschman et al., 2009) have all been shown to be involved in the localization of MMP14 to the invadopodia. However, the exact mechanism governing MMP14 targeting remains to be fully understood. Rab8 GTPase has been shown to be involved in mobilization of MMP14 from intracellular storage compartments allowing polarized recruitment of MMP14 to the invasive front of cells (Bravo-Cordero et al., 2007). Some of the exocytic components shown to regulate delivery of MMP14 to the invadopodia are cortactin, the Exocyst complex (consists of 8 subunits) and VAMP7. Cortactin has been shown to regulate the cell surface expression of MMP14 (Clark et al., 2007). Sec8, a subunit of the Exocyst complex, was shown to localize at the invadopodia and is required for MMP14 targeting to the invadopodia (Monteiro et al., 2013; **Figure 1A**). Active RhoA and Cdc42 trigger the interaction between the Exocyst subunits Sec3 and Sec8 and the polarity protein IQGAP1. This interaction has been shown to be required for the accumulation of MMP14 at the invadopodia (Sakurai-Yageta et al., 2008). The Exocyst complex has also been shown to interact with Arp2/3 activator Wiskott-Aldrich syndrome protein and Scar homolog (WASH) to ensure focal delivery of MMP14 to the invadopodia (Monteiro et al., 2013). Since exocytosis depends on SNAREs (soluble N-ethylmaleimide sensitive factor attachment protein receptors) which drive the fusion of transport vesicles with the plasma membrane, several recent studies have investigated the role of SNAREs in mediating MMP14 transport. Consequently, it was shown that VAMP7 (vesicle associated membrane protein 7) is required for trafficking of MMP14 to invadopodia (Steffen et al., 2008; **Figure 1A**).

Transport vesicle targeting and fusion with its destination membranes often relies on specific tethering factors that impart specificity to membrane transport. The tethering factors regulating MMP14 targeting remain to be identified. However, Tks4, a scaffolding factor related to Tks5, has been shown to be required for the formation and function of invadopodia (Buschman et al., 2009). In the absence of Tks4, recruitment of MMP14 to podosomes is inhibited, implicating the role of Tks4 in targeting of MMP14 to the invadopodia (Buschman et al., 2009; **Figure 1A**). Additionally, cortactin was reported to have a novel role in invadopodial maturation and invasion by regulating secretion of MMP14 at the invadopodia (Clark and Weaver, 2008). However, it remains to be tested whether Tks4 and cortactin function as MMP14-vesicle tethers, or whether they affect MMP14 targeting indirectly by regulating the actin cytoskeleton within during invadopodia formation and maturation. Recently, it was also shown that Orai1 calcium channels can also regulate MMP14 targeting, since Orail1-mediated Ca^{2+} oscillations regulate MMP14 recycling to the plasma membrane (Sun et al., 2014).

TARGETING OF MMP2 AND MMP9

MMP2 and MMP9 are gelatinases that possess fibronectin type II repeats that allow them to degrade collagen and gelatin. Gelatinolytic degradation can cause the release of signaling molecules from the ECM that aid cell migration and angiogenesis. A lot of effort has been focused on understanding the transport and targeting of gelatinases because they are overexpressed in a variety of tumors and are associated with tumor aggressiveness and poor patient prognosis (Pacheco et al., 1998; Egeblad and Werb, 2002; Hiratsuka et al., 2002; van 't Veer et al., 2002). Although MMP2 and MMP9 have been shown to be enriched at the invadopodia, there is not much known about how these proteases are transported and targeted to the invadopodia. It has been reported that MMP2 and MMP9 are stored and transported in small vesicles that move along microtubules powered by kinesin in human melanoma cells (Schnaeker et al., 2004; **Figure 1B**). Similar to MMP14, the secretion of MMP2 and MMP9 were also shown to depend on cortactin (Clark and Weaver, 2008). Interestingly, in contrast to MMP14, endocytic transport and the Exocyst complex do not appear to play a role in targeted transport of MMP2 and MMP9 (Jacob et al., 2013). Instead, MMP2 and MMP9 are transported via specialized secretory vesicles directly from the trans-Golgi Network (TGN) to the invadopodia (Jacob et al., 2013). While much of the machinery mediating the formation and transport of these secretory vesicles remains to be defined, it was recently shown that Rab40b GTPase plays an important role in targeting MMP2/9 to the invadopodia (Jacob et al., 2013; **Figure 1B**).

INVADOPODIA FORMATION AND FUNCTION *IN VIVO*

Although there is an increasing amount of evidence demonstrating the existence of invadopodia *in vitro*, the formation and function of invadopodia *in vivo* is less well understood due to challenges associated with visualizing and distinguishing these structures in animal models. Cancer invasion usually occurs deep in tissues and these events are highly dynamic and unpredictable making it difficult to visualize invadopodia during primary tumor metastasis. Though the majority of invadopodial studies have been conducted in 2D tissue culture systems, some groups have studied invadopodia formation in 3D matrices as they better simulate the physiological *in vivo* environment. Such studies of invadopodia in complex 3D matrices have shown that the matrix degrading activity is localized to the base rather than the tip of the invadopodia (Wolf et al., 2007; Tolde et al., 2010). These 3D studies have also helped establish criteria for the identification of invadopodia *in vivo* and provide a good model to study formation of invadopodia.

Despite the challenges mentioned above, there is some compelling evidence drawn from elegant *in vivo* experiments that confirm that invadopodia are not just *in vitro* artifacts. Recently, the chorioallantoic membrane of the chicken embryo was used to visualize the intravascular formation of invadopodia and the extravasation of tumor cells into the stroma (Leong et al., 2014). The same group also showed that knocking down invadopodial components like cortactin, Tks4 and Tks5 decreases extravasation of cells into the lung stroma in tail vein injected mice. Intravital live animal imaging has also allowed the visualization of MtLn3-GFP (a highly invasive rat mammary carcinoma line) invading into blood vessels using protrusions identified as invadopodia-like structures using invadopodia markers such as cortactin and N-WASP (Lohmer et al., 2014). Using live-cell imaging, it was shown that during the uterine-vulval development in *Caenorhabditis elegans*, the anchor cell breaches the uterine and vulval basement membranes by making an invadopodium (Hagedorn et al., 2013). The Src-Tks5 pathway was shown to be required for the migration of neural crest cells using actin-rich protrusions in zebrafish (Murphy et al., 2011). Finally, it has also been shown that the intestinal epithelium of the zebrafish mutant *meltdown* forms invadopodia-like protrusions that invade into the stromal tissue in response to cues from the surrounding smooth muscle layer (Seiler et al., 2012).

CONCLUSIONS AND FUTURE DIRECTIONS

Significant advances have been made in understanding the formation and function of invadopodia. However, there are still a lot more unknowns regarding this subcellular structure. While all of the above mentioned studies have helped to confirm the physiological role of invadopodia as a structure used by invasive cells to penetrate the basement membrane, more evidence is required to elucidate the functional role of invadopodia *in vivo* and understand how widespread the use of invadopodia by cells is. Many questions regarding the importance of invadopodia in cancer invasion and metastasis still exist. Future studies in the field of invadopodia will need to focus on detection of invadopodia in human cancer samples as well as to identify the role of invadopodia in the different steps of the metastatic cascade. The other areas that require focus are the identification of components specific to invadopodia that can be targeted and the development of compounds that can specifically inhibit invadopodia formation and function. These issues will need to be addressed before invadopodia can become a candidate for development of new cancer therapies.

ACKNOWLEDGMENTS

We are grateful to Drs. Traci Lyons and Paul Jedlicka for critical reading of the manuscript. We apologize to all colleagues whose work could not be cited due to space limitations. We also acknowledge the financial support by NIH (DK064380 to Rytis Prekeris), Cancer League of Colorado (to Rytis Prekeris) and Susan G. Komen for the Cure foundation (BCTR0706749 to Rytis Prekeris).

REFERENCES

Abram, C. L., Seals, D. F., Pass, I., Salinsky, D., Maurer, L., Roth, T. M., et al. (2003). The adaptor protein fish associates with members of the ADAMs family and localizes to podosomes of Src-transformed cells. *J. Biol. Chem.* 278, 16844–16851. doi: 10.1074/jbc.M300267200

Ala-aho, R., and Kahari, V. M. (2005). Collagenases in cancer. *Biochimie* 87, 273–286. doi: 10.1016/j.biochi.2004.12.009

Artym, V. V., Zhang, Y., Moiseiwitsch, F. S., Yamada, K. M., and Mueller, S. C. (2006). Dynamic interactions of cortactin and membrane type 1 matrix metalloproteinase at invadopodia:defining the stages of invadopodia formaton and function. *Cancer Res.* 66, 3034–3043. doi: 10.1158/0008-5472.CAN-05-2177

Beaty, B. T., Sharma, V. P., Bravo-Cordero, J. J., Simpson, M. A., Eddy, R. J., Koleske, A. J., et al. (2013). Beta1 integrin regulates Arg to promote invadopodial maturation and matrix degradation. *Mol. Biol. Cell* 24, 1661–1675, S1661–S1611. doi: 10.1091/mbc.E12-12-0908

Bergers, G., Brekken, R., McMahon, G., Vu, T. H., Itoh, T., Tamaki, K., et al. (2000). Matrix metalloproteinase-9 triggers the angiogenic switch during carcinogenesis. *Nat. Cell Biol.* 2, 737–744. doi: 10.1038/35036374

Bjorklund, M., and Koivunen, E. (2005). Gelatinase-mediated migration and invasion of cancer cells. *Biochim. Biophys. Acta* 1755, 37–69. doi: 10.1016/j.bbcan.2005.03.001

Bourguignon, L. Y., Gunja-Smith, Z., and Iida, N. (1998). CD44v(3,8-10) is involved in cytoskelton-mediated tumor cell migration and matrix metalloproteinase MMP-9 association in breast cancer cells. *J. Cell Physiol.* 176, 206–215.

Bravo-Cordero, J. J., Marrero-Diaz, R., Megias, D., Genis, L., Garcia-Grande, A., Garcia, M. A., et al. (2007). MT1-MMP proinvasive activity is regulated by a novel Rab8-dependent exocytic pathway. *EMBO J.* 26, 1499–1510. doi: 10.1038/sj.emboj.7601606

Buschman, M. D., Bromann, P. A., Cejudo-Martin, P., Wen, F., Pass, I., and Courtneidge, S. A. (2009). The novel adaptor protein Tks4 (SH3PXD2B) is required for functional podosome formation. *Mol. Biol. Cell* 20, 1302–1311. doi: 10.1091/mbc.E08-09-0949

Busco, G., Cardone, R. A., Greco, M. R., Bellizzi, A., Colella, M., Antelmi, E., et al. (2010). NHE1 promotes invadopodial ECM proteolysis through acidification of the peri-invadopodial space. *FASEB J.* 24, 3903–3915. doi: 10.1096/fj.09-149518

Chamber, A. F., and Matrisian, L. M. (1997). Changing views of the role of matrix metalloproteinases in metastasis. *J. Natl. Cancer Inst.* 89, 1260–1270. doi: 10.1093/jnci/89.17.1260

Chen, W. T. (1989). Proteolytic activity of specialized surface protrusions formed at rosette contact sites of transformed cells. *J. Exp. Zool.* 251, 167–185. doi: 10.1002/jez.1402510206

Chen, W. T., and Wang, J. Y. (1999). Specialized surface protrusions of invasive cells, invadopodia and lamellipodia, have differential MT1-MMP, MMP-2, and TIMP-2 localization. *Ann. N.Y. Acad. Sci.* 878, 361–371. doi: 10.1111/j.1749-6632.1999.tb07695.x

Chuang, Y., Xu, X., Kwiatkowska, A., Tsapraillis, G., Hwang, H., Petritis, K., et al. (2012). Regulation of synaptojanin 2 5'-phosphatase activity by Src. *Cell Adh. Migr.* 6, 518–525. doi: 10.4161/cam.22139

Clark, E. S., and Weaver, A. M. (2008). A new role for cortactin in invadopodia:regulation of protease secretion. *Eur. J. Cell Biol.* 87, 581–590. doi: 10.1016/j.ejcb.2008.01.008

Clark, E. S., Whigham, A. S., Yarbrough, W. G., and Weaver, A. M. (2007). Cortactin is an essential regulator of matrix metalloproteinase secretion and extracellular matrix degradation in invadopodia. *Cancer Res.* 67, 4227–4235. doi: 10.1158/0008-5472.CAN-06-3928

Coopman, P. J., Do, M. T., Thompson, E. W., and Mueller, S. C. (1998). Phagocytosis of cross-linked gelatin matrix by human breast carcinoma cells correlates with their invasive capacity. *Clin. Cancer Res.* 4, 507–515.

Courtneidge, S. A., Azucena, E. F., Pass, I., Seals, D. F., and Tesfay, L. (2005). Src substrate Tks5, podosomes (invadopodia), and cancer cell invasion. *Cold Spring Harb. Symp. Quant. Biol.* 70, 161–171. doi: 10.1101/sqb.2005.70.014

Coussens, L. M., Tinkle, C. L., Hanahan, D., and Werb, Z. (2000). MMP-9 supplied by bone marrow-derived cells contributes to skin carcinogenesis. *Cell* 103, 481–490. doi: 10.1016/S0092-8674(00)00139-2

David-Pfeuty, T., and Singer, S. J. (1980). Altered distributions of the cytoskeletal proteins vinculin and alpha-actinin in cultured fibroblasts transformed by Rous sarcoma virus. *Proc. Natl. Acad. Sci. U.S.A.* 77, 6687–6691. doi: 10.1073/pnas.77.11.6687

Deryugina, E. I., and Quigley, J. P. (2006). Matrix metalloproteinases and tumor metastasis. *Cancer Metastasis Rev.* 25, 9–34. doi: 10.1007/s10555-006-7886-9

Edwards, D. R., Handsley, M. M., and Pennington, C. J. (2008). The ADAM metalloproteinases. *Mol. Aspects Med.* 29, 258–289. doi: 10.1016/j.mam.2008.08.001

Egeblad, M., and Werb, Z. (2002). New functions of the matrix metalloproteinases in cancer progression. *Nat. Rev. Cancer* 2, 161–174. doi: 10.1038/nrc745

Fernandez-Catalan, C., Bode, W., Huber, R., Turk, D., Calvete, J. J., Lichte, A., et al. (1998). Crystal structure of the complex formed by the membrane type 1-matrix metalloproteinase with the tissue inhibitor of metalloproteinases-2, the soluble progelatinase A receptor. *EMBO J.* 17, 5238–5248. doi: 10.1093/emboj/17.17.5238

Fingleton, B. (2006). Matrix metalloproteinases: roles in cancer and metastasis. *Front. Biosci.* 11, 479–491. doi: 10.2741/1811

Frantz, C., Barreiro, G., Dominguez, L., Chen, X., Eddy, R., Condeelis, J., et al. (2008). Cofilin is a pH sensor for actin free barbed end formation: role of phosphoinositide binding. *J. Cell Biol.* 183, 865–879. doi: 10.1083/jcb.200804161

Gialeli, C., Theocharis, A. D., and Karamanos, N. K. (2011). Roles of matrix metalloproteinases in cancer progression and their pharmacological targeting. *FEBS J.* 278, 16–27. doi: 10.1111/j.1742-4658.2010.07919.x

Gimona, M., Kaverina, I., Resch, G. P., Vignal, E., and Burgstaller, G. (2003). Calponin repeats regulate actin filament stability and formation of podosomes in smooth muscle cells. *Mol. Biol. Cell* 14, 2482–2491. doi: 10.1091/mbc.E02-11-0743

Gomis-Ruth, F. X., Maskos, K., Betz, M., Bergner, A., Huber, R., Suzuki, K., et al. (1997). Mechanism of inhibition of the human matrix metalloproteinase stromelysin-1 by TIMP-1. *Nature* 389, 77–81. doi: 10.1038/37995

Greco, M. R., Antelmi, E., Busco, G., Guerra, L., Rubino, R., Casavola, V., et al. (2014). Protease activity at invadopodial focal digestive areas is dependent on NHE1-driven acidic pHe. *Oncol. Rep.* 31, 940–946. doi: 10.3892/or.2013.2923

Ha, H. Y., Moon, H. B., Nam, M. S., Lee, J. W., Ryoo, Z. Y., Lee, T. H., et al. (2001). Overexpression of membrane-type matrix metalloproteinase-1 gene induces mammary gland abnormalities and adenocarcinoma in transgenic mice. *Cancer Res.* 61, 984–990.

Hagedorn, E. J., Ziel, J. W., Morrissey, M. A., Linden, L. M., Wang, Z., Chi, Q., et al. (2013). The netrin receptor DCC focuses invadopodia-driven basement membrane transmigration *in vivo*. *J. Cell Biol.* 201, 903–913. doi: 10.1083/jcb.201301091

Hiratsuka, S., Nakamura, K., Iwai, S., Murakami, M., Itoh, T., Kijima, H., et al. (2002). MMP9 induction by vascular endothelial growth factor receptor-1 is involved in lung-specific metastasis. *Cancer Cell* 2, 289–300. doi: 10.1016/S1535-6108(02)00153-8

Hofmann, U. B., Houben, R., Brocker, E. B., and Becker, J. C. (2005). Role of matrix metalloproteinases in melanoma cell invasion. *Biochimie* 87, 307–314. doi: 10.1016/j.biochi.2005.01.013

Hoshino, D., Branch, K. M., and Weaver, A. M. (2013). Signaling inputs to invadopodia and podosomes. *J. Cell Sci.* 126, 2979–2989. doi: 10.1242/jcs.079475

Itoh, T., Tanioka, M., Matsuda, H., Nishimoto, H., Yoshioka, T., Suzuki, R., et al. (1999). Experimental metastasis is suppressed in MMP-9-deficient mice. *Clin. Exp. Metastasis* 17, 177–181. doi: 10.1023/A:1006603723759

Itoh, T., Tanioka, M., Yoshida, H., Yoshioka, T., Nishimoto, H., and Itohara, S. (1998). Reduced angiogenesis and tumor progression in gelatinase A- deficient mice. *Cancer Res.* 58, 1048–1051.

Jacob, A., Jing, J., Lee, J., Schedin, P., Peden, A. A., Junutula, J. R., et al. (2013). Rab40b regulates MMP2 and MMP9 targeting to the invadopodia during breast cancer cell invasion. *J. Cell Sci.* 126, 4647–4658. doi: 10.1242/jcs.126573

Jiang, A., Lehti, K., Wang, X., Weiss, S. J., Keski-Oja, J., and Pei, D. (2001). Regulation of membrane-type matrix metalloproteinase 1 activity by dynamin-mediated endocytosis. *Proc. Natl. Acad. Sci. U.S.A.* 98, 13693–13698. doi: 10.1073/pnas.241293698

Kalluri, R. (2003). Basement membranes: structure, assembly and role in tumour angiogenesis. *Nat. Rev. Cancer* 3, 422–433. doi: 10.1038/nrc1094

Kerkela, E., and Saarialho-Kere, U. (2003). Matrix metalloproteinases in tumor progression: focus on basal and squamous cell skin cancer. *Exp. Dermatol.* 12, 109–125. doi: 10.1034/j.1600-0625.2003.120201.x

Kohrmann, A., Kammerer, U., Kapp, M., Dietl, J., and Anacker, J. (2009). Expression of matrix metalloproteinases (MMPs) in primary human breast cancer and breast cancer cell lines: new findings and review of the literature. *BMC Cancer* 9:188. doi: 10.1186/1471-2407-9-188

Kruger, A., Fata, J. E., and Khokha, R. (1997). Altered tumor growth and metastasis of a T-cell lymphoma in Timp-1 transgenic mice. *Blood* 90, 1993–2000.

Kruger, A., Sanchez-Sweatman, O. H., Martin, D. C., Fata, J. E., Ho, A. T., Orr, F. W., et al. (1998). Host TIMP-1 overexpression confers resistance to experimental brain metastasis of a fibrosarcoma cell line. *Oncogene* 16, 2419–2423. doi: 10.1038/sj.onc.1201774

Lebeau, A., Nerlich, A. G., Sauer, U., Lichtinghagen, R., and Lohrs, U. (1999). Tissue distribution of major matrix metalloproteinases and their transcripts in human breast carcinomas. *Anticancer Res.* 19, 4257–4264.

Lehto, V. P., Hovi, T., Vartio, T., Badley, R. A., and Virtanen, I. (1982). Reorganization of cytoskeletal and contractile elements during transition of human monocytes into adherent macrophages. *Lab. Invest.* 47, 391–399.

Leong, H. S., Robertson, A. E., Stoletov, K., Leith, S. J., Chin, C. A., Chien, A. E., et al. (2014). Invadopodia are required for cancer cell extravasation and are a therapeutic target for metastasis. *Cell Rep.* 8, 1558–1570. doi: 10.1016/j.celrep.2014.07.050

Li, A., Dawson, J. C., Forero-Vargas, M., Spence, H. J., Yu, X., Konig, I., et al. (2010). The actin-bundling protein fascin stabilizes actin in invadopodia and potentiates protrusive invasion. *Curr. Biol.* 20, 339–345. doi: 10.1016/j.cub.2009.12.035

Linder, S., and Kopp, P. (2005). Podosomes at a glance. *J. Cell Sci.* 118, 2079–2082. doi: 10.1242/jcs.02390

Linder, S., Nelson, D., Weiss, M., and Aepfelbacher, M. (1999). Wiskott-Aldrich syndrome protein regulates podosomes in primary human macrophages. *Proc. Natl. Acad. Sci. U.S.A.* 96, 9648–9653. doi: 10.1073/pnas.96.17.9648

Linder, S., Wiesner, C., and Himmel, M. (2011). Degrading devices: invadosomes in proteolytic cell invasion. *Annu. Rev. Cell Dev. Biol.* 27, 185–211. doi: 10.1146/annurev-cellbio-092910-154216

Lizarraga, F., Poincloux, R., Romao, M., Montagnac, G., Le Dez, G., Bonne, I., et al. (2009). Diaphanous-related formins are required for invadopodia formation and invasion of breast tumor cells. *Cancer Res.* 69, 2792–2800. doi: 10.1158/0008-5472.CAN-08-3709

Lochter, A., Sternlicht, M. D., Werb, Z., and Bissell, M. J. (1998). The significance of matrix metalloproteinases during early stages of tumor progression. *Ann. N.Y. Acad. Sci.* 857, 180–193. doi: 10.1111/j.1749-6632.1998.tb10116.x

Lohmer, L. L., Kelley, L. C., Hagedorn, E. J., and Sherwood, D. R. (2014). Invadopodia and basement membrane invasion *in vivo*. *Cell Adh. Migr.* 8, 246–255. doi: 10.4161/cam.28406

Mader, C. C., Oser, M., Magalhaes, M. A., Bravo-Cordero, J. J., Condeelis, J., Koleske, A. J., et al. (2011). An EGFR-Src-Arg-cortactin pathway mediates functional maturation of invadopodia and breast cancer cell invasion. *Cancer Res.* 71, 1730–1741. doi: 10.1158/0008-5472.CAN-10-1432

Marchisio, P. C., Cirillo, D., Naldini, L., Primavera, M. V., Teti, A., and Zambonin-Zallone, A. (1984). Cell-substratum interaction of cultured avian osteoclasts is mediated by specific adhesion structures. *J. Cell Biol.* 99, 1696–1705. doi: 10.1083/jcb.99.5.1696

Mehner, C., Hockla, A., Miller, E., Ran, S., Radisky, D. C., and Radisky, E. S. (2014). Tumor cell-produced matrix metalloproteinase 9 (MMP-9) drives malignant progression and metastasis of basal-like triple negative breast cancer. *Oncotarget* 5, 2736–2749. doi: 10.1371/journal.pone.0013066

Monsky, W. L., Kelly, T., and Lin, C. Y. (1993). Binding and localization of M(r) 72,000 matrix metalloproteinase at cell surface invadopodia. *Cancer Res.* 53, 3159–3164.

Monteiro, P., Rosse, C., Castro-Castro, A., Irondelle, M., Lagoutte, E., Paul-Gilloteaux, P., et al. (2013). Endosomal WASH and exocyst complexes control exocytosis of MT1-MMP at invadopodia. *J. Cell Biol.* 203, 1063–1079. doi: 10.1083/jcb.201306162

Mook, O. R., Frederiks, W. M., and Van Noorden, C. J. (2004). The role of gelatinases in colorectal cancer progression and metastasis. *Biochim. Biophys. Acta* 1705, 69–89. doi: 10.1016/j.bbcan.2004.09.006

Moreau, V., Tatin, F., Varon, C., and Genot, E. (2003). Actin can reorganize into podosomes in aortic endothelial cells, a process controlled by Cdc42 and RhoA. *Mol. Cell. Biol.* 23, 6809–6822. doi: 10.1128/MCB.23.19.6809-6822.2003

Mori, H., Lo, A. T., Inman, J. L., Alcaraz, J., Ghajar, C. M., Mott, J. D., et al. (2013). Transmembrane/cytoplasmic, rather than catalytic, domains of Mmp14 signal to MAPK activation and mammary branching morphogenesis via binding to integrin beta1. *Development* 140, 343–352. doi: 10.1242/dev.084236

Mueller, S. C., Ghersi, G., Akiyama, S. K., Sang, Q. X., Howard, L., Pineiro-Sanchez, M., et al. (1999). A novel protease-docking function of integrin at invadopodia. *J. Biol. Chem.* 274, 24947–24952. doi: 10.1074/jbc.274.35.24947

Murphy, D. A., and Courtneidge, S. A. (2011). The 'ins' and 'outs' of podosomes and invadopodia:characteristics, formation and function. *Nat. Rev. Mol. Cell Biol.* 12, 413–426. doi: 10.1038/nrm3141

Murphy, G., and Crabbe, T. (1995). Gelatinases A and B. *Methods Enzymol.* 248, 470–484. doi: 10.1016/0076-6879(95)48030-7

Murphy, D. A., Diaz, B., Bromann, P. A., Tsai, J. H., Kawakami, Y., Maurer, J., et al. (2011). A Src-Tks5 pathway is required for neural crest cell migration during embryonic development. *PLoS ONE* 6:e22499. doi: 10.1371/journal.pone.0022499

Nakahara, H., Howard, L., Thompson, E. W., Sato, H., Seiki, M., Yeh, Y., et al. (1997). Transmembrane/cytoplasmic domain-mediated membrane type1-matrix metalloprotease docking to invadopodiais required for cell invasion. *Proc. Natl. Acad. Sci. U.S.A.* 94, 7959–7964. doi: 10.1073/pnas.94.15.7959

Oser, M., Yamaguchi, H., Mader, C. C., Bravo-Cordero, J. J., Arias, M., Chen, X., et al. (2009). Cortactin regulates cofilin and N-WASp activities to control the stages of invadopodium assembly and maturation. *J. Cell Biol.* 186, 571–587. doi: 10.1083/jcb.200812176

Osiak, A. E., Zenner, G., and Linder, S. (2005). Subconfluent endothelial cells form podosomes downstream of cytokine and RhoGTPase signaling. *Exp. Cell Res.* 307, 342–353. doi: 10.1016/j.yexcr.2005.03.035

Pacheco, M. M., Mourao, M., Mantovani, E. B., Nishimoto, I. N., and Brentani, M. M. (1998). Expression of gelatinases A and B, stromelysin-3 and matrilysin genes in breast carcinomas: clinico-pathological correlations. *Clin. Exp. Metastasis* 16, 577–585. doi: 10.1023/A:1006580415796

Poincloux, R., Lizarraga, F., and Chavrier, P. (2009). Matrix invasion by tumor cells: a focus on MT1-MMP trafficking to invadopodia. *J. Cell Sci.* 122, 3015–3024. doi: 10.1242/jcs.034561

Polette, M., Nawrocki-Raby, B., Gilles, C., Clavel, C., and Birembaut, P. (2004). Tumour invasion and matrix metalloproteinases. *Crit. Rev. Oncol. Hematol.* 49, 179–186. doi: 10.1016/j.critrevonc.2003.10.008

Pope, B. J., Zierler-Gould, K. M., Kuhne, R., Weeds, A. G., and Ball, L. J. (2004). Solution structure of human cofilin: actin binding, pH sensitivity, and relationship to actin-depolymerizing factor. *J. Biol. Chem.* 279, 4840–4848. doi: 10.1074/jbc.M310148200

Remacle, A., Murphy, G., and Roghi, C. (2003). Membrane type I-matrix metalloproteinase (MT1-MMP) is internalised by two different pathways and is recycled to the cell surface. *J. Cell Sci.* 116, 3905–3916. doi: 10.1242/jcs.00710

Sakamoto, T., and Seiki, M. (2010). A membrane protease regulates energy production in macrophages by activating hypoxia-inducible factor-1 via a non-proteolytic mechanism. *J. Biol. Chem.* 285, 29951–29964. doi: 10.1074/jbc.M110.132704

Sakurai-Yageta, M., Recchi, C., Le Dez, G., Sibarita, J. B., Daviet, L., Camonis, J., et al. (2008). The interaction of IQGAP1 with the exocyst complex is required for tumor cell invasion downstream of Cdc42 and RhoA. *J. Cell Biol.* 181, 985–998. doi: 10.1083/jcb.200709076

Sameni, M., Dosescu, J., Yamada, K. M., Sloane, B. F., and Cavallo-Medved, D. (2008). Functional live-cell imaging demonstrates that beta1-integrin promotes type IV collagen degradation by breast and prostate cancer cells. *Mol. Imaging* 7, 199–213.

Sang, Q. A., Bodden, M. K., and Windsor, L. J. (1996). Activation of human progelatinase A by collagenase and matrilysin: activation of procollagenase by matrilysin. *J. Protein Chem.* 15, 243–253. doi: 10.1007/BF01887112

Schnaeker, E. M., Ossig, R., Ludwig, T., Dreier, R., Oberleithner, H., Wilhelmi, M., et al. (2004). Microtubule-dependent matrix metalloproteinase-2/matrix metalloproteinase-9 exocytosis: prerequisite in human melanoma cell invasion. *Cancer Res.* 64, 8924–8931. doi: 10.1158/0008-5472.CAN-04-0324

Schoumacher, M., Goldman, R. D., Louvard, D., and Vignjevic, D. M. (2010). Actin, microtubules, and vimentin intermediate filaments cooperate for elongation of invadopodia. *J. Cell Biol.* 189, 541–556. doi: 10.1083/jcb.200909113

Scorilas, A., Karameris, A., Arnogiannaki, N., Ardavanis, A., Bassilopoulos, P., Trangas, T., et al. (2001). Overexpression of matrix-metalloproteinase-9 in human breast cancer: a potential favourable indicator in node-negative patients. *Br. J. Cancer* 84, 1488–1496. doi: 10.1054/bjoc.2001.1810

Seiler, C., Davuluri, G., Abrams, J., Byfield, F. J., Janmey, P. A., and Pack, M. (2012). Smooth muscle tension induces invasive remodeling of the zebrafish intestine. *PLoS Biol.* 10:e1001386. doi: 10.1371/journal.pbio.1001386

Sharma, V. P., Eddy, R., Entenberg, D., Kai, M., Gertler, F. B., and Condeelis, J. (2013). Tks5 and SHIP2 regulate invadopodium maturation, but not initiation, in breast carcinoma cells. *Curr. Biol.* 23, 2079–2089. doi: 10.1016/j.cub.2013.08.044

Soloway, P. D., Alexander, C. M., Werb, Z., and Jaenisch, R. (1996). Targeted mutagenesis of Timp-1 reveals that lung tumor invasion is influenced by Timp-1 genotype of the tumor but not by that of the host. *Oncogene* 13, 2307–2314.

Steffen, A., Le Dez, G., Poincloux, R., Recchi, C., Nassoy, P., Rottner, K., et al. (2008). MT1-MMP-dependent invasion is regulated by TI-VAMP/VAMP7. *Curr. Biol.* 18, 926–931. doi: 10.1016/j.cub.2008.05.044

Sternlicht, M. D., and Werb, Z. (2001). How matrix metalloproteinases regulate cell behavior. *Annu. Rev. Cell Dev. Biol.* 17, 463–516. doi: 10.1146/annurev.cellbio.17.1.463

Sternlicht, M. D., Lochter, A., Sympson, C. J., Huey, B., Rougier, J. P., Gray, J. W., et al. (1999). The stromal proteinase MMP3/stromelysin-1 promotes mammary carcinogenesis. *Cell* 98, 137–146. doi: 10.1016/S0092-8674(00)81009-0

Strongin, A. Y., Collier, I., Bannikov, G., Marmer, B. L., Grant, G. A., and Goldberg, G. I. (1995). Mechanism of cell surface activation of 72-kDa type IV collagenase. Isolation of the activated form of the membrane metalloprotease. *J. Biol. Chem.* 270, 5331–5338. doi: 10.1074/jbc.270.10.5331

Stylli, S. S., Kaye, A. H., and Lock, P. (2008). Invadopodia: at the cutting edge of tumour invasion. *J. Clin. Neurosci.* 15, 725–737. doi: 10.1016/j.jocn.2008.03.003

Sun, J., Lu, F., He, H., Shen, J., Messina, J., Mathew, R., et al. (2014). STIM1- and Orai1-mediated Ca2+ oscillation orchestrates invadopodium formation and melanoma invasion. *J. Cell Biol.* 207, 535–548. doi: 10.1083/jcb.201407082

Sutoh, M., Hashimoto, Y., Yoneyama, T., Yamamoto, H., Hatakeyama, S., Koie, T., et al. (2010). Invadopodia formation by bladder tumor cells. *Oncol. Res.* 19, 85–92. doi: 10.3727/096504010X12875107808008

Takeha, S., Fujiyama, Y., Bamba, T., Sorsa, T., Nagura, H., and Ohtani, H. (1997). Stromal expression of MMP-9 and urokinase receptor is inversely associated with liver metastasis and with infiltrating growth in human colorectal cancer: a novel approach from immune/inflammatory aspect. *Jpn. J. Cancer Res.* 88, 72–81. doi: 10.1111/j.1349-7006.1997.tb00304.x

Takkunen, M., Hukkanen, M., Liljestrom, M., Grenman, R., and Virtanen, I. (2010). Podosome-like structures of non-invasive carcinoma cells are replaced in epithelial-mesenchymal transition by actin comet-embedded invadopodia. *J. Cell. Mol. Med.* 14, 1569–1593. doi: 10.1111/j.1582-4934.2009.00868.x

Tarone, G., Cirillo, D., Giancotti, F. G., Comoglio, P. M., and Marchisio, P. C. (1985). Rous sarcoma virus-transformed fibroblasts adhere primarily at discrete protrusions of the ventral membrane called podosomes. *Exp. Cell Res.* 159, 141–157. doi: 10.1016/S0014-4827(85)80044-6

Tolde, O., Rosel, D., Vesely, P., and Brabek, J. (2010). The structure of invadopodia in a complex 3D environment. *Eur. J. Cell Biol.* 89, 674–680. doi: 10.1016/j.ejcb.2010.04.003

van den Oord, J. J., Paemen, L., Opdenakker, G., and de Wolf-Peeters, C. (1997). Expression of gelatinase B and the extracellular matrix metalloproteinase inducer EMMPRIN in benign and malignant pigment cell lesions of the skin. *Am. J. Pathol.* 151, 665–670.

van 't Veer, L. J., Dai, H., van de Vijver, M. J., He, Y. D., Hart, A. A., Mao, M., et al. (2002). Gene expression profiling predicts clinical outcome of breast cancer. *Nature* 415, 530–536. doi: 10.1038/415530a

Wagenaar-Miller, R. A., Gorden, L., and Matrisian, L. M. (2004). Matrix metalloproteinases in colorectal cancer: is it worth talking about? *Cancer Metastasis Rev.* 23, 119–135. doi: 10.1023/A:1025819214508

Weaver, A. M. (2008). Invadopodia. *Curr. Biol.* 18, R362–R364. doi: 10.1016/j.cub.2008.02.028

Woessner, J. F. Jr. (1991). Matrix metalloproteinases and their inhibitors in connective tissue remodeling. *FASEB J.* 5, 2145–2154.

Wolf, K., Wu, Y. I., Liu, Y., Geiger, J., Tam, E., Overall, C., et al. (2007). Multi-step pericellular proteolysis controls the transition from individual to collective cancer cell invasion. *Nat. Cell Biol.* 9, 893–904. doi: 10.1038/ncb1616

Yamaguchi, H., Lorenz, M., Kempiak, S., Sarmiento, C., Coniglio, S., Symons, M., et al. (2005a). Molecular mechanisms of invadopodium formation: the role of the N-WASP-Arp2/3 complex pathway and cofilin. *J. Cell Biol.* 168, 441–452. doi: 10.1083/jcb.200407076

Yamaguchi, H., Wyckoff, J., and Condeelis, J. (2005b). Cell migration in tumors. *Curr. Opin. Cell Biol.* 17, 559–564. doi: 10.1016/j.ceb.2005.08.002

Yamaguchi, H., Yoshida, S., Muroi, E., Yoshida, N., Kawamura, M., Kouchi, Z., et al. (2011). Phosphoinositide 3-kinase signaling pathway mediated by p110alpha regulates invadopodia formation. *J. Cell Biol.* 193, 1275–1288. doi: 10.1083/jcb.201009126

Yamamoto, H., Sutoh, M., Hatakeyama, S., Hashimoto, Y., Yoneyama, T., Koie, T., et al. (2011). Requirement for FBP17 in invadopodia formation by invasive bladder tumor cells. *J. Urol.* 185, 1930–1938. doi: 10.1016/j.juro.2010.12.027

Yang, W., Arii, S., Gorrin-Rivas, M. J., Mori, A., Onodera, H., and Imamura, M. (2001). Human macrophage metalloelastase gene expression in colorectal carcinoma and its clinicopathologic significance. *Cancer* 91, 1277–1283. doi: 10.1002/1097-0142(20010401)91:7<1277::AID-CNCR1129>3.0.CO;2-H

Yurchenco, P. D. (2011). Basement membranes: cell scaffoldings and signaling platforms. *Cold Spring Harb. Perspect. Biol.* 3:a004911. doi: 10.1101/cshperspect.a004911

Zeng, Z. S., Huang, Y., Cohen, A. M., and Guillem, J. G. (1996). Prediction of colorectal cancer relapse and survival via tissue RNA levels of matrix metalloproteinase-9. *J. Clin. Oncol.* 14, 3133–3140.

Conflict of Interest Statement: The authors declare that the research was conducted in the absence of any commercial or financial relationships that could be construed as a potential conflict of interest.

20

Common themes in PrP signaling: the Src remains the same

Katharina Ochs and Edward Málaga-Trillo *†

Department of Biology, University of Konstanz, Konstanz, Germany

Edited by:
Jean-Luc Vilotte, National Institute of Agronomical Research, France

Reviewed by:
Mario Cioce, NYU Langone Medical Center, USA
Husheng Ding, Mayo Clinic, USA

*Correspondence:
Edward Málaga-Trillo, Laboratorios de Investigación y Desarrollo, Facultad de Ciencias y Filosofía, Universidad Peruana Cayetano Heredia, Av. Honorio Delgado 430, Lima 31, Perú
e-mail: edward.malaga@upch.pe

†Present address:
Edward Málaga-Trillo, Laboratorios de Investigación y Desarrollo, Facultad de Ciencias y Filosofía, Universidad Peruana Cayetano Heredia, Lima, Perú

The ability of the cellular prion protein (PrP^C) to trigger intracellular signals appears central to neurodegeneration pathways, yet the physiological significance of such signals is rather puzzling. For instance, PrP^C deregulation disrupts phenomena as diverse as synaptic transmission in mammals and cell adhesion in zebrafish. Although unrelated, the key proteins in these events -the NMDA receptor (NMDAR) and E-cadherin, respectively- are similarly modulated by the Src family kinase (SFK) Fyn. These observations highlight the importance of PrP^C-mediated Fyn activation, a finding reported nearly two decades ago. Given their complex functions and regulation, SFKs may hold the key to intriguing aspects of PrP biology such as its seemingly promiscuous functions and the lack of strong phenotypes in knockout mice. Here we provide a mechanistic perspective on how SFKs might contribute to the uncertain molecular basis of neuronal PrP phenotypes affecting ion channel activity, axon myelination and olfactory function. In particular, we discuss SFK target proteins involved in these processes and the role of tyrosine phosphorylation in the regulation of their activity and cell surface expression.

Keywords: prion protein, Src family kinases, ion channels, myelination, tyrosine phosphorylation, endocytosis, knockout mice, zebrafish

Yes, there are two paths you can go by but in the long run
There's still time to change the road you're on
(R. Plant and J. Page)

INTRODUCTION

The abnormal accumulation of misfolded proteins is a defining molecular landmark of neurodegenerative conditions like prion, Alzheimer's and Parkinson's diseases. Interestingly, the identity of the pathogenic protein is different in each of these illnesses, suggesting that neurotoxicity may result from the excess of virtually any rogue protein in neuronal tissue. On the other hand, at least some of these proteins may play physiological roles in their non-aggregated state, which are key to neuronal survival and disease. For instance, the amyloid precursor protein and α-synuclein—the protein culprits of Alzheimer's and Parkinson's disease—are enriched in neuronal synapses and nerve terminals (Maroteaux et al., 1988; Schubert et al., 1991). Likewise, expression of PrP^C on neuronal cell surfaces is required for pathogenic prions and aβ oligomers to trigger cellular damage (Chesebro et al., 2005; Um et al., 2012). Beyond their propensity to misfold and aggregate, little is known about how exactly these proteins contribute to neuronal physiology and disease. For PrP, the quest for a gene/protein function has been particularly challenging because of the paucity of clear knockout phenotypes in mice (Steele et al., 2007). While this phenomenon may be explained by compensatory mechanisms (Málaga-Trillo and Sempou, 2009), the actual extent to which other genes can functionally replace PrP is only partly understood (Passet et al., 2013). More recent analyses of PrP knockout mice have revealed subtle defects in axon myelination and olfactory function as well as in the proliferation of neural precursors and self-renewal of hematopoietic stem cells (Steele et al., 2006; Zhang et al., 2006; Le Pichon et al., 2009; Bremer et al., 2010). Unfortunately, the precise molecular basis of these phenotypes remains unclear. Adding complexity to the matter, many of the dissimilar functions and interaction partners proposed for PrP^C cannot easily be accounted for by a single biological activity.

Among the various molecular roles ascribed to PrP^C, its ability to elicit intracellular signals is in good correspondence with its cell surface localization and involvement in physiological processes as diverse and complex as cell adhesion, lymphocyte activation, neuroprotection, and synaptic function (Aguzzi et al., 2008). That PrP signaling is pathophysiologically relevant was elegantly shown by Chesebro and colleagues, who demonstrated that without anchoring to the plasma membrane, PrP may misfold and aggregate but not induce neuronal damage (Chesebro et al., 2005). Presently, a number of transmembrane and intracellular molecules are known to help PrP^C transduce signals into the cell's interior (Linden et al., 2008). Among the latter, SFKs have gathered renewed interest because of their connection to prion- and aβ-induced neurotoxicity as well as for their contribution to PrP-mediated cell communication *in vivo*. The purpose of this article is to revisit putative functions of PrP in the context of SFK signaling.

SFKs, A MULTIFUNCTIONAL PROTEIN FAMILY

The Src family of non-receptor protein tyrosine kinases comprises nine members in vertebrates, five of which are active in mammalian CNS neurons, namely the ubiquitously expressed Src, Fyn, and Yes, and the tissue-restricted Lck and Lyn (Thomas and Brugge, 1997; Salter and Kalia, 2004). Their association to the cytoplasmic face of the plasma membrane via myristoylation and palmytoylation enables them to receive extracellular stimuli and transduce them into the cell interior (Silverman et al., 1993). SFK enzymatic activity is triggered by autophosphorylation at tyrosine residue 416 and inhibited by phosphorylation of tyrosine residue 527 (Roskoski, 2005). Owing to their interaction with a diverse array of transmembrane receptors and downstream targets, SFKs can regulate a broad range of cellular processes, including cell differentiation, proliferation, adhesion, migration, apoptosis, and immunity (Thomas and Brugge, 1997; Roskoski, 2004). Because of their strong structural conservation, SFK members are functionally redundant and can often compensate for each other, encumbering analysis of their individual roles via gene knockout approaches (Lowell and Soriano, 1996). Given their expression and multiple functions in neurons, it is not surprising that SFK deregulation should contribute to neurodegeneration. In fact, prion infection triggers increased levels of Src and phosphotyrosine proteins in cultured cells and mice (Nixon, 2005; Gyllberg et al., 2006). Moreover, Fyn is selectively upregulated in Alzheimer's brains (Shirazi and Wood, 1993), where it mediates hyperphosphorylation of the aggregation-prone, microtubule-associated protein Tau (Lee et al., 2004). Fittingly, Fyn overexpression triggers synaptic damage in mouse models of Alzheimer's disease (Chin et al., 2005) whereas Fyn knockdown leads to decreased Tau phosphorylation, increased aβ levels and impaired spatial learning (Minami et al., 2012). Together, these findings imply a direct involvement of SFKs in neurotoxic pathways that are common to multiple neurodegenerative conditions.

SFKs AS FUNCTIONAL HUBS OF PrP

Being tethered to the outer leaflet of the plasma membrane via a short glycosylphosphatidylinositol- (GPI) anchor, PrP is unlikely to physically associate with cytosolic SFKs. Nevertheless, co-immunoprecipitation data indicate that, in epithelial cells, PrP and SFKs interact at least indirectly within a larger protein complex (Morel et al., 2004, 2008). In addition, plenty of evidence supports the notion that PrP and SFKs are functional partners in neurons. Using antibody-mediated cross-linking, Mouillet-Richard and colleagues were the first to show that PrP clustering at neuronal cell surfaces triggers caveolin-dependent activation of Fyn (Mouillet-Richard et al., 2000). Related experiments in neuron-like PC12 cells corroborated the contribution of caveolin to PrP^C-mediated Fyn activation, along with the downstream activation of the Ras-Raf/ERK pathway and the negative regulation of the entire cascade by phospho-caveolin (Pantera et al., 2009). In cultured mouse hippocampal neurons, PrP^C was found to indirectly activate Fyn via *cis* or *trans* associations with the neural cell adhesion molecule (NCAM) within lipid rafts, thereby stimulating neurite outgrowth (Santuccione et al., 2005). Importantly, *in vivo* data from a wide range of animal models underscore the relevance of the PrP/SFK pathway in physiology and disease. For instance, expression of a pathogenic PrP mutant in *Caenorhabditis elegans* induced neuronal dysfunction in a Fyn-dependent manner (Bizat et al., 2010), a remarkable result since nematodes do not possess endogenous PrPs. In early zebrafish embryos, exogenously added mouse PrP mimicks the positive regulation of E-cadherin-mediated cell adhesion by endogenous PrPs, an effect that requires Fyn and Yes activity (Málaga-Trillo et al., 2009; Sempou et al., submitted). In undoubtedly one of the most sensational developments in the recent field of neurodegeneration, neuronal PrP^C was reported to act as a receptor for aβ oligomers. Notably, the resulting synaptic impairment was mediated by Fyn (Um et al., 2012).

At first glance, some of these findings have no apparent connection with each other, aside from the common involvement of SFKs. However, closer scrutiny reveals mechanistic similarities pertinent to our understanding of PrP function. This is particularly manifest in two of the aforementioned models, where the downstream targets of PrP turned out to be transmembrane proteins modulated by tyrosine phosphorylation. In the study by the Strittmatter lab, binding of aβ oligomers to PrP^C triggered Fyn activation and hyperphosphorylation of the NR2B subunit of the NMDAR at tyrosine residue 1472 (Um et al., 2012). As a result, NMDARs became overstabilized at the plasma membrane and glutamate excitotoxicity ensued. Under non-pathological conditions, phosphorylation of NR2B at this C-terminal regulatory site promotes its normal cell surface expression by preventing the binding of AP-2 adaptor complexes, which would otherwise initiate its internalization via clathrin- and dynamin-dependent endocytosis (Salter and Kalia, 2004). Similarly, postsynaptic density protein 95 (PSD-95) prevents NR2B endocytosis by binding a site adjacent to tyrosine 1472 and thereby blocking access to AP-2 (Lavezzari et al., 2003). Overall, the phosphorylation state of NMDAR subunits is controlled by the dynamic interplay between SFKs and protein tyrosine phosphatases (PTPs), as well as by the input from various upstream signaling molecules (Salter and Kalia, 2004). In addition, NMDAR function is further regulated via differential transcription, translation and trafficking of specific subunits as well as through protein-protein interactions (Sanz-Clemente et al., 2013). Notably, the regulation of protein endocytosis by molecules that competitively bind target motifs at cytoplasmic tails is also pivotal to the modulation of E-cadherin cell surface expression (Ishiyama et al., 2010). Thus, association of p120 catenin to the C-terminal juxtamembrane domain (JMD) of E-cadherin sterically blocks binding of AP-2 and the E3 ubiquitin ligase Hakai, thereby preventing its clathrin-mediated or ubiquitination-dependent endocytosis, respectively (Thoreson et al., 2000; Fujita et al., 2002; Miyashita and Ozawa, 2007; Ishiyama et al., 2010). Importantly, SFK-mediated phosphorylation of the JMD and/or p120 catenin triggers E-cadherin internalization (Reynolds, 2010). Additionally, SFKs phosphorylate and modulate components of the endocytic machinery like clathrin, dynamin and AP-2 (Ahn et al., 1999, 2002; Wilde et al., 1999; Zimmerman et al., 2009). In our zebrafish model, PrP^C promotes embryonic cell adhesion by stabilizing E-cadherin at the cell-surface (Málaga-Trillo et al., 2009). Crucially, this activity entails the SFK-dependent inhibition of E-cadherin endocytosis (Sempou et al., submitted). Therefore, the roles of PrP^C as a modulator of cell-cell adhesion in zebrafish and synaptic transmission in mammals converge at a common mechanism, namely the ability of SFKs to control the internalization of transmembrane proteins by phosphorylating their endocytic signals and/or the corresponding binding molecules. Given the functional diversity of SFK target proteins, we reasoned that altered SFK activity may account for at least some of the distinct neuronal phenotypes observed in PrP knockout and transgenic mice (**Figure 1**, **Table 1**). To address this possibility, we examined three compelling, yet poorly understood paradigms of PrP function.

PrP AND ION CHANNELS

The NMDAR-mediated glutamate excitotoxicity induced by aβ oligomers and PrP^C clearly shows that the latter can influence the activity of ion channels (Um et al., 2012). Intriguingly, while this effect relies on the PrP-dependent overactivation of NR2B subunits, a different study found that PrP^C suppresses NMDAR activity and glutamate excitotoxicity via the inhibition of NR2D subunits (Khosravani et al., 2008). How PrP may influence this pathway under physiological conditions remains to be established. Nonetheless, a general role of PrP as a modulator of ion channel activity and neurotransmission is consistent with its synaptic localization (Moya et al., 2000; Um et al., 2012) and the common occurrence of electrophysiological alterations in PrP knockout mice (Aguzzi et al., 2008; Biasini et al., 2012). Moreover, transgenic mice expressing neurotoxic PrP mutants show distinct neurological phenotypes related to ion channel dysfunction. One of these mutants, PrPΔCR, has 21 aa deleted from its central region and triggers neonatal lethality characterized by the spontaneous degeneration of cerebellar granule neurons (CGNs) (Li et al., 2007). Subsequent work in cultured cells and cerebellar slices revealed that PrPΔCR induces spontaneous currents that increase neuronal susceptibility to glutamate-dependent, Ca^{2+}-mediated excitotoxicity (Solomon et al., 2010; Biasini et al., 2013). Based on these data, the authors suggested that PrPΔCR might either activate an endogenous ion channel or form itself an ion channel. A second PrP mutant, PG14, causes age-related degeneration of CGNs owing to defective glutamate exocytosis and impaired calcium dynamics (Senatore et al., 2012). This phenotype was explained by the ability of PG14 to bind the voltage-gated calcium channel (VGCC) $\alpha_2\delta$-1 subunit and retain

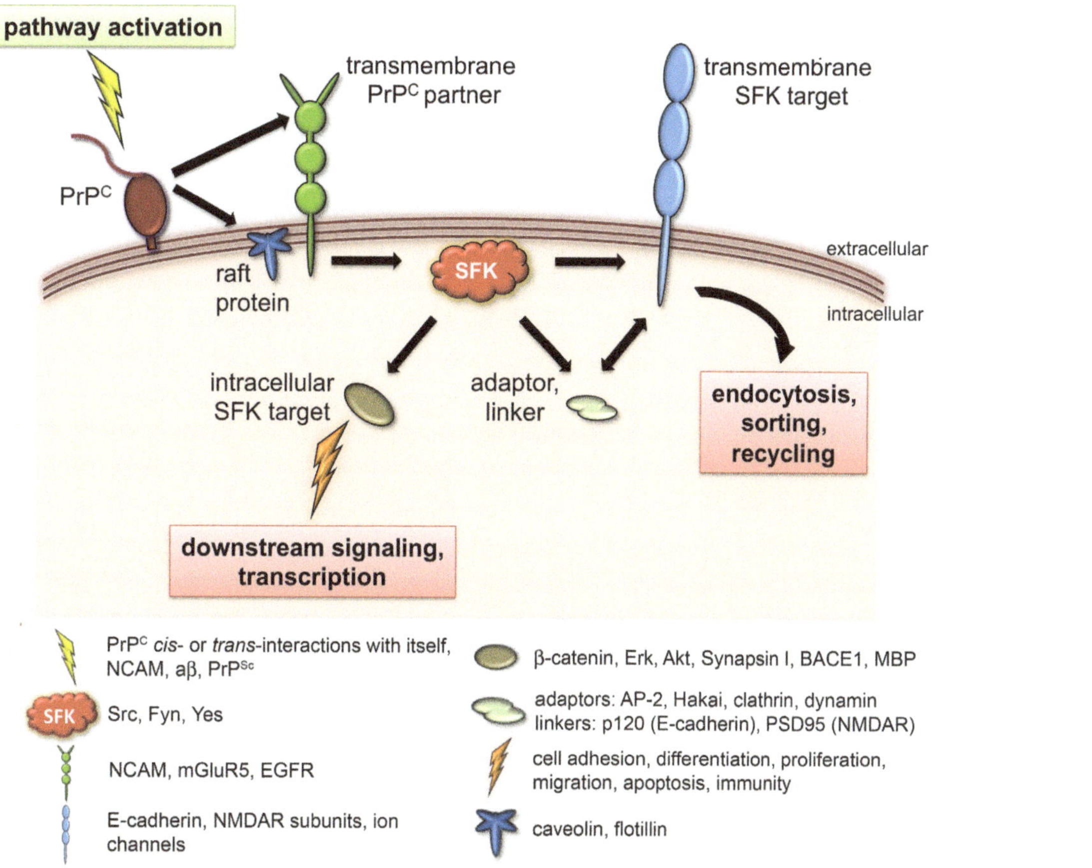

FIGURE 1 | PrP^C signaling via SFKs. Known physiological roles of PrP^C such as neurotransmission, embryonic cell adhesion, olfactory function, and myelination converge at the use of SFKs as intracellular signaling partners. In this model, engagement of PrP^C in various *cis-* or *trans-*interactions (depicted as "pathway activation") elicits the catalytic activity of SFKs, leading to the modulation of diverse downstream targets that include ion channels, adhesion complexes, and cytosolic signaling molecules. Phosphorylation of transmembrane SFK targets controls their cell surface expression and/or signaling properties, whereas phosphorylation of intracellular SFK targets regulates the activation of additional downstream pathways, gene transcription, and protein translation. The caption below provides concrete examples of documented components of this cascade.

Table 1 | SFKs and downstream events involved in PrP activities.

PrP activities reported	SFKs involved	SFK target events affected	References
aβ oligomer neurotoxicity	Fyn	Endocytosis of NMDA receptor subunit NR2B	Salter and Kalia, 2004
Regulation of embryonic cell adhesion (zebrafish)	Fyn, Yes	Endocytosis of E-cadherin	Ishiyama et al., 2010
PrPΔCR neurotoxicity	Fyn	Gating and/or endocytosis of NMDARs	Salter and Kalia, 2004
PrP PG14 neurotoxicity	Src	Gating and/or endocytosis of VGCCs	Davis et al., 2001
Regulation of olfactory physiology	Src	Binding of Synapsin Ib to synaptic vesicles and actin	Messa et al., 2010
	Src	Gating and/or trafficking of K_V1.3 channels	Fadool and Levitan, 1998; Fadool et al., 2000
	Fyn	Gating and/or endocytosis of NMDARs	Chen et al., 2000; Halabisky et al., 2000
Molecular control of myelination	Src	Activation of Akt/PKB	Chen et al., 2001; Jiang and Qiu, 2003
	Src	Endocytosis and activation of BACE1	Zou et al., 2007
	Src	mTORC1-induced transcription of the MBP gene	Vojtechova et al., 2008; Ondrusova et al., 2013
	Fyn	Localized translation of MBP mRNA at axon–glia contact sites via hnRNP A2	White et al., 2008
	unknown	Endocytosis of MAG and MOG	Bo et al., 1995; Kroepfl and Gardinier, 2001a,b
	Src, Fyn	Gating and/or endocytosis of K_V1.5 and 2.1 channels	Sobko et al., 1998; Peretz et al., 1999; Tiran et al., 2006

Summary of some previously described activities of PrP (see main text) and the relevant molecular targets/events regulated by SFKs. The actual involvement of these pathways in the corresponding PrP phenotypes is hypothetical, except for those under "aβ oligomer neurotoxicity" and "embryonic cell adhesion," which have been experimentally confirmed.

it at the ER, thus disrupting its delivery to the plasma membrane. Notably, while the molecular mechanisms of these two phenotypes seem unrelated, they are both consistent with the ability of SFKs to regulate the activity of ligand- and voltage-gated ion channels (Salter and Kalia, 2004). For instance, the spontaneous ionic currents induced by PrPΔCR may be explained by alterations in SFK-mediated phosphorylation and trafficking of NMDAR subunits. Detailed biochemical analyses will be needed to distinguish between this scenario and one proposed by the authors, namely, that the mutant PrP molecule activates NMDARs via direct interaction (Biasini et al., 2013). For the PG14 phenotype, the data show that intracellular retention of VGCC $\alpha_2\delta$-1 during biosynthesis is the most likely explanation for its reduced levels at the plasma membrane (Senatore et al., 2012). However, here it also might be worthwhile considering that VGCCs are modulated via tyrosine phosphorylation by, among others, Src (Davis et al., 2001). Whether SFK regulation of VGCCs targets their trafficking or only their gating properties is presently unknown.

PrP AND OLFACTION

Unique among mouse PrP knockout phenotypes was the finding of impaired olfactory behavior (Le Pichon et al., 2009). In electrophysiological terms, the phenotype was attributed to olfactory bulb-specific alterations in the patterns of oscillatory activity and in the plasticity of the dendrodendritic synapse. No molecular basis was reported for these defects but the authors suggested that due to its localization at olfactory synaptic membranes and its interaction with Synapsin Ib (Spielhaupter and Schatzl, 2001), PrP may be a regulatory component of the synaptic machinery. Interestingly, several lines of evidence argue for a contribution of SFKs and their targets to this phenotype. First, phosphorylation of Synapsin I by Src increases its binding to synaptic vesicles (SVs) and the actin cytoskeleton, thereby limiting the number of recycled SVs available for exocytosis and neurotransmitter release (Messa et al., 2010). Second, Fyn-deficient mice exhibit defects related to the olfactory dysfunction described by Le Pichon et al., such as impaired suckling behavior, altered synaptic transmission via GABA and NMDA receptors in the olfactory bulb, and abnormal axon guidance and fasciculation in the olfactory nerve (Yagi et al., 1993; Kitazawa et al., 1998; Morse et al., 1998). Third, K_V1.3 voltage-gated potassium channels are expressed along the dendrodendritic synapse and are negatively regulated by Src phosphorylation (Fadool and Levitan, 1998; Fadool et al., 2000). Remarkably, olfactory neurons from K_V1.3 null mice have modified action potentials and deregulated potassium currents, resulting in structurally altered olfactory bulbs and a heightened sense of smell (Fadool et al., 2004). Fourth, NMDARs play a crucial role at the dendrodendritic synapse by generating the Ca^{2+} influx that triggers inhibitory GABA release (Chen et al., 2000; Halabisky et al., 2000). This essential component of olfactory learning is therefore also subject to modulation by PrP and SFKs. Hence, the combined, SFK-mediated deregulation of Synapsin I, K_V1.3 channels, and NMDARs provides testable explanations for the olfactory defects of PrP knockout mice.

PrP AND MYELINATION

Re-examination of a mild, late-onset phenotype in PrP knockout mice led to the discovery of defects in the maintenance of peripheral myelin (Bremer et al., 2010). This chronic demyelinating polyneuropathy (CDP) was triggered by the loss of PrP in neurons but not in Schwann cells. The mechanistic basis of this phenotype remains unidentified, as the authors found no alterations in the phosphorylation of Akt, and Erk, two signaling molecules down the myelination pathway initiated with the cleavage of Neuregulin-1 (NRG1) by the BACE1 endopeptidase (Hu et al., 2006; Liu et al., 2007). Similarly, no differences were detected in the expression levels of NRG1, although, neither BACE1 activity nor BACE1-mediated cleavage of NRG1 was directly assessed in PrP knockout mice. Interestingly, Src is known to modulate several events along the BACE1/NRG1 pathway. Some of them, like Akt phosphorylation by Src (Chen et al., 2001; Jiang and Qiu, 2003), are probably not involved in the PrP phenotype, as implied by Bremer and colleagues. Others like BACE1 endocytosis and activation (Zou et al., 2007) or the mTORC1-induced transcription of the myelin basic protein (MBP) gene (Vojtechova et al., 2008; Ondrusova et al., 2013) would require further verification. Of note, BACE-null mice exhibit myelination defects reminiscent of those in PrP knockout mice, as well as increased levels of full-length NRG1 and—unlike PrP knockouts—reduced Akt phosphorylation (Hu et al., 2006, 2008).

As with the olfactory phenotype of Le Pichon et al., copious evidence suggests a potential involvement of SFKs in the PrP-CDP phenotype. Comprehensive reviews on the role of Fyn during myelination are available elsewhere (Kramer-Albers and White, 2011; White and Kramer-Albers, 2014) but for example, myelination requires Fyn activation by the myelin-associated glycoprotein (MAG), and activated Fyn promotes MBP gene transcription (Umemori et al., 1994, 1999). In addition, Fyn stimulates the localized translation of MBP mRNA at sites of axon–glia contact via phosphorylation of the heterogeneous nuclear ribonucleoprotein (hnRNP) A2 (White et al., 2008). Furthermore, Fyn-deficient mice show myelination defects similar to those of BACE- and PrP-null mice, albeit with some differences in the localization (PNS vs. CNS), onset (late vs. early) and severity of the phenotype (Umemori et al., 1994, 1999; Sperber et al., 2001). Interestingly, the presence of tyrosine-based endocytic motifs in MAG and the myelin/oligodendrocyte glycoprotein (MOG) strongly suggests that their cell surface expression is modulated by SFKs (Bo et al., 1995; Kroepfl and Gardinier, 2001a,b). In fact, the endocytic trafficking and sorting of myelin proteins is key to the establishment of membrane domains in oligodendroglial cells (Winterstein et al., 2008). Particularly relevant to the PrP-CDP phenotype, adult and aged MAG-null mice undergo degeneration of myelinated fibers in the PNS (Schachner and Bartsch, 2000).

Finally, voltage- and ligand-gated ion channels constitute further regulatory targets of SFKs during myelination. For instance, the phosphorylation state of $K_V1.5$ and $K_V2.1$ potassium channels in Schwann cells is controlled by the interplay between Src, Fyn, PTPα, and PTPε (Sobko et al., 1998; Peretz et al., 1999; Tiran et al., 2006). This complex mechanism modulates the gating properties of the channels, influencing Schwann cell development and the onset of myelination. Likewise, NMDARs promote the maturation of oligodendrocyte precursor cells (OPCs) by enhancing the expression of myelin proteins (Li et al., 2013). Nevertheless, experiments with NMDAR-deficient mice suggest that, *in vivo*, this regulation might be relevant for re-myelination but not for OPC development (De Biase et al., 2011; Li et al., 2013). Thus, although the molecular control of myelination is a rather intricate phenomenon, the PrP-CDP phenotype seems consistent with the alteration of multiple pathways downstream of SFKs.

Altogether, the common regulation of ion channels, olfaction and myelination by PrP and SFKs suggests that this phenomenon may extend to some of the other subtle PrP phenotypes mentioned above. For instance, it would be interesting to see if the role of PrP^C in the proliferation and differentiation of various stem cell lineages (reviewed in Martin-Lannerée et al., 2014) requires the differential expression of SFKs and their targets at specific stages of stem cell maturation. Along these lines, the contribution of tissue-restricted SFKs to PrP signaling deserves further investigation, as most of the available data on the PrP/SFK pathway so far involve only the ubiquitously expressed Src, Fyn, and Yes.

CONCLUSIONS

Several key observations emerge from the present analysis. First, the ability of PrP^C to elicit intracellular signals at the cell surface is central to its multiple roles in health and disease. Second, the specificity of these signals is largely determined by extrinsic factors such as the plasma membrane microenvironment and the availability of cell- and tissue-specific protein partners. Third, the ubiquitous expression and functional redundancy of SFKs, along with the diversity of their downstream targets offer plausible explanations for some of the striking facts of PrP biology, such as its functional promiscuity, the viability of PrP knockout mice, and the diversity of PrP transgenic/mutant phenotypes. Based on these considerations, we propose that PrP^C acts as a gatekeeper between neuronal survival and toxicity by controlling the activity and/or endocytic trafficking of ion channels, synaptic proteins and cell adhesion molecules via SFKs. Testing these complex scenarios should prove a formidable but rewarding endeavor.

ACKNOWLEDGMENTS

Our warm gratitude to Florence Kranitz and the CJD Foundation for their kind support. Special thanks to Chuck for his dogged encouragement. Funded by the Deutsche Forschungsgemeinschaft (Grant MA 2525/1-1) and the University of Konstanz AFF.

REFERENCES

Aguzzi, A., Baumann, F., and Bremer, J. (2008). The prion's elusive reason for being. *Annu. Rev. Neurosci.* 31, 439–477. doi: 10.1146/annurev.neuro.31.060407.125620

Ahn, S., Kim, J., Lucaveche, C. L., Reedy, M. C., Luttrell, L. M., Lefkowitz, R. J., et al. (2002). Src-dependent tyrosine phosphorylation regulates dynamin self-assembly and ligand-induced endocytosis of the epidermal growth factor receptor. *J. Biol. Chem.* 277, 26642–26651. doi: 10.1074/jbc.M201499200

Ahn, S., Maudsley, S., Luttrell, L. M., Lefkowitz, R. J., and Daaka, Y. (1999). Src-mediated tyrosine phosphorylation of dynamin is required for beta2-adrenergic receptor internalization and mitogen-activated protein kinase signaling. *J. Biol. Chem.* 274, 1185–1188. doi: 10.1074/jbc.274.3.1185

Biasini, E., Turnbaugh, J. A., Unterberger, U., and Harris, D. A. (2012). Prion protein at the crossroads of physiology and disease. *Trends Neurosci.* 35, 92–103. doi: 10.1016/j.tins.2011.10.002

Biasini, E., Unterberger, U., Solomon, I. H., Massignan, T., Senatore, A., Bian, H., et al. (2013). A mutant prion protein sensitizes neurons to glutamate-induced excitotoxicity. *J. Neurosci.* 33, 2408–2418. doi: 10.1523/JNEUROSCI.3406-12.2013

Bizat, N., Peyrin, J. M., Haik, S., Cochois, V., Beaudry, P., Laplanche, J. L., et al. (2010). Neuron dysfunction is induced by prion protein with an insertional mutation via a Fyn kinase and reversed by sirtuin activation in *Caenorhabditis elegans*. *J. Neurosci.* 30, 5394–5403. doi: 10.1523/JNEUROSCI.5831-09.2010

Bo, L., Quarles, R. H., Fujita, N., Bartoszewicz, Z., Sato, S., and Trapp, B. D. (1995). Endocytic depletion of L-MAG from CNS myelin in quaking mice. *J. Cell Biol.* 131, 1811–1820. doi: 10.1083/jcb.131.6.1811

Bremer, J., Baumann, F., Tiberi, C., Wessig, C., Fischer, H., Schwarz, P., et al. (2010). Axonal prion protein is required for peripheral myelin maintenance. *Nat. Neurosci.* 13, 310–318. doi: 10.1038/nn.2483

Chen, R., Kim, O., Yang, J., Sato, K., Eisenmann, K. M., Mccarthy, J., et al. (2001). Regulation of Akt/PKB activation by tyrosine phosphorylation. *J. Biol. Chem.* 276, 31858–31862. doi: 10.1074/jbc.C100271200

Chen, W. R., Xiong, W., and Shepherd, G. M. (2000). Analysis of relations between NMDA receptors and GABA release at olfactory bulb reciprocal synapses. *Neuron* 25, 625–633. doi: 10.1016/S0896-6273(00)81065-X

Chesebro, B., Trifilo, M., Race, R., Meade-White, K., Teng, C., Lacasse, R., et al. (2005). Anchorless prion protein results in infectious amyloid disease without clinical scrapie. *Science* 308, 1435–1439. doi: 10.1126/science.1110837

Chin, J., Palop, J. J., Puolivali, J., Massaro, C., Bien-Ly, N., Gerstein, H., et al. (2005). Fyn kinase induces synaptic and cognitive impairments in a transgenic mouse model of Alzheimer's disease. *J. Neurosci.* 25, 9694–9703. doi: 10.1523/JNEUROSCI.2980-05.2005

Davis, M. J., Wu, X., Nurkiewicz, T. R., Kawasaki, J., Gui, P., Hill, M. A., et al. (2001). Regulation of ion channels by protein tyrosine phosphorylation. *Am. J. Physiol. Heart Circ. Physiol.* 281, H1835–H1862.

De Biase, L. M., Kang, S. H., Baxi, E. G., Fukaya, M., Pucak, M. L., Mishina, M., et al. (2011). NMDA receptor signaling in oligodendrocyte progenitors is not required for oligodendrogenesis and myelination. *J. Neurosci.* 31, 12650–12662. doi: 10.1523/JNEUROSCI.2455-11.2011

Fadool, D. A., and Levitan, I. B. (1998). Modulation of olfactory bulb neuron potassium current by tyrosine phosphorylation. *J. Neurosci.* 18, 6126–6137.

Fadool, D. A., Tucker, K., Perkins, R., Fasciani, G., Thompson, R. N., Parsons, A. D., et al. (2004). Kv1.3 channel gene-targeted deletion produces "Super-Smeller Mice" with altered glomeruli, interacting scaffolding proteins, and biophysics. *Neuron* 41, 389–404. doi: 10.1016/S0896-6273(03)00844-4

Fadool, D. A., Tucker, K., Phillips, J. J., and Simmen, J. A. (2000). Brain insulin receptor causes activity-dependent current suppression in the olfactory bulb through multiple phosphorylation of Kv1.3. *J. Neurophysiol.* 83, 2332–2348.

Fujita, Y., Krause, G., Scheffner, M., Zechner, D., Leddy, H. E. M., Behrens, J., et al. (2002). Hakai, a c-Cbl-like protein, ubiquitinates and induces endocytosis of the E-cadherin complex. *Nat. Cell Biol.* 4, 222–231. doi: 10.1038/ncb758

Gyllberg, H., Löfgren, K., Lindegren, H., and Bedecs, K. (2006). Increased Src kinase level results in increased protein tyrosine phosphorylation in scrapie-infected neuronal cell lines. *FEBS Lett.* 580, 2603–2608. doi: 10.1016/j.febslet.2006.03.092

Halabisky, B., Friedman, D., Radojicic, M., and Strowbridge, B. W. (2000). Calcium influx through NMDA receptors directly evokes GABA release in olfactory bulb granule cells. *J. Neurosci.* 20, 5124–5134.

Hu, X., He, W., Diaconu, C., Tang, X., Kidd, G. J., Macklin, W. B., et al. (2008). Genetic deletion of BACE1 in mice affects remyelination of sciatic nerves. *FASEB J.* 22, 2970–2980. doi: 10.1096/fj.08-106666

Hu, X., Hicks, C. W., He, W., Wong, P., Macklin, W. B., Trapp, B. D., et al. (2006). Bace1 modulates myelination in the central and peripheral nervous system. *Nat. Neurosci.* 9, 1520–1525. doi: 10.1038/nn1797

Ishiyama, N., Lee, S.-H., Liu, S., Li, G.-Y., Smith, M. J., Reichardt, L. F., et al. (2010). Dynamic and static interactions between p120 catenin and E-cadherin regulate the stability of cell-cell adhesion. *Cell* 141, 117–128. doi: 10.1016/j.cell.2010.01.017

Jiang, T., and Qiu, Y. (2003). Interaction between Src and a C-terminal proline-rich motif of Akt is required for Akt activation. *J. Biol. Chem.* 278, 15789–15793. doi: 10.1074/jbc.M212525200

Khosravani, H., Zhang, Y., Tsutsui, S., Hameed, S., Altier, C., Hamid, J., et al. (2008). Prion protein attenuates excitotoxicity by inhibiting NMDA receptors. *J. Cell Biol.* 181, 551–565. doi: 10.1083/jcb.200711002

Kitazawa, H., Yagi, T., Miyakawa, T., Niki, H., and Kawai, N. (1998). Abnormal synaptic transmission in the olfactory bulb of Fyn-kinase-deficient mice. *J. Neurophysiol.* 79, 137–142.

Kramer-Albers, E. M., and White, R. (2011). From axon-glial signalling to myelination: the integrating role of oligodendroglial Fyn kinase. *Cell Mol. Life Sci.* 68, 2003–2012. doi: 10.1007/s00018-010-0616-z

Kroepfl, J. F., and Gardinier, M. V. (2001a). Identification of a basolateral membrane targeting signal within the cytoplasmic domain of myelin/oligodendrocyte glycoprotein. *J. Neurochem.* 77, 1301–1309. doi: 10.1046/j.1471-4159.2001.00343.x

Kroepfl, J. F., and Gardinier, M. V. (2001b). Mutually exclusive apicobasolateral sorting of two oligodendroglial membrane proteins, proteolipid protein and myelin/oligodendrocyte glycoprotein, in Madin-Darby canine kidney cells. *J. Neurosci. Res.* 66, 1140–1148. doi: 10.1002/jnr.10035

Lavezzari, G., Mccallum, J., Lee, R., and Roche, K. W. (2003). Differential binding of the AP-2 adaptor complex and PSD-95 to the C-terminus of the NMDA receptor subunit NR2B regulates surface expression. *Neuropharmacology* 45, 729–737. doi: 10.1016/S0028-3908(03)00308-3

Lee, G., Thangavel, R., Sharma, V. M., Litersky, J. M., Bhaskar, K., Fang, S. M., et al. (2004). Phosphorylation of tau by fyn: implications for Alzheimer's disease. *J. Neurosci.* 24, 2304–2312. doi: 10.1523/JNEUROSCI.4162-03.2004

Le Pichon, C. E., Valley, M. T., Polymenidou, M., Chesler, A. T., Sagdullaev, B. T., Aguzzi, A., et al. (2009). Olfactory behavior and physiology are disrupted in prion protein knockout mice. *Nat. Neurosci.* 12, 60–69. doi: 10.1038/nn.2238

Li, A., Christensen, H. M., Stewart, L. R., Roth, K. A., Chiesa, R., and Harris, D. A. (2007). Neonatal lethality in transgenic mice expressing prion protein with a deletion of residues 105-125. *EMBO J.* 26, 548–558. doi: 10.1038/sj.emboj.7601507

Li, C., Xiao, L., Liu, X., Yang, W., Shen, W., Hu, C., et al. (2013). A functional role of NMDA receptor in regulating the differentiation of oligodendrocyte precursor cells and remyelination. *Glia* 61, 732–749. doi: 10.1002/glia.22469

Linden, R., Martins, V. R., Prado, M. A., Cammarota, M., Izquierdo, I., and Brentani, R. R. (2008). Physiology of the prion protein. *Physiol. Rev.* 88, 673–728. doi: 10.1152/physrev.00007.2007

Liu, Y., Tao, Y. M., Woo, R. S., Xiong, W. C., and Mei, L. (2007). Stimulated ErbB4 internalization is necessary for neuregulin signaling in neurons. *Biochem. Biophys. Res. Commun.* 354, 505–510. doi: 10.1016/j.bbrc.2007.01.009

Lowell, C. A., and Soriano, P. (1996). Knockouts of Src-family kinases: stiff bones, wimpy T cells, and bad memories. *Genes Dev.* 10, 1845–1857. doi: 10.1101/gad.10.15.1845

Málaga-Trillo, E., and Sempou, E. (2009). PrPs: proteins with a purpose: lessons from the zebrafish. *Prion* 3, 129–133. doi: 10.4161/pri.3.3.9651

Málaga-Trillo, E., Solis, G. P., Schrock, Y., Geiss, C., Luncz, L., Thomanetz, V., et al. (2009). Regulation of embryonic cell adhesion by the prion protein. *PLoS Biol.* 7:e55. doi: 10.1371/journal.pbio.1000055

Maroteaux, L., Campanelli, J. T., and Scheller, R. H. (1988). Synuclein: a neuron-specific protein localized to the nucleus and presynaptic nerve terminal. *J. Neurosci.* 8, 2804–2815.

Martin-Lannerée, S., Hirsch, T. Z., Hernandez-Rapp, J., Halliez, S., Vilotte, J., Launay, J., et al. (2014). PrPC from stem cells to cancer. *Front. Cell Dev. Biol.* 2:55. doi: 10.3389/fcell.2014.00055

Messa, M., Congia, S., Defranchi, E., Valtorta, F., Fassio, A., Onofri, F., et al. (2010). Tyrosine phosphorylation of synapsin I by Src regulates synaptic-vesicle trafficking. *J. Cell Sci.* 123, 2256–2265. doi: 10.1242/jcs.068445

Minami, S. S., Clifford, T. G., Hoe, H. S., Matsuoka, Y., and Rebeck, G. W. (2012). Fyn knock-down increases Abeta, decreases phospho-tau, and worsens spatial learning in 3xTg-AD mice. *Neurobiol. Aging 33*, 825.e815–825.e824. doi: 10.1016/j.neurobiolaging.2011.05.014

Miyashita, Y., and Ozawa, M. (2007). Increased internalization of p120-uncoupled E-cadherin and a requirement for a dileucine motif in the cytoplasmic domain for endocytosis of the protein. *J. Biol. Chem.* 282, 11540–11548. doi: 10.1074/jbc.M608351200

Morel, E., Fouquet, S., Chateau, D., Yvernault, L., Frobert, Y., Pincon-Raymond, M., et al. (2004). The cellular prion protein PrPc is expressed in human enterocytes in cell-cell junctional domains. *J. Biol. Chem.* 279, 1499–1505. doi: 10.1074/jbc.M308578200

Morel, E., Fouquet, S., Strup-Perrot, C., Pichol Thievend, C., Petit, C., Loew, D., et al. (2008). The cellular prion protein PrP(c) is involved in the proliferation of epithelial cells and in the distribution of junction-associated proteins. *PLoS ONE* 3:e3000. doi: 10.1371/journal.pone.0003000

Morse, W. R., Whitesides, J. G. 3rd., Lamantia, A. S., and Maness, P. F. (1998). p59fyn and pp60c-src modulate axonal guidance in the developing mouse olfactory pathway. *J. Neurobiol.* 36, 53–63. doi: 10.1002/(SICI)1097-4695(199807)36:1<53::AID-NEU5>3.0.CO;2-9

Mouillet-Richard, S., Ermonval, M., Chebassier, C., Laplanche, J. L., Lehmann, S., Launay, J. M., et al. (2000). Signal transduction through prion protein. *Science* 289, 1925–1928. doi: 10.1126/science.289.5486.1925

Moya, K. L., Sales, N., Hassig, R., Creminon, C., Grassi, J., and Di Giamberardino, L. (2000). Immunolocalization of the cellular prion protein in normal brain. *Microsc. Res. Tech.* 50, 58–65. doi: 10.1002/1097-0029(20000701)50:1<58::AID-JEMT9>3.0.CO;2-5

Nixon, R. (2005). Prion-associated increases in Src-family kinases. *J. Biol. Chem.* 280, 2455. doi: 10.1074/jbc.M410883200

Ondrusova, L., Reda, J., Zakova, P., and Tuhackova, Z. (2013). Inhibition of mTORC1 by SU6656, the selective Src kinase inhibitor, is not accompanied by activation of Akt/PKB signalling in melanoma cells. *Folia Biol. (Praha)* 59, 162–167.

Pantera, B., Bini, C., Cirri, P., Paoli, P., Camici, G., Manao, G., et al. (2009). PrPc activation induces neurite outgrowth and differentiation in PC12 cells: role for caveolin-1 in the signal transduction pathway. *J. Neurochem.* 110, 194–207. doi: 10.1111/j.1471-4159.2009.06123.x

Passet, B., Halliez, S., Beringue, V., Laude, H., and Vilotte, J. L. (2013). The prion protein family: looking outside the central nervous system. *Prion* 7, 127–130. doi: 10.4161/pri.22851

Peretz, A., Sobko, A., and Attali, B. (1999). Tyrosine kinases modulate K+ channel gating in mouse Schwann cells. *J. Physiol.* 519(Pt 2), 373–384. doi: 10.1111/j.1469-7793.1999.0373m.x

Reynolds, A. B. (2010). Exposing p120 catenin's most intimate affair. *Cell* 141, 20–22. doi: 10.1016/j.cell.2010.03.022

Roskoski, R. Jr. (2004). Src protein-tyrosine kinase structure and regulation. *Biochem. Biophys. Res. Commun.* 324, 1155–1164. doi: 10.1016/j.bbrc.2004.09.171

Roskoski, R. Jr. (2005). Src kinase regulation by phosphorylation and dephosphorylation. *Biochem. Biophys. Res. Commun.* 331, 1–14. doi: 10.1016/j.bbrc.2005.03.012

Salter, M. W., and Kalia, L. V. (2004). Src kinases: a hub for NMDA receptor regulation. *Nat. Rev. Neurosci.* 5, 317–328. doi: 10.1038/nrn1368

Santuccione, A., Sytnyk, V., Leshchyns'ka, I., and Schachner, M. (2005). Prion protein recruits its neuronal receptor NCAM to lipid rafts to activate p59fyn and to enhance neurite outgrowth. *J. Cell Biol.* 169, 341–354. doi: 10.1083/jcb.200409127

Sanz-Clemente, A., Nicoll, R. A., and Roche, K. W. (2013). Diversity in NMDA receptor composition: many regulators, many consequences. *Neuroscientist* 19, 62–75. doi: 10.1177/1073858411435129

Schachner, M., and Bartsch, U. (2000). Multiple functions of the myelin-associated glycoprotein MAG (siglec-4a) in formation and maintenance of myelin. *Glia* 29, 154–165. doi: 10.1002/(SICI)1098-1136(20000115)29:2<154::AID-GLIA9>3.0.CO;2-3

Schubert, W., Prior, R., Weidemann, A., Dircksen, H., Multhaup, G., Masters, C. L., et al. (1991). Localization of Alzheimer beta A4 amyloid precursor protein at central and peripheral synaptic sites. *Brain Res.* 563, 184–194. doi: 10.1016/0006-8993(91)91532-6

Senatore, A., Colleoni, S., Verderio, C., Restelli, E., Morini, R., Condliffe, S. B., et al. (2012). Mutant PrP suppresses glutamatergic neurotransmission in cerebellar granule neurons by impairing membrane delivery of VGCC alpha(2)delta-1 Subunit. *Neuron* 74, 300–313. doi: 10.1016/j.neuron.2012.02.027

Shirazi, S. K., and Wood, J. G. (1993). The protein tyrosine kinase, fyn, in Alzheimer's disease pathology. *Neuroreport* 4, 435–437. doi: 10.1097/00001756-199304000-00024

Silverman, L., Sudol, M., and Resh, M. D. (1993). Members of the src family of nonreceptor tyrosine kinases share a common mechanism for membrane binding. *Cell Growth Differ.* 4, 475–482.

Sobko, A., Peretz, A., and Attali, B. (1998). Constitutive activation of delayed-rectifier potassium channels by a src family tyrosine kinase in Schwann cells. *EMBO J.* 17, 4723–4734. doi: 10.1093/emboj/17.16.4723

Solomon, I. H., Huettner, J. E., and Harris, D. A. (2010). Neurotoxic mutants of the prion protein induce spontaneous ionic currents in cultured cells. *J. Biol. Chem.* 285, 26719–26726. doi: 10.1074/jbc.M110.134619

Sperber, B. R., Boyle-Walsh, E. A., Engleka, M. J., Gadue, P., Peterson, A. C., Stein, P. L., et al. (2001). A unique role for Fyn in CNS myelination. *J. Neurosci.* 21, 2039–2047.

Spielhaupter, C., and Schatzl, H. M. (2001). PrPC directly interacts with proteins involved in signaling pathways. *J. Biol. Chem.* 276, 44604–44612. doi: 10.1074/jbc.M103289200

Steele, A. D., Emsley, J. G., Ozdinler, P. H., Lindquist, S., and Macklis, J. D. (2006). Prion protein (PrPc) positively regulates neural precursor proliferation during developmental and adult mammalian neurogenesis. *Proc. Natl. Acad. Sci. U.S.A.* 103, 3416–3421. doi: 10.1073/pnas.0511290103

Steele, A. D., Lindquist, S., and Aguzzi, A. (2007). The prion protein knockout mouse: a phenotype under challenge. *Prion* 1, 83–93. doi: 10.4161/pri.1.2.4346

Thomas, S. M., and Brugge, J. S. (1997). Cellular functions regulated by Src family kinases. *Annu. Rev. Cell Dev. Biol.* 13, 513–609. doi: 10.1146/annurev.cellbio.13.1.513

Thoreson, M. A., Anastasiadis, P. Z., Daniel, J. M., Ireton, R. C., Wheelock, M. J., Johnson, K. R., et al. (2000). Selective uncoupling of p120(ctn) from E-cadherin disrupts strong adhesion. *J. Cell Biol.* 148, 189–202. doi: 10.1083/jcb.148.1.189

Tiran, Z., Peretz, A., Sines, T., Shinder, V., Sap, J., Attali, B., et al. (2006). Tyrosine phosphatases epsilon and alpha perform specific and overlapping functions in regulation of voltage-gated potassium channels in Schwann cells. *Mol. Biol. Cell* 17, 4330–4342. doi: 10.1091/mbc.E06-02-0151

Um, J. W., Nygaard, H. B., Heiss, J. K., Kostylev, M. A., Stagi, M., Vortmeyer, A., et al. (2012). Alzheimer amyloid-beta oligomer bound to postsynaptic prion protein activates Fyn to impair neurons. *Nat. Neurosci.* 15, 1227–1235. doi: 10.1038/nn.3178

Umemori, H., Kadowaki, Y., Hirosawa, K., Yoshida, Y., Hironaka, K., Okano, H., et al. (1999). Stimulation of myelin basic protein gene transcription by Fyn tyrosine kinase for myelination. *J. Neurosci.* 19, 1393–1397.

Umemori, H., Sato, S., Yagi, T., Aizawa, S., and Yamamoto, T. (1994). Initial events of myelination involve Fyn tyrosine kinase signalling. *Nature* 367, 572–576. doi: 10.1038/367572a0

Vojtechova, M., Tureckova, J., Kucerova, D., Sloncova, E., Vachtenheim, J., and Tuhackova, Z. (2008). Regulation of mTORC1 signaling by Src kinase activity is Akt1-independent in RSV-transformed cells. *Neoplasia* 10, 99–107. doi: 10.1593/neo.07905

White, R., Gonsior, C., Kramer-Albers, E. M., Stohr, N., Huttelmaier, S., and Trotter, J. (2008). Activation of oligodendroglial Fyn kinase enhances translation of mRNAs transported in hnRNP A2-dependent RNA granules. *J. Cell Biol.* 181, 579–586. doi: 10.1083/jcb.200706164

White, R., and Kramer-Albers, E. M. (2014). Axon-glia interaction and membrane traffic in myelin formation. *Front. Cell. Neurosci.* 7:284. doi: 10.3389/fncel.2013.00284

Wilde, A., Beattie, E. C., Lem, L., Riethof, D. A., Liu, S. H., Mobley, W. C., et al. (1999). EGF receptor signaling stimulates SRC kinase phosphorylation of clathrin, influencing clathrin redistribution and EGF uptake. *Cell* 96, 677–687. doi: 10.1016/S0092-8674(00)80578-4

Winterstein, C., Trotter, J., and Kramer-Albers, E. M. (2008). Distinct endocytic recycling of myelin proteins promotes oligodendroglial membrane remodeling. *J. Cell. Sci.* 121, 834–842. doi: 10.1242/jcs.022731

Yagi, T., Aizawa, S., Tokunaga, T., Shigetani, Y., Takeda, N., and Ikawa, Y. (1993). A role for Fyn tyrosine kinase in the suckling behaviour of neonatal mice. *Nature* 366, 742–745. doi: 10.1038/366742a0

Zhang, C. C., Steele, A. D., Lindquist, S., and Lodish, H. F. (2006). Prion protein is expressed on long-term repopulating hematopoietic stem cells and is important for their self-renewal. *Proc. Natl. Acad. Sci. U.S.A.* 103, 2184–2189. doi: 10.1073/pnas.0510577103

Zimmerman, B., Simaan, M., Lee, M. H., Luttrell, L. M., and Laporte, S. A. (2009). c-Src-mediated phosphorylation of AP-2 reveals a general mechanism for receptors internalizing through the clathrin pathway. *Cell Signal* 21, 103–110. doi: 10.1016/j.cellsig.2008.09.013

Zou, L., Wang, Z., Shen, L., Bao, G. B., Wang, T., Kang, J. H., et al. (2007). Receptor tyrosine kinases positively regulate BACE activity and Amyloid-beta production through enhancing BACE internalization. *Cell Res.* 17, 389–401. doi: 10.1038/cr.2007.5

Conflict of Interest Statement: The authors declare that the research was conducted in the absence of any commercial or financial relationships that could be construed as a potential conflict of interest.

21

Distinct N-terminal regions of the exomer secretory vesicle cargo Chs3 regulate its trafficking itinerary

Amanda M. Weiskoff and J. Christopher Fromme*

Department of Molecular Biology and Genetics, Weill Institute for Cell and Molecular Biology, Cornell University, Ithaca, NY, USA

Edited by:
James E. Casanova, University of Virginia, USA

Reviewed by:
Vladimir Lupashin, University of Arkansas for Medical Sciences, USA
Christopher Stroupe, University of Virginia School of Medicine, USA

***Correspondence:**
J. Christopher Fromme, Department of Molecular Biology and Genetics, Weill Institute for Cell and Molecular Biology, Cornell University, 457 Weill Hall, Ithaca, NY 14853, USA
e-mail: jcf14@cornell.edu

Cells transport integral membrane proteins between organelles by sorting them into vesicles. Cargo adaptors act to recognize sorting signals in transmembrane cargos and to interact with coat complexes that aid in vesicle biogenesis. No coat proteins have yet been identified that generate secretory vesicles from the *trans*-Golgi network (TGN) to the plasma membrane, but the exomer complex has been identified as a cargo adaptor complex that mediates transport of several proteins in this pathway. Chs3, the most well-studied exomer cargo, cycles between the TGN and the plasma membrane in synchrony with the cell cycle, providing an opportunity to study regulation of proteins that cycle in response to signaling. Here we show that different segments of the Chs3 N-terminus mediate distinct trafficking steps. Residues 10–27, known to mediate retention, also appear to play a role in internalization. Residues 28–52 are involved in transport to the plasma membrane and recycling out of endosomes to prevent degradation in the vacuole. We also present the crystal structure of residues 10–27 bound to the exomer complex, suggesting different cargo adaptors could compete for binding to this segment, providing a potential mechanism for regulation.

Keywords: cargo sorting, Golgi, secretory vesicle, AP-1 complex, chitin synthase

INTRODUCTION

Eukaryotic cells must transport transmembrane proteins between different subcellular compartments, often in response to specific signals or conditions. This transport is mediated by coat complexes, which help to form the shape of transport vesicles (Bonifacino and Glick, 2004). These coat complexes also contain or interact with adaptor proteins that recognize sorting signals in the cytosolic domains of cargo proteins to sort them into the vesicles.

No coat complexes are known to mediate transport directly from the *trans*-Golgi network (TGN) to the apical plasma membrane (PM) of polarized cells, which corresponds to secretory vesicles in the model organism *Saccharomyces cerevisiae* (budding yeast). Therefore, it remains poorly understood how cells regulate this trafficking step. One of the many transmembrane proteins that follow this route of transport in yeast is the chitin synthase enzyme Chs3. Chs3 cycles between the TGN and the cell surface in a cell cycle dependent manner (Chuang, 1996; Ziman et al., 1998; Zanolari et al., 2011). This localization pattern is reminiscent of other proteins for which localization is regulated by signaling, such as the human Glut4 glucose transporter (Bryant et al., 2002). Chs3 is localized to the bud neck (junction of mother and daughter cells) through its interaction with its activator Chs4, which binds the septin-interacting protein Bni4 (DeMarini, 1997; Reyes et al., 2007).

The transport of Chs3 to the cell surface requires the exomer complex, which acts as a cargo adaptor for Chs3 and other cargos. Exomer consists of the core protein Chs5, and four paralagous adaptor proteins called Chs5-Arf1-binding Proteins (ChAPs): Chs6, Bud7, Bch1, and Bch2. Deletion of the *CHS5* or *CHS6* genes, or simultaneous deletion of the *BUD7* and *BCH1* genes, prevents Chs3 transport to the cell surface (Santos and Snyder, 1997; Sanchatjate and Schekman, 2006; Trautwein et al., 2006; Wang et al., 2006; Starr et al., 2012).

Two other proteins have been identified as exomer cargos: Fus1, a protein involved in mating, and Pin2, a prion-like domain-containing protein. Transport of Pin2, unlike Chs3, requires Bch1 or Bch2, and requires the C-terminus of Pin2, which has little similarity to any part of Chs3 (Ritz et al., 2014) Fus1 has a sorting signal not found in Chs3 and requires Bch1 and Bud7 for transport (Barfield et al., 2009). This suggests that exomer recognizes multiple motifs, possibly through interaction with different subsets of ChAPs.

Retention of Chs3 at the TGN requires AP-1, an adaptor protein complex that mediates trafficking between the TGN and endosomes (Valdivia et al., 2002). Disruption of the AP-1 complex partially rescues the phenotype of an exomer deletion, due to escape of Chs3 to the cell surface when both exomer and AP-1 pathways are blocked. Disruption of the Gga1/2 clathrin adaptors has a similar effect (Copic et al., 2007).

It was recently reported that both the exomer dependent transport and the AP-1 dependent retention of Chs3 are mediated by a motif near the N-terminus of the protein, $_{19}$DEESLL$_{24}$ (Starr et al., 2012). An interaction between exomer and the C-terminus of Chs3 was also found to be required for its transport out of the TGN (Rockenbauch et al., 2012).

In this study we examine the interaction between Chs3 and exomer. We find that different regions of Chs3 play different roles in balancing Chs3 traffic to and away from the PM. We also present a crystal structure of a portion of the Chs3 N-terminus bound to the exomer complex. We propose a role for the N-terminus in regulating both transport and retention of Chs3 by facilitating competition between the protein complexes required for these two processes.

MATERIALS AND METHODS

MICROSCOPY

Cells were grown to log phase ($OD_{600} \sim 0.5$) in synthetic dropout media, and imaged on a DeltaVision RT wide-field deconvolution microscope (Applied Precision). Images were deconvolved in SoftWoRx 3.5.0 software (Applied Precision) and min/max light levels adjusted for clarity in ImageJ (Abramoff et al., 2004) with levels kept consistent within each experiment.

EXOMER PURIFICATION

Recombinant "core" exomer complex (Chs5 residues 1–77 and Chs6-6xHis) was purified as described for the Chs5(1–299)/Chs6-6xHis construct (Paczkowski et al., 2012). Protein was concentrated to ~25 mg/ml for crystallography, or 5 mg/ml for the interaction assay.

INTERACTION ASSAY

GST-Chs3 fragment constructs were constructed in the pGEX-2T vector (GE Healthcare) and transformed into Rosetta2 (DE3) *E. coli* cells (Novagen) for expression. 1 L culture was grown to ~3 OD_{600} in TB media at 37°C, temperature lowered to 18°C, then expression induced with 240 uM IPTG. After overnight expression, cells were harvested by centrifugation, resuspended in 50 ml PBS buffer with 1 mM DTT, and lysed by sonication. GST fusion proteins were isolated by adding 100 μl equilibrated glutathione resin (G-Biosciences) to 5 ml cleared lysate and incubating with rotation at least 2 h at 4°C. Resin was washed 3 times with 1 ml PBS+DTT and resuspended in 500 μl PBS+DTT. 10 μl of 5 mg/ml exomer protein was added and mixture was incubated ~1 h at 4°C. Resin was washed 3 times with 1 ml PBS+DTT and analyzed by SDS-PAGE and Western blot with anti-6xHis antibody (Covance).

CRYSTALLOGRAPHY

The Chs5(1–77)/Chs6-6xHis exomer complex was co-crystallized with Chs3 peptides (Genscript, 95% purity) using the hanging drop vapor diffusion method. The peptides were resuspended in the precipitant solution (0.3 M ammonium sulfate, 0.1 M citric acid pH 4.0) to a concentration of 100 μM. 1 μl of the peptide solution was mixed with 1 μl of 25 mg/ml exomer, resulting in a molar ratio of 3.85:1 peptide:exomer, and this drop was placed on a cover slip above the precipitant solution. Hexagonal plate-shaped crystals appeared after 5–7 days. Crystals were cryoprotected in 0.3 M ammonium sulfate, 0.1 M citric acid pH 4.0, 30% glycerol, and 100 μM peptide. Diffraction data were collected at CHESS (Cornell High Energy Synchrotron Source) beamline A1 and processed using HKL-2000 (Otwinowski and Minor, 1997). The structure was solved by molecular replacement with Phaser in the PHENIX software suite (Adams et al., 2010) using residues 1–77 of Chs5 and all of Chs6 from the Chs5(1–299)/Chs6 exomer complex structure (PDB: 4GNS; Paczkowski et al., 2012). Density for the Chs3 peptide was clearly visible in initial difference maps (**Figure 6A**). The model was refined by manual adjustment in Coot (Emsley et al., 2010) and refinement in PHENIX. Our software is maintained by SBGrid (Morin et al., 2013).

The coordinates and structure factors have been deposited in the Protein Data Bank (www.rcsb.org) with accession number 4U9T.

RESULTS

THE N-TERMINUS OF Chs3 IS IMPORTANT FOR FUNCTION AND LOCALIZATION

Chs3 is a polytopic membrane protein with both termini exposed to the cytoplasm (**Figure 1A**). The cell cycle dependent localization of Chs3 (**Figure 1B**) depends upon its transit through multiple trafficking pathways. Several potential sorting signals reside in the N-terminus of Chs3 (**Figure 1C**).

To assess the importance of the N- and C-termini of Chs3 in its trafficking, we transformed truncated forms of the Chs3 protein into a strain in which Chs3 had been deleted. Chs3 transport defects were assayed by resistance to calcofluor white (CW), which indicates decreased levels of chitin in the cell wall. Increased resistance to CW arises through inactivation or mislocalization of Chs3. All mutations were made in a Chs3-GFP plasmid, which was found to be functional by its ability to rescue normal CW sensitivity levels. Deleting either residues 2–52 (ΔN52) or 1145–1165(ΔC21) of Chs3 caused significant CW resistance (**Figure 1D**).

Disrupting the AP-1 clathrin adaptor complex impairs Chs3 retention within the cell, and therefore will partially rescue mutants that are compromised specifically in TGN to PM trafficking (**Figure 1E**) (Valdivia et al., 2002). CW sensitivity of a strain containing Chs3(ΔN52)-GFP was rescued by deletion of the AP-1 component *APL2* (**Figure 1D**), suggesting this truncation disrupts transport to the PM. In contrast, CW sensitivity of the Chs3(ΔC21)-GFP mutant was not rescued by the AP-1 disruption (**Figure 1D**). This suggests an additional unknown role for the C-terminus of Chs3 in trafficking or activation, in addition to its identified role interacting with the exomer complex (Rockenbauch et al., 2012). Together, these results indicate that both the N-terminus and C-terminus of Chs3 are important for its trafficking, though they play different roles.

We examined the localization of the N-terminally truncated Chs3-GFP relative to that of the wild type Chs3-GFP. While Chs3 is seen at the bud neck in small buds and at the septum during cytokinesis in both the mutant and wild type strains (**Figure 1F**), the fluorescence intensity at these structures was lower for the Chs3(ΔN52)-GFP than the wild type (**Figures 1F,G**). Surprisingly, there are fewer visible internal punctae of GFP containing the truncated protein, suggesting an additional defect in retention at the TGN. To determine if this mutant protein has reduced expression relative to the wild-type, we performed α-GFP Western blot analysis and observed that the expression level of the mutant appears similar to that of the wild-type protein (**Figure 1H**, Figure S1). Deletion of residues 1-63 of Chs3 was previously reported to result in Chs3 being retained in the ER (Sacristan et al., 2013), but we did not observe any

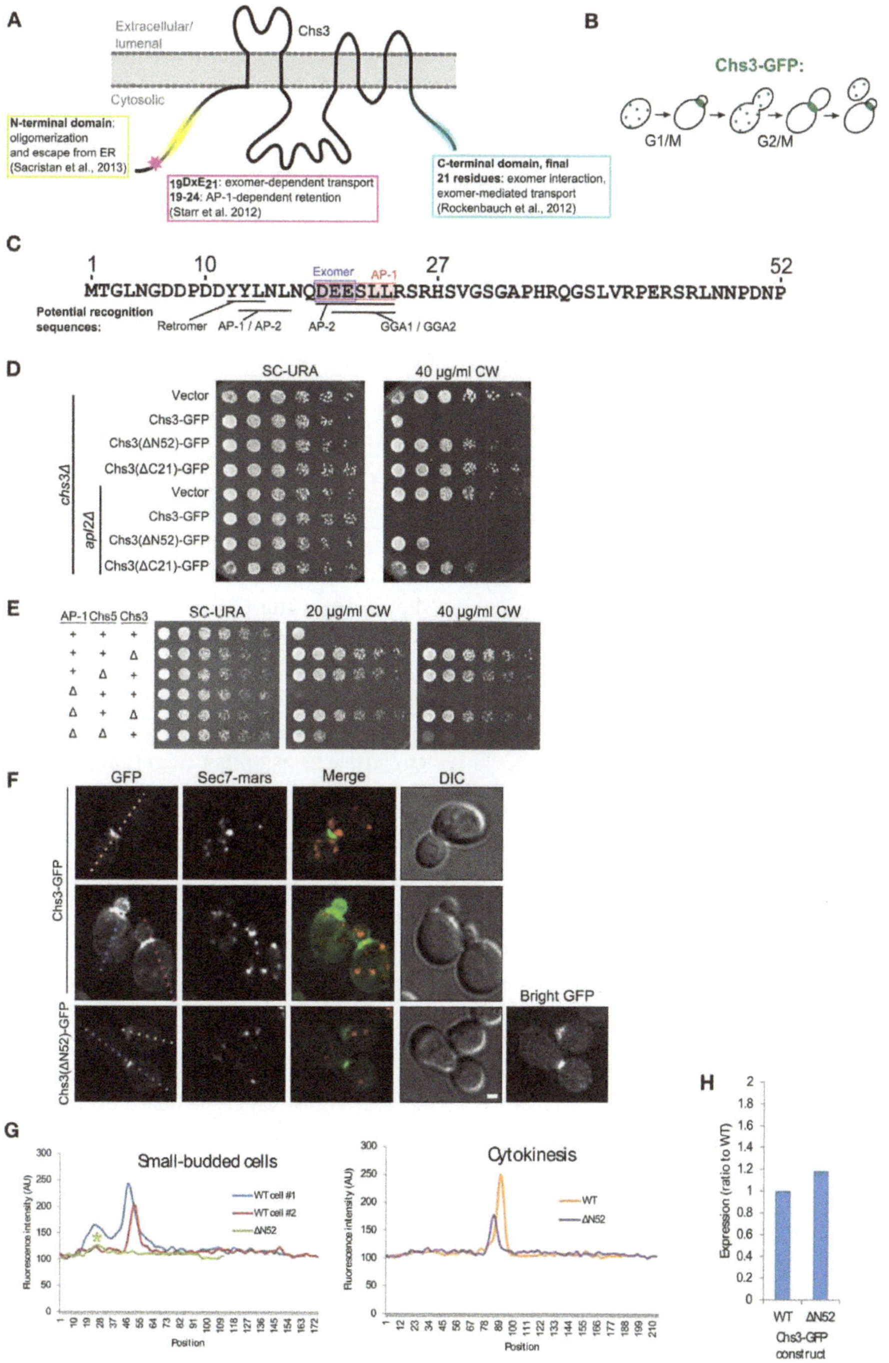

FIGURE 1 | The N-terminus of Chs3 is important for its function and localization. **(A)** The topology of Chs3 (Sacristan et al., 2013) and regions known to be important for its trafficking. **(B)** The itinerary of Chs3 throughout the cell cycle. **(C)** The N-terminus of Chs3 with sites known to be required for adaptor protein interaction (shaded rectangles) and predicted sorting signals for other cargo adaptors. **(D)** Chs3-GFP mutant plasmids were transformed into both a *chs3Δ SEC7-Mars* strain and a *chs3Δ SEC7-Mars apl2Δ* strain and a 10-fold dilution series was plated on indicated media. Plates were imaged after 2 days at 30°C. See also Figure S2 for additional CW concentrations. **(E)** Strains with indicated phenotypes were plated on indicated concentrations of CW. All strains are genomic *chs3Δ*. Chs3 "+" strains contain Chs3-GFP plasmid, and other strains contain empty expression vector to support growth on SC-URA. See also Figure S2 for additional CW concentrations. **(F)** Chs3-GFP and Sec7-Mars localization in *chs3Δ* cells. Wild type cells showing bud neck localization in small buds (first row) and at cytokinesis (second row) are represented. Chs3(ΔN52)-GFP intensity at bud neck is very low, so a "bright GFP" image with light levels scaled to increase visibility is included. Scale bar, 1 µm. **(G)** Line profile of intensity along dotted lines in (F) of cells with small buds (left) or cytokinesis (right). Asterisk marks significantly lower peak at the bud neck of Chs3(ΔN52)-GFP localization. **(H)** Protein expression level of Δ52 construct relative to wild type Chs3-GFP, as calculated from the α-GFP Western blots in Figure S1, normalized to loading control. Value shown is an average between the two experiments.

ER-retention of the ΔN52 mutant. It is unclear where in the cell the remaining ΔN52 protein is being retained when not transported to the bud neck, since it is not visibly enriched in any compartments. It is possible that it is not visible because it is spread out over multiple locations, perhaps partially retained in the TGN and endosomes and spread diffusely along the PM. Overall, residues 1–52 appear to play a role in efficient transport to the bud neck, but their deletion does not completely prevent transport to the PM.

DISTINCT REGIONS OF THE Chs3 N-TERMINUS MEDIATE DIFFERENT TRAFFICKING PATHWAYS

Since the deletion of residues 2–52 may remove multiple sorting signals (**Figure 1C**), shorter segments within residues 1–52 (**Figure 2A**) were deleted to more precisely map the function of the N-terminus. Deletions comprising residues 2–9, 10–27, or 28–52 all conferred some resistance to CW (**Figure 2B**). Deletion of residues 2-9 had the mildest phenotype. Deletion of residues 10–27, which contain residues reported to interact with exomer and AP-1 (Starr et al., 2012), had a moderate phenotype. The deletion of residues 28–52 had the greatest effect, nearly equivalent to that of a strain lacking Chs3. This Chs3(Δ28–52)-GFP construct also had somewhat lower expression than the wild type or Chs3(Δ10–27)-GFP, which may contribute to the strong CW growth phenotype (**Figure 2C**, Figure S1). These results suggest that residues 28–52 are the most important in this region for the transport of Chs3 to the PM.

The Chs3(Δ10–27)-GFP protein localized to small buds and the primary septum as well as the wild type (**Figure 2D**). However, it also mislocalized to large portions of the PM, especially in non-budded and large-budded cells (**Figures 2D,E**). This suggests a defect in internalization by endocytosis. Chs3 enzymatic activity requires activation by Chs4, which is held at the bud neck by an indirect interaction with septins (DeMarini, 1997). Therefore, the mutant Chs3 mislocalized throughout the PM is unlikely to be actively producing chitin, likely causing the increased CW resistance we observed.

The Chs3(Δ28–52)-GFP mutant exhibited a different pattern of localization (**Figure 2F**). It localized to the bud neck and septum, but with lower fluorescence intensity than the wild type protein (**Figure 2G**). We also observed GFP in the vacuole with this mutant, and less localization to TGN punctae. The inability to divert endocytosed Chs3(Δ28–52)-GFP back to the TGN and away from the vacuole could explain the lower level of this protein in cells (**Figure 2C**, Figure S1).

MUTATION OF RESIDUES 19–21 LEADS TO INCREASED PM LOCALIZATION

To identify the residues within the 10–27 segment most important for its function, alanine scanning mutagenesis was performed, with groups of three sequential amino acids mutated to alanine. In addition, the two serine residues were mutated, as serine is a potential phosphorylation target. The serines were mutated to either alanine or aspartate, to prevent or mimic phosphorylation, respectively. Two of these mutants had a mild CW resistance phenotype: ${}_{19}DEE_{21} \rightarrow AAA$, and S24D/S26D (**Figure 3A**). While the mutated constructs varied in expression level, the mutant constructs with the lowest levels were still sensitive to CW, indicating all mutants were expressed at levels sufficient to maintain normal levels of chitin in the cell wall if trafficked properly (**Figure 3B**, Figure S1).

Since a role for residues 19–21 in exomer-mediated transport is consistent with previous findings (Starr et al., 2012), we also looked at the localization of this mutant protein (**Figures 3C,D**). Its pattern of localization appeared the same as that of the deletion of residues 10–27 (**Figure 2D**): normal localization at the bud neck, and mislocalization along the entire PM. This suggests that residues 19–21 are important for internalization and possibly retention of Chs3. Previously these residues were shown to be important for the interaction between Chs3 and AP-1 (Starr et al., 2012), and it is possible they are also important for the interaction of Chs3 with AP-2 during endocytosis, as AP-1 and AP-2 recognize similar sorting signals (Bonifacino and Traub, 2003).

We next determined whether disruption of AP-1 function would rescue the CW sensitivity of the Chs3(Δ10–27)-GFP or Chs3(${}_{19}DEE_{21} \rightarrow AAA$)-GFP mutants. AP-1 disruption did increase the CW sensitivity of both mutants (**Figure 3E**). This result indicates that AP-1 contributes to retention of the Chs3(${}_{19}DEE_{21} \rightarrow AAA$)-GFP mutant, and implies AP-1 can still partially interact with this mutant despite disruption of this motif.

MUTATION OF RESIDUES 41–43 LEADS TO DECREASED PM LOCALIZATION

We used alanine scanning mutagenesis to determine the most important portions of the segment containing Chs3 residues 28–52. The mutation of residues 41–43 resulted in the strongest CW resistance, equivalent to that of the ΔN52 truncation (**Figure 4A**). Mutations in residues immediately preceding or following these residues (38–40 and 44–46) resulted in more modest CW resistance phenotypes. This indicates a region centered around residues 41–43 is required for normal transport or function of Chs3. Importantly, this region is highly conserved (**Figure 2A**). The defect of this mutant is not due to lower levels of the protein, as the ${}_{41}LVR_{43} \rightarrow AAA$ mutation did not decrease expression (**Figure 4B**, Figure S1). The ${}_{41}LVR_{43} \rightarrow AAA$ mutation affects Chs3 localization, with decreased levels at the bud neck and the septum (**Figures 4C,D**), with no apparent increase in retention at the TGN. This localization defect was slightly less severe than deleting residues 28–52. This could indicate the residues around 41–43 are still able to facilitate transport even when 41–43 are mutated, perhaps by interacting with exomer.

Surprisingly, the Chs3(${}_{41}LVR_{43} \rightarrow AAA$)-GFP mutant was more resistant to CW than the Chs3(Δ28–52)-GFP mutant (**Figure 4E**). This may suggest a more drastic local structural rearrangement when these residues are mutated to alanine vs. what occurs when the entire segment is deleted. Alternatively, more than one trafficking pathway may be disrupted by deletion of the entire segment. The Chs3(${}_{41}LVR_{43} \rightarrow AAA$)-GFP mutant was partially rescued by the *apl2* deletion (**Figure 4E**), indicating it is still able to interact with AP-1 and suggesting a role for this region in transport to the PM. In contrast, the CW sensitivity of the Chs3(Δ28–52)-GFP mutant

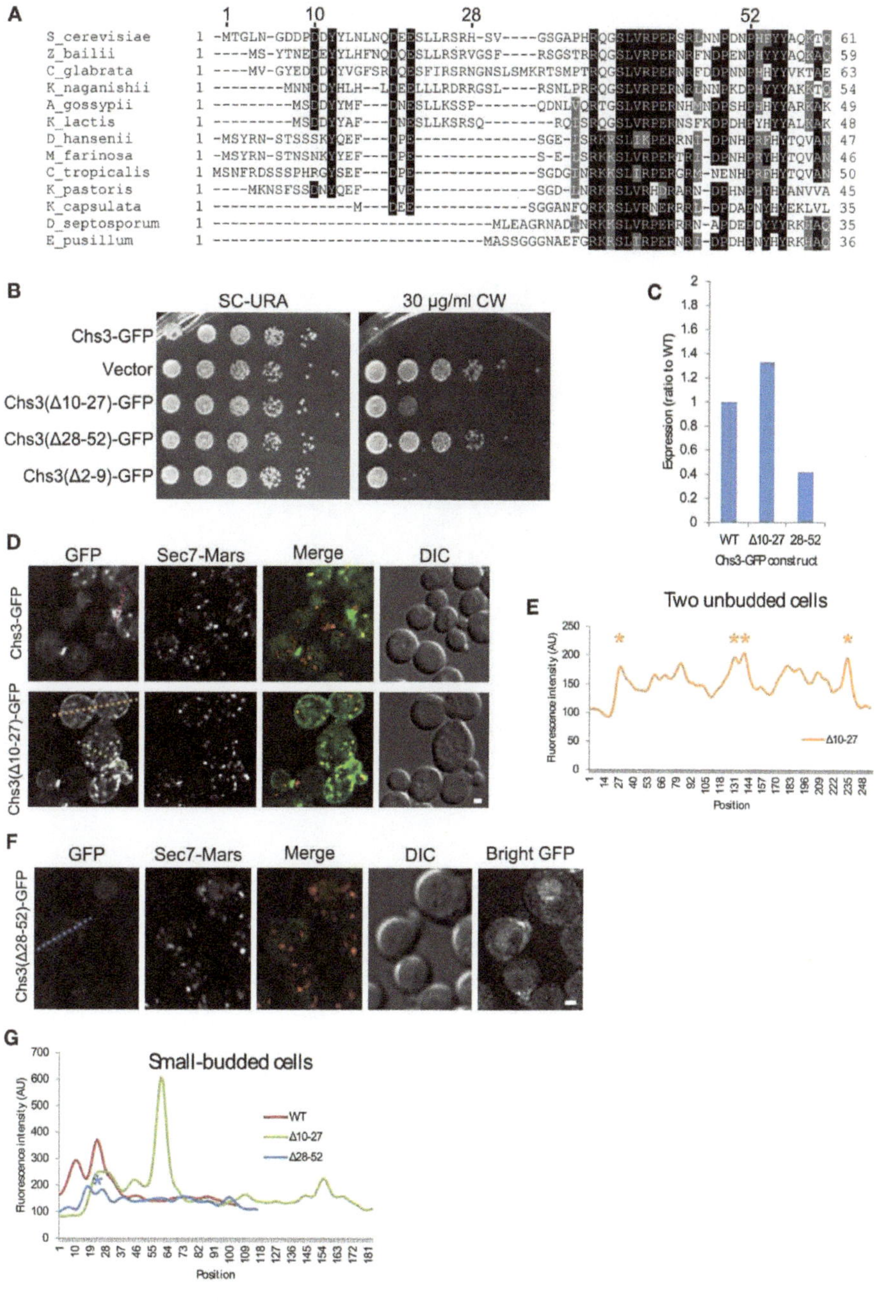

FIGURE 2 | Residues 10–27 and 28–52 are important for distinct trafficking steps. (A) Multiple sequence alignment of several Chs3 homologs. Residues are shaded to highlight similarity (gray) or identity (black). **(B)** The indicated plasmids were transformed into both a *chs3*Δ *SEC7-Mars* strain and a *chs3*Δ *SEC7-Mars apl2*Δ strain and plated on indicated media. Plates were imaged after 2 days at 30°C. See also Figure S2 for additional CW concentrations. **(C)** Protein expression level of mutated Chs3-GFP constructs relative to wild type as calculated from the α-GFP Western blots in Figure S1, normalized to loading control. **(D)** Chs3-GFP and Sec7-Mars localization in *chs3*Δ cells for WT Chs3-GFP and Chs3(Δ10–27)-GFP, at equivalent light levels. Scale bar, 1 μm. **(E)** Line profile of intensity along dotted lines in **(D)** of unbudded cells. Asterisks mark higher localization of Chs3(Δ10–27)-GFP to the plasma membrane at cell boundaries. **(F)** Chs3(Δ28–52)-GFP localization examined as in **(C)**. Additional "Bright GFP" image with light levels scaled to improve visibility is presented. **(G)** Line profile as in **(E)** for small-budded cells. Asterisk indicates lower bud neck localization of Chs3(Δ28–52)-GFP.

was not rescued by *apl2* deletion. This suggests this mutant is not efficiently retained by AP-1, and is consistent with our observation that this mutant is degraded in the vacuole (**Figure 2E**).

RESIDUES 28–52 INTERACT WITH EXOMER MORE STRONGLY THAN RESIDUES 10–27 OR THE C-TERMINUS

We created an *in vitro* assay to detect the direct interaction of Chs3 fragments with exomer. GST-fusion proteins were immobilized

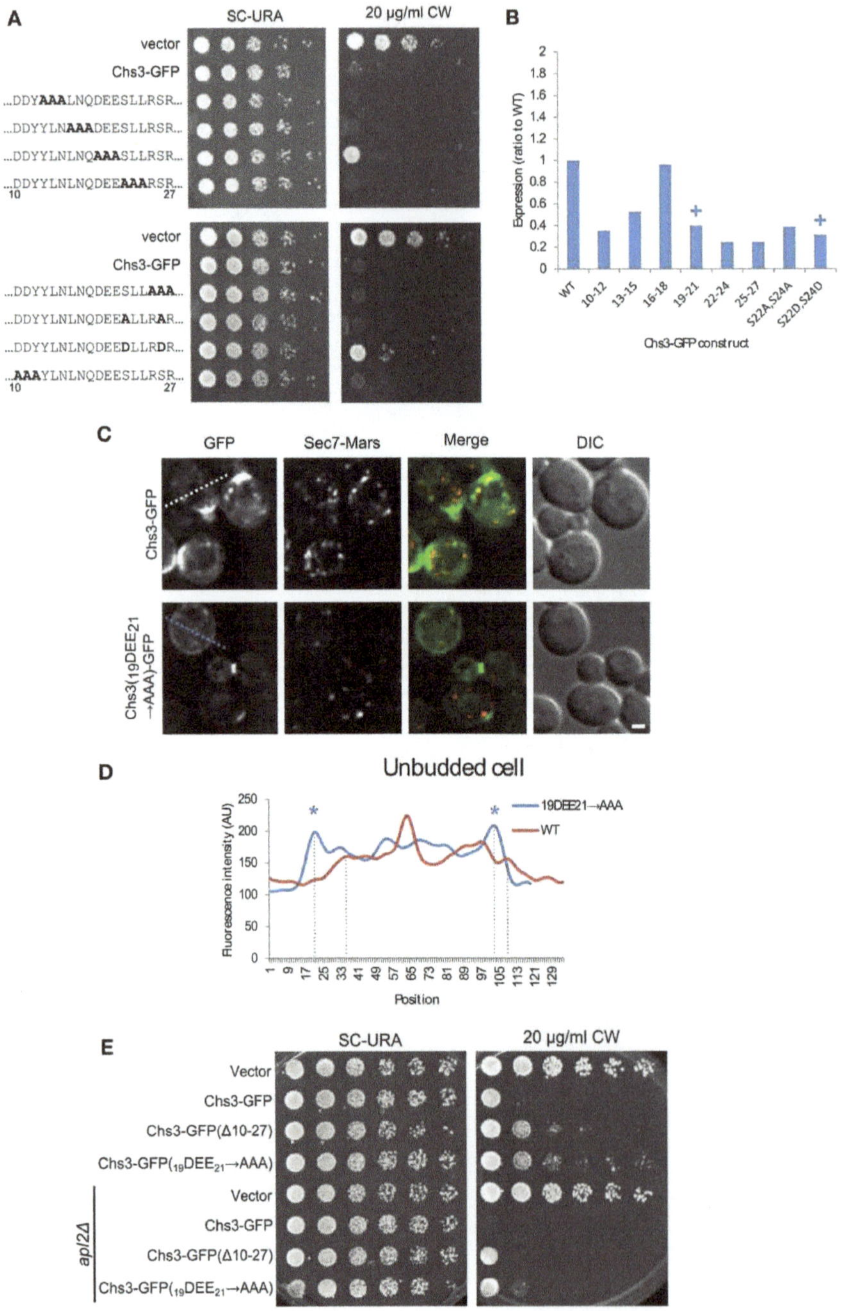

FIGURE 3 | Mutation of Chs3 residues 19–21 causes a defect in internalization. (A) Chs3-GFP plasmids containing alanine scanning mutants were transformed into a *chs3Δ SEC7-Mars* strain and plated on indicated concentrations of CW. Plates were imaged after 2 days at 30°C. See also Figure S2 for additional CW concentrations. **(B)** Protein expression level of alanine-scanning Chs3-GFP constructs relative to wild type as calculated from the α-GFP Western blots in Figure S1, normalized to loading control. Mutations that confer CW resistance phenotypes are marked with (+). **(C)** GFP and Sec7-Mars localization in *chs3Δ* cells for WT Chs3-GFP and Chs3($_{19}$DEE$_{21}$→AAA)-GFP, at equivalent light levels. Scale bar, 1 μm. **(D)** Line profile of intensity along dotted line in **(C)** of unbudded cell expressing mutant protein, or across non-bud section of cell expressing WT protein. Asterisks mark higher localization of Chs3($_{19}$DEE$_{21}$ →AAA)-GFP to the plasma membrane at cell boundaries (dotted lines). **(E)** Chs3(Δ10–27)-GFP and Chs3($_{19}$DEE$_{21}$ →AAA)-GFP were introduced into the *chs3Δ* and *chs3Δapl2Δ* strains and plated on indicated media. Plates were imaged after 2 days at 30°C. See also Figure S2 for additional CW concentrations.

on glutathione resin, which was incubated with exomer complex. Residues 1–52 were able to interact with exomer (**Figure 5A**). Residues 10–27 did not interact with exomer in this assay, and deleting residues 10–27 from the residue 1–52 segment only decreased exomer binding by about half. Together, these indicate residues 10–27 are neither necessary nor sufficient for interaction in this assay. Residues 28–52 were able to interact with exomer even more strongly than residues 1–52 (**Figure 5B**).

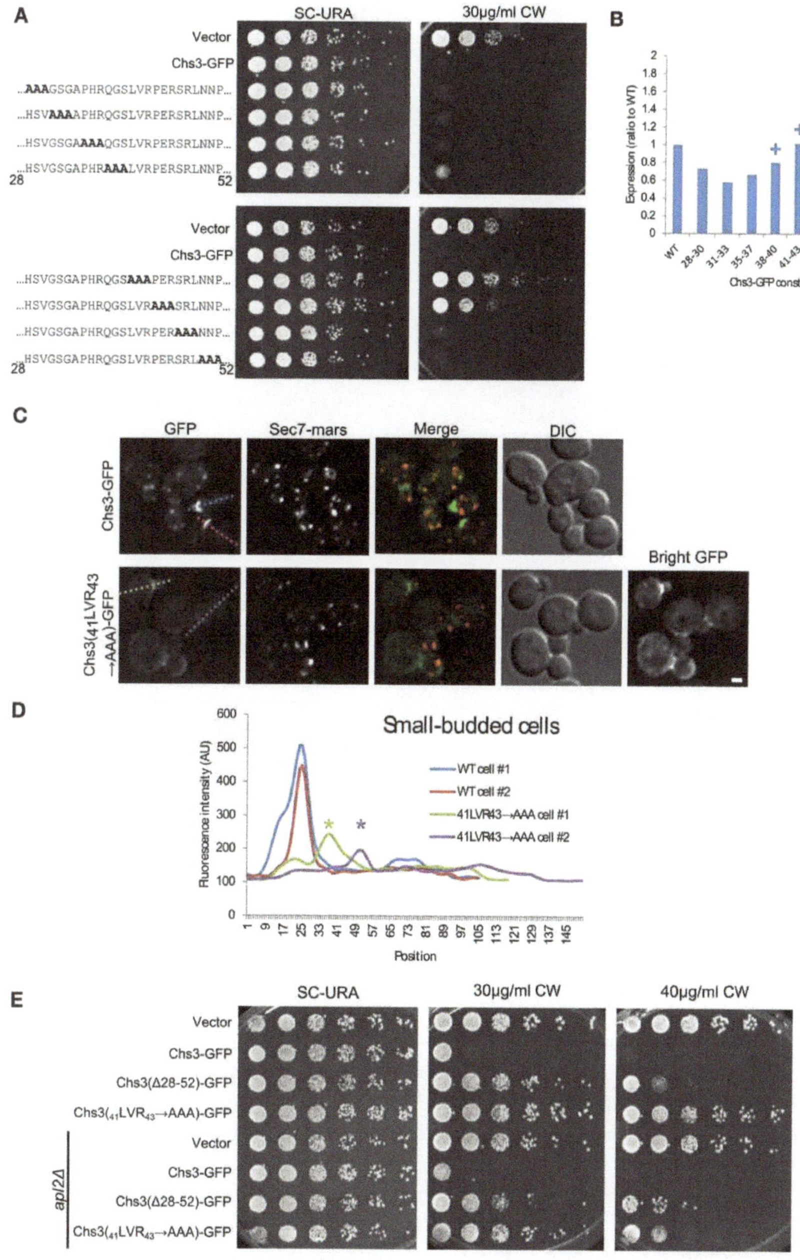

FIGURE 4 | Deletion of Chs3 residues 41–43 has a strong defect in transport. (A) Chs3-GFP plasmids containing alanine scanning mutants were transformed into a *chs3Δ SEC7-Mars* strain and plated on indicated concentrations of CW. Plates were imaged after 2 days at 30°C. See also Figure S2 for additional CW concentrations. **(B)** Protein expression level of alanine-scanning Chs3-GFP constructs relative to wild type as calculated from the α-GFP Western blots in Figure S1, normalized to loading control. Mutations that confer CW resistance phenotypes are marked with (+). **(C)** GFP and Sec7-Mars localization in *chs3Δ* cells for WT Chs3-GFP and Chs3($_{41}$LVR$_{43}$ →AAA)-GFP, at equivalent light levels. Scale bar, 1 μm. **(D)** Line profile of intensity along dotted lines in **(C)** of small-budded cells. Asterisks mark significantly lower peaks at the bud neck of Chs3($_{41}$LVR$_{43}$ →AAA)-GFP localization. **(E)** Chs3(Δ28–52)-GFP and Chs3($_{41}$LVR$_{43}$ →AAA)-GFP were introduced into the *chs3Δ* and *chs3Δapl2Δ* strains and plated on indicated media. Plates were imaged after 2 days at 30°C. See also Figure S2 for additional CW concentrations.

A segment at the C-terminus of Chs3 was unable to interact with exomer in this assay (**Figure 5C**), despite previously published evidence that it does interact with exomer (Rockenbauch et al., 2012). The previous report detected an interaction with exomer from cell lysates, rather than with purified exomer as used in our assay, so it is possible that the interaction requires another binding partner or a post-translational modification to occur. Alternatively, a more complete Chs5 protein may be required for the interaction, since the exomer complex used here contained only residues 1–77 of Chs5. This truncated Chs5 is not fully

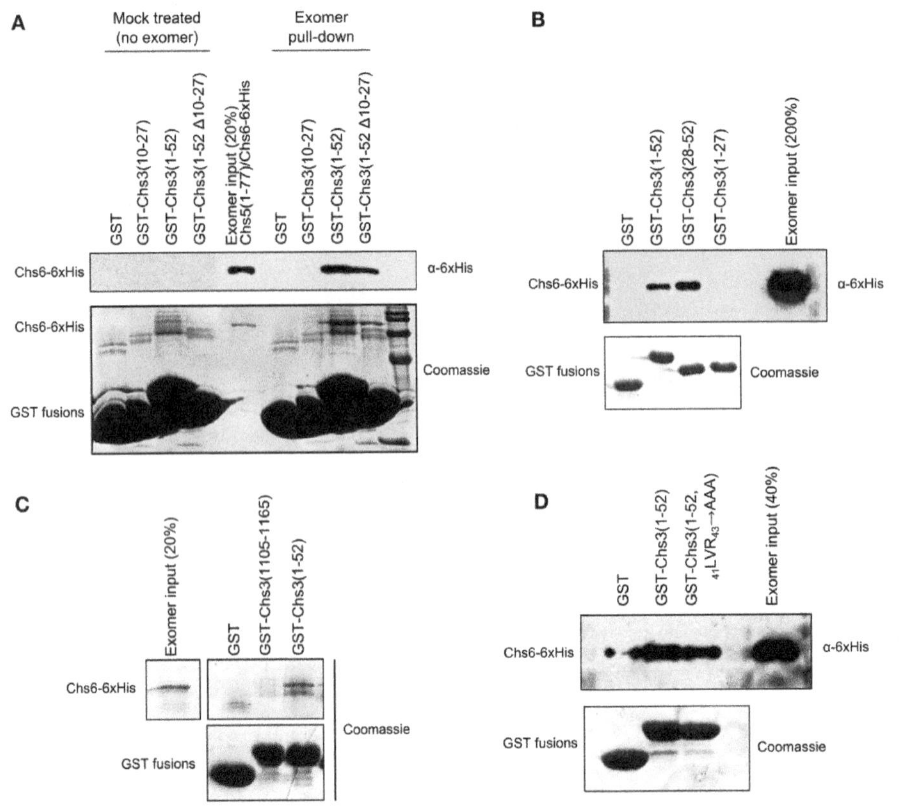

FIGURE 5 | Residues 28–52 interact with exomer *in vitro*. **(A)** *In vitro* interaction assay to measure ability of GST-tagged Chs3 fragments to interact with purified exomer complex. Exomer was detected by α-6xHis antibody which recognized His-tagged Chs6. GST-tagged constructs were detected by Coomassie. In the "mock treated" samples, the final exomer purification buffer was added in place of exomer. **(B)** Chs3 fragment pull-down assay as described in **(A)** comparing the two halves of the Chs3(1–52) segment. **(C)** Assay as described in **(A)** to compare interaction of N- and C-termini, but with bound exomer detected by Coomassie staining. **(D)** Assay as described in **(A)** to analyze effect of $_{41}$LVR$_{43}$ →AAA mutation on interaction.

functional, but it contains all the structural components necessary to homodimerize and interact with ChAP proteins (Paczkowski et al., 2012), which should allow the exomer complex to interact normally with cargo *in vitro*. Alternatively, post-translational modification of exomer or Chs3 may regulate the interaction.

Since the $_{19}$DEE$_{21}$ →AAA mutation had such a strong effect on Chs3 trafficking, we tested its effect on the *in vitro* interaction between Chs3(1–52) and purified exomer. Surprisingly, this mutated fragment interacted as well as the wild-type GST-Chs3(1–52) fragment (**Figure 5D**). The assay may not be sensitive enough to observe a slight reduction in binding affinity, which could strongly affect Chs3 trafficking *in vivo*.

RESIDUES 10–27 INTERACT WITH EXOMER BASED ON STRUCTURAL ANALYSIS

Obtaining a crystal structure of exomer bound to its sorting signal within Chs3 would elucidate both the location of binding within Chs6, and the structure of that segment of Chs3. Therefore, we attempted to co-crystallize exomer with several peptides from the N- and C-termini of Chs3. Most crystals diffracted well but did not contain the peptide. After screening many crystals, several were found that contained the peptide matching residues 10–27. This peptide formed an alpha-helical structure, and bound to a region on the surface of Chs6 roughly opposite where Chs5 binds (**Figures 6A,B; Table 1**). This observed helix is 14 residues in length, and we found residues 12–25 to be the best fit to the density (Figure S3). This indicates residues 10–27 can interact directly with exomer at high concentrations, despite the lack of binding seen in our *in vitro* assay (**Figure 5B**). It is possible other peptides that interact more strongly with exomer disrupted crystal contacts or destabilized the structure, leading to their exclusion from the crystals. This is supported by our observation that when a peptide containing residues 1–52 of Chs3 was added to the cryoprotectant solution, the exomer crystals dissolved. We also note that no additional electron density was visible when crystals were soaked with a peptide corresponding to residues 1128–1165 of the Chs3 C-terminus.

The Chs3 peptide makes direct contact with several residues on the surface of Chs6 (**Figures 6C–E**). Chs3 residue D19 appears to participate in a hydrogen bonding network with Chs6 residues D731 and R713. Chs3 residue E20 appears to hydrogen bond with the main chain carbonyl oxygen of Chs6 residue G31. Both of these Chs3 residues, D19 and E20, are part of the conserved DEESLL sequence that was previously found to be important for interaction with AP-1 (Starr et al., 2012) and that we found to be important for endocytosis of Chs3 (**Figure 3C**).

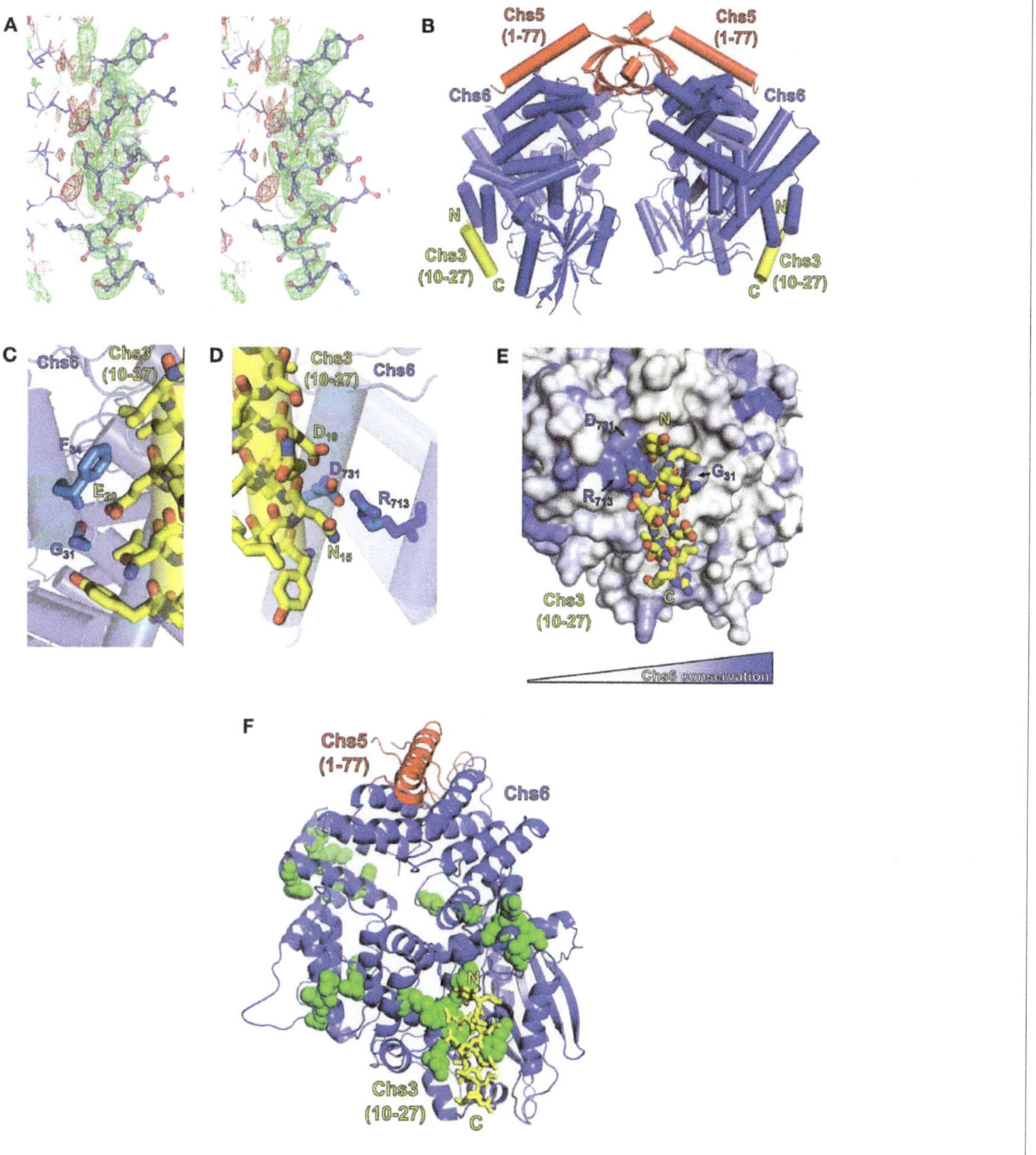

FIGURE 6 | Exomer co-crystalizes with Chs3 residues 10–27. (A) Stereo image (wall-eyed) of the original F_O-F_C difference map calculated after molecular replacement, used to model the Chs3 peptide (shown as ball-and-stick model), contoured at $+2.5\ \sigma$ (green) and $-2.5\ \sigma$ (red). The negative density to the left of the peptide is assumed to be due to series truncation errors arising from the modest resolution of the data. **(B)** Model of an exomer complex containing Chs5(1–77) (red), Chs6-6xHis (blue) and a peptide of Chs3 residues 10–27 (yellow, only residues 12–25 are visible). Model was created using Chs5(1–77)/Bch1 structure (PDB ID: 4IN3) (Richardson and Fromme, 2013) as a template, replacing each Bch1 subunit with the structure of Chs6 and peptide from this study. **(C,D)** Close-up views of the interaction of the Chs3 peptide with Chs6. **(E)** Chs6 colored by conservation within Chs6 and Bch2 ChAP proteins, white to blue representing low to high conservation. **(F)** Structure rotated 90° from **(A)**, showing location of Chs6 mutations that had no effect on Chs3 trafficking (green spheres), as listed in **Table 2**.This panel also depicts the crystallographic asymmetric unit.

Based on the crystal structure, several mutations were made in the Chs6 protein in an attempt to disrupt the interaction (**Figure 6F**, **Table 2**). These mutated proteins, in addition to many mutations made based on sequence conservation (**Figure 6E**), were still able to mediate transport of Chs3. This supports our hypothesis that residues 10–27 are not required for transport of Chs3 to the PM, but rather play a more important role in endocytosis of Chs3. Nevertheless, the structure reveals how this clathrin adaptor motif can be bound by exomer, perhaps providing a basis for competition between adaptors at the TGN.

DISCUSSION

We have shown the importance of the N-terminal 52 residues of Chs3 for its transport, as deletion of these residues decreases the amount of Chs3 at the bud neck as well as TGN punctae, and the

Table 1 | Crystallographic statistics.

Wavelength (Å)	0.9767
Resolution range (Å)	44.65–2.59 (2.683–2.59)
Space group	P 63 2 2
Unit cell	a = 218.613 Å, b = 218.613 Å, c = 137.848 Å α = 90°, β = 90°, γ = 120°
Total reflections	59,962
Unique reflections	59936 (5730)
Multiplicity	12.2 (10.0)
Completeness (%)	99.00 (96.33)
<I>/< σ(I) >	19.45 (5.06)
Wilson B-factor	50.18
R_{sym}	0.090 (0.717)
R_{cryst}	0.1796 (0.2202)
R_{free}	0.2107 (0.2534)
Number of atoms	6100
Protein (755 residues)	6008
Water	92
RMS (bonds) (Å)	0.006
RMS (angles) (°)	1.09
Ramachandran favored (%)	94
Ramachandran outliers (%)	1.5
Clash score	7.88
Average B-factor	68.20
Protein	68.40
Water	55.60

Values in parentheses represent the highest resolution shell.

Table 2 | Many Chs6 mutations have no phenotype.

Plasmid name	Residues mutated
MUTATIONS MADE BEFORE CRYSTAL STRUCTURE	
pAS62	Wild-type Chs6-Myc
pJC1	G540A W541A
pJC2	R548E F552A
pJC3	C582A W585A D587K
pAS70	S237A
pAS71	S253A
pAS72	T516A
pAS73	S612A
pAS92	C216A K217D K218D
MUTATIONS BASED ON CRYSTAL STRUCTURE	
pAS109	R713A
pAS110	D724A
pAS111	V728D
pAS112	D731A
pAS115	R713A V728D
pAS116	R713A D731A
pAS117	D724A D731A
pAS126	V728W
pAS127	A735W
pAS128	V728R
pAS129	A735R
pAS130	V728E
pAS131	A735E
pAS152	V728R A735R
pAS153	V728E A735E
pAS180	K210A
pAS183	R637A T638A
pAS184	D672A
pAS191	F34A
pAS192	F34A R713A
pAS193	F34A D731A

The indicated residues were mutated in a Chs6-Myc plasmid and transformed into a chs6Δ strain. The cells showed no CW resistance. Mutations in italics were made based on the crystal structure in (6A).

mutant cells are CW resistant. We also endeavored to more precisely identify sorting signals within this region that interact with exomer and potentially other cargo adaptor proteins. Residues 10–27 contribute to internalization, while residues 28–52 contribute to retrieval of Chs3 from the endosomal system and to exomer-mediated transport to the PM, and are sufficient to interact with exomer *in vitro*. While residues 10–27 did not interact in our *in vitro* pull-down assay, they did interact with exomer in the crystals, suggesting they can contribute to the interaction, though perhaps with low affinity.

The itinerary of Chs3 involves transport through several pathways in the cell (**Table 3**, **Figure 7**). The first step after synthesis of Chs3 is exit from the ER, which requires Chs7 (Trilla, 1999). This process requires the N-terminus of Chs3, since truncation of the first 126 residues prevents ER exit (Sacristan et al., 2013), but Chs3(ΔN52)-GFP exits the ER normally indicating the required residues are located between 52 and 126. After exit from the ER and transit through the Golgi, Chs3 can be transported to the PM by two pathways. The first is exomer-mediated, and requires both the N- and C-termini of Chs3. Our results indicated residues 28–52 are particularly important for this step. The second pathway is not exomer-mediated, and can allow normal levels of Chs3 to reach the PM when retention is disrupted. It is unknown whether this is mediated by any cargo adaptors or requires sorting signals, or whether it is a non-specific inclusion of Chs3 into secretory vesicles. Chs3 endocytosis is mediated by the AP-2 clathrin adaptor. The mislocalization of Chs3(Δ10–27)-GFP and Chs3($_{19}$DEE$_{21}$ →AAA)-GFP to the PM suggests AP-2 may bind the DEESLL motif at residues 19–24. Therefore, AP-1, AP-2, and exomer likely bind to this motif during different stages of Chs3 trafficking. Interestingly, the conformation of this motif when bound to Chs6 is alpha-helical, in contrast to the conformation (D/E)XXXLL motifs adopt when bound to AP-1 and AP-2 (Kelly et al., 2008; Jackson et al., 2010). In those interactions, the (D/E)XXXLL motif is in an extended conformation.

The TGN in yeast is a transient compartment that acts to sort proteins before distributing them to other locations within the cell. Therefore, retention of a protein at the TGN is not a static storage process, but a dynamic process of tubular and vesicular transport from the TGN to the endosomal system and then back to the TGN. Chs3 requires AP-1 and Gga1/2 to transport it into the endosomal system. AP-1 binds Chs3 residues 19–24, but also must interact elsewhere in Chs3, since AP-1 deletion can rescue the CW sensitivity even when residues 1–52 are deleted.

Table 3 | Summary of Chs3 mutations analyzed in this study and their effect on trafficking.

Mutant	CW growth phenotype	CW sensitivity rescued by *apl2*Δ	Localization phenotype	Likely pathways disrupted
*chs3*Δ	++++	No	N/A	N/A
ΔN52	++++	Strongly	Weak bud neck, few punctae	TGN→PM (exomer) TGN→Endosomes (AP-1,Gga1/2) Endosomes→TGN (Retromer, Snx4/41/42)
$_{19}$DEE$_{21}$ →AAA	+	Moderately	Normal bud neck and punctae, additional at PM	Endocytosis (AP-2) TGN→Endosomes (AP-1,Gga1/2)
Δ10-27	++	Moderately	Normal bud neck and punctae, additional at PM	Endocytosis (AP-2) TGN→Endosomes (AP-1,Gga1/2)
Δ28-52	+++	No	Weak bud neck, few punctae, vacuole	TGN→PM (exomer) Endosomes→TGN (Retromer, Snx4/41/42)
$_{41}$LVR$_{43}$ →AAA	++++	Moderately	Weak bud neck, few punctae	TGN→PM (exomer)
ΔC21	++++	No	TGN punctae only (Rockenbauch et al., 2012)	TGN→PM (exomer)

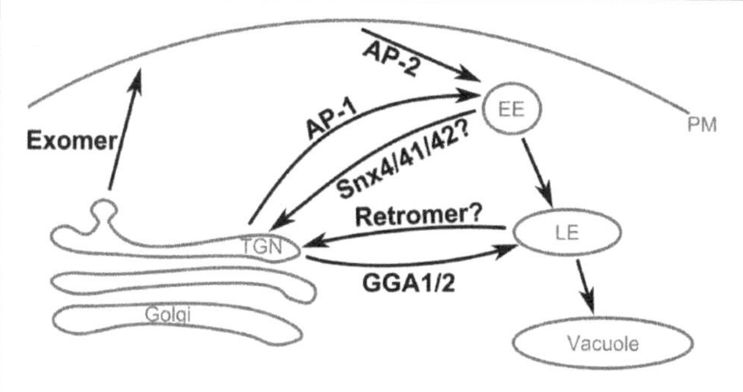

FIGURE 7 | Summary of Chs3 trafficking pathways. Chs3 is transported through several intracellular trafficking pathways throughout the cell cycle. As the yeast TGN is not a stable compartment, retention of cargo at the TGN appears to require traffic to endosomes by the AP-1 and Gga1/2 clathrin adaptors and subsequent retrieval back to the TGN, likely via the retromer and/or SNX4/41/42 pathways.

Two complexes that mediate transport from the endosomal system to the TGN are the retromer complex and the Snx4/41/42 complex. One or both of these complexes could be required for the retrieval of Chs3 back to the TGN. One of the complexes involved in this step may bind to residues 28–52 of Chs3, as we saw mutations in this region resulted in mislocalization to the vacuole. The lack of vacuolar localization in either larger (Δ52) or smaller ($_{41}$LVR$_{43}$ →AAA) disruptions suggests that a very specific set of interactions must be blocked, and others allowed, to result in this phenotype.

Our results suggest a competition of cargo adaptors for overlapping or closely adjacent sorting signals could be used to regulate and balance the pathways required for Chs3 transport. This form of regulation is unlikely to be specific to Chs3, since another exomer cargo, Pin2, contains a potential AP-1 sorting signal within the segment required for exomer-mediated transport (Ritz et al., 2014). An analogous mechanism has been demonstrated for the binding of the myosin V motor to different cargos, although the situation is reversed, in which multiple cargos compete for binding the same site on the motor (Eves et al., 2012).

This complex set of overlapping sorting signals and redundant transport pathways could allow for careful control of Chs3 localization throughout the cell cycle and efficient changes in response to signals. It is possible there are proteins in many organisms that use competitive binding of cargo adaptors to cycle between the cell surface and internal compartments.

ACKNOWLEDGMENTS

We thank Jon Paczkowski for assistance with exomer complex crystallography and reading the manuscript. We thank Brian Richardson for helpful discussions and reading the manuscript. We thank Trevor Starr and Randy Schekman, and Anne Spang and Cesar Roncero, for sharing unpublished data. We thank Robyn Barfield for plasmid pRB259. We thank the Emr lab for use of their microscope. We thank the staff of the CHESS for the use of beamline A1 and their technical assistance. CHESS is supported by the National Science Foundation and the National Institutes of Health/National Institute of General Medical Sciences under NSF award DMR-0936384, using the Macromolecular Diffraction at CHESS (MacCHESS) facility, which is supported by award GM-103485 from the National Institute of General Medical Sciences, National Institutes of Health. This work was funded by startup funds to J. Christopher Fromme from Cornell University, and by NIH Training Grant (T32GM07273) awards to Amanda M. Weiskoff.

REFERENCES

Abràmoff, M. D., Magalhães, P. J., and Ram, S. J. (2004). Image processing with ImageJ. *Biophotonics Intl.* 11, 36–42. doi: 10.1117/1.3589100

Adams, P. D., Afonine, P. V., Bunkóczi, G., Chen, V. B., Davis, I. W., Echols, N., et al. (2010). PHENIX: a comprehensive Python-based system for macromolecular structure solution. *Acta Crystallogr. D Biol. Crystallogr.* 66, 213–221. doi: 10.1107/S0907444909052925

Barfield, R. M., Fromme, J. C., and Schekman, R. (2009). The exomer coat complex transports Fus1p to the plasma membrane via a novel plasma membrane sorting signal in yeast. *Mol. Biol. Cell* 20, 4985–4996. doi: 10.1091/mbc.E09-04-0324

Bonifacino, J. S., and Glick, B. S. (2004). The mechanisms of vesicle budding and fusion. *Cell* 116, 153–166. doi: 10.1016/S0092-8674(03)01079-1

Bonifacino, J. S., and Traub, L. M. (2003). Signals for sorting of transmembrane proteins to endosomes and lysosomes. *Annu. Rev. Biochem.* 72, 395–447. doi: 10.1146/annurev.biochem.72.121801.161800

Bryant, N. J., Govers, R., and James, D. E. (2002). Regulated transport of the glucose transporter GLUT4. *Nat. Rev. Mol. Cell Biol.* 3, 267–277. doi: 10.1038/nrm782

Chuang, J. S. (1996). Differential trafficking and timed localization of two chitin synthase proteins, Chs2p and Chs3p [published erratum appears in J Cell Biol 1996 Dec;135(6 Pt 2):1925]. *J. Cell Biol.* 135, 597–610. doi: 10.1083/jcb.135.3.597

Copic, A., Starr, T. L., and Schekman, R. (2007). Ent3p and Ent5p exhibit cargo-specific functions in trafficking proteins between the trans-Golgi network and the endosomes in yeast. *Mol. Biol. Cell* 18, 1803–1815. doi: 10.1091/mbc.E06-11-1000

DeMarini, D. J. (1997). A septin-based hierarchy of proteins required for localized deposition of chitin in the *Saccharomyces cerevisiae* cell wall. *J. Cell Biol.* 139, 75–93. doi: 10.1083/jcb.139.1.75

Emsley, P., Lohkamp, B., Scott, W. G., and Cowtan, K. (2010). Features and development of Coot. *Acta Crystallogr. D Biol. Crystallogr.* 66, 486–501. doi: 10.1107/S0907444910007493

Eves, P. T., Jin, Y., Brunner, M., and Weisman, L. S. (2012). Overlap of cargo binding sites on myosin V coordinates the inheritance of diverse cargoes. *J. Cell Biol.* 198, 69–85. doi: 10.1083/jcb.201201024

Jackson, L. P., Kelly, B. T., McCoy, A. J., Gaffry, T., James, L. C., Collins, B. M., et al. (2010). A large-scale conformational change couples membrane recruitment to cargo binding in the AP2 clathrin adaptor complex. *Cell* 141, 1220–1229. doi: 10.1016/j.cell.2010.05.006

Kelly, B. T., McCoy, A. J., Späte, K., Miller, S. E., Evans, P. R., Höning, S., et al. (2008). A structural explanation for the binding of endocytic dileucine motifs by the AP2 complex. *Nature* 456, 976–979. doi: 10.1038/nature07422

Morin, A., Eisenbraun, B., Key, J., Sanschagrin, P. C., Timony, M. A., Ottaviano, M., et al. (2013). Collaboration gets the most out of software. *Elife* 2:e01456. doi: 10.7554/eLife.01456

Otwinowski, Z., and Minor, W. (1997). Macromolecular crystallography part A. *Meth. Enzymol.* 276, 307–326. doi: 10.1016/S0076-6879(97)76066-X

Paczkowski, J. E., Richardson, B. C., Strassner, A. M., and Fromme, J. C. (2012). The exomer cargo adaptor structure reveals a novel GTPase-binding domain. *EMBO J.* 31, 4191–4203. doi: 10.1038/emboj.2012.268

Reyes, A., Sanz, M., Duran, A., and Roncero, C. (2007). Chitin synthase III requires Chs4p-dependent translocation of Chs3p into the plasma membrane. *J. Cell Sci.* 120, 1998–2009. doi: 10.1242/jcs.005124

Richardson, B. C., and Fromme, J. C. (2013). The exomer cargo adaptor features a flexible hinge domain. *Structure* 21, 486–492. doi: 10.1016/j.str.2013.01.003

Ritz, A. M., Trautwein, M., Grassinger, F., and Spang, A. (2014). The prion-like domain in the exomer-dependent cargo Pin2 serves as a trans-Golgi retention motif. *Cell Rep.* 7, 249–260. doi: 10.1016/j.celrep.2014.02.026

Rockenbauch, U., Ritz, A. M., Sacristan, C., Roncero, C., and Spang, A. (2012). The complex interactions of Chs5p, the ChAPs, and the cargo Chs3p. *Mol. Biol. Cell* 23, 4402–4415. doi: 10.1091/mbc.E11-12-1015

Sacristan, C., Manzano-Lopez, J., Reyes, A., Spang, A., Muñiz, M., and Roncero, C. (2013). Oligomerization of the chitin synthase Chs3 is monitored at the Golgi and affects its endocytic recycling. *Mol. Microbiol.* 90, 252–266. doi: 10.1111/mmi.12360

Sanchatjate, S., and Schekman, R. (2006). Chs5/6 Complex: a multiprotein complex that interacts with and conveys chitin synthase III from the trans–Golgi network to the cell surface. *Mol. Biol. Cell* 17, 4157–4166. doi: 10.1091/mbc.E06-03-0210

Santos, B., and Snyder, M. (1997). Targeting of chitin synthase 3 to polarized growth sites in yeast requires Chs5p and Myo2p. *J. Cell Biol.* 136, 95–110.

Starr, T. L., Pagant, S., Wang, C.-W., and Schekman, R. (2012). Sorting signals that mediate traffic of chitin synthase III between the TGN/endosomes and to the plasma membrane in yeast. *PLoS ONE* 7:e46386. doi: 10.1371/journal.pone.0046386

Trautwein, M., Schindler, C., Gauss, R., Dengjel, J., Hartmann, E., and Spang, A. (2006). Arf1p, Chs5p and the ChAPs are required for export of specialized cargo from the Golgi. *EMBO J.* 25, 943–954. doi: 10.1038/sj.emboj.7601007

Trilla, J. A. (1999). Chs7p, a new protein involved in the control of protein export from the endoplasmic reticulum that is specifically engaged in the regulation of chitin synthesis in *Saccharomyces cerevisiae*. *J. Cell Biol.* 145, 1153–1163. doi: 10.1083/jcb.145.6.1153

Valdivia, R. H., Baggott, D., Chuang, J. S., and Schekman, R. W. (2002). The yeast clathrin adaptor protein complex 1 is required for the efficient retention of a subset of late Golgi membrane proteins. *Dev. Cell* 2, 283–294. doi: 10.1016/S1534-5807(02)00127-2

Wang, C.-W., Hamamoto, S., Orci, L., and Schekman, R. (2006). Exomer: a coat complex for transport of select membrane proteins from the trans-Golgi network to the plasma membrane in yeast. *J. Cell Biol.* 174, 973–983. doi: 10.1083/jcb.200605106

Zanolari, B., Rockenbauch, U., Trautwein, M., Clay, L., Barral, Y., and Spang, A. (2011). Transport to the plasma membrane is regulated differently early and late in the cell cycle in *Saccharomyces cerevisiae*. *J. Cell Sci.* 124, 1055–1066. doi: 10.1242/jcs.072371

Ziman, M., Chuang, J. S., Tsung, M., Hamamoto, S., and Schekman, R. (1998). Chs6p-dependent anterograde transport of Chs3p from the chitosome to the plasma membrane in *Saccharomyces cerevisiae*. *Mol. Biol. Cell* 9, 1565–1576.

Conflict of Interest Statement: The authors declare that the research was conducted in the absence of any commercial or financial relationships that could be construed as a potential conflict of interest.

22

Cellular prion protein and NMDA receptor modulation: protecting against excitotoxicity

Stefanie A. G. Black [1,2]***, Peter K. Stys*** [2,3]***, Gerald W. Zamponi*** [1,2]**** and Shigeki Tsutsui*** [2,3]

[1] *Department of Physiology and Pharmacology, University of Calgary, Calgary, AB, Canada*
[2] *Hotchkiss Brain Institute, University of Calgary, Calgary, AB, Canada*
[3] *Department of Clinical Neurosciences, University of Calgary, Calgary, AB, Canada*

Edited by:
Sophie Mouillet-Richard, INSERM Unit 1124, France

Reviewed by:
Flavio Beraldo, Western University, Canada
Federico Benetti, SISSA, Italy

****Correspondence:***
Gerald W. Zamponi, Department of Physiology and Pharmacology, Hotchkiss Brain Institute, University of Calgary, 3330 Hospital Drive NW, Calgary, AB T2N 4N1, Canada
e-mail: zamponi@ucalgary.ca

Although it is well established that misfolding of the cellular prion protein (PrP^C) into the β-sheet-rich, aggregated scrapie conformation (PrP^{Sc}) causes a variety of transmissible spongiform encephalopathies (TSEs), the physiological roles of PrP^C are still incompletely understood. There is accumulating evidence describing the roles of PrP^C in neurodegeneration and neuroinflammation. Recently, we identified a functional regulation of NMDA receptors by PrP^C that involves formation of a physical protein complex between these proteins. Excessive NMDA receptor activity during conditions such as ischemia mediates enhanced Ca^{2+} entry into cells and contributes to excitotoxic neuronal death. In addition, NMDA receptors and/or PrP^C play critical roles in neuroinflammation and glial cell toxicity. Inhibition of NMDA receptor activity protects against PrP^{Sc}-induced neuronal death. Moreover, in mice lacking PrP^C, infarct size is increased after focal cerebral ischemia, and absence of PrP^C increases susceptibility of neurons to NMDA receptor-dependent death. Recently, PrP^C was found to be a receptor for oligomeric beta-amyloid (Aβ) peptides, suggesting a role for PrP^C in Alzheimer's disease (AD). Our recent findings suggest that Aβ peptides enhance NMDA receptor current by perturbing the normal copper- and PrP^C-dependent regulation of these receptors. Here, we review evidence highlighting a role for PrP^C in preventing NMDA receptor-mediated excitotoxicity and inflammation. There is a need for more detailed molecular characterization of PrP^C-mediated regulation of NMDA receptors, such as determining which NMDA receptor subunits mediate pathogenic effects upon loss of PrP^C-mediated regulation and identifying PrP^C binding site(s) on the receptor. This knowledge will allow development of novel therapeutic interventions for not only TSEs, but also for AD and other neurodegenerative disorders involving dysfunction of PrP^C.

Keywords: NMDA receptor, cellular prion protein, excitotoxicity, neuroinflammation, ischemia, beta-amyloid, Alzheimer's disease

INTRODUCTION

Misfolding of the cellular prion protein (PrP^C) into the β-sheet-rich, aggregate-prone scrapie conformation (PrP^{Sc}) is well known to result in several progressive and often fatal diseases termed prionopathies. This group of disorders, also known as transmissible spongiform encephalopathies (TSEs), includes scrapie, bovine spongiform encephalopathy (BSE, or mad cow disease), and the human diseases Creutzfeld–Jakob disease, Gerstmann-Straussler-Scheinker syndrome, fatal familial insomnia, and Kuru (Aguzzi et al., 2008). PrP^{Sc} aggregation and formation of amyloid-like plaques disrupt neuronal physiology, and most TSEs involve eventual loss of neurons (Aguzzi et al., 2008). Although much work has been done investigating the deleterious effects of misfolded/aggregated prion proteins, the physiological roles of PrP^C remain incompletely understood.

Mature PrP^C is a glycoprotein that is anchored to the extracellular leaflet of the plasma membrane through a carboxyl (C)-terminal glycosylphosphatidylinositol (GPI) anchor. Initially, PrP^C is synthesized as a precursor polypeptide (254 amino acids for mouse PrP^C) containing an amino (N)-terminal signal sequence that directs PrP^C to the endoplasmic reticulum and is post-translationally cleaved, and a C-terminal GPI anchor sequence that is removed and replaced with a GPI anchor (**Figure 1**; Biasini et al., 2012). The N-terminal portion of PrP^C is intrinsically unstructured, contains the octarepeat region, and has multiple copper binding sites both within and outside of the octarepeat region (**Figure 1**; Klewpatinond et al., 2008; Stanyon et al., 2014). The PrP^C N-terminus also contains sites for binding of oligomeric β-amyloid (Aβ) peptides (**Figure 1**; Lauren et al., 2009; Chen et al., 2010). In contrast to the N-terminus, the C-terminus of PrP^C is globular in structure, with three α-helical and two short β-strand regions, and is the location of two asparagine residues that undergo glycosylation (**Figure 1**; Biasini et al., 2012). In addition, PrP^C can undergo physiological cleavage of different types: α-cleavage in the vicinity of residue 109 (mouse sequence) generates N1 and C1 fragments; β-cleavage in

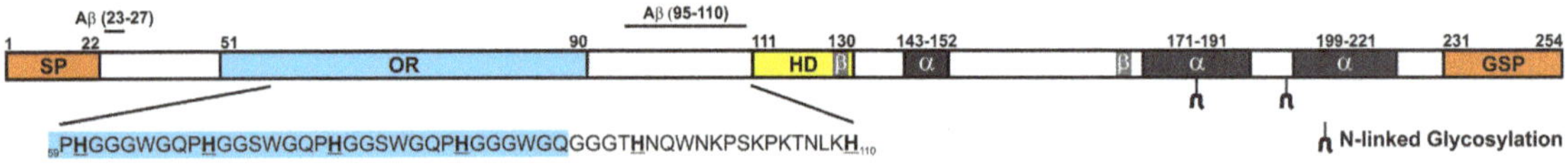

FIGURE 1 | Schematic representation of PrP^C. Numbering of amino acid residues corresponds to the mouse PrP^C sequence. Shown are the locations of histidine (H) residues that bind copper ions (residues 60, 68, 76, 84, 95, 110), the amino (N)-terminal signal peptide (SP; residues 1–22), β-amyloid oligomer (Aβ) binding regions (residues 23–27, 95–110), the octapeptide repeat region (OR; residues 51–90), the central hydrophobic domain (HD; residues 111–130), α-helical regions (α; residues 143–152, 171–191, 199–221), β-sheet regions (β; residues 127–129, 166–168), sites of asparagine (N)-linked glycosylation (residues 180, 196), and the carboxyl (C)-terminal glycosylphosphatidylinositol (GPI) signal peptide (GSP; residues 231–254).

the vicinity of residue 90 (mouse sequence) produces N2 and C2 fragments; and cleavage at the GPI anchor releases full-length PrP^C into the extracellular environment (Altmeppen et al., 2013; McDonald et al., 2014). However, recent data from McDonald and co-workers using recombinant mouse PrP^C reveal that cleavage of PrP^C is quite complex, generating a number of different N1, C1, N2, and C2 fragments (McDonald et al., 2014), although the biological functions of these cleavage products remain to be determined.

PrP^C interacts with and signals through many different cell surface proteins, including the α7 nicotinic acetylcholine receptor (Beraldo et al., 2010), metabotropic glutamate receptors mGluR1 (Beraldo et al., 2011) and mGluR5 (Beraldo et al., 2011; Um et al., 2013), kainate receptor GluR6/7 (Carulla et al., 2011), and α-amino-3-hydroxy-5-methyl-4-isoxazole propionic acid (AMPA) receptor subunits GluA1 (Watt et al., 2012) and GluA2 (Kleene et al., 2007; Watt et al., 2012). Furthermore, Aβ oligomers can bind to PrP^C and signal through mGluR5 to activate Fyn kinase (Um et al., 2013), and an Aβ-PrP^C-mGluR5 pathway is also involved in facilitation of long-term depression (LTD) (Hu et al., 2014). The diversity of signaling mediated by PrP^C highlights the need for a better understanding of how the interaction of PrP^C with its binding partners is regulated in both physiological and pathological conditions. Work from our laboratory revealed that PrP^C also interacts with N-methyl-D-aspartate (NMDA) receptors; PrP^C-deficient mice display increased NMDA receptor-dependent neuronal excitability and are more susceptible to NMDA-induced excitotoxicity (Khosravani et al., 2008). Here, we briefly review properties of the NMDA receptor and focus on evidence supporting a role for PrP^C in preventing NMDA receptor-mediated excitotoxicity, a process that may contribute to the pathogenesis of a variety of disorders including ischemic stroke, Alzheimer's disease (AD), Huntington's disease, and epilepsy (Lai et al., 2014; Parsons and Raymond, 2014). Similar mechanisms may be involved in PrP^{Sc}-induced toxicity (Müller et al., 1993; Riemer et al., 2008; Resenberger et al., 2011).

NMDA RECEPTOR CHARACTERISTICS

Glutamate is the major excitatory transmitter in the mammalian central nervous system (CNS), and it exerts its actions by binding to a variety of different receptor proteins. Glutamate receptors (GluRs) are divided into two families, ionotropic and metabotropic. Metabotropic GluRs are G protein-coupled receptors, while ionotropic GluRs are ion channels. There are three types of ionotropic GluRs, namely AMPA, kainate, and NMDA receptors (Mayer, 2005).

NMDA receptors are cation channels that mediate entry of Na^+ and Ca^{2+} ions and are activated by the co-agonists glutamate (or NMDA) and glycine or D-serine. Activation of NMDA receptors thus contributes to the excitatory postsynaptic potential (EPSP), as well as both long-term potentiation (LTP) and LTD (Bartlett and Wang, 2013). In addition to co-agonist binding, activation of NMDA receptors requires membrane depolarization to remove the Mg^{2+} ion block of the channel pore that occurs at resting membrane potentials. NMDA receptors are heterotetrameric channels formed by the assembly of two obligatory GluN1 and two GluN2/GluN3 subunits. GluN1 subunits contain the binding site for glycine, while glutamate binds to GluN2 subunits. To date, there are seven known NMDA receptor subunits: GluN1, GluN2A-GluN2D, and GluN3A-GluN3B, with alternative splicing of GluN1 and GluN3 subunits generating further diversity (Paoletti et al., 2013). GluN3 subunits also bind glycine, and thus NMDA receptors containing only GluN1 and GluN3 subunits give rise to "glycine-only" receptors that cannot be activated by glutamate or NMDA (Chatterton et al., 2002; Pina-Crespo et al., 2010). Identity of the GluN2 subunits that combine with GluN1 subunits to form the tetrameric channel dictates NMDA receptor properties, such as activation and deactivation kinetics, ion conductance, and affinity for glutamate (Cull-Candy and Leszkiewicz, 2004). Pharmacological characterization using ligands for specific GluN2 subunits allows differentiation of receptor subtypes (Wyllie et al., 2013).

NMDA receptor localization can be synaptic, perisynaptic, extrasynaptic, or even presynaptic, with receptor activity at each location coupling to specific cellular events (Corlew et al., 2008; Hardingham, 2009; Hardingham and Bading, 2010; Paoletti et al., 2013). In general, activation of synaptic NMDA receptors activates pro-survival signaling, while extrasynaptic NMDA receptor activity mediates pro-death signaling (Hardingham, 2009; Hardingham and Bading, 2010). Thus, while NMDA receptors mediate key physiological functions such as learning and memory under normal conditions, they also play a role in glutamate excitotoxicity, which can occur in response to ischemia (Lau and Tymianski, 2010) and is involved in many neurodegenerative conditions, including AD (Parsons and Raymond, 2014). In these abnormal situations, excessive glutamate can spill over from synaptic to extrasynaptic sites, thus activating not only the desired synaptically-localized NMDA receptors, but also those receptors located extrasynaptically. Because NMDA receptors are highly permeable to Ca^{2+} ions, this overstimulation leads to enhanced Ca^{2+} influx that ultimately can be fatal to cells (Lau and Tymianski, 2010). NMDA receptors are found not only in neurons, but also in oligodendrocytes and myelin, and ischemia can cause NMDA receptor-dependent damage to both myelin and oligodendrocyte processes (Karadottir et al., 2005; Salter and Fern, 2005; Micu et al., 2006). Hence, tight regulation of NMDA receptor activity is of extreme importance in maintaining physiological signaling while preventing excitotoxicity. NMDA receptor activity is regulated by a number of different mechanisms, one of which involves an intrinsic desensitization mechanism that results in termination of channel activity even in the prolonged presence of glutamate, a process that is potently regulated by the receptor co-agonist glycine (Mayer et al., 1989). Recently, we demonstrated that the absence of PrP^C enhances NMDA receptor glycine affinity, leading to persistent NMDA receptor activity upon prolonged agonist application (You et al., 2012), which likely underlies the greater susceptibility of PrP^C-null neurons to NMDA receptor-mediated damage and dysfunction (Rangel et al., 2007; Khosravani et al., 2008; Gadotti and Zamponi, 2011; Gadotti et al., 2012; You et al., 2012; Fleisch et al., 2013).

PrP^C AS A REGULATOR OF NMDA RECEPTORS

In support of a role for PrP^C in preventing NMDA receptor hyperactivity, data from our and other laboratories reveal that NMDA receptor activity is enhanced when PrP^C is absent. Our work using mouse hippocampal slices showed that the enhanced neuronal excitability in PrP^C-null neurons can be reduced by the NMDA receptor blocker aminophosphonovaleric acid (APV), and that NMDA receptor-mediated miniature excitatory postsynaptic currents (mEPSCs) recorded from cultured hippocampal neurons from PrP^C-null mice had significantly larger amplitudes and longer decay times than those recorded from wild-type neurons (Khosravani et al., 2008). In this study we also showed that NMDA application resulted in whole-cell currents with a prolonged time course of deactivation in cultured PrP^C-null hippocampal neurons that could be restored to wild-type levels by ectopic expression of mouse PrP^C and mimicked by reducing PrP^C expression in cultured wild-type hippocampal neurons. Co-immunoprecipitation experiments demonstrated that PrP^C and GluN2D subunits were found in the same protein complex, and immunofluorescent staining of hippocampal neurons showed co-localization of these two proteins. Indeed, the properties of the NMDA-evoked currents we observed in PrP^C-null neurons in this study appeared to resemble those of heterologously-expressed GluN2D-containing NMDA receptors (Cull-Candy and Leszkiewicz, 2004). Furthermore, siRNA knockdown of GluN2D expression dramatically accelerated NMDA receptor current deactivation kinetics and reduced current amplitude. Taken together, these observations point to an upregulation of GluN2D-containing NMDA receptor activity in the absence of PrP^C. However, it is also possible that activity of NMDA receptors containing other subunits is altered by the lack of PrP^C, producing GluN2D-like deactivation kinetics. Finally, in support of a role for PrP^C in protecting against NMDA receptor-mediated toxicity, we observed significantly more cell death in response to NMDA treatment of PrP^C-null hippocampal cultures than in wild-type cultures, and that focal injection of NMDA into the hippocampus resulted in a lesion of significantly greater size in PrP^C-null mice than in wild-type mice (Khosravani et al., 2008). This increase in NMDA receptor activity not only gives rise to an increased susceptibility of neurons to cell death, but also affects other physiological processes that are linked to NMDA receptor function. For example, NMDA receptors are important for the transmission of pain signals at the spinal level. Consequently, in models of inflammatory and neuropathic pain, we found enhanced nociceptive responses in PrP^C-null mice that could be restored to control levels by MK-801 (Gadotti and Zamponi, 2011). It is also known that NMDA receptors play a role in depressive-like behavior at the level of the hippocampus. Consistent with this notion, we found that PrP^C-null mice displayed increased depressive-like behavior that could be abrogated by NMDA receptor blockers (Gadotti et al., 2012).

It is well established that PrP^C is a copper-binding protein, with up to six copper ions binding to histidine residues in the N-terminal region both within (mouse PrP^C residues 60, 68, 76, and 84) and outside of (mouse PrP^C residues 95 and 110) the octarepeat region (**Figure 1**; Klewpatinond et al., 2008; Quintanar et al., 2013). In addition, copper may bind to histidine residues 176 and 186 (mouse sequence) in the C-terminus of PrP^C (Quintanar et al., 2013), and recent evidence shows that copper can bind to the N-terminal amino group of PrP^C (Stanyon et al., 2014). Occupancy of these copper sites is known to alter PrP^C structure (Qin et al., 2000; Thakur et al., 2011; Younan et al., 2011; Quintanar et al., 2013), and thus likely its interactions with other protein partners. The ability to examine NMDA receptor activity as an indirect readout of the interaction of the receptors with PrP^C has provided interesting insights into the roles of the copper binding sites for PrP^C function. Under normal circumstances, NMDA receptors undergo desensitization in the prolonged presence of glutamate or NMDA, and this process is modulated by the co-agonist glycine (Mayer et al., 1989). In cultured mouse hippocampal neurons lacking PrP^C, the desensitization kinetics appear to be slowed such that a non-desensitizing current is observed (You et al., 2012). In wild-type neurons, copper chelation using bathocuproine sulfonate (BCS) or cuprizone induced a persistent NMDA current similar to that seen in PrP^C-null neurons (You et al., 2012). Acute treatment with phosphatidylinositol-specific phospholipase C (PI-PLC), which enzymatically removes proteins such as PrP^C that are linked to the extracellular leaflet of the plasma membrane via GPI anchors, also induced a persistent NMDA-dependent current in rat hippocampal neurons that was similar to both the current seen with copper chelation and to that recorded from PrP^C-null mouse hippocampal neurons (You et al., 2012). Moreover, PI-PLC treatment did not alter the steady-state NMDA current in PrP^C-null neurons, indicating that the enzymatic cleavage did not measurably affect other potential regulatory proteins (You et al., 2012). Experiments investigating the effect of glycine concentration on steady-state NMDA current revealed that at a given glycine concentration, a higher level of steady-state current was seen in neurons lacking PrP^C than in wild-type neurons; a similar effect was seen upon copper chelation (You et al., 2012). In rat hippocampal cultures, copper chelation induced cell death that could be prevented by NMDA receptor inhibition or re-addition of excess copper (You et al., 2012). Taken together, these results indicate that copper ions modulate NMDA receptors by virtue of their interactions with

PrP^C. Furthermore, these findings suggests that copper ions are key regulators of NMDA receptor function, thus adding another layer of metal ion regulation of these receptors in addition to the well described role of zinc ions.

Further support for PrP^C-mediated regulation of NMDA receptor activity comes from a study using zebrafish with a targeted mutation of the *prp2* gene encoding the zebrafish prion protein PrP2, which has characteristics similar to mammalian PrP^C (Fleisch et al., 2013). Consistent with observations in mammalian brain, these mutants displayed increased convulsant-induced seizure activity compared to wild-type animals. Analysis of mEPSCs recorded *in vivo* from intact zebrafish hindbrain neurons showed that the frequency of NMDA receptor-mediated mEPSCs was reduced in *prp2*$^{-/-}$ fish compared to controls, and although NMDA receptor current amplitudes were the same in both groups, NMDA receptor currents had longer decay times in *prp2*$^{-/-}$ zebrafish. These recent observations agree with the earlier findings of Walz and colleagues showing that mice lacking PrP^C were more sensitive to convulsant-induced seizures, and significantly more PrP^C-null mice died than wild-type mice (Walz et al., 1999). However, a recent study found elevated thresholds for convulsant-induced epileptiform activity in hippocampal slices from PrP^C-null mice compared to wild-type controls (Ratte et al., 2011). The reason for this disparity in the literature remains to be determined.

The presence of misfolded prion proteins could induce NMDA receptor hyperfunction by preventing normal, PrP^C-mediated control of NMDA receptor activity. Indeed, there are several examples of augmented NMDA receptor function in the presence of mutant or scrapie forms of the prion protein. It has been demonstrated that NMDA receptor antagonists memantine and MK-801 prevented PrP^{Sc}-induced toxicity in cultured rat cortical neurons (Müller et al., 1993). Memantine also delayed death of scrapie-infected mice (Riemer et al., 2008) and blocked the PrP^{Sc}-induced increase in apoptosis of PrP^C-expressing SH-SY5Y neuroblastoma cells (Resenberger et al., 2011). Recently, Biasini and co-workers revealed a role for aberrant NMDA receptor activity in the toxic effects of a prion protein lacking residues 105–125 in the central region, termed ΔCR PrP (Biasini et al., 2013). In this study, organotypic cerebellar slice cultures from ΔCR PrP-expressing mice were more susceptible to glutamate-, kainate-, and NMDA-induced cell death compared to those from $PrP^{+/-}$ controls, an effect that was rescued by overexpression of wild-type PrP^C. Notably, NMDA had a more marked effect on cell death compared to the other treatments. These authors also observed that ΔCR PrP expression induces spontaneous inward currents, which may contribute to enhanced NMDA receptor activity via depolarization-induced relief of the Mg^{2+} ion block of these receptors. Importantly, the toxic NMDA receptor-dependent effects of ΔCR PrP in this study were prevented by overexpression of wild-type PrP^C, again supporting a neuroprotective role of PrP^C via prevention of NMDA receptor-mediated toxicity. In another recent study, the toxicity of the misfolded form of a prion protein fragment comprised of residues 90–231 was found to be mediated in part by NMDA receptor-dependent excitotoxicity (Thellung et al., 2013). In this study, treatment of cerebellar granule neuron cultures for 1–3 days with the misfolded prion protein 90–231 fragment augmented both intracellular Ca^{2+} levels and apoptosis compared to that measured in control cells, effects that were nearly completely abolished by NMDA receptor blockade.

Altogether, the data obtained from studies using PrP^C-null animals suggest that there is aberrant NMDA receptor activity in the absence of normal PrP^C function, which is further supported by the fact that exogenous expression of PrP^C can prevent the NMDA receptor-dependent toxic effects of misfolded or mutant prion proteins. This raises the possibility of restoring PrP^C-mediated regulation of NMDA receptors in pathologies where PrP^C function is perturbed as a therapeutic intervention to control NMDA receptor-mediated toxicity. This could in theory be accomplished by mimicking the interaction of PrP^C with NMDA receptors. To achieve this, more details regarding the molecular determinants governing the interaction of PrP^C and NMDA receptor proteins are required. Additionally, if loss of physiological PrP^C function results in augmented activity of certain NMDA receptor subtypes, such as those containing GluN2D subunits, a thorough molecular characterization of the PrP^C-NMDA receptor interaction will allow development of targeted pharmacotherapies to reduce hyperfunction of only those receptors affected whose activity is aberrantly upregulated. This approach could prevent undesired effects of blocking physiological NMDA receptor signaling.

ABSENCE OF PrP^C EXACERBATES ISCHEMIC INJURY—A ROLE FOR NMDA RECEPTOR HYPERACTIVITY?

Given the neuroprotective effects of PrP^C, it may not be surprising that absence of PrP^C can exacerbate the neuronal damage that ensues following an ischemic insult. Several studies have demonstrated increased ischemic brain damage in PrP^C-null mice (McLennan et al., 2004; Sakurai-Yamashita et al., 2005; Spudich et al., 2005; Weise et al., 2006; Mitteregger et al., 2007; Steele et al., 2009). However, ischemia can induce an upregulation of PrP^C mRNA (McLennan et al., 2004; Mitsios et al., 2007; Mitteregger et al., 2007) and protein (McLennan et al., 2004; Weise et al., 2004; Shyu et al., 2005; Adle-Biassette et al., 2006; Mitsios et al., 2007; Mitteregger et al., 2007), presumably as a compensatory protective mechanism. The idea of compensatory upregulation of PrP^C expression protecting against cell death in ischemia is supported by the findings that adenovirus-mediated overexpression of PrP^C reduced ischemia-induced infarct volume in rats (Shyu et al., 2005) and PrP^C-overexpressing mice showed smaller infarct volumes in response to ischemia than wild-type controls (Weise et al., 2008). However, another showed no difference in infarct sizes between wild-type and PrP^C-overexpressing mice exposed to ischemia (Spudich et al., 2005). Of note, these studies used different PrP^C-overexpressing mouse models and times of ischemia, which could account for the observed differences. Currently, the role of NMDA receptors in the enhanced ischemic cell death seen in the absence of PrP^C is unknown. However, because NMDA receptor hyperfunction contributes to cell death in ischemia (Lau and Tymianski, 2010; Lai et al., 2014), and since NMDA receptor desensitization is reduced in the absence of PrP^C (You et al., 2012), enhanced NMDA receptor-mediated Ca^{2+} influx may very well contribute to the increased ischemia-induced cell death that is observed in PrP^C-null mice.

Increases in extracellular glutamate concentration, such as those seen in ischemic conditions, cause activation of both synaptic and extrasynaptic receptors (Hardingham and Bading, 2010; Parsons and Raymond, 2014). Extrasynaptic NMDA receptor activity can induce apoptotic signaling by promoting nuclear import of FOXO transcription factor proteins (Hardingham and Bading, 2010). FOXO3a has recently been identified as a negative regulator of the gene encoding PrP^C, *prnp* (Liu et al., 2013), and in this study it was shown that insulin-like growth factor-1 (IGF-1)-induced PI3K/Akt activity promoted *prnp* expression by preventing nuclear import of FOXO proteins, thus inhibiting their negative effect on *prnp* expression. Therefore, it is possible that conditions where extrasynaptic NMDA receptor activity is upregulated may lead to a down-regulation of PrP^C protein expression, which may initiate a positive feedback loop of further dysregulation of NMDA receptor activity and thus toxicity. Hence, even if there is an initial protective upregulation of PrP^C, this may not be able to be sustained due to persistent activation of extrasynaptic NMDA receptors in the presence of prolonged elevation of glutamate levels.

Another important observation concerns increased shedding of PrP^C in response to neurotoxic conditions (Wang et al., 2012). PrP^C shedding results from cleavage of PrP^C near the C-terminal GPI anchor, releasing full-length, soluble PrP^C into the extracellular milieu (Altmeppen et al., 2013; McDonald et al., 2014). Wang and colleagues showed that upon treatment of cultured rat cortical neurons with NMDA, soluble PrP^C was released into the culture medium, while total PrP^C protein levels remained unchanged (Wang et al., 2012). As noted earlier, our data demonstrated that treatment with PI-PLC, which cleaves PrP^C from its GPI anchor thus releasing it from the cell surface, resulted in persistent NMDA receptor-mediated current in cultured rat hippocampal neurons that was similar to those seen in PrP^C-null hippocampal neurons (You et al., 2012). If loss of cell surface anchored-PrP^C results in aberrant NMDA receptor activity, one could envision a scenario where a positive feedback loop contributes to further shedding of PrP^C and enhanced NMDA receptor currents, which would be of detriment to cells. However, it has been suggested that PrP^C shedding could serve a protective role, especially in preventing Aβ-mediated toxicity (discussed in Altmeppen et al., 2013; Beland et al., 2014). A recent study by Beland and colleagues showed that shed PrP^C co-immunoprecipitated with Aβ (Beland et al., 2014), suggesting that extracellular Aβ can be sequestered by shed PrP^C to prevent Aβ-induced cellular damage. As discussed below, we found that $A\beta_{1-42}$ induced a PrP^C-dependent increase in steady-state NMDA receptor current (You et al., 2012). Shedding of PrP^C may thus aid in preventing $A\beta_{1-42}$-induced dysregulation of NMDA receptor function. Hence, PrP^C shedding may ultimately be protective in situations where PrP^C function is compromised, for example by binding of Aβ peptides. On the other hand, shedding of PrP^C may promote aberrant NMDA receptor activity by altering desensitization kinetics that could contribute to cell death in conditions such as ischemia where glutamate concentrations are elevated.

There is also evidence that ischemia induces cleavage of PrP^C into its N- and C-terminal fragments, and that these cleavage products can be neuroprotective. Mitteregger and coworkers also observed that ischemia caused increased cleavage of PrP^C, as detected by elevated levels of PrP^C C1 fragment on immunoblots of mouse brain homogenates (Mitteregger et al., 2007). In another study, oxygen glucose deprivation increased the secretion of the PrP^C N1 fragment from cultured retinal ganglion cells (Guillot-Sestier et al., 2009). Moreover, application of recombinant PrP^C N1 (PrP^C residues 23–110) was able to prevent ischemia-induced death of retinal ganglion cells (Guillot-Sestier et al., 2009). The details of how PrP^C N1 protects against ischemic cell death were not fully explored, but the mechanism did involve a reduction of ischemia-induced caspase-3 activation and p53 expression (Guillot-Sestier et al., 2009). An interesting possibility is that the PrP^C N1 fragment, which contains the octarepeat region that binds copper, binds to NMDA receptors, thus providing copper-dependent regulation of receptor desensitization. Of note, inhibition of NMDA receptor activity has been shown to attenuate death of retinal ganglion cells following ischemia-reperfusion injury (Li et al., 2014).

Further support of a role for PrP^C in protecting against ischemic cell death comes from the work of Mitteregger and coworkers where PrP^C-null mice and mice expressing a prion protein lacking amino acid residues 32–93 had infarcts of similar volumes (Mitteregger et al., 2007), suggesting that this region of PrP^C provides protection against ischemic injury. One reason for increased ischemic injury in mice expressing this form of prion protein could be that cleavage of this mutant would not produce the normal PrP^C N1 fragment. However, this deletion mutant lacks the copper-binding octarepeat region of PrP^C. As mentioned previously, we found a copper-dependent regulation of NMDA receptors that is mediated in large part through PrP^C, as demonstrated by no further augmentation of NMDA receptor currents by copper chelation in the absence of PrP^C (You et al., 2012). Moreover, exogenous copper can protect neurons from NMDA receptor-mediated death (Schlief et al., 2006), whereas copper chelation induced cell death that could be prevented by NMDA receptor inhibition (You et al., 2012). Recently, data from *in vitro* experiments have revealed that copper enhances one type of α-cleavage, termed α2-cleavage, of mouse PrP^C (McDonald et al., 2014). Copper induced α2-cleavage such that the resultant N-terminal PrP^C fragments terminate at alanine residues 116 or 117, in contrast to α1-cleavage observed in the absence of copper, which produces PrP^C N1 ending at lysine residue 109 (McDonald et al., 2014). These observations suggest that copper levels may regulate PrP^C processing by influencing the amount and identity of the N1 fragment produced, although the biological outcomes of this copper-dependent differential PrP^C cleavage are not yet known. Of further interest is the finding that the prion proteins carrying pathogenic mutations have decreased α2-cleavage upon addition of copper (McDonald et al., 2014), suggesting aberrant copper-dependent PrP^C processing could be detrimental to the health of cells. These findings reveal the possibility of copper-dependent processing of PrP^C as another mechanism contributing to regulation of NMDA receptor activity by copper. Thus, copper may alter the activity of NMDA receptors in multiple ways: PrP^C-independent control of receptor activity, as evidenced by restoration of NMDA receptor desensitization in PrP^C-null neurons by addition of excess copper ions (You et al., 2012); by mediating interaction of PrP^C and NMDA receptors, as shown by a decrease in the amount of GluN1 subunit that co-precipitated with PrP^C upon copper chelation (You et al., 2012); and by influencing PrP^C cleavage to generate PrP^C N1 fragments, which could bind to NMDA receptors to modulate their function. Although evidence exists for the former two mechanisms, the latter remains to be investigated.

As mentioned previously, the activity of myelinic NMDA receptors is increased by ischemic conditions, resulting in damage to myelin (Micu et al., 2006). Thus, a loss of PrP^C-mediated regulation of NMDA current may not only be neurotoxic, but may also contribute to demyelination that has been observed in PrP^C-null mice and in mice expressing mutant prion proteins (Radovanovic et al., 2005; Baumann et al., 2007; Bremer et al., 2010).

Aβ CAN DISRUPT PrP^C-MEDIATED REGULATION OF NMDA RECEPTOR ACTIVITY

PrP^C has been shown to be a receptor for oligomeric Aβ peptides (Lauren et al., 2009; Chen et al., 2010). Since these discoveries, many studies have investigated the role of PrP^C in the pathological effects of Aβ (reviewed in Lauren, 2014). Overall, these findings suggest a role for PrP^C in protecting from the pathological effects of Aβ peptides that occur in AD, although the requirement of PrP^C for Aβ-induced toxicity has been controversial, with some investigators finding that PrP^C was not required for the pathological effects of Aβ (Balducci et al., 2010; Calella et al., 2010; Kessels et al., 2010; Cisse et al., 2011). Recent findings by Nicoll and coworkers may help to reconcile these discrepancies. Their work showed that Aβ protofibrils, and not monomers or fibrils, bind with highest affinity to full-length PrP^C and the PrP^C N1 fragment (Nicoll et al., 2013). Furthermore, protofibril-rich Aβ preparations blocked LTP in a PrP^C-dependent manner, whereas LTP inhibition by fibrillar Aβ was independent of PrP^C. These observations underscore the need for careful characterization of Aβ preparations in order to determine the nature of the Aβ species used for experiments, which allow better interpretation of data resulting from these experiments.

Notably, many studies have demonstrated that Aβ can alter NMDA receptor function (reviewed in Rush and Buisson, 2014). Recently, we demonstrated that PrP^C is required for Aβ-induced alterations in NMDA receptor kinetics (You et al., 2012). Application of oligomerized $A\beta_{1-42}$, which is known to be a high-affinity copper chelator, to hippocampal neurons gave rise to a persistent (i.e., non-desensitizing) NMDA current, similar to what was observed in the presence of the copper chelator BCS or in the absence of PrP^C. Of note, the effects of $A\beta_{1-42}$ and copper chelation on NMDA current were not additive, suggesting that they acted via the same mechanism. Finally, in neurons isolated from 5XFAD mice (a mouse model of AD), NMDA current displayed a qualitatively similar persistent component. Altogether, these data then suggest that Aβ-mediated perturbation of NMDA receptor activity depends on both PrP^C and copper ions. Furthermore, the $A\beta_{1-42}$-induced upregulation of NMDA receptor activity may in part be responsible for the $A\beta_{1-42}$-mediated neurodegeneration observed in AD. If this were the case, then why is neurodegeneration observed in 5XFAD mice, but relatively absent in PrP^C-null mice even though they both display similar non-desensitizing NMDA receptor currents? Slowed receptor desensitization can only manifest itself pathologically under conditions where there is a prolonged excess of glutamate. It has been shown that $A\beta_{1-42}$ prevents glutamate re-uptake into neurons and glial cells (Danysz and Parsons, 2012), and induces release of glutamate from astrocytes (Talantova et al., 2013), which then would be expected to lead to just such an excess of glutamate. Then, together with the slowed desensitization kinetics of the receptor, this excess glutamate may damage neuronal structures such as spines. In PrP^C-null mice, this accumulation of glutamate would not be expected to occur, thus rendering the slowed desensitization kinetics physiologically inert (**Figure 2**). Nonetheless, as we discussed earlier, there are other pathophysiological events such as chemically-induced seizures or ischemia that are known to lead to an accumulation of glutamate that then causes increased brain damage in PrP^C-null mice (Walz et al., 1999; McLennan et al., 2004; Sakurai-Yamashita et al., 2005; Spudich et al., 2005; Weise et al., 2006; Mitteregger et al., 2007; Steele et al., 2009).

Work by Um and colleagues provides further evidence for the requirement of PrP^C on the effects of Aβ on NMDA receptor function (Um et al., 2012). In their study, they found that phosphorylation of the GluN2B subunit of the NMDA receptor at tyrosine residue 1472 was transiently increased by Aβ oligomers in Fyn-overexpressing neuroblastoma cells and in cortical cultures; longer treatment with Aβ oligomers (1–3 h) resulted in decreased phospho-GluN2B levels, which correlated with an increase in the level of STEP protein phosphatase. Moreover, the Aβ oligomer-induced increase in GluN2B phosphorylation was absent in PrP^C- and Fyn-null cultures. Indeed, they found that Aβ oligomers activated Fyn kinase in a PrP^C-dependent manner. Furthermore, brain extract from AD patients induced Fyn activation in cultured mouse cortical neurons in a PrP^C-dependent manner, whereas extract from healthy controls did not activate Fyn. The initial GluN2B phosphorylation was paralleled by an increase in surface levels of GluN2B, and with longer Aβ oligomer treatment a reduction in GluN2B surface expression was observed; again, both PrP^C and Fyn were required for these Aβ oligomer-mediated effects on GluN2B surface expression. In cultures from wild-type mice, pre-treatment for 15 min with Aβ oligomers significantly enhanced the NMDA-induced increase in intracellular Ca^{2+} levels, whereas a 60 min Aβ oligomer pre-treatment significantly reduced the NMDA-induced Ca^{2+} signal; in PrP^C-null cultures, Aβ oligomers had no effect on the Ca^{2+} signal induced by NMDA. Thus, Aβ oligomers, acting via PrP^C, induce a transient increase then a decrease in NMDA receptor-mediated Ca^{2+} influx into neurons. LDH release from cortical cultures was increased upon 90 min Aβ oligomer treatment and no further toxicity was observed with a 72 h treatment. This toxicity was NMDA receptor-dependent, as it was reduced by NMDA receptor inhibition with both APV and ifenprodil, the latter being a specific GluN2B antagonist. Thus, brief exposure to Aβ oligomers causes toxicity, likely by Ca^{2+} influx mediated by the transient increase in surface localization of GluN2B-containing

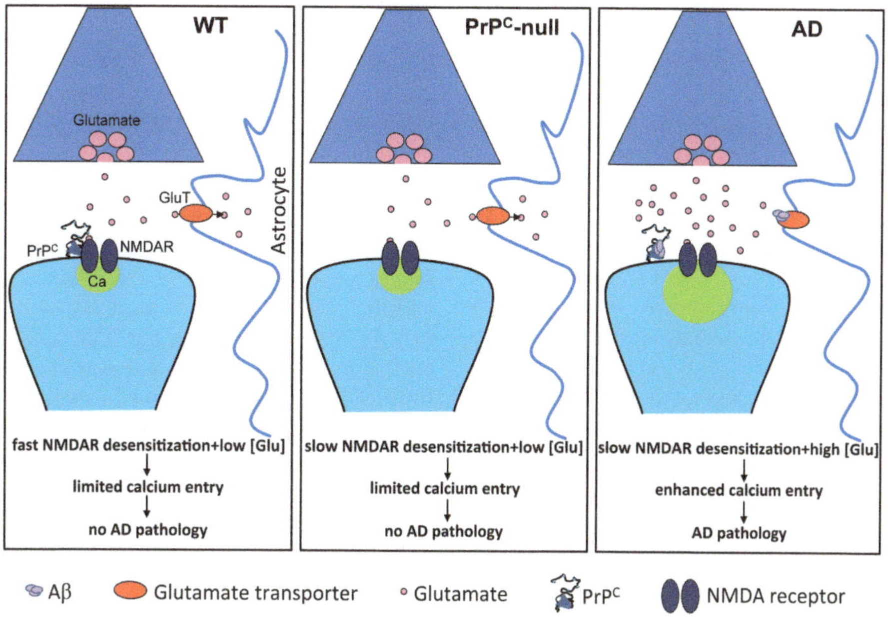

FIGURE 2 | Slowed NMDA receptor desensitization is pathological only in conditions of prolonged excess of glutamate. Under normal conditions (WT), glutamate homeostasis is unperturbed and prevents glutamate accumulation, and NMDA receptors (NMDAR) undergo fast desensitization, thus limiting calcium entry. In the absence of PrP^C (PrP^C-null), NMDA receptor desensitization is slowed but glutamate homeostasis remains unperturbed, thus limited calcium entry occurs. However, in Alzheimer's disease (AD), Aβ oligomers bind to PrP^C and cause slowed NMDA receptor desensitization. This, in combination with Aβ-induced elevation of glutamate levels, for example by inhibiting glutamate re-uptake by astrocytic glutamate transporters, leads to enhanced calcium entry and pathology.

NMDA receptors. In this study, it was also shown that Aβ oligomers induced spine loss in a PrP^C- and Fyn-dependent manner, and that the seizures and death due to status epilepticus in a mouse model of AD were prevented by genetic deletion of PrP^C. Taken together, these data suggest that Aβ oligomers signal through PrP^C to induce aberrant Fyn activation, leading to first an increase then later a decrease in activity of GluN2B-containing receptors, which is ultimately deleterious to neurons. Altogether, the recent data indicate that Aβ oligomers, by interacting with PrP^C, can alter both NMDA receptor kinetics (You et al., 2012) and the signaling pathways that modulate NMDA receptor function (Um et al., 2012).

A study by Resenberger and colleagues showed that Aβ- and PrP^{Sc}-induced cell death of PrP^C-expressing SH-SY5Y neuroblastoma cells were not additive, suggesting that these species act via the same pathway (Resenberger et al., 2011). Furthermore, pre-treatment with the NMDA receptor blocker memantine prevented apoptosis of PrP^C-expressing SH-SY5Y neuroblastoma cells induced by PrP^{Sc}, Aβ, and a synthetic peptide designed to be rich in β-sheet structure. This suggests that several types of β-sheet-containing proteins/peptides share a common mechanism of toxicity, aberrant NMDA receptor activity. Thus, PrP^C could contribute to toxicity in disorders other than TSEs and AD, possibly by mediating aberrant NMDA receptor activity in the presence of β-sheet-rich peptides.

Of further interest is the finding that the toxicity induced by β-sheet-rich conformers was dependent upon the N-terminal region of PrP^C and was prevented by a soluble N-terminal domain of PrP^C (Resenberger et al., 2011). Similar protective effects of the PrP^C N1 fragment against Aβ-induced toxicity have been observed in other studies (Beland et al., 2012, 2014; Guillot-Sestier et al., 2012; Nieznanski et al., 2012; Fluharty et al., 2013). This raises the possibility that increased PrP^C cleavage to produce the N1 fragment is neuroprotective. Interestingly, cleavage of PrP^C was increased in a mouse model of AD (Ostapchenko et al., 2013) and in post-mortem human AD brain (Beland et al., 2014). Furthermore, the recent findings from Beland and coworkers provide strong evidence for a protective role of the soluble N1 fragment that is produced by α-cleavage of PrP^C. In this study, the binding of PrP^C N1 and Aβ induced a conformational change that produced amorphous aggregates, the amount of PrP^C N1 in guanidine hydrochloride extracts from insoluble amyloid deposits from AD brain was increased compared to non-demented controls, and there was a significant correlation between the amount of α-cleavage, which generates the N1 fragment, and duration of AD (Beland et al., 2014). Altogether, these observations indicate that cleavage of PrP^C may reduce Aβ-mediated toxicity via N1-mediated neutralization of Aβ peptides, and N-terminal fragments may be useful as a therapy to prevent pathological effects of Aβ or other β-sheet-rich peptides, including alterations in NMDA receptor activity.

NMDA RECEPTORS, PrP^C, AND NEUROINFLAMMATION

Recent evidence has emerged that glutamatergic mechanisms operate not only in the nervous system, but also in a wide variety of non-neuronal cells (reviews: Gill and Pulido, 2001; Skerry and

Genever, 2001; Hinoi et al., 2004), including in the immune system (reviews: Boldyrev et al., 2005, 2012). GluRs, both ionotropic and metabotropic, are highly expressed in various immune cells, such as T cells, B cells, neutrophils, macrophages and dendritic cells, which are thought to subserve communication between the immune and nervous systems (review: Ganor and Levite, 2012). Interestingly, different GluRs, or different levels of certain GluRs, are expressed in resting and activated T cells (for review see Ganor and Levite, 2012). In particular, NMDA receptors expressed on T cells are involved in a wide variety of T cell functions, such as regulation of cytokine secretion (Kahlfuβ et al., 2014), proliferation (Kahlfuβ et al., 2014), apoptosis (Affaticati et al., 2011), and induction of reactive oxygen species (Tuneva et al., 2003).

As described above, PrP^C interacts with NMDA receptors on neurons and modulates NMDA receptor-dependent neuronal excitability and excitotoxicity (Khosravani et al., 2008). Although the precise functions of PrP^C in immune cells remain unclear, PrP^C expression is detected in the lymphoid system (Bendheim et al., 1992; Ford et al., 2002). Human T lymphocytes constitutively express PrP^C and its surface expression is influenced by the activation state of the cells (Cashman et al., 1990; Mabbott et al., 1997). In addition, PrP^C is upregulated on mouse T cells and dendritic cells after activation, and the lack of PrP^C increases T cell proliferation and causes T cell over-activation in mouse (Tsutsui et al., 2008). Therefore, PrP^C may play a role as a regulator of NMDA receptors in the immune system as we observed in the nervous system (Khosravani et al., 2008; You et al., 2012), with a common thread being "hyperfunction" when PrP^C is absent and cannot restrain normal physiological mechanisms.

The CNS is conventionally recognized as being "immunologically privileged" (Bailey et al., 2006) and is anatomically separated from the peripheral immune system by the presence of the blood-brain barrier (BBB) (Xiao and Link, 1998; Weller et al., 2009). Despite the presence of the BBB, which limits the entry of cells and pathogens to the brain, lymphocytes can traffic into the CNS to surveil the local environment (Xiao and Link, 1998; Kivisakk et al., 2003). It is increasingly recognized that the CNS is capable of shaping the immune response (for review see Xiao and Link, 1998); it is now clear that neuroinflammation is a well-established hallmark of a number of neurodegenerative diseases, including AD, and is frequently detrimental to neurological function (Akiyama et al., 2000; Rubio-Perez and Morillas-Ruiz, 2012; Solito and Sastre, 2012; Cappellano et al., 2013; Enciu and Popescu, 2013; Lynch, 2014). In the AD brain, the inflammatory response is thought to be a secondary response caused by an initial brain insult, which is provided by damaged neurons, neurites, insoluble Aβ aggregates, and neurofibrillary tangles as stimuli (Akiyama et al., 2000). However, recent evidence suggests that inflammatory mediators may stimulate amyloid precursor protein processing, and therefore, play a role as a driving force to establish a deleterious cycle to AD progression (Heneka et al., 2010). As described above, PrP^C is a receptor for oligomeric Aβ peptides (Lauren et al., 2009; Chen et al., 2010; Lauren, 2014), and Aβ oligomers, acting via PrP^C, modulate NMDA receptor-mediated Ca^{2+} influx into neurons. Perhaps, even soluble forms of small oligomers, may directly stimulate PrP^C and modulate NMDA receptor-mediated Ca^{2+} influx into immune cells to trigger signaling pathways to secrete cytokines, reactive oxygen species, and/or regulate T cell polarization at the early stage of AD. Gaining more insight into the role of PrP^C-NMDA receptor interactions in neuroinflammation may reveal novel approaches to visualize early AD pathogenesis and diagnosis.

CONCLUDING REMARKS

It is becoming clear that loss of PrP^C regulation of NMDA receptors can result in toxicity in a variety of pathological conditions, and that Aβ can cause aberrant activation of NMDA receptors in a PrP^C-dependent manner. The possibility of restoring the NMDA receptor-PrP^C interaction as a way to protect against excitotoxicity will need to be tested experimentally, and in order to accomplish this, elucidation of the molecular details of the interaction between PrP^C and NMDA receptors is needed. In light of the recent findings regarding cleavage of PrP^C and how this is altered in disease, there is a need for investigation of how physiological processing of PrP^C impacts NMDA receptor activity in both health and disease. Overall, a better understanding of the molecular mechanisms that determine how PrP^C regulates NMDA receptor function will open up new therapeutic avenues in situations where the regulation of NMDA receptors by PrP^C is perturbed, such as in AD and prion diseases.

ACKNOWLEDGMENTS

Stefanie A. G. Black is supported by a Postdoctoral Fellowship from the Multiple Sclerosis Society of Canada (MSSOC). Peter K. Stys is supported by an Alberta Innovates-Health Solutions (AI-HS) Scientist Award; work in the lab of Peter K. Stys is supported by the Canadian Institutes of Health Research (CIHR), MSSOC, the Alberta Prion Research Institute (APRI), Canada Research Chairs. Gerald W. Zamponi is a Canada Research Chair and an AI-HS Scientist; work in the Gerald W. Zamponi lab is supported by funds from APRI and CIHR. Work in the lab of Shigeki Tsutsui is supported by APRI/Alzheimer Society of Alberta and Northwest Territories (ASANT).

REFERENCES

Adle-Biassette, H., Verney, C., Peoc'h, K., Dauge, M. C., Razavi, F., Choudat, L., et al. (2006). Immunohistochemical expression of prion protein (PrPC) in the human forebrain during development. *J. Neuropathol. Exp. Neurol.* 65, 698–706. doi: 10.1097/01.jnen.0000228137.10531.72

Affaticati, P., Mignen, O., Jambou, F., Potier, M., Klingel-Schmitt, I., Degrouard, J., et al. (2011). Sustained calcium signalling and caspase-3 activation involve NMDA receptors in thymocytes in contact with dendritic cells. *Cell Death Differ.* 18, 99–108. doi: 10.1038/cdd.2010.79

Aguzzi, A., Baumann, F., and Bremer, J. (2008). The prion's elusive reason for being. *Annu. Rev. Neurosci.* 31, 439–477. doi: 10.1146/annurev.neuro.31.060407.125620

Akiyama, H., Barger, S., Barnum, S., Bradt, B., Bauer, J., Cole, G., et al. (2000). Inflammation and Alzheimer's disease. *Neurobiol. Aging* 21, 383–421. doi: 10.1016/S0197-4580(00)00124-X

Altmeppen, H. C., Prox, J., Puig, B., Dohler, F., Falker, C., Krasemann, S., et al. (2013). Roles of endoproteolytic alpha-cleavage and shedding of the prion protein in neurodegeneration. *FEBS J.* 280, 4338–4347. doi: 10.1111/febs.12196

Bailey, S., Carpentier, P., McMahon, E., Begolka, W., and Miller, S. (2006). Innate and adaptive immune responses of the central nervous system. *Crit. Rev. Immunol.* 26, 149–188. doi: 10.1615/CritRevImmunol.v26.i2.40

Balducci, C., Beeg, M., Stravalaci, M., Bastone, A., Sclip, A., Biasini, E., et al. (2010). Synthetic amyloid-beta oligomers impair long-term memory independently of cellular prion protein. *Proc. Natl. Acad. Sci. U.S.A.* 107, 2295–2300. doi: 10.1073/pnas.0911829107

Bartlett, T. E., and Wang, Y. T. (2013). The intersections of NMDAR-dependent synaptic plasticity and cell survival. *Neuropharmacology* 74, 59–68. doi: 10.1016/j.neuropharm.2013.01.012

Baumann, F., Tolnay, M., Brabeck, C., Pahnke, J., Kloz, U., Niemann, H. H., et al. (2007). Lethal recessive myelin toxicity of prion protein lacking its central domain. *EMBO J.* 26, 538–547. doi: 10.1038/sj.emboj.7601510

Beland, M., Bedard, M., Tremblay, G., Lavigne, P., and Roucou, X. (2014). Abeta induces its own prion protein N-terminal fragment (PrPN1)-mediated neutralization in amorphous aggregates. *Neurobiol. Aging* 35, 1537–1548. doi: 10.1016/j.neurobiolaging.2014.02.001

Beland, M., Motard, J., Barbarin, A., and Roucou, X. (2012). PrP(C) homodimerization stimulates the production of PrPC cleaved fragments PrPN1 and PrPC1. *J. Neurosci.* 32, 13255–13263. doi: 10.1523/JNEUROSCI.2236-12.2012

Bendheim, P., Brown, H., Rudelli, R., Scala, L., Goller, N., Wen, G., et al. (1992). Nearly ubiquitous tissue distribution of the scrapie agent precursor protein. *Neurology* 42, 149–156. doi: 10.1212/WNL.42.1.149

Beraldo, F. H., Arantes, C. P., Santos, T. G., Machado, C. F., Roffe, M., Hajj, G. N., et al. (2011). Metabotropic glutamate receptors transduce signals for neurite outgrowth after binding of the prion protein to laminin γ1 chain. *FASEB J.* 25, 265–279. doi: 10.1096/fj.10-161653

Beraldo, F. H., Arantes, C. P., Santos, T. G., Queiroz, N. G., Young, K., Rylett, R. J., et al. (2010). Role of alpha7 nicotinic acetylcholine receptor in calcium signaling induced by prion protein interaction with stress-inducible protein 1. *J. Biol. Chem.* 285, 36542–36550. doi: 10.1074/jbc.M110.157263

Biasini, E., Turnbaugh, J. A., Unterberger, U., and Harris, D. A. (2012). Prion protein at the crossroads of physiology and disease. *Trends Neurosci.* 35, 92–103. doi: 10.1016/j.tins.2011.10.002

Biasini, E., Unterberger, U., Solomon, I. H., Massignan, T., Senatore, A., Bian, H., et al. (2013). A mutant prion protein sensitizes neurons to glutamate-induced excitotoxicity. *J. Neurosci.* 33, 2408–2418. doi: 10.1523/JNEUROSCI.3406-12.2013

Boldyrev, A., Bryushkova, E., and Vladychenskaya, E. (2012). NMDA receptors in immune competent cells. *Biochemistry (Mosc.)* 77, 128–134. doi: 10.1134/S0006297912020022

Boldyrev, A., Carpenter, D., and Johnson, P. (2005). Emerging evidence for a similar role of glutamate receptors in the nervous and immune systems. *J. Neurochem.* 95, 913–918. doi: 10.1111/j.1471-4159.2005.03456.x

Bremer, J., Baumann, F., Tiberi, C., Wessig, C., Fischer, H., Schwarz, P., et al. (2010). Axonal prion protein is required for peripheral myelin maintenance. *Nat. Neurosci.* 13, 310–318. doi: 10.1038/nn.2483

Calella, A. M., Farinelli, M., Nuvolone, M., Mirante, O., Moos, R., Falsig, J., et al. (2010). Prion protein and Abeta-related synaptic toxicity impairment. *EMBO Mol. Med.* 2, 306–314. doi: 10.1002/emmm.201000082

Cappellano, G., Carecchio, M., Fleetwood, T., Magistrelli, L., Cantello, R., Dianzani, U., et al. (2013). Immunity and inflammation in neurodegenerative diseases. *Am. J. Neurodegener. Dis.* 2, 89–107.

Carulla, P., Bribián, A., Rangel, A., Gavín, R., Ferrer, I., Caelles, C., et al. (2011). Neuroprotective role of PrPC against kainate-induced epileptic seizures and cell death depends on the modulation of JNK3 activation by GluR6/7-PSD-95 binding. *Mol. Biol. Cell* 22, 3041–3054. doi: 10.1091/mbc.E11-04-0321

Cashman, N., Loertscher, R., Nalbantoglu, J., Shaw, I., Kascsak, R., Bolton, D., et al. (1990). Cellular isoform of the scrapie agent protein participates in lymphocyte activation. *Cell* 61, 185–192. doi: 10.1016/0092-8674(90)90225-4

Chatterton, J. E., Awobuluyi, M., Premkumar, L. S., Takahashi, H., Talantova, M., Shin, Y., et al. (2002). Excitatory glycine receptors containing the NR3 family of NMDA receptor subunits. *Nature* 415, 793–798. doi: 10.1038/nature715

Chen, S., Yadav, S. P., and Surewicz, W. K. (2010). Interaction between human prion protein and amyloid-beta (Abeta) oligomers: role of N-terminal residues. *J. Biol. Chem.* 285, 26377–26383. doi: 10.1074/jbc.M110.145516

Cisse, M., Sanchez, P. E., Kim, D. H., Ho, K., Yu, G. Q., and Mucke, L. (2011). Ablation of cellular prion protein does not ameliorate abnormal neural network activity or cognitive dysfunction in the J20 line of human amyloid precursor protein transgenic mice. *J. Neurosci.* 31, 10427–10431. doi: 10.1523/JNEUROSCI.1459-11.2011

Corlew, R., Brasier, D. J., Feldman, D. E., and Philpot, B. D. (2008). Presynaptic NMDA receptors: newly appreciated roles in cortical synaptic function and plasticity. *Neuroscientist* 14, 609–625. doi: 10.1177/10738584083 22675

Cull-Candy, S. G., and Leszkiewicz, D. N. (2004). Role of distinct NMDA receptor subtypes at central synapses. *Sci. STKE* 2004:re16. doi: 10.1126/stke.2552004re16

Danysz, W., and Parsons, C. G. (2012). Alzheimer's disease, beta-amyloid, glutamate, NMDA receptors and memantine–searching for the connections. *Br. J. Pharmacol.* 167, 324–352. doi: 10.1111/j.1476-5381.2012.02057.x

Enciu, A., and Popescu, B. (2013). Is there a causal link between inflammation and dementia? *Biomed. Res. Int.* 2013:316495. doi: 10.1155/2013/316495

Fleisch, V. C., Leighton, P. L., Wang, H., Pillay, L. M., Ritzel, R. G., Bhinder, G., et al. (2013). Targeted mutation of the gene encoding prion protein in zebrafish reveals a conserved role in neuron excitability. *Neurobiol. Dis.* 55, 11–25. doi: 10.1016/j.nbd.2013.03.007

Fluharty, B. R., Biasini, E., Stravalaci, M., Sclip, A., Diomede, L., Balducci, C., et al. (2013). An N-terminal fragment of the prion protein binds to amyloid-beta oligomers and inhibits their neurotoxicity *in vivo*. *J. Biol. Chem.* 288, 7857–7866. doi: 10.1074/jbc.M112.423954

Ford, M., Burton, L., Morris, R., and Hall, S. (2002). Selective expression of prion protein in peripheral tissues of the adult mouse. *Neuroscience* 113, 177–192. doi: 10.1016/S0306-4522(02)00155-0

Gadotti, V. M., Bonfield, S. P., and Zamponi, G. W. (2012). Depressive-like behaviour of mice lacking cellular prion protein. *Behav. Brain. Res.* 227, 319–323. doi: 10.1016/j.bbr.2011.03.012

Gadotti, V. M., and Zamponi, G. W. (2011). Cellular prion protein protects from inflammatory and neuropathic pain. *Mol. Pain* 7:59. doi: 10.1186/1744-8069-7-59

Ganor, Y., and Levite, M. (2012). "Glutamate in the immune system: glutamate receptors in immune cells, potent effects, endogenous production and involvement in disease," in *Nerve-Driven Immunity*, ed M. Levite (Wien: Springer-Verlag/Wien), 121–161. doi: 10.1007/978-3-7091-0888-8

Gill, S., and Pulido, O. (2001). Glutamate receptors in peripheral tissues: current knowledge, future research, and implications for toxicology. *Toxicol. Pathol.* 29, 208–223. doi: 10.1080/019262301317052486

Guillot-Sestier, M. V., Sunyach, C., Druon, C., Scarzello, S., and Checler, F. (2009). The alpha-secretase-derived N-terminal product of cellular prion, N1, displays neuroprotective function *in vitro* and *in vivo*. *J. Biol. Chem.* 284, 35973–35986. doi: 10.1074/jbc.M109.051086

Guillot-Sestier, M. V., Sunyach, C., Ferreira, S. T., Marzolo, M. P., Bauer, C., Thevenet, A., et al. (2012). alpha-Secretase-derived fragment of cellular prion, N1, protects against monomeric and oligomeric amyloid beta (Abeta)-associated cell death. *J. Biol. Chem.* 287, 5021–5032. doi: 10.1074/jbc.M111.323626

Hardingham, G. E. (2009). Coupling of the NMDA receptor to neuroprotective and neurodestructive events. *Biochem. Soc. Trans.* 37, 1147–1160. doi: 10.1042/BST0371147

Hardingham, G. E., and Bading, H. (2010). Synaptic versus extrasynaptic NMDA receptor signalling: implications for neurodegenerative disorders. *Nat. Rev. Neurosci.* 11, 682–696. doi: 10.1038/nrn2911

Heneka, M., O'banion, M., Terwel, D., and Kummer, M. (2010). Neuroinflammatory processes in Alzheimer's disease. *J. Neural Transm.* 117, 919–947. doi: 10.1007/s00702-010-0438-z

Hinoi, E., Takarada, T., Ueshima, T., Tsuchihashi, Y., and Yoneda, Y. (2004). Glutamate signaling in peripheral tissues. *Eur. J. Biochem.* 271, 1–13. doi: 10.1046/j.1432-1033.2003.03907.x

Hu, N. W., Nicoll, A. J., Zhang, D., Mably, A. J., O'Malley, T., Purro, S. A., et al. (2014). mGlu5 receptors and cellular prion protein mediate amyloid-β-facilitated synaptic long-term depression *in vivo*. *Nat. Commun.* 5, 3374. doi: 10.1038/ncomms4374

Kahlfuβ, S., Simma, N., Mankiewicz, J., Bose, T., Lowinus, T., Klein-Hessling, S., et al. (2014). Immunosuppression by N-methyl-D-aspartate receptor antagonists is mediated through inhibition of Kv1.3 and KCa3.1 channels in T cells. *Mol. Cell. Biol.* 34, 820–831. doi: 10.1128/MCB.01273-13

Karadottir, R., Cavelier, P., Bergersen, L. H., and Attwell, D. (2005). NMDA receptors are expressed in oligodendrocytes and activated in ischaemia. *Nature* 438, 1162–1166. doi: 10.1038/nature04302

Kessels, H. W., Nguyen, L. N., Nabavi, S., and Malinow, R. (2010). The prion protein as a receptor for amyloid-beta. *Nature* 466, E3–E4. discussion: E4–E5. doi: 10.1038/nature09217

Khosravani, H., Zhang, Y., Tsutsui, S., Hameed, S., Altier, C., Hamid, J., et al. (2008). Prion protein attenuates excitotoxicity by inhibiting NMDA receptors. *J. Cell Biol.* 181, 551–565. doi: 10.1083/jcb.200711002

Kivisakk, P., Mahad, D., Callahan, M., Trebst, C., Tucky, B., Wei, T., et al. (2003). Human cerebrospinal fluid central memory CD4+ T cells: evidence for trafficking through choroid plexus and meninges via P-selectin. *Proc. Natl. Acad. Sci. U.S.A.* 100, 8389–8394. doi: 10.1073/pnas.1433000100

Kleene, R., Loers, G., Langer, J., Frobert, Y., Buck, F., and Schachner, M. (2007). Prion protein regulates glutamate-dependent lactate transport of astrocytes. *J. Neurosci.* 27, 12331–12340. doi: 10.1523/JNEUROSCI.1358-07.2007

Klewpatinond, M., Davies, P., Bowen, S., Brown, D. R., and Viles, J. H. (2008). Deconvoluting the Cu2+ binding modes of full-length prion protein. *J. Biol. Chem.* 283, 1870–1881. doi: 10.1074/jbc.M708472200

Lai, T. W., Zhang, S., and Wang, Y. T. (2014). Excitotoxicity and stroke: identifying novel targets for neuroprotection. *Prog. Neurobiol.* 115, 157–188. doi: 10.1016/j.pneurobio.2013.11.006

Lau, A., and Tymianski, M. (2010). Glutamate receptors, neurotoxicity and neurodegeneration. *Pflugers Arch.* 460, 525–542. doi: 10.1007/s00424-010-0809-1

Lauren, J. (2014). Cellular prion protein as a therapeutic target in Alzheimer's disease. *J. Alzheimers Dis.* 38, 227–244. doi: 10.3233/JAD-130950

Lauren, J., Gimbel, D., Nygaard, H., Gilbert, J., and Strittmatter, S. (2009). Cellular prion protein mediates impairment of synaptic plasticity by amyloid-beta oligomers. *Nature* 457, 1128–1132. doi: 10.1038/nature07761

Li, J. B., Lu, Z. G., Xu, L., Wang, Q., Zhang, Z. H., and Fang, J. H. (2014). Neuroprotective effects of bis(7)-tacrine in a rat model of pressure-induced retinal ischemia. *Cell Biochem. Biophys.* 68, 275–282. doi: 10.1007/s12013-013-9707-4

Liu, T., Yi, W., Feng, B., Zhou, Z., and Xiao, G. (2013). IGF-1-induced enhancement of PRNP expression depends on the negative regulation of transcription factor FOXO3a. *PLoS ONE* 8:e71896. doi: 10.1371/journal.pone.0071896

Lynch, M. (2014). The impact of neuroimmune changes on development of amyloid pathology; relevance to Alzheimer's disease. *Immunology* 141, 292–301. doi: 10.1111/imm.12156

Mabbott, N., Brown, K., Manson, J., and Bruce, M. (1997). T-lymphocyte activation and the cellular form of the prion protein. *Immunology* 92, 161–165. doi: 10.1046/j.1365-2567.1997.00331.x

Mayer, M. L. (2005). Glutamate receptor ion channels. *Curr. Opin. Neurobiol.* 15, 282–288. doi: 10.1016/j.conb.2005.05.004

Mayer, M. L., Vyklicky, L. Jr., and Clements, J. (1989). Regulation of NMDA receptor desensitization in mouse hippocampal neurons by glycine. *Nature* 338, 425–427. doi: 10.1038/338425a0

McDonald, A. J., Dibble, J. P., Evans, E. G., and Millhauser, G. L. (2014). A new paradigm for enzymatic control of alpha-cleavage and beta-cleavage of the prion protein. *J. Biol. Chem.* 289, 803–813. doi: 10.1074/jbc.M113.502351

McLennan, N. F., Brennan, P. M., McNeill, A., Davies, I., Fotheringham, A., Rennison, K. A., et al. (2004). Prion protein accumulation and neuroprotection in hypoxic brain damage. *Am. J. Pathol.* 165, 227–235. doi: 10.1016/S0002-9440(10)63291-9

Micu, I., Jiang, Q., Coderre, E., Ridsdale, A., Zhang, L., Woulfe, J., et al. (2006). NMDA receptors mediate calcium accumulation in myelin during chemical ischaemia. *Nature* 439, 988–992. doi: 10.1038/nature04474

Mitsios, N., Saka, M., Krupinski, J., Pennucci, R., Sanfeliu, C., Miguel Turu, M., et al. (2007). Cellular prion protein is increased in the plasma and peri-infarcted brain tissue after acute stroke. *J. Neurosci. Res.* 85, 602–611. doi: 10.1002/jnr.21142

Mitteregger, G., Vosko, M., Krebs, B., Xiang, W., Kohlmannsperger, V., Nolting, S., et al. (2007). The role of the octarepeat region in neuroprotective function of the cellular prion protein. *Brain Pathol.* 17, 174–183. doi: 10.1111/j.1750-3639.2007.00061.x

Müller, W. E., Ushijima, H., Schroder, H. C., Forrest, J. M., Schatton, W. F., Rytik, P. G., et al. (1993). Cytoprotective effect of NMDA receptor antagonists on prion protein (PrionSc)-induced toxicity in rat cortical cell cultures. *Eur. J. Pharmacol.* 246, 261–267. doi: 10.1016/0922-4106(93)90040-G

Nicoll, A. J., Panico, S., Freir, D. B., Wright, D., Terry, C., Risse, E., et al. (2013). Amyloid-beta nanotubes are associated with prion protein-dependent synaptotoxicity. *Nat. Commun.* 4, 2416. doi: 10.1038/ncomms3416

Nieznanski, K., Choi, J. K., Chen, S., Surewicz, K., and Surewicz, W. K. (2012). Soluble prion protein inhibits amyloid-beta (Abeta) fibrillization and toxicity. *J. Biol. Chem.* 287, 33104–33108. doi: 10.1074/jbc.C112.400614

Ostapchenko, V. G., Beraldo, F. H., Guimaraes, A. L., Mishra, S., Guzman, M., Fan, J., et al. (2013). Increased prion protein processing and expression of metabotropic glutamate receptor 1 in a mouse model of Alzheimer's disease. *J. Neurochem.* 127, 415–425. doi: 10.1111/jnc.12296

Paoletti, P., Bellone, C., and Zhou, Q. (2013). NMDA receptor subunit diversity: impact on receptor properties, synaptic plasticity and disease. *Nat. Rev. Neurosci.* 14, 383–400. doi: 10.1038/nrn3504

Parsons, M. P., and Raymond, L. A. (2014). Extrasynaptic NMDA receptor involvement in central nervous system disorders. *Neuron* 82, 279–293. doi: 10.1016/j.neuron.2014.03.030

Pina-Crespo, J. C., Talantova, M., Micu, I., States, B., Chen, H. S., Tu, S., et al. (2010). Excitatory glycine responses of CNS myelin mediated by NR1/NR3 "NMDA" receptor subunits. *J. Neurosci.* 30, 11501–11505. doi: 10.1523/JNEUROSCI.1593-10.2010

Qin, K., Yang, D. S., Yang, Y., Chishti, M. A., Meng, L. J., Kretzschmar, H. A., et al. (2000). Copper(II)-induced conformational changes and protease resistance in recombinant and cellular PrP. Effect of protein age and deamidation. *J. Biol. Chem.* 275, 19121–19131. doi: 10.1074/jbc.275.25.19121

Quintanar, L., Rivillas-Acevedo, L., Grande-Aztatzi, R., Gómez-Castro, C. Z., Arcos-López, T., and Vela, A. (2013). Copper coordination to the prion protein: insights from theoretical studies. *Coordin. Chem. Rev.* 257, 429–444. doi: 10.1016/j.ccr.2012.06.026

Radovanovic, I., Braun, N., Giger, O. T., Mertz, K., Miele, G., Prinz, M., et al. (2005). Truncated prion protein and Doppel are myelinotoxic in the absence of oligodendrocytic PrPC. *J. Neurosci.* 25, 4879–4888. doi: 10.1523/JNEUROSCI.0328-05.2005

Rangel, A., Burgaya, F., Gavin, R., Soriano, E., Aguzzi, A., and Del Rio, J. A. (2007). Enhanced susceptibility of Prnp-deficient mice to kainate-induced seizures, neuronal apoptosis, and death: role of AMPA/kainate receptors. *J. Neurosci. Res.* 85, 2741–2755. doi: 10.1002/jnr.21215

Ratte, S., Vreugdenhil, M., Boult, J. K., Patel, A., Asante, E. A., Collinge, J., et al. (2011). Threshold for epileptiform activity is elevated in prion knockout mice. *Neuroscience* 179, 56–61. doi: 10.1016/j.neuroscience.2011.01.053

Resenberger, U. K., Harmeier, A., Woerner, A. C., Goodman, J. L., Muller, V., Krishnan, R., et al. (2011). The cellular prion protein mediates neurotoxic signalling of beta-sheet-rich conformers independent of prion replication. *EMBO J.* 30, 2057–2070. doi: 10.1038/emboj.2011.86

Riemer, C., Burwinkel, M., Schwarz, A., Gultner, S., Mok, S. W., Heise, I., et al. (2008). Evaluation of drugs for treatment of prion infections of the central nervous system. *J. Gen. Virol.* 89, 594–597. doi: 10.1099/vir.0.83281-0

Rubio-Perez, J., and Morillas-Ruiz, J. (2012). A review: inflammatory process in Alzheimer's disease, role of cytokines. *ScientificWorldJournal* 2012:756357. doi: 10.1100/2012/756357

Rush, T., and Buisson, A. (2014). Reciprocal disruption of neuronal signaling and Abeta production mediated by extrasynaptic NMDA receptors: a downward spiral. *Cell Tissue Res.* 356, 279–286. doi: 10.1007/s00441-013-1789-1

Sakurai-Yamashita, Y., Sakaguchi, S., Yoshikawa, D., Okimura, N., Masuda, Y., Katamine, S., et al. (2005). Female-specific neuroprotection against transient brain ischemia observed in mice devoid of prion protein is abolished by ectopic expression of prion protein-like protein. *Neuroscience* 136, 281–287. doi: 10.1016/j.neuroscience.2005.06.095

Salter, M. G., and Fern, R. (2005). NMDA receptors are expressed in developing oligodendrocyte processes and mediate injury. *Nature* 438, 1167–1171. doi: 10.1038/nature04301

Schlief, M. L., West, T., Craig, A. M., Holtzman, D. M., and Gitlin, J. D. (2006). Role of the Menkes copper-transporting ATPase in NMDA receptor-mediated neuronal toxicity. *Proc. Natl. Acad. Sci. U.S.A.* 103, 14919–14924. doi: 10.1073/pnas.0605390103

Shyu, W. C., Lin, S. Z., Chiang, M. F., Ding, D. C., Li, K. W., Chen, S. F., et al. (2005). Overexpression of PrPC by adenovirus-mediated gene targeting reduces ischemic injury in a stroke rat model. *J. Neurosci.* 25, 8967–8977. doi: 10.1523/JNEUROSCI.1115-05.2005

Skerry, T., and Genever, P. (2001). Glutamate signalling in non-neuronal tissues. *Trends Pharmacol. Sci.* 22, 174–181. doi: 10.1016/S0165-6147(00)01642-4

Solito, E., and Sastre, M. (2012). Microglia function in Alzheimer's disease. *Front. Pharmacol.* 3:14. doi: 10.3389/fphar.2012.00014

Spudich, A., Frigg, R., Kilic, E., Kilic, U., Oesch, B., Raeber, A., et al. (2005). Aggravation of ischemic brain injury by prion protein deficiency: role of ERK-1/-2 and STAT-1. *Neurobiol. Dis.* 20, 442–449. doi: 10.1016/j.nbd.2005.04.002

Stanyon, H. F., Patel, K., Begum, N., and Viles, J. H. (2014). Copper(II) sequentially loads onto the N-terminal amino group of the cellular prion protein before the individual octarepeats. *Biochemistry* 53, 3934–3939. doi: 10.1021/bi500643b

Steele, A. D., Zhou, Z., Jackson, W. S., Zhu, C., Auluck, P., Moskowitz, M. A., et al. (2009). Context dependent neuroprotective properties of prion protein (PrP). *Prion* 3, 240–249. doi: 10.4161/pri.3.4.10135

Talantova, M., Sanz-Blasco, S., Zhang, X., Xia, P., Akhtar, M. W., Okamoto, S., et al. (2013). Abeta induces astrocytic glutamate release, extrasynaptic NMDA receptor activation, and synaptic loss. *Proc. Natl. Acad. Sci. U.S.A.* 110, E2518–E2527. doi: 10.1073/pnas.1306832110

Thakur, A. K., Srivastava, A. K., Srinivas, V., Chary, K. V., and Rao, C. M. (2011). Copper alters aggregation behavior of prion protein and induces novel interactions between its N- and C-terminal regions. *J. Biol. Chem.* 286, 38533–38545. doi: 10.1074/jbc.M111.265645

Thellung, S., Gatta, E., Pellistri, F., Corsaro, A., Villa, V., Vassalli, M., et al. (2013). Excitotoxicity through NMDA receptors mediates cerebellar granule neuron apoptosis induced by prion protein 90-231 fragment. *Neurotox. Res.* 23, 301–314. doi: 10.1007/s12640-012-9340-9

Tsutsui, S., Hahn, J., Johnson, T., Ali, Z., and Jirik, F. (2008). Absence of the cellular prion protein exacerbates and prolongs neuroinflammation in experimental autoimmune encephalomyelitis. *Am. J. Pathol.* 173, 1029–1041. doi: 10.2353/ajpath.2008.071062

Tuneva, E., Bychkova, O., and Boldyrev, A. (2003). Effect of NMDA on production of reactive oxygen species by human lymphocytes. *Bull. Exp. Biol. Med.* 136, 159–161. doi: 10.1023/A:1026366907084

Um, J. W., Kaufman, A. C., Kostylev, M., Heiss, J. K., Stagi, M., Takahashi, H., et al. (2013). Metabotropic glutamate receptor 5 is a coreceptor for Alzheimer aβ oligomer bound to cellular prion protein. *Neuron* 79, 887–902. doi: 10.1016/j.neuron.2013.06.036

Um, J. W., Nygaard, H. B., Heiss, J. K., Kostylev, M. A., Stagi, M., Vortmeyer, A., et al. (2012). Alzheimer amyloid-beta oligomer bound to postsynaptic prion protein activates Fyn to impair neurons. *Nat. Neurosci.* 15, 1227–1235. doi: 10.1038/nn.3178

Walz, R., Amaral, O. B., Rockenbach, I. C., Roesler, R., Izquierdo, I., Cavalheiro, E. A., et al. (1999). Increased sensitivity to seizures in mice lacking cellular prion protein. *Epilepsia* 40, 1679–1682. doi: 10.1111/j.1528-1157.1999.tb01583.x

Wang, K. K., Zoltewicz, J. S., Chiu, A., Zhang, Z., and Rubenstein, R. (2012). Release of full-length PrP(C) from cultured neurons following neurotoxic challenges. *Front. Neurol.* 3:147. doi: 10.3389/fneur.2012.00147

Watt, N. T., Taylor, D. R., Kerrigan, T. L., Griffiths, H. H., Rushworth, J. V., Whitehouse, I. J., et al. (2012). Prion protein facilitates uptake of zinc into neuronal cells. *Nat. Commun.* 3, 1134. doi: 10.1038/ncomms2135

Weise, J., Crome, O., Sandau, R., Schulz-Schaeffer, W., Bahr, M., and Zerr, I. (2004). Upregulation of cellular prion protein (PrPc) after focal cerebral ischemia and influence of lesion severity. *Neurosci. Lett.* 372, 146–150. doi: 10.1016/j.neulet.2004.09.030

Weise, J., Doeppner, T. R., Muller, T., Wrede, A., Schulz-Schaeffer, W., Zerr, I., et al. (2008). Overexpression of cellular prion protein alters postischemic Erk1/2 phosphorylation but not Akt phosphorylation and protects against focal cerebral ischemia. *Restor. Neurol. Neurosci.* 26, 57–64.

Weise, J., Sandau, R., Schwarting, S., Crome, O., Wrede, A., Schulz-Schaeffer, W., et al. (2006). Deletion of cellular prion protein results in reduced Akt activation, enhanced postischemic caspase-3 activation, and exacerbation of ischemic brain injury. *Stroke* 37, 1296–1300. doi: 10.1161/01.STR.0000217262.03192.d4

Weller, R., Preston, S., Subash, M., and Carare, R. (2009). Cerebral amyloid angiopathy in the aetiology and immunotherapy of Alzheimer disease. *Alzheimers Res. Ther.* 1:6. doi: 10.1186/alzrt6

Wyllie, D. J., Livesey, M. R., and Hardingham, G. E. (2013). Influence of GluN2 subunit identity on NMDA receptor function. *Neuropharmacology* 74, 4–17. doi: 10.1016/j.neuropharm.2013.01.016

Xiao, B., and Link, H. (1998). Immune regulation within the central nervous system. *J. Neurol. Sci.* 157, 1–12. doi: 10.1016/S0022-510X(98)00049-5

You, H., Tsutsui, S., Hameed, S., Kannanayakal, T., Chen, L., Xia, P., et al. (2012). Abeta neurotoxicity depends on interactions between copper ions, prion protein, and N-methyl-D-aspartate receptors. *Proc. Natl. Acad. Sci. U.S.A.* 109, 1737–1742. doi: 10.1073/pnas.1110789109

Younan, N. D., Klewpatinond, M., Davies, P., Ruban, A. V., Brown, D. R., and Viles, J. H. (2011). Copper(II)-induced secondary structure changes and reduced folding stability of the prion protein. *J. Mol. Biol.* 410, 369–382. doi: 10.1016/j.jmb.2011.05.013

Conflict of Interest Statement: The authors declare that the research was conducted in the absence of any commercial or financial relationships that could be construed as a potential conflict of interest.

23

Prion protein and aging

Lisa Gasperini and Giuseppe Legname*

Laboratory of Prion Biology, Department of Neuroscience, Scuola Internazionale Superiore di Studi Avanzati, Trieste, Italy

Edited by:
Sophie Mouillet-Richard, INSERM Unit 1124, France

Reviewed by:
Mohammed Moudjou, Institut National de la Recherche Agronomique, France
Inga Zerr, University Medical Center Göttingen, Germany

***Correspondence:**
Giuseppe Legname, Laboratory of Prion Biology, Department of Neuroscience, Scuola Internazionale Superiore di Studi Avanzati, Via Bonomea 265, Trieste 34136, Italy
e-mail: legname@sissa.it

The cellular prion protein (PrP^C) has been widely investigated ever since its conformational isoform, the prion (or PrP^{Sc}), was identified as the etiological agent of prion disorders. The high homology shared by the PrP^C-encoding gene among mammals, its high turnover rate and expression in every tissue strongly suggest that PrP^C may possess key physiological functions. Therefore, defining PrP^C roles, properties and fate in the physiology of mammalian cells would be fundamental to understand its pathological involvement in prion diseases. Since the incidence of these neurodegenerative disorders is enhanced in aging, understanding PrP^C functions in this life phase may be of crucial importance. Indeed, a large body of evidence suggests that PrP^C plays a neuroprotective and antioxidant role. Moreover, it has been suggested that PrP^C is involved in Alzheimer disease, another neurodegenerative pathology that develops predominantly in the aging population. In prion diseases, PrP^C function is likely lost upon protein aggregation occurring in the course of the disease. Additionally, the aging process may alter PrP^C biochemical properties, thus influencing its propensity to convert into PrP^{Sc}. Both phenomena may contribute to the disease development and progression. In Alzheimer disease, PrP^C has a controversial role because its presence seems to mediate β-amyloid toxicity, while its down-regulation correlates with neuronal death. The role of PrP^C in aging has been investigated from different perspectives, often leading to contrasting results. The putative protein functions in aging have been studied in relation to memory, behavior and myelin maintenance. In aging mice, PrP^C changes in subcellular localization and post-translational modifications have been explored in an attempt to relate them to different protein roles and propensity to convert into PrP^{Sc}. Here we provide an overview of the most relevant studies attempting to delineate PrP^C functions and fate in aging.

Keywords: PrP, aging, Alzheimer disease, prion diseases

INTRODUCTION

The conformational conversion of the cellular form of the prion protein (PrP^C) into a β-sheet enriched isoform denoted as prion (PrP^{Sc}) is central in neurodegenerative pathologies collectively know as prion diseases (Prusiner, 2001).

The PrP^C is a sialoglycoprotein that is attached to the outer leaflet of the plasma membrane via a C-terminal glycosylphosphatidylinositol (GPI) anchor (Stahl et al., 1990). In the cell PrP^C may be expressed in different glycosylated forms, corresponding to the variable occupancy of the residues Asn180 and Asn196 (Haraguchi et al., 1989). While the C-terminal domain is folded in ordered and distinctive secondary structures, the N-terminal portion is flexible, unstructured and contains a unique octapeptide repeat (OR) region (Zahn et al., 2000). The OR confers PrP^C one of its most salient features, i.e., the ability to bind divalent cations, prominently copper and to a lesser extent zinc, nickel, iron, and manganese (Stockel et al., 1998; Jackson et al., 2001; Singh et al., 2010; Arena et al., 2012). Indeed, at physiological pH copper shows the highest binding affinity (Chattopadhyay et al., 2005; Liu et al., 2011). The importance of copper binding for PrP^C function is reflected in the high OR structural homology among different species (Hornshaw et al., 1995). Two of the most investigated putative functions for PrP^C are neuroprotection and defense against oxidative stress, both relevant in neurodegeneration and cell aging (Vassallo and Herms, 2003; Roucou and LeBlanc, 2005; Aguzzi et al., 2008). In agreement with these findings, PrP^C copper binding sites support the reduction of Cu^{2+} to Cu^+, thus preventing reactive oxygen species (ROS) formation (Liu et al., 2011).

Although contrasting results are likely due to the different experimental protocols employed, PrP^C is expressed in almost all body organs and tissues, from embryonic to aging stages (Fournier et al., 1998; Goldmann et al., 1999; Sales et al., 2002). However, the expression levels are differently regulated according to the age and tissue distribution. For instance, while PrP^C shows a high expression in the central nervous system (CNS), where it is differently regulated during development and aging, it is present at very low levels in the liver (Horiuchi et al., 1995; Moudjou et al., 2001; Peralta et al., 2012). The broad homology conservation of the PrP^C-encoding gene among mammalian and avian species (Hornshaw et al., 1995), the expression of the protein in many tissues during the entire lifespan, and its high turnover rate strongly indicate that PrP^C possesses key physiological roles. However, the definition of PrP^C's univocal function is still under debate.

The PrP^{Sc} isoform can form protein aggregates, known as prion deposits, often present as amyloid structures, which can propagate and possibly cause cell death. Despite continuous steps forward toward the definition of the conformational switch mechanism, the trigger of the posttranslational remodeling event is still obscure, especially in sporadic prion disorders. In the inherited forms of prion diseases, a genetic mutation in the open-reading frame of the gene leads to amino acid substitutions, which in turn destabilizes the protein structure over time, promoting PrP^C to PrP^{Sc} conversion. In the infective forms, a preformed PrP^{Sc}-aggregate triggers the conversion of endogenous PrP^C. In the sporadic forms of the disease, several unknown factors may perturb the protein structure, thus favoring PrP^{Sc} formation (Prusiner, 1991, 1994).

If a cause-effect relationship between prion deposits and cell death does not always exist (Lasmezas et al., 1997), a probable consequence of PrP^C conversion into PrP^{Sc} is the loss or alteration of its cellular function. Whether this impairment contributes or not to disease development is still under debate. Moreover, PrP^C is also involved in Alzheimer disease (AD) (Kellett and Hooper, 2009). Whether in AD PrP^C mediates neuroprotection or β-amyloid toxicity is still under investigation.

All these observations, together with the fact that the highest incidence of prion disorders and AD is in the elderly population, have renewed the interest of scientists in investigating of the role of PrP^C during aging. The two main topics under study have been: (i) PrP^C physiological functions in aging models, and (ii) consequent changes in PrP^C biochemical properties. The latter may be triggered by the cellular aging process and may promote prion aggregate formation. To date, studies into PrP^C role in aging are contrasting, mostly due to different animal/cellular systems employed and the lack of consistency in defining aging in experimental models.

Here we provide a comprehensive overview of studies attempting to understand PrP^C function and fate in aging and relate them to the neurodegenerative processes.

AGING

In mammals, aging is defined as the accumulation of changes in an individual over time and represents a multifaceted process of social, psychological, and physical alterations (Bowen and Atwood, 2004). Normal aging is associated with significant cognitive declines, such as decreased speed of information processing, working memory capacity, and long-term memory function (Hedden and Gabrieli, 2004). However, changes that occur in aging brains are less linked to chronological time than it was thought in the past. For instance, individual brains may grow old developing less alterations than others (Esiri, 2007).

The study of age-related changes in the human brain is challenging for two main reasons: (i) increasing improvement of nutrition during the last century limits our ability to directly compare human brain samples between generations; and (ii) most elderly brains show some pathological hallmarks, thus complicating the distinction between "normal aging" and "disease" (Esiri, 2007). In the past, it was thought that the main cause of age-related cognitive decline was massive neuron loss (Brody, 1955) and deterioration of dendritic branches (Scheibel et al., 1976; Scheibel, 1979). Now we know that alterations occurring in normal aging are more subtle and selective. In fact, it seems that most age-linked behavioral impairments are due to region-specific changes in dendritic morphology, cellular connectivity, calcium dysregulation, gene expression or other factors that affect plasticity and alter neuron network dynamics (Burke and Barnes, 2006). A crucial factor for brain aging is the enormous requirement that neurons have for oxidative metabolism required by energy consumption. Neurons need extraordinary amounts of energy because of: (i) the size and the ensuing energy-consumption system of transport for molecules and organelles; and (ii) the electrical activity for impulse transmission, implying ion gradient maintenance. The high degree of oxidative metabolism generates high amounts of ROS that damage proteins, nucleic acids and lipids, thus interfering with normal cell metabolism and resulting, for instance, in altered gene expression and abnormal protein generation (Esiri, 2007).

Mitochondrial efficiency decline, paralleled by oxidative stress, which occurs with aging, is connected to alterations in calcium homeostasis, in particular to higher calcium content in neuron cytosol after excitatory stimulation. Indeed, glutamate activation of N-methyl-D-aspartate (NMDA) receptors produces a transient elevation in intracellular calcium concentration, which is increased in normal aging. Disruption in calcium homeostasis predisposes aging neurons to more extended damage after stress and eventually leads to cell death by apoptosis (Cowan et al., 2001; Thibault et al., 2001; Esiri, 2007).

In this scenario, PrP^C may possess different functions: (i) neuron protection from oxidative stress through antioxidant activity, by sensing copper and/or free radical stimuli (Vassallo and Herms, 2003); (ii) modulation of calcium entry through NMDA receptor pore by inhibition of the channel activity (Cowan et al., 2001; Lo et al., 2007); (iii) anti-apoptotic effect on Bax-mediated cell death (Bounhar et al., 2006; Lo et al., 2007).

WHICH MOUSE AGE SHOULD BE CONSIDERED AS AN AGING MODEL?

For many reasons, researchers often use terms as *aged* or *old mice* that may not be proper by definition. A wild-type mouse goes through the following life phases (http://research.jax.org/faculty/harrison/ger1vLifespan1.html) (Flurkey et al., 2007):

- Mature adult [3–6 months old (mo) mouse]: this period is the reference for any age change; these mice are fully developed, but not senescent; after 6 months of age mice can show some age-related changes, for instance females are retired from breeding because litter size begins to diminish.
- Middle age (10–15 mo mouse): senescent changes can be detected in some, but not all, biomarkers of aging; these mice are generally used to determine if an age-related change is progressive or occurs only in old age.
- Old (18–24 mo mouse, or older): senescent changes can be detected in almost all biomarkers in all animals.

Taking into account these categories, few studies claiming the use of old mice indeed employ a properly aged model. For the purposes of this manuscript, this point has to be carefully taken into account, especially in respect of studies on PrP^C biochemical property changes in aging. On the other hand, studies on PrP^C knockout mice (also noted as $Prnp^{0/0}$), when not performed on old animals, as defined above, can nevertheless reveal progressive age-related changes that start earlier because of PrP^C absence. To let the reader critically consider reported results and conclusions, we will specify for each reviewed study the age of the animals used.

PrP^C ROLE IN AGING

As previously mentioned, PrP^C is highly conserved and expressed among mammals, and it may play important physiological roles. However, numerous efforts aimed at identifying PrP^C function have harbored contrasting results. Most of the work has focused on the nervous system, which is the organ with the highest PrP^C expression and the site of prion disease pathology. Several results have linked PrP^C to many cellular processes, such as neuronal survival, neurite outgrowth, synapse formation/maintenance/functionality, and myelinated fiber formation/maintenance (Aguzzi et al., 2008). However, its expression in many other tissues indicates that PrP^C has either many different functions or a function compatible with diverse cellular types.

A large body of evidence suggests that PrP^C plays a role in essential metal homeostasis, resulting in protection from oxidative stress. An overlap between systems controlling essential metal homeostasis and oxygen radical metabolism has been extensively documented (Avery, 2001). For instance, many antioxidant enzymes need metal ions as cofactors. PrP^C has been associated with cellular systems that control redox balance and protect against oxidative stress (Brown and Sassoon, 2002). Indeed, PrP^C-null models show: (i) increased neuronal sensitivity to oxidative stress (Brown et al., 1997; Rachidi et al., 2003); (ii) alterations of superoxide dismutase 1 (SOD1) activity due to impairments in copper incorporation (Brown and Besinger, 1998; Kralovicova et al., 2009); (iii) higher levels of lipids and protein oxidation (Wong et al., 2001). Moreover, PrP^C expression is increased by heat shock, hypoxia, ischemia, hypoglycemia, stroke, and knockdown of any SOD protein while shutting down PrP^C increases extracellular-SOD expression and SOD2 activity (Brown and Besinger, 1998; Shyu et al., 2000, 2004, 2005; McLennan et al., 2004; Mitsios et al., 2007; Kralovicova et al., 2009). Furthermore, *in vitro* PrP^C possesses SOD-like activity with a dismutation constant rate similar to SOD2 (Brown et al., 1999, 2001; Cui et al., 2003; Treiber et al., 2007). A possible link between SOD and PrP^C may be copper. Indeed, SOD activity depends on copper incorporation and probably involves Cu^{2+} reduction, characteristics shared with PrP^C (Brown et al., 2001). It should also be noted that PrP^C is cleaved at the end of the OR through the action of ROS, a process termed β-cleavage. β-cleavage is considered an early and critical event in the mechanism through which PrP^C protects cells against oxidative stress. If a PrP construct lacks the OR, the protein will fail to undergo ROS-mediated β-cleavage, as occurs with two mutant forms associated with prion diseases (Watt et al., 2005).

PrP^C ROLE IN BEHAVIOR AND LEARNING AS A FUNCTION OF AGE

Two main observations suggest that PrP^C plays a role in behavior and learning mechanisms in aging: (i) PrP^C is predominantly expressed in neurons, reaching the highest level in the hippocampus (DeArmond et al., 1987; Benvegnu et al., 2010); (ii) PrP^C brain expression is increased in aging (Williams et al., 2004). However, studies carried out on PrP^C knockout mice have sometimes produced contrasting results, likely due to differences in lines and age.

The behavioral characterization of the first PrP^C knockout (ZurichI) mouse generation was performed in 7 mo mice (Bueler et al., 1992). At this age these mice cannot be considered *aged*. Nevertheless, the latter study represents a good starting point toward analyzing alterations triggered by PrP^C ablation. In the swim test and in the Y-maze, PrP^C knockout mice did not reveal learning impairments, although there were large individual differences within test and control groups. In the two-way avoidance test, in which animals have to avoid an electrical shock announced by a light by running to the opposite chamber of the box, thus involving also motorial ability, PrP^C knockout mice showed a slightly lower performance, but the great variability in each group hid a possible significant difference. As Büeler and colleagues state, these results should be carefully considered because of the dissimilar loci surrounding $Prnp^{0/0}$ and wild-type ($Prnp^{+/+}$) alleles, due to different genetic background.

However, a follow-up study using the same mouse line showed that $Prnp^{0/0}$ mouse aging-related behavior alterations could not be ascribed to surrounding loci differences, but to PrP^C ablation itself (Coitinho et al., 2003). Differently from Büeler and colleagues data, here statistically significant results were obtained by applying other behavioral tests and by increasing the animal age: 9 and 3 mo $Prnp^{0/0}$ and $Prnp^{+/+}$ mice were compared in regards to fear-motivated learning, locomotor activity, exploratory behavior and anxiety. Nine mo PrP^C knockout mice showed impairments in short- and long-term memory and exploratory activity. With exception of the exploratory activity, 9 mo $Prnp^{0/0}$ mouse performances were recapitulated in 9 mo rats injected in the hippocampus with anti-PrP^C antibody, thus ruling out the involvement of the surrounding loci in memory alterations in the aging PrP^C knockout line.

Another study concerning aging-related PrP^C role in behavior control was performed on 3 and 11 mo mice from three lines (wild-type, $Prnp^{0/0}$ and PrP^C overexpressing Tga20 on a $Prnp^{0/0}$ genetic background) (Rial et al., 2009). Collectively the data showed that PrP^C overexpressing Tga20 mice were less susceptible to aging-caused alterations in locomotion, anxiety like responses and short-term social recognition memory in comparison with $Prnp^{+/+}$ and $Prnp^{0/0}$ mice. On the contrary, PrP^C knockout mice resulted more susceptible to age-related decline in comparison with wild-type mice. Biochemical analyses revealed that Tga20 mice have lower neuron caspase3 activation and serum acetylcholinesterase levels, and higher synaptic density. In light of these cognition-enhancing properties, Rial and colleagues suggest in particular the interaction of PrP^C with the stress-inducible protein 1 as a target for pharmaceutical intervention to attenuate age-related cognitive impairments.

More recent results confirmed age-dependent behavioral abnormalities in ZurichI PrP^C knockout mice (Schmitz et al., 2014). The following alterations were reported: (i) poor native nest building behavior in young (3 mo) and old (9 and 20 mo)

PrP^C knockout mice; (ii) higher latency in exploring a new environment only in young but not in old PrP^C knockout mice; (iii) more pronounced drop in anxiety during aging in PrP^C knockout mice; (iv) decline in associative learning in old PrP^C knockout mice compared to age-matched wild-type. Schmitz and colleagues wanted to relate the observed behavioral abnormalities to neuron structural alterations in PrP^C knockout mice. They found PrP^C-dependent alterations in cytoskeletal proteins that are responsible for morphology, structure, and stability of neurons, thus related to learning processes. Indeed, PrP^C was previously reported to functionally interact with cytoskeletal proteins (Dong et al., 2008; Malaga-Trillo et al., 2009).

Telling's group performed a work on a slightly different line, the ZurichI crossed with wild-type FVB animals (Nazor et al., 2007). On the rotarod, *Prnp*$^{0/0}$ on FVB background mice showed an age-dependent motor behavior deficit in comparison to wild-type, suggesting a function for PrP^C in maintaining sensorimotor coordination. This difference between wild-type and PrP^C knockout mice was detected starting from 3 to 8 mo. Moreover, these mice at 6 mo revealed vacuolation in different brain regions. Although vacuolation has been rarely reported in PrP^C knockout mice, Telling and coworkers parallel their findings, concerning both spongiosis and motor impairments, with results from a transgenic murine model expressing the PrP mutation linked to Gerstmann–Sträussler–Scheinker syndrome (GSS), thus suggesting the loss of PrP^C function as disease mechanism.

Another study was performed by Massimino and colleagues using three groups of congenic mice: wild-type FVB, PrP^C knockout on FVB background and Tg46 mice in which PrP^C expression was rescued over a PrP^C knockout background (Massimino et al., 2013). While young (3 mo) PrP^C knockout mice did not show any alteration, aged (15–18 mo) animals revealed perturbed behavioral pattern, in particular difficulties in adapting to new situations and in locomotor activity, likely due to a depressive syndrome. The authors suggest that PrP^C absence affects emotional reactivity. The authors did not carry out histological analysis of old *Prnp*$^{0/0}$ mice, therefore, it is not known whether the PrP^C knockout colony employed showed the vacuolation phenotype observed by Telling's group studying the same mouse line. To our knowledge, the result obtained by Telling's and coworkers is the only one showing vacuolation in *Prnp*$^{0/0}$ mice. For instance, Aguzzi and colleagues did not detect any morphological anomaly in 14 mo ZurichI *Prnp*$^{0/0}$ mice (Bremer et al., 2010).

Taken together, all these studies demonstrate that PrP^C absence affects learning, cognitive and behavioral skills in aged mice. A possible cause of these alterations is the lack of PrP^C-mediated neuroprotection and reduction of oxidative stress, particularly important to preserve neuron functions in the aging brain milieu.

PrP^C ROLE IN MYELIN MAINTENANCE

Alterations occurring during both normal aging and neurodegenerative pathologies largely involve maintenance and regeneration of myelin structure (Verdu et al., 2000; Peters, 2002; Bartzokis, 2011). Metal ions, in particular copper and iron, are necessary for both myelin formation and maintenance (Skripuletz et al., 2008; Benetti et al., 2010). Therefore, PrP^C may influence myelin by maintaining copper and iron homeostasis, pathways in which PrP^C has been involved (Pauly and Harris, 1998; Brown and Harris, 2003; Rachidi et al., 2003; Miura et al., 2005; Singh et al., 2009). By studying 14 mo (middle aged) mice, Aguzzi's group reported that PrP^C knockout mouse sciatic nerve contains more *digestion chambers*, i.e., macrophages ingesting myelin debris of degenerating nerve fibers, compared to wild-type mice. Moreover, they showed that PrP^C expression on neuronal cells, but not on Schwann's cells, is fundamental to preserve myelin fibers and prevent chronic demyelinating polyneuropathy (CDP) in the peripheral nervous system. This finding suggests that neuronal PrP^C *in trans* expression is required for a directional communication from axons to Schwann cells (Bremer et al., 2010).

As suggested in a study from our group (Benvegnu et al., 2011), PrP^C may exert a direct role in myelin sheath formation and structural preservation through its modulation of β-secretase 1 (BACE1) (Parkin et al., 2007). Results obtained in sciatic nerves from 15 mo mice highlight that PrP^C influences cleavage and processing of neuregulins. Neuregulins are a class of proteins crucial for myelin maintenance in the peripheral nervous system, and their cleavage by BACE1 is necessary for signaling functions. Hence, PrP^C positively regulates neuregulin processing, thus affecting their function in myelin homeostasis, maybe by modulating BACE1 activity. Interestingly, in the previously mentioned work, Aguzzi's group showed that CDP is prevented by PrP^C variants that are permissive for proteolytic amino-proximal cleavage, but not by variants that do not undergo cleavage. Indeed, all transgenic mice showing CDP lacked C1, i.e., the fragment generated by α-cleavage, while all transgenic mice in which the CDP was rescued produced high C1 levels. Hence, the cleavage of PrP^C seems to be functional for its myelinotrophic role (Bremer et al., 2010).

AGING-RELATED MODIFICATION OF PrP^C BIOCHEMICAL PROPERTIES

The most common occurrence of human prion diseases (about 80%) is sporadic. This means that a trigger of prion pathologies may be neither a genetic mutation nor an infectious seed, but most likely an unknown alteration that provokes the switch from PrP^C to PrP^{Sc} (Prusiner, 2001). The aging process itself may modify PrP^C biochemical properties, making the protein more prone to convert into PrP^{Sc}. For instance, loss of glycosylation in cell models favors PrP^C acquisition of PrP^{Sc}-like features (Lehmann and Harris, 1997). Indeed, glycans regulate PrP^C folding, intracellular trafficking, localization and function on the neuronal surface, thus their pattern modifications can be likely linked to the disease development. PrP^C biochemical properties that may be altered by cellular changes occurring during the aging process are: expression levels, post-translational modifications, such as glycosylation and phosphorylation, as well as localization in cellular domains, such as inside/outside membrane lipid rafts.

As mentioned in a previous paragraph, analysis of both parenchyma and microvessels revealed higher levels of both glycosylated and unglycosylated PrP^C in C57Bl/6J old (18 and 24 mo) mice (Williams et al., 2004). Concerning PrP^C diverse glycosylation isoforms, the three main bands appearing in one-dimensional immunoblots are commonly considered corresponding to di-glycosylated (~35 KDa), mono-glycosylated (~32 KDa) and un-glycosylated (~28 KDa) (Collinge, 2001; Pan et al., 2002). However, by applying a panel of monoclonal antibodies, the presence of additional lower molecular weight bands representing N-terminally truncated PrP^C isoforms has been shown (Pan et al., 2002). Moreover, two-dimensional immunoblot analysis revealed in human brain more than 50 PrP^C species which derive from several glycosylation and cleavage combinations. Interestingly, accumulation of aberrant full-length PrP^C bound to immature N-linked glycans is indeed a common feature of prion disease (Pan et al., 2005a,b,c). Spurred by these findings, Goh and coworkers investigated N-linked glycans on PrP^C during normal aging in mouse, providing the first glycan profile of full-length and truncated PrP^C isoforms (Goh et al., 2007). The oldest mice they used were 15 mo. First, they found that different mouse lines (i.e., CD-1 and FVB) have different PrP^C metabolism resulting in isoform heterogeneity. In general, they showed that the truncated form of PrP^C undergoes a simplification in its glycosylation process during aging, while the amount of complex oligosaccharides on the full-length PrP^C increases. This finding contrasts with what was observed in aging compared to young cattle (Yoshioka et al., 2010), but resembles what was previously detected in human brains affected by prion disorders (Pan et al., 2005b). It has been shown that PrP^C glycosylation state may modulate affinity for copper binding (Moudjou et al., 2007). In particular, the non-glycosylated species showed stronger binding to divalent cations (copper and cobalt) *in vitro*. Therefore, decreasing PrP^C glycosylation levels during aging may increase copper binding capacity, thus improving PrP^C antioxidant function. Interestingly, galactose, which has been detected in this study on full-length PrP^C in aging, is also highly present on PrP^{Sc} (Safar et al., 1990). Moreover, sialic acid content increases with aging on the truncated PrP^C but remains unchanged on the full-length PrP^C. As the authors observe, the presence of anionic residues may modulate PrP^{Sc} mobility, thus facilitating its propagation in prion disorders.

As previously mentioned, aging can modify PrP^C subcellular localization. Indeed, our group reported that PrP^C moves from detergent soluble membrane fractions to lipid rafts in aged (20–21 mo) mouse hippocampus (Agostini et al., 2013). No differences in the total PrP^C expression amount were detected, in contrast with what had been previously reported by another group that used the same mouse line and a very similar age but lacked repeated samples, normalization and statistical analysis (Williams et al., 2004). Changes in the cholesterol/sphingolipid ratio has been reported in normal brain aging, and in degenerative disorders, such as AD (Martin et al., 2010). As cholesterol and sphingolipids are the two main components of lipid rafts, their relative amount influences many cellular pathways, including protein localization (Martin et al., 2008; Trovo et al., 2011). PrP^C is a GPI-anchored protein, thus bound to lipid rafts and likely affected by aging-related alterations in membrane composition. PrP^C shift toward lipid rafts triggered by changes in lipid composition was confirmed using model systems of lipid manipulation. Moreover, decreasing sphingolipids, thus mimicking a juvenile condition, reduced the formation of PrP^{Sc} in a cell line model. This suggests that age-related changes influence PrP^C localization and ensuing propensity to PrP^{Sc} conversion (Agostini et al., 2013).

PrP^C AND AD

The most prevalent form of dementia is AD (Burns and Iliffe, 2009). Various factors are conducive to the risk of late-onset AD, in particular old age, genetic factors, family history, a history of head trauma, midlife hypertension, obesity, diabetes, and hypercholesterolemia (Bendlin et al., 2010). AD is characterized by the misprocessing of two proteins: intraneuronal tau and extracellular β-amyloid peptide (Aβ) (Querfurth and LaFerla, 2010). Aβ is generated by the aberrant proteolytic processing of the amyloid precursor protein by BACE1. Several studies have been performed to understand if a toxic interaction between Aβ and PrP exists, but the use of different *in vitro* or transgenic models has yielded contrasting results (Schwarze-Eicker et al., 2005; Lauren et al., 2009; Balducci et al., 2010; Calella et al., 2010; Chung et al., 2010; Kessels et al., 2010; Morales et al., 2010; Ordonez-Gutierrez et al., 2013). Thus, in this manuscript, we will consider only what has been found by analyzing AD patient brains. Whitehouse and colleagues found that PrP^C levels are decreased in AD patient hippocampus, normal aging hippocampus, and temporal lobe, but not in AD patient temporal lobe (Whitehouse et al., 2010). With exception of AD patient temporal lobe data, these results are in contrast with what was reported by Saijo et al. (2011), but correlate with the PrP^C decrease detected in the cerebrospinal fluid of patients affected by different neurological disorders including AD (Meyne et al., 2009). Since PrP^C inhibits BACE1 (Parkin et al., 2007), PrP^C level reduction may elevate Aβ production. Therefore, Whitehouse and colleagues suggest that PrP^C decrease is not a secondary consequence, but a primary cause of AD and, by occurring also in normal aging, increases the incidence of AD in old individuals. Interestingly, individuals with mutations in the PrP encoding gene generating a truncated form of the protein developed clinical AD symptoms at relatively young age (Kitamoto et al., 1993). In aging, BACE1 activity strongly increases, thus enhancing Aβ production and the possibility of deposit formation. The reduction of PrP^C-mediated BACE1 inhibition due to the decrease in PrP^C levels may be a cause of the enzyme activity elevation (Whitehouse et al., 2010). Additionally, since PrP^C is an antioxidant protein (Vassallo and Herms, 2003; Aguzzi et al., 2008), its downregulation likely increases neuron susceptibility to ROS, which rises in normal aging and in AD (Halliwell, 2006; Zhu et al., 2007), hence contributing to disease progression. However, a decrease in PrP^C levels in AD does not exclude that the residual protein mediates some Aβ toxic effects. Besides, it has been recently reported that Aβ and PrP^C do interact specifically in AD patient brains (Dohler et al., 2014). This result was obtained by means of a PrP^C-Aβ binding assay and size exclusion chromatography, and no binding was detected in non-demented age-matched controls.

Taken together, these results indicate that PrP^C may be involved in AD. Therefore, its fate in aging is likely related to the molecular mechanisms that induce neurodegeneration.

CONCLUSIONS

As presented in this review, relatively few studies have been carried out on the role of aging in the expression and regulation of PrP^C. Nevertheless, these experiments suggest that there may be a correlation between the physiology of the cells in which PrP^C is present and its localization and processing during aging. Particularly in the CNS, where it is abundantly expressed, PrP^C plays a prominent role as the precursor of PrP^{Sc} and thereby dictates the amount and extension of the conversion and accumulation process of prions. In aging, the physiology and the cellular localization of the protein may change concomitantly to different biochemical milieus in the cell membrane. Indeed, either membrane composition, in particular lipid raft composition, or additional protein complexes proximity to PrP^C, may influence its physiological functions. As indicated in the final paragraph, these changes may have a general relevance for more common causes of dementia such as AD. More work is necessary to define the precise role of PrP^C in the progression of AD and perhaps in other neurodegenerative diseases.

ACKNOWLEDGMENT

The authors would like to thank Kate Pischke and Erica Sarnataro for editing and proofreading the manuscript.

REFERENCES

Agostini, F., Dotti, C. G., Perez-Canamas, A., Ledesma, M. D., Benetti, F., and Legname, G. (2013). Prion protein accumulation in lipid rafts of mouse aging brain. *PLoS ONE* 8:e74244. doi: 10.1371/journal.pone.0074244

Aguzzi, A., Baumann, F., and Bremer, J. (2008). The prion's elusive reason for being. *Annu. Rev. Neurosci.* 31, 439–477. doi: 10.1146/annurev.neuro.31.060407.125620

Arena, G., La Mendola, D., Pappalard, G., Sóvágód, I., and Rizzarelli, E. (2012). Interactions of Cu2+ with prion family peptide fragments: considerations on affinity, speciation and coordination. *Coord. Chem. Rev.* 256, 2202–2218. doi: 10.1016/j.ccr.2012.03.038

Avery, S. V. (2001). Metal toxicity in yeasts and the role of oxidative stress. *Adv. Appl. Microbiol.* 49, 111–142. doi: 10.1016/S0065-2164(01)49011-3

Balducci, C., Beeg, M., Stravalaci, M., Bastone, A., Sclip, A., Biasini, E., et al. (2010). Synthetic amyloid-beta oligomers impair long-term memory independently of cellular prion protein. *Proc. Natl. Acad. Sci. U.S.A.* 107, 2295–2300. doi: 10.1073/pnas.0911829107

Bartzokis, G. (2011). Alzheimer's disease as homeostatic responses to age-related myelin breakdown. *Neurobiol. Aging* 32, 1341–1371. doi: 10.1016/j.neurobiolaging.2009.08.007

Bendlin, B. B., Carlsson, C. M., Gleason, C. E., Johnson, S. C., Sodhi, A., Gallagher, C. L., et al. (2010). Midlife predictors of Alzheimer's disease. *Maturitas* 65, 131–137. doi: 10.1016/j.maturitas.2009.12.014

Benetti, F., Ventura, M., Salmini, B., Ceola, S., Carbonera, D., Mammi, S., et al. (2010). Cuprizone neurotoxicity, copper deficiency and neurodegeneration. *Neurotoxicology* 31, 509–517. doi: 10.1016/j.neuro.2010.05.008

Benvegnu, S., Gasperini, L., and Legname, G. (2011). Aged PrP null mice show defective processing of neuregulins in the peripheral nervous system. *Mol. Cell. Neurosci.* 47, 28–35. doi: 10.1016/j.mcn.2011.02.005

Benvegnu, S., Poggiolini, I., and Legname, G. (2010). Neurodevelopmental expression and localization of the cellular prion protein in the central nervous system of the mouse. *J. Comp. Neurol.* 518, 1879–1891. doi: 10.1002/cne.22357

Bounhar, Y., Mann, K. K., Roucou, X., and LeBlanc, A. C. (2006). Prion protein prevents Bax-mediated cell death in the absence of other Bcl-2 family members in *Saccharomyces cerevisiae*. *FEMS Yeast Res.* 6, 1204–1212. doi: 10.1111/j.1567-1364.2006.00122.x

Bowen, R. L., and Atwood, C. S. (2004). Living and dying for sex. A theory of aging based on the modulation of cell cycle signaling by reproductive hormones. *Gerontology* 50, 265–290. doi: 10.1159/000079125

Bremer, J., Baumann, F., Tiberi, C., Wessig, C., Fischer, H., Schwarz, P., et al. (2010). Axonal prion protein is required for peripheral myelin maintenance. *Nat. Neurosci.* 13, 310–318. doi: 10.1038/nn.2483

Brody, H. (1955). Organization of the cerebral cortex. III. A study of aging in the human cerebral cortex. *J. Comp. Neurol.* 102, 511–516. doi: 10.1002/cne.901020206

Brown, D. R., and Besinger, A. (1998). Prion protein expression and superoxide dismutase activity. *Biochem. J.* 334(Pt 2), 423–429.

Brown, D. R., Clive, C., and Haswell, S. J. (2001). Antioxidant activity related to copper binding of native prion protein. *J. Neurochem.* 76, 69–76. doi: 10.1046/j.1471-4159.2001.00009.x

Brown, D. R., and Sassoon, J. (2002). Copper-dependent functions for the prion protein. *Mol. Biotechnol.* 22, 165–178. doi: 10.1385/MB:22:2:165

Brown, D. R., Schulz-Schaeffer, W. J., Schmidt, B., and Kretzschmar, H. A. (1997). Prion protein-deficient cells show altered response to oxidative stress due to decreased SOD-1 activity. *Exp. Neurol.* 146, 104–112. doi: 10.1006/exnr.1997.6505

Brown, D. R., Wong, B. S., Hafiz, F., Clive, C., Haswell, S. J., and Jones, I. M. (1999). Normal prion protein has an activity like that of superoxide dismutase. *Biochem. J.* 344(Pt 1), 1–5. doi: 10.1042/0264-6021:3440001

Brown, L. R., and Harris, D. A. (2003). Copper and zinc cause delivery of the prion protein from the plasma membrane to a subset of early endosomes and the Golgi. *J. Neurochem.* 87, 353–363. doi: 10.1046/j.1471-4159.2003.01996.x

Bueler, H., Fischer, M., Lang, Y., Bluethmann, H., Lipp, H. P., DeArmond, S. J., et al. (1992). Normal development and behaviour of mice lacking the neuronal cell-surface PrP protein. *Nature* 356, 577–582. doi: 10.1038/356577a0

Burke, S. N., and Barnes, C. A. (2006). Neural plasticity in the ageing brain. *Nat. Rev. Neurosci.* 7, 30–40. doi: 10.1038/nrn1809

Burns, A., and Iliffe, S. (2009). Alzheimer's disease. *BMJ* 338, b158. doi: 10.1136/bmj.b158

Calella, A. M., Farinelli, M., Nuvolone, M., Mirante, O., Moos, R., Falsig, J., et al. (2010). Prion protein and Abeta-related synaptic toxicity impairment. *EMBO Mol. Med.* 2, 306–314. doi: 10.1002/emmm.201000082

Chattopadhyay, M., Walter, E. D., Newell, D. J., Jackson, P. J., Aronoff-Spencer, E., Peisach, J., et al. (2005). The octarepeat domain of the prion protein binds Cu(II) with three distinct coordination modes at pH 7.4. *J. Am. Chem. Soc.* 127, 12647–12656. doi: 10.1021/ja053254z

Chung, E., Ji, Y., Sun, Y., Kascsak, R. J., Kascsak, R. B., Mehta, P. D., et al. (2010). Anti-PrPC monoclonal antibody infusion as a novel treatment for cognitive deficits in an Alzheimer's disease model mouse. *BMC Neurosci.* 11:130. doi: 10.1186/1471-2202-11-130

Coitinho, A. S., Roesler, R., Martins, V. R., Brentani, R. R., and Izquierdo, I. (2003). Cellular prion protein ablation impairs behavior as a function of age. *Neuroreport* 14, 1375–1379. doi: 10.1097/01.wnr.0000078541.07662.90

Collinge, J. (2001). Prion diseases of humans and animals: their causes and molecular basis. *Annu. Rev. Neurosci.* 24, 519–550. doi: 10.1146/annurev.neuro.24.1.519

Cowan, C. M., Thai, J., Krajewski, S., Reed, J. C., Nicholson, D. W., Kaufmann, S. H., et al. (2001). Caspases 3 and 9 send a pro-apoptotic signal from synapse to cell body in olfactory receptor neurons. *J. Neurosci.* 21, 7099–7109.

Cui, T., Daniels, M., Wong, B. S., Li, R., Sy, M. S., Sassoon, J., et al. (2003). Mapping the functional domain of the prion protein. *Eur. J. Biochem.* 270, 3368–3376. doi: 10.1046/j.1432-1033.2003.03717.x

DeArmond, S. J., Mobley, W. C., DeMott, D. L., Barry, R. A., Beckstead, J. H., and Prusiner, S. B. (1987). Changes in the localization of brain prion proteins during scrapie infection. *Neurology* 37, 1271–1280. doi: 10.1212/WNL.37.8.1271

Dohler, F., Sepulveda-Falla, D., Krasemann, S., Altmeppen, H., Schluter, H., Hildebrand, D., et al. (2014). High molecular mass assemblies of amyloid-beta oligomers bind prion protein in patients with Alzheimer's disease. *Brain* 137(Pt 3), 873–886. doi: 10.1093/brain/awt375

Dong, C. F., Shi, S., Wang, X. F., An, R., Li, P., Chen, J. M., et al. (2008). The N-terminus of PrP is responsible for interacting with tubulin and fCJD related PrP mutants possess stronger inhibitive effect on microtubule assembly *in vitro*. *Arch. Biochem. Biophys.* 470, 83–92. doi: 10.1016/j.abb.2007.11.007

Esiri, M. M. (2007). Ageing and the brain. *J. Pathol.* 211, 181–187. doi: 10.1002/path.2089

Flurkey, K., Currer, J. M., and Harrison, D. E. (2007). "The mouse in aging research," in *The Mouse in Biomedical Research*, Vol. 3, ed J. G. Fox (Burlington, MA: Elsevier), 637–672.

Fournier, J. G., Escaig-Haye, F., Billette de Villemeur, T., Robain, O., Lasmezas, C. I., Deslys, J. P., et al. (1998). Distribution and submicroscopic immunogold localization of cellular prion protein (PrPc) in extracerebral tissues. *Cell Tissue Res.* 292, 77–84. doi: 10.1007/s004410051036

Goh, A. X., Li, C., Sy, M. S., and Wong, B. S. (2007). Altered prion protein glycosylation in the aging mouse brain. *J. Neurochem.* 100, 841–854. doi: 10.1111/j.1471-4159.2006.04268.x

Goldmann, W., O'Neill, G., Cheung, F., Charleson, F., Ford, P., and Hunter, N. (1999). PrP (prion) gene expression in sheep may be modulated by alternative polyadenylation of its messenger RNA. *J. Gen. Virol.* 80(Pt 8), 2275–2283.

Halliwell, B. (2006). Oxidative stress and neurodegeneration: where are we now? *J. Neurochem.* 97, 1634–1658. doi: 10.1111/j.1471-4159.2006.03907.x

Haraguchi, T., Fisher, S., Olofsson, S., Endo, T., Groth, D., Tarentino, A., et al. (1989). Asparagine-linked glycosylation of the scrapie and cellular prion proteins. *Arch. Biochem. Biophys.* 274, 1–13. doi: 10.1016/0003-9861(89)90409-8

Hedden, T., and Gabrieli, J. D. (2004). Insights into the ageing mind: a view from cognitive neuroscience. *Nat. Rev. Neurosci.* 5, 87–96. doi: 10.1038/nrn1323

Horiuchi, M., Yamazaki, N., Ikeda, T., Ishiguro, N., and Shinagawa, M. (1995). A cellular form of prion protein (PrPC) exists in many non-neuronal tissues of sheep. *J. Gen. Virol.* 76(Pt 10), 2583–2587. doi: 10.1099/0022-1317-76-10-2583

Hornshaw, M. P., McDermott, J. R., and Candy, J. M. (1995). Copper binding to the N-terminal tandem repeat regions of mammalian and avian prion protein. *Biochem. Biophys. Res. Commun.* 207, 621–629. doi: 10.1006/bbrc.1995.1233

Jackson, G. S., Murray, I., Hosszu, L. L., Gibbs, N., Waltho, J. P., Clarke, A. R., et al. (2001). Location and properties of metal-binding sites on the human prion protein. *Proc. Natl. Acad. Sci. U.S.A.* 98, 8531–8535. doi: 10.1073/pnas.151038498

Kellett, K. A., and Hooper, N. M. (2009). Prion protein and Alzheimer disease. *Prion* 3, 190–194. doi: 10.4161/pri.3.4.9980

Kessels, H. W., Nguyen, L. N., Nabavi, S., and Malinow, R. (2010). The prion protein as a receptor for amyloid-beta. *Nature* 466, E3–E4. discussion: E-5. doi: 10.1038/nature09217

Kitamoto, T., Iizuka, R., and Tateishi, J. (1993). An amber mutation of prion protein in Gerstmann-Straussler syndrome with mutant PrP plaques. *Biochem. Biophys. Res. Commun.* 192, 525–531. doi: 10.1006/bbrc.1993.1447

Kralovicova, S., Fontaine, S. N., Alderton, A., Alderman, J., Ragnarsdottir, K. V., Collins, S. J., et al. (2009). The effects of prion protein expression on metal metabolism. *Mol. Cell. Neurosci.* 41, 135–147. doi: 10.1016/j.mcn.2009.02.002

Lasmezas, C. I., Deslys, J. P., Robain, O., Jaegly, A., Beringue, V., Peyrin, J. M., et al. (1997). Transmission of the BSE agent to mice in the absence of detectable abnormal prion protein. *Science* 275, 402–405. doi: 10.1126/science.275.5298.402

Lauren, J., Gimbel, D. A., Nygaard, H. B., Gilbert, J. W., and Strittmatter, S. M. (2009). Cellular prion protein mediates impairment of synaptic plasticity by amyloid-beta oligomers. *Nature* 457, 1128–1132. doi: 10.1038/nature 07761

Lehmann, S., and Harris, D. A. (1997). Blockade of glycosylation promotes acquisition of scrapie-like properties by the prion protein in cultured cells. *J. Biol. Chem.* 272, 21479–21487. doi: 10.1074/jbc.272.34.21479

Liu, L., Jiang, D., McDonald, A., Hao, Y., Millhauser, G. L., and Zhou, F. (2011). Copper redox cycling in the prion protein depends critically on binding mode. *J. Am. Chem. Soc.* 133, 12229–12237. doi: 10.1021/ja2045259

Lo, R. Y., Shyu, W. C., Lin, S. Z., Wang, H. J., Chen, S. S., and Li, H. (2007). New molecular insights into cellular survival and stress responses: neuroprotective role of cellular prion protein (PrPC). *Mol. Neurobiol.* 35, 236–244. doi: 10.1007/s12035-007-8003-y

Malaga-Trillo, E., Solis, G. P., Schrock, Y., Geiss, C., Luncz, L., Thomanetz, V., et al. (2009). Regulation of embryonic cell adhesion by the prion protein. *PLoS Biol.* 7:e55. doi: 10.1371/journal.pbio.1000055

Martin, M., Dotti, C. G., and Ledesma, M. D. (2010). Brain cholesterol in normal and pathological aging. *Biochim. Biophys. Acta* 1801, 934–944. doi: 10.1016/j.bbalip.2010.03.011

Martin, M. G., Perga, S., Trovo, L., Rasola, A., Holm, P., Rantamaki, T., et al. (2008). Cholesterol loss enhances TrkB signaling in hippocampal neurons aging *in vitro*. *Mol. Biol. Cell* 19, 2101–2112. doi: 10.1091/mbc.E07-09-0897

Massimino, M. L., Redaelli, M., Bertoli, A., Sorgato, M. C., and Mucignat-Caretta, C. (2013). Altered behavioral aspects of aged mice lacking the cellular prion protein. *Physiol. Behav.* 119, 86–91. doi: 10.1016/j.physbeh.2013.06.006

McLennan, N. F., Brennan, P. M., McNeill, A., Davies, I., Fotheringham, A., Rennison, K. A., et al. (2004). Prion protein accumulation and neuroprotection in hypoxic brain damage. *Am. J. Pathol.* 165, 227–235. doi: 10.1016/S0002-9440(10)63291-9

Meyne, F., Gloeckner, S. F., Ciesielczyk, B., Heinemann, U., Krasnianski, A., Meissner, B., et al. (2009). Total prion protein levels in the cerebrospinal fluid are reduced in patients with various neurological disorders. *J. Alzheimers. Dis.* 17, 863–873. doi: 10.3233/JAD-2009-1110

Mitsios, N., Saka, M., Krupinski, J., Pennucci, R., Sanfeliu, C., Miguel Turu, M., et al. (2007). Cellular prion protein is increased in the plasma and peri-infarcted brain tissue after acute stroke. *J. Neurosci. Res.* 85, 602–611. doi: 10.1002/jnr.21142

Miura, T., Sasaki, S., Toyama, A., and Takeuchi, H. (2005). Copper reduction by the octapeptide repeat region of prion protein: pH dependence and implications in cellular copper uptake. *Biochemistry* 44, 8712–8720. doi: 10.1021/bi0501784

Morales, R., Estrada, L. D., Diaz-Espinoza, R., Morales-Scheihing, D., Jara, M. C., Castilla, J., et al. (2010). Molecular cross talk between misfolded proteins in animal models of Alzheimer's and prion diseases. *J. Neurosci.* 30, 4528–4535. doi: 10.1523/JNEUROSCI.5924-09.2010

Moudjou, M., Bernard, J., Sabuncu, E., Langevin, C., and Laude, H. (2007). Glycan chains modulate prion protein binding to immobilized metal ions. *Neurochem. Int.* 50, 689–695. doi: 10.1016/j.neuint.2007.01.001

Moudjou, M., Frobert, Y., Grassi, J., and La Bonnardière, C. (2001). Cellular prion protein status in sheep: tissue-specific biochemical signatures. *J. Gen. Virol.* 82(Pt 8), 2017–2024. doi: 10.1099/vir.0.17776-0

Nazor, K. E., Seward, T., and Telling, G. C. (2007). Motor behavioral and neuropathological deficits in mice deficient for normal prion protein expression. *Biochim. Biophys. Acta* 1772, 645–653. doi: 10.1016/j.bbadis.2007.04.004

Ordonez-Gutierrez, L., Torres, J. M., Gavin, R., Anton, M., Arroba-Espinosa, A. I., Espinosa, J. C., et al. (2013). Cellular prion protein modulates beta-amyloid deposition in aged APP/PS1 transgenic mice. *Neurobiol. Aging* 34, 2793–2804. doi: 10.1016/j.neurobiolaging.2013.05.019

Pan, T., Li, R., Kang, S. C., Pastore, M., Wong, B. S., Ironside, J., et al. (2005a). Biochemical fingerprints of prion diseases: scrapie prion protein in human prion diseases that share prion genotype and type. *J. Neurochem.* 92, 132–142. doi: 10.1111/j.1471-4159.2004.02859.x

Pan, T., Li, R., Wong, B. S., Kang, S. C., Ironside, J., and Sy, M. S. (2005b). Novel antibody-lectin enzyme-linked immunosorbent assay that distinguishes prion proteins in sporadic and variant cases of Creutzfeldt-Jakob disease. *J. Clin. Microbiol.* 43, 1118–1126. doi: 10.1128/JCM.43.3.1118-1126.2005

Pan, T., Li, R., Wong, B. S., Liu, T., Gambetti, P., and Sy, M. S. (2002). Heterogeneity of normal prion protein in two- dimensional immunoblot: presence of various glycosylated and truncated forms. *J. Neurochem.* 81, 1092–1101. doi: 10.1046/j.1471-4159.2002.00909.x

Pan, T., Wong, P., Chang, B., Li, C., Li, R., Kang, S. C., et al. (2005c). Biochemical fingerprints of prion infection: accumulations of aberrant full-length and N-terminally truncated PrP species are common features in mouse prion disease. *J. Virol.* 79, 934–943. doi: 10.1128/JVI.79.2.934-943.2005

Parkin, E. T., Watt, N. T., Hussain, I., Eckman, E. A., Eckman, C. B., Manson, J. C., et al. (2007). Cellular prion protein regulates beta-secretase cleavage of the Alzheimer's amyloid precursor protein. *Proc. Natl. Acad. Sci. U.S.A.* 104, 11062–11067. doi: 10.1073/pnas.0609621104

Pauly, P. C., and Harris, D. A. (1998). Copper stimulates endocytosis of the prion protein. *J. Biol. Chem.* 273, 33107–33110. doi: 10.1074/jbc.273.50.33107

Peralta, O. A., Huckle, W. R., and Eyestone, W. H. (2012). Developmental expression of the cellular prion protein (PrP(C)) in bovine embryos. *Mol. Reprod. Dev.* 79, 488–498. doi: 10.1002/mrd.22057

Peters, A. (2002). The effects of normal aging on myelin and nerve fibers: a review. *J. Neurocytol.* 31, 581–593. doi: 10.1023/A:1025731309829

Prusiner, S. B. (1991). Molecular biology and transgenetics of prion diseases. *Crit. Rev. Biochem. Mol. Biol.* 26, 397–438. doi: 10.3109/10409239109086789

Prusiner, S. B. (1994). Biology and genetics of prion diseases. *Annu. Rev. Microbiol.* 48, 655–686. doi: 10.1146/annurev.mi.48.100194.003255

Prusiner, S. B. (2001). Shattuck lecture–neurodegenerative diseases and prions. *N. Engl. J. Med.* 344, 1516–1526. doi: 10.1056/NEJM200105173442006

Querfurth, H. W., and LaFerla, F. M. (2010). Alzheimer's disease. *N. Engl. J. Med.* 362, 329–344. doi: 10.1056/NEJMra0909142

Rachidi, W., Vilette, D., Guiraud, P., Arlotto, M., Riondel, J., Laude, H., et al. (2003). Expression of prion protein increases cellular copper binding and antioxidant enzyme activities but not copper delivery. *J. Biol. Chem.* 278, 9064–9072. doi: 10.1074/jbc.M211830200

Rial, D., Duarte, F. S., Xikota, J. C., Schmitz, A. E., Dafre, A. L., Figueiredo, C. P., et al. (2009). Cellular prion protein modulates age-related behavioral and neurochemical alterations in mice. *Neuroscience* 164, 896–907. doi: 10.1016/j.neuroscience.2009.09.005

Roucou, X., and LeBlanc, A. C. (2005). Cellular prion protein neuroprotective function: implications in prion diseases. *J. Mol. Med. (Berl.)* 83, 3–11. doi: 10.1007/s00109-004-0605-5

Safar, J., Wang, W., Padgett, M. P., Ceroni, M., Piccardo, P., Zopf, D., et al. (1990). Molecular mass, biochemical composition, and physicochemical behavior of the infectious form of the scrapie precursor protein monomer. *Proc. Natl. Acad. Sci. U.S.A.* 87, 6373–6377. doi: 10.1073/pnas.87.16.6373

Saijo, E., Scheff, S. W., and Telling, G. C. (2011). Unaltered prion protein expression in Alzheimer disease patients. *Prion* 5, 109–116. doi: 10.4161/pri.5.2.16355

Sales, N., Hassig, R., Rodolfo, K., Di Giamberardino, L., Traiffort, E., Ruat, M., et al. (2002). Developmental expression of the cellular prion protein in elongating axons. *Eur. J. Neurosci.* 15, 1163–1177. doi: 10.1046/j.1460-9568.2002.01953.x

Scheibel, A. B. (1979). Dendritic changes in senile and presenile dementias. *Res. Publ. Assoc. Res. Nerv. Ment. Dis.* 57, 107–124.

Scheibel, M. E., Lindsay, R. D., Tomiyasu, U., and Scheibel, A. B. (1976). Progressive dendritic changes in the aging human limbic system. *Exp. Neurol.* 53, 420–430. doi: 10.1016/0014-4886(76)90082-0

Schmitz, M., Greis, C., Ottis, P., Silva, C. J., Schulz-Schaeffer, W. J., Wrede, A., et al. (2014). Loss of prion protein leads to age-dependent behavioral abnormalities and changes in cytoskeletal protein expression. *Mol. Neurobiol.* doi: 10.1007/s12035-014-8655-3. [Epub ahead of print].

Schwarze-Eicker, K., Keyvani, K., Gortz, N., Westaway, D., Sachser, N., and Paulus, W. (2005). Prion protein (PrPc) promotes beta-amyloid plaque formation. *Neurobiol. Aging* 26, 1177–1182. doi: 10.1016/j.neurobiolaging.2004.10.004

Shyu, W. C., Chen, C. P., Saeki, K., Kubosaki, A., Matusmoto, Y., Onodera, T., et al. (2005). Hypoglycemia enhances the expression of prion protein and heat-shock protein 70 in a mouse neuroblastoma cell line. *J. Neurosci. Res.* 80, 887–894. doi: 10.1002/jnr.20509

Shyu, W. C., Kao, M. C., Chou, W. Y., Hsu, Y. D., and Soong, B. W. (2000). Heat shock modulates prion protein expression in human NT-2 cells. *Neuroreport* 11, 771–774. doi: 10.1097/00001756-200003200-00023

Shyu, W. C., Lin, S. Z., Saeki, K., Kubosaki, A., Matsumoto, Y., Onodera, T., et al. (2004). Hyperbaric oxygen enhances the expression of prion protein and heat shock protein 70 in a mouse neuroblastoma cell line. *Cell. Mol. Neurobiol.* 24, 257–268. doi: 10.1023/B:CEMN.0000018620.41913.d2

Singh, A., Kong, Q., Luo, X., Petersen, R. B., Meyerson, H., and Singh, N. (2009). Prion protein (PrP) knock-out mice show altered iron metabolism: a functional role for PrP in iron uptake and transport. *PLoS ONE* 4:e6115. doi: 10.1371/journal.pone.0006115

Singh, N., Das, D., Singh, A., and Mohan, M. L. (2010). Prion protein and metal interaction: physiological and pathological implications. *Curr. Issues Mol. Biol.* 12, 99–107.

Skripuletz, T., Lindner, M., Kotsiari, A., Garde, N., Fokuhl, J., Linsmeier, F., et al. (2008). Cortical demyelination is prominent in the murine cuprizone model and is strain-dependent. *Am. J. Pathol.* 172, 1053–1061. doi: 10.2353/ajpath.2008.070850

Stahl, N., Baldwin, M. A., Burlingame, A. L., and Prusiner, S. B. (1990). Identification of glycoinositol phospholipid linked and truncated forms of the scrapie prion protein. *Biochemistry* 29, 8879–8884. doi: 10.1021/bi00490a001

Stockel, J., Safar, J., Wallace, A. C., Cohen, F. E., and Prusiner, S. B. (1998). Prion protein selectively binds copper(II) ions. *Biochemistry* 37, 7185–7193. doi: 10.1021/bi972827k

Thibault, O., Hadley, R., and Landfield, P. W. (2001). Elevated postsynaptic [Ca2+]i and L-type calcium channel activity in aged hippocampal neurons: relationship to impaired synaptic plasticity. *J. Neurosci.* 21, 9744–9756.

Treiber, C., Pipkorn, R., Weise, C., Holland, G., and Multhaup, G. (2007). Copper is required for prion protein-associated superoxide dismutase-I activity in *Pichia pastoris*. *FEBS J.* 274, 1304–1311. doi: 10.1111/j.1742-4658.2007.05678.x

Trovo, L., Van Veldhoven, P. P., Martin, M. G., and Dotti, C. G. (2011). Sphingomyelin upregulation in mature neurons contributes to TrkB activity by Rac1 endocytosis. *J. Cell Sci.* 124(Pt 8), 1308–1315. doi: 10.1242/jcs.078766

Vassallo, N., and Herms, J. (2003). Cellular prion protein function in copper homeostasis and redox signalling at the synapse. *J. Neurochem.* 86, 538–544. doi: 10.1046/j.1471-4159.2003.01882.x

Verdu, E., Ceballos, D., Vilches, J. J., and Navarro, X. (2000). Influence of aging on peripheral nerve function and regeneration. *J. Peripher. Nerv. Syst.* 5, 191–208. doi: 10.1046/j.1529-8027.2000.00026.x

Watt, N. T., Taylor, D. R., Gillott, A., Thomas, D. A., Perera, W. S., and Hooper, N. M. (2005). Reactive oxygen species-mediated beta-cleavage of the prion protein in the cellular response to oxidative stress. *J. Biol. Chem.* 280, 35914–35921. doi: 10.1074/jbc.M507327200

Whitehouse, I. J., Jackson, C., Turner, A. J., and Hooper, N. M. (2010). Prion protein is reduced in aging and in sporadic but not in familial Alzheimer's disease. *J. Alzheimers. Dis.* 22, 1023–1031. doi: 10.3233/JAD-2010-101071

Williams, W. M., Stadtman, E. R., and Moskovitz, J. (2004). Ageing and exposure to oxidative stress *in vivo* differentially affect cellular levels of PrP in mouse cerebral microvessels and brain parenchyma. *Neuropathol. Appl. Neurobiol.* 30, 161–168. doi: 10.1111/j.1365-2990.2003.00523.x

Wong, B. S., Liu, T., Li, R., Pan, T., Petersen, R. B., Smith, M. A., et al. (2001). Increased levels of oxidative stress markers detected in the brains of mice devoid of prion protein. *J. Neurochem.* 76, 565–572. doi: 10.1046/j.1471-4159.2001.00028.x

Yoshioka, Y., Ishiguro, N., and Inoshima, Y. (2010). Proteasome activity and biological properties of normal prion protein: a comparison between young and aged cattle. *J. Vet. Med. Sci.* 72, 1583–1587. doi: 10.1292/jvms.10-0149

Zahn, R., Liu, A., Luhrs, T., Riek, R., von Schroetter, C., Lopez Garcia, F., et al. (2000). NMR solution structure of the human prion protein. *Proc. Natl. Acad. Sci. U.S.A.* 97, 145–150. doi: 10.1073/pnas.97.1.145

Zhu, X., Su, B., Wang, X., Smith, M. A., and Perry, G. (2007). Causes of oxidative stress in Alzheimer disease. *Cell. Mol. Life Sci.* 64, 2202–2210. doi: 10.1007/s00018-007-7218-4

Conflict of Interest Statement: The authors declare that the research was conducted in the absence of any commercial or financial relationships that could be construed as a potential conflict of interest.

24

Acquisition of lipid metabolic capability in hepatocyte-like cells directly induced from mouse fibroblasts

Shizuka Miura[1] and Atsushi Suzuki[1,2]*

[1] Division of Organogenesis and Regeneration, Medical Institute of Bioregulation, Kyushu University, Fukuoka, Japan
[2] Core Research for Evolutional Science and Technology, Japan Science and Technology Agency, Saitama, Japan

Edited by:
Alessandra Sacco, Sanford-Burnham Medical Research Institute, USA

Reviewed by:
In-Hyun Park, Yale University, USA
Akiyoshi Uezumi, Fujita Health University, Japan

***Correspondence:**
Atsushi Suzuki, Division of Organogenesis and Regeneration, Medical Institute of Bioregulation, Kyushu University, 3-1-1 Maidashi, Higashi-ku, Fukuoka 812-8582, Japan
e-mail: suzukicks@bioreg.kyushu-u.ac.jp

Recently, the numbers of patients with non-alcoholic fatty liver disease (NAFLD) and non-alcoholic steatohepatitis (NASH) have increased worldwide. NAFLD and NASH are known as risk factors for liver cirrhosis and hepatocellular carcinoma. Because many factors can promote the progression of NAFLD and NASH, the treatment of these patients involves various strategies. Thus, it is desired that drugs for patients with NAFLD and NASH should be developed more easily and rapidly using cultures of primary hepatocytes. However, it is difficult to use hepatocytes as a tool for drug screening, because these cells cannot be functionally maintained in culture. Thus, in this study, we sought to examine whether induced hepatocyte-like (iHep) cells, which were directly induced from mouse dermal fibroblasts by infection with a retrovirus expressing *Hnf4α* and *Foxa3*, possess the potential for lipid metabolism, similar to hepatocytes. Our data showed that iHep cells were capable of synthesizing lipids from a cis-unsaturated fatty acid, a trans-unsaturated fatty acid, and a saturated fatty acid, accumulating the synthesized lipids in cellular vesicles, and secreting the lipids into the culture medium. Moreover, the lipid synthesis in iHep cells was significantly inhibited in cultures with lipid metabolism improvers. These results demonstrate that iHep cells could be useful not only for screening of drugs for patients with NAFLD and NASH, but also for elucidation of the mechanisms underlying hereditary lipid metabolism disorders, as an alternative to hepatocytes.

Keywords: liver, hepatocyte, lipid metabolism, iHep cell, reprogramming, fibroblast

INTRODUCTION

The recent increase of the number of patients with non-alcoholic fatty liver disease (NAFLD), which often develops through obesity and lipid metabolism disorders, is an issue that has become problematic in liver diseases (Vuppalanchi and Chalasani, 2009). NAFLD can be characterized into simple steatosis and non-alcoholic steatohepatitis (NASH). In particular, NASH is associated with the risk of inducing not only liver cirrhosis and hepatocellular carcinoma, but also heart diseases (Hashimoto et al., 2009; Vuppalanchi and Chalasani, 2009). In general, as the cause of NASH developing from fatty liver disease, a "two-hit hypothesis" has been proposed (Day and James, 1998). The first hit is the onset of steatohepatitis, which is induced as a result of cellular lipid accumulation and steatosis. The second hit includes activation of inflammatory cytokines and hepatic oxidative stress (Day and James, 1998; Dowman et al., 2010). As the current therapeutic drugs for NASH, insulin-resistance improvers and antioxidants are frequently used. However, there are many factors that can promote the progression of NAFLD and NASH. Therefore, if we are able to develop many drugs for patients with NAFLD and NASH more easily and rapidly it will become possible to beat the causative factors of these diseases specifically.

Ideally, primary hepatocytes should be used in the screening of drugs for patients with liver diseases. However, it is difficult to maintain the function of hepatocytes in culture. Thus, liver cancer cell lines are often substituted for hepatocytes, although the nature of these cell lines differs from that of normal hepatocytes. To obtain functional hepatocytes for the screening of drugs, hepatocyte-like cells can be induced from embryonic stem (ES) cells and induced pluripotent stem (iPS) cells (Basma et al., 2009; Rashid et al., 2010; Si-Tayeb et al., 2010; Cayo et al., 2012; Choi et al., 2013). Moreover, recent advances in the induction of cellular reprogramming have enabled the generation of hepatocyte-like cells directly from fibroblasts, which is termed "direct reprogramming" (Huang et al., 2011; Sekiya and Suzuki, 2011). In our previous paper, we reported that three specific combinations of two transcription factors, comprising Hnf4α plus Foxa1, Foxa2, or Foxa3, can be used to convert mouse fibroblasts into hepatocyte-like cells (Sekiya and Suzuki, 2011). The induced hepatocyte-like (iHep) cells exhibited multiple hepatocyte-specific functions and restored hepatic tissues after transplantation. Thus, it is expected that iHep cells could be used for developing regenerative therapies and examining the pharmacological effects of drugs. Although iHep cells may be useful in the screening of drugs for patients with NAFLD and NASH, the capability for lipid metabolism in iHep cells remains to be clarified. Thus, in this study, we examined whether iHep cells possess the potential for lipid metabolism, similar to hepatocytes, and further attempted to clarify whether iHep cells could be useful not only for the screening of drugs but also for the progression of basic research for patients with liver diseases involving lipid metabolism.

MATERIALS AND METHODS

GENERATION OF iHEP CELLS

We generated iHep cells from mouse dermal fibroblasts (MDFs) and conducted co-immunofluorescence staining of albumin with E-cadherin, as described previously (Sekiya and Suzuki, 2011). Some modifications that were added to the methods are specifically described below. In the preparation of MDFs, we incubated small pieces of skin tissue obtained from C57BL/6 adult mice (8–10 weeks of age) in Hanks' balanced salt solution containing 0.05% collagenase (Wako, Osaka, Japan) and 0.01% trypsin inhibitor (Nacalai Tesque, Kyoto, Japan) for 20 min at 37°C. The treated tissues were then collected by centrifugation (200 g for 3 min), plated on gelatin-coated six-well plates, and grown until expanded MDFs reached confluency. In the production of recombinant retroviruses, we used PLAT-E cells (Morita et al., 2000) for transfection of plasmid DNA. In this study, we used iHep cells at passage 9–12 for examination.

INDUCTION OF LIPID SYNTHESIS AND SECRETION

To induce lipid synthesis, iHep cells were cultured in our hepato-medium (Sekiya and Suzuki, 2011) containing 20 ng/ml hepatocyte growth factor (HGF) (Sigma-Aldrich, St. Louis, MO, USA), 20 ng/ml epithelial growth factor (EGF) (Sigma-Aldrich), and 5% bovine serum albumin (BSA) (Wako) for 3 days with or without 1 mM oleic acid (Sigma-Aldrich), 1 mM elaidic acid (Tokyo Chemical Industry, Tokyo, Japan), or 1 mM palmitic acid (Wako). Hepatocytes isolated from adult mice by two-step collagenase digestion (Seglen, 1979) were also cultured under the same conditions used for iHep cells, except that HGF and EGF were omitted. MDFs were cultured in Dulbecco's modified Eagle's medium containing 10% fetal bovine serum, 2 mM L-glutamine (Nacalai Tesque), penicillin/streptomycin (Nacalai Tesque), and 5% BSA for 3 days with or without 1 mM oleic acid. To induce lipid secretion, iHep cells were cultured in phenol red-free hepato-medium containing 20 ng/ml HGF, 20 ng/ml EGF, and 5% BSA for 48 h with or without 1 mM oleic acid. The culture medium was then changed to remove the oleic acid, and the supernatants were collected at 12, 24, and 36 h after changing the culture medium.

OIL RED O STAINING

Cultured cells were fixed in 10% formalin for 10 min at room temperature, washed with phosphate-buffered saline (PBS), and treated with 60% isopropanol for 1 min at room temperature. The cells were then incubated with 60% Oil red O (Muto Pure Chemicals, Tokyo, Japan) in water for 20 min at room temperature. After washing with 60% isopropanol for 1 min and additional washing with PBS, the cells were incubated with hematoxylin (Muto Pure Chemicals) for 5 min.

MEASUREMENT OF TRIGLYCERIDE AND APOLIPOPROTEIN B (APO B)

Cultured cells were washed with PBS, collected using a cell scraper into a triglyceride assay buffer [5% Triton X-100 (Nacalai Tesque), 1 mM EDTA (Nacalai Tesque), and 25 mM Tris-HCl pH 7.5], and lysed by sonication. The triglyceride contents in the cell lysates and supernatants and the apo B contents in the supernatants were detected using a Triglyceride Assay Kit (BioAssay Systems, Hayward, CA, USA) and a Total Human Apo B ELISA Assay Kit (ALerCHEK Inc., Portland, ME, USA), respectively, according to the manufacturer's instructions. The absorbance signals were measured with a Multiskan FC microplate reader (Thermo Fisher Scientific, Waltham, MA, USA).

ANALYSIS OF THE EFFECTS OF LIPID METABOLISM IMPROVERS

iHep cells were cultured in hepato-medium containing 20 ng/ml HGF, 20 ng/ml EGF, and 5% BSA for 3 days with or without 1 mM oleic acid, 15 μM 5-(tetradecyloxy)-2-furoic acid (TOFA) (Sigma-Aldrich), and 10 μM clofibrate (Tokyo Chemical Industry). Hepatocytes were also cultured under the same conditions used for iHep cells, except that HGF and EGF were omitted. We added clofibrate to the culture media every day.

STATISTICAL ANALYSIS

Data are shown as mean ± standard deviation (SD), and statistical significance was analyzed using an unpaired Student's *t*-test or a One-Way ANOVA. $P < 0.05$ was considered statistically significant.

STUDY APPROVAL

The experiments were approved by the Kyushu University Animal Experiment Committee, and the care of the animals was in accordance with institutional guidelines.

RESULTS

iHEP CELLS CAN SYNTHESIZE LIPIDS FROM A CIS-UNSATURATED FATTY ACID, SIMILAR TO HEPATOCYTES

In a previous study, we reported a method for the conversion of MDFs into iHep cells (Sekiya and Suzuki, 2011). In accordance with this method, we generated iHep cells by introducing *Hnf4α* and *Foxa3* into dermal fibroblasts obtained from adult mice. As shown in **Figure 1A**, iHep cells had features of epithelial cells and hepatocytes, including the expression of E-cadherin and albumin, respectively. iHep cells became attached to each other through intercellular adhesion molecules, and the size of the iHep cells was smaller than that of MDFs. To investigate the function of lipid metabolism in iHep cells, we used iHep cells that were generated in three independent experiments ($n = 3$). In the presence of oleic acid, a cis-unsaturated fatty acid, in the culture medium, many lipid droplets were observed in the cytoplasm of iHep cells and primary hepatocytes, but not in the cytoplasm of MDFs (**Figure 1B**). Moreover, the amounts of triglyceride in iHep cells and primary hepatocytes were significantly increased in cultures with oleic acid (**Figure 1C**). These data demonstrated that iHep cells acquired the ability to synthesize lipids from a cis-unsaturated fatty acid and accumulate these lipids into many cellular vesicles, similar to hepatocytes, by reprogramming of the original fate of fibroblasts into the hepatic fate.

iHEP CELLS CAN SYNTHESIZE LIPIDS FROM A TRANS-UNSATURATED FATTY ACID AND A SATURATED FATTY ACID

In addition to a cis-unsaturated fatty acid, we examined whether iHep cells could synthesize lipids from a trans-unsaturated fatty

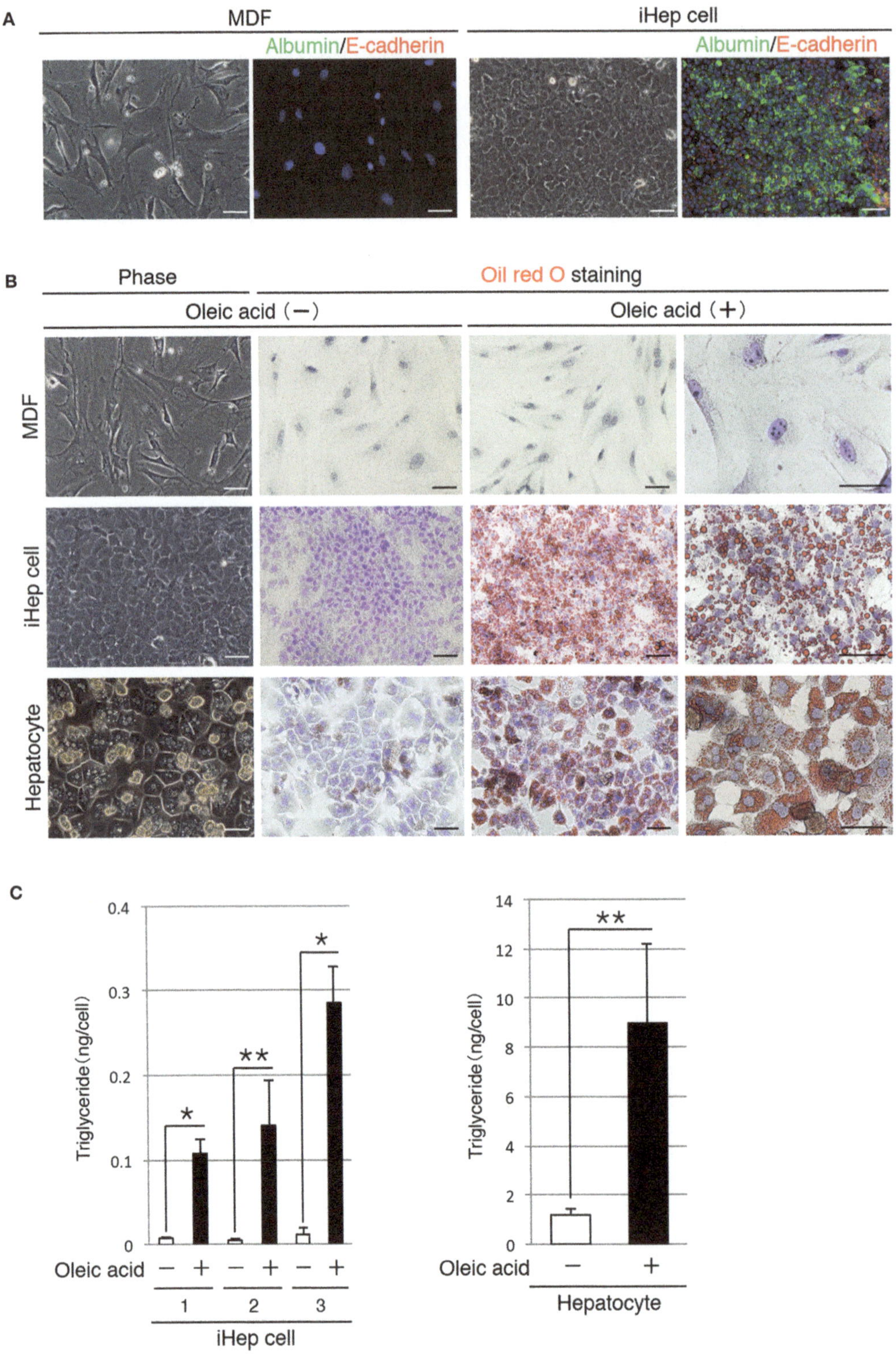

FIGURE 1 | iHep cells can synthesize lipids from a cis-unsaturated fatty acid. (A) Morphologies of MDFs and MDF-derived iHep cells. Co-immunofluorescence staining of albumin and E-cadherin was conducted for MDFs and iHep cells. DNA was stained with DAPI. **(B)** iHep cells can synthesize and store abundant lipids in cultures containing oleic acid, a cis-unsaturated fatty acid. Phase-contrast images of MDFs, iHep cells, and primary hepatocytes in cultures without oleic acid are shown. Oil red O staining was conducted for MDFs, iHep cells, and primary hepatocytes cultured with (+) or without (−) oleic acid. DNA was stained with hematoxylin. **(C)** Quantification of triglyceride contents in iHep cells and primary hepatocytes cultured with (+) or without (−) oleic acid. We chose three independent iHep cells for examination. The graphs show the average of three independent experiments (mean ± SD). $*P < 0.01$, $**P < 0.05$ (Student's *t*-test). Scale bars, 50 μm.

acid and a saturated fatty acid, both of which are often present in foods. iHep cells and primary hepatocytes were each cultured with elaidic acid, a trans-unsaturated fatty acid, or palmitic acid, a saturated fatty acid, for 3 days, and then subjected to oil red O staining to investigate their lipid synthesis and accumulation. The data showed that many lipid droplets were formed in both types of cells, and that the sizes of the droplets in cells cultured with elaidic acid were larger than those in cells cultured with palmitic acid (**Figure 2A**). In addition, the amounts of triglyceride in iHep cells and primary hepatocytes were increased in cultures with elaidic acid or palmitic acid (**Figure 2B**). Thus, similar to hepatocytes, iHep cells can synthesize lipids from a trans-unsaturated fatty acid and a saturated fatty acid, and accumulate these lipids into many droplets in cells, similar to the case for the culture with a cis-unsaturated fatty acid.

iHEP CELLS POSSESS THE CAPABILITY FOR LIPID SECRETION

Next, we examined whether iHep cells could secrete lipids to the culture medium, similar to hepatocytes. To this end, we maintained iHep cells in culture with oleic acid for 48 h, and then changed the culture medium to remove the oleic acid. The supernatants were collected at 12, 24, and 36 h after changing the culture medium for analysis (**Figure 3A**). Quantification of the triglyceride contents showed that the amounts of triglyceride in the supernatants increased after removal of oleic acid from the culture medium (**Figure 3B**). In addition to triglyceride, the amounts of apo B, which is the major protein component of triglyceride-rich lipoproteins, in the supernatants also increased after removal of oleic acid from the culture medium (**Figure 3C**). Moreover, the number of lipid droplets in iHep cells was decreased at 36 h after changing the culture medium,

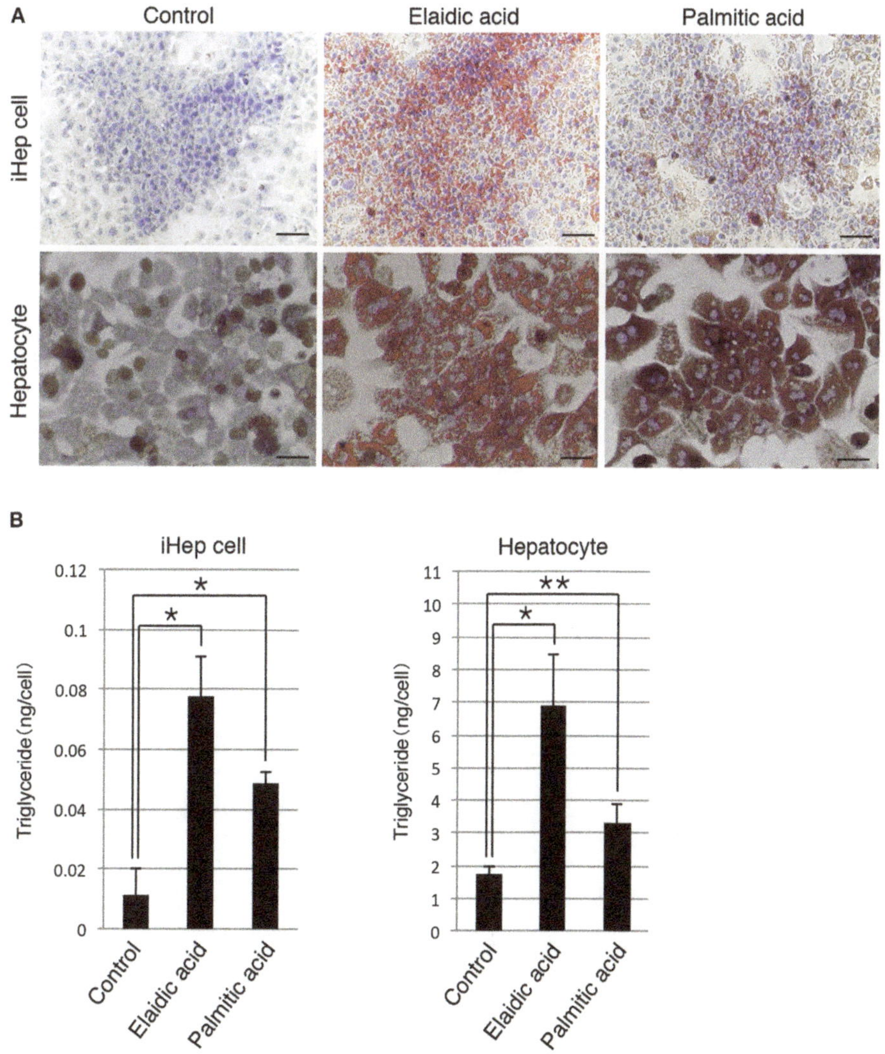

FIGURE 2 | iHep cells are capable of synthesizing lipids from a trans-unsaturated fatty acid and a saturated fatty acid. (A) Oil red O staining was conducted for iHep cells and primary hepatocytes cultured with elaidic acid, a trans-unsaturated fatty acid, or palmitic acid, a saturated fatty acid, and without any fatty acids (control). DNA was stained with hematoxylin. Scale bars, 50 μm. **(B)** Quantification of triglyceride contents in iHep cells and primary hepatocytes cultured with elaidic acid or palmitic acid and without any fatty acids (control). The graphs show the average of three independent experiments (mean ± SD). $^{*}P < 0.01$, $^{**}P < 0.05$ (One-Way ANOVA).

suggesting that the lipids synthesized from oleic acid in iHep cells were secreted from the cells into the supernatant (**Figure 3D**). These data demonstrated that iHep cells possessed the ability to not only synthesize lipids from fatty acids, but also secrete these lipids by exocytosis, similar to hepatocytes.

INHIBITION OF LIPID SYNTHESIS IN iHEP CELLS BY LIPID METABOLISM IMPROVERS

As shown above, iHep cells had the potential to synthesize, accumulate, and secrete lipids in culture, similar to hepatocytes, suggesting that iHep cells could be used in the screening of lipid metabolism improvers for patients with liver diseases involving lipid metabolism. Thus, we further examined whether the lipid synthesis in iHep cells could be inhibited by the effects of well-known lipid metabolism improvers. To this end, we cultured iHep cells with TOFA or clofibrate, both of which are known to act as lipid metabolism improvers, in addition to oleic acid. TOFA blocks lipid synthesis by inhibiting acetyl-CoA carboxylase in the fatty acid biosynthetic pathway, as shown in rat hepatocytes *in vitro* (Panek et al., 1977; McCune and Harris, 1979). Clofibrate is an agonist of peroxisome proliferator-activated receptor α and stimulates peroxisomal β-oxidation (Scotto et al., 1995; Wheelock et al., 2007). Both TOFA and clofibrate are used for the treatment of fatty liver diseases in humans. Our data showed that the amounts of triglyceride in iHep cells and primary hepatocytes were decreased in cultures with oleic acid and TOFA or clofibrate, indicating that iHep cells could respond to the lipid metabolism improvers and reduce lipid synthesis, similar to hepatocytes (**Figure 4**). Thus, iHep cells could be used for evaluating the effects of lipid metabolism improvers and screening of drugs for patients with lipid metabolism disorders, as an alternative to hepatocytes.

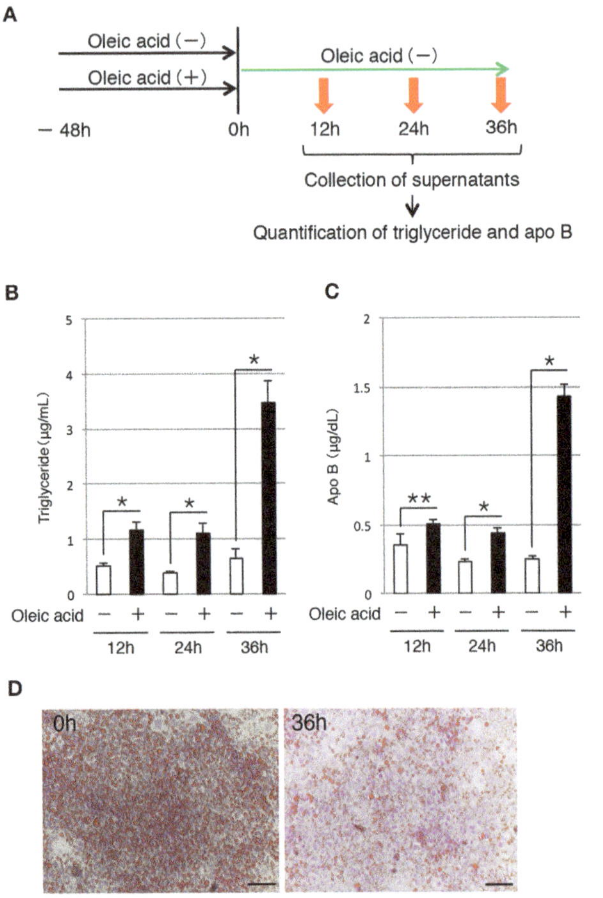

FIGURE 3 | iHep cells have the ability to secrete triglyceride and apo B. (A) Schematic diagram of the experimental procedure. iHep cells were cultured with or without oleic acid for 48 h, and then cultured for 36 h after changing the culture medium. The supernatants were collected at 12, 24, and 36 h after changing the culture medium, and the amounts of triglyceride and apo B in the supernatants were measured. **(B,C)** Quantification of triglyceride **(B)** and apo B **(C)** contents in the supernatants collected at 12, 24, and 36 h after changing the culture medium. The amounts of triglyceride and apo B in the supernatants were normalized to the values per 1 × 10^4 cultured cells. The graphs show the average of three independent experiments (mean ± SD). *$P < 0.01$, **$P < 0.05$ (Student's *t*-test). **(D)** Oil red O staining was conducted for iHep cells at 0 and 36 h after changing the culture medium. DNA was stained with hematoxylin. Scale bars, 50 μm.

DISCUSSION

In this study, we investigated the potential for lipid metabolism in iHep cells that were directly induced from MDFs by infection with a retrovirus expressing *Hnf4α* and *Foxa3*. We found that iHep cells were not only able to synthesize lipids from a cis-unsaturated fatty acid, but also able to accumulate and secrete the synthesized lipids, similar to hepatocytes. In addition, iHep cells were capable of synthesizing lipids from a trans-unsaturated fatty acid and a saturated fatty acid, as well as from a cis-unsaturated fatty acid. Moreover, both iHep cells and primary hepatocytes were capable of responding to the well-known lipid metabolism improvers TOFA and clofibrate. Thus, it is possible that iHep cells will also respond to other lipid metabolism improvers, in addition to TOFA and clofibrate. Taken together, our data demonstrate that iHep cells have some of the important functions in hepatic

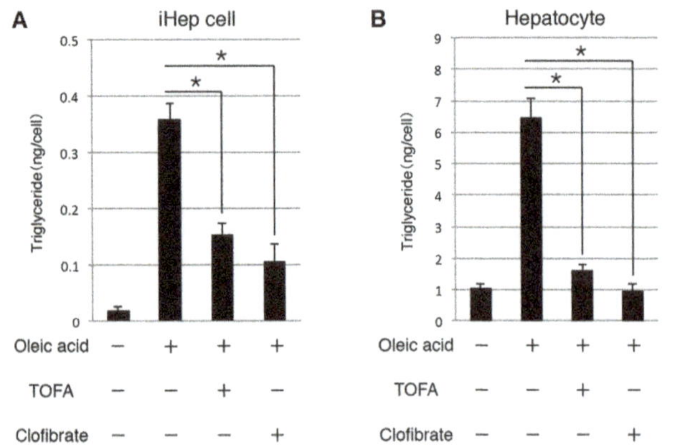

FIGURE 4 | iHep cells can respond to lipid metabolism improvers. (A,B) Quantification of triglyceride contents in iHep cells **(A)** and primary hepatocytes **(B)** cultured with or without oleic acid, and supplemented with or without the well-known lipid metabolism improvers TOFA or clofibrate. The graphs show the average of three independent experiments (mean ± SD). *$P < 0.01$ (One-Way ANOVA).

lipid metabolism and could be useful in the screening of drugs for patients with lipid metabolism disorders.

In the present study, we measured the amounts of triglyceride to quantify the cellular lipid contents within iHep cells and primary hepatocytes. However, hepatocytes can also synthesize other types of lipid, including cholesterol. In our previous study, microarray data demonstrated that the expressions of genes involved in cholesterol metabolism in iHep cells became close to those in hepatocytes, suggesting that iHep cells are also able to synthesize cholesterol, similar to hepatocytes (Sekiya and Suzuki, 2011). In addition, iHep cells expressed genes involved in glucose and xenobiotic metabolism (Sekiya and Suzuki, 2011). Thus, it is possible to speculate that iHep cells possess the potential for glucose and xenobiotic metabolism, as well as that the potential for lipid metabolism that was demonstrated in this study.

Although the amount of triglyceride in iHep cells was less than that in primary hepatocytes, the function and number of iHep cells could be maintained in culture, unlike the case for hepatocytes. Currently, particular cell lines derived from liver cancer cells, which have some hepatic functions and can be maintained in culture, are used as alternatives to hepatocytes in the screening of drugs. Thus, these liver cancer cell lines would appear to be more useful than iHep cells. However, the nature of liver cancer cell lines is different from that of normal hepatocytes, and it has thus been desirable to identify a new type of cells that can be maintained in culture with the properties of normal hepatocytes to examine the pharmacological effects of drugs. Recently, in addition to iHep cells, hepatocyte-like cells induced from ES cells and iPS cells have been expected to be useful in the screening of drugs (Basma et al., 2009; Rashid et al., 2010; Si-Tayeb et al., 2010; Cayo et al., 2012; Choi et al., 2013). However, it has remained unclear whether these iHep cells and stem cell-derived hepatocyte-like cells have the potential for lipid metabolism, similar to hepatocytes. In this study, we clearly showed that iHep cells possess the properties of hepatocytes in lipid metabolism, by synthesizing lipids from fatty acids, accumulating the lipids in cellular vesicles, secreting the lipids into the culture medium, and responding to lipid metabolism improvers, even after expansion in culture for a relatively long period. Thus, iHep cells will be useful in the screening of drugs for patients with liver diseases involving lipid metabolism, as an alternative to hepatocytes and liver cancer cell lines. Moreover, it is expected that iHep cells will be useful in basic research to analyze the mechanisms underlying hereditary lipid metabolism disorders.

ACKNOWLEDGMENTS

We thank Dr. Toshio Kitamura for sharing reagents, and Momoko Hamano, Mayumi Yamamoto, and Chiaki Kaieda for technical assistance. This work was supported in part by Grants-in-Aid for Scientific Research from the Ministry of Education, Culture, Sports, Science and Technology (MEXT) of Japan, a Health Labour Sciences Research Grant in Japan, The Core Research for Evolutional Science and Technology Program of the Japan Science and Technology Agency, The Suzuken Memorial Foundation, The ONO Medical Research Foundation, and The Uehara Memorial Foundation.

REFERENCES

Basma, H., Soto-Gutiérrez, A., Yannam, G. R., Liu, L., Ito, R., Yamamoto, T., et al. (2009). Differentiation and transplantation of human embryonic stem cell-derived hepatocytes. *Gastroenterology* 136, 990–999. doi: 10.1053/j.gastro.2008.10.047

Cayo, M. A., Cai, J., DeLaForest, A., Noto, F. K., Nagaoka, M., Clark, B. S., et al. (2012). JD induced pluripotent stem cell-derived hepatocytes faithfully recapitulate the pathophysiology of familial hypercholesterolemia. *Hepatology* 56, 2163–2171. doi: 10.1002/hep.25871

Choi, S. M., Kim, Y., Shim, J. S., Park, J. T., Wang, R. H., Leach, S. D., et al. (2013). Efficient drug screening and gene correction for treating liver disease using patient-specific stem cells. *Hepatology* 57, 2458–2468. doi: 10.1002/hep.26237

Day, C. P., and James, O. F. (1998). Steatohepatitis: a tale of two "hits"? *Gastroenterology* 114, 842–845. doi: 10.1016/S0016-5085(98)70599-2

Dowman, J. K., Tomlinson, J. W., and Newsome, P. N. (2010). Pathogenesis of non-alcoholic fatty liver disease. *QJM* 103, 71–83. doi: 10.1093/qjmed/hcp158

Hashimoto, E., Yatsuji, S., Tobari, M., Taniai, M., Torii, N., Tokushige, K., et al. (2009). Hepatocellular carcinoma in patients with nonalcoholic steatohepatitis. *J. Gastroenterol.* 44(Suppl. 19), 89–95. doi: 10.1007/s00535-008-2262-x

Huang, P., He, Z., Ji, S., Sun, H., Xiang, D., Liu, C., et al. (2011). Induction of functional hepatocyte-like cells from mouse fibroblasts by defined factors. *Nature* 475, 386–389. doi: 10.1038/nature10116

McCune, S. A., and Harris, R. A. (1979). Mechanism responsible for 5-(tetradecyloxy)-2-furoic acid inhibition of hepatic lipogenesis. *J. Biol. Chem.* 254, 10095–10101.

Morita, S., Kojima, T., and Kitamura, T. (2000). Plat-E: an efficient and stable system for transient packaging of retroviruses. *Gene Ther.* 7, 1063–1066. doi: 10.1038/sj.gt.3301206

Panek, E., Cook, G. A., and Cornell, N. W. (1977). Inhibition by 5-(tetradecyloxy)-2-furoic acid of fatty acid and cholesterol synthesis in isolated rat hepatocytes. *Lipids* 12, 814–818. doi: 10.1007/BF02533270

Rashid, S. T., Corbineau, S., Hannan, N., Marciniak, S. J., Miranda, E., Alexander, G., et al. (2010). Modeling inherited metabolic disorders of the liver using human induced pluripotent stem cells. *J. Clin. Invest.* 120, 3127–3136. doi: 10.1172/JCI43122

Scotto, C., Keller, J. M., Schohn, H., and Dauça, M. (1995). Comparative effects of clofibrate on peroxisomal enzymes of human (Hep EBNA2) and rat (FaO) hepatoma cell lines. *Eur. J. Cell Biol.* 66, 375–381.

Seglen, P. O. (1979). Hepatocyte suspensions and cultures as tools in experimental carcinogenesis. *J. Toxicol. Environ. Health* 5, 551–560. doi: 10.1080/15287397909529766

Sekiya, S., and Suzuki, A. (2011). Direct conversion of mouse fibroblasts to hepatocyte-like cells by defined factors. *Nature* 475, 390–393. doi: 10.1038/nature10263

Si-Tayeb, K., Noto, F. K., Nagaoka, M., Li, J., Battle, M. A., Duris, C., et al. (2010). Highly efficient generation of human hepatocyte-like cells from induced pluripotent stem cells. *Hepatology* 51, 297–305. doi: 10.1002/hep.23354

Vuppalanchi, R., and Chalasani, N. (2009). Nonalcoholic fatty liver disease and nonalcoholic steatohepatitis: selected practical issues in their evaluation and management. *Hepatology* 49, 306–317. doi: 10.1002/hep.22603

Wheelock, C. E., Goto, S., Hammock, B. D., and Newman, J. W. (2007). Clofibrate-induced changes in the liver, heart, brain and white adipose lipid metabolome of Swiss-Webster mice. *Metabolomics* 3, 137–145. doi: 10.1007/s11306-007-0052-8

Conflict of Interest Statement: The authors declare that the research was conducted in the absence of any commercial or financial relationships that could be construed as a potential conflict of interest.

25

Mouse adipose tissue stromal cells give rise to skeletal and cardiomyogenic cell sub-populations

Cécile Dromard[1,2,3,4], Corinne Barreau[1,2,3,4], Mireille André[1,3,4], Sandra Berger-Müller[1,3,4], Louis Casteilla[1,2,3,4] and Valerie Planat-Benard[1,2,3,4]*

[1] *Centre National de la Recherche Scientifique (CNRS) UMR 5273, STROMALab, Toulouse, France*
[2] *Université Paul Sabatier de Toulouse, Toulouse, France*
[3] *Institut National de la Santé et de la Recherche Médical (INSERM) U1031, STROMALab, Toulouse, France*
[4] *Etablissement Français du Sang (EFS) Pyrénées-Méditerranée, STROMALab, Toulouse, France*

Edited by:
Anthony Scime, York University, Canada

Reviewed by:
Atsushi Asakura, University of Minnesota, USA
Guillaume Grenier, Université de Sherbrooke, Canada

****Correspondence:***
Valerie Planat-Benard, UMR 5273 CNRS, UPS, EFS, INSERM U1031, STROMALab, BP 84 225, 31 432 Toulouse, France
e-mail: valerie.planat@inserm.fr

We previously reported that adipose tissue could generate cardiomyocyte-like cells from crude stromal vascular fraction (SVF) *in vitro* that improved cardiac function in a myocardial infarction context. However, it is not clear whether these adipose-derived cardiomyogenic cells (AD-CMG) constitute a homogenous population and if AD-CMG progenitors could be isolated as a pure population from the SVF of adipose tissue. This study aims to characterize the different cell types that constitute myogenic clusters and identify the earliest AD-CMG progenitors *in vitro* for establishing a complete phenotype and use it to sort AD-CMG progenitors from crude SVF. Here, we report cell heterogeneity among adipose-derived clusters during their course of maturation and highlighted sub-populations that exhibit original mixed cardiac/skeletal muscle phenotypes with a progressive loss of cardiac phenotype with time in liquid culture conditions. Moreover, we completed the phenotype of AD-CMG progenitors but we failed to sort them from the SVF. We demonstrated that micro-environment is required for the maturation of myogenic phenotype by co-culture experiments. These findings bring complementary data on AD-CMG and suggest that their emergence results from *in vitro* events.

Keywords: Adipose tissue stem cells, cardiac and skeletal myogenesis, *in vitro* differentiation, cell plasticity

INTRODUCTION

The inexorability of heart failure has prompted the development of cell therapy as a new strategy to replenish the pool of dead cardiomyocytes. A prerequisite in such efforts is to determine which progenitor cells can be differentiated into a functional cardiac phenotype. Resident cardiac stem and progenitor cells have been identified in adult mammalian myocardium (Askari et al., 2003; Beltrami et al., 2003; Martin et al., 2004; Matsuura et al., 2004; Messina et al., 2004; Laugwitz et al., 2005) and a phase 1 clinical trial showed that autologous cardiac stem cells may be efficient in patients with ischemic cardiomyopathy (Bolli et al., 2011). However, their use requires a previous isolation step for *in vitro* expansion, which is invasive for the heart to be treated.

Currently, a variety of other autologous adult progenitor cells that could generate differentiated cells beyond their own tissue boundaries are of great interest. Skeletal muscle myoblasts and bone marrow-derived cell subsets (hematopoietic stem cells, mesenchymal stem cells) were tested as potential sources of cardiac progenitors for cell replacement therapy and yielded positive results in infarcted myocardium of various animal models.

Abbreviations: SVF, stromal vascular fraction; AD-CMG, adipose-derived cardiomyogenic cells; cTNT, cardiac troponin T; BAT, brown adipose tissue; WAT-ING, white inguinal adipose tissue; WAT-P, white peritoneal adipose tissue; ASC, adipose stromal cells.

However, despite integration and survival their predominant *in vivo* effect may be related to neoangiogenesis or supportive effect (for review, see Menasche, 2011). Indeed, with the exception of Spoc cells (Skeletal-based precursors of cardiomyocytes) isolated on the basis of several surface markers from adult mice skeletal muscles, that have demonstrated their potential to differentiate into beating and functional cardiomyocytes both *in vitro* and *in vivo* (Winitsky et al., 2005), it is now recognized that both skeletal myoblasts and bone marrow cells lack the degree of plasticity allowing them to widely convert into cardiomyocytes *in vivo* (Reinecke et al., 2002; Scherschel et al., 2008).

Previous studies highlighted the existence of adipose tissue derived progenitor cells possessing cardiogenic potential and being able to promote myocardial regeneration (Planat-Benard et al., 2004; Yamada et al., 2006; Leobon et al., 2009). Indeed, clusters of myogenic cells spontaneously emerge from culture of the crude stromal vascular fraction (SVF) of adipose tissue in semi-solid medium. The clusters contain cells that exhibit pace-maker contractile activity, are responsive to chronotropic agents and express different cardiac markers such as transcription factors and specific contractile proteins (Planat-Benard et al., 2004). Until now, the origin of adipose derived-cardiomyogenic cells (AD-CMG) has not been clearly determined. Indeed, we and other teams have tried to identify in adipose tissue, progenitor cells owning the potential of cardiomyogenic differentiation, but only partial phenotypes were established from cells freshly prepared from SVF and the progenitors of AD-CMG have still never been identified *in situ* (Yamada et al., 2006, 2007).

The present works aims to identify and characterize the earliest AD-CMG progenitors in myogenic clusters *in vitro* for subsequently use these hallmarks in order to prospectively isolate such progenitors from SVF of adipose tissue.

MATERIALS AND METHODS

ETHICAL APPROVAL

Eight-week-old male C57Bl6J mice (Harlan) were housed in a controlled environment (12-h light/dark cycle at 21°C) with free access to water and a standard chow diet (UAR). All procedures were performed in accordance with the European Community guidelines for the care and use of laboratory animals (EEC/No. 07430).

ISOLATION AND CULTURE OF ADIPOSE DERIVED CELLS

Mice were euthanized by cervical dislocation. Brown interscapular and white peritoneal or inguinal adipose tissues were withdrawn and subjected to mechanical dissociation and digestion in DMEM-F12 medium (Invitrogen, Carlsbad, USA) supplemented with bovine serum albumin (BSA) (2%) and 2 mg/ml collagenase A (Roche Diagnostics), for 30 min at 37°C. After elimination of undigested fragments by filtration through 25-μm filters, cell suspension was centrifuged at 486 g for 10 min to separate floating mature adipocytes from the SVF. SVF was incubated in erythrocytes lysis buffer (ammonium chloride solution) (StemCell Technologies) for 5 min at 4°C and washed in PBS. SVF cells were counted and used for further analysis (flow cytometry) or cultured in semi-solid medium as previously described (Leobon et al., 2009). Briefly, freshly prepared SVF cells were plated (27000 cells/mL) and maintained for 2 weeks in methylcellulose (Methocult M3534, StemCell Technologies), corresponding to semi-solid culture condition.

For some experiments, myogenic clusters were transferred from semi-solid to liquid culture condition (Leobon et al., 2009). Indeed myogenic clusters of cells were dissected under an inverted phase-contrast microscope and were plated (1500 cells/cm^2) into 30-mm culture dishes coated with 0.1% gelatin and cultured in BHK21 medium containing fetal bovine serum (10%), supplemented with b-mercaptoethanol (1024 M), glutamine (2 mM), pyruvate (1 mM), non-essential amino acid (0.1 mM), and a solution of amphotericin (0.25 mg/mL), penicillin G (100 U/mL), and streptomycin (100 mg/mL), corresponding to liquid culture condition. In that case passaging was performed by collecting non adherent cells from the culture medium that were successively plated to reach passage 15 (P15). For flow cytometry, myogenic clusters were picked from methylcellulose under an inverted phase-contrast microscope, washed with PBS and mechanically dissociated.

CELL PHENOTYPING

Freshly-isolated SVF cells or AD-CMG were stained in PBS supplemented with 0.5% new calf serum and FcR Block reagent (StemCell Technologies). Sextuple stainings were performed by direct immunofluorescence with conjugated rat anti-mouse monoclonal antibodies (listed in **Table 1**) as compared with their matched immunoglobulin isotype controls (BD Biosciences). Cells were washed in PBS and analyzed on a fluorescence-activated cell sorter (FACS Canto II, Becton Dickinson). Data acquisition and analysis were performed using FACS Diva software (Becton Dickinson).

Table 1 | List of conjugated monoclonal antibodies used for cell phenotyping.

Antibodies	Clone	Conjugates
CD29[a]	Ha2/5	FITC
CD31[a]	MEC13.3	APC
CD34[a]	RAM34	A647
CD44[a]	IM7	PerCp-Cy5.5
CD45[a]	30-F11	APC-Cy7
CD73[a]	TY/23	PE
CD81[a]	Eat2	PE
CD90[a]	53-2.1	FITC
CD117[a]	2B8	PerCp-Cy5.5
CD133[b]	13A4	APC
CMH1[a]	AF6-88.5	PE
CMH2[a]	AF6-120.1	FITC
Flk1[a]	AVAS 12α1	PerCp-Cy5.5
Sca1[b]	D7	PE-Cy7

[a]Antibodies obtained from BD Biosciences, San Jose, CA.
[b]Antibodies obtained from eBiosciences, San Diego, CA.
FITC, fluorescein isothiocyanate; PE, phycoerythrin; PerCP, peridinin chlorophyll protein; PE-Cy7, phycoerythrin cyanine 7; APC, allophycocyanin; APC-Cy7, allophycocyanin cyanine 7.

IMMUNOCHEMISTRY

Cultured cells were washed out from methylcellulose with PBS then fixed with paraformaldehyde (3,7%). Cells were stained with antibodies against cardiac troponin T (c-TnT clone 13-11), myogenin (Santa Cruz, clone M-225), and Numb (Abcam) as compared with their controls (matched immunoglobulin isotype for monoclonal antibody and primary antibody omission for polyclonal antibodies). After washing, conjugated secondary antibodies were added for 1 h. Nuclei were stained with DAPI (1/10,000). Cell staining was observed with a fluorescence microscope (DMRB Leica) and analyzed with the image analyzer software NIS (Nikon).

WESTERN BLOT ANALYSIS

Extensor digitorum longus muscles and hearts were dissected from C57Bl6J mice, minced with scissors, washed in PBS, and ground in RIPA lysis buffer containing protease inhibitors. Samples were incubated on ice for 1 h and vortexed every 20 min. Protein extracts were then sonicated, centrifuged at 15000 g for 15 min, and protein concentration of the supernatant was determined using the BCA assay (Pierce). 25 μg proteins were loaded on a 12% acrylamide gel. Proteins were transferred on a PVDF membrane using a semi-wet transfer system. Membranes were incubated with mouse anti-cTnT antibody (clone 13-11) and with HRP-conjugated anti-mouse antibody. Membranes were then stripped and probed with rabbit anti-β-tubulin antibody (Abcam) and HRP-conjugated anti-rabbit antibody.

DETECTION OF CELL FUSION

SVF cells were labeled with PKH26 using the PKH26-Red Fluorescence Cell Linker kit (Sigma–Aldrich) according to the manufacturer's instructions. In parallel, SVF cells were isolated from transgenic GFP-expressing mice [C57Bl/6 TgN(act-EGFP)OsbC15-001-FJ001] kindly provided by Prof. M. Okabe (Okabe et al., 1997). Equal numbers of GFP- and PKH26-labeled cells were mixed and plated in methylcellulose (5000 cells/cm^2). As controls, GFP- and PKH26-labeled cells plated separately in methylcellulose. After 5 days of culture, cells were distinguished by fluorescence microscopy (green and red fluorescence respectively) directly in methylcellulose.

DATA ANALYSIS

Results were expressed as means ± s.e.m. Data from at least three independent preparations of AD-CMG derived from BAT, WAT-ING, WAT-P, or inguinal derived adipose stromal cells (ASC) were statistically processed with the Wilcoxon test using the Prism 4 software (GraphPad, San Diego, CA). Significance was defined as $^{*}p \leq 0.05$, $^{**}p \leq 0.01$, and $^{***}p \leq 0.001$.

RESULTS

MORPHOLOGICAL HETEROGENEITY IN CONTRACTILE MYOGENIC CLUSTERS

Mascroscopic analysis of contractile myogenic clusters obtained in semi-solid medium revealed cell heterogeneity in size, shape, and nucleus number as initially described (**Figures 1A,B**). Distinct cell morphologies appeared sequentially, some of them being maintained over time. Small round or bipolar mono-nucleated cells (15 μm diameter) appeared during the first 3 days of culture (**Figures 1C,F**). From day 5, large round multi-nucleated cells (30 μm diameter) constituted the main cell type in the clusters and some of them displayed spontaneous contractile activity (**Figures 1D,F**). From day 8, no more mono-nucleated cells could be observed and from day 10, multi-nucleated tube-shaped cells constituted the main cell type in the clusters (**Figures 1E,F**). Two types of tube-shaped cells could be distinguished; thin elongated multi-nucleated cells, and thick elongated multi-nucleated cells (reaching 17 μm in diameter). The latters displayed spontaneous beating activity, in part in a synchronous way. Our previous data had demonstrated their cardiogenic phenotype sustained by electrophysiological characteristics of cardiomyocytes (Planat-Benard et al., 2004); here, we provided additional functional evidence. Considering that electrical stimuli cannot elicit contractions in cardiomyocytes in the absence of external calcium ions but are capable of doing so in skeletal muscle (Nabauer et al., 1989), we tested the ability of beating cells to develop or not contractions in the absence of extracellular calcium. Replacement of the culture medium for a medium containing the selective calcium scavenger EGTA stopped spontaneous beating and confirmed the dependence of beating cells on extracellular calcium. This effect was reversed by reintroduction of the calcium containing-control culture medium that induced once again spontaneous contractions (Supplemental video online). Thus, spontaneously contracting thick elongated multi-nucleated cells are a population of adipose-derived cells exhibiting a cardiac rather than skeletal-like contractile phenotype.

CHARACTERIZATION OF CLUSTER-INITIATING SUB-POPULATIONS

In order to evaluate the emergence and persistence of cardiogenic phenotype, expression of cardiac proteins was assessed in each cell type present in the clusters obtained in semi-solid medium or passaged in liquid medium.

Many proteins initially considered as cardiac specific have actually been demonstrated to be expressed in skeletal muscle as well, such as the transcription factor Myocyte Enhancer factor 2C (MEF2C) (Chen et al., 1998; Le Bihan et al., 2006; Potthoff et al., 2007). So, we first verified by western blot the specificity of the tools used as markers of cardiomyogeneity. Among the proteins we tested, the cardiac contractile protein troponin C (cTnT) only, appeared to be exclusively expressed in the mouse heart (Figure S1).

Then we performed double immunocytochemical stainings against the cardiac troponin C and the skeletal transcription factor myogenin, respectively located in cell cytoplasm and nucleus to clearly distinguish specific labeling. In early formed clusters appearing in semi-solid medium, we obtained distinct profiles of expression for these markers associated with distinct cell morphologies (**Table 2**, **Figure 2**). cTNT and myogenin were present in cluster-initiating cells, i.e., the small mono-nucleated cells, with various combinations. Some cells appeared exclusively stained with cTNT or with myogenin and somewhat surprisingly, some expressed both markers simultaneously (**Figures 2A,B**). Whereas thin tube-shaped cells always co-expressed cTNT and myogenin, large multi-nucleated round cells and thick tube-shaped cells exclusively expressed cTNT (**Figures 2C,D**). Based on these protein expressions and on the functional evidences of their cardiogenic phenotype as described above, we restricted the AD-CMG designation name to the contractile thick tube-shaped cells. Our results suggested that cluster-initiating cells may be intrinsically heterogeneous. During the time course of cluster maturation, it is likely that cTnT(+)myogenin(−) cell populations initially represented in small round cells, then large multi-nucleated cells and later on thick contractile tube-shaped cells (AD-CMG) were cardiomyogenic cells that could be distinguished from cTnT(+/−) but clearly myogenin(+) skeletal muscle cell populations represented in small round-shaped then thin tube-shaped cells. In order to target the progenitor of AD-CMG, we focused on the large round multi-nucleated cell population displaying a cTnT(+)myogenin(−) phenotype that emerged before myotube-shaped cells in semi-solid medium, thus representing a good candidate for AD-CMG progenitors identification.

While emergence of myogenic clusters required a first step of culture in semi-solid medium, for long term expansion myogenic clusters were picked from the semi-solid medium and transferred to liquid medium as reported by Leobon et al. (2009). In liquid medium, differentiated cells adhered to the plastic dishes while large round multi-nucleated cells (the cluster-initiating cells), were non adherent so; subsequent cell passages were performed by harvesting cells in suspension in the liquid medium and replating.

Clusters that emerged in liquid medium contained cells that displayed the same morphologies as in semi-solid medium but the cardiac markers cTnT and MLC2v stainings declined

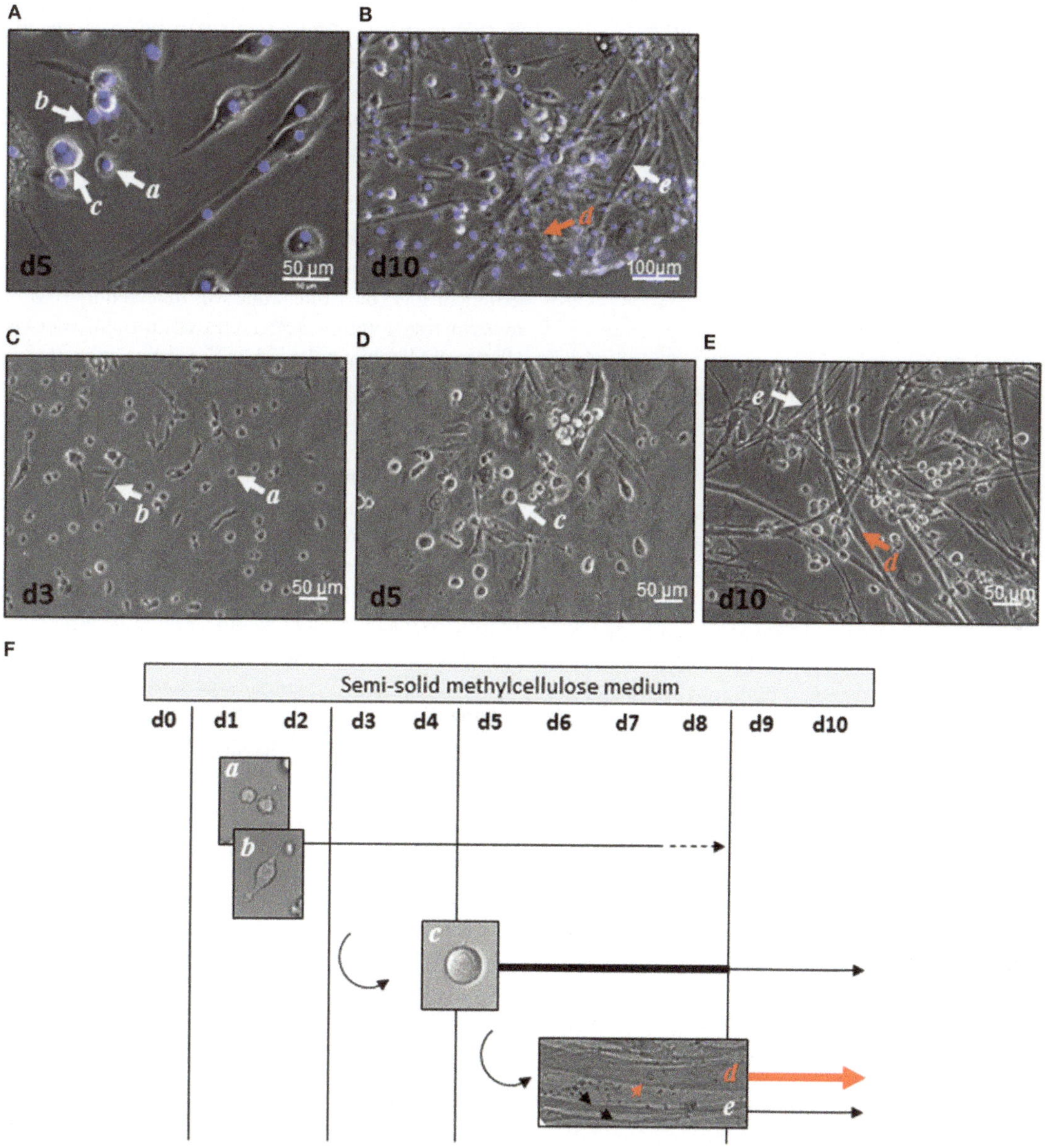

FIGURE 1 | Cell heterogeneity in myogenic clusters. **(A,B)** After 5 days **(A)** and 10 days **(B)** of culture, AD-CMG displayed mono and multi-nucleation associated to various morphologies: small round mono-nucleated cells (*a* cells), bipolar mono-nucleated cells (*b* cells), large round multi-nucleated cells (*c* cells), thick multi-nucleated tube-shaped cells (*d* cells), and thin multi-nucleated tube-shaped cells (*e* cells). Nuclei were stained with Dapi (blue). **(C–E)** Distinct cell morphologies appeared along maturation of myogenic clusters 3 days **(C)**, 5 days **(D)**, and 10 days **(E)** after plating of SVF cells in semi-solid medium (scale bars 50 μm). **(F)** Time course of distinct cell morphologies emergence in myogenic clusters after plating of SVF in semi-solid. Thick lines represent the prevalent cell type, whereas dotted lines represent the minor cell type.

progressively and at the end of passage 1 nearly all cells in the clusters expressed myogenin exclusively (**Figure 3**). Of note, from passage 9, skeletal muscle markers also began to decline (**Figure 3B**) and after passage 15, skeletal markers could not be detected either. Same results were obtained with the skeletal muscle marker MyoD (data not shown). Taken together, these data pointed out that adipose-derived myogenic cluster displayed important cell heterogeneity and original mixed cardiogenic/skeletal phenotypes during the course of their differentiation. Moreover, it appeared that cardiomyogenic phenotype was transient and progressively declined in favor to skeletal-like phenotype upon large expansion in liquid medium.

CHARACTERIZATION OF LARGE ROUND MULTI-NUCLEATED CELLS AS POTENT PROGENITORS OF AD-CMG

First, we determined that the large round cells multi-nucleation resulted from cell fusion rather than endomitosis events (Figure S2). As large round multi-nucleated cells were mingled with other cells in cardiomiogenic clusters, they could hardly be set apart for phenotyping. To overcome this limit, individual

Table 2 | Combinations of cardiac and skeletal myogenic protein expression associated with distinct cell morphologies in cardiomyogenic clusters.

Cell phenotype	Troponin T	Myogenin
Small round mono-nucleated (a)	+/–	+/–
Small bipolar mono-nucleated (b)	+	+
Large round multi-nucleated (c)	+	–
Thick multi-nucleated tube-shaped contractile (d)	++	–
Thin multi-nucleated tube-shaped (e)	+	+

clusters were selectively picked up among surrounding cells present in the culture plate at day 5, when large round multi-nucleated cells were numerous and predominant. The phenotype of selected cells was analyzed and compared to the phenotype of the remaining surrounding cells in order to bring out a specific phenotype that would be exclusively representative of myogenic clusters and in this case, mainly of large round multi-nucleated cells. Among $CD45^-$ cells collected, a differential population of $CD29^+/Sca1^-/CD44^+$ could be exclusively identified in the cells coming from the clusters, that represented 26,8 % (± 7.8) of the myogenic clusters cells and 1,7 % (± 0.3) of the environment cells in culture. This sub-population was positive for CD81 and CD133 (**Figure 4A**) and negative for CD31, CD34, CD73, CD90, CD117, CMH1, CMH2, and Flk1 (**Figure 4B**).

AD-CMG PROGENITORS REQUIRE ACCESSORY CELLS TO GENERATE MYOGENIC CLUSTERS

A cell population with the antigenic profile described above was identified in SVF, where it represented 0.1% (± 0.7) of total cells. To test whether this population has the potential to differentiate toward cardiomyocytes, we cultured a native cells subpopulation sorted by $CD45^-$ $CD44^+$, $CD81^+$, or $Sca1^-$; unfortunately, whatever the surface marker combination and the sorting technique (flow cytometry or magnetic cell sorting); sorted cells survived but no myogenic cluster emerged. Reconstitution of the crude SVF by mixing again cell populations after sorting also failed to induce contractile cluster formation in semi-solid medium, suggesting that the sorting process itself may impair cell properties. We hypothesized that cell separation before culture may have a negative impact on cardiogenic commitment by depleting specific surface proteins crucial for inducing proliferation and differentiation.

Alternatively to the sorting approach, clonal experiments were set up. Individual 5 days old clusters (when large round multi-nucleated cells were in majority) were picked up and seeded after limiting dilution to plate one cell in every 3 well. In these conditions, clonal cells survived but remained in a stationary state without proliferating and none of them gave rise to a myogenic cluster. Thus, AD-CMG could not be purified from crude SVF or myogenic clusters. However, these experiments revealed that AD-CMG progenitors required a specific cell environment to evolve into myogenic clusters.

In order to better understand the role of microenvironment on AD-CMG emergence and to develop strategies to improve their yield, we investigated the influence of surrounding cells. To that purpose, adipose tissue derived SVF cells were cultured under our classical conditions in semi-solid medium to generate myogenic clusters and in some experiments, SVF cells from other adipose pads were added to provide different cellular environments. First, the intrinsic cardiomyogenic potential of SVF cells from brown (BAT), white inguinal (WAT-ING), and white peritoneal (WAT-P) adipose tissues as well as cultured ASC from inguinal fat pad was evaluated independently. It appeared that only SVF cells from BAT and WAT-ING could give rise to myogenic clusters (only one clone was detected in WAT-P among six experiments), with a better efficacy for BAT-derived SVF cells (**Figure 5A**); while cultured ASC did not generate any myogenic clusters by themselves. Then, we comparatively tested the addition of SVF cells from WAT-ING and WAT-P as well as ASC to BAT SVF cells. When co-cultured with BAT SVF cells, neither WAT-ING nor ASC did affect the potential of BAT SVF cells to generate myogenic clusters while WAT-P did potentiate the number of BAT SVF derived myogenic clusters (**Figure 5B**). Elsewhere, co-culture of BAT SVF cells with inguinal WAT-ING SVF induced striking acceleration in myogenic clusters maturation, characterized by beating events occurring as soon as 5 days after plating (**Figure 5C**), whereas equivalent beating activity was not observed before 12 days in control BAT SVF cultures (**Figure 5D**). This results suggested that some microenvironment were prone to support myogenic clusters emergence (i.e., SVF of WAT-P) while some other promote cardiogenic differentiation (i.e., SVF of WAT-ING).

DISCUSSION

We previously reported that cardiomyogenic cells (AD-CMG) could be obtained from white and brown adipose tissue *in vitro* and improved cardiac function in a myocardial infarction context (Planat-Benard et al., 2004; Leobon et al., 2009). Here, we reported complementary data on adipose-derived myogenic cluster formation and shed light on AD-CMG progenitors.

Kinetic analysis revealed morphological cell heterogeneity among clusters during their course of maturation. Molecular characterization confirmed that some sub-populations did exhibit cardiomyogenic phenotype and some of them also expressed markers of skeletal muscle, corresponding to an original mixed cardiac/skeletal muscle phenotype. Such a bi-phenotype has never been described before, excepted in the P19CL6 cell-line, a mouse embryonal carcinoma cell-line extensively used as an *in vitro* model of cardiomyocytes (Khodadadi et al., 2010).

In addition, cardiac phenotype appeared to be transient. We demonstrated that transfer of myogenic clusters into liquid medium allowed culture expansion. However the cardiac protein expression diminished with increasing passage numbers leading to a complete loss of their expression and beating activity after nine passages, while cells expressing skeletal transcript markers continued until passage 15. These findings suggest that AD-CMG phenotype was linked to specific culture conditions and was lost with time and expansion rate in liquid culture conditions.

In order to characterize AD-CMG progenitors, we focused on large round multi-nucleated cells strictly expressing marker

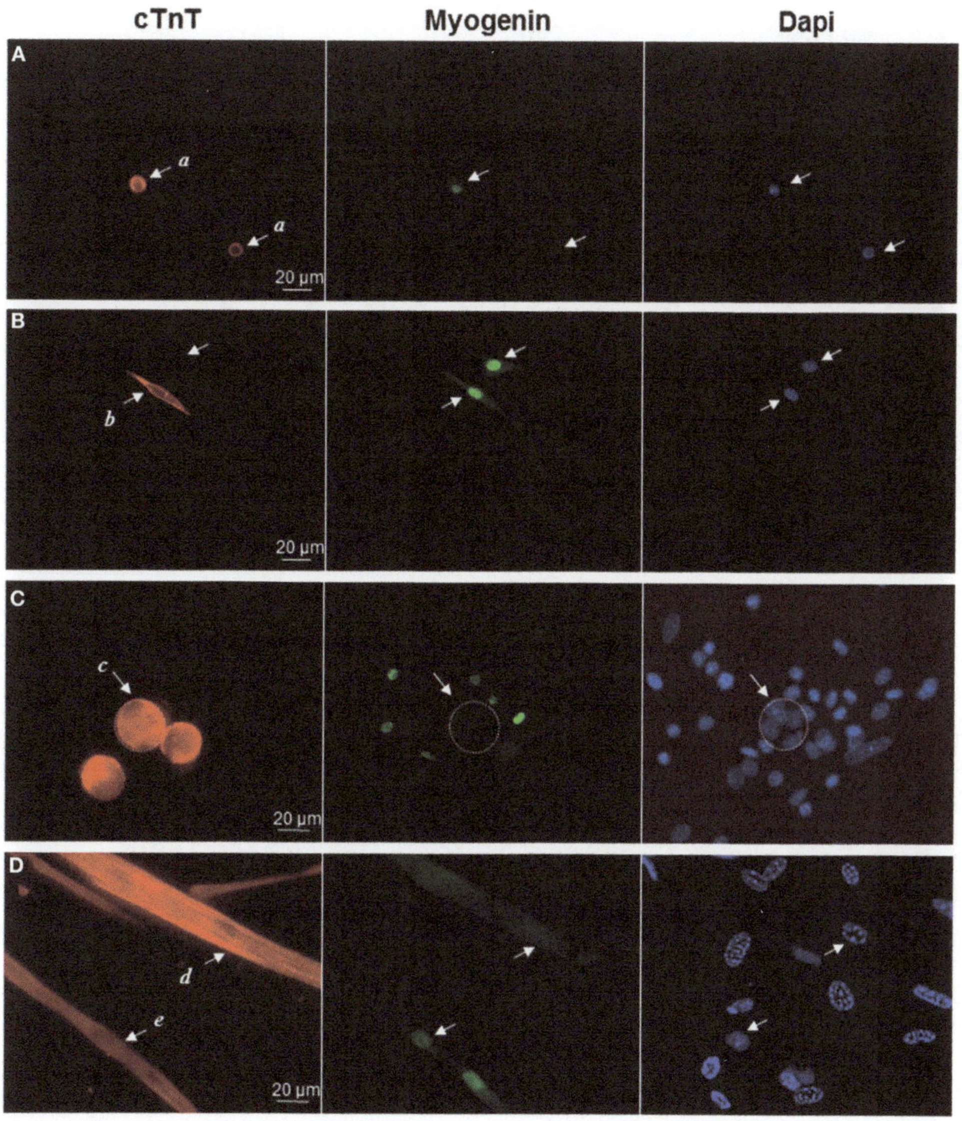

FIGURE 2 | Differential expression of cardiac and skeletal markers reflect cell heterogeneity in myogenic clusters. Immunolabeling of cardiac TnC (red) and myogenin (green) in **(A)** small round cells (*a* cells), **(B)** bipolar cells (*b* cells), **(C)** large round cells (*c* cells), **(D)** thick tube-shaped cells (*d* cells), and thin tube-shaped cells (*e* cells). Nuclei were stained with Dapi (blue).

of cardiomyocytes that emerged in semi-solid medium before myotube-like beating structures. We showed that multi-nucleation events in myogenic clusters resulted mostly from cell fusion events. Cell fusion is generally associated with skeletal muscle myoblasts forming myotubes but multinucleation in postnatal ventricular myocytes can occur by uncoupling karyokinesis from cytokinesis (Oh et al., 2003) and cardiomyocytes polyploidization was also described (Meckert et al., 2005; Liu et al., 2010), showing that multi-nucleation of AD-CMG (mature or not) is not incompatible with their cardiomyogeneity.

Distinct from hematopoietic stem cells (based on CD45+, CD117+) and endothelial progenitor cells (based on CD45−, CD34+, Flk−1+), large multi-nucleated AD-CMG progenitors displayed the following profile: CD45−/CD29+/Sca1−/CD44+/CD81+/CD133+/CD31−/CD34−/CD73−/CD90−/CD117−/Flk1−/CMH1−/CMH2−. They may be related to the muscle "Spoc" cells (Sca−1−,CD34−, CD45−, CD117−) (Winitsky et al., 2005), though the authors describe that no beating cells emerged when SVF cells were cultured under standard Spoc cell conditions. Moreover, AD-CMG progenitors differed from multipotential muscle cells (Sca−1+, CD34+) (Torrente et al., 2001) and resident cardiac stem cells (CD117+, Lin−) (Beltrami et al., 2003)/(CD117−, Sca1+) (Oh et al., 2003).

Previous studies could isolate an enriched cardiogenic progenitors population based on expression of CD133 in brown adipose tissue SVF among $CD45^{-}$/$ter119^{-}$/$CD31^{-}$ cells (non hematopoietic, non endothelial cells) (Yamada et al., 2007). They differentiated into cardiomyocytes with a higher incidence

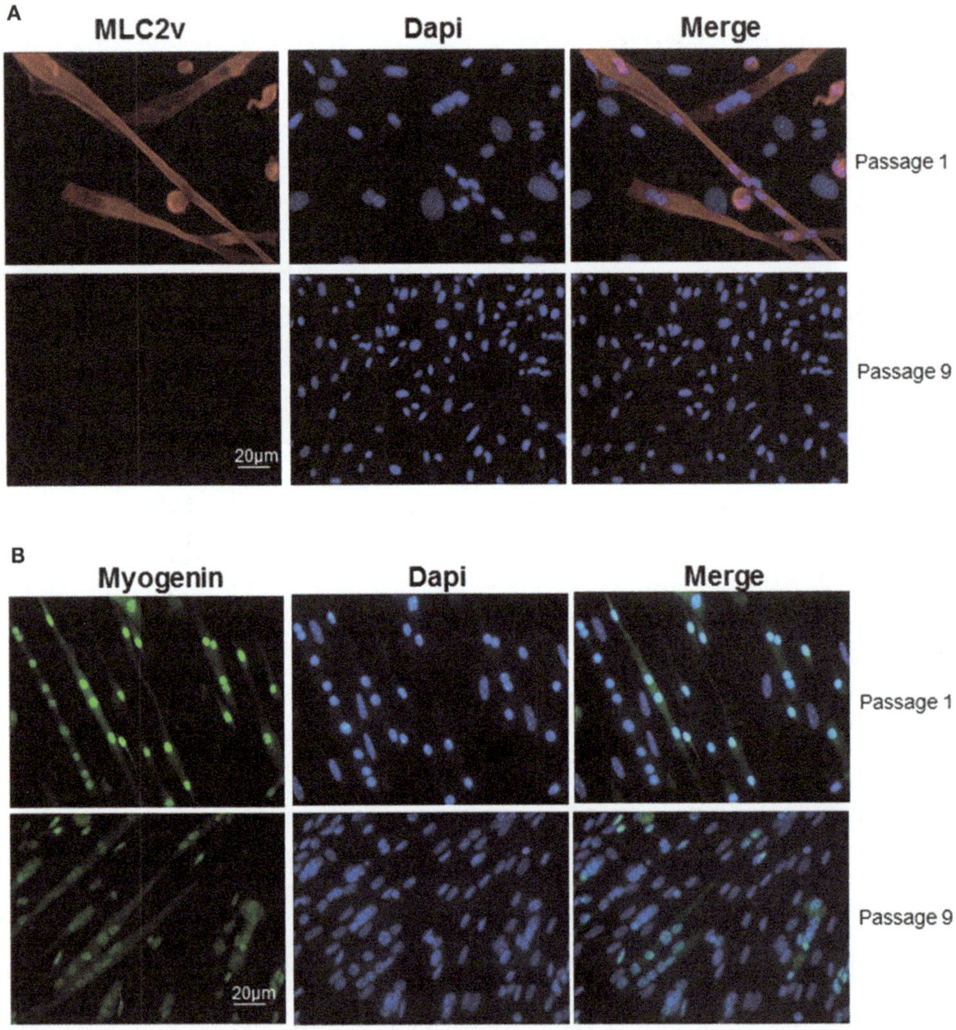

FIGURE 3 | Decline of cardiac/skeletal myogenic phenotype in liquid culture conditions. Immunolabeling of **(A)** myogenin (green) and **(B)** MLC2v (red) in myogenic clusters at passages 1 and 9 in liquid medium. Nuclei were stained with Dapi (blue).

than in other fractions and $CD29^+$ ($CD45^-$/$ter119^-$/$CD31^-$) cells, which had been previously described as a cardiomyogenic enriched population (Yamada et al., 2006). However, if sorted cells could differentiate into cardiomyocytes with a higher frequency than other fractions, the selected sub-population was not restricted to AD-CMG progenitors as >50% of them didn't generate cardiogenic cells.

A complete profile is still needed for purification but we showed that myogenic clusters emergence was largely impaired when AD-CMG progenitors were separated from supporting cells by cell sorting on the basis of markers such as CD44 CD81 or Sca1 or by clonal analysis, highlighting the important role of accessory cells for acquisition of cardiogenic phenotype. This requirement of sorted cells for accessory cells has been observed earlier: Zangiacomi et al showed that $CD133^+$ cord blood stem/progenitor cells required co-culture with $CD133^-$ cells to undergo neuronal differentiation (Zangiacomi et al., 2008).

This influence of surrounding cells on cardiogenic phenotype was underlined by our co-culture experiments showing that inguinal WAT SVF cells accelerated maturation of brown adipose tissue-derived cardiomyogenic cells. The fact that the emergence of myogenic phenotypes depends on time and culture conditions, raised some concerns about the role of the *in vitro* microenvironment regarding to secreted factors and cell contacts. According to our study, the myogenic potential of some adipose tissue cells requires surrounding cells and is silenced at the clonal level. The culture medium was optimized to contractile clusters formation however the subtle combination provided by the presence of cells couldn't be reproduced in the absence of supporting cells (data not shown). Thus, the initial step consisting in heterogenous SVF cells plating in semi-solid medium remains mandatory.

Requirement of surrounding cells evoked the stem cell "niche" concept described to condition and control stem cell fate. An abundant set of literature lead to the notion that adult stem

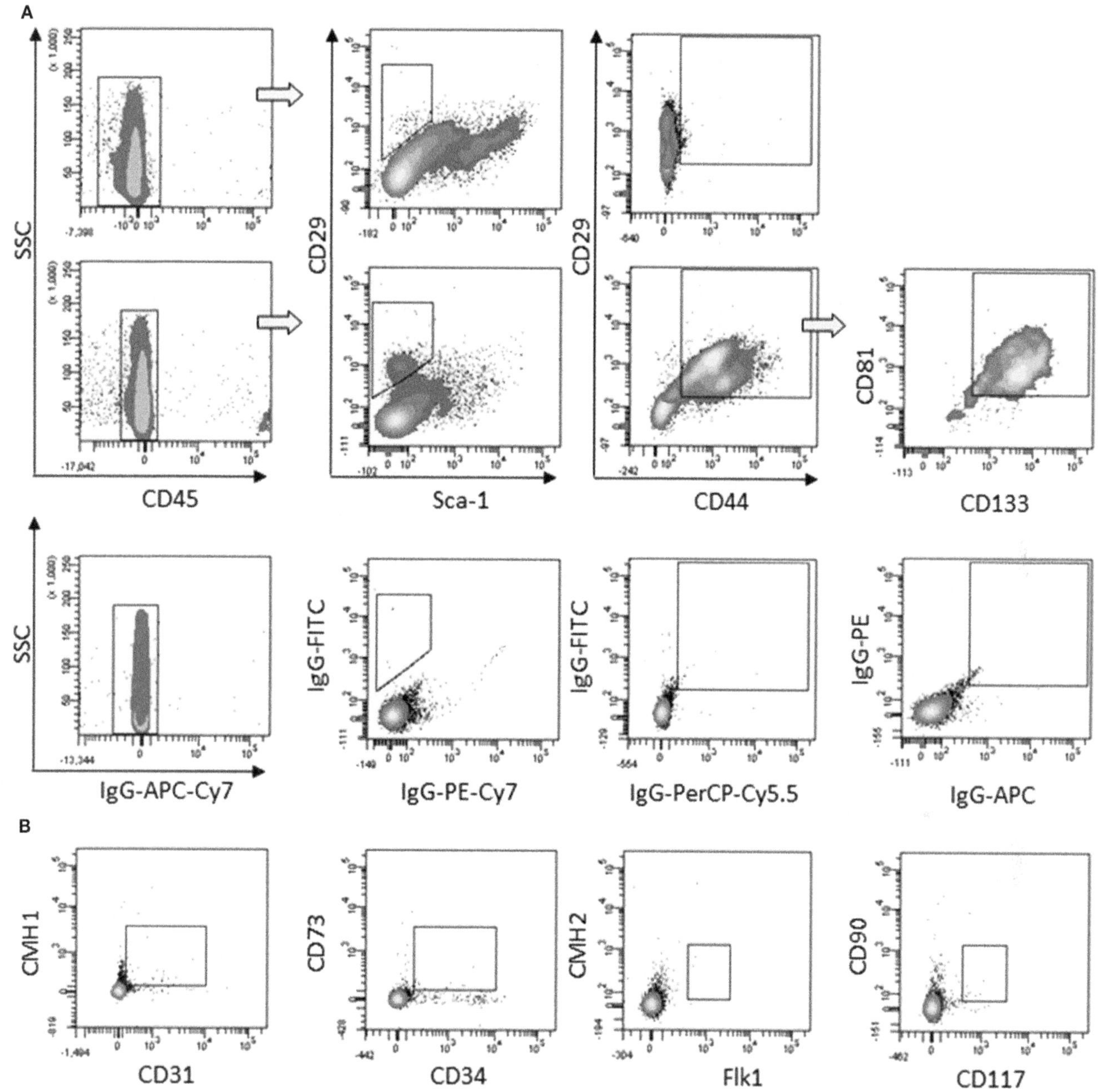

FIGURE 4 | Characterization of large round multi-nucleated cells (*c* cells). **(A)** Flow cytometry analysis of surrounding cells (upper panels) and myogenic clusters (medium panels) picked off 8 days after plating. Cells were gated on $CD45^-$ population and analyzed for CD29/Sca-1/CD44 expression. CD81/CD133 cells were gated on $CD29^+/CD44^+$ cells. Representative dot plots of specific (upper and medium panels) or isotype (lower panels) staining obtained in three separate experiments are shown. **(B)** Expression of CD31, CD34, CD73, CD90, CD117, CMH1, CMH2, and Flk-1 on $CD45^-$ gated cells among picked myogenic clusters 8 days after plating. Abbreviation: SSC, side scatter.

cells are actually highly plastic and have the ability to cross lineage boundaries (Clarke and Frisen, 2001). It could be hypothesized that apart the well described mesenchymal progenitors (ASC), adult adipose tissue contained distinct highly plastic cells with skeletal/cardiac myogenic potentials. Considering that cardiac and skeletal muscle cells do not developmentally originate from a common progenitor, the obtained phenotypes could rather result from gain of potency through fusion events and/or differential distribution of potency in their progeny through asymmetric division. Indeed, multinucleated cells in adipose-derived myogenic clusters resulting from fusion events were observed (Figure S2). In addition, based on Numb repartition we highlighted asymmetric divisions as early events that occurred in small round mono-nucleated cells before emergence of other cell phenotypes (Figure S3). Numb is known to be asymmetrically distributed between mother and progeny cells in myogenic

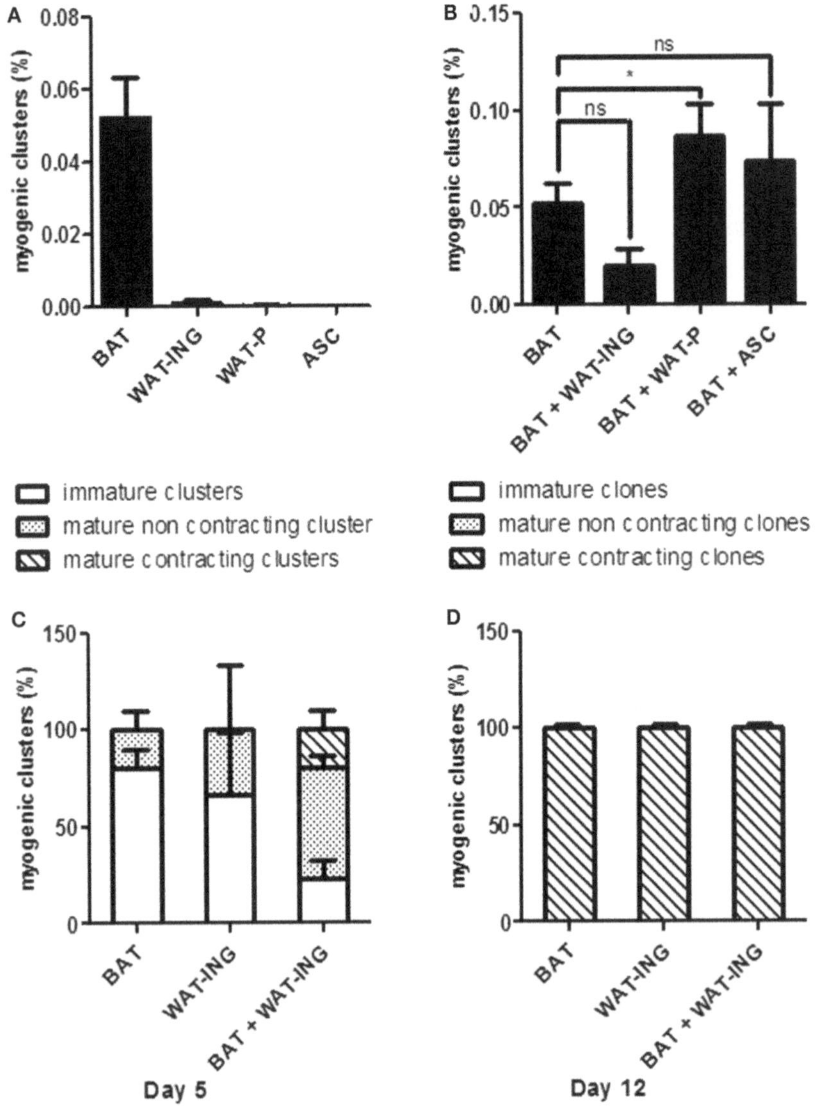

FIGURE 5 | Microenvironment influence on myogenic clusters emergence. (A) Quantification of myogenic clusters generated 8 days after independent plating of BAT, WAT-ING, and WAT-P SVF cells and inguinal pad derived ASC. Results are expressed as percentage of total plated cells. **(B)** Quantification of myogenic clusters generated 8 days after co-culture of BAT SVF cells with WAT-ING or WAT-P SVF cells or inguinal pad derived ASC. Results are expressed as percentage of total plated BAT SVF cells. **(C,D)** Quantification of immature clusters, mature but not contractile clusters and mature and contractile clusters, 5 days **(C)** and 12 days **(D)** post plating. Results are expressed as percentage myogenic clusters. Abbreviations: BAT, brown adipose tissue; WAT-ING, inguinal white adipose tissue; WAT-P: peritoneal white adipose tissue. $*p < 0.05$.

progenitors where it acts to block Notch-mediated repression of genes expressed in myogenic progenitor cells (Ruiz Gomez and Bate, 1997). In brown adipose tissue, asymmetric cell divisions may lead to a diversification of progenitors that would be subdivided to produce distinct populations of myogenic cells (likely cardiogenic and skeletal).

Much more beating cardiomyocytes are obtained from SVF of brown than inguinal adipose tissue. This can be linked to the vicinity of interscapular brown adipose tissue with muscle and to a common developmental origin. Indeed, Seale et al. described a common progenitor for brown fat and skeletal muscle (Seale et al., 2008). Thus, generation of cardiomyogenic cells from inguinal SVF is coherent with the emergence of ectopic brown adipocytes in such white adipose depots during cold acclimation in animals (Cousin et al., 1996) and it seems now very likely that, at least in some WAT pads, brown adipocytes can emerge from differentiation of brown preadipocytes or transdifferentiation of existing white adipocytes (for review, see Boss and Farmer, 2012).

Taken together, our data show that some non-adherent cells from adipose tissue SVF, when trapped in semi-solid culture medium undergo fusion and asymmetric division events leading to a cardiomyogenic phenotype. Cardiomyogenic phenotype can be influenced by accessory cells present in adipose tissue and may progressively be lost over cell expansion, suggesting that cardiogenic differentiation of such adipose-derived cells is merely acquired in culture and not a reflection of native biology.

ACKNOWLEDGMENTS

We thank Pascale Guillou for technical assistance in fat pads removal. This work was supported by the Etablissement Français du Sang – Pyrénées Méditerranée. Sandra Berger-Müller was financially supported by SANOFI.

REFERENCES

Askari, A. T., Unzek, S., Popovic, Z. B., Goldman, C. K., Forudi, F., Kiedrowski, M., et al. (2003). Effect of stromal-cell-derived factor 1 on stem-cell homing and tissue regeneration in ischaemic cardiomyopathy. *Lancet* 362, 697–703. doi: 10.1016/S0140-6736(03)14232-8

Beltrami, A. P., Barlucchi, L., Torella, D., Baker, M., Limana, F., Chimenti, S., et al. (2003). Adult cardiac stem cells are multipotent and support myocardial regeneration. *Cell* 114, 763–776. doi: 10.1016/S0092-8674(03)00687-1

Bolli, R., Chugh, A. R., D'Amario, D., Loughran, J. H., Stoddard, M. F., Ikram, S., et al. (2011). Cardiac stem cells in patients with ischaemic cardiomyopathy (SCIPIO): initial results of a randomised phase 1 trial. *Lancet* 378, 1847–1857. doi: 10.1016/S0140-6736(11)61590-0

Boss, O., and Farmer, S. R. (2012). Recruitment of brown adipose tissue as a therapy for obesity-associated diseases. *Front. Endocrinol. (Lausanne)* 3:14. doi: 10.3389/fendo.2012.00014

Chen, J., Kubalak, S. W., and Chien, K. R. (1998). Ventricular muscle-restricted targeting of the RXRalpha gene reveals a non-cell-autonomous requirement in cardiac chamber morphogenesis. *Development* 125, 1943–1949.

Clarke, D., and Frisen, J. (2001). Differentiation potential of adult stem cells. *Curr. Opin. Genet. Dev.* 11, 575–580. doi: 10.1016/S0959-437X(00)00235-5

Cousin, B., Bascands-Viguerie, N., Kassis, N., Nibbelink, M., Ambid, L., Casteilla, L., et al. (1996). Cellular changes during cold acclimatation in adipose tissues. *J. Cell Physiol.* 167, 285–289.

Khodadadi, I. J., Plant, N., Mersinias, V., and Thumser, A. (2010). Applicability of the P19CL6 cells as a model of cardiomyocytes – a transcriptome analysis. *Health* 2, 8. doi: 10.4236/health.2010.21005

Laugwitz, K. L., Moretti, A., Lam, J., Gruber, P., Chen, Y., Woodard, S., et al. (2005). Postnatal isl1+ cardioblasts enter fully differentiated cardiomyocyte lineages. *Nature* 433, 647–653. doi: 10.1038/nature03215

Le Bihan, M. C., Hou, Y., Harris, N., Tarelli, E., and Coulton, G. R. (2006). Proteomic analysis of fast and slow muscles from normal and kyphoscoliotic mice using protein arrays, 2-DE and MS. *Proteomics* 6, 4646–4661. doi: 10.1002/pmic.200500746

Leobon, B., Roncalli, J., Joffre, C., Mazo, M., Boisson, M., Barreau, C., et al. (2009). Adipose-derived cardiomyogenic cells: *in vitro* expansion and functional improvement in a mouse model of myocardial infarction. *Cardiovasc. Res.* 83, 757–767. doi: 10.1093/cvr/cvp167

Liu, Z., Yue, S., Chen, X., Kubin, T., and Braun, T. (2010). Regulation of cardiomyocyte polyploidy and multinucleation by CyclinG1. *Circ. Res.* 106, 1498–1506. doi: 10.1161/CIRCRESAHA.109.211888

Martin, C. M., Meeson, A. P., Robertson, S. M., Hawke, T. J., Richardson, J. A., Bates, S., et al. (2004). Persistent expression of the ATP-binding cassette transporter, Abcg2, identifies cardiac SP cells in the developing and adult heart. *Dev. Biol.* 265, 262–275. doi: 10.1016/j.ydbio.2003.09.028

Matsuura, K., Nagai, T., Nishigaki, N., Oyama, T., Nishi, J., Wada, H., et al. (2004). Adult cardiac Sca-1-positive cells differentiate into beating cardiomyocytes. *J. Biol. Chem.* 279, 11384–11391. doi: 10.1074/jbc.M310822200

Meckert, P. C., Rivello, H. G., Vigliano, C., Gonzalez, P., Favaloro, R., and Laguens, R. (2005). Endomitosis and polyploidization of myocardial cells in the periphery of human acute myocardial infarction. *Cardiovasc. Res.* 67, 116–123. doi: 10.1016/j.cardiores.2005.02.017

Menasche, P. (2011). Stem cell therapy for chronic heart failure: lessons from a 15-year experience. *C. R. Biol* 334, 489–496. doi: 10.1016/j.crvi.2011.03.006

Messina, E., De Angelis, L., Frati, G., Morrone, S., Chimenti, S., Fiordaliso, F., et al. (2004). Isolation and expansion of adult cardiac stem cells from human and murine heart. *Circ. Res.* 95, 911–921. doi: 10.1161/01.RES.0000147315.71699.51

Nabauer, M., Callewaert, G., Cleemann, L., and Morad, M. (1989). Regulation of calcium release is gated by calcium current, not gating charge, in cardiac myocytes. *Science* 244, 800–803. doi: 10.1126/science.2543067

Oh, H., Bradfute, S. B., Gallardo, T. D., Nakamura, T., Gaussin, V., Mishina, Y., et al. (2003). Cardiac progenitor cells from adult myocardium: homing, differentiation, and fusion after infarction. *Proc. Natl. Acad. Sci. U.S.A.* 100, 12313–12318. doi: 10.1073/pnas.2132126100

Okabe, M., Ikawa, M., Kominami, K., Nakanishi, T., and Nishimune, Y. (1997). 'Green mice' as a source of ubiquitous green cells. *FEBS Lett.* 407, 313–319. doi: 10.1016/S0014-5793(97)00313-X

Planat-Benard, V., Menard, C., Andre, M., Puceat, M., Perez, A., Garcia-Verdugo, J. M., et al. (2004). Spontaneous cardiomyocyte differentiation from adipose tissue stroma cells. *Circ. Res.* 94, 223–229. doi: 10.1161/01.RES.0000109792.43271.47

Potthoff, M. J., Arnold, M. A., McAnally, J., Richardson, J. A., Bassel-Duby, R., and Olson, E. N. (2007). Regulation of skeletal muscle sarcomere integrity and postnatal muscle function by Mef2c. *Mol. Cell. Biol.* 27, 8143–8151. doi: 10.1128/MCB.01187-07

Reinecke, H., Poppa, V., and Murry, C. E. (2002). Skeletal muscle stem cells do not transdifferentiate into cardiomyocytes after cardiac grafting. *J. Mol. Cell. Cardiol.* 34, 241–249. doi: 10.1006/jmcc.2001.1507

Ruiz Gomez, M., and Bate, M. (1997). Segregation of myogenic lineages in Drosophila requires numb. *Development* 124, 4857–4866.

Scherschel, J. A., Soonpaa, M. H., Srour, E. F., Field, L. J., and Rubart, M. (2008). Adult bone marrow-derived cells do not acquire functional attributes of cardiomyocytes when transplanted into peri-infarct myocardium. *Mol. Ther.* 16, 1129–1137. doi: 10.1038/mt.2008.64

Seale, P., Bjork, B., Yang, W., Kajimura, S., Chin, S., Kuang, S., et al. (2008). PRDM16 controls a brown fat/skeletal muscle switch. *Nature* 454, 961–967. doi: 10.1038/nature07182

Torrente, Y., Tremblay, J. P., Pisati, F., Belicchi, M., Rossi, B., Sironi, M., et al. (2001). Intraarterial injection of muscle-derived CD34(+)Sca-1(+) stem cells restores dystrophin in mdx mice. *J. Cell Biol.* 152, 335–348. doi: 10.1083/jcb.152.2.335

Winitsky, S. O., Gopal, T. V., Hassanzadeh, S., Takahashi, H., Gryder, D., Rogawski, M. A., et al. (2005). Adult murine skeletal muscle contains cells that can differentiate into beating cardiomyocytes *in vitro*. *PLoS Biol* 3:e87. doi: 10.1371/journal.pbio.0030087

Yamada, Y., Wang, X. D., Yokoyama, S., Fukuda, N., and Takakura, N. (2006). Cardiac progenitor cells in brown adipose tissue repaired damaged myocardium. *Biochem. Biophys. Res. Commun.* 342, 662–670. doi: 10.1016/j.bbrc.2006.01.181

Yamada, Y., Yokoyama, S., Wang, X. D., Fukuda, N., and Takakura, N. (2007). Cardiac stem cells in brown adipose tissue express CD133 and induce bone marrow nonhematopoietic cells to differentiate into cardiomyocytes. *Stem Cells* 25, 1326–1333. doi: 10.1634/stemcells.2006-0588

Zangiacomi, V., Balon, N., Maddens, S., Lapierre, V., Tiberghien, P., Schlichter, R., et al. (2008). Cord blood-derived neurons are originated from CD133+/CD34 stem/progenitor cells in a cell-to-cell contact dependent manner. *Stem Cells Dev.* 17, 1005–1016. doi: 10.1089/scd.2007.0248

Conflict of Interest Statement: The authors declare that the research was conducted in the absence of any commercial or financial relationships that could be construed as a potential conflict of interest.

26

Lipid rafts: linking prion protein to zinc transport and amyloid-β toxicity in Alzheimer's disease

Nicole T. Watt[1], Heledd H. Griffiths[2] and Nigel M. Hooper[2]*

[1] Division of Cardiovascular and Diabetes Research, Multidisciplinary Cardiovascular Research Centre, University of Leeds, Leeds, UK
[2] Faculty of Medical and Human Sciences, Institute of Brain, Behaviour and Mental Health, University of Manchester, Manchester, UK

Edited by:
Sophie Mouillet-Richard, INSERM Unit 1124, France

Reviewed by:
NADIA ALFAIDY, INSERM, France
Gerold Schmitt-Ulms, University of Toronto, Canada

***Correspondence:**
Nigel M. Hooper, Faculty of Medical and Human Sciences, Institute of Brain, Behaviour and Mental Health, University of Manchester, Rm 2.015 AV Hill Building, Oxford Road, Manchester M13 9PT, UK
e-mail: nigel.hooper@manchester.ac.uk

Dysregulation of neuronal zinc homeostasis plays a major role in many processes related to brain aging and neurodegenerative diseases, including Alzheimer's disease (AD). Yet, despite the critical role of zinc in neuronal function, the cellular mechanisms underpinning its homeostatic control are far from clear. We reported that the cellular prion protein (PrP^C) is involved in the uptake of zinc into neurons. This PrP^C-mediated zinc influx required the metal-binding octapeptide repeats in PrP^C and the presence of the zinc permeable AMPA channel with which PrP^C directly interacted. Together with the observation that PrP^C is evolutionarily related to the ZIP family of zinc transporters, these studies indicate that PrP^C plays a key role in neuronal zinc homeostasis. Therefore, PrP^C could contribute to cognitive health and protect against age-related zinc dyshomeostasis but PrP^C has also been identified as a receptor for amyloid-β oligomers which accumulate in the brains of those with AD. We propose that the different roles that PrP^C has are due to its interaction with different ligands and/or co-receptors in lipid raft-based signaling/transport complexes.

Keywords: Alzheimer's disease, AMPA receptor, amyloid, cholesterol, prion protein, zinc

ZINC HOMEOSTASIS AND ALZHEIMER'S DISEASE

Zinc is a trace element that is essential for life and whose importance to the function of the central nervous system is increasingly being appreciated. Zinc serves as a cofactor for >300 enzymes that regulate a variety of cellular processes and signaling pathways, and is also a key structural component of numerous other proteins (Frederickson et al., 2005; Sensi et al., 2009). In the brain, which has one of the highest zinc contents with respect to other organs, zinc-containing axons are particularly abundant in the hippocampus and cortex (Toth, 2011). Zinc is predominantly, but not exclusively, localized within synaptic vesicles at glutamatergic nerve terminals [sometimes referred to as gluzinergic neurons (Mocchegiani et al., 2005)]. During neuronal activity, zinc is released along with glutamate into the synaptic cleft where it affects the activity of various receptors. In addition, zinc is itself a signaling molecule, and within the neuron it regulates the activity of multiple enzymes and plays a critical role in the formation and stabilization of the postsynaptic density (Beyersmann and Haase, 2001; Grabrucker et al., 2011; Wilson et al., 2012). Both intracellular and extracellular zinc concentrations must be tightly maintained within narrow optimal ranges for the correct functioning of the nervous system. An excess influx of zinc can damage postsynaptic neurons (Plum et al., 2010) and zinc deficiency affects neurogenesis and increases neuronal apoptosis, which can lead to learning and memory deficits (Szewczyk, 2013). Under normal circumstances, zinc homeostasis is maintained by the coordinated actions of a range of different proteins involved in its uptake, efflux, and intracellular storage and trafficking. Zinc enters neurons through members of the ZIP (Zrt/Irt-like Protein; SLC39) family of zinc transporters, as well as through activated voltage-gated Ca^{2+} channels, α-amino-3-hydroxy-5-methyl-4-isoxazoleproprionate (AMPA) receptors, and N-methyl-D-aspartate (NMDA) receptors (Cousins et al., 2006; Sensi et al., 2009). Zinc exporters, members of the zinc transporter (ZnT; SLC30) family transport zinc from the cytosol to the lumen of intracellular organelles or out of the cell (Sensi et al., 2009).

Alzheimer's disease (AD) is the most prevalent form of dementia, affecting millions of individuals world-wide. With currently no cure, the economic and social costs associated with the disease are set to increase dramatically with our aging population. AD is characterized by the deposition in the brain of extracellular plaques of amyloid-β (Aβ) peptide and intracellular inclusions of tau protein. Aβ is proteolytically cleaved from the larger amyloid precursor protein (APP), and both Aβ and APP have binding sites for zinc (Watt et al., 2010; Wong et al., 2014). Zinc, particularly that released from glutamatergic nerve terminals, has a crucial role to play in the aggregation of Aβ into neurotoxic oligomers and fibrils (Bush et al., 1994; Esler et al., 1996; Deshpande et al., 2009), and is also co-localized with Aβ in amyloid plaques (Dong et al., 2003). The activity of the zinc transporter ZnT-3 is required for the transport of zinc into the glutamate-containing presynaptic vesicles. Key evidence for a link between zinc and amyloid pathology in AD comes from the crossing of ZnT-3 knockout mice with Tg2576 mice, a commonly used transgenic mouse model of AD. The crossed mice had minimal synaptic zinc and as a consequence both brain plaque load and amyloid angiopathy were significantly reduced (Lee et al., 2002; Friedlich et al., 2004). Furthermore, in a separate study of aged ZnT-3 knockout mice, there were marked differences in learning and memory observed compared to wild type mice, supporting a requirement for zinc in memory function and the maintenance of synaptic health upon aging (Adlard et al., 2010).

PrP^{C} AND NEURONAL ZINC UPTAKE

Recently we reported that PrP^{C} facilitates the uptake of zinc into neuronal cells (Watt et al., 2012). PrP^{C} is infamous because of its conformational conversion into PrP^{Sc} being responsible for the fatal transmissible spongiform encephalopathies, such as Creutzfeldt-Jakob disease. PrP^{C} is a glycosyl-phosphatidylinositol (GPI)-anchored protein located on the surface of neurons, at both pre- and post-synaptic sites. It is found throughout the central nervous system and is particularly abundant in the hippocampus and frontal cortex (Sales et al., 1998). Using two zinc specific fluorescent dyes (Newport green and Zinpyr-1) we showed that PrP^{C} enhanced the uptake of zinc into human SH-SY5Y neuroblastoma cells and rat primary hippocampal neurons (Watt et al., 2012). This PrP^{C}-mediated zinc influx required the metal-binding octapeptide repeat region in PrP^{C} but not its endocytosis. We then used selective channel antagonists to identify that AMPA receptors were involved in the PrP^{C}-mediated zinc uptake, and that PrP^{C} interacted with both GluA1 and GluA2 AMPA receptor subunits as shown by co-immunoprecipitation. Intracellular protein tyrosine phosphatase activity, which is potently inhibited by zinc (Brautigan et al., 1981; Wilson et al., 2012), was increased in the brains of PrP^{C} null mice, providing evidence of a physiological consequence of altered zinc uptake in the absence of PrP^{C}. These observations provided the first mechanistic explanation for the reduced zinc in the hippocampus and other brain regions of PrP^{C} null mice (Pushie et al., 2011) and indicate that PrP^{C} is a key player in neuronal zinc uptake.

Interestingly, during the course of our work showing that PrP^{C} mediates neuronal zinc uptake, the protein was reported to be evolutionarily related to a subset of the ZIP family of zinc transporters (Schmitt-Ulms et al., 2009). Bioinformatic analysis revealed that the N-terminal extracellular domain of a distinct sub-branch of the LIV-1 subfamily of ZIPs that includes ZIPs 5, 6, and 10 had sequence similarity to the C-terminal globular domain of PrP^{C}. These three ZIPs are also equipped with histidine-rich sequences N-terminal to their "prion-like" domain, capable of divalent metal binding, which is reminiscent of the octapeptide repeat domain of PrP^{C} (**Figure 1A**). Interestingly, the orientation and distance of the "prion-like" domain to the respective membrane attachment sites in both PrP^{C} and the ZIPs are similar (**Figure 1A**), and the primary sequence of the first transmembrane domain in the ZIPs and the GPI anchor attachment sequence of PrP^{C} are also comparable, providing further evidence that PrP^{C} is evolutionarily related to members of the ZIP family (Schmitt-Ulms et al., 2009).

As PrP^{C} is a GPI-anchored protein residing in the outer leaflet of the plasma membrane, in order to correlate a function for PrP^{C} in the transport of zinc across the lipid bilayer, we proposed that PrP^{C} requires the assistance of the channel properties of another protein(s), such as AMPA receptors, in order to achieve zinc transport (**Figure 1B**) (Watt et al., 2012). The octapeptide repeats in the GPI-anchored PrP^{C} act as a zinc sensor/scavenger in the extracellular environment which then presents the metal to the transmembrane channel for uptake of the zinc into the cell (Watt et al., 2013). Interestingly, not all ZIP family members have an extracellular N-terminal extension that may function in zinc sensing: for example ZIPs 1, 2, 3, 9, and 11 have minimal N-terminal extensions on the first transmembrane domain. This raises the intriguing possibility that the two properties (zinc

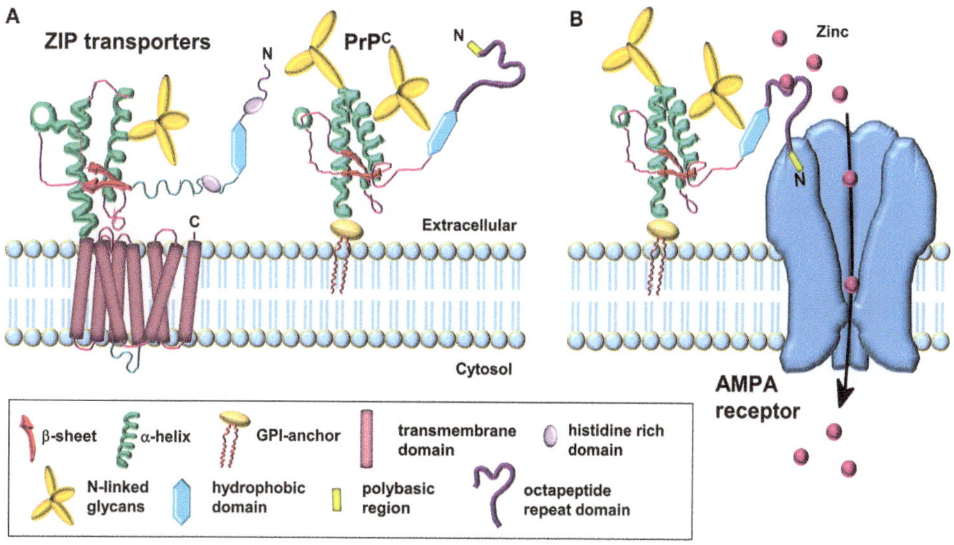

FIGURE 1 | Schematic comparison of ZIP transporters and the PrP^{C}-AMPA receptor zinc transporter. (A) Model of the LIV-1 subfamily of ZIPs with a prion-like ectodomain which is coupled to the C-terminal transmembrane domain. PrP^{C} is structurally similar to the ectodomain of the ZIP transporter. **(B)** PrP^{C} acts as a sensor for zinc in the extracellular space and coordinates the low affinity binding of zinc to the octapeptide repeats. The N-terminal polybasic region of PrP^{C} interacts with the AMPA receptor subunits; this interaction facilitates the transport of zinc through the AMPA receptor which forms a channel across the membrane for the uptake of zinc in a manner similar to the C-terminal region of the ZIP transporter. Modified from Watt et al. (2013).

sensing and zinc channel) present together in certain ZIPs (e.g., ZIPs 6 and 10) reside in separate proteins in those ZIPs with minimal N-terminal extensions. It will be interesting to see if PrP^C acts as the zinc sensor in combination with certain ZIPs, similar to the situation we have proposed for PrP^C and the AMPA receptor (**Figure 1B**) (Watt et al., 2012, 2013). Such PrP^C-based zinc sensing modules may be restricted to higher eukaryotes in which PrP^C with the octapeptide metal binding repeats is present.

In addition to its role in zinc metabolism, PrP^C also plays a role in the homeostasis of other metals such as iron and copper (Kozlowski et al., 2010; Singh et al., 2014). PrP^C expression alters regional zinc, copper and iron content in the mouse brain (Pushie et al., 2011) and there appears to be crosstalk between the different metals, with zinc modulating copper coordination to the octapeptide repeats (Stellato et al., 2011). In addition, PrP^C modulates NMDA receptor activity in a copper-dependent manner (Stys et al., 2012), suggesting that multiple receptors/channels may be regulated by PrP^C in a ligand/metal-dependent manner.

A CELL-SURFACE, LIPID RAFT-BASED COMPLEX INVOLVED IN THE REGULATION OF ZINC UPTAKE

The GPI-anchored PrP^C is localized in cholesterol-rich, detergent-resistant lipid rafts at the cell surface (Taylor et al., 2005) and it has been proposed that PrP^C functions as a key scaffolding protein for the dynamic assembly of cell surface signaling modules (Linden et al., 2008). PrP^C, along with the microdomain-forming flotillin or caveolin proteins, may lead to the local assembly of membrane protein complexes at sites involved in cellular communication, such as cell-cell contacts, focal adhesions, the T-cell cap and synapses (Solis et al., 2010). Lipid rafts are essential for synapse development, stabilization and maintenance and caveolin-1 organizes and targets synaptic components to rafts (Hering et al., 2003; Willmann et al., 2006; Guirland and Zheng, 2007). Interestingly, when ZIP1 and ZIP3 were stably expressed in HEK293 cells, the punctate cell surface staining observed led the authors to suggest that they were localized to lipid rafts (although no experimental evidence for this was provided) (Wang et al., 2004). Furthermore, treatment of the HEK293 cells with methyl-β-cyclodextrin, which sequesters cholesterol and disrupts cholesterol-rich lipid rafts, resulted in a more diffuse surface staining (Wang et al., 2004), reminiscent of what is observed for PrP^C and other GPI-anchored proteins (Parkin et al., 2003; Taylor et al., 2005). We also have disrupted rafts in neuronal cells using methyl-β-cyclodextrin and observed a reduction in zinc uptake, an effect exacerbated when the cells also expressed PrP^C (Watt and Hooper, unpublished). In contrast, it has been reported that ZIP10 only partially colocalizes with PrP^C in N2a cells and is not detected in detergent-resistant rafts (Ehsani et al., 2012). These data raise the intriguing possibility that a cell-surface, lipid raft-based complex, possibly stabilized by PrP^C, regulates the cell surface expression of certain zinc transporters and thus zinc uptake.

PrP^C IS A CELL SURFACE RECEPTOR FOR Aβ OLIGOMERS

In 2009, PrP^C was identified as a high-affinity receptor for Aβ oligomers, the primary neurotoxic species in AD (Lauren et al., 2009). The presence of PrP^C in hippocampal slices was shown to be responsible for the Aβ oligomer-mediated inhibition of long-term potentiation (LTP) (Lauren et al., 2009). PrP^C was also required for the manifestation of memory impairments in an AD mouse model (Gimbel et al., 2010), which were reversed by intracerebral infusion of an anti-PrP^C monoclonal antibody (Chung et al., 2010). Immuno-targeting of PrP^C was shown to block completely the LTP impairments caused by Aβ oligomers derived from human AD brain extracts (Barry et al., 2011; Freir et al., 2011). Aβ oligomers bound to PrP^C activate the non-receptor tyrosine kinase Fyn (Um et al., 2012; Rushworth et al., 2013) and results in pathological changes in tau (Larson et al., 2012). Although there is general consensus that PrP^C can bind oligomeric forms of Aβ, some studies dispute a role for PrP^C in mediating Aβ toxicity (Calella et al., 2010; Kessels et al., 2010).

It has been hypothesized that a putative transmembrane co-receptor is required to connect the binding of Aβ to the GPI-anchored PrP^C on the outer surface of the plasma membrane with downstream effects inside the cell (Cisse and Mucke, 2009; Lauren et al., 2009). The transmembrane low-density lipoprotein-receptor related protein-1 (LRP1) is highly expressed in neuronal cells (Nykjaer and Willnow, 2002), facilitates the endocytosis of PrP^C (Taylor and Hooper, 2007) and has been implicated in the neuronal uptake of Aβ (Fuentealba et al., 2010; Kanekiyo et al., 2011). We hypothesized, therefore, that LRP1 may play a role in the PrP^C-mediated action of Aβ oligomers (**Figure 2**). We showed that LRP1 is required for the binding of Aβ oligomers to cells, as well as for their subsequent internalization and cytotoxicity (Rushworth et al., 2013). In parallel with our work on LRP1, Strittmatter and colleagues identified the metabotropic glutamate receptor, mGluR5, as a co-receptor with PrP^C for Aβ oligomers and was required to activate Fyn (Um et al., 2013). Furthermore, antagonists of mGluR5 reversed the deficits in learning, memory and synapse density in AD transgenic mice (Um et al., 2013). In addition, the Aβ-PrP^C-mGluR5 interplay is involved in mediating both long-term depression facilitation and LTP inhibition (Hu et al., 2014). Thus, both LRP1 and mGluR5 appear to be transmembrane co-receptors in the Aβ-PrP^C interaction, which are required for the PrP^C-mediated action of Aβ oligomers. Whether both LRP1 and mGluR5 reside in the same complex with PrP^C or in separate signaling complexes is currently unclear.

A CELL-SURFACE, LIPID RAFT-BASED SIGNALING COMPLEX INVOLVED IN Aβ OLIGOMER ACTION

Cell surface cholesterol-rich, detergent-resistant lipid rafts are intimately involved in the production, aggregation and toxicity of Aβ (Rushworth and Hooper, 2010; Di Paolo and Kim, 2011). For example, these membrane microdomains have been linked with Aβ toxicity via Fyn, and rafts are involved in the neuronal internalization of Aβ (Williamson et al., 2008; Lai and McLaurin, 2010). We investigated whether the integrity of lipid rafts is required for the binding of Aβ oligomers and the subsequent activation of Fyn. Treatment of cells with methyl-β-cyclodextrin caused the re-localization of PrP^C and Fyn from detergent-resistant rafts to detergent-soluble, non-raft regions of the membrane as analyzed by sucrose density gradient centrifugation in the presence of Triton X-100. Surprisingly, disruption of the rafts with methyl-β-cyclodextrin significantly reduced (by

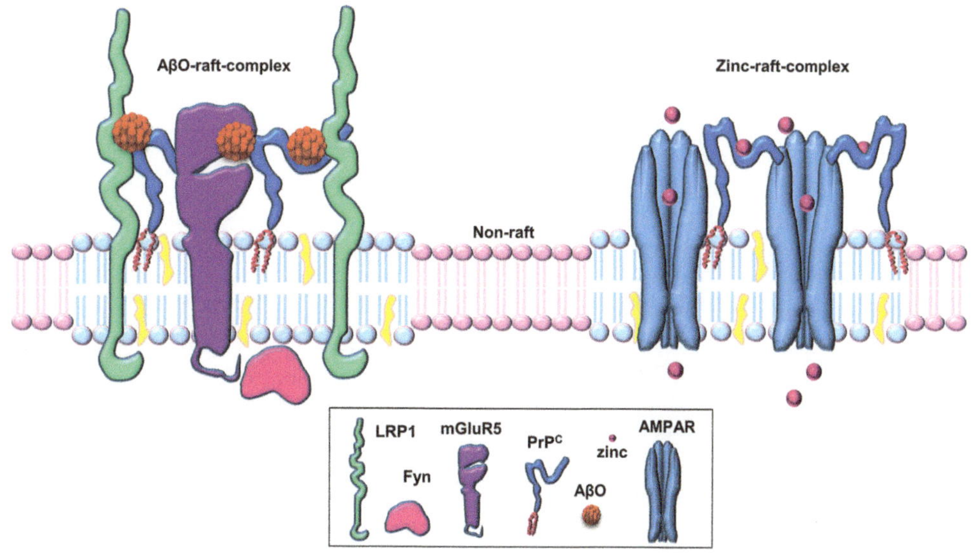

FIGURE 2 | Schematic of PrP^C-based cell-surface, raft-based complexes involved in zinc uptake and Aβ binding. PrP^C acts as a hub for cell surface, lipid raft-based signaling/transport complexes. PrP^C associates with various transmembrane proteins (such as AMPA receptors, LRP1, mGluR5, etc.) in these multi-protein complexes in a ligand-dependent manner (here shown for the ligands zinc and Aβ oligomers within separate raft-based complexes).

80.6%) the cell surface binding of the Aβ oligomers, even though the cell surface expression of PrP^C was unaffected, and prevented the Aβ oligomers from activating Fyn (Rushworth et al., 2013). Thus, raft localization of PrP^C is required for the binding of Aβ oligomers, and the integrity of rafts and/or other raft-localized proteins are required for the Aβ-mediated activation of Fyn. These observations suggest that there is a cell-surface, raft-based signaling complex that is key to the binding of Aβ oligomers and the subsequent generation of cellular responses.

Aβ oligomers caused mGluR5 receptors to manifest reduced lateral diffusion as they became aberrantly clustered (Renner et al., 2010). Antibodies against PrP^C and the NR1 subunit of NMDA receptors had a similar inhibitory effect on Aβ oligomer binding as the antibody against mGluR5, but there was not an additive effect of the individual antibodies, suggesting that mGluR5, PrP^C, and NR1 may be in proximity to each other as well as the Aβ oligomer binding site (Renner et al., 2010). Interestingly, addition of Aβ oligomers to neurons caused a large increase of mGluR5 in the Triton-resistant fraction (Renner et al., 2010), although no link to its possible redistribution into detergent-resistant lipid rafts was made. In another study, disruption of lipid rafts by cholesterol depletion reduced the interaction of Aβ with α-7-nicotinic acetylcholine receptors (Khan et al., 2010). These observations are consistent with multiple (co)receptor proteins for Aβ oligomers residing in cell-surface, raft-based complex(es).

PrP^C AND RAFTS ARE DYSREGULATED IN AGING

Zinc homeostasis plays a major role in many processes related to brain aging. For example, mouse models of accelerated aging, such as senescence-accelerated mouse prone 10 (SAMP10) mice display a low total zinc concentration in synaptic vesicles that is associated with brain atrophy and defects in learning and memory (Saito et al., 2000), and dietary zinc deficiency influences hippocampal learning and memory in an age-dependent manner (Mocchegiani et al., 2005; Szewczyk, 2013). ZIP6 expression has been reported to be decreased in an age-dependent manner (Wong et al., 2013), although nothing has been reported about the expression, subcellular localization or function of any other ZIP in the aged brain. Caveolin-1 has been identified as a novel control point for healthy neuronal aging, with the localization of caveolin-1, PSD95, and AMPA receptors in lipid rafts being decreased in aged (> 18 months) mice compared with young (3–6 month) mice (Head et al., 2010). Furthermore, in an aging series (age 20–88 years) of human brains we reported that PrP^C was reduced in the hippocampus with increasing age (Whitehouse et al., 2010). Preliminary data indicate that there is a significant reduction of both ZIP1 and ZIP3 in the hippocampus of old (79–88 years) compared with young (20–26 years) individuals (Watt and Hooper, unpublished). These observations suggest that alterations to the structure and function of the cell-surface, raft-based zinc transporter complex may occur in aging and contribute to the age-related dysregulated zinc homeostasis.

Aging is the greatest risk factor for AD and recently the REST protein was reported to have a central role in protecting aging neurons from degeneration (Lu et al., 2014). There is a close relationship between lipid rafts, cholesterol, and the age-associated decline and dysregulation of cellular signaling pathways (Ohno-Iwashita et al., 2010). A key component of a subset of rafts, caveolin-1, was identified as a novel control point for both Aβ-based neurodegeneration and healthy neuronal aging (Head et al., 2010), and caveolin-1 expression is altered in AD (Gaudreault et al., 2004). In addition, the level of PrP^C is altered in both the AD and aging brain (Whitehouse et al., 2010, 2013; Larson et al., 2012). Thus, dysregulation of the structure and function of cell-surface, raft-based signaling complexes may occur in, and contribute to, both AD and aging.

CONCLUSION

PrP^C appears to be involved in multiple physiological and pathological processes, including as highlighted here, neuronal zinc uptake and Aβ oligomer binding and toxicity. We propose that PrP^C acts as a hub for cell surface, lipid raft-based signaling/transport complexes, and that the different roles that the protein has are due to the selective interaction of PrP^C with different ligands (such as zinc or Aβ oligomers) and/or co-receptors (such as AMPA receptors, LRP1, mGluR5) in these multi-protein complexes (**Figure 2**). It is likely that these complexes are relatively transient in nature, being stabilized upon binding of a particular ligand for a limited time period before dissociation or endocytosis terminates the signaling or transport process. The use of more sophisticated techniques, such as super resolution light microscopy, should enable the molecular details of these different lipid raft-based signaling/transport complexes to be determined and provide a clearer picture of the role of PrP^C and lipid rafts in neuronal zinc transport and Aβ action in AD. It will also be interesting to determine how the interplay between PrP^C, zinc, and Aβ may underlie aging and age-related diseases.

ACKNOWLEDGMENTS

The authors acknowledge the financial support of the Medical Research Council of Great Britain (G0802189).

REFERENCES

Adlard, P. A., Parncutt, J. M., Finkelstein, D. I., and Bush, A. I. (2010). Cognitive loss in zinc transporter-3 knock-out mice: a phenocopy for the synaptic and memory deficits of Alzheimer's disease? *J. Neurosci.* 30, 1631–1636. doi: 10.1523/JNEUROSCI.5255-09.2010

Barry, A. E., Klyubin, I., McDonald, J. M., Mably, A. J., Farrell, M. A., Scott, M., et al. (2011). Alzheimer's disease brain-derived amyloid-β-mediated inhibition of LTP *in vivo* is prevented by immunotargeting cellular prion protein. *J. Neurosci.* 31, 7259–7263. doi: 10.1523/JNEUROSCI.6500-10.2011

Beyersmann, D., and Haase, H. (2001). Functions of zinc in signaling, proliferation and differentiation of mammalian cells. *Biometals* 14, 331–341. doi: 10.1023/A:1012905406548

Brautigan, D. L., Bornstein, P., and Gallis, B. (1981). Phosphotyrosyl-protein phosphatase. Specific inhibition by Zn. *J. Biol. Chem.* 256, 6519–6522.

Bush, A. I., Pettingell, W. H., Multhaup, G., d Paradis, M., Vonsattel, J. P., Gusella, J. F., et al. (1994). Rapid induction of Alzheimer Aβ amyloid formation by zinc. *Science* 265, 1464–1467. doi: 10.1126/science.8073293

Calella, A. M., Farinelli, M., Nuvolone, M., Mirante, O., Moos, R., Falsig, J., et al. (2010). Prion protein and Aβ-related synaptic toxicity impairment. *EMBO Mol. Med.* 2, 306–314. doi: 10.1002/emmm.201000082

Chung, E., Ji, Y., Sun, Y., Kascsak, R. J., Kascsak, R. B., Mehta, P. D., et al. (2010). Anti-PrPC monoclonal antibody infusion as a novel treatment for cognitive deficits in an Alzheimer's disease model mouse. *BMC Neurosci.* 11:130. doi: 10.1186/1471-2202-11-130

Cisse, M., and Mucke, L. (2009). Alzheimer's disease: a prion protein connection. *Nature* 457, 1090–1091. doi: 10.1038/4571090a

Cousins, R. J., Liuzzi, J. P., and Lichten, L. A. (2006). Mammalian zinc transport, trafficking, and signals. *J. Biol. Chem.* 281, 24085–24089. doi: 10.1074/jbc.R600011200

Deshpande, A., Kawai, H., Metherate, R., Glabe, C. G., and Busciglio, J. (2009). A role for synaptic zinc in activity-dependent Aβ oligomer formation and accumulation at excitatory synapses. *J. Neurosci.* 29, 4004–4015. doi: 10.1523/JNEUROSCI.5980-08.2009

Di Paolo, G., and Kim, T.-W. (2011). Linking lipids to Alzheimer's disease: cholesterol and beyond. *Nat. Rev. Neurosci.* 12, 284–296. doi: 10.1038/nrn3012

Dong, J., Atwood, C. S., Anderson, V. E., Siedlak, S. L., Smith, M. A., Perry, G., et al. (2003). Metal binding and oxidation of amyloid-β within isolated senile plaque cores: Raman microscopic evidence. *Biochemistry* 42, 2768–2773. doi: 10.1021/bi0272151

Ehsani, S., Salehzadeh, A., Huo, H., Reginold, W., Pocanschi, C. L., Ren, H., et al. (2012). LIV-1 ZIP ectodomain shedding in prion-infected mice resembles cellular response to transition metal starvation. *J. Mol. Biol.* 422, 556–574. doi: 10.1016/j.jmb.2012.06.003

Esler, W. P., Stimson, E. R., Jennings, J. M., Ghilardi, J. R., Mantyh, P. W., and Maggio, J. E. (1996). Zinc-induced aggregation of human and rat β-amyloid peptides *in vitro*. *J. Neurochem.* 66, 723–732. doi: 10.1046/j.1471-4159.1996.66020723.x

Frederickson, C. J., Koh, J. Y., and Bush, A. I. (2005). The neurobiology of zinc in health and disease. *Nat. Rev. Neurosci.* 6, 449–462. doi: 10.1038/nrn1671

Freir, D. B., Nicoll, A. J., Klyubin, I., Panico, S., McDonald, J. M., Risse, E., et al. (2011). Interaction between prion protein and toxic amyloid β assemblies can be therapeutically targeted at multiple sites. *Nat. Commun.* 2, 336. doi: 10.1038/ncomms1341

Friedlich, A. L., Lee, J. Y., van Groen, T., Cherny, R. A., Volitakis, I., Cole, T. B., et al. (2004). Neuronal zinc exchange with the blood vessel wall promotes cerebral amyloid angiopathy in an animal model of Alzheimer's disease. *J. Neurosci.* 24, 3453–3459. doi: 10.1523/JNEUROSCI.0297-04.2004

Fuentealba, R. A., Liu, Q., Zhang, J., Kanekiyo, T., Hu, X., Lee, J. M., et al. (2010). Low-density lipoprotein receptor-related protein 1 (LRP1) mediates neuronal Aβ42 uptake and lysosomal trafficking. *PLoS ONE* 5:e11884. doi: 10.1371/journal.pone.0011884

Gaudreault, S. B., Dea, D., and Poirier, J. (2004). Increased caveolin-1 expression in Alzheimer's disease brain. *Neurobiol. Aging* 25, 753–759. doi: 10.1016/j.neurobiolaging.2003.07.004

Gimbel, D. A., Nygaard, H. B., Coffey, E. E., Gunther, E. C., Lauren, J., Gimbel, Z. A., et al. (2010). Memory impairment in transgenic Alzheimer mice requires cellular prion protein. *J. Neurosci.* 30, 6367–6374. doi: 10.1523/JNEUROSCI.0395-10.2010

Grabrucker, A. M., Knight, M. J., Proepper, C., Bockmann, J., Joubert, M., Rowan, M., et al. (2011). Concerted action of zinc and ProSAP/Shank in synaptogenesis and synapse maturation. *EMBO J.* 30, 569–581. doi: 10.1038/emboj.2010.336

Guirland, C., and Zheng, J. Q. (2007). Membrane lipid rafts and their role in axon guidance. *Adv. Exp. Med. Biol.* 621, 144–155. doi: 10.1007/978-0-387-76715-4_11

Head, B. P., Peart, J. N., Panneerselvam, M., Yokoyama, T., Pearn, M. L., Niesman, I. R., et al. (2010). Loss of caveolin-1 accelerates neurodegeneration and aging. *PloS ONE* 5:e15697. doi: 10.1371/journal.pone.0015697

Hering, H., Lin, C. C., and Sheng, M. (2003). Lipid rafts in the maintenance of synapses, dendritic spines, and surface AMPA receptor stability. *J. Neurosci.* 23, 3262–3271.

Hu, N.-W., Nicoll, A. J., Zhang, D., Mably, A. J., O'Malley, T., Purro, S. A., et al. (2014). mGlu5 receptors and cellular prion protein mediate amyloid-β-facilitated synaptic long-term depression *in vivo*. *Nat. Commun.* 5:3374. doi: 10.1038/ncomms4374

Kanekiyo, T., Zhang, J., Liu, Q., Liu, C.-C., Zhang, L., and Bu, G. (2011). Heparan sulphate proteoglycan and the low-density lipoprotein receptor-related protein 1 constitute major pathways for neuronal amyloid-β uptake. *J. Neurosci.* 31, 1644–1651. doi: 10.1523/JNEUROSCI.5491-10.2011

Kessels, H. W., Nguyen, L. N., Nabavi, S., and Malinow, R. (2010). The prion protein as a receptor for amyloid-β. *Nature* 466, E3–E4; discussion: E4–E5. doi: 10.1038/nature09217

Khan, G. M., Tong, M., Jhun, M., Arora, K., and Nichols, R. A. (2010). β-Amyloid activates presynaptic α7 nicotinic acetylcholine receptors reconstituted into a model nerve cell system: involvement of lipid rafts. *Eur. J. Neurosci.* 31, 788–796. doi: 10.1111/j.1460-9568.2010.07116.x

Kozlowski, H., Luczkowski, M., and Remelli, M. (2010). Prion proteins and copper ions. Biological and chemical controversies. *Dalton Trans.* 39, 6371–6385. doi: 10.1039/c001267j

Lai, A. Y., and McLaurin, J. (2010). Mechanisms of amyloid-β peptide uptake by neurons: the role of lipid rafts and lipid raft-associated proteins. *Int. J. Alzheimers Dis.* 2011:548380. doi: 10.4061/2011/548380

Larson, M., Sherman, M. A., Amar, F., Nuvolone, M., Schneider, J. A., Bennett, D. A., et al. (2012). The complex PrPc-Fyn couples human oligomeric Aβ with pathological Tau changes in Alzheimer's disease. *J. Neurosci.* 32, 16857–16871. doi: 10.1523/JNEUROSCI.1858-12.2012

Lauren, J., Gimbel, D. A., Nygaard, H. B., Gilbert, J. W., and Strittmatter, S. M. (2009). Cellular prion protein mediates impairment of synaptic plasticity by amyloid-β oligomers. *Nature* 457, 1128–1132. doi: 10.1038/nature07761

Lee, J. Y., Cole, T. B., Palmiter, R. D., Suh, S. W., and Koh, J. Y. (2002). Contribution by synaptic zinc to the gender-disparate plaque formation in human Swedish mutant APP transgenic mice. *Proc. Natl. Acad. Sci. U.S.A.* 99, 7705–7710. doi: 10.1073/pnas.092034699

Linden, R., Martins, V. R., Prado, M. A., Cammarota, M., Izquierdo, I., and Brentani, R. R. (2008). Physiology of the prion protein. *Physiol. Rev.* 88, 673–728. doi: 10.1152/physrev.00007.2007

Lu, T., Aron, L., Zullo, J., Pan, Y., Kim, H., Chen, Y., et al. (2014). REST and stress resistance in ageing and Alzheimer's disease. *Nature* 507, 448–454. doi: 10.1038/nature13163

Mocchegiani, E., Bertoni-Freddari, C., Marcellini, F., and Malavolta, M. (2005). Brain, aging and neurodegeneration: role of zinc ion availability. *Prog. Neurobiol.* 75, 367–390. doi: 10.1016/j.pneurobio.2005.04.005

Nykjaer, A., and Willnow, T. E. (2002). The low-density lipoprotein receptor gene family: a cellular Swiss army knife? *Trends Cell Biol.* 12, 273–280. doi: 10.1016/S0962-8924(02)02282-1

Ohno-Iwashita, Y., Shimada, Y., Hayashi, M., and Inomata, M. (2010). Plasma membrane microdomains in aging and disease. *Geriatr. Gerontol. Int.* 10 Suppl. 1, S41–S52. doi: 10.1111/j.1447-0594.2010.00600.x

Parkin, E. T., Tan, F., Skidgel, R. A., Turner, A. J., and Hooper, N. M. (2003). The ectodomain shedding of angiotensin-converting enzyme is independent of its localisation in lipid rafts. *J. Cell Sci.* 116, 3079–3087. doi: 10.1242/jcs.00626

Plum, L. M., Rink, L., and Haase, H. (2010). The essential toxin: impact of zinc on human health. *Int. J. Environ. Res. Public health* 7, 1342–1365. doi: 10.3390/ijerph7041342

Pushie, M. J., Pickering, I. J., Martin, G. R., Tsutsui, S., Jirik, F. R., and George, G. N. (2011). Prion protein expression level alters regional copper, iron and zinc content in the mouse brain. *Metallomics* 3, 206–214. doi: 10.1039/c0mt00037j

Renner, M., Lacor, P. N., Velasco, P. T., Xu, J., Contractor, A., Klein, W. L., et al. (2010). Deleterious effects of amyloid beta oligomers acting as an extracellular scaffold for mGluR5. *Neuron* 66, 739–754. doi: 10.1016/j.neuron.2010.04.029

Rushworth, J. V., Griffiths, H. H., Watt, N. T., and Hooper, N. M. (2013). Prion protein-mediated toxicity of amyloid-β oligomers requires lipid rafts and the transmembrane LRP1. *J. Biol. Chem.* 288, 8935–8951. doi: 10.1074/jbc.M112.400358

Rushworth, J. V., and Hooper, N. M. (2010). Lipid rafts: linking Alzheimer's amyloid-β production, aggregation, and toxicity at neuronal membranes. *Int. J. Alzheimers Dis.* 2011, 603052. doi: 10.4061/2011/603052

Saito, T., Takahashi, K., Nakagawa, N., Hosokawa, T., Kurasaki, M., Yamanoshita, O., et al. (2000). Deficiencies of hippocampal Zn and ZnT3 accelerate brain aging of Rat. *Biochem. Biophys. Res. Commun.* 279, 505–511. doi: 10.1006/bbrc.2000.3946

Sales, N., Rodolfo, K., Hassig, R., Faucheux, B., Di Giamberardino, L., and Moya, K. L. (1998). Cellular prion protein localization in rodent and primate brain. *Eur. J. Neurosci.* 10, 2464–2471. doi: 10.1046/j.1460-9568.1998.00258.x

Schmitt-Ulms, G., Ehsani, S., Watts, J. C., Westaway, D., and Wille, H. (2009). Evolutionary descent of prion genes from the ZIP family of metal ion transporters. *PLoS ONE* 4:e7208. doi: 10.1371/journal.pone.0007208

Sensi, S. L., Paoletti, P., Bush, A. I., and Sekler, I. (2009). Zinc in the physiology and pathology of the CNS. *Nat. Rev. Neurosci.* 10, 780–791. doi: 10.1038/nrn2734

Singh, N., Haldar, S., Tripathi, A. K., McElwee, M. K., Horback, K., and Beserra, A. (2014). Iron in neurodegenerative disorders of protein misfolding: a case of prion disorders and Parkinson's disease. *Antioxid. Redox Signal.* 21, 471–484. doi: 10.1089/ars.2014.5874

Solis, G. P., Malaga-Trillo, E., Plattner, H., and Stuermer, C. A. (2010). Cellular roles of the prion protein in association with reggie/flotillin microdomains. *Front. Biosci.* 15, 1075–1085. doi: 10.2741/3662

Stellato, F., Spevacek, A., Proux, O., Minicozzi, V., Millhauser, G., and Morante, S. (2011). Zinc modulates copper coordination mode in prion protein octa-repeat subdomains. *Eur. Biophys. J.* 40, 1259–1270. doi: 10.1007/s00249-011-0713-4

Stys, P. K., You, H., and Zamponi, G. W. (2012). Copper-dependent regulation of NMDA receptors by cellular prion protein: implications for neurodegenerative disorders. *J. Physiol.* 590, 1357–1368. doi: 10.1113/jphysiol.2011.225276

Szewczyk, B. (2013). Zinc homeostasis and neurodegenerative disorders. *Front. Aging Neurosci.* 5:33. doi: 10.3389/fnagi.2013.00033

Taylor, D. R., and Hooper, N. M. (2007). The low-density lipoprotein receptor-related protein 1 (LRP1) mediates the endocytosis of the cellular prion protein. *Biochem. J.* 402, 17–23. doi: 10.1042/BJ20061736

Taylor, D. R., Watt, N. T., Perera, W. S., and Hooper, N. M. (2005). Assigning functions to distinct regions of the N-terminus of the prion protein that are involved in its copper-stimulated, clathrin-dependent endocytosis. *J. Cell Sci.* 118, 5141–5153. doi: 10.1242/jcs.02627

Toth, K. (2011). Zinc in neurotransmission. *Annu. Rev. Nutr.* 31, 139–153. doi: 10.1146/annurev-nutr-072610-145218

Um, J. W., Kaufman, A. C., Kostylev, M., Heiss, J. K., Stagi, M., Takahashi, H., et al. (2013). Metabotropic glutamate receptor 5 is a coreceptor for Alzheimer Aβ oligomer bound to cellular prion protein. *Neuron* 79, 887–902. doi: 10.1016/j.neuron.2013.06.036

Um, J. W., Nygaard, H. B., Heiss, J. K., Kostylev, M. A., Stagi, M., Vortmeyer, A., et al. (2012). Alzheimer amyloid-β oligomer bound to postsynaptic prion protein activates Fyn to impair neurons. *Nat. Neurosci.* 15, 1227–1235. doi: 10.1038/nn.3178

Wang, F., Dufner-Beattie, J., Kim, B. E., Petris, M. J., Andrews, G., and Eide, D. J. (2004). Zinc-stimulated endocytosis controls activity of the mouse ZIP1 and ZIP3 zinc uptake transporters. *J. Biol. Chem.* 279, 24631–24639. doi: 10.1074/jbc.M400680200

Watt, N. T., Griffiths, H. H., and Hooper, N. M. (2013). Neuronal zinc regulation and the prion protein. *Prion* 7, 203–208. doi: 10.4161/pri.24503

Watt, N. T., Taylor, D. R., Kerrigan, T. L., Griffiths, H. H., Rushworth, J. V., Whitehouse, I. J., et al. (2012). Prion protein facilitates uptake of zinc into neuronal cells. *Nat. Commun.* 3, 1134. doi: 10.1038/ncomms2135

Watt, N. T., Whitehouse, I. J., and Hooper, N. M. (2010). The role of zinc in Alzheimer's disease. *Int. J. Alzheimers Dis.* 2011:971021. doi: 10.4061/2011/971021

Whitehouse, I. J., Jackson, C. D., Turner, A. J., and Hooper, N. M. (2010). Prion protein is reduced in aging and in sporadic but not in familial Alzheimer's disease. *J. Alzheimers Dis.* 22, 1023–1031. doi: 10.3233/JAD-2010-101071

Whitehouse, I. J., Miners, J. S., Glennon, E. B., Kehoe, P. G., Love, S., Kellett, K. A., et al. (2013). Prion protein is decreased in Alzheimer's brain and inversely correlates with BACE1 activity, amyloid-β levels and Braak stage. *PLoS ONE* 8:e59554. doi: 10.1371/journal.pone.0059554

Williamson, R., Usardi, A., Hanger, D. P., and Anderton, B. H. (2008). Membrane-bound β-amyloid oligomers are recruited into lipid rafts by a Fyn-dependent mechanism. *FASEB J.* 22, 1552–1559. doi: 10.1096/fj.07-9766com

Willmann, R., Pun, S., Stallmach, L., Sadasivam, G., Santos, A. F., Caroni, P., et al. (2006). Cholesterol and lipid microdomains stabilize the post-synapse at the neuromuscular junction. *EMBO J.* 25, 4050–4060. doi: 10.1038/sj.emboj.7601288

Wilson, M., Hogstrand, C., and Maret, W. (2012). Picomolar concentrations of free zinc(II) ions regulate receptor protein-tyrosine phosphatase β activity. *J. Biol. Chem.* 287, 9322–9326. doi: 10.1074/jbc.C111.320796

Wong, B. X., Hung, Y. H., Bush, A. I., and Duce, J. A. (2014). Metals and cholesterol: two sides of the same coin in Alzheimer's disease pathology. *Front. Aging Neurosci.* 6:91. doi: 10.3389/fnagi.2014.00091

Wong, C. P., Magnusson, K. R., and Ho, E. (2013). Increased inflammatory response in aged mice is associated with age-related zinc deficiency and zinc transporter dysregulation. *J. Nutr. Biochem.* 24, 353–359. doi: 10.1016/j.jnutbio.2012.07.005

Conflict of Interest Statement: The authors declare that the research was conducted in the absence of any commercial or financial relationships that could be construed as a potential conflict of interest.

27

The prion protein family: a view from the placenta

Samira Makzhami[1], Bruno Passet[1], Sophie Halliez[2], Johan Castille[1], Katayoun Moazami-Goudarzi[1], Amandine Duchesne[1], Marthe Vilotte[1], Hubert Laude[2], Sophie Mouillet-Richard[3], Vincent Béringue[2], Daniel Vaiman[4] and Jean-Luc Vilotte[1]*

[1] *INRA, UMR1313 Génétique Animale et Biologie Intégrative, Jouy-en-Josas, France*
[2] *INRA, U892 Virologie et Immunologie Moléculaires, Jouy-en-Josas, France*
[3] *INSERM, UMR-S1124 Signalisation et Physiopathologie Neurologique, Université Paris Descartes, Paris, France*
[4] *Faculté Paris Descartes, UMR8104 CNRS, U1016 INSERM, Institut Cochin, Paris, France*

Edited by:
Craig Michael Walsh, University of California, Irvine, USA

Reviewed by:
Chunying Li, Wayne State University School of Medicine, USA
Panagiotis N. Moschou, Swedish University of Agricultural Sciences, Sweden

Correspondence:
Jean-Luc Vilotte, INRA, UMR1313 Génétique Animale et Biologie Intégrative, Equipe MoDiT- Batiment 440, 78350 Jouy-en-Josas, France
e-mail: jean-luc.vilotte@jouy.inra.fr

Based on its developmental pattern of expression, early studies suggested the implication of the mammalian Prion protein PrP, a glycosylphosphatidylinositol-anchored ubiquitously expressed and evolutionary conserved glycoprotein encoded by the *Prnp* gene, in early embryogenesis. However, gene invalidation in several species did not result in obvious developmental abnormalities and it was only recently that it was associated in mice with intra-uterine growth retardation and placental dysfunction. A proposed explanation for this lack of easily detectable developmental-related phenotype is the existence in the genome of one or more gene (s) able to compensate for the absence of PrP. Indeed, two other members of the *Prnp* gene family have been recently described, Doppel and Shadoo, and the consequences of their invalidation alongside that of PrP tested in mice. No embryonic defect was observed in mice depleted for Doppel and PrP. Interestingly, the co-invalidation of PrP and Shadoo in two independent studies led to apparently conflicting observations, with no apparent consequences in one report and the observation of a developmental defect of the ectoplacental cone that leads to early embryonic lethality in the other. This short review aims at summarizing these recent, apparently conflicting data highlighting the related biological questions and associated implications in terms of animal and human health.

Keywords: prion, shadoo, doppel, placenta, development, mammals

INTRODUCTION

Studies on the neuropathology of transmissible spongiform encephalopathy (TSE) revealed that misfolded, aggregated conformers of the host-encoded cellular protein PrP are the major if not the sole constituent of the infectious agent termed prion (Prusiner, 1982). Prion pathologies affect humans, with the Creutzfeldt-Jakob and Kuru diseases for example, and animals, with for instance the bovine spongiform encephalopathy and sheep scrapie. PrP is a glycosylphosphatidylinositol-anchored, ubiquitously expressed, glycoprotein encoded by the *Prnp* gene. The PrP primary and tertiary structures are well conserved among mammals. Although many studies focused on PrP potential role in the central nervous system in association with prion neurotoxicity, the evolutionary conservation of this protein and the *Prnp* gene expression pattern suggested that PrP may exert important biological roles. Numerous functions were attributed to PrP such as its implication in various signal transductions, cell adhesion, neuroprotection, basic biology of embryonic and tissue-specific stem cells, T-cell regulation and immune function, oxidative stress homeostasis and synaptic function (Westergard et al., 2007; Linden et al., 2008; Zomosa-Signoret et al., 2008; Haigh et al., 2010; Resenberger et al., 2011; Schneider et al., 2011; Lopes and Santos, 2012, for recent reviews as well as associated reviews from this special issue). However, the precise PrP biological role remains rather elusive. It may relate to the multiplicity of partners with which PrP appears to interact, with noticeable differences according to the cell type considered and/or the physiological condition assessed. These diverse cellular proteins include transmembrane receptors, metal ion transporters, signaling molecules, cytoskeleton-associated proteins which could suggest that PrP is a key component of a versatile signaling scaffold complex that allows the activation of various biological pathways.

One further difficulty in understanding PrP physiological function relates to the lack of obvious phenotype in *Prnp*-knockout mice, apart from resistance to TSE and the disappearance of the neurotoxicity associated with prions (Büeler et al., 1992, 1993; Manson et al., 1994). This absence of a strong phenotype could have various, non-mutually exclusive, origins such as (i) the lack of a crucial function of this protein in non-challenging breeding conditions, (ii) genetic adaptation of the animals to the lack of this protein, (iii) biological functional redundancy with other host-encoded proteins able to take over at least some of the PrP roles, and (iv) genetic robustness with PrP being involved in biological functions for which several pathways naturally exist. As mentioned above, the evolutionary conservation of the PrP protein during evolution, at least in mammals, and its involvement in the homeostasis of various stem cells appear somehow contradictory with an absence of important function(s). However, such function could be crucial only to allow adaptation of the animal against stressful conditions, situations that have not been yet really tested, or only in a few occasions, for *Prnp*-knockout mice. The second hypothesis has been challenged by the observation that invalidation only in adult neurons of the *Prnp* locus did not induce histopathological changes but resulted in a modulation of neuronal excitability (Mallucci et al., 2002), with very limited associated transcriptional alteration at a global brain level (Chadi et al., 2010). However, despite its high level of expression in the adult nervous system, PrP biological role might be crucial at earlier developmental stages in this and/or other tissues of the animal. Two other members of the prion gene family, that share with PrP some structural features (**Figure 1**), were recently described in the mammalian genomes, *Prnd* which encodes Doppel and *Sprn* which encodes Shadoo. These three genes are supposedly evolutionary derived from the retro-insertion event of a gene belonging to one of the four subfamilies of the Zinc transporter containing protein-encoding genes, the ZIP LIV-1 branch (Schmitt-Ulms et al., 2009; Ehsani et al., 2011). The existence of these two related proteins could be seen as being in favor of the existence of a biological redundant mechanism. Investigations toward testing this hypothesis led to apparently contradictory observations that have yet to be conciliated (Young et al., 2009; Daude et al., 2012; Passet et al., 2012).

Besides the understanding of the precise mechanism leading to the lack of strong phenotypes in *Prnp*-knockout mammals, a related and yet unsolved question is the tissue where and the developmental stage at which this phenotype would be the more likely to occur. Because of its involvement in TSE and of its higher level of expression in the adult nervous system, many studies first focused on this tissue. However, accumulating evidences associate PrP with stem cell biology and early developmental and regenerative processes. Of specific interest is thus the potential involvement of these proteins in the regulation of the extra-embryonic annexes and more precisely the placenta, based on the observation of the developmental regulation of this gene family in such normal or pathological tissues and on the association of alleles, including experimental gene knockouts, with phenotypes involving or at least recalling placental functional failures. This review summarizes our current knowledge on this specific topic.

PRION PROTEIN GENE FAMILY EXPRESSION IN EXTRA-EMBRYONIC TISSUES

Available data on the regulation of the prion protein gene family in extra-embryonic tissues are limited and mostly related to the *Prnp* gene. Both *in situ* hybridization (Manson et al., 1992; Alfaidy et al., 2013) and uses of a *Prnp*-promoter/LacZ reporter transgene (Tremblay et al., 2007) allowed to identify mouse *Prnp*-gene expression in extra-embryonic membranes as early as 6.5 days post-coïtum, E6.5, the earliest analyzed developmental stage in these studies. This stage corresponds to a transition toward oxidative metabolism and is associated with an increase of the *Prnp* gene expression level (Miele et al., 2003). A similar early extra-embryonic expression of *Prnp* was reported in human (Donadio et al., 2007), rat (Tanji et al., 1995), and ruminant (Kubosaki et al., 2000; O'Rourke et al., 2011) placentas. However, this gene developmental regulation might differ between species, as observed for example in human where the high level of *Prnp* expression was restricted to the first trimester of pregnancy.

Transcription of members of the prion protein gene family was also detected at earlier developmental stages (**Table 1**). Mouse embryonic stem cells were reported to express the three genes (Miranda et al., 2011) and human embryonic stem cells to express at least the *Prnp* one (Krejciova et al., 2011; Lee and Baskakov, 2013). *Prnp* gene expression was also observed in human trophoblast cells (Alfaidy et al., 2013) and that of *Sprn* indirectly suspected to occur in mouse trophectoderm (Passet et al., 2012). We have detected *Sprn* and *Prnd* gene expressions in extra-embryonic tissues of mouse embryos at E10.5 and E13.5 (our unpublished observation and Young et al., 2011). In sheep, *Prnp*

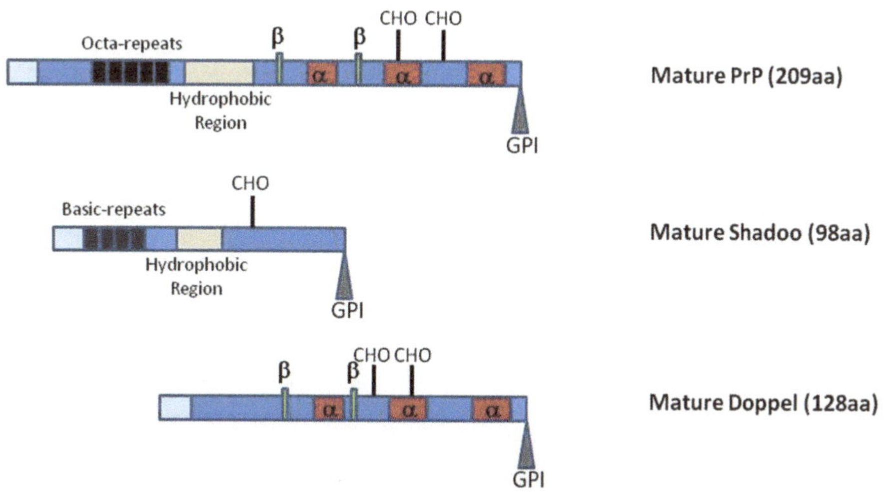

FIGURE 1 | Schematic representation of PrP, Shadoo and Doppel mature proteins. Representations are not at scale. Light blue boxes: N-terminal charged regions of the mature proteins. α, α-helices; β, β-strands; CHO, N-glycosylation sites. GPI glycosylphosphatidylinositol. The PrP octa-repeat region and the Shadoo basic repeat region are indicated as well as the two protein hydrophobic domains.

Table 1 | Summary on prion protein family expression in placenta and knockout phenotypes.

Gene (Protein)	Expression in embryonic stem cells	Expression in placenta	Placental and related phenotype in knockout mice
Prnp (PrP)	Published	Published	Fetal and post-natal growth retardation. Placental abnormalities (weight, structure). Transcriptomic alterations. (Khalifé et al., 2011; Alfaidy et al., 2013)
Sprn (Shadoo)	Published	Published	Post-natal growth retardation (Daude et al., 2012). Transcriptomic alterations to be confirmed (knockdown experiments only, Passet et al., 2012)
Prnd (Doppel)	Published	Our observation	No described effect but no reported specific analysis.

Data on gene expression analyses are derived from various sources, origin of which can be found in the article.

transcripts were also detected in mature and immature oocytes and at the morula stage (Thumdee et al., 2007) while in mouse, the recent sequencing of single oocytes revealed that the three genes are potentially expressed at very low amounts in such cells (Reich et al., 2012).

Overall, these studies suggest that in most mammalian species, the prion family genes are transcribed very early in embryogenesis, in totipotent cells, and that this expression remains both in embryonic and extra-embryonic tissues at early differentiation stages (**Table 1**). At later developmental times, more restricted and cell-specific expression patterns have been reported, but only few studies focused on these loci. Nevertheless, these observations suggest a potential involvement of this gene family in early mammalian developmental stages and highlight its potential implication in the differentiation of the extra-embryonic tissues.

POTENTIAL INCIDENCE OF SINGLE PRION PROTEIN FAMILY GENE ON PLACENTA

THE *Prnp* LOCUS

In sheep, PrP variants are strongly associated with differences in animal susceptibility to prion. The observation that some of these PrP variants confer a stronger resistance of the animal to some prion strains resulted in selection breeding strategies aiming at reducing the prevalence of the disease in the sheep population through the selection of so-called resistant *Prnp* alleles, such as ARR vs. VRQ at codons 136, 154, and 171, respectively (Clouscard et al., 1995; Jeffrey et al., 2014). While selecting ARR animals seems efficient against scrapie development, it could have negative consequences should *Prnp* or closely related genes be involved in important economical traits, since it would force the breeder not to consider selecting for interesting traits if they are associated to a sensitive (VRQ) genotype. Indeed, several articles described a potential association between the *Prnp* genotype and specific traits such as lamb birth weight (Sawalha et al., 2007b) and (post-natal) survival (Sawalha et al., 2007a; Gubbins et al., 2009). Although these are complex traits with several potential origins, a correlation between these observations and a role of PrP during placentation/early development could be suspected.

Potential implication of PrP in the placental physiology was also suggested by the observation of the deregulation of its gene expression in pathological situations. In human, one of the most studied albeit still mysterious gestational disease is preeclampsia. Preeclampsia is characterized by a *de novo* gestational hypertension and proteinuria. It is a major source of maternal and neonatal morbidity and mortality (Sibai et al., 2005). In one third of the cases, preeclampsia is accompanied with Intra-Uterine Growth Restriction (IUGR), a disease where the fetus fails to reach its genetic growth potential, often through placental dysfunction. Oxidative stress is associated with preeclampsia (Shaker and Sadik, 2013) and abnormal placenta levels of zinc and copper with preterm gestations and IUGR both in humans (Zadrozna et al., 2009; Uriu-Adams and Keen, 2010; Kambe et al., 2014) and mouse (Andrews et al., 2004; Tian et al., 2014). Placental tissues deriving from preeclamptic pregnancies were found to over-express both the *Prnp* gene mRNA and the PrP protein (Nishizawa et al., 2007; Hwang et al., 2010). However, this over-expression was not equally distributed within the tissue and appeared to be restricted to the syncytiotrophoblast (the external layer of the placental villi) and to the cytotrophoblasts (the layer of cells that underline the syncytiotrophoblast). Both in human (Bilodeau, 2014) and in a recently published preeclamptic mouse model (Doridot et al., 2013, 2014), strong links between preeclampsia and oxidative stress were described. The octapeptide region and adjacent sites, around histidines 96 and 111, of the PrP protein localized in its amino-terminal region bind several divalent metals (Jackson et al., 2001; Kramer et al., 2001; Walter et al., 2009). This ability has been associated with putative PrP functions in the binding and internalizations of ions, ferrireductase and superoxide dismutase-like activities, signal transduction and protection against oxidative stress (Brown et al., 2001; Choi et al., 2007; Bertuchi et al., 2012; Watt et al., 2012; Singh et al., 2013). Concurrently, a recent study by Alfaidy et al. depicted an involvement of PrP in the response to oxidative stress in the placenta (Alfaidy et al., 2013). It was thus hypothesized that these increased protein and mRNA PrP levels could be part of a compensatory mechanism induced by a preeclamptic status. However, beside this protective function, PrP could also play an active role in the establishment of preeclampsia as signaling pathways involving PrP have been proposed to participate or to worsen this disease. The TGFβ pathway, modulated by the prion protein (Wurm and Wechselberger, 2006), is suspected to contribute to the preeclamptic etiopathology (Stanczuk et al., 2007; Feizollahzadeh et al., 2012; Ozkan et al., 2013), possibly by altering Treg cells equilibrium (Laresgoiti-Servitje et al., 2010; Robertson et al., 2013). Alteration of the Notch-signaling pathway leads to a defect of trophoblast invasion that contributes to the pathogenesis of preeclampsia (Hunkapiller et al., 2011). Interestingly, invalidation of the PrP appears to modulate the placental Notch signaling pathway through alteration of the expression of the Jagged1 ligand (SMR et al., unpublished observation). Thus, PrP could coordinate various signal pathways in response to an oxidative stress that lead to the development of preeclampsia.

Another indirect evidence of PrP implication in placentation came from comparative transcriptomic studies of E6.5 and E7.5 FVB/N and FVB/N *Prnp*$^{-/-}$ embryos (Khalifé et al., 2011). These data revealed that in *Prnp* gene-invalidated embryos several major metabolic pathways such as angiogenesis, cell proliferation, adhesion and movement were affected at these early developmental stages. Such pathways are important in placental physiology. Furthermore, some of the identified deregulated genes were previously described as key actors of mammalian placentation, such as Adam12 (Huppertz et al., 2006) and activin receptors (Munir et al., 2004).

A direct implication of PrP in regulating placental function and pregnancy outcomes was recently documented (Alfaidy et al., 2013). This study comparatively analyzed the reproductive performances of transgenic mice either over-expressing mouse PrP (Tga20) or knockout for *Prnp* (*Prnp*$^{-/-}$) with those of their wild type (WT) counterpart. The litter size was decreased in Tga20 comparatively to that of *Prnp*$^{-/-}$ or WT mice. At E17.5, the fetal and placental weights of the *Prnp*$^{-/-}$ mice were lower than that of the two other genotypes and in their adulthood, both the *Prnp*$^{-/-}$ and Tga20 mice had lower body weights compared to their WT counterparts. This study further highlighted a role of PrP in placental Cu homeostasis and protection against oxidative stress as well as its involvement in placental angiogenesis, two of the altered metabolic pathways identified in the above-mentioned embryonic transcriptomic study (Khalifé et al., 2011). Phenotypic analysis of *Prnp*$^{-/-}$ mice revealed a compacted placental labyrinth structure and a disorganization of the vascular tree, disorganization also observed in Tga20 placentas. Alteration of the expression of genes specific of spongiotrophoblasts, invasive trophoblasts or related to the labyrinth branching were noticed. Trophoblastic cell mobility is an important physiological process that allows placental cells to invade uterine tissue, contributing to the anchorage of the placenta, to the regulation of decidual angiogenesis and to the remodeling of the maternal spiral arteries. This mechanism is complex and involves different cell types and signaling pathways (Gasperowicz and Otto, 2008; Knofler, 2010, for recent reviews). PrP might also be directly involved in angiogenesis (Turu et al., 2008) and hematopoiesis (Palmqvist et al., 2007). This might be related to its localization in caveolae and interaction with Caveolin-1, known to be involved in angiogenesis (Griffoni et al., 2000; Massimino et al., 2002).

Overall these studies highlighted a role of PrP during placentation. Deregulation of the expression of the gene and differences in PrP genotypes were associated with placental perturbations (**Table 1**). However, these events did not result in overt phenotypes, suggesting either that PrP function in this tissue is not crucial enough for its development to induce embryonic distress or that compensatory mechanisms exist.

THE *Prnd* AND *Sprn* LOCI

The *Prnd* gene has been invalidated in two different mouse genetic backgrounds, 129/Ola (Paisley et al., 2004) and a mixed C57BL6/CBA one (Behrens et al., 2002). In both experiments, knockout males suffered from severe subfertility involving a defect in acrosome biogenesis. These males also suffered from other sperm abnormalities that varied between the two reports probably in relation with the different genetic background. Because of this male phenotype, a knockout transgenic line could not be propagated by classical breeding and homozygous knockout animals were only routinely obtained using heterozygous males. In both articles, reproduction of the knockout females were reported to be normal in terms of fertility and litter sizes. Although not specifically studied, none of these reports described other associated phenotypes, suggesting either that Doppel has no role during early embryogenesis or that, should it be involved in placentation, its gene invalidation does not induce drastic alterations (**Table 1**). However, as Doppel is expressed in mouse extra-embryonic tissues (our unpublished observation), more dedicated analyses should probably be conducted before formally excluding it from the list of proteins involved in the physiology of the placenta. The α-helical region of the Doppel protein binds copper (Qin et al., 2003), but this interaction does not appear to induce the protein internalization, as it is observed for PrP (Cereghetti et al., 2004). It was also reported that Doppel expression might exacerbate oxidative damage (Wong et al., 2001), but this property is discussed (Qin et al., 2003). Thus, analysis of a putative role of Doppel in oxidative stress homeostasis during placental development would be worth looking at.

The *Sprn* gene invalidation was also recently obtained by homologous recombination in 129Pas ES cells (Daude et al., 2012). These cells were used to derive C57BL/129 and FVB/NCr/129 *Sprn*$^{-/-}$ mice that were reported to have no gross morphological alterations and to be fertile (Daude et al., 2012). However, 8–16 days old knockout females from the first genetic background were found to have slightly lighter body weights compared to their wild-type counterparts, while knockout males of both genetic backgrounds had a similar weight phenotype at older ages, from day 28 till 42 for C57BL/129 and from day 22 to 50 for FVB/NCr/129 *Sprn*$^{-/-}$ mice, respectively. Overall this subtle alteration resulted in a weight decrement of about 9%. In a following up review, the authors suggested that the origin of this phenotype could be related to the natural expression of Shadoo in hypothalamic neurons, a structure associated with the control of feeding behavior (Daude and Westaway, 2012). Several other explanations could also be proposed. A mammary defect resulting in modified lactation ability of the knockout females would also induce growth retardations; however such a phenotype would probably affect all pups similarly. Interestingly, uteroplacental insufficiency and altered maternal nutrition during gestation have also been associated with sex-specific post-natal development defects of rat offsprings (Howie et al., 2012; Wadley et al., 2013). It could thus be also hypothesized that the body weight phenotype described for the *Sprn*$^{-/-}$ pups arises from a defect in the placentation process. In line with this suggestion is the observation that the mean litter size of these *Sprn*$^{-/-}$ females is smaller than that of their *Sprn*$^{+/-}$ counterparts (Table S1 in Daude et al., 2012). Transcriptomic analyses performed on FVB/N *Sprn*-knockdown embryos at E6.5 and E7.5 also revealed the differential expression of around 60 transcripts associated with cellular movement and development and with differentiation of the hematological system. Interestingly, some of them, such as the prolactin-related genes, pointed to a transcriptomic deregulation affecting precisely the trophectoderm-derived compartment (Passet et al., 2012, 2013). Regulation of cellular metal uptake by PrP is probably regulated through the interaction of this protein with other membrane-bound partners such as the lipoprotein-receptor-related protein LRP1 (Taylor and Hooper, 2007) or the glutamate receptor AMPA (Watt et al., 2012) and, as already discussed, this function was shown to be important for placentation (Alfaidy et al., 2013). Although not known to bind divalent metals, Shadoo might also indirectly regulate oxidative stress homeostasis regulatory processes through interaction with PrP and/or other proteins from their shared interactome (Watts et al., 2009). This hypothesis has yet to be substantiated by experimental evidences. Overall, these observations are compatible with a gene dosage effect of *Sprn* expression on the ectoplacental cone differentiation and of a potential incidence of the genetic background on the resulting phenotype, as already mentioned for *Prnd* (**Table 1**).

CO-INVALIDATION OF SEVERAL MEMBERS OF THE PRION PROTEIN GENE FAMILY

As already mentioned, the absence of a strong phenotype in mammals invalidated for a single member of the prion protein gene family has led to the hypothesis of the existence of a biological functional redundancy with another host-encoded protein. This model was first developed for the implication of PrP in the central nervous system (Shmerling et al., 1998). Such proteins should share with PrP an overlapping pattern of expression, at least at some developmental time points, and related biological properties. The two PrP related proteins, Shadoo and Doppel, were suspected to play this role.

Prnd AND *Prnp* CO-INVALIDATION

Doppel is unlikely to compensate for the lack of PrP based on its adult pattern of expression, mainly restricted to non-nervous tissues and the male gonads, and on its biological properties in terms of neurotoxicity. Indeed, induced ectopic activation of the *Prnd* gene expression in the adult nervous system of some *Prnp*-knockout mice was shown to be neurotoxic (Moore et al., 1999, 2001; Rossi et al., 2001; Anderson et al., 2004) and antagonized by full length PrP expression (Yamaguchi et al., 2004). However, because Doppel and PrP are both expressed during embryogenesis, it remained possible that these two proteins share crucial properties at earlier developmental stages. To investigate a potential compensation by Doppel for the absence of PrP, both loci were co-invalidated in ES cells, taking advantage of their physical close linkage (Paisley et al., 2004). The resulting animals were reported to develop normally into adulthood with no overt phenotype beside an infertility syndrome in the double-knockout males indistinguishable from that observed in their single *Prnd*-knockout counterparts. Thus, this experiment strongly suggests that Doppel and PrP share no obvious compensatory mechanisms. The reproductive performances of these mice, apart that of the males, were however not specifically assessed.

Sprn AND *Prnp* CO-INVALIDATION

The co-invalidation of *Sprn* and *Prnp* was achieved using transgenic lines derived from targeted 129Sv/Pas ES cells. Double knockout animals were obtained by crossing *Sprn*$^{-/-}$ with FVB.129-*Prnp*$^{-/-}$ mice (Daude et al., 2012). The resulting mice were viable, fertile and produced at expected Mendelian ratios. No overt phenotype was reported, even in aged individuals, and the above-discussed lighter body-weight of the single *Sprn*$^{-/-}$ mice not mentioned in these animals (Daude et al., 2012). Altogether, this work strongly argued against compensation or redundancy between Shadoo and PrP in the adult nervous system and during embryogenesis (Daude and Westaway, 2012).

However, this conclusion drastically contrasts with previous experiments aimed at lowering the expression of the *Sprn* gene in various genetic backgrounds by RNA interference, RNAi (Young et al., 2009; Passet et al., 2012). Indeed, using lentiviral ShRNA-delivery in mouse embryos, *Sprn* gene knockdown resulted in early embryonic lethality, between E8 and E11, in an FVB/N.129-*Prnp*$^{-/-}$ genetic background that was not observed in FVB/N mice (Young et al., 2009). This lethality was associated with a developmental defect of the ectoplacental cone and with important hemorrhage surrounding the embryos (Passet et al., 2012). The restriction of the lentiviral infection to the trophoblastic cell lineage resulted in a similar lethal phenotype (Passet et al., 2012). Consistently, embryos invalidated for *Prnp* and/or knockdown for *Sprn* (Khalifé et al., 2011; Passet et al., 2013) display alterations of the expression of genes involved in placental angiogenesis, such as angiopoietin genes (Geva et al., 2005), cathepsins (Screen et al., 2008) and matrix metalloproteinases (Fontana et al., 2012). It would be of interest to assess in which extent, inactivation of Shadoo and co-inactivation with PrP affects Notch signaling in regards of PrP invalidation alone and consistently with the crucial role of this pathway for mouse placental fetal angiogenesis and proper trophoblast cell type specification (Gasperowicz and Otto, 2008). PrP expression is positively regulated by MEK and p38 MAPK kinases (Wang et al., 2005) and it is noteworthy that Mek-1 deficient mice die at early embryonic stages, around E10.5, from an angiogenesis defect of the placenta (Giroux et al., 1999). This age is reminiscent of that of the lethality occurring in *Prnp*$^{-/-}$*Sprn*-knockdown embryos (Young et al., 2009) and it would be of interest to also assess the effect of Mek-1 on the transcriptional regulation of *Sprn* in a regular context and in the absence of PrP.

It remains that this RNAi strategy yielded seemingly contradictory result compared to the classical knockout approach. Several explanations were proposed to conciliate these observations (Daude and Westaway, 2012; Passet et al., 2013). It was suggested that the RNAi approach may induce a potential artifact resulting from either an off target effect of the two used shRNA only phenotypically visible in *Prnp*-knockout embryos or a toxicity associated with an shRNA over expression in single *Prnp* or double *Prnp*/*Sprn* knockout genotypes, linked to the role of PrP and potentially Shadoo in the regulation of the RISC complex (Gibbings et al., 2012). Alternative hypotheses may also be proposed, such as a higher susceptibility of the double knockout genotype to external stresses, such as lentiviral infection. Furthermore, even if the knockdown of a gene can result in phenotypic outcomes comparable to that of the knockout (De Souza et al., 2006), haplo-insufficiency may on the opposite induce stronger phenotypes than the complete gene loss, as described for Dicer1, a key regulator of microRNA biogenesis (Lambertz et al., 2010).

The use of different genetic backgrounds is also known to greatly influence the phenotype of genetically engineered mouse (Doetschman, 2009). Although cautions were taken in that respect in the knockout experiments (see comments in Daude and Westaway, 2012), this strategy involves inbreeding crosses over several generations. Doing so, starting from different genetic backgrounds [129Pas for the ES cells, C57BL6 for the first established knockout line that was then backcrossed in FVB/NCr (Daude et al., 2012)], a selection process might take place that progressively eliminates unfavorable allelic combinations. This could also lead to potential miss-association between (absence of) a phenotype and the targeted gene, as recently exemplified in the case of PrP where abnormalities suspected to result from this gene invalidation actually originated from co-segregating 129Sv-derived loci (Nuvolone et al., 2013; Striebel et al., 2013). Such a mechanism will not occur using an RNAi approach, potentially leading to stronger associated phenotypes. One way to avoid these potential artifacts would be to use the new custom-designed DNA scissors to invalidate these loci in pure genetic backgrounds (Gaj et al., 2013).

On this basis, we have engineered *Sprn* knockout mice through the use of the Zinc Finger Nuclease (ZFN) technology. Our preliminary results indicate that, as reported by Daude et al. (2012), FVB/N *Sprn*$^{-/-}$ and FVB/N.129 *Prnp*$^{-/-}$/*Sprn*$^{-/-}$ mice are viable. However, these transgenic mice suffer from various defects that are under scrutiny. In both cases, pups exhibit a growth retardation that can be attributed, for part of it, to a mammary defect of the knockout females, probably associated with *Sprn* invalidation. The FVB/N.129 *Prnp*$^{-/-}$/*Sprn*$^{-/-}$ mice are also characterized by increased embryonic and perinatal lethality rates compared to FVB/N or FVB/N.129 *Prnp*$^{-/-}$ controls, the extent of which is under analysis. Finally, some phenotypic differences in the survival of the mice at the past weaning stage between FVB/N *Sprn*$^{-/-}$ and FVB/N.129 *Prnp*$^{-/-}$/*Sprn*$^{-/-}$ mice appear to result from the 129 genetic contribution in the double knockout mice. Altogether, these preliminary and unpublished observations are consistent with a role of these two related proteins in placentation. They further reassert the importance of the influence of the genetic background and suggest the contribution of other yet unknown loci in the phenotypic consequences of the *Prnp* and *Sprn* invalidations.

A remaining issue concerns the discrepancy observed between the knockout and the knockdown experiments. We could exclude a specific susceptibility of the FVB/N.129 *Prnp*$^{-/-}$/*Sprn*$^{-/-}$ mice toward lentiviral infection *per se*. Three other hypothesis are currently tested: (i) an impact of ShRNA over-expression in FVB/N.129 *Prnp*$^{-/-}$/*Sprn*$^{-/-}$ embryos in relation with a potential function of these proteins in miRNA regulation, (ii) an off-target effect of the anti-*Sprn* ShRNAs used in Young et al. (2009) that would down-regulate the expression of a protein leading to a detrimental effect on the development of the ecoplacental cone in FVB/N.129 *Prnp*$^{-/-}$ and/or FVB/N.129 *Prnp*$^{-/-}$/*Sprn*$^{-/-}$ mice or (iii) a genetic adaptation of the double-knockout embryos to the lack of these two proteins highlighting other loci able to compensate for the lack of them. Pursuing this analysis will thus undoubtedly increase our knowledge on early mammalian embryogenesis.

SIMILARITIES OF THE SIGNALING PATHWAYS ASSOCIATED WITH THE PRION PROTEIN FAMILY IN PLACENTATION AND IN ZEBRAFISH EARLY EMBRYOGENESIS

As mentioned before, the biological roles of the PrP, of the related Shadoo and to a lesser extend of the Doppel proteins still remain elusive in mammals. Investigations on the PrP biological role were also performed on more distantly species such as Zebrafish. Again, the outcome of these studies surprisingly differed according to the experimental approaches used, morpholinos or knockout, somehow recalling the situation described in mouse. Morpholino-induced downregulation of PrP1 or PrP2 resulted in high mortality of the depleted early embryos. Lethality occurred at developmental stages related to the respective two genes' spatio-temporal patterns of expression, and was associated with deficient morphogenic cell movements and induced apoptotic cell death (Malaga-Trillo et al., 2009; Nourizadeh-Lillabadi et al., 2010). Surviving PrP2- (Nourizadeh-Lillabadi et al., 2010) or PrP1- (Kaiser et al., 2012) depleted embryos also highlighted a neuroprotective role of these proteins. In contrast to PrP-2 morpholino experiments, the knockout of this gene by a ZFN-based approach resulted in no overt developmental phenotypes (Fleisch et al., 2013). This apparent discrepancy was attributed in part to an off target effect of the used morpholinos. In contrast, PrP1-knockdown by morpholinos can be considered as specific as the phenotype can be partially rescued by the injection of PrP1 or PrP2 mRNA.

Interestingly, the function of PrP1 in early embryonic development of the Zebrafish could be partly rescued in depleted embryos by the injection of mammalian PrP-encoding mRNAs, suggesting evolutionary conservation of some of the protein functions (Chiesa and Harris, 2009; Malaga-Trillo and Sempou, 2009; Malaga-Trillo et al., 2009; Solis et al., 2013). Besides, in mouse *Prnp*$^{-/-}$ embryos, transcriptomic analyses revealed similarities with PrP1-depleted Zebrafish eggs by highlighting the over-expression of genes encoding proteins with proteolysis activities that could modulate cellular adhesion and proliferation as well as extracellular matrix deposition (Khalifé et al., 2011). The same pathways were also deregulated in Shadoo-depleted embryos, suggesting that this protein shares some of these PrP-functions at least in mammals (Passet et al., 2012). In Zebrafish, the PrP1-depleted phenotype is associated with an abnormal intracellular processing and/or transport of E-cadherin, potentially resulting from a modulation of the Fyn signaling pathway (Malaga-Trillo et al., 2009). Similarly in mammals, PrP has been linked to the activation of Src kinases, including Fyn (Mouillet-Richard et al., 2000), while in early mouse embryos, both *Prnp* and *Sprn* down-regulations induced protocadherin and cadherin transcriptomic alterations (Khalifé et al., 2011; Passet et al., 2012). In mouse embryos, the trophectoderm is the first differentiated tissue to form, with cells requiring complex adhesive structures and invasive capacities. Migrating trophoblast cells are first observed at

E6.5 (Croy et al., 2012). Thus, the alterations of cell adhesion pathways in *Sprn* and *Prnp* invalidated mice, similar to the alterations exhibited by PrP1-depleted zebrafish, are very likely to affect this specialized tissue at these early embryonic stages.

CONCLUSION—ONGOING OR POTENTIAL FUTURE DIRECTIONS

Current data clearly indicate that the PrP protein has an important role in early mammalian embryogenesis through its implication in placental physiology. The mechanistic behind these observations has yet to be clearly deciphered and could involve several, not mutually exclusive, processes such as an involvement in cell adhesion and angiogenesis as well as a role in the control and response to oxidative stress. A similar implication of Shadoo has also been suggested by several studies but whether this protein acts through identical, complementary or unrelated biological pathways with those of PrP remains a debatable question. The properties of Doppel and its pattern of expression make it an interesting potential partner, but no evidence so far has pointed to this protein as being a key actor of placental physiology. Taking into account the phenotypes observed through the modulation of the expression of these genes in placenta and the affected pathways, the knowledge accumulated on the function of these proteins in this tissue could be of particular interest in regards with their more and more recognized implication in cancers. A crosstalk between these so far independent studies would be of great interest with potential benefits for human and animal health.

A yet poorly investigated question is the relation between these proteins and the biology of mitochondria. PrP depletion was shown to result in reduced mitochondrial numbers and unusual mitochondrial morphology with increased diameters and poorly defined and sparse cristae in several mouse tissues (Miele et al., 2002). PrP depletion induced elevated mitochondrial manganese superoxide dismutase activity, manganese-induced mitochondrial depolarization and reactive oxygen species generation. Brains of Prion-infected mice have reduced cytochrome C oxidase and manganese superoxide dismutase activities and contain swelling mitochondria. Oxidative stress conditions can induce PrP localization in raft-like microdomains of the mitochondria membranes, resulting in loss of this membrane potential. Impaired mitochondrial function is associated with perturbations of mouse placental (Wakefield et al., 2011) and potentially with preeclampsia (Shi et al., 2013; Doridot et al., 2014) developments. Thus, it is tempting to suggest that oxidative stress conditions and preeclampsia through its induced over-expression of PrP, might trigger the delocalization of some PrP proteins in sub-domains of the mitochondria membrane resulting in induced mitochondria dysfunctions and placentation defects. Similar implication of the other prion protein family members is a yet-to-be explored direction for future experimentations.

One difficulty to delineate the role of these proteins might be related with their mode of action. As nicely highlighted by Alfaidy et al. (2013), these proteins may act as gatekeeper of cellular integrity and thus, in the absence of specific insults such as oxidative stress conditions, their absence might not yield strong phenotypes. Challenging through diverse approaches the existing panel of created animal models with altered expression of these genes would reveal such roles.

It was also suggested that these proteins, which have diverse and large interactomes, may participate to versatile signaling scaffold complexes, allowing activation of various biological pathways. As such, one might expect this function to be revealed by invalidation of their encoding genes unless (i) biological redundancy exists, (ii) the involved biological pathways are associated with non-essential functions or related to response to various stress as discussed just before, (iii) their absence may be, at least partially, compensated through selection of specific genetic environment. This latter hypothesis is attractive as it would explain the different outcomes of the knockout of these genes according to the genetic background, as observed for Doppel, and/or to the methodology used, as observed for Shadoo. It would also reconcile the discrepancy between *in vitro* data, suggesting that these proteins have key biological functions in crucial pathways such as stem cell homeostasis, with *in vivo* observations, which appear to deny or limit their importance. The classical knockout process involves the establishment of transgenic lines in mixed genetic backgrounds, allowing the selection process of the less detrimental genetic combinations. New approaches now exist to assess gene invalidation in controlled genetic backgrounds, such as the use of endonucleases (Wijshake et al., 2014). The implementation of this approach to target the prion gene family in different mouse strains might help us to better understand the role of each of these proteins and their potential biological redundancy. Such experiments are currently ongoing and their conclusions much awaited.

AUTHOR CONTRIBUTIONS

All authors contributed to the conception of this review, revised the manuscript and approved the submitted version.

REFERENCES

Alfaidy, N., Chauvet, S., Donadio-Andrei, S., Salomon, A., Saoudi, Y., Richaud, P., et al. (2013). Prion protein expression and functional importance in developmental angiogenesis: role in oxidative stress and copper homeostasis. *Antioxid. Redox Signal.* 18, 400–411. doi: 10.1089/ars.2012.4637

Anderson, L., Rossi, D., Linehan, J., Brandner, S., and Weissmann, C. (2004). Transgene-driven expression of the Doppel protein in Purkinje cells causes Purkinje cell degeneration and motor impairment. *Proc. Natl. Acad. Sci. U.S.A.* 101, 3644–3649. doi: 10.1073/pnas.0308681101

Andrews, G. K., Wang, H., Dey, S. K., and Palmiter, R. D. (2004). Mouse zinc transporter 1 gene provides an essential function during early embryonic development. *Genesis* 40, 74–81. doi: 10.1002/gene.20067

Behrens, A., Genoud, N., Naumann, H., Rulicke, T., Janett, F., Heppner, F. L., et al. (2002). Absence of the prion protein homologue Doppel causes male sterility. *EMBO J.* 21, 3652–3658. doi: 10.1093/emboj/cdf386

Bertuchi, F. R., Bourgeon, D. M., Landemberger, M. C., Martins, V. R., and Cerchiaro, G. (2012). PrPC displays an essential protective role from oxidative stress in an astrocyte cell line derived from PrPC knockout mice. *Biochem. Biophys. Res. Commun.* 418, 27–32. doi: 10.1016/j.bbrc.2011.12.098

Bilodeau, J. F. (2014). Review: maternal and placental antioxidant response to preeclampsia —impact on vasoactive eicosanoids. *Placenta* 35(Suppl.), S32–S38. doi: 10.1016/j.placenta.2013.11.013

Brown, D. R., Clive, C., and Haswell, S. J. (2001). Antioxidant activity related to copper binding of native prion protein. *J. Neurochem.* 76, 69–76. doi: 10.1046/j.1471-4159.2001.00009.x

Büeler, H., Aguzzi, A., Sailer, A., Greiner, R. A., Autenried, P., Aguet, M., et al. (1993). Mice devoid of PrP are resistant to scrapie. *Cell* 73, 1339–1347. doi: 10.1016/0092-8674(93)90360-3

Büeler, H., Fischer, M., Lang, Y., Bluethmann, H., Lipp, H. P., Dearmond, S. J., et al. (1992). Normal development and behaviour of mice lacking the neuronal cell-surface PrP protein. *Nature* 356, 577–582. doi: 10.1038/356577a0

Cereghetti, G. M., Negro, A., Vinck, E., Massimino, M. L., Sorgato, M. C., and Van Doorslaer, S. (2004). Copper(II) binding to the human Doppel protein may mark its functional diversity from the prion protein. *J. Biol. Chem.* 279, 36497–36503. doi: 10.1074/jbc.M404341200

Chadi, S., Young, R., Le Guillou, S., Tilly, G., Bitton, F., Martin-Magniette, M. L., et al. (2010). Brain transcriptional stability upon prion protein-encoding gene invalidation in zygotic or adult mouse. *BMC Genomics* 11:448. doi: 10.1186/1471-2164-11-448

Chiesa, R., and Harris, D. A. (2009). Fishing for prion protein function. *PLoS Biol.* 7:e75. doi: 10.1371/journal.pbio.1000075

Choi, C. J., Anantharam, V., Saetveit, N. J., Houk, R. S., Kanthasamy, A., and Kanthasamy, A. G. (2007). Normal cellular prion protein protects against manganese-induced oxidative stress and apoptotic cell death. *Toxicol. Sci.* 98, 495–509. doi: 10.1093/toxsci/kfm099

Clouscard, C., Beaudry, P., Elsen, J. M., Milan, D., Dussaucy, M., Bounneau, C., et al. (1995). Different allelic effects of the codons 136 and 171 of the prion protein gene in sheep with natural scrapie. *J. Gen. Virol.* 76(Pt 8), 2097–20101. doi: 10.1099/0022-1317-76-8-2097

Croy, B. A., Chen, Z., Hofmann, A. P., Lord, E. M., Sedlacek, A. L., and Gerber, S. A. (2012). Imaging of vascular development in early mouse decidua and its association with leukocytes and trophoblasts. *Biol. Reprod.* 87, 125. doi: 10.1095/biolreprod.112.102830

Daude, N., and Westaway, D. (2012). Shadoo/PrP (*Sprn*(0/0) /*Prnp*(0/0)) double knockout mice: more than zeroes. *Prion* 6, 420–424. doi: 10.4161/pri.21867

Daude, N., Wohlgemuth, S., Brown, R., Pitstick, R., Gapeshina, H., Yang, J., et al. (2012). Knockout of the prion protein (PrP)-like *Sprn* gene does not produce embryonic lethality in combination with PrP(C)-deficiency. *Proc. Natl. Acad. Sci. U.S.A.* 109, 9035–9040. doi: 10.1073/pnas.1202130109

De Souza, A. T., Dai, X., Spencer, A. G., Reppen, T., Menzie, A., Roesch, P. L., et al. (2006). Transcriptional and phenotypic comparisons of Ppara knockout and siRNA knockdown mice. *Nucleic Acids Res.* 34, 4486–4494. doi: 10.1093/nar/gkl609

Doetschman, T. (2009). Influence of genetic background on genetically engineered mouse phenotypes. *Methods Mol. Biol.* 530, 423–433. doi: 10.1007/978-1-59745-471-1_23

Donadio, S., Alfaidy, N., De Keukeleire, B., Micoud, J., Feige, J. J., Challis, J. R., et al. (2007). Expression and localization of cellular prion and COMMD1 proteins in human placenta throughout pregnancy. *Placenta* 28, 907–911. doi: 10.1016/j.placenta.2006.11.006

Doridot, L. D., Chatre, L. D., Ducat, A., Vilotte, J. L., Lombes, A., Mehats, C., et al. (2014). Nitroso-redox balance and mitochondrial homeostasis are regulated by Stox1, a pre-eclampsia associated gene. *Antioxid. Redox Signal.* 21, 819–834. doi: 10.1089/ars.2013.5661

Doridot, L., Passet, B., Méhats, C., Rigourd, V., Barbaux, S., Ducat, A., et al. (2013). Preeclampsia-like symptoms induced in mice by fetoplacental expression of STOX1 are reversed by aspirin treatment. *Hypertension* 61, 662–668. doi: 10.1161/HYPERTENSIONAHA.111.202994

Ehsani, S., Tao, R., Pocanschi, C. L., Ren, H., Harrison, P. M., and Schmitt-Ulms, G. (2011). Evidence for retrogene origins of the prion gene family. *PLoS ONE* 6:e26800. doi: 10.1371/journal.pone.0026800

Feizollahzadeh, S., Taheripanah, R., Khani, M., Farokhi, B., and Amani, D. (2012). Promoter region polymorphisms in the transforming growth factor beta-1 (TGFbeta1) gene and serum TGFbeta1 concentration in preeclamptic and control Iranian women. *J. Reprod. Immunol.* 94, 216–221. doi: 10.1016/j.jri.2012.02.006

Fleisch, V. C., Leighton, P. L., Wang, H., Pillay, L. M., Ritzel, R. G., Bhinder, G., et al. (2013). Targeted mutation of the gene encoding prion protein in zebrafish reveals a conserved role in neuron excitability. *Neurobiol. Dis.* 55, 11–25. doi: 10.1016/j.nbd.2013.03.007

Fontana, V., Coll, T. A., Sobarzo, C. M., Tito, L. P., Calvo, J. C., and Cebral, E. (2012). Matrix metalloproteinase expression and activity in trophoblast-decidual tissues at organogenesis in CF-1 mouse. *J. Mol. Histol.* 43, 487–496. doi: 10.1007/s10735-012-9429-8

Gaj, T., Gersbach, C. A., and Barbas, C. F. 3rd. (2013). ZFN, TALEN, and CRISPR/Cas-based methods for genome engineering. *Trends Biotechnol.* 31, 397–405. doi: 10.1016/j.tibtech.2013.04.004

Gasperowicz, M., and Otto, F. (2008). The notch signalling pathway in the development of the mouse placenta. *Placenta* 29, 651–659. doi: 10.1016/j.placenta.2008.06.004

Geva, E., Ginzinger, D. G., Moore, D. H. 2nd., Ursell, P. C., and Jaffe, R. B. (2005). *In utero* angiopoietin-2 gene delivery remodels placental blood vessel phenotype: a murine model for studying placental angiogenesis. *Mol. Hum. Reprod.* 11, 253–260. doi: 10.1093/molehr/gah159

Gibbings, D., Leblanc, P., Jay, F., Pontier, D., Michel, F., Schwab, Y., et al. (2012). Human prion protein binds Argonaute and promotes accumulation of microRNA effector complexes. *Nat. Struct. Mol. Biol.* 19, 517–524. doi: 10.1038/nsmb.2273

Giroux, S., Tremblay, M., Bernard, D., Cardin-Girard, J. F., Aubry, S., Larouche, L., et al. (1999). Embryonic death of Mek1-deficient mice reveals a role for this kinase in angiogenesis in the labyrinthine region of the placenta. *Curr. Biol.* 9, 369–372. doi: 10.1016/S0960-9822(99)80164-X

Griffoni, C., Spisni, E., Santi, S., Riccio, M., Guarnieri, T., and Tomasi, V. (2000). Knockdown of caveolin-1 by antisense oligonucleotides impairs angiogenesis *in vitro* and *in vivo*. *Biochem. Biophys. Res. Commun.* 276, 756–761. doi: 10.1006/bbrc.2000.3484

Gubbins, S., Cook, C. J., Hyder, K., Boulton, K., Davis, C., Thomas, E., et al. (2009). Associations between lamb survival and prion protein genotype: analysis of data for ten sheep breeds in Great Britain. *BMC Vet. Res.* 5:3. doi: 10.1186/1746-6148-5-3

Haigh, C. L., Marom, S. Y., and Collins, S. J. (2010). Copper, endoproteolytic processing of the prion protein and cell signalling. *Front. Biosci. (Landmark Ed).* 15, 1086–1104. doi: 10.2741/3663

Howie, G. J., Sloboda, D. M., and Vickers, M. H. (2012). Maternal undernutrition during critical windows of development results in differential and sex-specific effects on postnatal adiposity and related metabolic profiles in adult rat offspring. *Br. J. Nutr.* 108, 298–307. doi: 10.1017/S000711451100554X

Hunkapiller, N. M., Gasperowicz, M., Kapidzic, M., Plaks, V., Maltepe, E., Kitajewski, J., et al. (2011). A role for Notch signaling in trophoblast endovascular invasion and in the pathogenesis of pre-eclampsia. *Development* 138, 2987–2998. doi: 10.1242/dev.066589

Huppertz, B., Bartz, C., and Kokozidou, M. (2006). Trophoblast fusion: fusogenic proteins, syncytins and ADAMs, and other prerequisites for syncytial fusion. *Micron* 37, 509–517. doi: 10.1016/j.micron.2005.12.011

Hwang, H. S., Park, S. H., Park, Y. W., Kwon, H. S., and Sohn, I. S. (2010). Expression of cellular prion protein in the placentas of women with normal and preeclamptic pregnancies. *Acta Obstet. Gynecol. Scand.* 89, 1155–1161. doi: 10.3109/00016349.2010.498497

Jackson, G. S., Murray, I., Hosszu, L. L., Gibbs, N., Waltho, J. P., Clarke, A. R., et al. (2001). Location and properties of metal-binding sites on the human prion protein. *Proc. Natl. Acad. Sci. U.S.A.* 98, 8531–8535. doi: 10.1073/pnas.151038498

Jeffrey, M., Martin, S., Chianini, F., Eaton, S., Dagleish, M. P., and González, L. (2014). Incidence of infection in *Prnp* ARR/ARR sheep following experimental inoculation with or natural exposure to classical scrapie. *PLoS ONE* 9:e91026. doi: 10.1371/journal.pone.0091026

Kaiser, D. M., Acharya, M., Leighton, P. L., Wang, H., Daude, N., Wohlgemuth, S., et al. (2012). Amyloid beta precursor protein and prion protein have a conserved interaction affecting cell adhesion and CNS development. *PLoS ONE* 7:e51305. doi: 10.1371/journal.pone.0051305

Kambe, T., Hashimoto, A., and Fujimoto, S. (2014). Current understanding of ZIP and ZnT zinc transporters in human health and diseases. *Cell. Mol. Life Sci.* doi: 10.1007/s00018-014-1617-0. [Epub ahead of print].

Khalifé, M., Young, R., Passet, B., Halliez, S., Vilotte, M., Jaffrezic, F., et al. (2011). Transcriptomic analysis brings new insight into the biological role of the prion protein during mouse embryogenesis. *PLoS ONE* 6:e23253. doi: 10.1371/journal.pone.0023253

Knofler, M. (2010). Critical growth factors and signalling pathways controlling human trophoblast invasion. *Int. J. Dev. Biol.* 54, 269–280. doi: 10.1387/ijdb.082769mk

Kramer, M. L., Kratzin, H. D., Schmidt, B., Romer, A., Windl, O., Liemann, S., et al. (2001). Prion protein binds copper within the physiological concentration range. *J. Biol. Chem.* 276, 16711–16719. doi: 10.1074/jbc.M006554200

Krejciova, Z., Pells, S., Cancellotti, E., Freile, P., Bishop, M., Samuel, K., et al. (2011). Human embryonic stem cells rapidly take up and then clear exogenous human and animal prions *in vitro*. *J. Pathol.* 223, 635–645. doi: 10.1002/path.2832

Kubosaki, A., Ueno, A., Matsumoto, Y., Doi, K., Saeki, K., and Onodera, T. (2000). Analysis of prion protein mRNA by *in situ* hybridization in brain and placenta of sheep. *Biochem. Biophys. Res. Commun.* 273, 890–893. doi: 10.1006/bbrc.2000.3035

Lambertz, I., Nittner, D., Mestdagh, P., Denecker, G., Vandesompele, J., Dyer, M. A., et al. (2010). Monoallelic but not biallelic loss of Dicer1 promotes tumorigenesis *in vivo*. *Cell Death Differ.* 17, 633–641. doi: 10.1038/cdd.2009.202

Laresgoiti-Servitje, E., Gómez-López, N., and Olson, D. M. (2010). An immunological insight into the origins of pre-eclampsia. *Hum. Reprod. Update* 16, 510–524. doi: 10.1093/humupd/dmq007

Lee, Y. J., and Baskakov, I. V. (2013). The cellular form of the prion protein is involved in controlling cell cycle dynamics, self-renewal, and the fate of human embryonic stem cell differentiation. *J. Neurochem.* 124, 310–322. doi: 10.1111/j.1471-4159.2012.07913.x

Linden, R., Martins, V. R., Prado, M. A., Cammarota, M., Izquierdo, I., and Brentani, R. R. (2008). Physiology of the prion protein. *Physiol. Rev.* 88, 673–728. doi: 10.1152/physrev.00007.2007

Lopes, M. H., and Santos, T. G. (2012). Prion potency in stem cells biology. *Prion* 6, 142–146. doi: 10.4161/pri.19035

Malaga-Trillo, E., and Sempou, E. (2009). PrPs: proteins with a purpose: lessons from the zebrafish. *Prion* 3, 129–133. doi: 10.4161/pri.3.3.9651

Malaga-Trillo, E., Solis, G. P., Schrock, Y., Geiss, C., Luncz, L., Thomanetz, V., et al. (2009). Regulation of embryonic cell adhesion by the prion protein. *PLoS Biol.* 7:e55. doi: 10.1371/journal.pbio.1000055

Mallucci, G. R., Ratte, S., Asante, E. A., Linehan, J., Gowland, I., Jefferys, J. G., et al. (2002). Post-natal knockout of prion protein alters hippocampal CA1 properties, but does not result in neurodegeneration. *EMBO J.* 21, 202–210. doi: 10.1093/emboj/21.3.202

Manson, J. C., Clarke, A. R., McBride, P. A., Mcconnell, I., and Hope, J. (1994). PrP gene dosage determines the timing but not the final intensity or distribution of lesions in scrapie pathology. *Neurodegeneration* 3, 331–340.

Manson, J., West, J. D., Thomson, V., McBride, P., Kaufman, M. H., and Hope, J. (1992). The prion protein gene: a role in mouse embryogenesis? *Development* 115, 117–122.

Massimino, M. L., Griffoni, C., Spisni, E., Toni, M., and Tomasi, V. (2002). Involvement of caveolae and caveolae-like domains in signalling, cell survival and angiogenesis. *Cell. Signal.* 14, 93–98. doi: 10.1016/S0898-6568(01) 00232-7

Miele, G., Alejo Blanco, A. R., Baybutt, H., Horvat, S., Manson, J., and Clinton, M. (2003). Embryonic activation and developmental expression of the murine prion protein gene. *Gene Expr.* 11, 1–12. doi: 10.3727/000000003783992324

Miele, G., Jeffrey, M., Turnbull, D., Manson, J., and Clinton, M. (2002). Ablation of cellular prion protein expression affects mitochondrial numbers and morphology. *Biochem. Biophys. Res. Commun.* 291, 372–377. doi: 10.1006/bbrc.2002.6460

Miranda, A., Pericuesta, E., Ramirez, M. A., and Gutierrez-Adan, A. (2011). Prion protein expression regulates embryonic stem cell pluripotency and differentiation. *PLoS ONE* 6:e18422. doi: 10.1371/journal.pone.0018422

Moore, R. C., Lee, I. Y., Silverman, G. L., Harrison, P. M., Strome, R., Heinrich, C., et al. (1999). Ataxia in prion protein (PrP)-deficient mice is associated with upregulation of the novel PrP-like protein doppel. *J. Mol. Biol.* 292, 797–817. doi: 10.1006/jmbi.1999.3108

Moore, R. C., Xiang, F., Monaghan, J., Han, D., Zhang, Z., Edstrom, L., et al. (2001). Huntington disease phenocopy is a familial prion disease. *Am. J. Hum. Genet.* 69, 1385–1388. doi: 10.1086/324414

Mouillet-Richard, S., Ermonval, M., Chebassier, C., Laplanche, J. L., Lehmann, S., Launay, J. M., et al. (2000). Signal transduction through prion protein. *Science* 289, 1925–1928. doi: 10.1126/science.289.5486.1925

Munir, S., Xu, G., Wu, Y., Yang, B., Lala, P. K., and Peng, C. (2004). Nodal and ALK7 inhibit proliferation and induce apoptosis in human trophoblast cells. *J. Biol. Chem.* 279, 31277–31286. doi: 10.1074/jbc.M400641200

Nishizawa, H., Pryor-Koishi, K., Kato, T., Kowa, H., Kurahashi, H., and Udagawa, Y. (2007). Microarray analysis of differentially expressed fetal genes in placental tissue derived from early and late onset severe pre-eclampsia. *Placenta* 28, 487–497. doi: 10.1016/j.placenta.2006.05.010

Nourizadeh-Lillabadi, R., Seilo Torgersen, J., Vestrheim, O., Konig, M., Alestrom, P., and Syed, M. (2010). Early embryonic gene expression profiling of zebrafish prion protein (Prp2) morphants. *PLoS ONE* 5:e13573. doi: 10.1371/journal.pone.0013573

Nuvolone, M., Kana, V., Hutter, G., Sakata, D., Mortin-Toth, S. M., Russo, G., et al. (2013). SIRPalpha polymorphisms, but not the prion protein, control phagocytosis of apoptotic cells. *J. Exp. Med.* 210, 2539–2552. doi: 10.1084/jem.20131274

O'Rourke, K. I., Zhuang, D., Truscott, T. C., Yan, H., and Schneider, D. A. (2011). Sparce PrP^{Sc} accumulation in the placentas of goats with naturally acquired scrapie. *BMC Vet. Res.* 7:7. doi: 10.1186/1746-6148-7-7

Ozkan, Z. S., Simsek, M., Ilhan, F., Deveci, D., Godekmerdan, A., and Sapmaz, E. (2013). Plasma IL-17, IL-35, interferon-gamma, SOCS3 and TGF-beta levels in pregnant women with preeclampsia, and their relation with severity of disease. *J. Matern. Fetal Neonatal Med.* doi: 10.3109/14767058.2013.861415. [Epub ahead of print].

Paisley, D., Banks, S., Selfridge, J., McLennan, N. F., Ritchie, A. M., McEwan, C., et al. (2004). Male infertility and DNA damage in Doppel knockout and prion protein/Doppel double-knockout mice. *Am. J. Pathol.* 164, 2279–2288. doi: 10.1016/S0002-9440(10)63784-4

Palmqvist, L., Pineault, N., Wasslavik, C., and Humphries, R. K. (2007). Candidate genes for expansion and transformation of hematopoietic stem cells by NUP98-HOX fusion genes. *PLoS ONE* 2:e768. doi: 10.1371/journal.pone.0000768

Passet, B., Halliez, S., Beringue, V., Laude, H., and Vilotte, J. L. (2013). The prion protein family: looking outside the central nervous system. *Prion* 7, 127–130. doi: 10.4161/pri.22851

Passet, B., Young, R., Makhzami, S., Vilotte, M., Jaffrezic, F., Halliez, S., et al. (2012). Prion protein and Shadoo are involved in overlapping embryonic pathways and trophoblastic development. *PLoS ONE* 7:e41959. doi: 10.1371/journal.pone.0041959

Prusiner, S. B. (1982). Novel proteinaceous infectious particles cause scrapie. *Science* 216, 136–144. doi: 10.1126/science.6801762

Qin, K., Coomaraswamy, J., Mastrangelo, P., Yang, Y., Lugowski, S., Petromilli, C., et al. (2003). The PrP-like protein Doppel binds copper. *J. Biol. Chem.* 278, 8888–8896. doi: 10.1074/jbc.M210875200

Reich, A., Neretti, N., Freiman, R. N., and Wessel, G. M. (2012). Transcriptome variance in single oocytes within, and between, genotypes. *Mol. Reprod. Dev.* 79, 502–503. doi: 10.1002/mrd.22061

Resenberger, U. K., Harmeier, A., Woerner, A. C., Goodman, J. L., Muller, V., Krishnan, R., et al. (2011). The cellular prion protein mediates neurotoxic signalling of beta-sheet-rich conformers independent of prion replication. *EMBO J.* 30, 2057–2070. doi: 10.1038/emboj.2011.86

Robertson, S. A., Prins, J. R., Sharkey, D. J., and Moldenhauer, L. M. (2013). Seminal fluid and the generation of regulatory T cells for embryo implantation. *Am. J. Reprod. Immunol.* 69, 315–330. doi: 10.1111/aji.12107

Rossi, D., Cozzio, A., Flechsig, E., Klein, M. A., Rulicke, T., Aguzzi, A., et al. (2001). Onset of ataxia and Purkinje cell loss in PrP null mice inversely correlated with Dpl level in brain. *EMBO J.* 20, 694–702. doi: 10.1093/emboj/20.4.694

Sawalha, R. M., Brotherstone, S., Conington, J., and Villanueva, B. (2007a). Lambs with scrapie susceptible genotypes have higher postnatal survival. *PLoS ONE* 2:e1236. doi: 10.1371/journal.pone.0001236

Sawalha, R. M., Brotherstone, S., Man, W. Y., Conington, J., Bunger, L., Simm, G., et al. (2007b). Associations of polymorphisms of the ovine prion protein gene with growth, carcass, and computerized tomography traits in Scottish Blackface lambs. *J. Anim. Sci.* 85, 632–640. doi: 10.2527/jas.2006-372

Schmitt-Ulms, G., Ehsani, S., Watts, J. C., Westaway, D., and Wille, H. (2009). Evolutionary descent of prion genes from the ZIP family of metal ion transporters. *PLoS ONE* 4:e7208. doi: 10.1371/journal.pone.0007208

Schneider, B., Pietri, M., Pradines, E., Loubet, D., Launay, J. M., Kellermann, O., et al. (2011). Understanding the neurospecificity of Prion protein signaling. *Front. Biosci. (Landmark Ed).* 16:169–186. doi: 10.2741/3682

Screen, M., Dean, W., Cross, J. C., and Hemberger, M. (2008). Cathepsin proteases have distinct roles in trophoblast function and vascular remodelling. *Development* 135, 3311–3320. doi: 10.1242/dev.025627

Shaker, O. G., and Sadik, N. A. (2013). Pathogenesis of preeclampsia: implications of apoptotic markers and oxidative stress. *Hum. Exp. Toxicol.* 32, 1170–1178. doi: 10.1177/0960327112472998

Shi, J., Feng, H., Lee, J., and Ning Chen, W. (2013). Comparative proteomics profile of lipid-cumulating oleaginous yeast: an iTRAQ-coupled 2-D LC-MS/MS analysis. *PLoS ONE* 8:e85532. doi: 10.1371/journal.pone.0085532

Shmerling, D., Hegyi, I., Fischer, M., Blattler, T., Brandner, S., Gotz, J., et al. (1998). Expression of amino-terminally truncated PrP in the mouse leading to ataxia and specific cerebellar lesions. *Cell* 93, 203–214. doi: 10.1016/S0092-8674(00)81572-X

Sibai, B., Dekker, G., and Kupferminc, M. (2005). Pre-eclampsia. *Lancet* 365, 785–799. doi: 10.1016/S0140-6736(05)17987-2

Singh, A., Haldar, S., Horback, K., Tom, C., Zhou, L., Meyerson, H., et al. (2013). Prion protein regulates iron transport by functioning as a ferrireductase. *J. Alzheimers. Dis.* 35, 541–552. doi: 10.3233/JAD-130218

Solis, G. P., Radon, Y., Sempou, E., Jechow, K., Stuermer, C. A., and Malaga-Trillo, E. (2013). Conserved roles of the prion protein domains on subcellular localization and cell-cell adhesion. *PLoS ONE* 8:e70327. doi: 10.1371/journal.pone.0070327

Stanczuk, G. A., McCoy, M. J., Hutchinson, I. V., and Sibanda, E. N. (2007). The genetic predisposition to produce high levels of TGF-beta1 impacts on the severity of eclampsia/pre-eclampsia. *Acta Obstet. Gynecol. Scand.* 86, 903–908. doi: 10.1080/00016340701416945

Striebel, J. F., Race, B., Pathmajeyan, M., Rangel, A., and Chesebro, B. (2013). Lack of influence of prion protein gene expression on kainate-induced seizures in mice: studies using congenic, coisogenic and transgenic strains. *Neuroscience* 238, 11–18. doi: 10.1016/j.neuroscience.2013.02.004

Tanji, K., Saeki, K., Matsumoto, Y., Takeda, M., Hirasawa, K., Doi, K., et al. (1995). Analysis of PrPc mRNA by *in situ* hybridization in brain, placenta, uterus and testis of rats. *Intervirology* 38, 309–315.

Taylor, D. R., and Hooper, N. M. (2007). The low-density lipoprotein receptor-related protein 1 (LRP1) mediates the endocytosis of the cellular prion protein. *Biochem. J.* 402, 17–23. doi: 10.1042/BJ20061736

Thumdee, P., Ponsuksili, S., Murani, E., Nganvongpanit, K., Gehrig, B., Tesfaye, D., et al. (2007). Expression of the prion protein gene (*PRNP*) and cellular prion protein (PrPc) in cattle and sheep fetuses and maternal tissues during pregnancy. *Gene Expr.* 13, 283–297. doi: 10.3727/000000006780666984

Tian, X., Anthony, K., Neuberger, T., and Diaz, F. J. (2014). Preconception zinc deficiency disrupts postimplantation fetal and placental development in mice. *Biol. Reprod.* 90, 83. doi: 10.1095/biolreprod.113.113910

Tremblay, P., Bouzamondo-Bernstein, E., Heinrich, C., Prusiner, S. B., and Dearmond, S. J. (2007). Developmental expression of PrP in the post-implantation embryo. *Brain Res.* 1139, 60–67. doi: 10.1016/j.brainres.2006.12.055

Turu, M., Slevin, M., Ethirajan, P., Luque, A., Elasbali, A., Font, A., et al. (2008). The normal cellular prion protein and its possible role in angiogenesis. *Front. Biosci.* 13, 6491–6500. doi: 10.2741/3169

Uriu-Adams, J. Y., and Keen, C. L. (2010). Zinc and reproduction: effects of zinc deficiency on prenatal and early postnatal development. *Birth Defects Res. B Dev. Reprod. Toxicol.* 89, 313–325. doi: 10.1002/bdrb.20264

Wadley, G. D., McConell, G. K., Goodman, C. A., Siebel, A. L., Westcott, K. T., and Wlodek, M. E. (2013). Growth restriction in the rat alters expression of metabolic genes during postnatal cardiac development in a sex-specific manner. *Physiol. Genomics* 45, 99–105. doi: 10.1152/physiolgenomics.00095.2012

Wakefield, S. L., Lane, M., and Mitchell, M. (2011). Impaired mitochondrial function in the preimplantation embryo perturbs fetal and placental development in the mouse. *Biol. Reprod.* 84, 572–580. doi: 10.1095/biolreprod.110.087262

Walter, E. D., Stevens, D. J., Spevacek, A. R., Visconte, M. P., Dei Rossi, A., and Millhauser, G. L. (2009). Copper binding extrinsic to the octarepeat region in the prion protein. *Curr. Protein Pept. Sci.* 10, 529–535. doi: 10.2174/138920309789352056

Wang, V., Chuang, T. C., Hsu, Y. D., Chou, W. Y., and Kao, M. C. (2005). Nitric oxide induces prion protein via MEK and p38 MAPK signaling. *Biochem. Biophys. Res. Commun.* 333, 95–100. doi: 10.1016/j.bbrc.2005.05.091

Watt, N. T., Taylor, D. R., Kerrigan, T. L., Griffiths, H. H., Rushworth, J. V., Whitehouse, I. J., et al. (2012). Prion protein facilitates uptake of zinc into neuronal cells. *Nat. Commun.* 3, 1134. doi: 10.1038/ncomms2135

Watts, J. C., Huo, H., Bai, Y., Ehsani, S., Jeon, A. H., Shi, T., et al. (2009). Interactome analyses identify ties of PrP and its mammalian paralogs to oligomannosidic N-glycans and endoplasmic reticulum-derived chaperones. *PLoS Pathog.* 5:e1000608. doi: 10.1371/annotation/9eb11869-6acb-49b0-978e-abedc3cc545a

Westergard, L., Christensen, H. M., and Harris, D. A. (2007). The cellular prion protein (PrP(C)): its physiological function and role in disease. *Biochim. Biophys. Acta* 1772, 629–644. doi: 10.1016/j.bbadis.2007.02.011

Wijshake, T., Baker, D. J., and Van De Sluis, B. (2014). Endonucleases: new tools to edit the mouse genome. *Biochim. Biophys. Acta.* doi: 10.1016/j.bbadis.2014.04.020. [Epub ahead of print].

Wong, B. S., Liu, T., Paisley, D., Li, R., Pan, T., Chen, S. G., et al. (2001). Induction of HO-1 and NOS in doppel-expressing mice devoid of PrP: implications for doppel function. *Mol. Cell. Neurosci.* 17, 768–775. doi: 10.1006/mcne.2001.0963

Wurm, S., and Wechselberger, C. (2006). Prion protein modifies TGF-beta induced signal transduction. *Biochem. Biophys. Res. Commun.* 349, 525–532. doi: 10.1016/j.bbrc.2006.08.074

Yamaguchi, N., Sakaguchi, S., Shigematsu, K., Okimura, N., and Katamine, S. (2004). Doppel-induced Purkinje cell death is stoichiometrically abrogated by prion protein. *Biochem. Biophys. Res. Commun.* 319, 1247–1252. doi: 10.1016/j.bbrc.2004.05.115

Young, R., Bouet, S., Polyte, J., Le Guillou, S., Passet, B., Vilotte, M., et al. (2011). Expression of the prion-like protein Shadoo in the developing mouse embryo. *Biochem. Biophys. Res. Commun.* 416, 184–187. doi: 10.1016/j.bbrc.2011.11.021

Young, R., Passet, B., Vilotte, M., Cribiu, E. P., Beringue, V., Le Provost, F., et al. (2009). The prion or the related Shadoo protein is required for early mouse embryogenesis. *FEBS Lett.* 583, 3296–3300. doi: 10.1016/j.febslet.2009.09.027

Zadrozna, M., Gawlik, M., Nowak, B., Marcinek, A., Mrowiec, H., Walas, S., et al. (2009). Antioxidants activities and concentration of selenium, zinc and copper in preterm and IUGR human placentas. *J. Trace Elem. Med. Biol.* 23, 144–148. doi: 10.1016/j.jtemb.2009.02.005

Zomosa-Signoret, V., Arnaud, J. D., Fontes, P., Alvarez-Martinez, M. T., and Liautard, J. P. (2008). Physiological role of the cellular prion protein. *Vet. Res.* 39, 9. doi: 10.1051/vetres:2007048

Conflict of Interest Statement: The authors declare that the research was conducted in the absence of any commercial or financial relationships that could be construed as a potential conflict of interest.

Recruitment of bone marrow-derived cells to periodontal tissue defects

Yasuyuki Kimura[1], Motohiro Komaki[2]*, Kengo Iwasaki[2], Masataka Sata[3], Yuichi Izumi[1,4] and Ikuo Morita[4,5]

[1] Department of Periodontology, Graduate School of Medical and Dental Science, Tokyo Medical and Dental University, Tokyo, Japan
[2] Department of Nanomedicine (DNP), Graduate School of Medical and Dental Science, Tokyo Medical and Dental University, Tokyo, Japan
[3] Department of Cardiovascular Medicine, Institute of Health Biosciences, The University of Tokushima Graduate School, Tokushima, Japan
[4] Global Center of Excellence Program, International Research Center for Molecular Science in Tooth and Bone Diseases, Tokyo Medical and Dental University, Tokyo, Japan
[5] Department of Cellular Physiological Chemistry, Graduate School of Medical and Dental Science, Tokyo Medical and Dental University, Tokyo, Japan

Edited by:
Mario Petrini, University of Pisa, Italy

Reviewed by:
Hiroyuki Moriyama, Kinki University, Japan
Jyotsna Dhawan, Institute for Stem Cell biology and Regenerative Medicine, India

***Correspondence:**
Motohiro Komaki, Department of Nanomedicine (DNP), Graduate School of Medical and Dental Science, Tokyo Medical and Dental University, Room#N607, M&D tower, 1-5-45 Yushima, Bunkyo-ku, 113-8549 Tokyo, Japan
e-mail: komaki.peri@tmd.ac.jp

Bone marrow-derived cells (BMCs) are considered to be a major source of mesenchymal stem cells (MSCs) in adults and are known to be effective in periodontal tissue regeneration. However, whether endogenous BMCs are involved in periodontal tissue repair process is uncertain. We therefore created periodontal tissue defects in the buccal alveolar bone of mandibular first molars in bone marrow chimeric mice, and immunohistochemically examined the expression of stromal cell derived factor-1 (SDF-1) and the mobilization of BMCs. We found that SDF-1 expression was increased around the defects at as early as 1 week after injury and that BMCs were mobilized to the defects, while GFP+/CD45+ were rarely observed. Fluorescence-activated cell sorting (FACS) analysis demonstrated that the number of platelet-derived growth factor receptor (pdgfr) α+/Sca-1+ (PαS) cells in the bone marrow decreased after injury. Taken together, these results suggest that BMCs are mobilized to the periodontal tissue defects. Recruitment of BMCs, including a subset of MSCs could be a new target of periodontal treatment.

Keywords: bone marrow chimeric mice, periodontal defects, mesenchymal stem cells, recruitment, SDF-1

INTRODUCTION

Periodontal disease is a bacterially induced chronic inflammatory disease that destroys the tooth-supporting tissue and is one of the main causes of tooth loss. The inflammation and breakdown of tissue can be prevented by conventional periodontal treatment such as scaling and root plaining (tissue debridement). However, the conventional treatment does not regain the tissue that has been lost during the disease process. Recently MSC-like cells have been discovered in periodontal ligament (PDLSCs) (Seo et al., 2004) and extensive studies have been carried out to investigate the potential use of PDLSCs as a therapeutic agent for periodontal regeneration.

Mesenchymal stem cells (MSCs) show multi-differentiation capability and self-renewability *in vitro*. Bone marrow is considered as the one of main source of MSCs. Bone marrow MSCs have attracted attention as donor cells for regenerative therapy, and the efficacy of bone marrow MSCs has been also shown by experiments in the periodontal tissue regeneration. For example, it has been reported that canine periodontal defects can be regenerated from bone marrow MSCs mixed with atelocollagen (Kawaguchi et al., 2004) and that transplanted bone marrow MSCs were observed in regenerated periodontal tissue (Hasegawa et al., 2006). Yang et al. demonstrated that engraftment of bone marrow-derived MSCs with gelatin beads successfully regenerated periodontal tissue in rats (Yang et al., 2010).

Endogenous bone marrow-derived cells (BMCs) including MSCs have been reported to promote repair of the remote tissue by mobilizing into peripheral blood by injury signals, and homing to injured tissues (Mansilla et al., 2006; Alm et al., 2010). Recently, SDF-1 has been reported to involve in the recruitment and engraftment of stem cells in wound sites (Yin et al., 2010). C-X-C chemokine receptor type 4 (CXCR4) is a unique receptor of SDF-1, and is known to express in various cell types, including stem cells (Honczarenko et al., 2006). However, whether the SDF-1/CXCR4 axis is involved in recruitment of BMCs to periodontal wound is not clarified. Particularly, the contribution of endogenous bone marrow MSCs during periodontal tissue repair is not fully understood due to lack of appropriate detection system of MSCs *in vivo*.

A lack of the unique marker is a major obstacle for detection of MSCs *in vivo*. The characteristics of MSCs were confirmed by combination expression of cell surface markers such as CD44, CD73, CD90, and CD105 in a single cell. Besides it must be shown that there is not expression of hematopoietic stem cell marker, CD34 and hematopoietic progeny markers such as CD11b, and CD45. Recently, Morikawa et al. reported the method to isolate MSCs from murine bone marrow without cell culture by cell sorting of pdgfrα (+)/Sca-1 (+) cells (Morikawa et al., 2009).

In this study, we created periodontal defects in the buccal alveolar bone of mandibular first molars in bone marrow chimeric mice and investigated the expression of SDF-1 and the recruitment of bone marrow MSCs during periodontal tissue repair process.

MATERIALS AND METHODS

PREPARATION OF BONE MARROW CHIMERIC MICE

BMCs were isolated from femurs and tibias of GFP transgenic mice as reported previously (Okabe et al., 1997; Sata et al., 2002; Fukuda et al., 2009). In brief, BMCs were hemolyzed with ACK lysing buffer (Lonza, Basel, Switzerland). C57/BL/6 mice (age, 8 weeks; male) were lethally irradiated 9.5 Gy (MBR-1520RB; Hitachi, Tokyo, Japan). Two days after irradiation, unfractionated BMCs (1×10^6 cells/0.3 ml D-PBS) from GFP transgenic mice were intravenously injected into irradiated mice by tail vein puncture. Eight weeks after the transplantation, peripheral bloods were collected from retro-orbital plexus. Replacement ratio of bone marrow was confirmed by fluorescence-activated cell sorting (FACS) aria (BD, Franklin Lakes, NJ, USA). Chimeric mice with a bone marrow substitution rate of over 83% were used in this experiment. All procedures involving the experimental animals were performed in accordance with protocols approved by the local institutional guidelines for animal care of The University of Tokyo and Tokyo Medical and Dental University (0120218A) and complied with the *Guide for the Care and Use of Laboratory Animals* (NIH guidelines 32).

EXPERIMENTAL PERIODONTAL TISSUE INJURY

Under general anesthesia with sodium pentobarbital (40–50 mg/kg, IP), we produced 2.0×1.5 mm periodontal tissue defects in the buccal alveolar bone of mandibular first molars in bone marrow chimeric mice, by removing alveolar bone, periodontal ligament and cementum using a round bar with water cooling under a stereoscopic microscope.

IMMUNOHISTOCHEMICAL STAINING

Before or at 1, 2, 4, 5, and 10 weeks after injury, tissue was fixed with 4% paraformaldehyde, followed by decalcification in 10% ethylenediaminetetraacetic acid (EDTA) solution at 4°C. Tissue sections (5 μ) were immunostained for GFP and SDF-1. Briefly, after deparaffinization, sections were washed with PBS, and were treated with 1% hydrogen peroxide (Wako, Osaka, Japan) in methanol. After washing with PBS, sections were blocked with blocking solution (0.5% goat serum in PBS) at room temperature for 30 min.

Before anti mouse SDF-1 antibody treatment, sections were treated with a M.O.M. Immunodetection Kit (Vector Laboratories Inc., Burlingame, CA, USA). Sections were then treated with either anti-mouse GFP antibody (1: 500 dilution, #47894A; Molecular Probes, Eugene, OR) or anti mouse SDF-1 antibody (1: 1000 dilution, #79014; R&D systems, Minneapolis, MN, USA) or anti-mouse CD45 antibody (1: 200 dilution #550539; BD) for 2 h. After washing with PBS three times, sections were treated with secondary antibodies, either anti-rabbit conjugated with biotin (Dako Japan, Kyoto, Japan) or anti-mouse conjugated with biotin (Dako Japan) or anti-rat Cy3 (1: 200 diluted, #56021; Jackson, West Grove, PA, USA) at room temperature for an hour. After washing with PBS three times, sections were treated with ABC-AP mix (Vector Laboratories Inc.) at room temperature for an hour. Detection was performed using Vector Red at room temperature for 10–20 min. Histological examination was performed under a fluorescence microscope (BZ-8000; Keyence, Osaka, Japan)

FLUORESCENCE-ACTIVATED CELL SORTER (FACS) ANALYSIS FOR IDENTIFICATION OF BONE MARROW MSCs

In the experimental group, BMCs were obtained from murine femurs and tibias 1 week after preparation of periodontal defects. Untreated mice were used as controls. To evaluate the ratio of bone marrow MSCs for all collected cell counts, we conducted at least three experiments using two mice in the experimental and control groups. BMCs were obtained as described previously (Morikawa et al., 2009). Briefly, femurs and tibias were aseptically removed from two mice and bones were crushed with an ice-cold pestle and mortar. Bone chips, including marrow were rinsed with HBSS+ and were digested using collagenase (#032-22364; Wako) for an hour at 37°C. Collected BMCs were hemolyzed and FcR was blocked with anti-mouse CD16/32 (#553142; BD) on ice for 5 min. BMCs ($2–5 \times 10^5$) were multi-stained with CD45.2-APC-eFlour780 (#47-0454; eBioscience, San Diego, CA, USA), TER119-PECy7 (#25-5921; eBioscience), pdgfrα-PE (#12-1401-81; eBioscience), Sca-1-APC (#17-5981-81; eBioscience) (1: 100 dilution, 30 min, on ice) and 7AAD (#559925; BD) (1: 100 dilution, 10 min, on ice). The ratio of MSCs [CD45.2 (−), TER119 (−), 7AAD (−), pdgfrα (+), Sca-1 (+)] in BMCs was evaluated by FACS analysis.

RESULTS

LOCALIZATION OF GFP-POSITIVE CELLS IN PERIODONTAL TISSUE

In order to investigate the localization of BMCs, we prepared bone marrow chimeric mice and examined GFP-positive cells in the periodontium. Both osteoclasts-like multinucleated giant cell in resorption pits of alveolar bone and macrophages in gingival epithelium, which are known to be derived from bone marrow, were positive for GFP (**Figures 1A,B**). In control mice, GFP-positive cells were observed in both periodontal ligament and dental pulp (**Figures 1C–E**). GFP-positive cells in periodontal ligament were mainly observed around blood vessels (**Figure 1D**). GFP-positive cells were rarely observed in alveolar bone or dentin.

TIME-COURSE OBSERVATIONS OF BMCs AROUND PERIODONTAL TISSUE DEFECTS

Experimental periodontal tissue defects were created in bone marrow chimeric mice. Five weeks after injury, GFP-positive cells were observed in the periodontal ligament in both control and experimental groups. In the experimental group, the number of GFP-positive cells increased significantly around the periodontal tissue defects (**Figure 2**). Ten weeks after injury, GFP-positive cells around the periodontal tissue defects decreased to control levels. Meanwhile, no changes were seen during the course of experiment in control tissue. In addition, we performed double staining for SDF-1 and GFP in order to check co-localization of SDF-1 and GFP. Five weeks after injury, SDF-1 expression was ubiquitously seen in periodontal ligament and the number of BMCs in experimental tissues was higher than in controls (**Figure 3**). In control tissue, SDF-1 expression was dominantly observed around blood vessels. A double staining for GFP and CD45 showed limited co-localization of GFP (green) and CD45

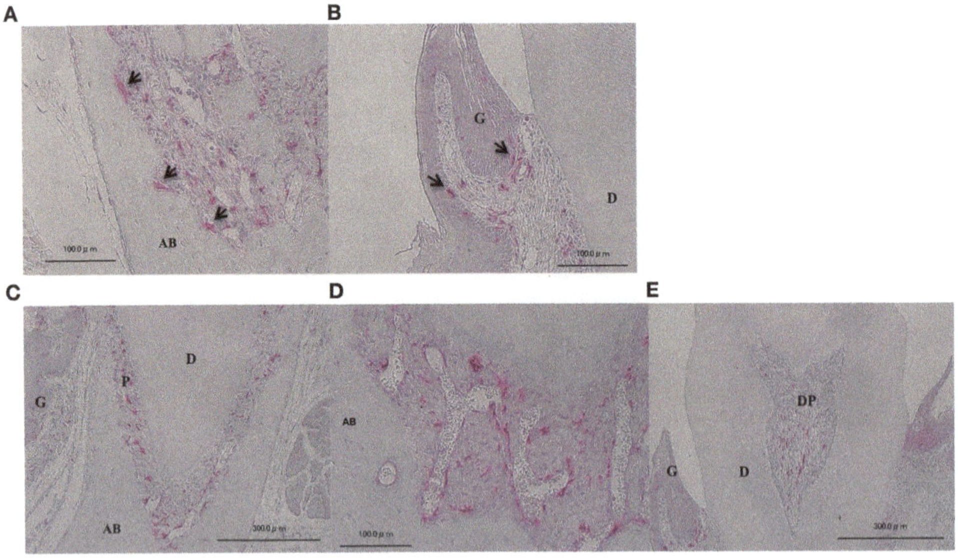

FIGURE 1 | Localization of GFP-positive cells in intact periodontal tissue. Visualization of BMCs *in vivo* on periodontal tissue. Arrows indicate osteoclast-like multinucleated giant cells **(A)** and macrophage-like cells in gingival epithelium **(B)** were GFP-positive, implying that GFP-positive images were confirmed to match BMCs. GFP-positive cells were also observed at the periodontal ligament **(C)**, blood vessels **(D)**, and dental pulp **(E)**. A, multinucleated giant cells in resorption lacunae of alveolar bone; B, gingival epithelium; C, periodontal ligament; D, blood vessels in periodontal ligament; E, dental pulp. G, gingiva; AB, alveolar bone; P, periodontal ligament; DP, dental pulp; D, dentin.

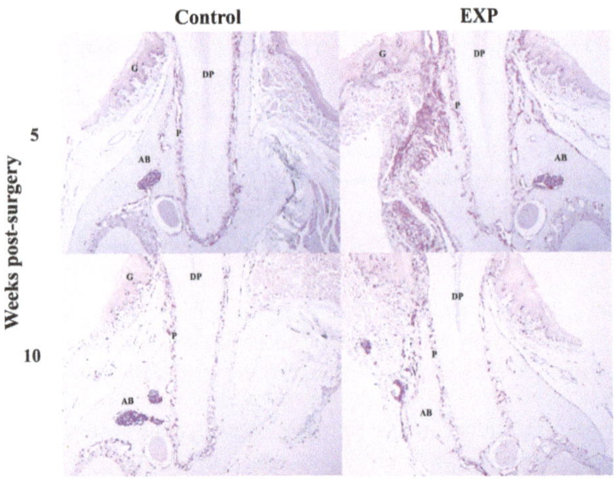

FIGURE 2 | Time-course observations of BMCs around periodontal tissue defects. Experimental periodontal tissue defects were introduced at the buccal surface of mandibular first molar roots in bone marrow chimeric mice. Five weeks after surgery, the number of GFP-positive cells was significantly elevated around the defects. Ten weeks after surgery, GFP-positive cells around the periodontal tissue defects decreased to control levels. G, gingiva; AB, alveolar bone; P, periodontal ligament; DP, dental pulp; D, dentin.

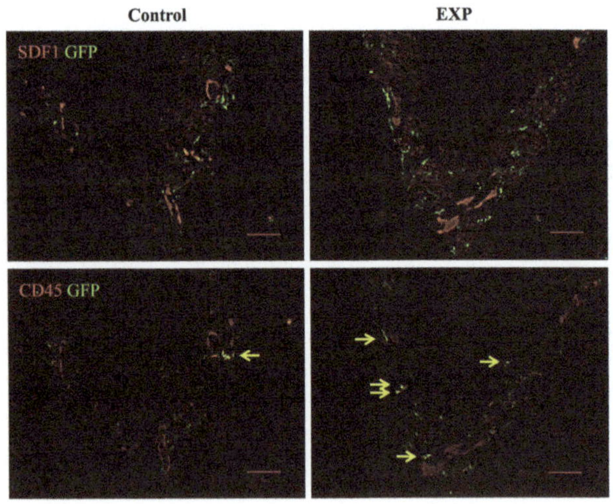

FIGURE 3 | Double staining of GFP and SDF-1/CD45 in periodontal ligaments. In control sites, blood vessel-like structures showed intense staining of SDF-1. GFP-positive cells were observed around SDF-1-positive cells. In experimental sites, SDF-1 expression was ubiquitously detected in periodontal ligament 5 weeks after surgery. The number of GFP-positive cells in experimental tissues was higher than in controls. In experimental tissue, the number of GFP(+)/CD45(+) cells (yellow arrows) was elevated, accounting for 13% of GFP-positive cells ($n = 4$).

(red). In experimental tissue the number of GFP(+)/CD45(+) cells peaked at 13% GFP-positive cells.

TIME-COURSE OBSERVATIONS OF SDF-1 EXPRESSION AROUND PERIODONTAL TISSUE DEFECTS

At 1, 2, and 4 weeks after injury, SDF-1 expression was immunohistochemically evaluated (**Figure 4**). One week after injury, some SDF-1 was observed within the blood vessels of periodontal ligaments in control tissue. In experimental tissue, weak and diffuse SDF-1 staining was observed in the defects and periodontal ligament. After 2 weeks, SDF-1 staining increased markedly and was attenuated by 4 weeks after injury.

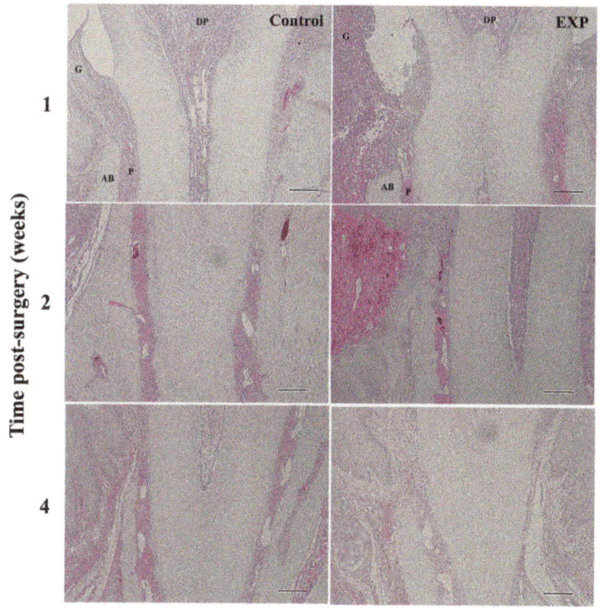

FIGURE 4 | Time-course observations of SDF-1 expression around periodontal tissue defects. In experimental sites, SDF-1 expression was slightly higher at 1 week after surgery when compared with controls. SDF-1 expression peaked at 2 weeks after surgery in experimental tissue. SDF-1 expression decreased to control levels by 4 weeks. G, gingiva; AB, alveolar bone; P, periodontal ligament; DP, dental pulp.

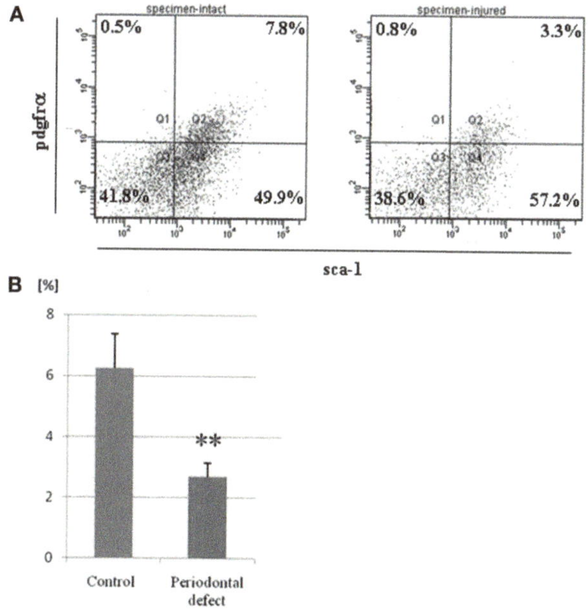

FIGURE 5 | Postoperative changes in MSC population in bone marrow. BMCs were collected from murine femurs and tibias in both the control and experimental groups. **(A)** A representative data from FACS analyses. **(B)** FACS analysis showed that the percentage of MSCs in bone marrow of mice with periodontal defects was significantly lower than in controls. (** $p = 0.01$) Independent experiments were repeated at least five times.

POST-OPERATIVE CHANGES IN MSC POPULATION IN BONE MARROW

One week after injury, BMCs were collected from murine femurs and tibias. Immediately after collecting BMCs, MSC population in bone marrow was evaluated by FACS analysis. FACS analysis confirmed that the percentage of pdgfrα (+)/Sca-1(+) cells in bone marrow of mice with periodontal defects was significantly lower when compared to controls (**Figure 5**).

DISCUSSION

Involvement of BMCs in the physiology and pathology of various tissues, including heart, lung, liver, kidney, skeletal muscle, bone, and blood vessels, has been reported. Some of these studies used bone marrow chimeric mice and successfully tracked BMCs. In order to determine roles of endogenous BMCs in periodontal tissue, we created bone marrow chimeric mice. Both osteoclast-like multinucleated giant cells at the alveolar bone and macrophages in gingival connective tissue, known to be of bone marrow origin, were positive for GFP staining (**Figures 1D,E**), implying BMCs were successfully tracked. However, we should be aware of possibility that recipient bone marrow cells may fuse with donor cells, acquiring such phenotypes in periodontal tissue. GFP-positive cells were observed in both the periodontal ligament and dental pulp (**Figures 1A,B**). To our knowledge, this is the first time that the presence of endogenous BMCs has been demonstrated in naive murine periodontal ligament, at least by observation of bone marrow chimeric mice. In addition, these cells were mainly observed around blood vessels in periodontal ligament (**Figure 1C**). McCulloch reported that stem cell-like cells with a slow rate of cell proliferation were located in paravascular sites in murine periodontal ligament (McCulloch, 1985). Chen et al. reported that putative stem cells that showed cross-reactivity with at least one among STRO-1, CD146, and CD44 antibodies, mainly in the paravascular region of human periodontal ligament (Chen et al., 2006). It remains elusive whether BMCs observed in our study were the same as the MSCs/progenitor cells residing in periodontal tissue. It is of interest to clarify the physiological role of BMCs in periodontium and to determine whether BMCs are the origin of MSCs/progenitor cells around blood vessels residing in the periodontal ligament. Further studies are thus necessary.

Next, in order to determine whether BMCs contribute to periodontal wound healing, we created periodontal defects in the bone marrow chimeric mice. Zhou et al. have reported engraftment and differentiation of BMCs to periodontal tissue-forming cells when periodontal defects were treated by regenerative procedure, a grafting of ceramic bovine bone (Zhou et al., 2011). In contrast, Ohta et al. have reported that bone marrow-derived MSCs were not detected in repairing periodontal ligament by a single staining of either STRO-1 or CD44 (Ohta et al., 2008). Our result clearly demonstrated that the number of BMCs around periodontal defects increased as a result of tissue injury in accordance with data reported by Zhou et al. The result that Zhou et al. demonstrated is different from ours in that they investigated the participation of BMCs after periodontal regenerative treatment. In contrast, we investigated whether BMCs are involved in tissue repair without treatment. We observed neither localization of BMCs in alveolar bone 10 weeks post-surgery nor restoration of alveolar bone. It is well known that the tissue debridement abrogates tissue degradation, but does not regain the lost tissue, implying that our model mimics this clinical situation. It is also possible that periodontal tissue healing in irradiated recipient mice may not represent a physiological process that occurs naturally in non-irradiated mice in response to injury. Thus, it remains to be elucidated whether mobilized BMCs actually contribute to periodontal tissue healing.

Kitaori et al. reported that SDF-1 increased during the repair of bone grafts at both the messenger RNA and protein levels, and that anti-SDF-1 antibody inhibited new bone formation (Kitaori et al., 2009). Moreover, Jones et al. demonstrated that migration of MSCs to bone and bone marrow is CXCR4/SDF-1 dependent in a murine osteogenesis imperfecta model and that SDF-1 up-regulates CXCR4 demonstrating chemotaxis *in vitro* and enhancing engraftment *in vivo* (Jones et al., 2012). It has been reported that inflammation and hypoxia play an important role in regulating the SDF-1/CXCR4 axis. First we confirmed that hypoxia around periodontal defects after periodontal injury (supplemental figure 1). We then checked the expression of SDF-1 in periodontium, and found that SDF-1 expression increased around periodontal defects and in periodontal ligament prior to the recruitment of BMCs (**Figure 4**). It has been reported that SDF-1 levels increase in periodontal disease. (Havens et al., 2008) We confirmed that SDF-1 stimulated the migration of plastic dish-adhered BMCs and that CXCR4-positive BMCs migrated to SDF-1 more promptly than CXCR4-negative BMCs in transwell assay (supplemental figure 2). It has been reported that in BMCs hematopoietic cells, hematopoietic stem cells, MSCs, and endothelial progenitor cells express CXCR4 and that other molecules such as inflammatory factors and cytokines are involved in mobilization of BMCs. (Salem and Thiemermann, 2010) Further experiments are necessary to examine the direct involvement of the SDF-1/CXCR4 axis in recruitment of BMCs *in vivo*. It is also interesting to compare recruitment potential among BMCs.

Adult bone marrow contains several distinct populations of stem cells, such as hematopoietic stem cells, MSCs, endothelial progenitor cells. We observed the number of GFP-positive cells increased in periodontal ligament 5 weeks after injury and about 13% of them are CD45 positive (**Figure 3**). This result suggested that most of cells detected in periodontal ligament were not hematopoietic cells at least at time point we evaluated (5 weeks after injury). It is necessary to directly determine what types of cells are recruited to periodontal defects.

In FACS analyses, we used CD45 (−), TER119 (−), PDGFRα (+), and Sca-1 (+) cells to detect PαS cells, a subset of MSCs in bone marrow. Based on the results of FACS analyses, we demonstrated for the first time that PαS cells in bone marrow decreased after periodontal injury (**Figure 5**). However, it was not clarified whether PαS cells contribute to periodontal tissue repair process. Morikawa et al. reported that PaS-derived clones with multi-differentiation potential are also positive for CD105 and CD90 *in vitro* (Morikawa et al., 2009). It is very important to analyze whether PαS cells that migrated to periodontal defects also express CD90 and CD105.

In summary, our data demonstrate that bone marrow cells are mobilized to periodontal defects. Further experiments are needed to determine cells to be mobilized from bone marrow to periodontal defects and regulatory mechanism involved in this process. Recruitment of BMCs, including a subset of MSCs could be a new target of periodontal treatment.

REFERENCES

Alm, J. J., Koivu, H. M., Heino, T. J., Hentunen, T. A., Laitinen, S., and Aro, H. T. (2010). Circulating plastic adherent mesenchymal stem cells in aged hip fracture patients. *J. Orthop. Res.* 28, 1634–1642. doi: 10.1002/jor.21167

Chen, S. C., Marino, V., Gronthos, S., and Bartold, P. M. (2006). Location of putative stem cells in human periodontal ligament. *J. Periodontal Res.* 41, 547–553. doi: 10.1111/j.1600-0765.2006.00904.x

Fukuda, D., Enomoto, S., Shirakawa, I., Nagai, R., and Sata, M. (2009). Fluvastatin accelerates re-endothelialization impaired by local sirolimus treatment. *Eur. J. Pharmacol.* 612, 87–92. doi: 10.1016/j.ejphar.2009.04.006

Hasegawa, N., Kawaguchi, H., Hirachi, A., Takeda, K., Mizuno, N., Nishimura, M., et al. (2006). Behavior of transplanted bone marrow-derived mesenchymal stem cells in periodontal defects. *J. Periodontol.* 77, 1003–1007. doi: 10.1902/jop.2006.050341

Havensm A. M., Chiu, E., Taba, M., Wang, J., Shiozawa, Y., Jung, Y., et al. (2008). Stromal-derived factor-1 alpha (CXCL12) levels increase in periodontal disease. *J. Periodontol.* 79, 845–853. doi: 10.1902/jop.2008.070514

Honczarenko, M., Le, Y., Swierkowski, M., Ghiran, I., Glodek, A. M., and Silberstein, L. E. (2006). Human bone marrow stromal cells express a distinct set of biologically functional chemokine receptors. *Stem Cells* 24, 1030–1041. doi: 10.1634/stemcells.2005-0319

Jones, G. N., Moschidou, D., Lay, K., Abdulrazzak, H., Vanleene, M., Shefelbine, S. J., et al. (2012). Upregulating CXCR4 in human fetal mesenchymal stem cells enhances engraftment and bone mechanics in a mouse model of osteogenesis imperfecta. *Stem Cells Transl. Med.* 1, 70–78. doi: 10.5966/sctm.2011-0007

Kawaguchi, H., Hirachi, A., Hasegawa, N., Iwata, T., Hamaguchi, H., Shiba, H., et al. (2004). Enhancement of periodontal tissue regeneration by transplantation of bone marrow mesenchymal stem cells. *J Periodontol.* 75, 1281–1287. doi: 10.1902/jop.2004.75.9.1281

Kitaori, T., Ito, H., Schwarz, E. M., Tsutsumi, R., Yoshitomi, H., Oishi, S., et al. (2009). Stromal cell-derived factor 1/CXCR4 signaling is critical for the recruitment of mesenchymal stem cells to the fracture site during skeletal repair in a mouse model. *Arthritis Rheum.* 60, 813–823. doi: 10.1002/art.24330

Mansilla, E., Marín, G. H., Drago, H., Sturla, F., Salas, E., Gardiner, C., et al. (2006). Bloodstream cells phenotypically identical to human mesenchymal bone marrow stem cells circulate in large amounts under the influence of acute large skin damage: new evidence for their use in regenerative medicine. *Transplant. Proc.* 38, 967–969. doi: 10.1016/j.transproceed.2006.02.053

McCulloch, C. A. G. (1985). Progenitor-cell populations in the periodontal-ligament of mice. *Anat. Rec.* 211, 258–262. doi: 10.1002/ar.1092110305

Morikawa, S., Mabuchi, Y., Kubota, Y., Nagai, Y., Niibe, K., Hiratsu, E., et al. (2009). Prospective identification, isolation, and systemic transplantation of multipotent mesenchymal stem cells in murine bone marrow. *J. Exp. Med.* 206, 2483–2496. doi: 10.1084/jem.20091046

Ohta, S., Yamada, S., Matuzaka, K., and Inoue, T. (2008). The behavior of stem cells and progenitor cells in the periodontal ligament during wound healing as observed using immunohistochemical methods. *J. Periodontal. Res.* 43, 595–603. doi: 10.1111/j.1600-0765.2007.01002.x

Okabe, M., Ikawa, M., Kominami, K., Nakanishi, T., and Nishimune, Y. (1997). Green mice as a source of ubiquitous green cells. *FEBS Lett.* 407, 313–319. doi: 10.1016/S0014-5793(97)00313-X

Salem, H. K., and Thiemermann, C. (2010). Mesenchymal stromal cells: current understanding and clinical status. *Stem Cells* 28, 585–596. doi: 10.1002/stem.269

Sata, M., Saiura, A., Kunisato, A., Tojo, A., Okada, S., Tokuhisa, T., et al. (2002). Hematopoietic stem cells differentiate into vascular cells that participate in the pathogenesis of atherosclerosis. *Nat. Med.* 8, 403–409. doi: 10.1038/nm0402-403

Seo, B. M., Miura, M., Gronthos, S., Bartold, P. M., Batouli, S., Brahim, J., et al. (2004). Investigation of multipotent postnatal stem cells from human periodontal ligament. *Lancet* 364, 149–155. doi: 10.1016/S0140-6736(04)16627-0

Yang, Y., Rossi, F. M. V., and Putnins, E. E. (2010). Periodontal regeneration using engineered bone marrow mesenchymal stromal cells. *Biomaterials* 31, 8574–8582. doi: 10.1016/j.biomaterials.2010.06.026

Yin, Y., Zhao, X., Fang, Y., Yu, S., Zhao, J., Song, M., et al. (2010). SDF-1 alpha involved in mobilization and recruitment of endothelial progenitor cells after arterial injury in mice. *Cardiovasc. Pathol.* 19, 218–227. doi: 10.1016/j.carpath.2009.04.002

Zhou, J., Shi, S., Shi, Y., Xie, H., Chen, L., He, Y., et al. (2011). Role of bone marrow-derived progenitor cells in the maintenance and regeneration of dental mesenchymal tissues. *J. Cell. Physiol.* 226, 2081–2090. doi: 10.1002/jcp.22538

Conflict of Interest Statement: The authors declare that the research was conducted in the absence of any commercial or financial relationships that could be construed as a potential conflict of interest.

Permissions

All chapters in this book were first published in FCELL, by Frontiers; hereby published with permission under the Creative Commons Attribution License or equivalent. Every chapter published in this book has been scrutinized by our experts. Their significance has been extensively debated. The topics covered herein carry significant findings which will fuel the growth of the discipline. They may even be implemented as practical applications or may be referred to as a beginning point for another development.

The contributors of this book come from diverse backgrounds, making this book a truly international effort. This book will bring forth new frontiers with its revolutionizing research information and detailed analysis of the nascent developments around the world.

We would like to thank all the contributing authors for lending their expertise to make the book truly unique. They have played a crucial role in the development of this book. Without their invaluable contributions this book wouldn't have been possible. They have made vital efforts to compile up to date information on the varied aspects of this subject to make this book a valuable addition to the collection of many professionals and students.

This book was conceptualized with the vision of imparting up-to-date information and advanced data in this field. To ensure the same, a matchless editorial board was set up. Every individual on the board went through rigorous rounds of assessment to prove their worth. After which they invested a large part of their time researching and compiling the most relevant data for our readers.

The editorial board has been involved in producing this book since its inception. They have spent rigorous hours researching and exploring the diverse topics which have resulted in the successful publishing of this book. They have passed on their knowledge of decades through this book. To expedite this challenging task, the publisher supported the team at every step. A small team of assistant editors was also appointed to further simplify the editing procedure and attain best results for the readers.

Apart from the editorial board, the designing team has also invested a significant amount of their time in understanding the subject and creating the most relevant covers. They scrutinized every image to scout for the most suitable representation of the subject and create an appropriate cover for the book.

The publishing team has been an ardent support to the editorial, designing and production team. Their endless efforts to recruit the best for this project, has resulted in the accomplishment of this book. They are a veteran in the field of academics and their pool of knowledge is as vast as their experience in printing. Their expertise and guidance has proved useful at every step. Their uncompromising quality standards have made this book an exceptional effort. Their encouragement from time to time has been an inspiration for everyone.

The publisher and the editorial board hope that this book will prove to be a valuable piece of knowledge for researchers, students, practitioners and scholars across the globe.

List of Contributors

Simone Pacini and Iacopo Petrini
Department of Clinicaland Experimental Medicine, University of Pisa, Pisa, Italy

Yoshinori Makino
Laboratory of Pathology and Development ,Institute of Molecular and Cellular Biosciences, The University of Tokyo,Tokyo,Japan
Career-Path Promotion Unit for Young Life Scientists, Center for the Promotion of Interdisciplinary Education and Research, Kyoto University, Kyoto, Japan

Erina Inoue
Laboratory of Pathology and Development, Institute of Molecular and Cellular Biosciences, The University of Tokyo, Tokyo, Japan

Masashi Hada
Laboratory of Pathology and Development, Institute of Molecular and Cellular Biosciences, The University of Tokyo, Tokyo, Japan
Department of Biotechnology, Graduate School of Agricultural and Life Sciences, The University of Tokyo, Tokyo, Japan

Keisuke Aoshima
Division of Molecular Pathobiology, Center for Zoonosis Control, Hokkaido University, Sapporo, Japan

Satsuki Kitano and Hitoshi Miyachi
Reproductive Engineering Team, Institute for Virus Research, Kyoto University, Kyoto, Japan

Yuki Okada
Laboratory of Pathology and Development, Institute of Molecular and Cellular Biosciences, The University of Tokyo, Tokyo, Japan
Career Path Promotion Unit for Young Life Scientists, Center for the Promotion of Interdisciplinary Education and Research, Kyoto University, Kyoto, Japan
PRESTO, Japan Science and Technology Agency, Saitama, Japan

John P. Dowling, Anirudh Nair and JiankeZhang
Department of Microbiology and Immunology, Sidney Kimmel Cancer Center, Sidney Kimmel Medical College, Thomas Jefferson University, Philadelphia, PA,USA

Bishuang Cai, Shuwei Xie, Steve Caplan and Naava Naslavsky
Department of Biochemistry and Molecular Biology and the Fredand Pamela Buffett Cancer Center, The University of Nebraska Medical Center, Omaha, NE, USA

Ming Li and Susumu Ikehara
Department of Stem Cell Disorders, Kansai Medical University, Hirakata City, Japan

Kuquan Guo
Department of Cardiac Surgery, Beijing Institute of Heart, Lung and Blood Vessel Disease, Beijing Anzhen Hospital Affiliated to Capital Medical University, Beijing, China

Mohadeseh Mehrabian, Sepehr Ehsani and Gerold Schmitt-Ulms
Tanz Centre for Research in Neurodegenerative Diseases, University of Toronto, Toronto, ON, Canada
Department of Laboratory Medicine and Pathobiology, University of Toronto, Toronto, ON, Canada

Séverine Martin-Lannerée, Théo Z. Hirsch and Sophie Mouillet-Richard
Toxicology, Pharmacology and Cellular Signaling, INSERMUMR-S1124, Paris, France
Toxicology, Pharmacology and Cellular Signaling, Université Paris Descartes, Sorbonne Paris Cité, UMR-S1124, Paris, France

Julia Hernandez-Rapp
Toxicology, Pharmacology and Cellular Signaling, INSERMUMR-S1124, Paris, France
Toxicology, Pharmacology and Cellular Signaling, Université Paris Descartes, Sorbonne Paris Cité,UMR-S1124, Paris, France
Université Paris Sud11, ED419Biosigne, Orsay, France

Sophie Halliez
U892 Virologieet Immunologie Moléculaires, INRA, Jouy-en-Josas, France

Jean-Luc Vilotte
UMR1313 Génétique Animaleet Biologie Intégrative, INRA, Jouy-en-Josas, France

Jean-Marie Launay
AP-HP Service de Biochimie, Fondation Fonda Mental, INSERMU942 Hôpital Lariboisière, Paris, France 7 Pharma Research Department, F. Hoffmann-La-Roche Ltd., Basel, Switzerland

Aurélie Allais-Bonnet and Eric Pailhoux
Institut National dela Recherche Agronomique, UMR1198, Biologiedu Développementet Reproduction, Jouy-en-Josas, France

Xavier Roucou
Department of Biochemistry, Faculty of Medicine, Universitéde Sherbrooke, Sherbrooke, QC, Canada

Sophie Halliez, Hubert Laude and Vincent Béringue
Institut Nationaldela Recherche Agronomique, U892 Virologieet Immunologie Moléculaires, Jouy-en-Josas, France

Bruno Passet and Jean-Luc Vilotte
Institut Nationaldela Recherche Agronomique,UMR1313 Génétique Animaleet Biologie Intégrative, Jouy-en-Josas, France

Séverine Martin-Lannerée, Julia Hernandez-Rapp and Sophie Mouillet-Richard
Institut Nationaldela Santéetdela Recherche Médicale, UMR-S1124, Paris, France
Université Paris Descartes, Sorbonne Paris Cité, UMR-S1124, Paris, France

VimalK.Singh
INSPIRE Faculty, Stem Cell Research Laboratory, Department of Biotechnology, Delhi Technological University, Delhi, India

Manisha Kalsan, Neeraj Kumar and Abhishek Saini
Stem Cell Research Laboratory, Department of Biotechnology, Delhi Technological University, Delhi, India

Ramesh Chandra
B. R. Ambedkar Centre for Biomedical Research, University of Delhi, Delhi, India

Akikazu Sakudo
Laboratory of Biometabolic Chemistry, Faculty of Medicine, School of Health Sciences, University of the Ryukyus, Nishihara, Japan

Takashi Onodera
Research Center for Food Safety, School of Agricultural and Life Sciences, University of Tokyo, Tokyo, Japan

Monika Sramkova, Laura Parente, Timothy Wigand, Myo-Pale'Aye, Akiko Shitara and RobertoWeigert
Intracellular Membrane Trafficking Unit, Oraland Pharyngeal Cancer Branch, National Institute of Dental and Craniofacial Research, National Institutes of Health, Bethesda, MD, USA

Sean P. Delaney
Sprott Centre for Stem Cell Research, Regenerative Medicine Program, Ottawa Hospital Research Institute, Ottawa, ON, Canada
Faculty of Graduate and Postdoctoral Studies, University of Ottawa, Ottawa, ON, Canada
Department of Cellular and Molecular Medicine, University of Ottawa, Ottawa, ON, Canada

Lisa M. Julian
Sprott Centre for Stem Cell Research, Regenerative Medicine Program, Ottawa Hospital Research Institute, Ottawa, ON, Canada
Faculty of Graduate and Postdoctoral Studies, University of Ottawa, Ottawa, ON, Canada

William L. Stanford
Sprott Centre for Stem Cell Research, Regenerative Medicine Program, Ottawa Hospital Research Institute, Ottawa, ON, Canada
Faculty of Graduate and Postdoctoral Studies, University of Ottawa, Ottawa, ON, Canada
Department of Cellular and Molecular Medicine, University of Ottawa, Ottawa, ON, Canada
Department of Biochemistry, Microbiology, and Immunology, University of Ottawa, Ottawa, ON, Canada

Riya R. Kanherkar, Naina Bhatia-Dey and Antonei B. Csoka
Epigenetics Laboratory, Department of Anatomy, Howard University, Washington, DC, USA

Evgeny Makarev
InSilico Medicine,Emerging Technology Center, Johns Hopkins University Eastern, Baltimore, MD, USA

Mitsuo Tagaya, Kohei Arasaki, HirokiInoue and Hana Kimura
Department of Molecular Life Sciences, School of Life Sciences, Tokyo University of Pharmacy and Life Sciences, Hachioji, Japan

B. Sophia Winkler, Franziska Oltmer, Julia Richter, Joachim Bischof, Pengfei Xu, Timo Burster, Frank Leithäuser and Uwe Knippschild
Department of Pathology, Ulm University Hospital, Ulm, Germany
Department of General and Visceral Surgery, Surgery Center,Ulm University Hospital, Ulm, Germany
Department of Neurosurgery, Ulm University Hospital, Ulm, Germany

Danica Ciric and Human Rezaei
Virologieet Immunologie Moléculaires, Institut Nationaldela Recherche Agronomique, Jouy-en-Josas, France

Abitha Jacob and Rytis Prekeris
Department of Cell and Developmental Biology, School of Medicine, Anschutz Medical Campus, University of Colorado Denver, Aurora, CO, USA

Katharina Ochs and Edward Málaga-Trillo
Department of Biology, University of Konstanz, Konstanz, Germany

Amanda M. Weiskoff and J. Christopher Fromme
Department of Molecular Biology and Genetics, Weill Institute for Cell and Molecular Biology, Cornell University, Ithaca, NY, USA

Stefanie A. G. Black, Peter K. Stys, Gerald W. Zamponi and Shigeki Tsutsui
Department of Physiology and Pharmacology, University of Calgary, Calgary, AB, Canada
Hotchkiss Brain Institute, University of Calgary, Calgary, AB, Canada
Department of Clinical Neurosciences, University of Calgary, Calgary, AB, Canada

Lisa Gasperini and Giuseppe Legname
Laboratory of Prion Biology, Department of Neuroscience, Scuola Internazionale Superioredi Studi Avanzati, Trieste, Italy

Shizuka Miura and Atsushi Suzuki
Division of Organogenesis and Regeneration, Medical Institute of Bioregulation, Kyushu University, Fukuoka, Japan
Core Research for Evolutional Science and Technology, Japan Science and Technology Agency, Saitama, Japan

Cécile Dromard, Corinne Barreau, Mireille André, Sandra Berger-Müller, Louis Casteilla and Valerie Planat-Benard
Centre Nationaldela Recherche Scientifique (CNRS) UMR5273, STROMA Lab, Toulouse, France
Université Paul Sabatierde Toulouse, Toulouse, France
Institut Nationaldela Santéetdela Recherche Médical (INSERM) U1031, STROMA Lab, Toulouse, France
Etablissement Franc¸ais du Sang (EFS) Pyrénées-Méditerranée, STROMA Lab, Toulouse, France

Nicole T. Watt, Heledd H. Griffiths and Nigel M. Hooper
Division of Cardiovascular and Diabetes Research, Multidisciplinary Cardiovascular Research Centre, University of Leeds, Leeds, UK
Faculty of Medical and Human Sciences, Institute of Brain, Behaviour and Mental Health, University of Manchester, Manchester, UK

Samira Makzhami, Bruno Passet, Johan Castille, Katayoun Moazami-Goudarzi, Amandine Duchesne, Marthe Vilotte and Jean-Luc Vilotte
INRA, UMR1313 Génétique Animaleet Biologie Intégrative, Jouy-en-Josas, France

Sophie Halliez, Hubert Laude and Vincent Béringue
INRA, U892 Virologieet Immunologie Moléculaires, Jouy-en-Josas, France

Sophie Mouillet-Richard
INSERM, UMR-S1124 Signalisationet Physiopathologie Neurologique, Université Paris Descartes, Paris, France

Daniel Vaiman
Faculté Paris Descartes, UMR8104CNRS, U1016 INSERM, Institut Cochin, Paris, France

Yasuyuki Kimura
Department of Periodontology, Graduate School of Medical and Dental Science, Tokyo Medical and Dental University, Tokyo, Japan

Motohiro Komaki and KengoIwasak
Department of Nanomedicine (DNP), Graduate School of Medical and Dental Science, Tokyo Medical and Dental University, Tokyo, Japan

Masataka Sata
Department of Cardiovascular Medicine, Institute of Health Biosciences, The University of Tokushima Graduate School, Tokushima, Japan

YuichiI zumi
Department of Periodontology, Graduate School of Medical and Dental Science, Tokyo Medical and Dental University, Tokyo, Japan
Global Center of Excellence Program, International Research Center for Molecular Science in Tooth and Bone Diseases, Tokyo Medical and Dental University, Tokyo, Japan

Ikuo Morita
Global Center of Excellence Program, International Research Center for Molecular Science in Tooth and Bone Diseases, Tokyo Medical and Dental University, Tokyo, Japan
Department of Cellular Physiological Chemistry, Graduate School of Medical and Dental Science, Tokyo Medical and Dental University, Tokyo, Japan